HANDBOOK OF BIODEGRADABLE POLYMERS

Drug Targeting and Delivery
A series of books discussing all aspects of the targeting and delivery of drugs.
Edited by A.T. Florence and G. Gregoriadis, The School of Pharmacy,
University of London, UK

HANDBOOK OF BIODEGRADABLE POLYMERS

Edited by

Abraham J. Domb

School of Pharmacy, Hebrew University of Jerusalem, Israel

Joseph Kost

Department of Chemical Engineering, Ben-Gurion University
Beer Sheva, Israel

and

David M. Wiseman

Synechion, Inc., Dallas, Texas, USA

CRC Press
Taylor & Francis Group
Boca Raton London New York

CRC Press is an imprint of the
Taylor & Francis Group, an **informa** business

CRC Press
Taylor & Francis Group
6000 Broken Sound Parkway NW, Suite 300
Boca Raton, FL 33487-2742

First issued in paperback 2019

© 1997 by Taylor & Francis Group, LLC
CRC Press is an imprint of Taylor & Francis Group, an Informa business

No claim to original U.S. Government works

ISBN-13: 978-90-5702-153-4 (hbk)
ISBN-13: 978-0-367-40064-4 (pbk)

Library of Congress Cataloging-in-Publication Data

Catalog record is available from the Library of Congress

Visit the Taylor & Francis Web site at
http://www.taylorandfrancis.com

and the CRC Press Web site at
http://www.crcpress.com

This book is dedicated to
Professor Lewin on his 78th birthday and to our families

CONTENTS

PREFACE TO THE SERIES

This book series aims to provide a comprehensive survey of the many facets of drug delivery and targeting for senior undergraduates, graduates and established workers carrying out research in this area. Both drug delivery technologies and targeting grow in scope and potential, as well as complexity, almost daily. New opportunities arise through the development of new materials for the design and fabrication of drug delivery vehicles and carriers; new challenges are posed by the discovery and development of new therapeutic agents, which include not only small organic and inorganic molecules but also macromolecules that frequently have a natural propensity for transport across biological barriers. The series has to date covered delivery technologies in microencapsulation[1], liposomal delivery[2,3], the promotion of drug absorption[4], the important issues surrounding peptide and protein delivery[5], and interfacial phenomena in drug delivery and targeting[6]. The technology of drug delivery can never be dealt with in isolation, but always in the context of the biological environment in which delivery vehicles will operate *in vivo*.

This volume in the series deals in a timely and comprehensive manner with the crucial topic of biodegradable polymers of natural synthetic or semi-synthetic origin. All the key classes of absorbable polymers used in the fabrication of microspheres and nanospheres, membranes, reservoirs and other components of injected or implanted delivery systems are covered. A better understanding of the chemical and physical factors affecting biodegradability will allow more precise prediction of the lifetime of systems *in vivo* and the more informed choice of polymeric vehicle.

Drexler[7] talks of conventional drug delivery as somewhat haphazard. "Drugs are dumped into the body, tumble and bump around in solution haphazardly until they bump a target molecule, fit and stick." He proposed that the future of therapy will consist of nanomachines which will gain access to diseased sites, recognize the target, disassemble damaged parts, rebuild and reassemble. We are perhaps some way from that vision, but the progress that is made step by step is the stuff of research in drug delivery and targeting. It is the role of a book series to record and analyze that progress, and we hope that this series will achieve those aims and provide some stimulus for further discovery.

Alexander T. Florence and Gregory Gregoriadis
Series Editors

REFERENCES

1. T.L. Whateley (Ed.) (1992) *Microencapsulation of Drugs*, Harwood Academic Publishers, Chur.
2. G. Gregoriadis, A.T. Florence and H.M. Patel (Eds) (1993) *Liposomes in Drug Delivery*, Harwood Academic Publishers, Chur.
3. P.N. Shek (Ed.) (1996) *Liposomes in Biomedical Applications*, Harwood Academic Publishers, Chur.
4. A.(Bert) G. de Boer (Ed) (1994) *Drug Absorption Enhancement: Concepts, Possibilities, Limitations and Trends*, Harwood Academic Publishers, Chur.

5. V.H.L. Lee, M. Hashida and Y. Mizushima (Eds) (1995) *Trends and Future Perspectives in Peptide and Protein Drug Delivery*, Harwood Academic Publishers, Chur.
6. G. Buckton (1995) *Interfacial Phenomena in Drug Delivery and Targeting*, Harwood Academic Publishers, Chur.
7. K.E. Drexler (1990) *Engines of Creation*, Fourth Estate, London.

PREFACE

There has been an explosion of interest recently in the topic of biodegradable polymers for medical applications. The reason is clear: these materials are versatile building blocks of devices that can achieve targeted delivery of drugs and/or tissue augmentation.

In its genesis, inventions in this field related to the production of suture materials. Fibers could be engineered with varying absorption rates to match the healing profiles of the tissues they helped to repair. The synthetic materials offered clear advantages over conventional gut and silk sutures in terms of versatility and low tissue reactivity. Synthetic bioabsorbable orthopedic fixation devices too offered clear advantages over their steel counterparts since they would ultimately be degradable where an indefinite material was not truly needed or might require removal. Further advances have centered around the development of degradable materials for vascular grafts, ligament reconstruction, adhesion prevention and organ regeneration. The development of endoscopic surgical techniques has widened the opportunity for placement of drug delivery devices at specific anatomic locations. A number of these devices are now at late stages in their development.

After reviewing the available books in the field we felt that a reference work that provides technical and practical information on biodegradable polymers would be useful. The purpose of this volume is to provide a source manual for synthetic procedures, properties and applications of bioerodible polymers. In pursuing this goal, we have included descriptions of widely available materials such as polylactides, collagen and gelatin as well as polymers of emerging importance such as the genetically engineered and elastin based polymers which are as yet either proprietary and/or in early stages of development. The book is organized into three sections and contains 23 chapters, one for each type of polymer. Section 1 deals with synthetic absorbable polymers, Section 2 contains seven chapters focusing on natural, semi-synthetic and biosynthetic polymers, and Section 3 provides information on the characterization of biopolymers and a chapter on non-medical polymers. In this way we hope that this book will have utility for many years to come.

We are honored to have received contributions from such a distinguished panel of investigators and experts in this field. We thank all our contributors for their time and expertise in preparing this work.

Abraham J. Domb
Joseph Kost
David M. Wiseman

CONTRIBUTORS

Mary Ann Accavitti
Department of Medicine
School of Medicine
The University of Alabama
at Birmingham
THT 466
Birmingham, AL 35294
USA

Rao S. Bezwada
Johnson & Johnson
Corporate Biomaterials Center
P.O. Box 151, Somerville, NJ 08876
USA

Joseph Cappello
Protein Polymer Technologies
10655 Sorrento Valley Road
San Diego, CA 92121
USA

Stewart A. Cederholm-Williams
Oxford Bioresearch Laboratory
The Magdalen Centre
Oxford Science Park
Oxford OX4 4GA
UK

Jun Chen
School of Pharmacy
Purdue University
West Lafayette, IN 47907
USA

Kevin Cooper
Johnson & Johnson
Corporate Biomaterials Center
P.O. Box 151, Somerville, NJ 08876
USA

Patrick Couvreur
Laboratoire de Physico-Chimie
URA CNRS 1218
5 rue Jean-Baptiste Clement
92296 Chatenay-Malabry Cedex
France

Martyn C. Davies
Laboratory of Biophysics and
Surface Analysis
Department of Pharmaceutical Sciences
University of Nottingham
University Park
Nottingham NG7 2RD
UK

S. Dejardin
Polymer Materials Research Group
Department of Organic Chemistry
University of Ghent
Krijgslaan 281 S4bis
9000 Ghent
Belgium

Yoshiharu Doi
Riken, The Institute of Physical and
Chemical Research
Head of Polymer Chemistry Laboratory
2-1 Hirosawa Wako-shi, Saitama 351-01
Japan

Abraham J. Domb
The Hebrew University of Jerusalem
School of Pharmacy-Faculty of Medicine
The Adolph Weinberger Building
Jerusalem 91120
Israel

Omar Elmalak
The Hebrew University of Jerusalem
School of Pharmacy-Faculty of Medicine
The Adolph Weinberger Building
Jerusalem 91120
Israel

James P. English
Consulting and Technology Assessment
Pharmaceutical Formulations and
Medical Devices
2440 County Road 39
Chelsea, AL 35043
USA

Elias Fattal
Laboratoire de Physico-Chimie
URA CNRS 1218
5 rue Jean-Baptiste Clement
92296 Chatenay-Malabry Cedex
France

Atul K. Garg
Integra LifeSciences, Co.
14 Currey Lane,
West Orange, NJ 07052
USA

Riki Goldbart
Ben Gurion Uiversity of the Negev
Department of Chemical Engineering
Center for Biomedical Engineering
and Program for Biotechnology
Beer Sheva 84105
Israel

Achim Göpferich
Department of Pharmaceutical
Technology
University of Erlangen-Nürnberg
Cauerstrasse 4
91058 Erlangen
Germany

D. Channe Gowda
Laboratory of Molecular Biophysics
School of Medicine
The University of Alabama
at Birmingham
1670 University Boulevard
VH 300
Birmingham, AL 35294-0019
USA

Cynthia M. Harris
Bioelastics Research, Ltd.
1075 South 13th Street
Birmingham, AL 35205
USA

Jorge Heller
Executive Director
APS Research Institute
3696 Haven Avenue
Redwood City, CA 94063
USA

Patrice P. Hildgen
Department of Chemical Engineering
Massachusetts Institute of Technology
Cambridge, MA 02139
USA

Dennis D. Jamiolkowski
Johnson & Johnson
Corporate Biomaterials Center
P.O. Box 151, Somerville, NJ 08876
USA

Naijie Jing
Laboratory of Molecular Biophysics
School of Medicine
The University of Alabama
at Birmingham
1670 University Boulevard
VH 300
Birmingham, AL 35294-0019
USA

Seongbong Jo
School of Pharmacy
Purdue University
West Lafayette, IN 47907
USA

Thomas R. Keenan
Kind and Knox Gelatin Inc.
P.O. Box 927
Sioux City, IA 51102
USA

John Kemnitzer
Department of Chemistry
Rutgers University
Taylor Road/Busch Campus
Piscataway, NJ 08855-0939
USA

Joachim Kohn
Department of Chemistry
Rutgers University
Taylor Road/Busch Campus
Piscataway, NJ 08855-0939
USA

Joseph Kost
Ben Gurion University of the Negev
Department of Chemical Engineering
Center for Biomedical Engineering
and Program for Biotechnology
Beer Sheva 84105
Israel

Robert Langer
Department of Chemical Engineering
Massachusetts Institute of Technology
Cambridge, MA 02139
USA

Y. Lemmouchi
Polymer Materials Research Group
Department of Organic Chemistry
University of Ghent
Krijgslaan 281 S4bis, 9000 Ghent
Belgium

Chi-Xiang Luan
Laboratory of Molecular Biophysics
School of Medicine
The University of Alabama
at Birmingham
1670 University Boulevard
VH 300, Birmingham, AL 35294-0019
USA

Michael G. Marks
Johnson and Johnson Medical Inc.
2500 Arbrook Boulevard
Arlington, TX 76014
USA

David M. Masters
Mayo Clinic and Foundation
Anesthesia Research
Rochester, MN 55905
USA

David T. McPherson
The UAB AIDS Center
The University of Alabama
at Birmingham
BBRB 346
Birmingham, AL 35294-2170
USA

Antonios G. Mikos
Cox Laboratory for Biomedical
Engineering
Institute of Biosciences and
Bioengineering
Department of Chemical Engineering
Rice University
P.O. Box 1892
Houston, TX 77251-1892
USA

Michael J. Miller
Department of Plastic and
Reconstructive Surgery
M.D. Anderson Cancer Center
The University of Texas
1515 Holcombe Boulevard
Box 62
Houston, TX 77030
USA

Yousef Najajrah
The Hebrew University of Jerusalem
School of Pharmacy-Faculty of Medicine
The Adolph Weinberger Building
Jerusalem 91120
Israel

Kinam Park
School of Pharmacy
Purdue University
West Lafayette, IN 47907
USA

Timothy M. Parker
Bioelastics Research, Ltd.
1075 South 13th Street
Birmingham, AL 35205
USA

Asima Pattanaik
Laboratory of Molecular Biophysics
School of Medicine
The University of Alabama
at Birmingham
1670 University Boulevard
VH 300
Birmingham, AL 35294-0019
USA

M. Teresa Peracchia
Laboratoire de Physico-Chimie
URA CNRS 1218
5 rue Jean-Baptiste Clement
92296 Chatenay-Malabry Cedex
France

Dana E. Perrin
Manager – Technical Services
Linvatec Corp.
11311 Concept Boulevard
Largo, FL 34643
USA

Susan J. Peter
Cox Laboratory for Biomedical
Engineering
Institute of Biosciences and
Bioengineering
Department of Chemical Engineering
Rice University
P.O. Box 1892
Houston, TX 77251-1892
USA

Lorraine E. Reeve
Alliance Pharmaceutical Corp.
3040 Science Park Road
San Diego, CA 92121
USA

Israel Ringel
The Hebrew University of Jerusalem
School of Pharmacy-Faculty of Medicine
The Adolph Weinberger Building
Jerusalem 91120
Israel

Clive J. Roberts
Laboratory of Biophysics and Surface
Analysis
Department of Pharmaceutical Sciences
University of Nottingham
University Park
Nottingham NG7 2RD
UK

Lowell Saferstein
Johnson and Johnson Medical Inc.
14 Currey Lane,
West Orange, NJ 07052
USA

Etienne Schacht
Polymer Materials Research Group
Department of Organic Chemistry
University of Ghent
Krijgslaan 281 S4bis
9000 Ghent
Belgium

Kevin M. Shakesheff
Laboratory of Biophysics and Surface
Analysis
Department of Pharmaceutical Sciences
University of Nottingham
University Park
Nottingham NG7 2RD
UK

Alex G. Shard
Laboratory of Biophysics and Surface
Analysis
Department of Pharmaceutical Sciences
University of Nottingham
University Park
Nottingham NG7 2RD
UK

Venkatram R. Shastri
Department of Chemical Engineering
Massachusetts Institute of Technology
Cambridge, MA 02139
USA

Frederick H. Silver
Division of Biomaterials
Department of Pathology
Robert Wood Johnson Medical School
University of Medicine and
Dentistry of New Jersey
675 Hoes Lane
Piscataway, NJ 08854-5635
USA

Reginald L. Stilwell
Johnson and Johnson Medical Inc.
2500 Arbrook Boulevard
Arlington, TX 76014
USA

Graham Swift
Rohm and Haas Company
727 Norristown Road
Spring House, PA 19477
USA

Zeev Ta-Shma
Israel Defense Forces (IDF)

Saul J.B. Tendler
Laboratory of Biophysics and
Surface Analysis
Department of Pharmaceutical Sciences
University of Nottingham
University Park
Nottingham NG7 2RD
UK

Doron Teomim
The Hebrew University of Jerusalem
School of Pharmacy-Faculty of Medicine
The Adolph Weinberger Building
Jerusalem 91120
Israel

Dan W. Urry
Laboratory of Molecular Biophysics
School of Medicine
The University of Alabama
at Birmingham
1670 University Boulevard
VH 300, Birmingham, AL 35294-0019
USA

J. Vandorpe
Polymer Materials Research Group
Department of Organic Chemistry
University of Ghent
Krijgslaan 281 S4bis
9000 Ghent
Belgium

John W. Weisel
Department of Cell and Developmental
Biology
School of Medicine
University of Pennsylvania
245 Anatomy-Chemistry Building
36th Street and Hamilton Walk
Philadelphia, PA 19104-6058
USA

David M. Wiseman
Synechion Inc.
6757 Arapaho,
Suite 711
Dallas, TX 75248
USA

Jie Xu
Laboratory of Molecular Biophysics
School of Medicine
The University of Alabama
at Birmingham
1670 University Boulevard
VH 300, Birmingham, AL 35294-0019
USA

Michael J. Yaszemski
Department of Orthopaedic Surgery
Wilford Hall Medical Center
Lackland Air Force Base, TX 78236
USA

SECTION 1:
SYNTHETIC ABSORBABLE POLYMERS

1. POLYGLYCOLIDE AND POLYLACTIDE

DANA E. PERRIN[1] and JAMES P. ENGLISH[2]

[1]*Linvatec Corporation, a division of Zimmer, a Bristol-Myers Squibb Company, 11311 Concept Boulevard, Largo, Florida 33773, USA*
[2]*Absorbable Polymer Technologies, 115B Hilltop Business Drive, Pelham, Alabama 35124, USA*

INTRODUCTION

The biodegradable polyesters, polyglycolide and polylactide, are often commonly referred to as poly(glycolic acid) (PGA) and poly(lactic acid) (PLA). However, high-molecular-weight polymers of glycolic and lactic acid are not possible to obtain by direct condensation of the related carboxylic acids because of the reversibility of the condensation reaction, backbiting reactions, and the high extent of reaction required. Therefore, polyglycolide and polylactide are typically made by ring-opening polymerization of their respective cyclic diester dimers, glycolide and lactide. Since the repeating unit of these polymers is actually glycolide or lactide, the nomenclature of these polymers should more correctly be polyglycolide (PG) and polylactide (PL) and copolymers should more correctly be named as poly(lactide-co-glycolide)s (PLG)s or poly(glycolide-co-lactide)s (PGL)s.

The monomers glycolide and lactide are prepared by first condensing glycolic or lactic acid into their respective low-molecular-weight condensation polymers. These low-molecular-weight polymers are then thermally cracked, preferentially forming the six-membered cyclic diesters. The crystalline cyclic diesters are highly purified by distillation, recrystallization, or both and then polymerized by ring-opening, addition polymerization to form the high-molecular-weight polymers.

PG is a crystalline, biodegradable polymer having a melting point (T_m) of ~225°C and a glass transition temperature (T_g) of ~35°C. The heat of fusion of 100% crystalline PG is 45.7 cal/gram. The repeating molecular structure of PG consists of one nonpolar methylene group and a single relatively polar ester group having the empirical formula of $(C_2H_2O_2)_n$. Relative to other biodegradable polymers, PG is a highly crystalline polymer, with crystallinity typically reported in the range of 35–75%. The molecular and subsequent crystalline structure of PG allow very tight chain packing and thus afford some very unique chemical, physical, and mechanical properties to the material. For example, the specific gravity is typically in the range of about 1.5–1.7 which is extremely high for a polymeric material. The polymer is very insoluble in most organic solvents with the only solvents of utility being hexafluoroisopropanol (HFIP) and hexafluoroacetone sesquihydrate (HFASH). In its highly-crystalline form, PG has a very high tensile strength (10,000–20,000 psi) and modulus of elasticity (~1,000,000 psi) (BPI 1995).

PG biodegrades by hydrolysis of the readily accessible and hydrolytically unstable aliphatic-ester linkages. The degradation time is just a few weeks depending on the molecular weight, degree of crystallinity, crystal morphology, physical geometry of the specimen, and the physico-chemical environment (Ginde and Gupta, 1987, Browning and Chu, 1986, and Katz and Turner, 1970). Without the benefit of its unique crystalline behavior, PG would degrade much more rapidly. Additional information regarding the *in-vitro* and *in-vivo* degradation of PG, PL, and PGL is given later in this chapter.

Although structurally very similar to PG, the polylactides (PL)s are quite different in chemical, physical and mechanical properties because of the presence of a pendant methyl group on the alpha carbon. This structure causes chirality at the alpha carbon of PL; and thus, L, D, and DL isomers are possible. L-PL is made from L(–)-lactide and D-PL is made from D(+)-lactide while DL-PL is made from DL-lactide which is a racemic mixture of the L(–) and D(+) isomers and the meso form having both the D(+) and L(–) configuration on the same dimer molecule.

To date, L-PL, DL-PL, and their copolymers have received the most attention and achieved the greatest commercial success; therefore, the remainder of this discussion will be limited to L-PL, DL-PL, and their copolymers.

L-PL is a crystalline, biodegradable polymer having a melting point (T_m) of approximately 175°C and a glass transition temperature (T_g) of approximately 65°C (Schindler *et al.*, 1977). Fisher, *et al.* (1973) has reported a calculated value for the heat of fusion of 100% crystalline PL as 93.7 J/gram, while the heat of fusion for commercial products has been reported as low as 30 J/gram (PURAC 1996). The empirical formula of the polylactides is $(C_3H_4O_2)_n$.

L-PL is generally less crystalline than PG, with crystallinity reported in the range of 35% (Gogolewski *et al.*, 1983). The specific gravity is approximately 1.2–1.3. The polymer is very soluble in common organic solvents such as chloroform. In its more crystalline form, L-PL has a very high tensile strength slightly lower, although similar to PG (~10,000–15,000 psi) but a much lower modulus of elasticity (~500,000 psi) (BPI 1995).

On the other hand, DL-PL is a completely amorphous polymer having a T_g of ~57°C (Hollinger and Battistone, 1986, Schindler *et al.*, 1977). The specific gravity is ~1.2–1.3, similar to L-PL. Because of its lack of crystallinity, DL-PL has a much lower tensile strength (~5,000 psi) and modulus of elasticity (~250,000 psi) (BPI 1995).

The methyl group in PL causes the carbonyl of the ester linkage to be sterically less accessible to hydrolytic attack; and depending on the type of PL, its molecular weight, degree of crystallinity, the physical geometry of the specimen, and the physico-chemical environment the PLs are typically more hydrolytically stable than PG. However, the lack of crystallinity in DL-PL causes this polymer to degrade faster than L-PL.

As stated above, high-molecular-weight polymers and copolymers of glycolide and L- and DL-lactides are prepared by ring-opening addition polymerization of their respective cyclic dimers. Copolymers having a wide range of physical and mechanical properties with varying rates of biodegradation can be prepared with glycolide and lactide and a variety of lactones, other lactides, cyclic carbonates, and lactams (Dijkstra *et al.*, 1991; Feng *et al.*, 1983; Kricheldorf *et al.*, 1985; Zhu *et al.*, 1986).

In making copolymers, attention to monomer reactivity differences is very important and depending on the specific conditions for copolymerization, wide differences in copolymer microstructure (monomer sequencing) are possible (Dunn *et al.*, 1988).

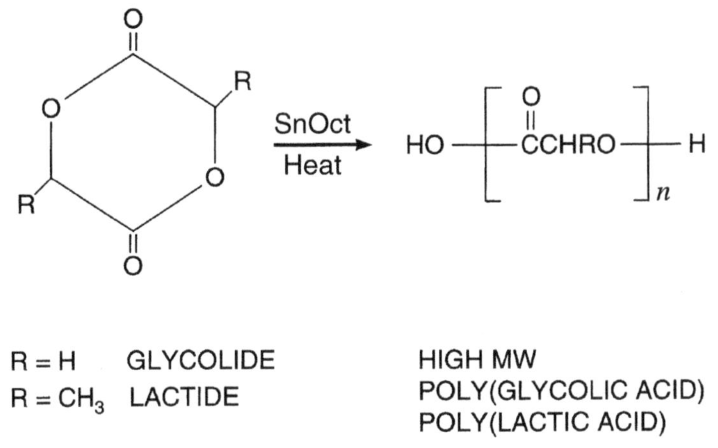

R = H   GLYCOLIDE

R = CH$_3$ LACTIDE

HIGH MW
POLY(GLYCOLIC ACID)
POLY(LACTIC ACID)

Figure 1 Polymerization of Polyglycolide, Polylactide, and Copolymers.

As with PG and PL homopolymers, the copolymers of lactide and glycolide are also subject to biodegradation because of the susceptibility of the aliphatic ester linkage to hydrolysis. However, biodegradation of the copolymers is normally faster than the homopolymers because copolymerization reduces the overall crystallinity of the polymer, thus giving the polymer a more open macrostructure for easier moisture penetration.

POLYMERIZATION OF POLYMERS AND COPOLYMERS OF POLYGLYCOLIDE AND POLYLACTIDE

The synthesis of PG, PL and PGL is depicted in Figure 1.

Polyglycolide

In the polymerization of glycolide, temperatures in the range of 140–235°C are typical, although lower temperatures can be used. When polymerization temperatures are less than the melting point of the polymer (~225°C), crystallization of the polymerizing polymer occurs resulting in what is known as solid-state polymerization. Solid-state polymerization can be a useful tool in forming very high-molecular-weight polymers.

The PG polymerization reaction is normally catalyzed by stannous 2-ethyl hexanoate (stannous octoate) or stannic chloride dihydrate because of their very low toxicity. Other catalysts which can be used include various Lewis acids, organometallic compounds and organic acids.

Poly(L-lactide)

In the polymerization of L-lactide, temperatures in the range of 105–185°C are typical. When polymerization temperatures are less than the melting point of the polymer (~175°C), crystallization of the polymerizing polymer occurs, resulting in solid-state polymerization as with PG. Solid-state polymerization has been useful in forming very high-molecular-weight polymers with molecular weights of ~1,000,000 Daltons (Bergsma *et al.*, 1993).

As with PG, the L-PL polymerization reaction is normally catalyzed by stannous octoate or stannic chloride dihydrate, however other catalysts can also be used.

Poly(DL-lactide)

In the polymerization of DL-lactide, temperatures in the range of 135–155°C are typical. Since DL-PL is an amorphous polymer, solid-state polymerization does not occur as with PG and L-PL. The DL-PL polymerization reaction is also normally catalyzed by stannous octoate or stannic chloride dihydrate.

Molecular weights of PG, PL, and PGL copolymers are controlled by the addition of chain-control agents. These chain-control agents are usually water or primary alcohols; however, amines or other active hydrogen compounds can be used. Linear polymers are typically prepared using alcohols having functionalities of less than or equal to two, such as 1-dodecanol, or 1,6-hexanediol. Branched polymers can also be prepared by using tri- or poly- functional chain-control agents (Shalaby and Jamiolkowski, 1985; Cowsar *et al.*, 1985).

The polymerization details are given below for small scale laboratory polymerizations of PG, L-PL, and DL-PL homopolymers and a pilot-scale polymerization of 75/25 DL-PLG copolymer.

Acquisition, Characterization, and Qualification of Glycolide and Lactide Monomers

High-purity glycolide and lactide monomers are manufactured by Purac and Boehringer Ingelheim and may be obtained from either PURAC America of Lincolnshire, Illinois or BI Chemicals of Montvale, New Jersey. Upon receipt, the monomer is sampled for identity and purity testing. During sampling, the monomer is blanketed with dry, high-purity nitrogen to avoid moisture contamination. The monomer is then stored inside a freezer at <0°C under a dry, high-purity nitrogen atmosphere in the original container until qualified for use.

The sample is first characterized by infrared spectroscopy (IR) and compared to a standard IR spectrum for identification. Monomer purity may be determined in a variety of ways such as melting point, DSC, etc., but is best determined by preparation of a small test sample of the homopolymer. A 10 gram test polymer is made by adding an appropriate quantity (0.03–0.10 wt %) of stannous octoate, available from Sigma Chemical Company of St. Louis Missouri, as a solution in dry toluene, with no chain control agent, to 10 grams of the monomer in a 20 mL glass vial under a dry, high-purity nitrogen atmosphere. The vial is sealed and submerged in an oil bath at 135–155°C for ~18–24 hours. After partial melting, the monomer/catalyst mixture is agitated well initially and every 5 minutes thereafter until the melt becomes too viscous to agitate. After ~18–24 hours the test polymer is removed from the oil bath, cooled to room temperature and a sample of the polymer is removed from the vial for determination of its molecular weight. The relative molecular weight for PG, L-PL, and DL-PL is determined by dilute solution viscosity. L-PL and DL-PL are also analyzed by gel permeation chromatography (GPC) as described below. GPC is not routinely done in most laboratories on PG because it requires the use of HFIP as the solvent and the mobile phase, which is both expensive and hazardous.

PG prepared in this manner should have an inherent viscosity of 2.0 dL/g or greater in HFIP to be assured of high-purity monomer. L-PL and DL-PL should have an inherent viscosity greater than 2.0 dL/g in chloroform and a weight-average molecular weight (M_W) by GPC of greater than 250,000 Daltons. If glycolide, L-lactide, or DL-lactide monomer purity is shown to be insufficient by these tests, they may be improved by recrystallization of the monomer.

Small-scale Polymerization of Glycolide, L-lactide, and DL-lactide

Small-scale polymerization of glycolide, L-lactide, or DL-lactide is conveniently carried out in a 1 Liter, stainless steel resin kettle equipped with a glass top, a stainless steel mechanical stirrer, a thermometer, and a gas inlet tube. The resin kettle, mechanical stirrer, and all glassware is predried overnight in an oven at 150°C and cooled inside a glove box in a dry, high-purity nitrogen atmosphere. Both the loading of the monomer and assembly of the apparatus are also conducted inside the glove box in a dry, high-purity nitrogen atmosphere. Qualified, high-purity glycolide, L-lactide, or DL-lactide monomer (300 grams) and an appropriate amount of chain control agent for the desired molecular weight is charged to the flask. The resin kettle is then closed, removed from the glove box, and immediately connected to a low flow source of dry, high-purity nitrogen as a continuous purge. The flask is then submerged in an oil bath at 133–155°C and slow stirring is begun. After the monomer/chain control agent mixture has reached the reaction temperature, stirring is discontinued. At this time, an appropriate amount of stannous octoate catalyst is added as a solution in dry toluene. Mechanical stirring is immediately resumed and maintained until the polymerizate becomes viscous or in the case of PG and L-PL begins to show signs of crystallization. Stirring is then discontinued, the stirrer and thermometer are raised from the melt, and heating is continued for a total of ~18–24 hours after which the resin kettle assembly is removed from the oil bath and cooled to room temperature under a continual nitrogen purge. When the polymer has cooled, the resin kettle top, stirrer, and thermometer are removed from the resin kettle. The polymer is removed by supercooling the resin kettle. When thoroughly cooled the polymer mass will separate readily from the resin kettle. While still cold, the polymer is removed and broken into several smaller pieces. The polymer pieces are allowed to warm to room temperature overnight inside a vacuum oven to avoid condensation of moisture from the air. At this point L-PL and DL-PL can be purified by methanol precipitation from a dichloromethane solution as described by Eling *et al.* (1982).

After precipitation, the purified polymer is then redried inside a vacuum oven under high vacuum at room temperature (~24 hours), cooled with liquid nitrogen, and ground on a Wiley Mill. The ground polymer is finally vacuum dried an additional 24 hours at room temperature. A 5 gram sample of the polymer is removed for characterization, and the remainder is stored inside a polyethylene bag evacuated and backfilled three times with dry, high-purity nitrogen prior to sealing. The polyethylene bag is then placed inside an outer polylined foil bag containing a bag of DRIERITE® desiccant, sealed, and then stored in a freezer at <0°C. The yield is typically 75–95%. The molecular weight and thus the inherent viscosity of a typical polymer prepared by this process vary depending on the quantity of catalyst, the type and quantity of chain control agent, and the specific reaction conditions chosen.

Pilot-scale Polymerization of 75/25 DL-PLG

A very useful reactor assembly for pilot-scale polymerization of lactide and glycolide polymers and copolymers is an 8-CV mixer from Design Integrated Technologies (DIT) of Warrington, Virginia. The 8-CV is a dual cone mixer (8 qt. capacity) equipped with double helical mixer blades, a gas inlet line, and a large vacuum port.

The reactor is first cleaned and dried thoroughly. Appropriate amounts of qualified, high-purity DL-lactide and glycolide monomers to yield a 75:25 copolymer ratio are then charged to the reactor. The reactor is heated under positive nitrogen pressure. When the monomers have partially melted, stirring is begun and continued until the monomer mixture reaches 135–155°C. Stirring is then discontinued and the reactor is depressurized. At this time, an appropriate amount of the desired chain-control agent is added, the reactor is repressurized and stirring is resumed. Next, a sufficient quantity (0.03–0.10 wt %) of stannous octoate catalyst is added, the reactor is repressurized, and stirring is resumed and continued until the polymerizate becomes very viscous. Stirring is then discontinued and heating of the polymer melt is continued under pressure for a total elapsed heating time of 18–24 hours.

After the polymerization is complete, a vacuum is slowly applied to the polymer melt with continual slow stirring to remove residual monomer. The reactor is then pressured with dry, high-purity nitrogen, and the discharge valve is opened slowly until a steady stream of polymer flows from the reactor. The polymer extrudate is quenched while continuously pelletizing the cooled polymer strand. The pelletized polymer is then dried under high vacuum at room temperature for 24 hours in a vacuum dryer, and packaged inside a polyethylene bag under dry, high-purity nitrogen. This bag is then placed inside an outer polylined foil bag containing desiccant, sealed, and finally stored in a freezer at <0°C. The final yield is typically 75–95%. The molecular weight and thus the inherent viscosity of a typical polymer prepared by this process vary depending on the quantity of catalyst, the type and quantity of chain control agent, and the specific reaction conditions chosen.

POLYMER CHARACTERIZATION

Molecular Weight and Molecular-Weight Distribution

PG, PL, and PLG copolymers are characterized by first determining their inherent viscosity (η_{inh}) and/or intrinsic viscosity, $[\eta]$, in a suitable solvent (benzene, toluene, chloroform or HFIP). A method is given in ASTM D-2857. Intrinsic viscosity is also a convenient means of determining the molecular weight of the homopolymers using the Mark-Houwink relationship between intrinsic viscosity $[\eta]$ and molecular weight.

$$[\eta] = K \times M_w^a$$

Relative molecular weight and molecular weight distribution (MWD) or polydispersity of PL and PLG copolymers is determined by GPC. GPC may be carried out in a variety of solvents. In a typical method for PL and PLG both the dissolution solvent and mobile phase solvent of choice is chloroform. A suitable flow rate for the mobile phase is 1 mL/min. Three 5 micron particle size, microstyrogel columns (1,000 Å, 10,000 Å, and 100,000 Å) arranged in series are convenient sizes to use. The polymer is dissolved in chloroform at a concentration of 0.5 g/dL and a 100 μL injection is made into the mobile phase. Data is collected using a refractive index (RI) detector. Comparison of the chromatographic data is made against chromatographic data obtained from a series of known M_w polystyrene standards in the same solvent. A similar method can be devised for PG using HFIP as the solvent and mobile phase.

The MWD or polydispersity of PL, and PLG polymers prepared as described above and determined in this manner is normally about 1.5–2.0.

Determination of the Final Comonomer Ratio in the Polymer

As stated, glycolide and lactide may be copolymerized or copolymerized with a variety of other monomers. The final ratio of monomers incorporated into a copolymer of lactide and glycolide is determined by proton nuclear magnetic resonance spectroscopy (^{1}H NMR) in deuterated chloroform. The ratio of the methylene protons adjacent to the oxygen in glycolide is compared against that of the methyne protons in the lactide. The ratio is expressed as lactide to glycolide.

Numerous other papers and patents have been published regarding the polymerization and physical properties of PG, PL, and PLG copolymers (Eenink, 1987; Dunn *et al.*, 1988; BIKG, 1994; PURAC, 1996).

PROCESSING

PL polymers with inherent viscosities less than about 4.0 g/dL can be injection molded using conventional reciprocating screw thermoplastic molding machines provided the polymers are dried thoroughly prior to molding (von Oepen. 1995). Even when residual moisture is minimized, a substantial reduction in the molecular weight is observed (von Oepen & Michaeli, 1992). Ellis and Tipton (1994) have shown that the tensile strength and molecular weight of PL test bars can be affected by processing temperature and injection speed. Thermal degradation of PL and PG polymers would not be unexpected since their respective cyclic dimers are made from thermal cracking of the low molecular weight polymers.

Typical injection molding conditions for PL are shown below (BIKG, 1995):

Feed Zone Temperature:	45°C
Barrel Zone 1:	200°C
Barrel Zone 2:	215°C
Barrel Zone 3:	220°C
Nozzle:	200°C
Injection Pressure:	1.5 bar
Mold Temperature:	ambient

Using similar techniques and lower processing temperatures, it is possible to mold DL PL, and PLG copolymers.

PG and PL have been successfully spun into fibers by various authors using both melt and solvent spinning techniques (Ginde & Gupta, 1987; Eling *et al.*, 1982; Gogolewski & Pennings, 1983; Frazza & Schmitt, 1971; Horacek & Kalisek, 1994; Fambri *et al.*, 1994). Eling, *et al* (1982) reported spinning fibers from a 6% solution of PL with a viscosity average molecular weight of 5.3×10^5 in toluene at $110°C$ through a conical capillary die with an opening of 1 mm. Wind up speeds were 25–35 cm/minute. The fibers were then drawn through a $196.5°C$ oven having a length of 500 mm at take up speeds of 25–35 mm/minute to yield fibers with draw ratios between 4 and 17, having tensile strengths ranging from 0.28 and 0.80 GPa, and elongations of 16 to 18 percent.

The research groups of Törmälä (Tampere University, Finland), Tunc (Johnson and Johnson, New Jersey, USA) and Ikada (Kyoto University, Japan) have developed methods for producing what are termed "self-reinforced" PL and PG devices. The process published by Tunc and Jadhav for PL devices has been identified as an "orientrusion" process. These technologies involve the molding and orientation of a "preformed" shape. The oriented preform is then machined or further processed during the manufacture of the device. This orientation results in a finished device with significantly improved tensile and flexural strength (Lautiainen *et al.*, 1994; Vainionpaa *et al.*, 1987; Tunc and Jadhav, 1988; Shimamoto *et al.*, 1995; Ikada *et al.*, 1990).

DEGRADATION

From a review of the literature, which includes both animal and human clinical study data, it is possible to construct a composite model of how absorbable devices degrade in living tissues. One version of such a model is presented below which we believe accounts for the majority of the published observations with PL devices. With the exception of degradation rate, this same model can also be generally applicable to devices made from PG and PLG copolymers. For additional information specific to PG and PLG polymers, the reader is referred to Chu (1985) and Katz & Turner (1970). Published observations of inflammation associated with devices made from PG and PLG copolymers may or may not be pertinent to the model described below for PL. Observations of inflammation will be discussed in the section on HUMAN CLINICAL STUDIES.

From a chemical standpoint, absorbable devices are thought to undergo five general stages of degradation (Kronenthal, 1975). These stages are not discrete and may overlap. First, hydration of the implant begins when the device is placed in the body. During this stage, the device absorbs water from the surrounding environment. Depending upon the mass and surface area of the implant, this diffusion process occurs over the course of days or months. Hydration of the amorphous segments of the polymer occurs faster than with the crystalline segments. Because of their hydrophilic nature, PG, PL, and their copolymers will absorb water. In a three dimensional polymer matrix, water penetrates deeply into the interior areas eventually resulting in a fragmentation degradation mechanism as opposed to the surface erosion mechanism observed with more hydrophobic polymer matrices.

Immediately after soft tissue implantation, a histological examination of the implant site shows a cellular response typical of acute trauma (hematoma), followed by the formation of a fibrous capsule around the implant (Chawla *et al.*, 1985). In osseous tissues, new bone formation begins to form periosteally and endosteally. Granulation tissue can be observed between the bone and the implant. A few giant cells are present, but no other inflammatory cells are observed (Majola *et al.*, 1991; Matsusue *et al.*, 1991).

The second stage of degradation is depolymerization or chemical cleavage of the polymer backbone which results in a reduction in mechanical properties (strength). In this process, water reacts with the polymer in a hydrolytic fashion resulting in cleavage of covalent chemical bonds with a commensurate reduction in average molecular weight and physical strength. The kinetics of the loss of strength depend upon a great many factors such as implant size, surface area, polymer type, polymer purity, polymer crystallinity, surrounding pH, sterilization method, and initial molecular weight (Matsusue *et al.*, 1991; Cutright *et al.*, 1974; Miller *et al.*, 1977; Rozema *et al.*, 1991; Nakamura *et al.*, 1989). The degradation of amorphous zones of the polymer matrix occurs first, followed by the degradation of the more crystalline zones. It has been demonstrated that bacterial contamination does not influence the rate of PL degradation *in-vitro* (Hoffman *et al.*, 1990).

In an *in-vitro* experiment performed at Linvatec, molded PL devices were soaked in buffered saline at 37 degrees Centigrade for a number of months and the loss in tensile strength as a function of time was measured. Figure 2 shows the change in tensile strength (as a percentage of the initial value) as a function of total elapsed degradation time. Initially, the tensile strength starts out at 100% and remains relatively high (above 75%) for approximately 20 weeks. Published *in-vivo* reports of PG (DEXON® Suture) degradation show a much faster degradation rate. After 1 week, approximately 80% of the original strength is retained. The tensile strength falls to about 10% after 3 weeks (Chu 1983). Figure 3 shows *the in-vitro* degradation curve for PG fibers (Ginde and Gupta 1987). The crystallinity and annealing conditions used during the production of the PG suture can affect the degradation rate (Hollinger & Battistone, 1986; Browning & Chu, 1986). For PGL copolymer (VICRYL® Suture), approximately 90% of the original strength is retained after 1 week, 55% is retained at 2 weeks, with essentially no strength remaining after 4–5 weeks (Chu, 1983).

Also during this degradation phase, the mass of the device remains essentially unchanged. As this depolymerization process occurs, histologically in soft tissue, the fibrous tissue capsule surrounding the implant begins to become thinner with a few foreign body cells observed at the tissue-implant interface. The capsule appears to consist mainly of collagen fibers with a few sporadic fibrocytes and mononuclear macrophages (Gogolewski *et al.*, 1993). In osseous tissue, the amount of granulation tissue around the implant decreases and the thickness of the new bone formation around the implant increases. No inflammatory reaction is observed. Cracks can be seen on the surface of the implant (Majola *et al.*, 1991). A transitory and slight infiltration of lymphocytes in the vicinity of the marrow cavity can be observed. More new bone forms around PL implants than around stainless steel controls (Matsusue *et al.*, 1991) It is during this depolymerization stage that osseous healing occurs.Consolidation and complete union of osteotomies of rat femora and rabbit tibiae are observed during this stage (Majola *et al.*, 1991; Matsusue *et al.*, 1991).

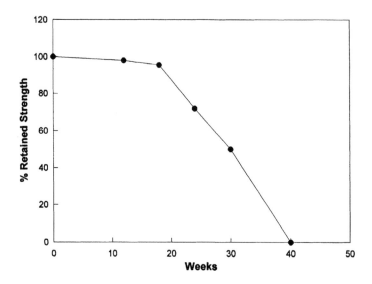

Figure 2 *In vitro* Degradation of Poly(L-lactide): Retained Tensile Strength vs Time.

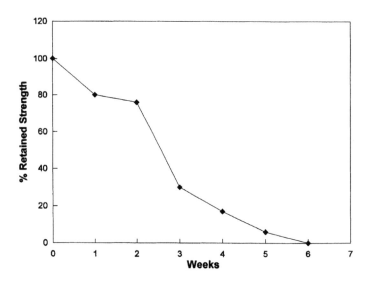

Figure 3 *In vitro* Degradation of Polyglycolide: Retained Tensile Strength vs Time. Adapted from Ginde and Gupta (1987).

The role of enzyme-mediated hydrolysis (depolymerization) of PL has been investigated (Schakenraad *et al.*, 1990). This study concluded that the major route of degradation of PL is most likely via simple (non-enzymatic) hydrolysis. However, the results of the study showed that the possibility of some enzyme-mediated hydrolysis could not be ruled out as a minor pathway. The conclusion that the process is not enzyme mediated is supported by another study which showed that the degradation rates measured *in-vivo* (sheep, dogs, and rats) were essentially the same as measured *in-vitro* (Leenslag *et al.*, 1987).

The third stage in the degradation process is loss of mass integrity which occurs when the implant has essentially no cohesive strength and begins to fragment into pieces of low molecular weight PL. At this stage histologically, in soft tissue, a collagen capsule around the implant can be observed and depending upon the extent of degradation, a mild inflammatory response is also observed. In osseous tissue, the thickness of the bony layer around the implant continues to increase without an observable foreign body response. The size of the implant is noticeably reduced. The amount of granulation tissue appears to be reduced (Majola *et al.*, 1991). In some instances, phagocytic cells and giant cells with villous projections are observed (Cutright & Hunsuck, 1972).

The fourth stage of degradation is absorption which occurs when further hydrolysis causes the fragment size to be suitable for assimilation by phagocytes or when further hydrolysis simply leads to soluble monomeric (lactate or glycolate) anions which dissolve into the intercellular fluid. One report suggests that the presence of hydrolyzing lactide-containing implants can modify and enhance the tissue regeneration process via modulation of lactate dehydrogenase present in macrophages (Salthouse & Matagla, 1984). It is during this stage that the implant undergoes an observed loss in mass which corresponds to the physical absorption process. Several published studies with PL devices indicate that the absorption of PL devices may require from 20 months to 5 years (Pihlajmaki *et al.*, 1992; Bergsma *et al.*, 1995). A long term sheep study sponsored by Linvatec revealed almost complete absorption of PL screws placed intraosseously after approximately 3 years.

Unlike PL, extracellular enzymes are thought to have a role in the *in-vivo* degradation of PG outside the cell. This suggests that PG degradation at the cellular level may occur via two possible pathways. First, extracellular degradation occurs via a combination of simple hydrolysis and enzyme-mediated hydrolysis. The fragments thus generated would then be brought into the cell via phagocytosis. Once inside the cell, the fragments would then degrade intracellularly via simple hydrolysis. Glycolate is metabolized along a slightly different route of the Krebs cycle than lactate. The absorption time for PG is 6–17 weeks (Hollinger and Battistone, 1986).

The final stage of degradation is elimination. During this stage, L-lactate is converted into carbon dioxide and pyruvate. Pyruvate then enters Krebs cycle via the acetylation of Coenzyme A. During this final stage, the majority of the elimination of PL degradation products occurs through respiration (carbon dioxide) with only minor elimination via urine and/or feces (Brady *et al.*, 1973). Final hydrolysis of PG results in glycolate, some of which is excreted directly in the urine. Some glycolate may also be oxidized to glyoxylate which is then converted to glycine, serine, and pyruvate. Pyruvate then can enter the Krebs cycle as before with lactate (Hollinger and Battistone, 1986).

A similar model is proposed by Grizzi and others (1995) to describe the hydrolysis of Poly(DL-Lactide). Recently, Cordewener *et al.* (1995) have begun to investigate copolymers of 96% L-co-4% D Lactic Acid. The presence of the 4% D Lactic Acid segment in the polymer yields a lower crystallinity and result in a faster degradation rate.

BIOCOMPATIBILITY

Biocompatibility of PL

Numerous animal studies have examined the biocompatibility of PL in soft tissue (Kulkarni *et al.*, 1966; Chawla *et al.*, 1985; Gogolewski *et al.*, 1993; Salthouse & Matagla 1984; Brady *et al.*, 1973; Kulkarni *et al.*, 1971; Pistner *et al.*, 1993a, 1993b, 1994). Generally, good biocompatibility is observed with normal histological changes occurring during the healing and degradation processes.

Several notable exceptions to this generalization were published recently.

In a 1993 study, particles of PTFE (control) and predegraded particles of PL having diameters of less than 38 microns were injected intraperitoneally into mice. After up to 7 days, cells were harvested from the abdominal cavity. Microscopic examination of cell morphology revealed evidence of cell damage and death caused by phagocytosed PL particles. Cell death was not observed with the phagocytosed PTFE particles (Lam *et al.*, 1993). In a similar study with PL and PTFE films, Lam and coworkers (1995) found a more pronounced inflammatory response with PL films than with PTFE films in subcutaneous tissues of rats. The response was more pronounced if the films were porous and wettable.

In a 1991 study, PL plates were implanted subcutaneously in 35 rats. At 104 weeks, no adverse reactions were noted. At 143 weeks; however, one animal exhibited a histologically indisputable late foreign body reaction to the PL implant (Bos *et al.*, 1991).

Bergsma *et al.* (1995d) reported the use of a copolymer made of 96% L and 4% D Lactic Acid in subcutaneous tissues of rats for up to 52 weeks. Prior to implantation, discs made from the copolymer and homopolymer PL were predegraded *in-vitro* (Bergsma *et al.*, 1995e). While the degradation of the copolymer was faster than the homopolymer, inflammation was observed with both.

As in the above soft tissue studies, osseous studies also yielded inconsistent observations. In the 1970s, three studies appeared wherein PL devices were used to repair mandibular defects and orbital blowout defects in monkeys and mandibular defects in dogs. Healing was observed without inflammation at 38 weeks. Follow-up of the dog study at 40 weeks revealed that the fracture sites were histologically similar to contiguous bone (Cutright & Hunsuck 1972; Cutright *et al.*, 1971; Getter *et al.*, 1972). In a 1974 article which contained a review of the past studies, the authors concluded that PL exhibited good tissue receptivity, slow resorption, attended by varying numbers of phagocytic cells, and without inflammatory reaction after the wound had healed (Cutright *et al.*, 1974).

PL and stainless steel screws were compared in a canine calcaneus osteotomy model over 12 weeks. Radiographically, all sites healed without any note of inflammation. Also, more bone per unit area was observed with the PL screws than with the stainless steel screws (Tunc *et al.*, 1986).

A similar study evaluated the use of PL screws versus stainless steel screws in a rabbit tibia osteotomy model. Healing of the osteotomy was observed within 4 to 8 weeks. Histomorphometric analysis of the osteotomy sites at 8 weeks revealed more bone present between the PL screw threads than for the stainless steel screw threads. Histologically no inflammation was observed with either screw type during the course of the 16 week study (Matsusue *et al.*, 1991).

The use of fixation screws made from PL in the repair of rabbit proximal femoral osteotomies has been reported. Histologically, at the maximum 48 week follow-up time, healing was observed without inflammation. The implants appeared to be completely surrounded by a layer of new bone. This report also references three studies performed by other investigators wherein the incidence of localized osteitis following dental extractions was reduced by 12.7% through the use of a PL surgical dressing (Majola *et al.*, 1991).

PL screws (4.5 mm diameter) have been implanted in canine femora and compared to stainless steel screws and "empty hole" bone defects. After 3 years, histologic evaluations showed that PL screws exhibited good biocompatibility. A "gel-like" PL residue was observed in the defect area, with bone repair and reorganization in progress, however, a filling-in of the bone defect was not observed (Chen *et al.*, 1992).

Porous PL cylinders have been used as substrates upon which rabbit periosteal grafts are implanted as a means to effect repair of articular defects. Two articular defects were made in both the left and right medial femoral condyles in 18 rabbits. In the right leg, the defect was filled with a PL cylinder only, while in the left leg, the defect was filled with a periosteal graft supported on a PL cylinder. After 12 weeks, it appeared that the PL had allowed for *de novo* growth of neocartilage in all grafted specimens. Also, small amounts of PL were observed underneath the neocartilage with the majority of the PL being replaced by bone. The overall graft survival rate was 89% (von Schroeder *et al.*, 1991).

There are three additional published animal studies with dogs and sheep wherein PL devices were evaluated in fracture fixation applications without the observation of any adverse effects (Raiha *et al.*, 1990; Manninen *et al.*, 1991; Suuronen 1991).

In contrast to the above studies where adverse effects were not observed, we are aware of only two published animal studies where adverse effects associated with the use of PL devices were observed in osseous applications. In the first study, the diaphyseal femora of a single goat were implanted with six different 5 mm × 7 mm plugs (two PL, two hydroxylapatite/PL composites, one PL implant plasma coated with hydroxylapatite, and one 316 stainless steel). After two years, the deep inguinal lymph nodes were harvested. Visual examination revealed that the nodes were swollen. Light microscopy revealed no evidence of chronic inflammation. Histology revealed the abundant presence of phagocytes with a morphology consistent with active sinus histiocytosis. Because this observation was not made with goats implanted with only hydroxylapatite or other non-PL-containing implants, the authors inferred that the histiocytosis observed in this one goat was related to the degradation of the PL-containing implants (Verheyen *et al.*, 1993).

The only other published animal study wherein adverse reactions were observed is in a canine study. In this study, PL and stainless steel coupons were implanted within an ultra high molecular weight polyethylene chamber in the right femora of eight dogs. While the intramedullary area around the PL coupons showed more bone growth than the stainless steel coupons at 3, and 6 weeks, the amount of bone present around both coupons at 12 weeks was markedly reduced. At 12 weeks, an inflammatory reaction was observed histologically in the area near the PL coupons, but not in the area of the stainless steel coupons (Suganuma & Alexander, 1993). The observation of reduced bone formation for rigid stainless steel coupons at 12 weeks was not unexpected since a previous paper reported similar findings with titanium coupons in the identical canine model. The authors attributed this resorption to stress shielding resulting from the use of the coupon chamber (Spivak *et al.*, 1990).

When considering the long term suitability of implantable materials in general, one must always consider the potential for material-induced tumor formation. Three reports of sarcoma formation in rats having soft tissue PL implants have been published (Gutwald *et al.*, 1992, Pistner *et al.*, 1993, Nakamura*et al.*, 1994). The authors indicated that the sarcoma formation was probably not material related due to the fact that many non reactive implantable materials are observed to induce tumors in rats, a phenomenon referred to as solid-state tumorigenesis or the Oppenheimer effect.

Biocompatibility of PG and PGL

Numerous reports regarding the biocompatibility of PG and PGL have appeared in the literature since the introduction of Vicryl suture in the 1970s. The reader is referred to four representative papers which give details of the biocompatibility of PG and PGL in rabbits, dogs and rats (Winet *et al.*, 1995; Riddick *et al.*, 1977; Case *et al.*, 1976; Craig *et al.*, 1975; Lautiainen *et al.*, 1994).

HUMAN CLINICAL STUDIES

Because the clinical results vary according the type of polymer used to make the device, we will summarize the results in a separate fashion.

PG and PGL Copolymers

The successful clinical use of VICRYL (PG 90%-co-PL 10%) and DEXON (PG homopolymer) suture has proven that PG containing polymers can be used safely in soft tissue applications. Recently, severe late-stage foreign body reactions have been associated with the use of PG rod implants used for fracture fixation (Bostman *et al.*, 1990). Additional reports of late-stage severe foreign body reactions in approximately 5–8% of patients with PG rod implants are summarized in two excellent reviews (Bostman *et al.*, 1991, Litsky, 1993). These authors infer that since PG and PL are in the same general polyester chemical category, that all PL devices can be expected to elicit the same severe foreign body reactions that are well known for PG devices. It is important to note that while PL and PG are chemically similar, they differ in two ways. First, it is well known that PG degrades (*in-vivo* and *in-vitro*) at a faster rate than PL (Kronenthal, 1975). The fast degradation rate of PG may challenge the body's ability to remove and assimilate the degradation products from the implant site.

Secondly, glycolic acid is metabolized by the cell along a metabolic pathway which is slightly different from the metabolic pathway for lactic acid (Hollinger *et al.*, 1986, Bostman 1991). Thus it would not be unreasonable to suspect that the body would assimilate PG differently than PL.

Svensson and coworkers (1944) reported the use of PG rod fracture fixation devices in 50 pediatric patients having transphyseal or osteochondral fractures. All of the fractures healed without incident during the one year study. In two cases followed for beyond one year, nonunions of intraarticular fractures were observed. The authors attributed the nonunions to the probable occurrence of late stage foreign body reactions.

Edwards and coworkers (1994) reported a high incidence of intracapsular synovitis in patients having shoulder (Bankart) repair procedures wherein the SURETAC® device was used. This device was made from a copolymer of glycolide and trimethylene carbonate. In a similar study, Paganini and others reported no device-related complications with 22 SURETAC patients followed for two years (Paganini *et al.*, 1995).

Toljan and Orthner (1995) published details of the clinical use of an interference screw (for reconstruction of the Anterior Cruciate Ligament) made from a PGL copolymer. After 6 months, MRI data show the presence of local reactions at the implant site (edema and defined cysts) and a "liquidization" of the screw in 41% of the patients. Clinically, no febrile episodes or sterile effusions were observed.

PL Homopolymers

There have been apparently conflicting reports in the literature regarding the incidence of late stage foreign body reactions in PL devices. In general, we are aware of eight clinical studies involving the implantation of PL devices in a total of 412 patients. In all, 20 patients (4.9%) experienced adverse outcomes which could generally be described as late stage foreign body reactions. As one looks closely at the studies, the devices, and the polymer morphologies used, it seems prudent to begin to make some basic distinctions. Clearly, the factors which contribute to late stage reactions are numerous and complex, and many of the details of the polymer characterizations are not specifically defined in the literature because of the proprietary nature of the products. Nonetheless, it appears that at least three distinctly different morphologic forms of PL have been used clinically:

1. Bulk-polymerizate: Several early papers reported the use of devices which were machined from a bulk polymerized block of PL. This material had an extremely high average molecular weight (approximately 1,300,000 Daltons), a very high crystallinity, and because the polymer was machined from the bulk polymerizate, it may have contained a relatively high level of residual free lactide dimer. We will refer to this material as bulk-PL.

2. Self-reinforced: The work of Törmälä and Tunc, as mentioned previously (see PROCESSING above) has resulted in several clinical trials with devices made using self-reinforced PL material. We will refer to this material as SRPL.

3. Melt-processed: Another set of studies have been conducted with devices made via a melt process from purified PL. The purification process was designed to

minimize the level of residual lactide dimer prior to injection molding. After injection molding, the devices had an average molecular weight (from GPC using Polystyrene standards) of approximately 500,000 daltons. We will refer to this material as MPPL.

Bulk-PL

Eitenmuller *et al* reported the use of bulk-PL screws and plates to repair ankle fractures in 25 patients. In four of these patients, remnants of the plates were observed to extrude through the skin and in one case, a partially degraded plate had to be removed after about 12 months (Eitenmueller *et al.*, 1990).

In a series of papers, Bergsma, *et al.* (1993, 1995c, 1995d, 1995e), Bos, et al. (1987), and Rozema, *et al.* (1992, 1994) reported that after three years post-operatively, a late foreign body reaction was observed in four of ten patients who originally had zygomatic fractures repaired with bulk-PL plates and screws (Rozema *et al.*, 1991). A follow up study with the same patient group revealed that five additional patients experienced late stage foreign body reactions after up to 68 months. One patient in this study died for unrelated reasons.

From the above, one would estimate that the incidence of late stage foreign body reactions associated with the use of bulk-PL devices would be approximately 13 in 39 or 33%.

SRPL

SRPL screws (made via the "orientrusion process" or other orientation processes) and stainless steel screws were used to repair ankle fractures in 60 patients followed over the course of 9 months. Union was achieved in each case. No soft tissue reactions to the PL screws were noted. The incidence of chronic irritation from the screw heads under the medial malleolus was greater in the metal screw group (29 patients) than in the PL screw group (31 patients) (Bucholz, 1992). In what is most probably a continuation of the 1992 study, a second paper reported results with 83 patients followed for up to 59 months. No cases of late stage foreign body reactions were observed, although three PL screws were removed due to chronic tissue irritation from the screw heads (Bucholz *et al.*, 1994).

SRPL screws (Tampere process) were used in 32 patients to repair small fragment fractures and osteotomies of hand, foot, elbow, and patella. With a follow-up time of up to 37 months, no inflammatory reactions were observed. Biopsies in two patients at 20 and 37 months revealed that no polymeric material remained, that the implant channel had become filled with fibrous connective tissue without evidence of restoration of bony architecture (Pihlajmake *et al.*, 1992).

SRPL screws (probably made via the "orientrusion" process) have been used in acetabular osteotomy repair of 28 hips followed for one year. Union occurred in all cases with no observation of inflammatory reactions (Nakamura *et al.*, 1993).

BIOFIX® pins made from a self-reinforced PL (Tampere process) were used to repair osteochondritis dissecans defects in 10 knees of 9 patients. In 5 knees, severe inflammatory reactions were observed between 3 days and 10 months post op. The authors also report circumstantial evidence that PL may activate the compliment system. Because the authors did not report compliment test results for control devices (like metal screws), it is difficult to draw firm conclusions regarding the role that PL may or may not play in compliment activation (Tegnander *et al.*, 1994). Mainil-Varlet (1995) commented that the Tegnander study was not well-controlled and that the compliment test methodology was not appropriately applied. Further testing must be performed to answer the questions raised by Tegnander and Mainil-Varlet.

From the above, one would estimate the incidence of late stage foreign body reactions associated with SRPL devices would be 5 in 152 or 3.3%.

Melt-processed PL

MPPL tacks have been used to repair glenohumeral defects in 127 patients. Of these 127, 37 underwent secondary procedures to remove loose tack fragments or for various other reasons. Of these 37, 12 developed some degree of synovitis presumably due to mechanical irritation of the joint from the device fragments. In only one case was there a clear histopathological indication of a foreign body reaction with birefringent material being observed under polarized light at 18 months post-op. In all cases, the synovitis resolved after the second surgery. Selected clinical results have been published (Snyder *et al.*, 1993, Elrod 1993).

In another human clinical study, 100 patients received BioScrew® MPPL interference screws as a means of providing fixation of bone-patellar tendon-bone autografts during anterior cruciate ligament reconstruction. The control group consisted of 99 patients who received standard titanium alloy interference screws. Patients were initially followed for up to 1 year after the surgery. In both study groups, the postoperative device related complication rates were low (0% for the BioScrew group and 1% for the metal screw group). In the BioScrew group, one patient experienced mild synovitis after 101 days post-op, however it is unlikely that this complication was device related. In no cases were late stage foreign body or other inflammatory reactions observed. Comparison of the study groups after 1 year (42 BioScrew, 38 metal screw) revealed no significant differences in Lysholm scores, Tegner activity levels, KT 1000/2000 Arthrometer results or range of motion reacquisition (McGuire *et al.*, 1994). Similar results were reported in a two year follow up (McGuire *et al.*, 1995, Barber *et al.*, 1995). At the time of this writing, many of the patients in this study had progressed without any adverse outcomes up to four years post implantation.

From the above studies with purified and melt processed PL, one observes a much lower incidence of late stage foreign body reactions (1 in 227 patients or 0.4%) than with bulk-PL (13 in 39 patients or 33%) or with SRPL (5 in 152 or 3.3%). The role of polymer purity has been considered as a mitigating factor in at least one paper (Pihlajmaki *et al.*, 1992).

A recent report from Dawes and Rushton (1994) gives evidence that prostaglandin E_2 is released by human synovial fibroblasts *in-vitro* upon exposure to lactic acid. The authors hypothesize that this may explain the isolated inflammatory responses exhibited by some individuals with PL implants.

APPLICATIONS

The first types of devices to be introduced were sutures, due largely to the fact that the initial absorbable polymer technology yielded polymers with low strength and molecular weight. As further advances were made however, it became possible to produce polymers with sufficient strength for some orthopaedic applications. While the list below shows some of the devices which are commercially available, other devices are currently under investigation.

DEXON® PG Suture[1]
VICRYL® PGL copolymer Suture[2]
MAXON® TMC/PG copolymer Suture[1]
MONOCRYL® PG/Polycaprolactone copolymer Suture[2]
POLYSORB™ PGL Suture[3]
BIOSYN™ PDS/PGA/TMC Suture[3]
BioScrew® PL Interference Screws[4]
BIOFIX® PG and PL Pins[5]
PL-FIX PL Pins[6]
SURETAC® TMC/PG copolymer Suture Anchor[7]
Biologically Quiet™ PGL copolymer Interference Screw[8]
Biologically Quiet™ PGL copolymer Staple[8]
Bio-Interference Screw[9]
ENDOFIX® TMC/PG Interference Screw[10]
SYSORB® DL-PL Interference Screw[11]
PHUSILINE 98%L-PL/2%D-PL Interference Screw[12]
Biostop® Cement Restrictor[13]

In addition to the above, investigators have reported the use of PG, PL, and their copolymers to produce suture anchors (Barber *et al.*, 1993), replacements for bone grafts (Coombes & Meilkle 1994), in tissue regeneration (Dunn *et al.*, 1994), a scaffold for treatment of bone and cartilage defects (Boyan and Walter 1996), subcuticular staples (Zachman *et al.*, 1994), for protein drug delivery (Agrawal *et al.*, 1995), and for numerous other drug delivery applications (Dunn *et al.*, 1988; Cowsar *et al.*, 1985; Heya *et al.*, 1991).

CONCLUSION

Considering the clinical observations of late stage foreign body reactions in osseous applications, it would seem beneficial if future research could be directed at understanding the role of the polymers in the etiology of such reactions. Even though such reactions have been reported to occasionally occur, PG, PL, and related copolymers have been extensively used in numerous applications over the past 25 years to improve the quality of life of large numbers of patients. In 1995 and 1996, the number of absorbable devices introduced into the US market nearly doubled compared with the number released in 1993 and 1994. This trend is expected to continue, particularly with regard to the potential suitability of these polymers in drug delivery and tissue engineering applications.

ENDNOTES

DEXON and MAXON are registered trademarks of Davis & Geck. For DEXON, see Herrmann *et al.*, 1970. For MAXON, see Metz *et al.*, 1989.

VICRYL and MONOCRYL are registered trademarks of Ethicon. For VICRYL see Conn *et al.*, 1974. For MONOCRYL, see Bezwada *et al.*, 1995.

POLYSORB and BIOSYN are trademarks of U.S. Surgical.

BioScrew is a registered trademark of the Linvatec Corporation. See McGuire *et al.*, 1994, 1995; Barber, 1995; Ross & Bassetti, 1995.

BIOFIX is a registered trademark of Bioscience, Ltd.

PL-FIX rods are sold by Zimmer Japan, a Bristol-Myers Squibb Company.

SURETAC is a registered trademark of Acufex Microsurgical, Inc. See Paganini *et al.*, 1995; Warner & Warren, 1991.

Biologically Quiet is a trademark of Instrument Makar. See Johnson, 1995.

Bio-Interference Screws are sold by Arthrex, Inc.

ENDOFIX is a registered trademark of Smith and Nephew.

SYSORB is a registered trademark of SYNOS.

PHUSILINE is a trademark of PHUSIS.

Biostop is a registered trademark of BIOLAND.

REFERENCES

Agrawal, C.M., Best, D., Heckmann, J.D., Boyan, B.D. (1995) Protein release kinetics of a biodegradable implant for fracture non-unions. *Biomaterials*, **16**, 1255–1260.

Barber, F.A., Cawley, P., Prudich, J.F. (1993) Suture anchor failure strength — an *in-vivo* study. *Arthroscopy*, **9**, 647–652.

Barber, F.A., Elrod, B.F., McGuire, D.A., Paulos, L.E.(1995) Preliminary results of an absorbable interference screw. *Arthroscopy*, **11**, 537–548.

Bergsma, E.J, Rozema, F.R., Bos, R.R.M., de Bruijn, W.C. (1993) Foreign body reactions to resorbable poly(l-lactide) bone plates and screws used for the fixation of unstable zygomatic fractures. *J. Oral Maxillofacial Surg.*, **51**, 666–670.

Bergsma, J.E., Rozema, F.R., Bos, R.R.M., Van Rozendaal, A.W.M., De Jong, W.H., Teppema, J.S., Joziasse, C.A.P. (1995a) Biocompatibility and degradation mechanisms of predegraded and non-predegraded poly(lactide) implants: an animal study. *Journal of Materials Science: Materials In Medicine*, **6**, 715–723.

Bergsma, J.E., Rozema, F.R., Bos, R.R.M., Boering, G., Joziasse, C.A.P., Pennings, A.J. (1995b) *In-vitro* predegradation at elevated temperatures of poly(lactide). *Journal of Materials Science: Materials In Medicine*, **6**, 642–646.

Bergsma, J.E., de Bruijn, Rozema, F.R., Bos, R.R.M., Boering, G. (1995c) Late degradation tissue response to poly(l-lactide)bone plates and screws. *Biomaterials*, **16**, 25–31.

Bergsma, J.E., Rozema, F.R., Bos, R.R.M., Boering, G., de Bruijn, W.C., and Pennings, A.J. (1995d) *In-vivo* degradation and biocompatibility study of *in-vitro* pre-degraded as-polymerized polylactide particles. *Biomaterials*, **16**, 267–274.

Bergsma, J.E. (1995e) Late complications using poly(lactide) osteosyntheses *in vivo* and *in vitro* tests, dissertation, University of Groningen, The Netherlands.

Bezwada, R.S., Jamiolkowski, D.D., Lee, I.L., Vishvaroop, A., Persivale, J., Trenka-Benthin, S., Erneta, E., Surydevara, A.Y., Liu, S. (1995) Monocryl suture, a new ultrapliable absorbable monofilament suture. *Biomaterials*, **16**, 1141–1148.

BIKG (1994) Technical Brochure: RESOMER® Resorbable Polyesters. Boehringer Ingelheim KG, Ingelheim am Rhein, Germany and BI Chemicals, Inc., Montvale, NJ, USA. RESOMER is a registered trademark of Boehringer Ingelheim KG.

BIKG (1995) Technical Report: Guidance for the injection moulding of poly-l-lactide. Boehringer Ingelheim KG, Ingelheim am Rhein, Germany.

Bos, R.R.M., Boering, G., Rozema, F.R., and Leenslag, J.W. (1987) Resorbable poly(l-lactide) plates and screws for the fixation of zygomatic fractures, *J. Oral Maxillofacial Surg.*, **45**, 751–753.

Bos, R.R., Rozema, F. Boering, G., Nijenhuis, A.J., Pennings, A.J., Verwey, A.B., Nieuwenhuis, P., Jansen, H.W.B. (1991) Degradation of and tissue reaction to biodegradable poly(l-lactide) for use as internal fixation of fractures: a study in rats. *Biomaterials*, **12**, 32–36.

Bostman, O., Hirvensalo, E., Makinen, J., Rokkanen, P. (1990) Foreign-Body Reaction to Fracture Fixation Implants of Biodegradable synthetic polymers. *J. Bone and Joint Surg.* [Br], **72–B**, 592–596.

Bostman, O.M. (1991) Current concepts review: absorbable implants for the fixation of fractures. *J. Bone and Joint Surg.*, **73–A**, 148–153.

Boyan, B.D. and Walter, M.A. (1996) Tissue Engineering Scientific Exhibit, Fifth World Biomaterials Congress, Toronto, Ontario, Canada.

BPI (1995) Technical Brochure: Lactel® Polymers. Birmingham Polymers, Inc., Birmingham, AL, USA. Lactel is a trademark of Birmingham Polymers, Inc.

Brady, J.M., Cutright, D.E., Miller, R.A., Battistone, G.C. (1973) Resorption rate, route of elimination, and ultrastructure of the implant site of polylactic acid in the abdominal wall of the rat. *J. Biomed. Mater. Res.*, **7**, 155–166.

Browning A., and Chu, C.C. (1986) The effect of annealing treatments on the tensile properties and hydrolytic degradation properties of polyglycolic acid sutures. *J. Biomed. Mater. Res.*, **20**, 613–632.

Bucholz, R. (1992) Prospective study of bioabsorbable screw fixation of ankle fractures. *J. Orthop. Trauma (Abstracts)*, **6**, 505.

Bucholz, R.W., Henry, S., Henley, M.B. (1994) Fixation with Bioabsorbable screws for the treatment of fractures of the ankle. *Journal of Bone and Joint Surgery*, **76–A**, 319–324.

Case, G.D., Glenn, J.F., Postlethwait, R.W. (1976) Comparison of absorbable sutures in urinary bladder. *Urology*, **7**, 165–168.

Chawla, A.S., Chang, T.M.S. (1985) *In vivo* Degradation of poly(lactic acid) of different molecular weights. *Biomat. Med. Dev. Art. Org.*, **13**, 153–162.

Chen, E., Clemow, A., Jadhav, B., Lancaster, R., Sommerich, R., Tunc, D. (1992) A three year study of implantation of absorbable screws in canine femora and tibiae, *The 38th Meeting of the Orthopaedic Research Society*, Washington, D.C.

Chu, C.C. (1983) Survey of Clinically Important Wound Closure. In M. Szycher, (ed.), *Biomaterials in BioCompatible Polymers, Metals, and Composites*, Technomic Publishing, Lancaster, PA, USA, pp. 477–523.

Chu, C.C. (1985) The degradation and biocompatibility of suture materials. In D.F. Williams, (ed.), *CRC Critical Reviews in Biocompatibility, volume 1*, CRC Press, Boca Raton, FL, USA, pp. 261–322.

Chujo, K, Kobayashi, H., Suzuki, J., Tokuhara, S., Tanabe, M. (1967) Ring-opening polymerization of glycolid. *Die Makromol. Chemie*, **100**, 262–266.

Coombes, A.G.A. and Meikle (1994) Resorbable synthetic polymers as replacements for bone graft. *Clinical Materials*, **17**, 35–67.

Conn, J., Oysasu, R., Welsh, M., Beal, J. (1974) Vicryl (polyglactin 910) synthetic absorbable sutures. *Am. J. Surg.*, **128**, 19–23.

Cowsar, D.R., Tice, T.R., Gilley, R.M., English, J.P. (1985) Poly(lactide-co-glycolide) for controlled release of steroids. *Methods in Enzymology*, **112**, 101–116.

Craig, P.H., Williams, J.A., Davis, K.W., Magoun, A.D., Levy, A.J., Bogdansky, S., Jones, J.P. (1975) A biologic comparison of polyglactin 910 and polyglycolic acid synthetic sutures. *Surgery, Gynecology, and Obstetrics*, **141**, 1–10.

Cutright, D.E., Hunsuck, E.E., Beasley, J.D. III (1971) Fracture fixation using a biodegradable material. Polylactic acid. *J. Oral Maxillofacial Surg.*, **29**, 393–397.

Cutright, D.E., Hunsuck, E.E. (1972) The repair of fractures of the orbital floor. *Oral Surg.*, **33**, 28–34.

Cutright, D.E., Perez, B., Beasley, J., Larson, W., Posey, W. (1974) Degradation rates of polymers and copolymers of polylactic acid and polyglycolic acid. *Oral Surgery*, **37**, 142–152.

Dawes, E. and Rushton, N. (1994) The effects of lactic acid on PGE_2 production by macrophages and human synovial fibroblasts: a possible explanation for problems associated with the degradation of poly(lactide) implants? *Clinical Materials*, **17**, 157–163.

Dijkstra, P.J., Bulte, A., Feijen, J. (1991) Block copolymers of L-lactide, D-lactide, and ϵ-caprolactone. *The 17th Annual Meeting of the Society for Biomaterials*, May, p. 184.

Dunn, R.L., English, J.P., Strobel, J.D., Cowsar, D.R., Tice, T.R. (1988) Preparation and evaluation of lactide/glycolide copolymers for drug delivery. In C. Migliarisi, L. Nicolaisi, P. Giusti, E. Chiellini, (eds.), *Polymers in Medicine III, volume 5*, Elsevier, Science Publishers, Amsterdam.

Dunn, R.L., Yewey, G.L., Duysen, E.G., Polson, A.M., Southard, G.L. (1994) *In-situ* forming biodegradable polymeric implants for tissue regeneration. *Polymer Reprints*, **35**, 437.

Edwards, D.J., Hoy, G., Saies, A.D., Hayes, M.G. (1994) Adverse reactions to an absorbable shoulder fixation device. *Journal of Elbow and Shoulder Surgery*, **3**, 230–233.

Eenink, M.J.D. (1987) Synthesis of biodegradable polymers and development of biodegradable hollow fibers for the controlled release of drugs. Dissertation, University of Twente, The Netherlands.

Eitenmuller, J., Muhr, G., David, J. (1990) Treatment of ankle fractures with completely degradable plates, and screws of high molecular weight polylactide. *Proced. 92nd Cong. Francais de Chirgie*, Paris, France.

Eling, B., Gogolewski, S., and Pennings, A.J. (1982) Biodegradable materials of poly(l-lactic acid): 1. Melt-spun and solution-spun fibers. *Polymer*, **23**, 1587–1593.

Ellis, D.N. and Tipton, A. (1994) Influence of injection rate on tensile strength and molecular weight of molded poly(l-lactide test pieces. *20th Annual Meeting of the Society for Biomaterials*, Boston, MA, USA, abstracts p. 479.

Elrod, B.F. (1993) Arthroscopic shoulder stabilization with a bioabsorbable tak. *12th Annual Meeting of the Arthroscopy Association of North America*, Palm Desert, CA, USA, April, abstracts pp. 23–25.

Fambri, L., Pegoretti, A., Mazzurana, M., and Miliagressi, C. (1994) Biodegradable fibres part I. Poly-l-lactic acid fibres produced by solution spinning. *Journal of Materials Science: Materials in Medicine*, **5**, 679–683.

Feng, X.D., Voong, S.T., Song, C.X., Chen, W.Y. (1983) Synthesis and evaluation of biodegradable block copolymers of ϵ-caprolactone and DL-lactide. *Journal of Polymer Science: Polymer Letters Edition*, **21**, 593–600.

Frazza, E.J., and Schmitt, E.E. (1971) A new absorbable suture. *J. Biomed Mater. Res. Symposium*, **1**, 43–58.

Getter, L., Cutright, D., Bhaskar, S., Augsburg, J. (1972) A biodegradable intraosseous appliance in the treatment of mandibular fractures. *J. Oral Surg.*, **30**, 344–348.

Ginde, R. M. and Gupta, R.K. (1987) *In-vitro* chemical degradation of poly(glycolic acid) pellets and fibers. Journal of Applied Polymer Science, **33**, 2411–2429.

Gogolewski, S., and Pennings, A.J. (1983) Resorbable Materials of poly(l-lactide). II. fibers spun from solutions of poly(l-lactide) in good solvents. *Journal of Applied Polymer Science*, **28**, 1045–1061.

Gogolewski, S., Jovanovic, M., Perren, S.M., Dillon, J.G., Hughes, M.K. (1993) Tissue response and *in vivo* degradation of selected polyhydroxyacids: polylactides (pla), poly(3–hydroxybutyrate) (phb), and poly(3–hydroxybutyrate-co-3–hydroxyvalerate) (phb/va). *J. Biomed. Mater. Res.*, **27**, 1135–1148.

Grizzi, I., Garreau, H., Li, S., Vert, M. (1995) Hydrolytic degradation of devices based on poly(dl-lactic acid) size dependence. *Biomaterials*, **16**, 305–511.

Gutwald, R., Pistner, H., Hoppert, T., Muhling, J. (1992) Unexpected malignant soft-tissue-reaction in a long-term biodegradation study. *Sixth Biennial Congress of the International Association of Oral Pathologists*, Hamburg, Germany.

Herrmann, J.B. (1970) Polyglycolic Acid sutures; laboratory and clinical evaluation of a new absorbable suture material. *Arch. Surg.*, **100**, 486.

Heya, T., Okada, H., Ogawa, Y., Toguchi, H. (1991) Factors influencing the profiles of TRH release from poly(D,L-lactic acid/glycolic acid) microspheres. *International Journal of Pharmaceutics*, **72**, 199–205.

Hofmann, G.O., Liedtke, H., Ruckdeschel, G., Lob, G. (1990) The Influence of bacterial contamination on biodegradation of pla implants. *Clinical Materials*, **6**, 137–150.

Hollinger, J.O., and Battistone, G.C. (1986) Biodegradable bone repair materials. *Clinical Orthopaedics and Related Research*, **207**, 290–305.

Horacek, I. and Kalisek, V. (1994) Polylactide. II. discontinuous dry spinning-hot drawing preparation of fibers. *J. Appl. Polymer Science*, **54**, 1759–1765.

Ikada, Y., Suong, H.H., Shimizu, Y, Watanabe, S., Nakamura, T., Suzuki, M., Shimamoto, T. (1990) Osteosynthetic pin. U.S. Patent 4,898,186, Gunze Limited.

Johnson, L.L. (1995) Comparison of Bioabsorbable and metal interference screw in anterior cruciate reconstruction (a clinical trial), *14th Annual Meeting of the Arthroscopy Association of North America*, San Francisco, CA, USA.

Katz, A.R., and Turner, R.J. (1970) Evaluation of tensile and absorption properties of polyglycolic acid sutures. *Surgery, Gynecology, and Obstetrics*, **146**, October, 701–716.

Kricheldorf, H.R., Jonte, J.M., Berl, M. (1985) Polylactone 3. copolymerization of glycolide with DL-lactide and other lactones. *Makromol. Chem., Suppl.*, **12**, 25–38.

Kronenthal. R.L. (1975) Biodegradable polymers in medicine and surgery. In R.L. Kronenthal, Z. Oser, and E. Martin, (eds.), *Polymers in Medicine and Surgery*, Plenum Publishing, New York, NY, USA, pp. 119–137.

Kulkarni R.K., Pani, K.C., Neuman, C., Leonard, F. (1966) Polylactic acid for surgical implants. *Arch. Surg.*, **93**, 839–843.

Kulkarni, R.K., Moore, E.G., Hegyeli, A.F., Leonard, F. (1971) Biodegradable poly(lactic acid) polymers. *J. Biomed. Mater. Res.*,**5**, 169–181.

Lam, K.H., Schakenraad, J., Esselbrugge, H., Feijen, J., Nieuwenhuis, P. (1993) The effect of phagocytosis of poly(l-lactic acid) fragments on cellular morphology and viability. *J. Biomed. Mater. Res.*, **27**, 1569–1577.

Lam, K.H., Schakenraad, J.M., Groen, H., Esselbrugge, H., Dijkstra, P.J., Feijen, J., Nieuwenhuis, P. (1995) The influence of surface morphology and wettability on the inflammatory response against poly(l-lactic acid): a semiquantitative study with monoclonal antibodies. *J. Biomed. Mater. Res.*,**29**, 929–942.

Lautiainen. I., Miettinen, H, Makela, A., Rokkanen, Tormala, P. (1994) Early effects of the self-reinforced pga implant on a growing bone: an experimental study in rats. *Clinical Materials*, **17**, 197–201.

Litsky, A.S. (1993) Clinical reviews: bioabsorbable implants for orthopaedic fracture fixation. *J. App. Biomat.*, **4**, 109–111.

Leenslag, J. W., Pennings, A., Bos, R., Klzema, F., Boering, G. (1987) Resorbable materials of poly(l-lactide) VII. *In-vivo* and *in vitro* degradation. *Biomaterials*, **8**, 311–314.

Mainil-Varlet, P. (1995) Polylactic acid pins. *Acta Orthop. Scand.* **88**, 573.

Majola, A., Vainionpaa, S., Vihtonen, K., Mero, M., Vasonius, J., Tormala, P., Rokkanen, P. (1991) Absorption, biocompatibility, and fixation properties of polylactic acid in bone tissue: an experimental study in rats. *Clinical Orthopaedics and Related Research*, **268**, 260–269.

Manninen, M.J. (1991) Absorbable poly-l-lactide screws in the fixation of olecranon osteotomy in sheep. *Proceedings of the 9th European Conference on Biomaterials*, Chester, UK, p. 213.

Matsusue, Y., Yamamuro, T., Yoshii, S., Oka, M., Ikada, Y., Hyon, S., Jhikinami, Y. (1991) Biodegradable screw fixation of rabbit tibia proximal osteotomies. *J. App. Biomat.*,**2**, 1–12.

McGuire, D.A., Hendricks, S., Barber, F.A., Elrod, B.F., Paulos, L.E. (1994) The use of bioabsorbable interference screws in anterior cruciate ligament reconstruction: mid-term follow-up results. *The 61st Meeting of the American Academy of Orthopaedic Surgeons*, New Orleans, LA, USA.

McGuire, D.A., Barber, F.A., Elrod, B.F., and Paulos, L.E. (1995) The BioScrew bioabsorbable interference screw in ACL reconstruction. *The 62nd Meeting of the American Academy of Orthopaedic Surgeons*, Orlando, FL, USA.

Metz, S.A., Chegini, N., Masterson, B. (1989) *In vivo* tissue reactivity and degradation of suture materials: a comparison of MAXON and PDS. *J. Gynecol. Surg.*, **5**, 37–43.

Miller, R.A., Brady, J., Cutright, D. (1977) Degradation rates of oral resorbable implants (polylactates and polyglycolates): rate modification with changes in pla/pga copolymer ratios. *J. Biomed. Mater. Res.*, **11**, 711–719.

Nakamura, T., Hitomi, S., Watanabe, S., Shimizu, Y., Jamshidi, K., Hyon, S., Ikada, Y. (1989) Bioabsorption of polylactides with different molecular properties. *J. Biomed. Mater. Res.*, **23**, 1115–1130.

Nakamura, T., Ninomiya, S., Takatori, Y., Morimoto, S., Kusaba, I., Kurokawa, T. (1993) Polylactide screws in acetabular osteotomy. *Acta Ortho. Scand.*, **64**, 301–302.

Nakamura, T., Shimizu, Y., Okumura, N., Matsui, T., Hyon, S.H., Shimamoto, T. (1994) T. Tumorigenicity of poly-l-lactide (PLLA) plates compared with medical-grade polyethylene.*J. Biomed. Mater. Res.*, **28**, 17–25.

Paganini, M.J., Speer, K.P., Altchek, D.W., Warren, R.F., and Dines, D.M. (1995) Arthroscopic fixation of superior labral lesions using a biodegradable implant: a preliminary report. *Arthroscopy*, **11**, 194–198.

Pihlajmaki, H., Bostman, O., Hirvensalo, E., Tormala, P., Rokkanen, P. (1992) Absorbable pins for self-reinforced poly-l-lactic acid for fixation of fracture and osteotomies,. *J. Bone and Joint Surg.* [Br], **74–B**, 853–857.

Pistner, H., Bendix, D., Muhling, J., Reuther, J.F. (1993a) Poly(l-lactide): A long-term degradation study *in-vivo*. part III: analytical characterization. *Biomaterials*, **14**, 291–298.

Pistner, H., Gutwald, R., Ordung, R., Reuther, J., Muhling, J. (1993b) Poly(L-lactide): A long-term degradation study *in-vivo*. part I: biological results, *Biomaterials*, **14**, 671–677.

Pistner, H., Stallforth, H., Gutwald, R., Muhling, J., Reuther, J., and Michel, C. (1994) Poly(l-lactide): a long-term degradation study *in-vivo*. part II: physico-mechanical behaviour of implants. *Biomaterials*, **15**, 439–449.

PURAC (1996) Technical Brochure: PURASORB® Monomers and Biodegradable Polymers, Gorinchem, Holland and Lincolnshire, IL, USA. PURASORB is a registered trademark of PURAC biochem bv.

Raiha, J.E. (1990) Fixation of trochanteric osteotomies in laboratory beagles with absorbable screws of polylactic acid. *VOCT*, **3**, 123–129.

Riddick, D.H., DeGrazia, C.T., Maenza, R.M. (1977) Comparison of polylactic and polyglycolic acid sutures in reproductive tissue. *Fertility and Sterility*, **28**, 1220–1225.

Ross, R.D., Bassetti, K.J (1995) Bioabsorbable interference bone fixation screw. US Patent 5,470,344, Linvatec Corporation.

Rozema, F.R., Bos, R., Boering, G., Van Asten, J., Nijenhuis, A., Pennings, A. (1991) The effects of different steam-sterilization programs on materials properties of poly(l-lactide). *J. App. Biomat.*, **2**, 23–28.

Rozema, R.M., (1991) Late tissue response of bone-plates and screws of poly(l-lactide) used for fixation of zygomatic fractures, *Trans. 9th Europ. Conf. on Biomat.*, Chester, UK, p. 154.

Rozema, F.R., de Bruijn, W.C., Bos, R.R.M., Boering, G., Nijenhuis, A.J., Pennings, A.J. (1992) Late Tissue response to bone-plates and screws of poly(l-lactide) used for fracture fixation of the zygomatic bone, biomaterial-tissue interfaces. In P.J. Dougherty, *et al.* (eds). *Advances in Biomaterials*, **10**, 349–335.

Rozema, F.R., Bergsma, J.E., Bos, R.R.M., Boering, G., Nijenhuis, A.J., Pennings, A.J., de Bruijn, W.C. (1994) Late degradation simulation of poly(l-lactide). *Journal of Materials Science: Materials in Medicine*, **5**, 575–581.

Salthouse, T.N., Matagla, B. (1984) Tissue regeneration associated with lactide containing implants. *Second World Congress on Biomaterials, 10th Annual Meeting of the Society of Biomaterials*, Washington, D.C., USA, abstracts p. 272.

Schakenraad, J.M., Hardonk, M., Feijen, Jr., Molenaar, I., Nieuwenhuis, P. (1990) Enzymatic activity toward poly(l-lactic acid) implants. *J. Biomed. Mater. Res.*, **24**, 529–545.

Schnidler, A., Jeffcoat, R., Kimmel, G.L., Pitt, C.G., Wall, M.E., Zweidinger, R. (1977) Biodegradable polymers for sustained drug delivery, in E.M. Pearce and J.R. Schaefgen, (eds.), *Contemporary Topics in Polymer Science volume 2*, Plenum Press, New York, pp. 251–289.

Shalaby, S.W., Jamiolkowski, D.D. (1985) Synthesis and intrinsic properties of crystalline copolymers of ε-caprolactone and glycolide. *Polymer Preprints, American Chemical Society Division of Polymer Chemistry*, **26**, 190.

Shimamoto, T., Oka, T., Adachi, M., Hyon, S.H., Nakayama, K., Kaito, A. (1995) Bone-treating devices and their manufacturing method. U.S. Patent 5,431,652, Gunze Limited and Agency of Industrial Science and Technology.

Snyder, S.J., Strafford, B. (1993) Arthroscopic management of instability of the shoulder,. *Orthopaedics*, **16**, 993–1002.

Spivak, J.M., Ricci, J., Blumenthal, N., Alexander, H. (1990) A new canine model to evaluate the biological response of intramedullary bone to implant materials and surfaces. *J. Biomed. Mater. Res.*, **24**, 1121–1149.

Suganuma, J., and Alexander, H. (1993) Biological Response of Intramedullary Bone to Poly-L-Lactic Acid, *J. App. Biomat.*, **4**, 13–27.

Suuronen, R. (1991) Comparison of absorbable self-reinforced poly-l-lactide screws and metallic screws in the fixation of mandibular condyle osteotomies: an experimental study in sheep. *J. Oral Maxillofac. Surg.*, **49**, 989–995.

Svensson, P-J., Janarv, P-M., Hirsch, G. (1994) Internal fixation with biodegradable rods in pediatric fractures: one-year follow-up of fifty patients. *J. Pediatric Orthop. Surg.*, **2**, 220–224.

Tegnander, A., Engebretsen, L., Bergh, K., Eide, E., Holen, K.J., and Iversen, O.J. (1994) Activation of the compliment system and adverse effects of biodegradable pins of poly-(lactic acid), (biofix) in osteochondritis dissecans. *Acta Orthop. Scand.*, **65**, 472–475.

Toljan, M.A., and Orthner, E. (1995) Bioresorbable interference screws in ACL surgery. *Arthroscopy*, **11**, 381.

Tunc, D.C., Rohousky, M., Zadwadsky, J., Spieker, J., Strauss, E. (1986) Evaluation of body absorbable screw in avulsion type fractures. *The 12th Annual Meeting of the Society of Biomaterials*, St. Paul, MN, USA.

Tunc, D.C., Jadhav, B. (1988) Development of absorbable, ultra high strength polylactide. *ACS Polymeric Materials Science and Engineering*, **59**, 383–387.

Vainionpaa, S., Kilpikari, J., Laiho, J., Helvirta, P., Rokkanen, P., Tormala, P. (1987) Strength and strength retention *in vitro*, of absorbable, self-reinforced polyglycolide (pga) rods for fracture fixation. *Biomaterials*, **8**, 46–48.

Verheyen, C.C.P.M., de Wijn, J., Van Blitterswijk, C.A., Rozing, P.M., de Groot, K. (1993) Examination of efferent lymph nodes after 2 years of transcortical implantation of poly(l-lactide) containing plug: a case report. *J. Biomed Mater. Res.*, **27**, 1115–1118.

Vert, M. (1986) Biomedical Polymers from chiral lactides and functional lactones: properties and applications. Makromol. *Chem. Macromol. Symp.*, **6**, 109–122.

Von Oepen, R. (1995) Development of special machine and processing techniques for producing resorbable implants. Dissertation, Institut fur Kunstoffverarbeitung (IKV), RWTH Aachen.

von Oepen, R. and Michaeli, W. (1992) Injection moulding of biodegradable implants. *Clinical Materials*, **10**, 21–28.

Von Schroeder, H.P., Kwan, M., Amiel, D., Coutts, R. (1991) The use of polylactic acid matrix and periosteal grafts for the reconstruction of rabbit knee articular defects. *J. Biomed. Mat. Res.*, **25**, 329–339.

Warner, J.P. and Warren, R.F. (1991) Arthroscopic bankart repair using a cannulated, absorbable fixation device. *Operative Techniques in Orthopaedics*, **1**, 192–198.

Winet, H., Hollinger, J.O., Stevanovic, M. (1995) Incorporation of polylactide-polyglycolide in a cortical defect: neoangiogenesis and blood supply in a bone chamber. *Journal of Orthopaedic Research*, **13**, 679–689.

Zachmann, G.C., Foresmanm, P.A., Bill, T.J., Bentrem, D.J., Rodeheaver, G.T., Edlich, R.F. (1994) Evaluation of a new absorbable Lactomer subcuticular staple. *Journal of Applied Biomaterials*, **5**, 221–226.

Zhu, K.J., Xiangzhou, L., Shilin, Y. (1986) Preparation and properties of DL-lactide and ethylene oxide copolymers: a modifying biodegradable material. *Journal of Polymer Science Part C: Polymer Letters*, **24**, 331–337.

APPENDIX

Polymer Name	Structure	Precursor/polymer	Source
Poly(L-lactide)	$(CH(CH_3)COO)n$	L-lactide	PURAC biochem bv Boehringer Ingelheim KG
		Stannous Octoate (catalyst)	Sigma Chemical Company
		1,6-Hexanediol (& other chain control agents)	Aldrich Chemical Company
		Poly(L-Lactide) (various Mw)	Birmingham Polymers, Inc. Boehringer Ingelheim KG PURAC biochem bv
Poly(D-lactide)	$(CH(CH_3)COO)n$	D-lactide	PURAC biochem bv
		Poly(D-lactide)	PURAC biochem bv Birmingham Polymers, Inc. Boehringer Ingelheim KG
Poly(DL-lactide)	$(CH(CH_3)COO)n$	DL-lactide	PURAC biochem bv Birmingham Polymers, Inc. Boehringer Ingelheim KG
		Poly(DL-lactide)	PURAC biochem bv Birmingham Polymers, Inc. Boehringer Ingelheim KG
Polyglycolide	$(CH_2COO)n$	Glycolide	PURAC biochem bv Boehringer Ingelheim KG
		Polyglycolide	PURAC biochem bv Birmingham Polymers, Inc. Boehringer Ingelheim KG
(Lactide/Glycolide) Copolymers		(various Mw)	PURAC biochem bv Birmingham Polymers, Inc. Boehringer Ingelheim KG

2. POLY(p-DIOXANONE) AND ITS COPOLYMERS

RAO S. BEZWADA, DENNIS D. JAMIOLKOWSKI and KEVIN COOPER

Johnson & Johnson, Inc., The J&J Corporate Biomaterials Center,
P.O. Box 151, Somerville, NJ 08876, USA

INTRODUCTION

Since the development in the early 1970's of the first synthetic absorbable polymer, polyglycolic acid, biodegradable polymers have become increasingly important in biomedical applications (Kronenthal, 1975; Frazza, 1976; Chu, 1983, 1985, 1990, 1995; Barrows, 1986; Heller, 1985; Shalaby and Johnson, 1994; Amecke *et al.*, 1995; Wu, 1995; Hollinger *et al.*, 1995; Gilding). Five cyclic lactone monomers (Figure 1) have been used alone or in combination to produce the synthetic absorbable sutures which are currently marketed: VICRYL™, DEXON™, POLYSORB™, PDS™, MAXON™, MONOCRYL™, and BIOSYN™ (Figure 2). VICRYL is a braided suture produced from a copolymer of glycolide/lactide at a 90/10 mol/mol composition (Craig *et al.*, 1975; Reed and Gilding, 1981). POLYSORB is also based upon a combination of glycolide and lactide but at a different composition. DEXON(Craig *et al.*, 1975; Reed and Gilding, 1981) is a braided suture based upon the homopolymer of polyglycolic acid (PGA), or more accurately, polyglycolide. MAXON (Katz *et al.*, 1985) is a monofilament suture produced from a segmented block copolymer of glycolide and trimethylene carbonate, while MONOCRYL (Bezwada *et al.*, 1992, 1995), the most pliable monofilament suture to date, is formed from a segmented block copolymer of glycolide and caprolactone. Recently, BIOSYN (Roby *et al.*, 1995), a monofilament suture based on a non-random terpolymer of p-dioxanone, trimethylene carbonate and glycolide, was introduced.

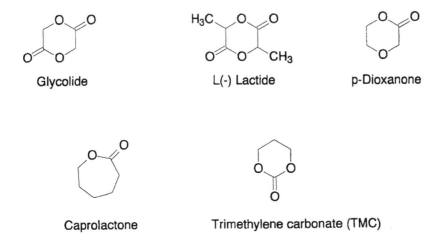

Glycolide L(-) Lactide p-Dioxanone

Caprolactone Trimethylene carbonate (TMC)

Figure 1 Cyclic monomers.

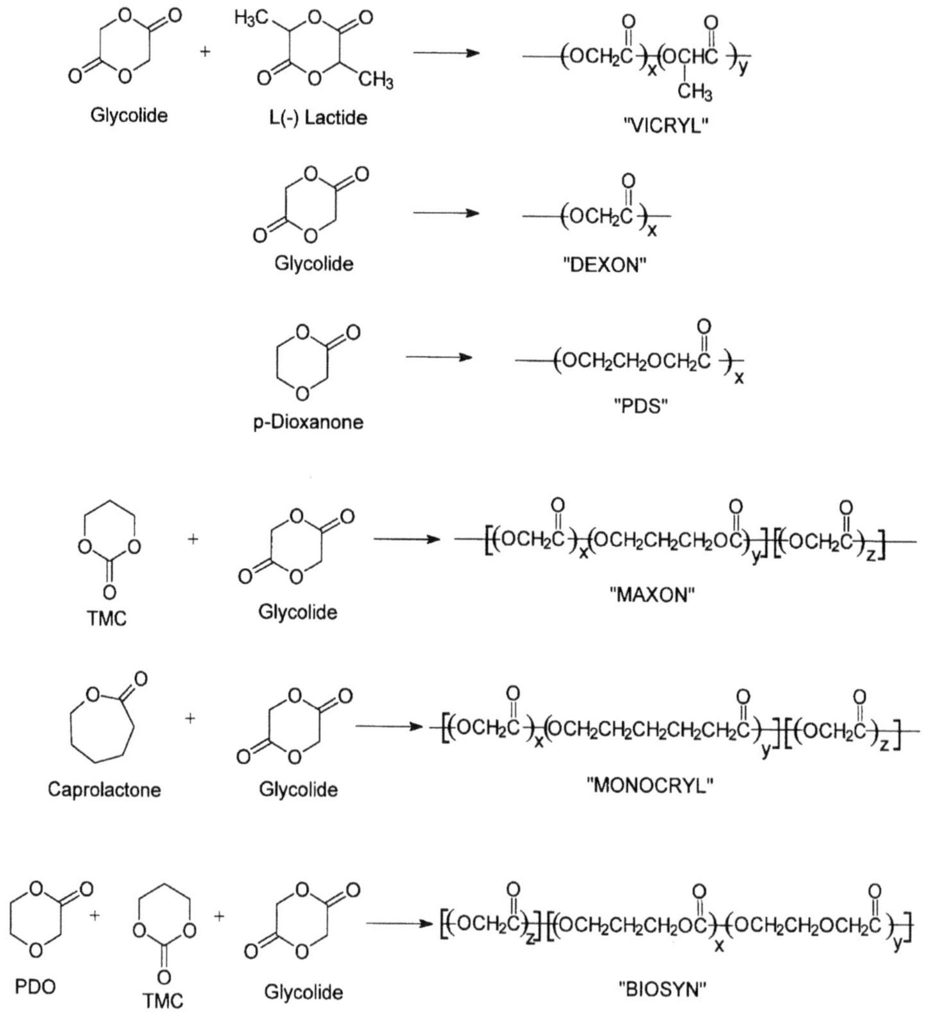

Figure 2 Synthetic absorbable sutures.

PDS suture, the very first synthetic absorbable monofilament, is formed from poly(p-dioxanone) polymer which is, of course, based on the homopolymerization of p-dioxanone monomer (Doddi *et al.*, 1977). Polymers and copolymers based on p-dioxanone have gained increasing interest in the medical device and pharmaceutical fields due to their degradability *in vivo*, low toxicity (Doddi *et al.*, 1977; Ray *et al.*, 1981), softness and flexibility. The low moduli of poly(p-dioxanone) and its copolymers distinguish them from glycolide and lactide based absorbable polymers which have a high degree of stiffness, and therefore, when used as sutures, must be used in a multifilament braided construction to achieve the handling characteristics needed by the surgeon.

PDS (polydioxanone suture) can thus be utilized in applications which require good handling, and tissue pass-through (Shalaby and Koelmel, 1984; Bezwada, Shalaby, and Newman, 1987, 1991; Bezwada, *et al.*, 1987, 1990; Bezwada, Shalaby, and Erneta, 1991; Bezwada and Shalaby, 1991; Bezwada, Shalaby, and Hunter, 1991). Although PDS resin is inherently soft, PDS II suture, formed by novel and proprietary processes, was recently introduced with improved handling properties (Broyer, 1995). In addition, the extended strength retention profile of PDS sutures makes them useful in comparison to other absorbable surgical sutures when extended healing times are encountered.

Although the softness and flexibility inherent in poly(p-dioxanone) has allowed researchers to develop this material for monofilament suture applications, poly(p-dioxanone) has also been injection molded into a number of non-filamentous surgical devices, namely ABSOLOK™ and LAPRA-TY™.

Commercial products which utilize poly(p-dioxanone) are:

Product	Device	Processing Method	Polymer type
PDS II	Suture	Extrusion	PDS homopolymer
BIOSYN	Suture	Extrusion	Terpolymer
ABSOLOK	Ligating clip	Injection molded	PDS homopolymer
LAPRA-TY	Suture clip	Injection molded	PDS homopolymer
ORTHOSORB	Pin	Extrusion	PDS homopolymer

The following describes the synthesis, physical characteristics and *in vitro/in vivo* properties of poly(p-dioxanone) and its copolymers with other lactone derived absorbable polyesters and related structures, as well as the synthesis of the monomer, p-dioxanone.

SYNTHESIS OF p-DIOXANONE MONOMER

One synthetic route to p-dioxanone monomer (Doddi *et al.*, 1977) is by first reacting the monosodium salt of ethylene glycol with chloroacetic acid. Sodium metal is dissolved in a large excess of ethylene glycol under a stream of nitrogen to form the monosodium salt of ethylene glycol.

This is then reacted with 0.5 moles of chloroacetic acid per mole of sodium to form sodium hydroxyethoxyacetate. (See Figure 3) Excess ethylene glycol and by-products of the reaction are removed by distillation and by washing with acetone. The sodium hydroxyethoxyacetate is converted to the free hydroxy acid by the addition of hydrochloric acid. The resulting sodium chloride is removed by precipitation with ethanol, followed by filtration. In the presence of $MgCO_3$, the hydroxyacid is then heated in a distillation apparatus to about 200°C. Upon further heating, crude p-dioxanone is formed and distilled over at 200–220°C. The crude p-dioxanone is purified to over 99% by multiple recrystallizations and/or distillations.

An alternative approach (Figure 4) to p-dioxanone utilizes the dehydrogenation of diethylene glycol (DEG) and leads to improved yields (Jiang, 1995a, 1995b). Thus, to a two neck 3L round bottom flask equipped with an overhead stirrer and nitrogen inlet, 2 Kg of diethylene glycol and a copper-chromium catalyst is added. The flask is placed in an oil bath at 220–240°C and stirred for several hours.

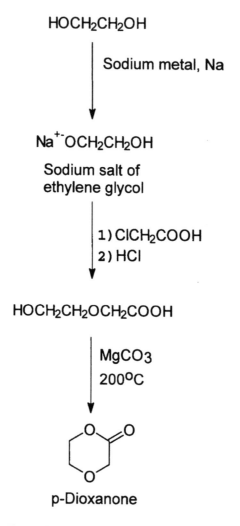

HOCH₂CH₂OH

→ Sodium metal, Na

Na⁺⁻OCH₂CH₂OH

Sodium salt of
ethylene glycol

1) ClCH₂COOH
2) HCl

HOCH₂CH₂OCH₂COOH

MgCO₃
200°C

p-Dioxanone

Figure 3 Preparation of p-dioxanone monomer.

The resulting crude reaction product of p-dioxanone and unreacted DEG is dried over molecular sieves (4 Angstroms) overnight and then filtered utilizing a Buchner funnel.

The dried mother liquor is then placed in a two neck flask equipped with a distillation head and stopper. Benzyl bromide (100 grams) and pyridine (50 grams) are then added. The mixture is then distilled under reduced pressure at a rate of 60 drops per minute. Approximately 1.6 Kg of crude p-dioxanone is recovered.

Two hundred grams of the crude reaction product is then placed in an Erlenmeyer flask along with 200 ml of ethyl acetate at room temperature. The stirred, clear, yellow solution is then cooled to −20°C. After 10 minutes, stirring is stopped and 2 grams of pure p-dioxanone monomer crystal seeds are added. After 1 hour, the solution is cooled to −34°C for 2 hours.

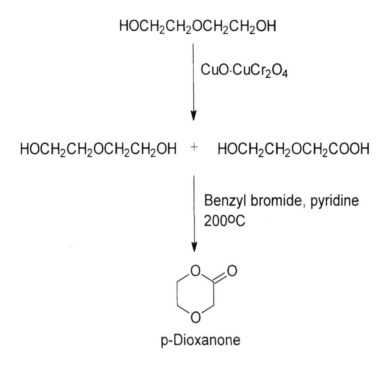

Figure 4 Preparation of p-dioxanone monomer via dehydrogenation.

The partially purified crystalline p-dioxanone monomer (100 grams) is then filtered using a Buchner funnel and then added to a Erlenmeyer flask along with 75 grams of ethyl acetate. The mixture is stirred until the monomer dissolves. Stirring is then stopped; the solution is cooled to 0°C and 1 gram of pure p-dioxanone seed crystals are added. The mixture is then allowed to stand at –30°C for 12 hours, then filtered and dried. Sixty grams of pure p-dioxanone are obtained (30% yield).

SYNTHESIS OF POLY(p-DIOXANONE)

Several polymerization methods, including melt, solution and emulsion have been applied to lactone monomers to form high molecular weight polymers. Typically, a catalyst such as a metal oxide or metal salt, stannous 2-ethylhexanoate or stannous chloride, is used along with a free hydroxy containing initiator such as water, an alcohol, hydroxy acid or ester to activate the lactone ring to initiate polymerization.

Furthermore, since impurities in the monomer feed can limit molecular weight, monomers must be highly pure and great care must be taken to use dry glassware and glove box procedures when setting up a polymerization run. Additionally, the formed crude polymer, especially poly(p-dioxanone) and its copolymers, will contain unreacted monomer in amounts typically greater than 1%. Vacuum or other extraction techniques must be utilized to remove residual monomer and catalyst in order to obtain polymers with optimized properties.

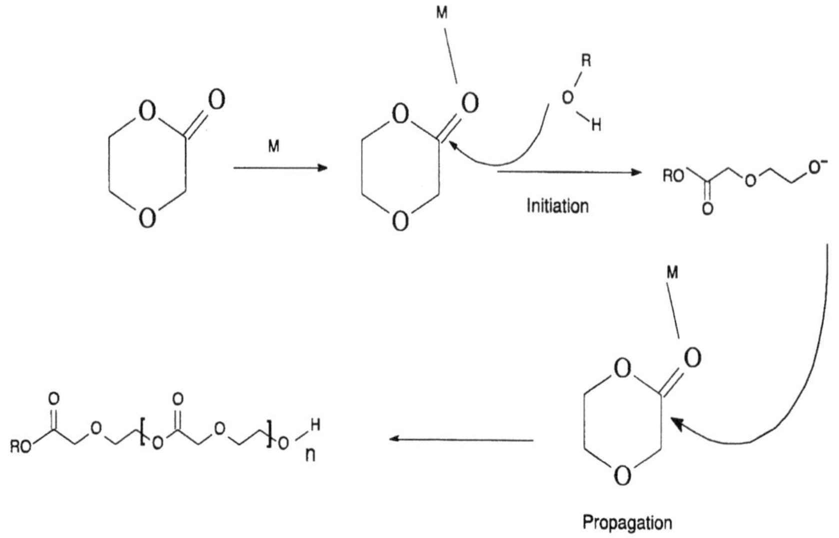

Figure 5 Polymerization mechanisms of p-dioxanone to poly(dioxanone).

The polymerization of p-dioxanone has been described in several references (Doddi *et al.*, 1977; Shalaby and Koelmel, 1984; Jamiolkowski *et al.*, 1989). As described above, highly purified p-dioxanone monomer is polymerized in the presence of an organometallic catalyst such as diethyl zinc or zirconium acetylacetone to obtain high molecular weight, fiber forming polymer (Figure 5).

Thus, in a nitrogen purged glove box, 1.0 mole (102.8 g) of highly pure (99.99+%) p-dioxanone monomer, 1–dodecanol (0.358 grams; 0.192 mole percent based on monomer) and a catalytic amount of stannous octoate in a toluene solution (0.0025 mole% based on monomer) is added to a two neck round bottom flask, equipped with an overhead stirrer and nitrogen inlet. The mixture is heated under an inert dry nitrogen at 90°C for one hour. The viscous polymer is discharged into trays, and post cured at 80°C for 96 hours under nitrogen. The polymer is isolated, ground, and dried *in vacuo* at room temperature for 10 hours and then at 80°C for 32 hours. A weight loss of approximately 4% (predominately unreacted monomer) is obtained during the vacuum drying. The polymer thus formed has an inherent viscosity ("IV") of about 1.72 dL/g, as determined at a concentration of 0.1 g/dL in hexafluoroisopropyl alcohol at 25°C. Poly(p-dioxanone), so produced, has a glass transition temperature (Tg) of −15°C, and a melting temperature (Tm) of about 115°C as measured by differential scanning calorimetry (DSC), and about 37 percent crystallinity as measured by X-ray diffraction.

Polymer Extrusion

In the preparation of fibers, PDS can be melt extruded through a spinnerette in a conventional manner using a single screw extruder to form one or more filaments. The following is a general procedure which can be used for laboratory scale experiments.

Extrusion of PDS can be accomplished using an INSTRON Capillary Rheometer. The resin is packed in the preheated (80 to 90°C) chamber (3/8" diameter) and extruded through a 40 mil diameter round hole die (L/D = 24.1) using a ram speed of 2 cm/min after a dwell time of about 11 minutes. Extrusion is typically done at temperatures of about 10 to 85°C above the Tm, ranging from 125 to 200°C, but temperatures from about 140 to 160°C work best. At lower temperatures, the melt viscosity may be too high, while at higher temperatures the rate of degradation is too high. The extrudate is taken up through an ice water quench bath. A take-off speed of 24 feet/minute provides an extrudate with a diameter of approximately 0.019 inches. Other take-up speeds can be used to vary the diameter of the extrudate.

The extrudate, which preferably should be allowed to partially crystallize (conveniently achieved by allowing it to stand at room temperature for 1 to 24 hours) are subsequently drawn about 6× to 7.5× in a one or multistage drawing process in order to achieve molecular orientation and improve tensile properties.

Drawing of the extrudate (diameter range, usually 18–20 mils) is accomplished by first passing it through rollers at an input speed of four feet per minute and into a heated draw bath of glycerine. The temperature of the draw bath can vary from 25 to 90°C, but is typically between 49 and 60°C. The draw ratio in this first stage of drawing can vary from 3× to 7×.

The partially drawn fibers are then placed over a second set of rollers into a glycerine bath and are kept at temperatures ranging from 67 to 73°C with draw ratios of up to 2×. The fiber is passed through a water-wash, taken up on a spool and dried. A set of hot rollers can be substituted for a portion or all of the glycerine draw bath.

The resulting oriented filaments have good straight and knot tensile strengths. Dimensional stability and *in vivo* tensile strength retention of the oriented filaments may be enhanced by subjecting the filaments to an annealing treatment. This optional treatment consists of heating the drawn filaments to a temperature of 60 to 90°C while restraining the filaments to prevent any substantial shrinkage or at least to control it. Restraining may begin with the filaments initially under tension or with up to 20% shrinkage allowed prior to restraint. The latter is referred to as relaxation and usually results in a softer, but somewhat slightly weaker, fiber. The filaments are held at the annealing temperature for a few minutes to several days, or longer, depending on the temperature. In general, annealing at 60 to 90°C for up to 24 hours is satisfactory. Optimum annealing time and temperature for maximizing a particular property or combination of properties, such as fiber *in vivo* strength retention and dimensional stability, is readily determined by simple experimentation.

The characteristic properties of the filaments such as tensile properties (i.e., straight and knot tensile strengths, Young's Modulus, and elongation) are generally determined with an INSTRON tensile tester. Typical fiber properties of PDS monofilament (size 2/0) are given below in Table 1. Improved mechanical properties and toughness can also be obtained by using higher molecular weight polymer (Jamiolkowski *et al.*, 1995; Datta *et al.*, 1995).

PDS monofilament sutures with improved flexibility and handling characteristics have been reported. These were achieved by additional processing improvements in a proprietary (Broyer, 1995) melt spinning process which includes the step of drawing the filaments in a heated zone maintained at a temperature above the melting temperature of the filament. The resulting monofilament sutures have a higher

Table 1 Fiber properties of PDS monofilament (size 2/0) (Ray *et al.*, 1981)

Property	Value
Diameter	13 mils
Straight tensile strength	80,000 psi
Knot tensile strength	50,000 psi
Elongation to break	30%
Young's modulus	250,000 psi

elongation and lower modulus than comparable sutures obtained without the heated drawing step, and are characterized by a crystalline structure which is more highly ordered in the core of the monofilament suture than in the surrounding annualar area. Some of the improved fiber properties (PDS II) with this new process are described in Broyer (Broyer, 1995).

Breaking Strength Retention *In Vivo* (BSR)

The mechanical properties of a suture should ensure adequate apposition of tissues until the wound heals. The initial strength of the suture should be such that the wound strength should be that of normal tissue. If the suture is absorbable, the reduction in breaking strength should, ideally, not be greater than the gain in wound strength that healing provides.

The *in vivo* breaking strength retention (BSR) of sutures is commonly evaluated by implanting the suture in laboratory animals, usually subcutaneously. The suture strands are recovered after various periods of *in vivo* residence and their breaking strength is determined using an appropriate tensiometer.

Quite commonly, the BSR of a fiber is determined by implanting two strands of the fiber in the dorsal subcutis of each of a number of Long-Evans rats. Typically eight (8) strands are used for each period requiring four rats per period. Thus 16, 24, or 32 segments of each fiber are implanted corresponding to two, three, or four implantation periods. Typical periods of *in vivo* residence for a study can be 5, 7, 14, 21, and/or 28 days. The ratio of the mean value of 8 determinations of the breaking strength at each period to the mean value for the fiber prior to implantation constitutes its breaking strength retention for that period.

In an important 1981 paper (Ray *et al.*, 1981), Ray *et al.* report their work on poly(p-dioxanone). PDS sutures, sizes 2/0 and 6/0, were implanted in the posterior dorsal subcutis of female Long Evans rats for periods of 14, 21 and 28 days. The sutures were recovered at the designated periods and tested for straight tensile strength (Table 2). Monofilament sutures, sizes 2/0 and 6/0, retained an average of 74 per cent of the unimplanted strength following two weeks of *in vivo* residence in the rat subcutis. At four weeks, an average of 58 per cent of the original strength of both dyed and undyed poly(p-dioxanone) suture remained. After six weeks, the strength remaining averaged 41 per cent. At eight weeks, both sizes (2/0 and 6/0) of monofilament poly(p-dioxanone) suture still retained an average of 14 percent of its original strength.

Table 2 *In-vivo* BSR of PDS suture (Ray *et al.*, 1981)

Suture size	Suture strength (%)			
	2 wks	4 wks	6 wks	8 wks
2/0	82	71	49	13
6/0	66	46	34	14
Avg.	74	58	41	14

In Vivo Absorption/Tissue Reaction

The term absorbable suture implies that absorption eventually will cause disappearance of the suture from the site of implantation, thereby limiting the duration of the tissue response. Absorption of synthetic absorbable sutures occurs following hydrolytic or enzymatic breakdown of the polymer chain and resultant molecular weight reduction. With properly produced filaments, this proceeds in a regular and predictable manner in tissue. Like breaking strength, the rate of absorption is assessed following implantation, often intramuscularly, of suture strands in laboratory animals. It is judged by molecular weight degradation, radiotracer studies utilizing labeled suture or, more commonly, by histologic assessment. Absorption of dyed, PDS monofilament suture following *in vivo* residence in rat muscle, judged histologically, is summarized below in Table 3.

Values in Table 3 are expressed in average percent suture remaining. Size 1, 4/0 and 8/0 are based on the data from one lot, whereas size 7/0 is based on data from two lots and size 2/0 data from three lots.

During the healing period, absorbable sutures are replaced by healthy tissue. In general, the tissue response to synthetic absorbable sutures is foreign body in nature. Further, the inflammatory response has been reported by Blomstedt (Blomstedt and Osterberg, 1978) to be less pronounced around suture materials with low capillarity.

Poly(p-dioxanone) is a polyester which degrades by chemical hydrolysis in the body; there is no evidence of chain cleavage by enzymatic degradation.

Ray and his coworkers, besides studying the BSR profiles of PDS monofilament sutures, and their absorption rates, also studied their *in vivo* performance by evaluating tissue reaction (Ray *et al.*, 1981).

Table 3 *In vivo* absorption of PDS monofilament sutures (Ray *et al.*, 1981)

Suture size	Average remaining (%)					
	5 days	91	140	154	168	182
1	100	88.3	—	30.7	—	1.25
2/0	100	85.2	79.4	77.7	43.9	2.0
4/0	100	87.7	—	43.9	—	0
7/0	100	61.4	30.5	—	19.6	—
8/0	100	12.0	—	4.6	—	0

Two, 2 cm, segments of monofilament fiber having a diameter corresponding to size 2–0 suture were implanted aseptically into the left gluteal muscles of 24 female Long Evans rats. The implant sites were recovered after periods of 60, 90, 120 and 180 days and examined microscopically to determine the extent of absorption. After 60 days the suture cross-sections were still transparent and intact. The tissue reactions were slight and most sutures were encapsulated with fibrous tissue.

At 90 days, the sutures were becoming translucent and had lost some of their birefringent properties. A few of the suture cross-sections stained pink (eosinophilic) around the periphery and the edges were indistinct, indicating the onset of absorption. The tissue reactions generally consisted of a fibrous capsule and a layer of macrophages interposed between it and the suture surface.

At 120 days the sutures were translucent, most cross-sections had taken on an eosinophilic stain, and the sutures appeared to be in the process of active absorption. The few reactions consisted of an outer layer of fibroblasts with an interface of several cell layers thick. Absorption at 120 days was estimated to be approximately 70 percent complete. At 180 days, absorption of the suture was substantially complete. The incision healed with minimal adverse tissue reaction.

Several histologic changes accompanied the degradation of the polymer. At the five and 91 day postimplantation intervals, the cross sections of poly(p-dioxanone) were unstained and retractile. At these intervals, the polymer was strongly anisotropic. By the 168th postimplantation day, anisotropy became much less pronounced, and the suture sections appeared to be blue gray. Occasionally, when filaments remained in sites at the 168th day interval, they were extensively fissured and were eosinophilic.

The tissue responses were judged to be slight to minimal for all sizes of poly(p-dioxanone) suture tested. The cellular responses were found only in the vicinity of the implant site and were predominately mononuclear in character. Such tissue responses were similar to those previously reported by Craig and colleagues (Craig *et al.*, 1975) following implantation of polyglactin 910 suture and polyglycolic acid suture in gluteal muscles of rats.

Poly(p-dioxanone) suture elicited foreign body reactions which were judged to be minimal or slight for all periods that the suture remained. At five days, the reaction consisted primarily of small numbers of macrophages and proliferating fibroblasts. Neutrophils, foreign body, giant cells, eosinophils and lymphoctyes were rarely seen in this interval. At 91 days and later periods, no neutrophils were seen and only macrophages and fibroblasts remained consistently present in implant sites until the suture completely absorbed. After absorption, reactions were either absent or identified by the presence of a few enlarged macrophages or fibroblasts localized between otherwise normal muscle cells. Occasionally foci of fat cells were noted to occupy the implant site after absorption of the suture.

The prolonged retention of breaking strength of poly(p-dioxanone) suture is accompanied by a somewhat slower rate of absorption than that reported for polyglactin 910 suture or polyglycolic acid suture. Poly(p-dioxanone) suture was found to be essentially completely absorbed from the rat muscle by 180 days versus 60 to 90 days for polyglactin 910 and more than 120 days for polyglycolic acid suture, when similarly tested.

From the study of ^{13}C-labeled poly(p-dioxanone) suture, absorption was also judged complete by approximately 180 days. Furthermore, ^{13}C was not accumulated in any organ or tissue following suture absorption, indicating that the degradation products of poly(dioxanone) suture are rapidly excreted by the body. Confirmatory evidence of this comes from the excretion profile which closely represents the absorption profile.

The direct correlation of molecular weight and breaking strength loss with both *in vivo* and *in vitro* incubation implies a similar mechanism of degradation. Because *in vitro* incubation provides only a buffered aqueous environment, the chemical degradation of the suture appears, in both instances, to be by nonenzymatic hydrolysis of the ester-bonds. Hydrolysis would be expected to proceed until small, soluble products are formed which then dissolve and are removed from the suture and the implant site.

CONCLUSIONS

Special grade poly(p-dioxanone) synthetic absorbable polymer has been developed into monofilament PDS™ suture, formulated to provide wound support through an extended healing period, as well as to minimize the variability of breaking strength retention and absorption and to invoke minimal tissue reaction. These features are particularly beneficial in critical applications, such as those involving slowly healing tissues.

The inherent flexibility of poly(p-dioxanone) allows it to be fabricated into a monofilament fiber useful for all sizes of sutures. Poly(p-dioxanone) suture has greater pliability than poly(propylene) suture and can provide substantial strength when compared to most other monofilament sutures.

In the body, poly(p-dioxanone) suture retains its strength for longer periods than other synthetic absorbable sutures. It elicits a low order of tissue response and is absorbed by simple hydrolysis.

POLY(p-DIOXANONE) COPOLYMERS

Although great commercial successes have been generated with poly(p-dioxanone) as shown by the development of PDS sutures, as well as ABSOLOK and LAPRA-TY clips, investigators have also prepared poly(p-dioxanone) copolymers (i.e., BIOSYN) with interesting physical properties that can be utilized for various medical device applications. The following describes the polymerization, properties and potential uses of poly(p-dioxanone) copolymers.

POLY(p-DIOXANONE-CO-L(–)LACTIDE) SEGMENTED COPOLYMERS

Like poly(p-dioxanone) homopolymer, copolymers of PDO and lactide can be prepared by several conventional polymerization means. The most commercially viable method relies upon melt polymerization. Since a thermodynamic equilibrium between the monomer (p-dioxanone) and the polymer [poly(p-dioxanone)], results during polymerization (Bezwada *et al.*, 1987, 1990) in conversions ranging from

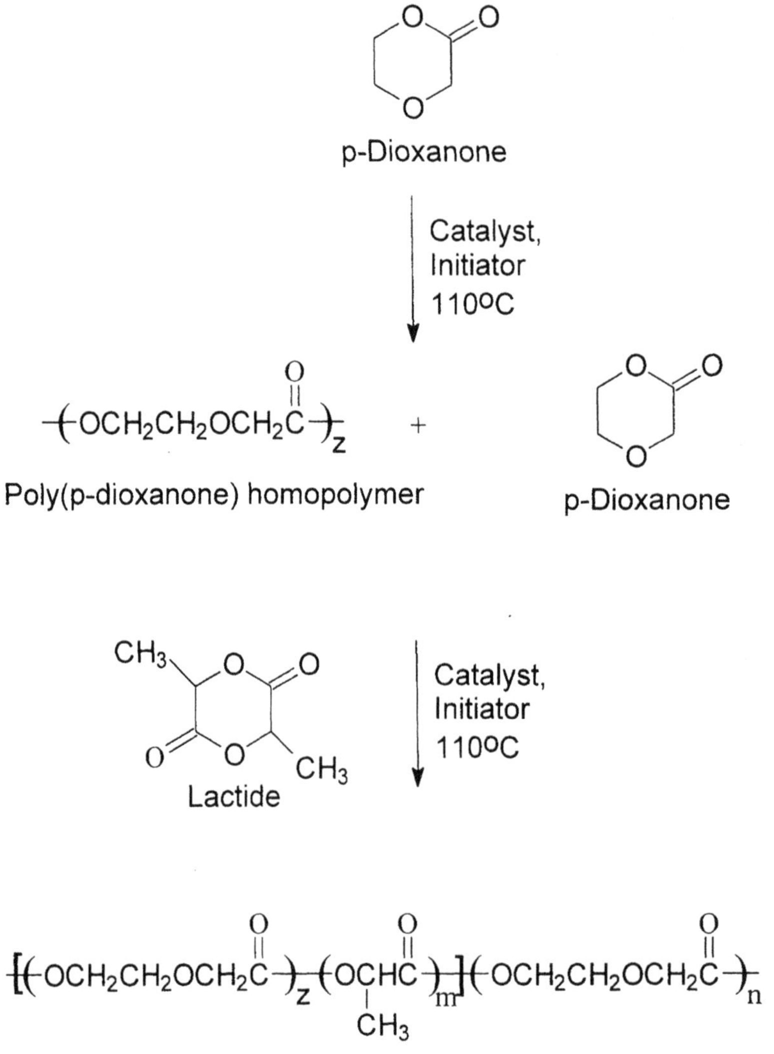

p-Dioxanone

Catalyst,
Initiator
110°C

$$-(OCH_2CH_2OCH_2\overset{O}{\overset{\|}{C}})_z$$ +

Poly(p-dioxanone) homopolymer   p-Dioxanone

Lactide

Catalyst,
Initiator
110°C

$$\{(OCH_2CH_2OCH_2\overset{O}{\overset{\|}{C}})_z (O\overset{}{\underset{CH_3}{C}}H\overset{O}{\overset{\|}{C}})_m\}(OCH_2CH_2OCH_2\overset{O}{\overset{\|}{C}})_n$$

Figure 6 Poly(p-dioxanone-co-(L-) lactide) segmented copolymers.

about 95% at reaction temperatures of 80–85°C, to about 75% at 110°C, to about 50% at 150°C, a convenient way to carry out copolymerization of PDO and other lactones is in a two step process. One first conducts a melt polymerization of p-dioxanone to produce a mixture of PDO homopolymer and monomer (Figure 6).

This homopolymerization is carried out in the presence of a catalytically effective amount of a suitable metal-containing catalyst such as stannous octoate (stannous-2-ethylhexoate — $Sn(C_8H_{15}O_2)_2$) or stannous oxalate. Typically the catalyst levels are reported as monomer:catalyst molar ratios. Level vary widely depending on the catalyst but stannous octoate is preferably used from 15,000:1 to 40,000:1. The polymerization is carried out in the presence of an initiator such as an alkanol, a glycol, a hydroxyacid, or an amine. Specific initiators that have been used include 1-dodecanol, diethylene glycol, glycolic acid, lactic acid, and ethanol amine. Typical proportions of the initiator, reported as monomer:initiator molar ratios, are 500:1 to 1800:1. The polymerization of p-dioxanone is carried out at elevated temperatures under an inert atmosphere for a period of time sufficient to produce a mixture of p-dioxanone homopolymer and p-dioxanone monomer. Typical polymerization reaction temperatures are within the range of 100°C to 130°C.

The polymerization reaction is normally carried out until an equilibrium is reached between polymer and monomer. Depending on the temperature and catalyst concentration, this reaction usually takes from 4 to 8 hours. At a preferred temperature of 110°C, the usual reaction time is 5 to 6 hours.

The resulting mixture is then reacted with lactide and subjected to an elevated temperature for a period of time sufficient to produce a copolymer. As a general rule, the reaction temperature for this polymerization will be within the range of 110°C to 160°C. At reaction temperatures within this range, the polymerization will be complete within a period of about 1 to 4 hours.

Thus, to a dry, round bottom, two neck flask, 95 grams of p-dioxanone, 0.197 ml of 1–dodecanol, and 0.0975 ml of stannous octoate (0.33 molar in toluene) are added. The stirred reaction mixture is heated under nitrogen for 6 hours at 100°C. Five grams of L(–)lactide are added to the reaction mixture, and the temperature is raised to 140°C and maintained there for 2 hours. The copolymer is isolated, ground, and vacuum dried for 48 hours at 80°C.

Using similar procedures, Bezwada and coworkers (Bezwada *et al.*, 1987, 1990) prepared copolymers of PDS-melt/L(–)lactide at 90/10, and 80/20 (by weight). A series of PDO/L(-)lactide random copolymers at 95/5 and 90/10 (by weight) were also prepared for comparative purposes. To a dry round bottom two neck flask, equipped with an overhead stirrer and nitrogen inlet, 95 grams (0.93 mole) of p-dioxanone, 0.197 ml of 1-dodecanol, 0.0975 ml of stannous octoate (0.33 molar in toluene) and 5 grams (0.347 mole) of L(–)lactide were added. The reaction mixture was heated at 110°C for 8 hours. The resulting copolymer was isolated, ground and vacuum dried for 64 hours at 80°C. The melting points of these copolymers were measured utilizing hot stage microscopy; the copolymer compositions were determined by NMR. Inherent viscosity values were measured at a concentration of 0.1 grams of polymer per dL of hexafluoroisopropyl alcohol at 25°C.

Extrusion of the poly(p-dioxanone-co-L(–)lactide) segmented and random copolymers was performed in a similar fashion as that of poly(p-dioxanone). The tensile properties were determined with an INSTRON tensile tester. The breaking strength retention (BSR) *in vivo* of the fibers was determined by implanting two strands of the fiber in the dorsal subcutis of each of a number of Long-Evans rats.

RESULTS AND DISCUSSION

The physical and biological properties of the segmented copolymers are summarized in Table 4. Compared with p-dioxanone homopolymer, the segmented copolymers displayed a gradual increase in compliance (the reciprocal of Young's modulus) with an increase in the concentration of lactide moieties. From a glass transition standpoint alone, one might think that adding lactide would increase the modulus of the copolymers. A small amount of lactide, however, is morphologically disruptive enough to slightly lower the crystallinity to result in a softer material. (Also note in Table 4 the drop in the melting point with increasing lactide content, indicative of increasingly poorer crystal perfection.) When considering segmented copolymers low in lactide, this disruption is still not great enough to significantly lower tensile strength.

Similar properties could not be realized for the random copolymers of the same overall composition because of the lack of adequate crystallinity in the fibers. (See Table 5; the melting transion for the 90–10 random copolymer is 12°C lower than the corresponding 90–10 segmented copolymer shown in Table 4.) The absorption profile and breaking strength retention of the segmented fibers reflected an interesting balance between the decreasing crystallinity of fibers and decreasing rate of hydrolysis of the copolymers with incorporation of lactide moieties in the chain. This situation is more noticeable in the BSR profile of the monofilaments than for their absorption.

Table 4 Physical properties of PDO-PLA segmented copolymers

	Composition (by wt.) of PDO-PLA			
	100–0 (PDS)	95–5	90–10	80–20
I.V. (dL/g)	1.8	2.36	2.15	2.53
Tm (°C)	107	100	102	98
Fiber properties				
Diameter (mils)	8.7	7.4	7.9	7.0
Ten. strength (Kpsi)	75	95	91	72
Knot strength (Kpsi)	45	60	48	54
Elongation (%)	28	32	38	48
Young's modulus (Kpsi)	278	222	143	152
In vitro BSR (%) @ 50°C/pH 7.27				
4 days	88	81	70	67
7 days	—	71	—	45
In vivo BSR (%)				
3 wks	60	—	64	—
4 wks	50	—	48	—
6 wks	25	—	—	—
8 wks	10	—	9	—

Table 5 Physical properties of PDO-PLA random copolymers

	Composition (by wt.) PDO-PLA	
	95–5	90–10
I.V. (dL/g)	1.5	1.75
Tm (°C)	—	90
Fiber properties		
Diameter (mils)	7.7	7.0
Ten. strength (Kpsi)	59	64
Knot strength (Kpsi)	39	41
Elongation (%)	43	46
Young's modulus (Kpsi)	217	95

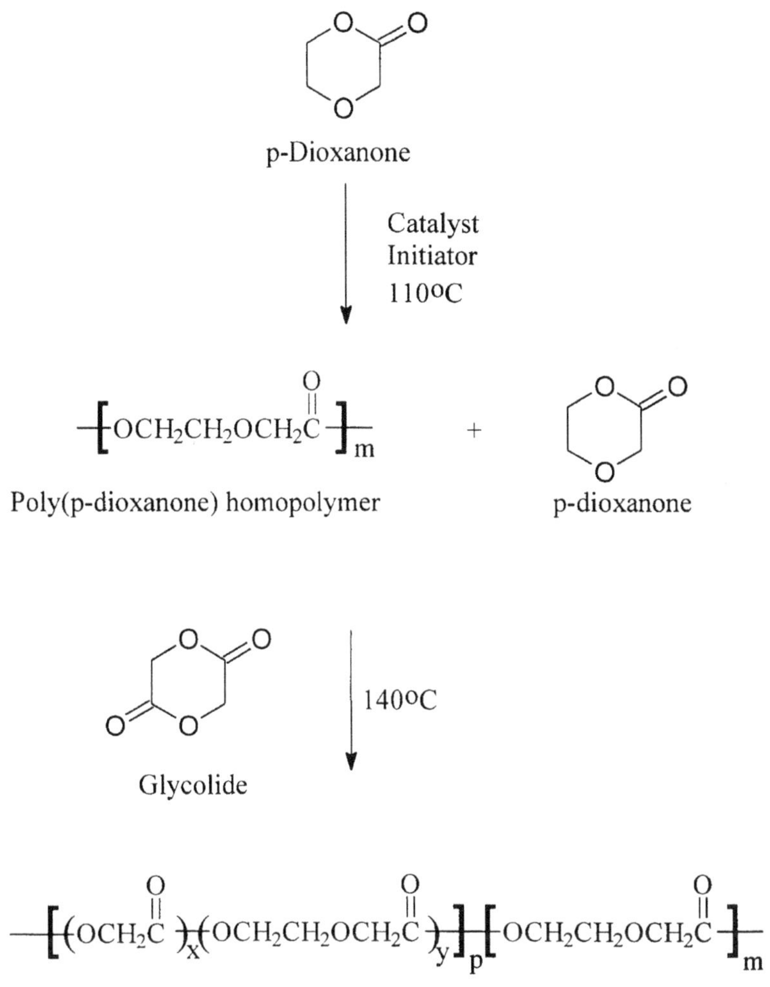

Figure 7 Poly(p-dioxanone-co-glycolide) segmented copolymers.

POLY(P-DIOXANONE-CO-GLYCOLIDE) SEGMENTED COPOLYMERS

Segmented copolymers of p-dioxanone/glycolide (Bezwada, Shalaby and Newman, 1987) can be prepared by a similar method as segmented copolymers of p-dioxanone/L(–)lactide. Firstly, p-dioxanone (PDO) monomer is melt polymerized to produce a mixture of poly(p-dioxanone) homopolymer and p-dioxanone monomer.

The polymerization is carried out in the presence of a catalyst, such as stannous octoate and an initiator such as 1-dodecanol, or diethylene glycol. Typical polymerization reaction temperatures are from 100°C to 130°C, preferably 110°C for 4 to 8 hours. The polymerization is carried out until an equilibrium is reached between polymer and monomer. This is usually obtained in about 5 to 6 hours.

Next, glycolide is added to the mixture, and the resulting reaction mixture's temperature is elevated to 120 to 180°C for 1 to 4 hours (Figure 7).

The proportion of glycolide that is added to the mixture of p-dioxanone homopolymer and monomer can be varied, and in the examples that follow, it is from 3 to 25 weight percent, based on total weight of the reaction mixture (i.e., total weight of glycolide, p-dioxanone homopolymer, and p-dioxanone monomer).

Thus, to a flame dried, 250 milliliter, round bottom, three-neck flask fitted with a flame dried overhead stirrer, nitrogen inlet and stopper, 95 grams (0.9306 mole) of p-dioxanone, 0.266 milliliter of 1–dodecanol, and 0.0984 milliliter of stannous octoate (0.33 molar solution in toluene) were added. The contents of the reaction flask were held under high vacuum at room temperature for about 16 hours. The reaction mixture was then heated to 110°C, and maintained there for 5 hours with stirring. A small sample of the resulting polymer was removed for analysis (inherent viscosity = 1.21 dL/g), and 50.0g (0.04308 mole) of glycolide was added to the reaction mixture. The temperature was raised to 140°C, and maintained there for 1 hour. The temperature was lowered to 90°C, and maintained there for 65 hours.

The resulting copolymer was isolated, ground, and dried for 48 hours at 80°C under 0.1 mm Hg to remove any unreacted monomer. A weight loss of 13.7% was observed.

Using this general scheme, two more copolymers of PDO/glycolide at 90/10 and 80/20 initial weight composition were prepared. It was found that the polymerization of the first stage occurred best at 6 hours at 100°C, with a second stage of 1 to 2 hours at 140°C.

The copolymers were characterized using Varian XL-300H and C^{13} NMR, DuPont DSC, Waters-Model 150C GPC, and X-ray diffractometers. Inherent viscosities (I.V.) were obtained in hexafluoroisopropyl alcohol (HFIP) at 25°C and 0.1 g/dL concentration using a Ubbhlode viscometer. Properties of these polymers are summarized in Table 6.

The fibers were extruded in a fashion similar to that used for the poly(p-dioxanone-co-L(–)lactide) copolymers. The resulting oriented filaments have good straight and knot strength. Oriented and rack annealed (12 hours/60°C, constrained length/no relaxation) fiber properties of these copolymers are summarized in Tables 7 and 8. The effect of initiators 1-dodecanol and diethylene glycol on the copolymer properties were also studied, and the properties are summarized in Table 9. *In vivo* absorption and breaking strength retention is compared with PDS homopolymer (Table 10).

Table 6 Properties of PDO-PGA segmented copolymers

Feed composition (by wt.) PDO-PGA	100–0	95–5	90–10	80–20
Final comp. (by mole %) PDO-PGA	100–0	—	87–13	83–17
I.V. (dL/g)	1.8	1.63	1.44	1.64
Conversion (%)	95	86	86	93
Crystallinity (%)	50	—	45	40
Tm (°C)	–12	—	–8	–2
Tg (°C)	112	—	97	125

Table 7 Drawing conditions and properties of unannealed PDO-PGA copolymers

PDO-PGA (wt. %)	95–5	90–10	80–20
Drawing Conditions	4× at 58°C followed by 2× at 75°C	5× at 52°C followed by 2× at 72°C	5× at 50°C followed by 2× at 71°C
Dia.(mil)	7.1	7.3	7.5
Tensile strength (Kpsi)	88	87	65
Knot strength (Kpsi)	53	49	43
Yield(%)	49	61	94
Modulus (Kpsi)	211	143	81

Table 8 Tensile properties of drawn, annealed monofilaments of PDO-PGA segmented copolymers

Composition (by wt.) PDO-PGA	100–0 (PDS)	95–5	90–10	80–20
Fiber properties				
Diameter (mils)	8.65	7.5	7.3	7.5
Str. strength (Kpsi)	75	79	85	61
Knot strength (Kpsi)	45	50	62	51
Elongation (%)	28	34	39	55
Young's modulus (Kpsi)	278	281	283	201
In vitro BSR (%)				
4 days/50°C	88	79	49	43

Table 9 Effect initiator on fiber properties

Composition (by wt.) PDO-PGA	90–10	90–10
Initiator type	1–dodecanol	diethylene glycol
I.V. (dL/g)	1.44	1.88
Fiber Properties (annealed 12hrs./60°C)		
Diameter (mils)	7.3	7.4
Str. strength (Kpsi)	85	88
Knot strength (Kpsi)	62	55
Elongation (%)	39	30
Young's Modulus (Kpsi)	283	204
In-vitro BSR (%)		
4 days/50°C	49	34

Table 10 Comparative properties of PDS and a PDO-PGA
90–10 segmented copolymer

Composition (PDO-PGA)	100–0 (PDS)	90–10
I.V. (dL/g)	1.8	1.82
Tm (°C)	112	100
Fiber properties		
Diameter (mils)	8.7	7.8
Str. strength (Kpsi)	75	86
Knot strength (Kpsi)	45	53
Elongation (%)	28	48
Young's modulus (Kpsi)	278	221
In-vivo BSR (%)		
3 wks	80	30
4 wks	70	12
6 wks	50	0
In vivo absorption (wt % remaining)		
91 days	96	23
119 days	83	0
182 days	43	—
210 days	0	—

RESULTS AND DISCUSSION

Melt polymerization of p-dioxanone proceeds to about 75% conversion in 5–6
hours at 110°C when using stannous octoate as a catalyst at a level as described
above. Addition of glycolide at this stage to the p-dioxanone polymerization
mixture allows for the copolymerization of the unreacted p-dioxanone with glycolide to produce amorphous glycolide/p-dioxanone segments at the end of the
poly(p-dioxanone) chains. Incorporation of such segments provides much more
compliant monofilaments as compared with those made from poly(p-dioxanone).
The segmented copolymers provided monofilaments with higher strength and
improved absorbability and more rapid loss of *in vivo* breaking strength than
poly(p-dioxanone).

With an increase of glycolide content from 0 to 20%, crystallinity appeared to
drop from 50% to 40%. The Tg of the copolymers increased with the increase in
glycolide ratio, and a slight increase of Tm could be observed. Copolymer with
20% glycolide displayed a Tm of 125°C. It appears that diethylene glycol (DEG)
initiated 90/10 copolymer yielded fiber possessing similar tensile strength to that
derived from 1-dodecanol initiated polymer; these fibers however lost strength *in vitro*
slightly faster and had a slightly lower modulus. This may suggest that the copolymer
has a different microstructure; this is important as certain key properties can be
effected by such microstructures. The segmented copolymers of PDO/glycolide at
90/10 exhibited a much faster BSR profile, with complete absorption in 119 days,
whereas complete absorption takes 210 days for poly(p-dioxanone) monofilament.
The absorption and BSR profile of the segmented copolymer of PDO/glycolide
monofilaments are similar to PGA braided sutures.

POLY(p-DIOXANONE-CO-GLYCOLIDE) BLOCK COPOLYMERS

Unlike poly(p-dioxanone-co-glycolide) segmented copolymers, poly(p-dioxanone-co-glycolide) block copolymers are prepared by reacting p-dioxanone homopolymer that is essentially free of unreacted monomer (Jamiolkowski *et al.*, 1989). That is, the homopolymer, prior to the copolymerization reaction, contains not more than about 3 or 4 weight percent of unreacted p-dioxanone monomer. The reaction of the p-dioxanone homopolymer with glycolide is preferably carried out in such a manner that the homopolymer is first dissolved (or at the very least, intimately mixed) in the glycolide monomer before significant polymerization of the glycolide occurs. This is done in order to minimize the presence of homopolymeric species in the final product. This leads to copolymers that are blocky, tending towards an ABA structure. Furthermore, by reacting substantially monomer-free poly(p-dioxanone) with glycolide, the copolymers will contain poly(glycolide) blocks or sequences that are capable of developing a significant degree of crystallinity (Figure 8).

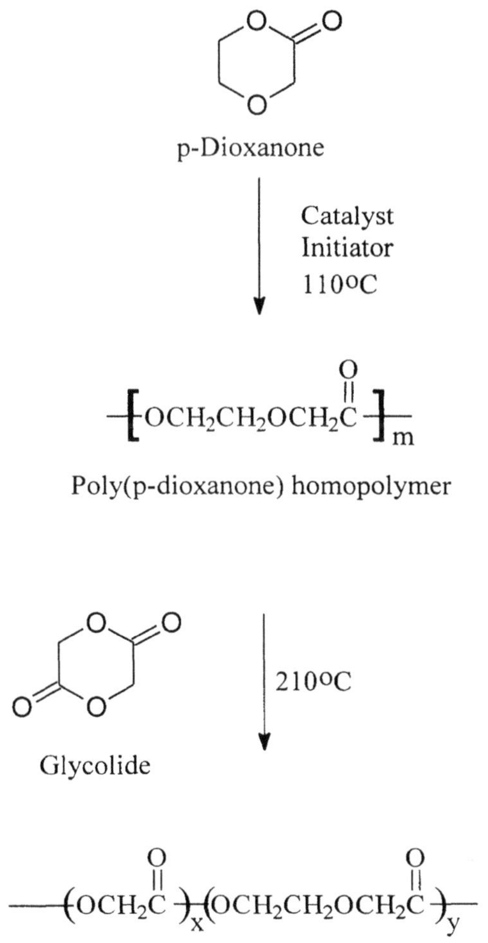

p-Dioxanone

Catalyst
Initiator
110°C

Poly(p-dioxanone) homopolymer

210°C

Glycolide

Figure 8 Poly(p-dioxanone-co-glycolide) block copolymers.

As a general rule, the reaction temperature for the reaction of poly(p-dioxanone) with glycolide is 140 to 240°C. When the glycolide content of the reaction mass is less than about 50 weight percent, the reaction temperature can be 140 to 180°C. When glycolide is the predominant component in the reaction mass, the reaction temperatures are usually raised to 200 to 235°C. By using as low of a reaction temperature as possible, the incidence of transesterification reactions is lowered, and therefore, randomization is diminished.

Thus, p-dioxanone homopolymer is prepared, as described previously, by charging pure p-dioxanone, dodecanol, and a catalytic amount of stannous octoate in toluene solution (0.0025 mole percent based on monomer) to an appropriate reactor, and heating under an inert dry nitrogen atmosphere at 90°C for one hour.

Then, to a flame dried, 250 ml round bottom three-neck flask, equipped with a overhead stirrer, nitrogen inlet and stopper, 25.0 g of the poly(p-dioxanone) homopolymer was added. The flask was equipped with a vacuum adapter. Vacuum was applied and the flask was lowered into a silicone oil bath heated at 80°C. Heating at 80°C under high vacuum was maintained for 16 hours to remove any residual water and to remove as much residual monomer as possible (i.e., 2 wt%).

The vessel was removed from the oil bath and allowed to cool. Then, one hundred grams of pure glycolide monomer was introduced into the flask under dry nitrogen. The flask was then placed in a preheated oil bath at 120°C. Within 10 minutes at 120°C the stirred reaction mass was noted to be clear and not very viscous. After 15 minutes, the temperature was increased to 140°C and maintained there for 10 minutes upon which it was noted that the PDS appeared to be completely dissolved. The reaction temperature was then raised to 215°C. Stirring of the highly viscous melt was continued for an additional 2 hours. At this stage, the vessel was removed from the oil bath and allowed to cool under a stream of nitrogen. The copolymer was isolated, ground, and dried in vacuo at room temperature and at 110°C for 1.5 and 16 hours, respectively, to remove any unreacted monomer.

The resulting poly(p-dioxanone-b-glycolide) block copolymer (20/80 initial weight ratio) had a melting point of 210°C with an IV of 1.59 dL/g. A weight loss of 11.1% was observed.

Using a similar procedure, poly(p-dioxanone-b-glycolide) block copolymers at 30/70, 40/60, 50/50, and 60/40 initial weight ratios were prepared, isolated and characterized.

The copolymers were extruded into monofilaments using an INSTRON capillary rheometer. The extrudates were drawn in two stages at about 54°C and then about 74°C using draw ratios of about 4.5 and 1.3X, respectively. After annealing, tensile properties were measured (Table 11).

Table 11 Properties of PDO-PGA block copolymers

Composition (by wt.) PDO-PGA	20–80	30–70	40–60	50–50	60–40
I.V. (dL/g)	1.59	1.94	1.68	1.60	1.53
Tm (°C)	210	205	200	200	170
Fiber properties					
Diameter (mils)	6.7	6.3	7.7	7.5	7.5
Str. strength (Kpsi)	47	58	115	99	104
Knot strength (Kpsi)	37	53	82	74	55
Elongation (%)	40	53	45	56	67
Young's Modulus (Kpsi)	627	551	591	230	93

RESULTS AND DISCUSSION

The block copolymers of PDS and PGA with 20 to 60% PDS moieties can be made into crystalline materials which may be converted to strong, complaint monofilaments. Copolymers based on 50 to 60% PDS moieties produce very high strength and highly compliant monofilaments. The melting data indicates that the crystalline phase is primarily composed of PGA moieties, while those of PDS dominate the amorphous phase. Therefore, copolymers with about 50–60% PDS moieties exhibit moduli equal to or less than those of p-dioxanone homopolymer, with melting temperatures closer to that of PGA. Monofilaments made from these copolymers, because of the presence of the rapidly hydrolyzable glycolide moieties and morphological considerations, display an accelerated loss of their *in vivo* breaking strength profile as compared with PDS sutures.

POLY(p-DIOXANONE-CO-ε-CAPROLACTONE) COPOLYMERS

Both blocky and random poly(p-dioxanone-co-ε-caprolactone) copolymers were prepared and extruded into monofilaments to study their physical and biological properties (Bezwada, Shalaby, and Erneta, 1991). Random copolymers were prepared by polymerizing the desired proportions of p-dioxanone and ε-caprolactone in a single step process in the presence of an organometallic catalyst and an initiator at elevated temperatures as described above for p-dioxanone.

Thus, to a flame dried 250 ml, round bottom three-neck flask, equipped with a flame dried overhead stirrer, nitrogen inlet and stopper, 90 grams (0.7885 mole) of distilled ε-caprolactone, 10 grams (0.0980 mole) of p-dioxanone, 0.0253 ml of distilled diethylene glycol, and 0.108 ml of stannous octoate (0.33 molar solution in toluene) were added. The flask was held under high vacuum (0.1 mm Hg) for about 16 hours then purged with nitrogen three times before being vented with nitrogen. The stirred reaction mixture was then placed in an oil bath at 160°C for 24 hours, then cooled to 110°C and stirred for an additional 24 hours. The polymer was cooled to room temperature under a stream of nitrogen, isolated, ground and devolatilized under vacuum (0.1 mm Hg) at 50°C for 16 hours, then at 80°C for an additional 48 hrs.

The polymer was melt spun, drawn and annealed to prepare oriented, dimensionally stable filaments using extrusion techniques previously described.

For the preparation of block copolymers, ε-caprolactone is polymerized into a "prepolymer" and then reacted with p-dioxanone monomer. This results in a copolymer with a center block of ε-caprolactone and one or two end blocks of p-dioxanone. This forms block copolymers with a diblock (AB) or triblock (ABA) architecture (Figure 9). Diblock copolymers can be prepared by first prepolymerizing ε-caprolactone with a monofunctional initiator such as 1-dodecanol, followed by polymerization with p-dioxanone monomer.

Thus, a diblock copolymer is formed by adding to a flame dried, 250 ml round bottom three-neck flask equipped with a flame dried overhead mechanical stirrer, nitrogen inlet and stopper, 10 grams (8.9 wt. percent) of monohydroxy-terminated poly(ε-caprolactone) prepolymer with a weight average molecular weight of 10,000 (gel permeation chromatography, Scientific Polymer Products, Inc). The reaction

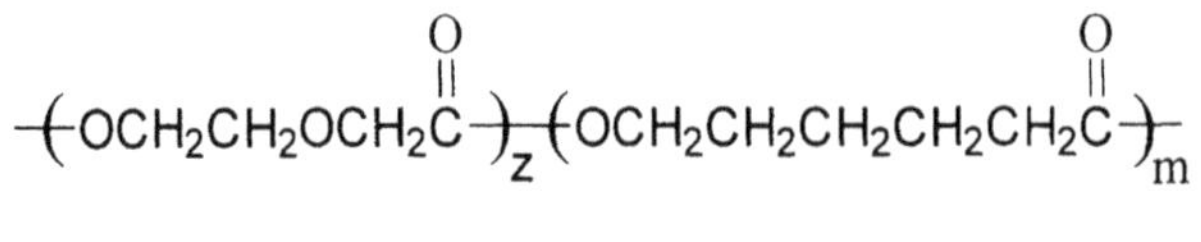

$$\text{-}\!\left(\text{OCH}_2\text{CH}_2\text{OCH}_2\overset{\overset{\text{O}}{\|}}{\text{C}}\right)_{\!z}\!\left(\text{OCH}_2\text{CH}_2\text{CH}_2\text{CH}_2\text{CH}_2\overset{\overset{\text{O}}{\|}}{\text{C}}\right)_{\!m}$$

Diblock (A-B) copolymer

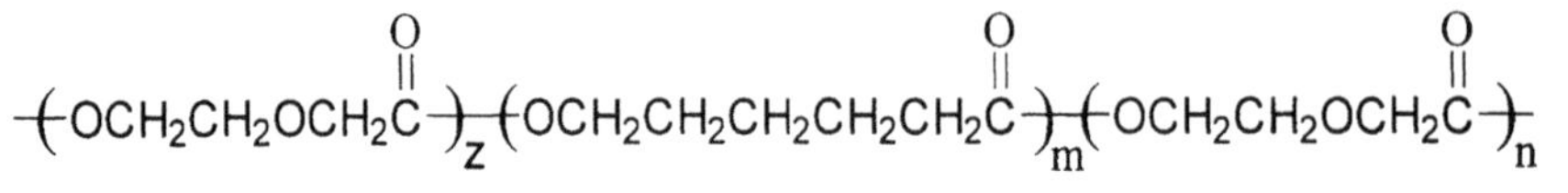

$$\text{-}\!\left(\text{OCH}_2\text{CH}_2\text{OCH}_2\overset{\overset{\text{O}}{\|}}{\text{C}}\right)_{\!z}\!\left(\text{OCH}_2\text{CH}_2\text{CH}_2\text{CH}_2\text{CH}_2\overset{\overset{\text{O}}{\|}}{\text{C}}\right)_{\!m}\!\left(\text{OCH}_2\text{CH}_2\text{OCH}_2\overset{\overset{\text{O}}{\|}}{\text{C}}\right)_{\!n}$$

Triblock (A-B-A) copolymer

Figure 9 Poly(p-dioxanone-co-caprolactone) copolymers.

flask was then held under high vacuum at 80°C for about 64 hours. After cooling to room temperature, the flask was charged with 102.1 gm (1.0 mole, 91.1 wt. percent) of p-dioxanone, and 0.101 ml of stannous octoate (0.33 molar solution in toluene). The contents of the reaction flask were then held under high vacuum at room temperature for about 16 hours, followed by purging with nitrogen three times before being vented with nitrogen.

The flask was then placed in an oil bath at 90°C. After 15 minutes, the stirred, low viscosity solution became clear. The viscosity began to increase and the temperature was raised to 100°C. Stirring of the high viscosity melt was continued for 4 hours. The temperature was then lowered to 90°C and stirring was continued for an additional 24 hours. Then, the temperature was lowered to 80°C. Stirring was continued at 80°C for three more days. The resulting copolymer was cooled to room temperature under a stream of nitrogen, isolated, ground and dried for 8 hrs. at 60°C, then 8 hrs. at 70°C under high vacuum (0.1 mm Hg) to remove residual monomers (about 15 percent).

The poly(p-dioxanone-co-ϵ-caprolactone) block copolymer had an inherent viscosity of 2.18 dL/g in hexafluoroisopropyl alcohol (HFIP) at 25°C, and a melting point by hot stage microscopy of 110°C.

Triblock copolymers can be prepared in a similar manner, except that the initiator used for prepolymerization of ϵ-caprolactone is difunctional (i.e., diethylene glycol). This will yield a triblock copolymer, with a center block of poly(ϵ-caprolactone) and end blocks of poly(p-dioxanone).

The copolymers were melt spun, drawn and annealed to prepare oriented, dimensionally stable filaments using conventional extrusion techniques.

The mechanical and biological properties are described in the literature (Bezwada, Shalaby and Erneta, 1991) and show that a surgical filament prepared from a block copolymer of ϵ-caprolactone and p-dioxanone has equivalent straight and knot tensile strength to p-dioxanone homopolymer, but has significantly enhanced flexibility as demonstrated by the reduction in Young's Modulus.

The block copolymer filaments are semi-crystalline and have mechanical properties substantially equivalent to the mechanical properties of p-dioxanone

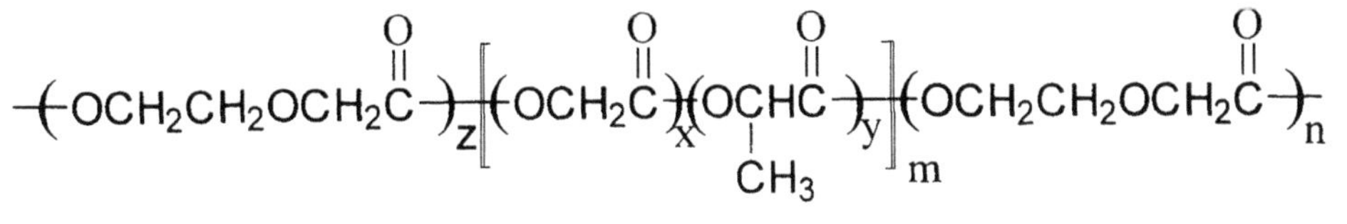

$$\text{+(OCH}_2\text{CH}_2\text{OCH}_2\overset{\overset{\displaystyle O}{\|}}{C}\text{+)}_z\left[\text{(OCH}_2\overset{\overset{\displaystyle O}{\|}}{C}\text{)}_x\text{(O}\underset{\underset{\displaystyle CH_3}{|}}{C}H\overset{\overset{\displaystyle O}{\|}}{C}\text{+)}_y\right]_m\text{(OCH}_2\text{CH}_2\text{OCH}_2\overset{\overset{\displaystyle O}{\|}}{C}\text{+)}_n$$

Triblock (A-B-A) terpolymer

Figure 10 Poly(p-dioxanone-co-glycolide/lactide) block terpolymers.

homopolymer, with a straight tensile strength of 60,000 psi, and a knot tensile strength of 40,000 psi. The Young's Modulus was 300,000 to 100,000 psi, with elongation-to-break of 30 to 80 percent. Furthermore, the block copolymer filaments exhibit an *in vivo* absorption profile comparable to poly(p-dioxanone), with complete absorption occurring not more than 210 days after implantation.

POLY(p-DIOXANONE-CO-GLYCOLIDE-CO-LACTIDE) TERPOLYMERS

Terpolymers comprising glycolide, lactide and p-dioxanone have also been prepared (Bezwada, Newman and Shalaby, 1991). Polymerization consists of reacting glycolide and lactide to form a low molecular weight "prepolymer", which is then further polymerized with p-dioxanone to form the terpolymer (Figure 10).

The ratio of lactide to glycolide used to prepare the prepolymer must be adjusted so that the resulting prepolymer structure is amorphous. The amount of amorphous lactide/glycolide prepolymer used to prepare the terpolymer can vary over a wide range, and will depend to a great extent on the physical and absorption properties desired. Typically, an amount from 5 to 50 percent by weight of the composition of the polyester is acceptable. This is important to facilitate the successful copolymerization of the prepolymer with p-dioxanone.

Amorphous prepolymers exhibiting the properties desired for copolymerization with p-dioxanone can generally be prepared at a molar ratio of lactide to glycolide as low as 50:50, but a ratio of 60:40 to 70:30 is preferred. Generally, if the amount of glycolide exceeds 50 mole percent of the prepolymer, then the prepolymer would not only crystallize, but its melting temperature would be greater than the desired temperature for copolymerizing with p-dioxanone. Prepolymers prepared from greater than 70 mole percent lactide typically provide a polyester exhibiting a slower rate of absorption and a larger retention of breaking strength. For most applications, this is not desired.

The prepolymer of lactide and glycolide is prepared by conventional polymerization techniques. The terpolymer is then prepared by polymerizing the desired proportions of prepolymer and p-dioxanone in the presence of an organometallic catalyst and an initiator at elevated temperatures from 100 to 160°C for approximately 16 hours. Alternatively, the polymerization can be carried out in 2 or more successive stages — for example, for 1–2 hours at 100–140°C, followed by 2 to 5 days at 80°C.

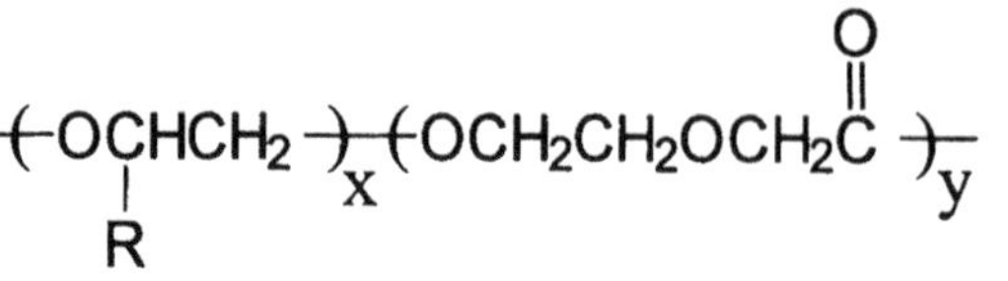

$$\text{+OCHCH}_2\text{)}_x\text{(OCH}_2\text{CH}_2\text{OCH}_2\overset{\overset{\displaystyle O}{\displaystyle \|}}{C}\text{)}_y$$
$$\text{R}$$

R = hydrogen or alkyl

Figure 11 Poly(p-dioxanone-co-alkylene oxide) copolymers.

These terpolymers have a degree of crystallinity and an intrinsic viscosity which render them suitable for extrusion into fibers or films and for injection molding into surgical devices such as staples. *In vivo* BSR and rate of absorption of the formed filaments were determined. The straight tensile strength of monofilaments is greater than 50,000 psi, while the knot tensile strength is greater than 40,000 psi. Additionally, the Young's Modulus was less than 500,000 psi, with some displaying moduli less than 300,000 psi, while the elongation was less than 80%.

POLY(p-DIOXANONE-CO-ALKYLENE OXIDE) COPOLYMERS

Copolymers of p-dioxanone and alkylene oxides have also been prepared (Bezwada and Shalaby, 1991) (Figure 11). These crystalline copolymers can be either blocky or branched. Diblock (AB) and triblock copolymers (ABA) are derived from mono-functional and difunctional poly(alkylene oxides), respectively, while branched copolymers, are formed from polyfunctional poly(alkylene oxides). Polymerization occurs by reacting p-dioxanone with the desired type and amount of poly(alkylene oxide) in the presence of an organometallic catalyst at elevated temperatures.

Thus, to a flame dried, 250 ml round bottom, three-neck flask, equipped with an overhead stirrer, nitrogen inlet and stopper, 4.18 grams of a propylene-ethylene oxide (Pluronic, Tm 68°C, Mn = 8350 g/mol) block copolymer was added. The reaction flask was held under high vacuum at 80°C for about 18 hours. After cooling to room temperature, the reaction flask was charged with 102.1 gm (1.0 mole, 96.1 wt. percent) of PDO, and 0.101 ml of stannous octoate (0.33 molar solution in toluene). The contents of the reaction flask were held under high vacuum at room temperature for about 16 hours. The flask was then fitted with a flame dried mechanical stirrer and an adapter, and purged with nitrogen three times before being vented with nitrogen. The flask was then placed in an oil bath at 110°C and maintained there for about one hour, lowered to 90°C for an additional 24 hours, and then lowered further to 80°C for three more days. The copolymer was cooled to room temperature, isolated, ground and dried for 16 hours at 60°C, 16 hours at 70°C, and finally for 32 hours at 80°C under high vacuum (0.1 mm Hg) to remove any unreacted monomer (16%).

The resulting poly(p-dioxanone-co-ethylene propylene oxide) copolymer had an inherent viscosity of 3.43 dL/g in hexafluoroisopropanol (HFIP), and a melting point of 120°C.

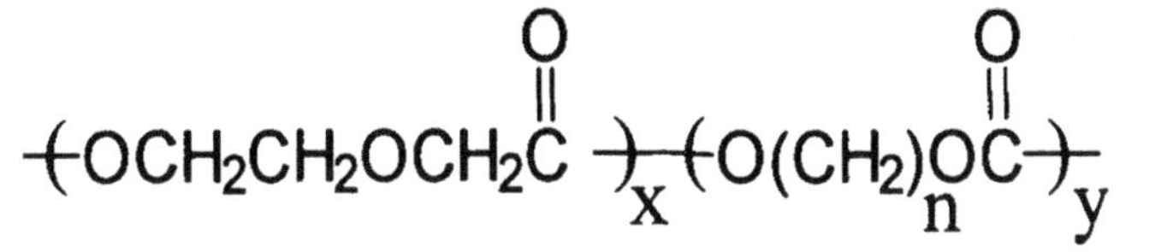

$$\left(OCH_2CH_2OCH_2\overset{\displaystyle O}{\overset{\displaystyle \|}{C}}\right)_x\left(O(CH_2)_n\overset{\displaystyle O}{\overset{\displaystyle \|}{OC}}\right)_y$$

n = 1 to 10

Figure 12 Poly(p-dioxanone-co-alkylene carbonate) copolymers.

Using methods previously described, poly(p-dioxanone-co-ethylene propylene oxide) copolymers were melt extruded into monofilaments. Tensile strengths were found to be equivalent to PDS homopolymer, but handling properties were improved (Bezwada and Shalaby, 1991).

POLY(p-DIOXANONE-CO-CARBONATE) COPOLYMERS

Such as the previously described, poly(p-dioxanone-co-alkylene oxide) copolymers, bioabsorbable copolymers comprising linkages other than esters have been prepared. Unless one views carbonates as esters of carbonic acid, this would also include the family of polymers based upon p-dioxanone and AA-BB type poly(alkylene carbonate)s such as poly(hexamethylene carbonate) (Bezwada, Shalaby and Hunter, 1991) (Figure 12).

Synthesis involves the formation of a poly(alkylene carbonate) prepolymer formed from intermediates of organic carbonate moieties by reaction of a diol and organic carbonate monomer at elevated temperatures in the presence of a catalyst such as stannous octoate. Diols include 1,6-hexanediol, 1,4-butanediol and 1,8-octanediol, while carbonates include diphenyl carbonate and dibutyl carbonate. p-Dioxanone is then added to form the poly(p-dioxanone-co-alkylene carbonate) copolymer.

The molecular weight of the final copolymer depends upon the molecular weight of the particular prepolymer employed in its synthesis. Low molecular weight prepolymers will yield low molecular weight bioabsorbable copolymers and high molecular weight prepolymers will yield high molecular weight polymers. Additionally, the ratio of reactants will also control the molecular weight of the copolymer.

As is typical in a polycondensation reaction employing transesterification, the molecular weights of the formed prepolymers are controlled by in large measure by reaction conditions. The ratios of the respective reactants can influence the time needed for reaction - the prepolymers monomer feed is typically an excess of 0.5 to 0.6 mole percent diol and a slight deficiency of 0.4 to 0.5 mole percent organic carbonate.

Hence, a prepolymer of poly(hexamethylene carbonate) was formed by adding to a flame dried 250 ml single neck flask 36.6 g of 1,6-hexanediol and 69.2 g of diphenyl carbonate. The flask was fitted with a mechanical stirrer, a distillation adapter and receiving flask. The reaction flask was held under high vacuum at room temperature for about 18 hours. Then, the flask was placed in an oil bath, under a stream of nitrogen, at 180°C for 1.5 hours. The stirred solution became clear after 15 minutes and began to increase in viscosity as volatiles were distilled off. The temperature of the reaction was then raised to 200°C for 1 hour, followed by polymerization at 220°C for 3 hours. Then, the temperature was lowered to 200°C and a strong vacuum (0.01 mm Hg) was slowly applied to remove residual monomers. Stirring under vacuum was continued for an additional 18 hours.

The resulting poly(hexamethylene carbonate) was cooled to room temperature under a stream of nitrogen, isolated, ground, and dried under vacuum (0.1 mm Hg). The polymer had an I.V. of 0.38 dL/g.

Five grams of the poly(hexamethylene carbonate) was then added to a flame dried 100 ml two-neck flask, equipped with an overhead stirrer and nitrogen inlet, and dried at 60°C and 0.1 mm Hg for 24 hours. p-Dioxanone (20 g (0.196 mole)) and 0.02 ml of stannous octoate (0.33 molar in toluene) were then added to the flask and held under high vacuum (0.1 mm Hg) at room temperature for an additional 24 hours. The flask was then placed in an oil bath at 110°C. After 15 minutes, the stirred solution became homogenous and began to increase in viscosity. Stirring was continued for 8 hours under nitrogen.

The resulting poly(p-dioxanone-co-hexamethylene carbonate) copolymer was cooled to room temperature, isolated, ground, and dried at 80°C under high vacuum to remove any unreacted monomer. The copolymer had an I.V. of 0.38 dL/g.

These low molecular weight copolymers have been found useful in the fabrication of surgical articles, particularly in bioabsorbable coatings for sutures due to their very soft, pliable characteristics.

Their softness allows penetration deeply into the coated substrate, and consequently, leaves less coating build-up on the substrate's surface. For coated sutures, such build-up is manifested by flaking which occurs when the suture is tied into a knot or secured otherwise. Thus, when sutures are coated with poly(p-dioxanone-co-alkylene carbonate), flaking is reduced. Furthermore, because the low molecular weight copolymers penetrate into the suture, the coatings exhibit enhanced conformability. Additionally, tactile smoothness is imparted to surgical articles. Such smoothness serves, for example, to reduce the tissue drag. Lower molecular weight coatings also lessen the degree of fibrillation during the formation of both wet and dry knots. In addition, the low molecular weight copolymers are more readily absorbed into bodily tissues than higher molecular weight polymers.

POLY(p-DIOXANONE-GLYCOLIDE-LACTIDE) RANDOM TERPOLYMERS

Random terpolymers of p-dioxanone, glycolide and lactide have also been prepared (Bezwada and Kronenthal, 1991) and have exhibited excellent coating properties due to their softness and pliability. Polymerization occurs by reacting p-dioxanone, lactide and glycolide with a mono- or polyhydric alcohol initiator such as diethylene glycol, glycerol, 1-dodecanol, and mannitol, or a hydroxy acid such as lactic or glycolic acid.

Molecular weights and viscosities can be controlled by the type of initiator and by the ratio of initiator to monomer. Handling properties of the coating copolymers can be controlled by changing the mole ratios of p-dioxanone, lactide and glycolide.

Type and proportion of initiator control the molecular structure and molecular weight of the copolymer. The proportion and type of initiator is selected such that the polymer will have an inherent viscosity of 0.05 to 0.5 dL/g as determined in hexafluoroispropanol at 25°C and a concentration of 0.1 g/dL.

The exact amount and type of initiator required to achieve the desired molecular weight can be determined by routine experimentation. For instance, a polyhydric alcohol initiator having three or more hydroxy groups will result in a branched chain polymeric structure, whereas a mono- or dihydric alcohol or a hydroxy acid (having only one carboxy group and one hydroxy group) will result in a linear polymeric structure. The monomers are used in proportions such that the resulting polymer is either a liquid or a waxy solid at room temperature (25°C). As a general rule, the ratio of the monomers is 0.1 to 0.5 moles of lactide and/or glycolide, with the remainder being PDO.

The copolymers are prepared by synthesis techniques that are analogous to processes described previously. For instance, the initiator, monomers, and a suitable esterification or ester exchange catalyst are charged to a suitable vessel and the contents of the vessel are heated to a temperature within the range of 80 to 180°C for a period of four hours to four days.

Hence, to a flame dried, 250 ml round bottom, two-neck flask, equipped with an overhead stirrer and nitrogen inlet, 12 milliliters of distilled diethylene glycol, 80 grams (0.7836 moles) of p-dioxanone, 20.0 grams (0.1388 moles) of L(–) lactide, and 0.093 milliliters of stannous octoate (0.33 molar solution in toluene) were added. The flask was held under high vacuum at room temperature for about 16 hours and then purged with nitrogen three times followed by venting with nitrogen. The flask was placed in an oil bath at 170°C and stirred for 3 hours. The temperature was lowered to 120°C and stirring of the low viscosity melt was continued for an additional 5 hours.

The resulting poly(p-dioxanone-co-lactide) copolymer was cooled to room temperature under a stream of nitrogen, isolated, and dried for 96 hours at 80°C under vacuum (0.1 mm Hg) to remove any unreacted monomers. The copolymer had an inherent viscosity of 0.11 dL/g.

RESULTS AND DISCUSSION

Low molecular weight absorbable copolymers of p-dioxanone ("PDO") with lactide and/or glycolide can be used as coatings for surgical filaments to improve their tie-down and tactile smoothness. The copolymers can be used as a coating on braided or monofilament sutures and ligatures. The improvements imparted by the copolymers are most pronounced on braided sutures and ligatures. The filaments can be absorbable or non-absorbable. Among the types of filaments which can be coated by the copolymers are absorbable materials such as polyglycolide, poly(lactide-co-glycolide), poly(p-dioxanone), and other absorbable materials, and non-absorbable materials such as aromatic polyesters, nylon, silk, and poly(propylene).

The copolymers are coated on the surgical filaments by procedures such as passing the filaments through an organic solvent solution of the copolymer, and then through a drying oven to evaporate the solvent. As a general rule, coating levels are from about 1 to 10%, based on the weight of the filament.

POLY(p-DIOXANONE-CO-ε-CAPROLACTONE) LIQUID/LOW MELT COPOLYMERS

Liquid and low melting p-dioxanone/ε-caprolactone copolymers were prepared by methods described previously (Bezwada, 1995; Roller and Bezwada, 1996; Cooper). Several specific compositions of p-(dioxanone) and ε-caprolactone, especially molar ratios of 40/60, 50/50, and 60/40 can form liquids at room temperature. This is caused in part by the low Tg's of the homopolymers, –60°C for PCL and –15°C for PDS. At the compositions described above, the copolymers exhibit no crystallinity. Consequently, at low molecular weights (I.V. of 0.1–0.5 dl/g), these copolymers are fluid at room and body temperatures. Low melt polymers (40–55°C) were prepared with compositions of PDO/caprolactone at molar ratios of 30/70 to 10/90.

By blending the liquid and low melt polymers, semi-solid fluids with yield points, can be prepared that form dispersions. A unique balance of physical, thermal and rheological properties was obtained, yielding fluids they can be administered as spreadable lotions, injectables or coatings for implantable or non-implantable devices. If properly designed and formulated, the fluids behave like semisolids that stay where they are placed, retain their shape and do not migrate like simple low viscosity liquids (Roller and Bezwada, 1996).

RADIATION STERILIZABLE ABSORBABLE PDO COPOLYMERS

All implantable surgical devices, including sutures, ligating clips, bone plates and screws, need to be sterilized. In most cases, they are sterilized by heat, ethylene oxide, or gamma radiation (employing Co^{60} source). In many cases, gamma irradiation is the most convenient and effective method of sterilization. However, not all polymeric materials, in particular, absorbable polymers, retain their properties, and therefore, their usefulness upon irradiation. This is due to free radical formation which causes branching, crosslinking and chain scission, and leads to chemical structure changes and physical property detriments.

However, polymers which contain free radical scavengers such as benzene rings can preserve their physical characteristics. Poly(ethylene 1,4-phenylene-bis-oxy-acetate) [PG-2] was investigated and found to retain useful properties (Jamiolkowski and Shalaby, 1990a). In order to increase the absorption rates, however, copolymerization with glycolide was attempted (Jamiolkowski and Shalaby, 1990a; 199b). Multifilament braids derived from the block copolymers of PG-2/PGA exhibited excellent physical and biological properties, even after gamma radiation (Jamiolkowski and Shalaby, 1990b). This concept was extended by replacing the hydroquinone moiety with a p-hydroxybenzoic acid moiety with similar results (Bezwada, Jamiolkowski, and Shalaby, 1985a, 1985b). Fibers derived from all these materials were too stiff for use as monofilament sutures. For certain procedures, surgeons prefer monofilament sutures with good handling (low modulus) properties.

In order to improve the handling properties, copolymerization with p-dioxanone was attempted (Koelmel *et al.*, 1985; 1991). These copolymers, like their PGA analogs, were also found to be radiation sterilizable with little loss of physical properties (Koelmel *et al.*, 1985; 1991).

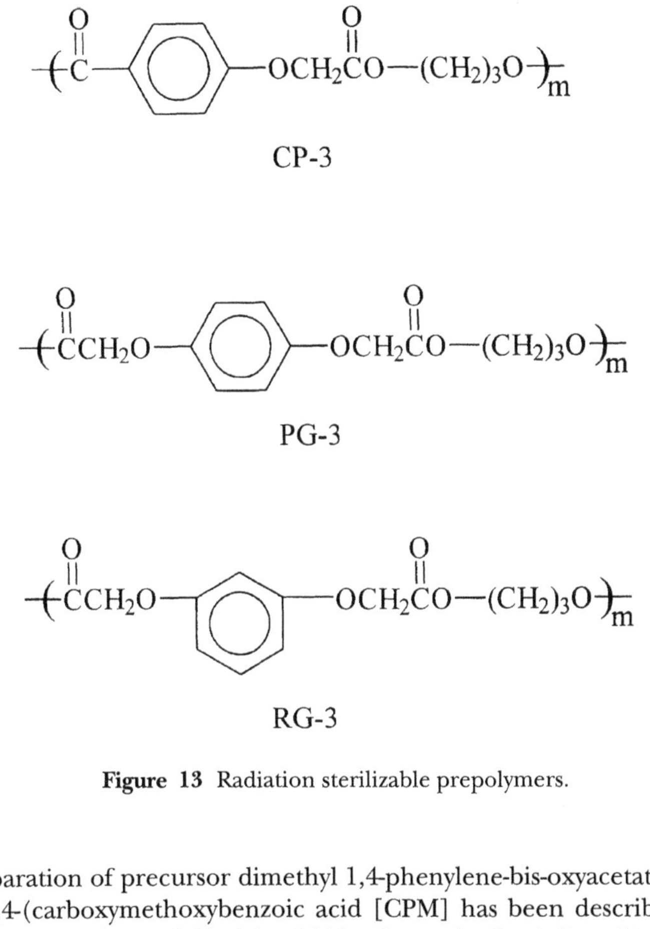

CP-3

PG-3

RG-3

Figure 13 Radiation sterilizable prepolymers.

The preparation of precursor dimethyl 1,4-phenylene-bis-oxyacetate [PGM] and dimethyl 1,4-(carboxymethoxybenzoic acid [CPM] has been described in the literature (Jamiolkowski, and Shalaby, 1990a; Bezwada, Jamiolkowski, and Shalaby, 1985a). The preparation of the prepolymers, poly(trimethylene 1,4-phenylene-bis-oxyacetate) [PG-3], poly(trimethylene 1,3-phenylene-bis-oxyacetate) [RG-3], and poly(trimethylene 4-(carboxymethoxy) benzoate) [CP-3] are also described in several patents (Koelmel *et al.*, 1985) and displayed in Figure 13.

Thus, copolymers of PG-3, RG-3, and CP-3 with p-dioxanone were prepared using the following general procedure; a flame dried, 250 ml, round bottom, three neck flask equipped with a overhead stirrer, nitrogen inlet and stopper, was charged under nitrogen with 15 grams of the prepolymer. The flask was held for 16 hours at 50°C under 0.1 mm Hg. After drying, 85 grams of p-dioxanone and 0.126 ml of 0.33M stannous octoate in toluene were added. The mixture was placed in an oil bath at 75°C for one hour to melt the p-dioxanone monomer and to dissolve the prepolymer in p-dioxanone. The temperature of the bath was raised to 90°C and maintained there for 24 hours. Stirring was stopped after 3 to 4 hours at 90°C. The temperature of the oil bath was lowered to 80°C and maintained there for 72 hours. The copolymer was isolated, ground and dried for 18 hours at 80°C under 0.1 mm Hg to remove any unreacted monomer. Inherent viscosities (IV), melting points by hot stage microscopy, and composition via NMR were determined.

Table 12 Properties of radiation sterilizable PDO copolymers

Polymer type	PDS	PG-3/PDS	RG-3/PDS	CP-3/PDS
Final composion (mole %)	100	13/87	17/83	16/84
I.V. (dL/g)	1.79	1.99	1.82	1.55
Tm (°C)	112	—	107	108
Extrusion conditions				
Extrusion Temperature(°C)	155	170	150	155
First stage (draw ratio/temp., °C)	4×/54	4×/55	4×/49	5×/60
Second stage (draw ratio/temp., °C)	1.5×/73	1.5×/82	1.5×/78	1.2×/80
Overall draw ratio	6×	6×	6×	6×
Fiber Properties*				
Straight strength (Kpsi)	71/58	80/70	85/69	59/54
Knot strength (Kpsi)	—	49/47	60/50	51/50
Elongation (%)	39/37	47/42	51/53	60/55
Young's modulus (Kpsi)	310/301	124/165	153/144	191/189

*fiber properties represent annealed, non-sterilized vs. annealed, Co-60 sterilized, respectively.

Table 13 *In vivo* breaking strength of cobalt sterilized PDO copolymers

Polymer type	PDS	PG-3/PDS	RG-3/PDS	CP-3/PDS
Zero day strength (lbs.)	3.1	3.6	3.1	2.7
BSR (%)				
1 wk	—	—	82	81
2 wks.	43	75	76	75
3 wks.	30	68	70	62
4 wks.	25	58	56	54

Table 14 *In vivo* absorption of cobalt sterilized PDO copolymers

Copolymer type	PDS	PG-3/PDS	RG-3/PDS	CP-3/PDS
Post-implantation time				
5 days	100	100	100	—
119 days	83	—	87	—
154 days	—	0	0	—
182 days	43	0	0	—
210 days	0	0	0	—

The copolymers were extruded using an INSTRON capillary rheometer. The extrudate filaments were annealed at room temperature from 2 hours to about one week prior to orientation. Fibers were drawn in a two stage glycerine bath, and annealed at 80° for 6 hours at 5% relaxation. The tables above summarize the data for these copolymers (Tables 12, 13 and 14).

All copolymers exhibited a nonrandom sequence distribution.

The properties of the monofilaments exhibited equivalent or higher strength than PDS homopolymer with much lower modulus (better handling properties).

Even after irradiation, significant strength was retained out to four weeks post-implantation. The absorption data indicate that absorption occurs at least as fast as PDS. Thus, these systems can provide radiation sterilizable absorbable monofilament sutures with good handling properties.

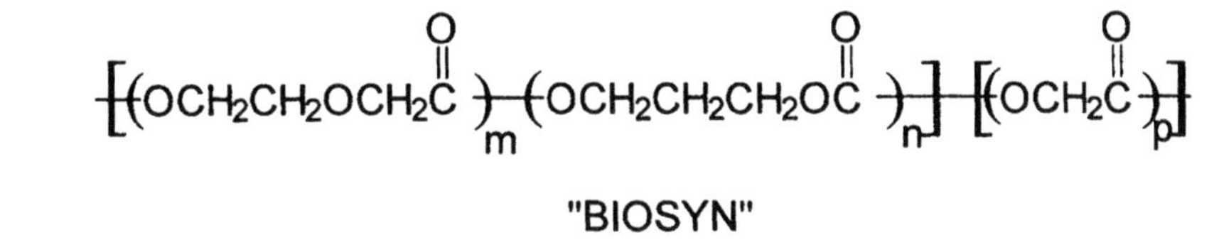

$$\left[\left(OCH_2CH_2OCH_2\overset{O}{\overset{\|}{C}}\right)_m\left(OCH_2CH_2CH_2O\overset{O}{\overset{\|}{C}}\right)_n\right]\left[\left(OCH_2\overset{O}{\overset{\|}{C}}\right)_p\right]$$

"BIOSYN"

Figure 14 Poly(p-dioxanone-trimethylenecarbonate-glycolide) terpolymers.

POLY(p-DIOXANONE-TRIMETHYLENE CARBONATE-GLYCOLIDE) BLOCK TERPOLYMERS

Block terpolymers containing a random soft middle block copolymer of p-dioxanone and trimethylene carbonate (TMC) and polyglycolic acid (PGA) end blocks were prepared (Roby *et al.*, 1995) as shown in Figure 14.

Monofilament sutures (size 3/0) developed (trade name BIOSYN) from this block terpolymer of [PDO/TMC]/glycolide at [16/26]/58 weight composition exhibit good properties: straight tensile strength of 80 Kpsi, knot tensile of 50 Kpsi, and Young's modulus of 145 Kpsi.

SUMMARY

As DEXON and VICRYL sutures are generally based on braided constructions, PDS suture was the first absorbable monofilament. PDS homopolymer has a glass transition of about −15° and melting point of about 115°. Most recently, PDS II suture was introduced with improved handling properties.

Over the years, polymers and copolymers based on p-dioxanone have gained increasing interest in the medical device and pharmaceutical fields due to their biodegradability, low toxicity, softness and flexibility. By combining p-dioxanone with certain other lactone monomers such as glycolide, lactide, ε-caprolactone, and trimethylene carbonate, polymers with a wide range of physical properties can be obtained without compromising absorbability or increasing tissue reaction. This allows p-dioxanone (co-)polymers to be utilized in many medical device applications such as monofilament sutures, ligating clips, orthopedic pins, meshes, bone waxes, coatings for sutures, soft tissue augmentation, and drug delivery vehicles.

For example, by copolymerizing with small amounts of glycolide, absorption of PDS can be reduced from 210 days to almost 100 days without compromising tensile properties. Similarly, by copolymerizing with small amounts of lactide, polymer morphology can be altered decreasing the modulus or stiffness, and thereby increasing flexibility.

Furthermore, due to the low Tg of PDS, low molecular weight copolymers of PDS with glycolide and/or lactide can be liquids or low melt polymers. Such copolymers can be used as absorbable coatings, or bone wax. Liquid copolymers of p-dioxanone and trimethylene carbonate, and p-dioxanone and caprolactone have potential for use in drug delivery and soft tissue augmentation.

REFERENCES

Kronenthal, R.L. (1975) Biodegradable polymers in medicine and surgery, *Polymer. Sci. Technol.*, **8** (Polym. Med. Surgery), 119–137.

Frazza, E.J. (1976) *Sutures, Encycl. Polymer Science & Technology*, Suppl. 1, 587–97.

Chu, C.C. (1983) *Survey of clinically important wound closure biomaterials, in Biocompatability of Polymers, Metals and Composites*, ed. M. Szycher. Technomic Publ. pp. 477–523.

Barrows, T.H. (1986) Degradable implant materials: A review of synthetic absorbable polymers and their applications, *Clinical Materials*, **1**, 233–257.

Heller, J. (1985) *Biodegradable polymers in controlled drug delivery. in CRC Critical Reviews in therapeutic Drug Carrier Systems*, ed. S.D. Bruck. Boca Raton, Ann Arbor, Boston, London, **1**, 39–90.

Shalaby, S.W. and Johnson, R.A. (1994) Synthetic absorbable polyesters, pp 1–34, ed. by Shalaby, S.W., *Biomedical Polymers: Designed-to-Degrade Systems*, Hanser/Gardner Publications, Cincinnati OH, 1994.

Amecke, B., Bendix, D. and Entenmann, G. (1995) *Synthetic resorbable polymers based on glycolide, lactide and similar monomers, Encylopedic Handbook of Biomaterials and Bioengineering, Part A: Materials*, ed. D.L. Wise, *et al.*, pp. 977–1007.

Wu, S.X. (1995) *Synthesis and properties of biodegradable Lactic/Glycolic acid Polymers, Encylopedic Handbook of Biomaterials and Bioengineering, Part A: Materials*, ed. D.L. Wise, *et al.*, pp. 1015–1054.

Chu, C.C. (1995) *Biodegradable suture materials: Intrinsic and Extrinsic factors affecting biodegradation phenomena, Encylopedic Handbook of Biomaterials and Bioengineering, Part A: Materials*, ed. D.L. Wise, *et al.*, pp. 543–687.

Hollinger, J.O., Jamiolkowski, D.D. and Shalaby, S.W. (1995) Bone repair and a unique class of biodegradable polymers: The poly(alfa-esters), pp. 197–233, ed. by Hollinger J.O., *Biomedical Applications of Synthetic Biodegradable Polymers*, CRC Press.

Chu, C.C. (1983) in *Biocompatible Polymer, Metals, and Composites*, M. Szycher, editor, Technomic Press, Lancaster, PA.

Chu, C.C. (1985) *CRC Critical Reviews in Biocompatability*, **1**(3), 261– 322.

Chu, C.C. (1990) *Suture materials, in Concise Encyclopedia of Medical and Dental Materials*, D.F. Williams, editor, Pergamon Press, London, pp. 345–353.

Gilding, D.K. (1981) Biodegradable polymers, in CRC Biocompatibility of clinical implant materials, Vol. 2, D.F. Williams, editor CRC Press, Boca Raton, FL., 209–232.

Craig, P.H., Williams, J.A., Davis, K.W. *et al.*, (1975) A biological comparison of polyglactin 910 and polyglycolic acid synthetic absorbable sutures, *Surg. Gynecol. Obstet.*, **141**(1), 1–10.

Reed, A.M. and Gilding A.K. (1981) Biodegradable polymers for use in surgery-Poly(glycolic)/poly(lactic acid) homo- and copolymers: 2. *In vitro* degradation, *Polymer*, **22**(4), 494–98.

Katz, A., Mukherjee, D.P., Kaganov, A.L. and Gordon, S. (1985) A new synthetic monofilament absorbable suture made from polytrimethylene carbonate, *Surg. Gynecol. Obstet.*, **161**, 213–22.

Bezwada, R.S., Jamiolkowski, D.D., Lee, I. *et al.* (1995) MONOCRYL suture, a new ultrapliable absorbable monofilament suture, *Biomaterials*, **16**, 1141– 1148.

Bezwada, R.S., Jamiolkowski, D.D. and Shalaby, S.W. (1992) Segmented copolymers of caprolactone and glycolide, U.S. patent (to Ethicon, Inc.) 5,133,739.

Roby, M.S., Bennett, S.L. and Liu, C.K. (1995) *Absorbable block copolymers and surgical articles fabricated from them*, U.S. Patent (to United States Surgical Corp.) 5,403,347.

Doddi, N., Versfelt, C. and Wasserman, W. (1977) Synthetic absorbable surgical devices of poly(p-dioxanone), U.S. Patent (to Ethicon, Inc.) 4,052,988.

Ray, J.A, Doddi, N., Regula, D. *et al.* (1981) *Surgery, Gynecology and Obstetrics*, **153**, 497–507.

Shalaby, S.W. and Koelmel, D.F. (1984) *Copolymers of p-dioxanone and 2,5-morpholinediones and surgical devices formed therefrom having accelerated absorption characteristics*, U.S. Patent (to Ethicon, Inc.) 4,441,496.

Bezwada, R.S., Shalaby, S.W. and Newman, H.D. (1987) *Crystalline p-dioxanone/glycolide copolymers and surgical devices made therefrom*, U.S. Patent (to Ethicon, Inc.) 4,653,497.

Bezwada, R.S., Shalaby, S.W., Newman, H.D. and Kafrauy A. (1987) *Crystalline copolymers of p-dioxanone and lactide and surgical devices made therefrom*, U.S. Patent (to Ethicon, Inc.) 4,643,191.

Bezwada, R.S., Shalaby, S.W., Newman, H.D. and Kafrauy A. (1990) Bioabsorbable copolymers of p-dioxanone and lactide for surgical devices, *Trans. Soc. Biomater.*, **13**, 194.

Bezwada, R.S., Shalaby, S.W. and Erneta M. (1991) *Crystalline copolymers of p-dioxanone and caprolactone*, U.S. Patent (to Ethicon, Inc.) 5,047,048.

Bezwada, R.S., Newman, H.D. and Shalaby S.W. (1991) *Crystalline copolyesters of amorphous (lactide/glycolide) and p-dioxanone*, U.S. Patent (to Ethicon, Inc.) 5,007,923.

Bezwada, R.S. and Shalaby, S.W. (1991) *Crystalline copolymers of p-dioxanone and poly(alkyleneoxide)*, U.S. Patent (to Ethicon, Inc.) 5,019,094 (May 1991).

Bezwada, R.S., Shalaby S.W. and Hunter, A. (1991) Bioabsorbable copolymers of polyalkylene carbonate/p-dioxanone for sutures and coatings, U.S. Patent (to Ethicon, Inc.) 5,037,950.

Broyer, E. (1995) *Thermal treatment of thermoplastic filaments for the preparation of surgical sutures*, U.S. Patents 5,294,395 (March, 1994) and 5,451,461 (September, 1995).

Jiang, Y. (1995a) *Process for the production of dioxanone*, U.S. Patent (to United States Surgical Corp.) 5,391,707.

Jiang, Y. (1995b) *Purification of 1,4-dioxan-2-one by crystallization*, U.S. Patent (to United States Surgical Corp.) 5,391,768.

Jamiolkowski, D.D., Shalaby S.W., Bezwada R.S. and Newman H.D. (1989) *Glycolide/p-dioxanone block copolymers*, U.S. Patent (to Ethicon, Inc.) 4,838,267 (January 1989).

Bezwada R.S. and Kronenthal, R. (1991) Random copolymers of p- dioxanone, lactide, and/or glycolide as coating polymers for surgical filaments, U.S. Patent (to Ethicon, Inc.) 5,076,807.

Bezwada, R.S. (1995) *Liquid copolymers of caprolactone and lactide*, U.S. Patent (to Ethicon, Inc.) 5,442,033.

Roller, M. and Bezwada R.S. (1996) Liquid and low melt absorbable copolymers and their blends-synthesis and rheological characterization, *Preprints of SPE ANTEC'96*, **2**, 2848–2951.

Koelmel, D.F., Jamiolkowski, D.D., Shalaby, S.W. and Bezwada R.S., Poly(p-dioxanone) polymers having improved radiation resistance, U.S. Patent (to Ethicon, Inc.) 4,546,152 (1985) and U.S. Patent (to Ethicon, Inc.) 4,649,921 (1987).

Koelmel, D.F., Jamiolkowski, D.D., Shalaby, S.W. and Bezwada R.S. (1991) Sterilizable monofilament sutures, *Polym. Prepr.*, **32**, 235.

Jamiolkowski, D.D. and Shalaby, S.W. (1990a) *Polymer Preprints*, **31**(2), 327.

Jamiolkowski, D.D. and Shalaby, S.W. (1990b) *Polymer Preprints*, **31**(2), 329.

Bezwada, R.S., Jamiolkowski, D.D. and Shalaby S.W. (1985a) *Absorbable polymers of substituted benzoic acid*, U.S. Patent (to Ethicon, Inc.) 4,510,295.

Bezwada, R.S., Jamiolkowski, D.D. and Shalaby S.W. (1985b) *Surgical sutures made from absorbable polymers of substituted benzoic acid*, U.S. Patent (to Ethicon, Inc.) 4,532,928.

Blomstedt, B. and Osterberg, B. (1978). Suture materials and wound infection, *Acta Chir. Scand.*, **144**, 269.

Jamiolkowski, D.D., Newman, H.D., Datta, A. *et al.* (1995) *European patent application* (to Ethicon, Inc.) 691359.

Datta, A., Jamiolkowski, D.D. and Roller, M.B. (1995) Mechanical properties of poly(p-dioxanone) as a function of annealing conditions and molecular weight, *Transactions of the 21th Annual Meeting of the Society for Biomaterials*, **XVIII**, 437.

Cooper, K., Bezwada, R. Roller, M.B. (1997) Absorbable liquid and low melt copolymers: Synthesis and hydrolysis characteristics, *Transactions of the 23rd Annual Meeting of the Society for Biomaterials*, **XX**, 361.

3. POLYCAPROLACTONE

DANA E. PERRIN[1] and JAMES P. ENGLISH[2]

[1]*Linvatec Corporation, a division of Zimmer, a Bristol-Myers Squibb Company,
11311 Concept Boulevard, Largo, Florida 33773, USA*
[2]*Absorbable Polymer Technologies, 115B Hilltop Business Drive,
Pelham, Alabama 35124, USA*

INTRODUCTION

Poly(ϵ-caprolactone) (PCL) is a semicrystalline, biodegradable polymer having a melting point (T_m) of ~60°C and a glass transition temperature (T_g) of ~ −60°C (Brode & Koeleske, 1973; TONE© P700, 1983).[1] The repeating molecular structure of PCL homopolymer consists of five nonpolar methylene groups and a single relatively polar ester group. This structure gives PCL some unique properties. The mechanical properties are similar to polyolefin because of its high olefinic content, while the presence of the hydrolytically unstable aliphatic-ester linkage causes the polymer to be biodegradable. This combination also gives PCL the unusual property of being compatible with numerous other polymers and PCL polymer blends having unique properties have been prepared (Brode & Koleske, 1972; Kalfogu, 1983; Olabisi *et al.*, 1979; TONE Polymers, 1983).

POLYMERIZATION OF POLYMERS AND COPOLYMERS OF POLY(ϵ-CAPROLACTONE)

High molecular weight polymers and copolymers of ϵ-caprolactone are prepared by ring-opening, addition polymerization over a wide range of temperatures. Copolymers can also be prepared with a variety of other lactones, lactides (Pitt *et al.*, 1981a), and lactams (TONE Polymers, 1983; Interox, 1977). In making copolymers, attention to monomer reactivity differences is important and depending on the specific conditions for copolymerization wide differences in copolymer microstructure (sequencing) are possible.

Polymerization temperatures in the range of 140–150°C are typical. The polymerization is normally catalyzed by stannous 2-ethyl hexanoate (stannous octoate) or stannic chloride dihydrate. Other catalysts which have been used include various Lewis acids, alkyl metals and organic acids (Schindler, *et al.*, 1982; Interox, 1977). Molecular weight is controlled by addition of chain control agents. These chain control agents are usually water, primary alcohols, amines, or some other active hydrogen compound. For example, linear polymers can be prepared using alcohols having functionalities less than or equal to two, such as 1-dodecanol or 1,6-hexanediol, and branched polymers can be prepared using alcohols having functionalities greater than two, such as various, sugars, pentaerythritol, and trimethylolpropane (Schindler, *et al.*, 1982).

---

[1]TONE© is a registered trademark of the Union Carbide Corporation, Danbury, Connecticut.

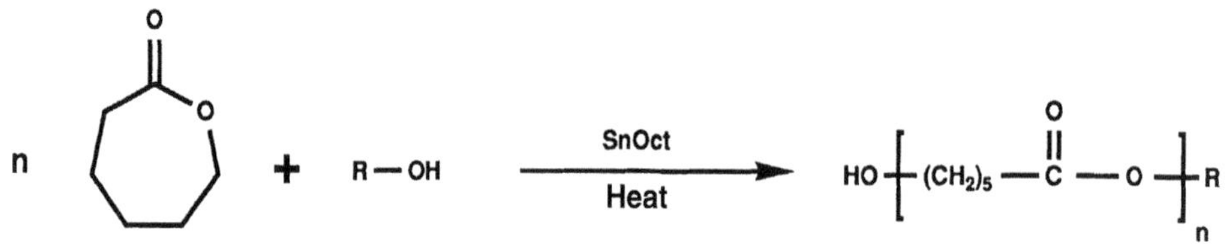

Figure 1 Polymerization of PCL.

While PCL and its copolymers are subject to biodegradation because of the susceptibility of its aliphatic ester linkage to hydrolysis, biodegradation of the PCL homopolymer is considerably slower than the poly(α-hydroxy acids) such as polylactide because of the combination of its crystallinity and high olefinic character. On the other hand, copolymers with lactides and blends with other degradable polymers and copolymers reduce overall crystallinity increasing the accessibility of ester linkage, and thus significantly enhance the rate of hydrolysis. Therefore, PCL copolymers and polyblends can be prepared having a wide range of physical and mechanical properties with varying rates of biodegradation. The synthesis of PCL is shown in Figure 1.

The polymerization details are given for a small scale laboratory polymerization and a larger pilot scale polymerization of PCL homopolymer. A general description for laboratory copolymerization with lactide or glycolide is also given.

Acquisition and Characterization of ϵ-Caprolactone Monomer

High purity ϵ-caprolactone monomer is readily available from several commercial sources. Upon receipt, the monomer is sampled for identity and purity testing. During sampling, the monomer is blanketed with dry, high purity nitrogen to avoid moisture contamination. The monomer is then stored under a dry, high purity nitrogen atmosphere in the original container at room temperature until qualified for use.

The sample is first characterized by infrared spectroscopy (IR) and compared to a standard IR spectrum of ϵ-caprolactone for identification. Monomer purity may be determined in a variety of ways such as boiling point, DSC, etc., but is best determined by preparation of a small test sample of the homopolymer. A 10 gram test polymer is made by adding an appropriate quantity (0.03–0.10 wt %) of stannous octoate, available from Sigma Chemical Company of St. Louis Missouri, as a solution in dry toluene, to 10 grams of the monomer sealed in a 20 mL scintillation vial under a dry, high-purity nitrogen atmosphere. The vial is sealed and submerged in an oil bath at 140–150°C for 18–24 hours. The monomer/catalyst mixture is well agitated until the melt becomes too viscous to agitate. After 18–24 hours, the test polymer is removed from the oil bath, cooled to room temperature, and a sample of the polymer is removed from the vial for molecular weight determination. The molecular weight is determined by dilute solution viscosity and gel permeation chromatography (GPC) as described below. PCL prepared in this manner should have a molecular weight of 200,000 Daltons or greater by GPC to be assured of high purity monomer.

If monomer purity is shown to be insufficient by this test, it may be improved by vacuum fractional distillation of the monomer from CaH_2 and collection over dry molecular sieves.

Lactide and Glycolide Monomers for Copolymerization

A wide variety of high-purity monomers are available commercially for copolymerization with ε-caprolactone. high-purity L-lactide, D-lactide, DL-lactide, and glycolide monomers may be obtained from either PURAC America of Lincolnshire, Illinois or BI Chemicals of Montvale, New Jersey.

Upon receipt the monomer is sampled for identity and purity testing. As with ε-caprolactone, the monomer is blanketed with dry, high-purity nitrogen to avoid moisture contamination during sampling. The monomer is then stored in the original container under dry, high-purity nitrogen in a freezer at $<0°C$. The monomer sample is characterized by IR comparing the test sample spectrum to a standard spectrum for identification; and similar to the ε-caprolactone, a 10 gram test polymer is made to determine monomer purity.

POLYMER SYNTHESIS

Small-scale Polymerization of ε-Caprolactone

Small-scale polymerization of ε-caprolactone is conveniently carried out in a 1 Liter, stainless steel resin kettle equipped with a glass top, a stainless steel mechanical stirrer, a thermometer, and a gas inlet tube. The resin kettle, mechanical stirrer, and all glassware are predried for 18 hours in an oven at $150°C$ and cooled inside a glove box in a dry, high-purity nitrogen atmosphere. Both the loading of the monomer and assembly of the apparatus are also conducted inside the glove box in a dry, high-purity nitrogen atmosphere. Qualified, high-purity ε-caprolactone monomer and an appropriate amount of the desired chain control agent are charged to the flask. The resin kettle is then closed, removed from the glove box, and immediately connected to a low flow source of dry, high-purity nitrogen as a continuous purge. The flask is then submerged in an oil bath at $140–150°C$ and slow stirring is begun. After the monomer/chain control mixture has reached the reaction temperature, stirring is discontinued. At this time, a sufficient quantity (0.03–0.10 wt %) of tin octoate catalyst is added as a solution in dry toluene. Mechanical stirring is immediately resumed and maintained until the polymerizate becomes very viscous. Stirring is then discontinued and the stirrer and thermometer are raised from the melt, and heating is continued for a total of 18–24 hours after which the resin kettle assembly is removed from the oil bath and cooled to room temperature under a continual nitrogen purge. When the polymer has cooled, the resin kettle top containing the stirrer and thermometer are removed from the resin kettle. The polymer is removed by submerging the resin kettle into liquid nitrogen in an insulated metal container (metal Dewar flask). The insulated container should be quickly covered with a loose metal cap. This procedure should be done quickly and carefully behind a safety shield. When thoroughly cooled, the polymer mass will separate readily from the resin kettle. While still cold, the polymer is removed and cut into several smaller pieces. The polymer pieces are allowed to warm to room temperature overnight inside a vacuum oven under high vacuum to avoid condensation of moisture from the air, and then dissolved in an excess of dichloromethane in a sealed glass jar on a laboratory shaker. The solution is then filtered through a sintered glass filter using aspirator vacuum, and finally precipitated into excess methanol.

After precipitation from dichloromethane into methanol, the purified polymer is then redried inside a vacuum oven under high vacuum at room temperature ($\sim$24 hours), cooled with liquid nitrogen, and ground through a 6 mm screen on a Wiley Mill. The ground polymer is finally vacuum dried an additional 24 hours at room temperature. A 5 gram sample of the polymer is removed for characterization, and the remainder is stored inside a polyethylene bag evacuated and backfilled three times with dry, high-purity nitrogen prior to sealing. The polyethylene bag is then placed inside an outer polylined foil bag containing a bag of DRIERITE® desiccant, sealed, and then stored in a freezer at <0°C.[2] The yield is typically 75–95%. The molecular weight and thus the inherent viscosity of a typical polymer prepared by the process vary depending on the quantity of catalyst, the type and quantity of chain control agent, and the specific reaction conditions chosen.

Small-scale Copolymerization of ε-Caprolactone

In preparation of small scale copolymers of ε-caprolactone, a convenient polymerization apparatus is again a 1 Liter stainless steel resin kettle equipped with a glass top, a stainless steel mechanical stirrer, a thermometer, and a gas inlet tube. As mentioned above, wide variations in monomer sequencing can be obtained depending on comonomer reactivity and the specific reaction conditions. For example, block copolymers of ε-caprolactone and glycolide have been prepared by first prepolymerizing a mixture of the two monomers rich in ε-caprolactone using stannous octoate as the catalyst and 1,6-hexanediol as the chain control agent. During the course of the polymerization, an additional quantity of a monomer mixture rich in glycolide is added. More random copolymers can be prepared by copolymerizing the entire monomer mixture at once (Shalaby & Jamiolkowski, 1985).

Pilot-scale Polymerization of ε-Caprolactone

A very useful reactor assembly for pilot-scale polymerization of ε-caprolactone is an 8-CV mixer from Design Integrated Technologies (DIT) of Warrington, Virginia. The 8-CV is a dual cone mixer (8 qt. capacity) equipped with double helical mixer blades, a gas inlet line, and a large vacuum port.

The reactor is first cleaned and dried thoroughly. The desired amount of high-purity ε-caprolactone monomer, and an appropriate amount of the desired chain-control agent are then charged to the reactor. The reactor is heated under positive nitrogen pressure until the monomer mixture reaches 140–150°C. At this time, a sufficient quantity of stannous octoate catalyst is added. Stirring is continued until the polymerizate becomes very viscous. Stirring is then discontinued and heating of the polymer is continued for a total elapsed heating time of 12–18 hours.

---

[2]DRIERITE® is a registered trademark of W.A. Hammond DRIERITE Company, Xenia, Ohio.

After the polymerization is complete, a vacuum is applied to the polymer melt with continual slow stirring to remove residual monomer. The reactor is then cooled, pressurized with dry, high-purity nitrogen, and the discharge valve is opened slowly until a steady stream of polymer flows from the reactor. The polymer extrudate is quenched by passing the molten strand through a cooling bath of deionized, reverse osmosis purified water at room temperature while continuously pelletizing the cooled strand. The pelletized polymer is then dried under high vacuum at room temperature for 48 hours in a rotary-cone dryer, and packaged inside a polyethylene bag under dry, high-purity nitrogen. This inner bag is then placed inside an outer polylined foil bag containing desiccant, sealed, and finally stored in a freezer at <0°C. The final yield is typically 75–95%. The molecular weight and thus the inherent viscocity of a typical polymer prepared by this process vary depending on the quantity of catalyst, the type and quantity of chain control agent, and the specific reaction condition chosen.

POLYMER CHARACTERIZATION

Molecular Weight and Molecular-Weight Distribution

PCL and PCL copolymers are characterized by first determining their inherent viscosity (η_{inh}) and/or intrinsic viscosity, $[\eta]$, in a suitable solvent (benzene, toluene, or chloroform). A convenient method is given in ASTM D-2857 using chloroform as the solvent at 30°C. Intrinsic viscosity is a convenient means of determining the molecular weight of the PCL homopolymer. The Mark-Houwink relationship between intrinsic viscosity $[\eta]$ in benzene and molecular weight is (TONE P700, 1983):

$$[\eta] = 1.96 \times 10^{-4} \, M_w^{0.76}$$

Relative molecular weight and molecular weight distribution (MWD) or polydispersity of PCL and PCL copolymers may be determined by GPC. GPC may also be carried out in a variety of solvents; however, in a typical method for PCL both the dissolution solvent and mobile phase solvent of choice is chloroform. A suitable flow rate for the mobile phase is 1 mL/min. Three 5 micron particle size, microstyrogel columns (1,000 Å, 10,000 Å, and 100,000 Å) arranged in series are convenient sizes to use. The polymer is dissolved in chloroform at a concentration of 0.5 g/dL and a 100 μL injection is made into the mobile phase. Data is collected using a refractive index (RI) detector. Comparison of the chromatographic data is made against chromatographic data obtained from a series of known M_w polystyrene standards in the same solvent. The MWD or polydispersity of polycaprolactone homopolymers prepared as described above and determined in this manner is normally about 1.5–2.0 (Schindler, *et al.*, 1982).

Caprolactone Comonomer Ratio

As stated, ϵ-caprolactone may be copolymerized with a variety of monomers. A method is given for determination of the monomer ratio in a copolymer with DL-lactide.

The final ratio of monomers incorporated into a copolymer of DL-lactide and ε-caprolactone is determined by proton nuclear magnetic resonance spectroscopy (^{1}H NMR) in deuterated chloroform. The ratio of the methylene protons adjacent to the oxygen in ε-caprolactone is compared against that of the methyne protons in the lactide dimer. The ratio is normally expressed as lactide dimer to ε-caprolactone; however it can be expressed as lactic acid to ε-caprolactone and care should be taken to not confuse the two.

Numerous other papers and patents have been published regarding the polymerization and physical properties of polycaprolactone (Pitt & Schindler, 1983a; Schindler, *et al.*, 1977; Brode & Koleske, 1973; Schindler, *et al.*, 1982; Hostettler, *et al.*, 1966). A comparison of the various physico-mechanical properties of a number of biodegradable polymers was published by Engelberg and Kohn (1991).

PROCESSING

PCL homopolymers have processing characteristics which are very similar to polyolefins with the exception that the overall processing temperatures are somewhat lower. Since hydrolytic degradation can occur during processing, it is usually prudent to dry the polymer prior to processing. While the possible range of conditions is actually quite large, injection molded articles can be made using the following conditions (TONE Polymers, 1990):

Screw Speed	90–290 rpm
Zone 1	160°F
Zone 2	180°F
Zone 3	200°F
Nozzle	200°F
Melt Temperature	230°F
Injection Pressure	2000–3000 psi
Mold Temperature	ambient

PCL may also be extruded into various profiles and can also be drawn into fibers having tensile strengths of over 60,000 psi. Such monofilaments will meet USP suture requirements after sterilization with an elongation of 20–30% (Perrin, 1989). Typical extrusion conditions are shown below (TONE Polymers, 1990):

Extruder Diameter	1.5 inches
Screw Length/Diameter	24/1
Zone 1	120°F
Zone 2	140°F
Zone 3	145°F
Adapter	135°F
Die	170°F
Melt Temperature	173°F

Additional information regarding the processing of PCL has been published by Liu and others (1991).

PCL HOMOPOLYMER DEGRADATION

The *in-vitro* and *in-vivo* degradation characteristics of PCL homopolymer have been characterized by Pitt and coworkers (1981b). In this study, the rates of degradation were measured both in saline at 40°C and in a rabbit model. The authors found the two hydrolytic rates were essentially similar and concluded that enzymatic involvement was not a factor in the degradation process. The observed rate constant for the initial part of the degradation (from 2–110 weeks) was found to be 3.07×10^{-3} day^{-1}. After 110 weeks, the crystallinity of the polymer increased from about 50% at 2 weeks to around 70% at 110 weeks. The reported crystallinity fractions were based upon an enthalpy of fusion value of 139.5 J/g for 100% crystalline PCL as reported by Crescenzi and others (1972). The degradation study also reported the observation that the chain scission of PCL was not accompanied by the loss of low molecular weight PCL fragments until the molecular weight had decreased to around 5,000.

Woodward and others (1985) have extensively studied the intracellular degradation of PCL. Their work provides perhaps the most comprehensive details on the biocompatibility of PCL throughout the degradation process using an animal model (Sprague-Dawley rats). During the first stage (nonenzymatic bulk hydrolysis), the implant became encapsulated by collagen filaments containing only occasional giant cells. The connective tissue capsule was observed to be avascular with a thickness of less than 100 micrometers. This capsule was well developed after two weeks post implantation. Due to the tissue trauma associated with implantation, a transient initial inflammatory response was observed during the first two weeks. This inflammation subsided as the initial stages of the wound healing process occured. Significant weight loss of the implant was not observed during this first stage which lasted approximately nine months. After nine months, when the molecular weight decreased to about 5,000, the second stage of the degradation process was observed. During this second stage, the rate of chain scission slowed, the hydrolytic process began to produce short chain oligomers, and weight loss was observed. Eventually the implant was observed to fragment into a powder. Predegraded PCL powders were produced with a molecular weight of about 3,000 and having particle sizes in two ranges: 53–106 micrometers and 221–500 micrometers. Approximately 25 mg of each powder were placed in gelatin capsules, sterilized with gamma radiation, and then implanted under the dorsal abdominal panniculus carnosus. Light and scanning electron microscopy techniques were used to evaluate the histologic response at various time intervals. The authors found that the histologic response was the same for both particle sizes. The degradation of fragmented PCL was observed inside the phagosomes of macrophage and giant cells. Inside these cells, the observed degradation rate was rapid, requiring only 13 days for complete absorption in some cases. PCL fragments were also observed inside some fibroblasts. The authors concluded that intracellular degradation was the primary degradative pathway for the absorption of PCL. Unpublished studies referenced by these authors specify ε-hydroxy caproic acid as the sole metabolite.

More recently, hydroxyl free radicals have been implicated as a major factor in the degradation of PCL (Ali, *et al.*, 1993; Williams, 1994). In addition, certain bacteria, yeast, and fungi have been shown to possess the ability to degrade PCL (Ookuma, *et al.*, 1993; Benedict, *et al.*, 1983a & 1983b; Cook, *et al.*, 1981).

APPLICATIONS

A review of the literature regarding PCL polymers reveals extensive research activity mainly in the areas of controlled release drug delivery and in suture applications. The reader is referred to two excellent review articles which cover PCL polymers in detail (Pitt & Schindler, 1983b; Gilbert *et al.*, 1982).

Drug Delivery Applications

The early drug delivery work was performed by Schindler and others who reported *in-vitro* values for diffusivities, solubilities, and normalized fluxes for five steroids in homopolymeric PCL capsules. The release kinetics showed initial rapid release during the first 20 days followed by a slower rate over the next 20–150 days. Observed variations in kinetics were attributed to the wall thickness of the device (Schindler, *et al.*, 1977).

In a related study, Pitt and others (1979a) reported release rates of various steroids from monolithic films and capsules made from homopolymeric PCL, poly (D,L-lactide), and copolymers of caprolactone and D,L-lactide (in both 60/40 and 90/10 caprolactone/D,L-lactide acid ratios), and copolymers of glycolic acid with D,L-lactide. The release rates of the steroids from the caprolactone-co-lactide polymers were similar to the homopolymeric PCL. However, the release rates from the glycolide-co-lactide polymers were much slower than those observed with PCL. Further data analysis revealed that the observed release kinetics depended upon the steroid dissolution rate (in the polymer) and the polymer crystallinity. While it is outside the scope of the current discussion, a good background reference on PCL thermodynamics and crystallinity is given by Lebedev (1979) and Chynoweth & Stachurscki (1986).

In another study, Pitt and others compared the steroidal release kinetics of caprolactone homopolymer, D,L-lactide homopolymer, and their related copolymers to silicone rubber. While the observed release rates from the poly (D,L-lactide) were much slower than any of the other materials studied, incorporation of a plasticizer such as tributyl citrate into the poly(D,L-lactide) seemed to increase the release rates. The other caprolactone-based materials showed release kinetics similar to that of silicone rubber (Pitt *et al.*, 1979b).

The *in-vivo* controlled release of norgestrel from PCL capsules implanted in rats was reported by Pitt and Schindler (1980a). This study showed that a constant release rate of 8–10 micrograms/day/cm capsule could be attained over a period of 292 days. This paper gives a valuable comprehensive review of the related literature concerning numerous other polymeric drug delivery systems.

In a related paper, Pitt, Marks, and Schindler (1980) reported similar results with naltrexone delivery pertinent to narcotic antagonist therapy. This study used radiolabelled naltrexone delivered to monkeys via implanted capsules made from a copolymer of caprolactone and D,L-lactide. Acute stage efficacy was observed over the course of 33 days with this system, even though the delivery rate was not quite zero order as with the norgestrel study.

The work of Pitt and Schindler culminated with the development of Capronor®, a biodegradable capsular delivery system for delivery of levonorgestrel contraceptive.[3]

In one of many papers dealing with this delivery system, the authors describe a general model for the *in-vivo* degradation rate of PCL capsules in rabbits. The polymer used in this study had an initial number average molecular weight of around 50,000 and a crystallinity of about 45%. The degradation of PCL *in-vivo* was observed to occur in two stages. The initial stage appeared to involve random hydrolytic cleavage of ester bonds. After 4 weeks, the crystallinity increased to 50% apparently due to the annealing effect of the 40°C *in-vivo* environment. During this first stage, the reduction in molecular weight as a function of time was essentially linear. Mass loss was negligible during this stage. A second stage in the degradation process was observed after 120 weeks. At this time, the number average molecular weight dropped to 4,600 and the crystallinity increased to about 80%. The authors attributed the late stage increase in crystallinity to the crystallization of tie segments made possible by chain cleavage in the amorphous phase (the glass transition of PCL is around −60°C, well below the *in-vivo* temperature). Due to the increase in crystallinity after 120 weeks, the rate of degradation dramatically slowed after this time.

The authors further investigated this late stage degradation of the PCL in detail using tritium labelled PCL which had been predegraded *in-vitro* using an accelerated hydrolytic technique to yield a polymer with a number average molecular weight of around 3,000. Both predegraded capsules and powder (of defined particle sizes) were implanted in rats contained in Roth metabolism cages. The degradation of the PCL was followed by measuring the radioactivity levels in urine, feces and expired water. In powder form, only 50% of the initial radioactivity was present at the implant site after 60 days and only 9% remained after 120 days. With the predegraded capsules, 22% of the original radioactivity was found at the implant site after 180 days. Additional kinetic data for the release of testosterone from PCL capsules are also reported in this study (Pitt & Schindler (1980b).

Ory and others (1983) reported the first human clinical use of Capronor implants in 8 women. This preliminary study investigated the feasibility of using a PCL implant to modulate the release of levonorgestrel as a means of contraception. Rods containing 15.9 mg of levonorgestrel were implanted in the hip of each patient. Ovulation was suppressed in seven of the eight women. Serum levels of the levonorgestrel were monitored. The authors concluded that the Capronor implant was worthy of a larger clinical trial.

The second report of the human clinical use of Capronor with 48 women followed over approximately 40 weeks was published by Darney and coworkers (1989). The study evaluated the contraceptive efficacy of subdermal rod implants which were 2.5 cm long (containing 12 mg of levonorgestrel) and 4.0 mm long (containing 21.6 mg of levonorgestrel). While the devices were found to deliver the levonorgestrel as expected, a few pregnancies did occur. The authors concluded that the 4.0 mm device was probably the minimum design for partial clinical efficacy and that further studies were necessary. The devices were well tolerated by the patients and no adverse systemic side effects were observed. At the beginning of the study, four patients experienced a localized swelling or itching at the implant site which resolved after treatment with topical corticosteroids.

---

[3]Capronor® is a registered trademark of the Research Triangle Institute, Research Triangle Park, North Carolina.

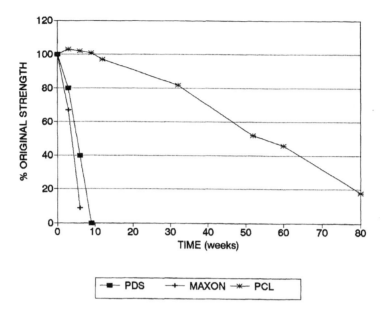

Figure 2 *In-vivo* Degradation of Absorbable Sutures.

Other drug delivery applications include delivery of 5-fluorouracil for anticancer therapy (Gebelein, *et al.*, 1990), delivery of neuroactive drugs (Menei, *et al.*, 1994), ketoprofen (Giunchedi, *et al.*, 1994), phenobarbitone (Berrabah, *et al.*, 1994), cyclosporin, (Guzman, *et al.*, 1993), cisplatin (Wada, *et al.*, 1991), carteolol (Marchal-Heussler, *et al.*, 1993), flurbiprofen (Pandey, *et al.*, 1994), chlorpromazine (Chang, *et al.*, 1994), povidone-iodine (Dunn, *et al.*, 1986), and minocycline (Kyun, *et al.*, 1990).

Sutures

Most absorbable sutures such as VICRYL®, PDS®, DEXON®, and MAXON®[4] are designed to retain strength so as to provide secure soft tissue approximation for about two to six weeks. For orthopaedic applications, where soft tissue is reattached to bone, the healing time may be as long as three to six months. Sutures made from homopolymeric PCL were evaluated in a rabbit model by Barber and Click (1992) to determine their associated degradation rate. The results of this study are shown in Figure 2. Included in Figure 2 are data from another rabbit study with these same PCL sutures conducted over a longer time frame (Perrin, 1989).

While ethylene oxide is an acceptable means of sterilizing homopolymeric PCL sutures, electron beam and gamma irradiation at 2.5 MRad causes only a slight loss in physical properties (Perrin, 1989). Narkis and others (1985) have published a study detailing the effects of radiation on PCL.

---

[4]DEXON® and MAXON® are registered trademarks of Davis & Geck (American Cyanamid Corporation), Danbury, Connecticut.

Sutures can also be made from a block copolymer of polyglycolide (PG) and PCL. Extensive *in-vitro* and *in-vivo* biocompatibility and efficacy studies with this copolymer have been performed and have lead to the introduction of MONOCRYL® sutures.[5] Because of the presence of PG in this copolymer, the degradation of MONOCRYL is rapid compared to the homopolymer. Using size 2/0 monofilament sutures implanted subdermally in rats, MONOCRYL sutures retain 50% of their original strength after about one week, whereas homopolymeric PCL monofilament size 2/0 sutures also implanted subdermally in rats retain 50% of their original strength after about 52 weeks (Bezwada, *et al.*, 1995; Perrin, 1989).

Other Applications

Copolymers of PCL with other comonomers (such as glycolide, lactide, and 1,5-dioxepan-2-one, and blends of PCL with other polymers (such as polyhydroxyvalerate and polyhydroxybutyrate) have been the subject of numerous papers and patents (Yasin & Tighe, 1992). Applications include suture coatings (Bezwada, 1995; Bennett & Liu, 1995; Bennett, *et al.*, 1995; Shalaby, 1980), as well as a wide variety of absorbable medical devices (Shalaby & Jamiolkowski, 1986; Jamiolkowski & Shalaby, 1987; Lin, 1987), including microporous intravascular stents (Rajasubramanianm, *et al.*, 1994), as a support to autologous vein grafts (Hinrichs, *et al.*, (1994), and as a nerve guide device (Perego, *et al.*, 1994).

In addition to the above, PCL and other copolymers have been evaluated for other medical uses such as an external casting material for broken bones (Phillips, 1972), as a material for use in making custom dental impression trays (Pelerin, 1993), and as a material used to fabricate a radiation therapy mask (Radiology, 1992).

CONCLUSION

In comparison to the other commercially available absorbable materials, PCL is one of the most flexible and easy to process. However, PCL has one of the slowest degradation rates of all such materials. This unique blend of characteristics fills a gap in the property spectrum unmatched by any other absorbable material.

REFERENCES

Ali, S.A., Ahong, S.P., Doherty, P.J. and Williams, D.F. (1993) Mechanisms of polymer degradation in implantable devices. *Biomaterials*, **14**, 648–656.

Barber, F.A. and Click, J.N. (1992) The effect of inflammatory synovial fluid on the breaking strength of new "long lasting" sutures. *Journal of Arthroscopy and Related Surgery*, **8**, 437–441.

Benedict, C.V., Cameron, J.A. and Huang, S.J. (1983a) Polycaprolactone degradation by mixed and pure cultures of bacteria and a yeast. *Journal of Applied Polymer Science*, **28**, 335–342.

Benedict, C.V., Cook, W.J., Jarrett, P., Cameron, J.A., Huang, S.J. and Bell, J.P. (1983b) Fungal degradation of polycaprolactones. *Journal of Applied Polymer Science*, **28**, 327–334.

---

[5]VICRYL®, PDS®, and MONOCRYL® are registered trademarks of Ethicon Incorporated, Somerville, New Jersey.

Bennett, S.L. and Liu, C.K. (1995) Absorbable block copolymers and surgical articles fabricated therefrom. US Patent #5,431,679, United States Surgical.

Bennett, S.L., Roby, M.S. and Muth, R.R. (1995) Bioabsorbable copolymer and coating composition containing same. US Patent #5,425,949, United States Surgical.

Berrabah, M., Andre, D., Prevot, F., Orecchiono, A.M. and Lafont, O. (1994) CG-MS Determination of phenobarbitone entrapped in poly (ϵ-caprolactone) nanocapsules. *Journal of Pharmaceutical and Biomedical Analysis*, **12**, 373–378.

Bezwada, R., (1995) Liquid copolymers of epsilon-caprolactone and lactide. US Patent #5,442,033, Ethicon.

Bezwada, R.S., Jamiolkowski, D.D., Lee, I., L., Vishvaroop, A., Persivale, J., Trenka-Benthin, S., Erneta, E., Surydevara, A.Y. and Liu, S. (1995) Monocryl® suture, a new ultra-pliable absorbable monofilament suture. *Biomaterials*, **16**, 1141–1148.

Brode, G.L. and Koleske, J.V. (1972) Lactone polymerization and polymer properties. *Journal of Macromolecular Science, Chemistry*, **A6**, 1109–1144.

Brode, G.L. and Koleske, J.V. (1973) Lactone polymerization and polymer properties", In , O. Vogl and J. Furukawa, (eds.), *Polymerization of Heterocycles*, Marcell Dekker, New York, pp. 97–116.

Chang, R.K., Price, J. and Whitworth, C.W. (1994) Enhancement of dissolution rate by incorporation into a water insoluble polymer, polycaprolactone. *Drug Development and Industrial Pharmacy*, **13**, 249–256.

Chynoweth, K.R. and Stachurski, Z.H. (1986) Crystallization of poly(ϵ-caprolactone). *Polymer*, **27**, 1912–1916.

Cook, W.J., Cameron, J.A., Bell, J.P. and Huang, S.J. (1981) Scanning electron microscopic visualization of biodegradation of polycaprolactones by fungi. *Journal of Polymer Science, Polymer Letters Edition*, **19**, 159–165.

Crescenze, V., Manzini, G., Calzolari, G. and Borri, C. (1972) Thermodynamics of fusion of poly-β-propiolactone and poly-ϵ-caprolactone. comparative analysis of the melting of aliphatic polylactone and polyester chains. *European Polymer Journal*, **8**, 449–463.

Darney, P.D., Monroe, S.E., Klaisle, C.M. and Alvarado, A. (1989) Clinical evaluation of the Capronor contraceptive implant: preliminary report. *American Journal of Obstetrics & Gynecology*, **160**, 1292–1295.

Dunn, R.L., Lewis, DH. and Laufe, L.E. (1986) Povidone iodine dispensing fiber. US Patent #4,582,052, Repromed.

Engelberg, I. and Kohn, J. (1991) Physico-mechanical properties of degradable polymers used in medical applications: a comparative study. *Biomaterials*, **12**, 292–304.

Gebelein, C.G., Chapman, M., Davison, M.K. and Gober, T. (1990) The controlled release of 5-fluorouracil from annealed monolithic systems. In C.G. Gebelein and R.L. Dunn, (eds.), *Progress in Biomedical Polymers*, Plenum Press, New York, pp. 321–333.

Gilbert,R.D., Stannett, V., Pitt, C.G. and Schindler, A. (1982) The design of biodegradable polymers: two approaches. In N. Grassie, (ed.), *Developments in Polymer Degradation, volume IV*, Elsevier, New York, pp. 259–293.

Giunchedi, P., Conti, B., Maggi, L. and Conte, U. (1994) Cellulose acetate butyrate and polycaprolactone for ketoprofen spray-dried microsphere preparation. *Journal of Microencapsulation*, **11**, 381–393.

Guzman, M., Molpeceres, J., Garcia, F., Aberturas, M.R. and Rodriguez, M. (1993) Formation and characterization of cyclosporin-loaded nanoparticles. *Journal of Pharmaceutical Sciences*, **82**, 498–502.

Hinrichs, W.L., Zweep, H.P., Satoh, S., Feijen, J. and Wildavuur, C.R. (1994) Supporting, microporous, degradable protheses to improve the arterialization of autologous vein grafts. *Biomaterials*, **15**, 83–91.

Hostettler, F, Magnus, G. and Vineyard, H. (1966) Process for the preparation of lactone polyesters. US Patent 3,284,417, Union Carbide Corporation.

Interox (1977) Caprolactone and its polymers. *Paint and Colour Journal*, **August 10/24**, A20–104.

Jamiolkowski, D.D. and Shalaby, S.W. (1987) Surgical articles of copolymers of glycolide and ϵ-caprolactone and methods of producing the same. US Patent #4,700,704, Ethicon.

Kalfogu, N.K. (1983) Compatibility of low density polyethylene-poly(ϵ-caprolactone) blends. *Journal of Applied Polymer Science*, **28**, 2541–2551.

Kyun, K.D., Yun, K.S., Young, J.S., Pyoung, C.C. and Heuf, S.S. (1990) Development of minocycline containing polycaprolactone films as a local drug delivery, *Taehan Chikkwa Uisa Hyphoe Chi*, **28**, 279–290.

Lebedev, B.V., Yevstropov, A.A., Lebedev, N.K., Karpova, Y.A., Lyudvig, Y.B. and Belen'kaya, B.G. (1979) The thermodynamics of ϵ-caprolactone, its polymer and of ϵ-caprolcatone in the 0-350°K range. *Polymer Science USSR*, **20**, 2219–2226.

Lin, S. (1987) Lactide/caprolactone polymer, method of making the same, composites thereof, and prostheses produced therefrom. US Patent #4,643,734, Hexcel Corporation.

Liu, Q., Donabedian, D., Lisuardi, A., Wang, X., Gross, R.A. and McCarthy, S.P. (1991) Melt processing and reactive extrusion of biodegradable materials. *201st National Meeting of the American Chemical Society*, Atlanta, April 1994.

Marchal-Heussler, L., Sirbat, D., Hoffman, M. and Maincent, P. (1993) Poly (ε-caprolactone) nanocapsules in carteolol ophthalmic delivery. *Pharmaceutical Research*, **10**, 386–390.

Menei, P., Croue, A, Daniel, V., Pouplard-Barthelaix, A. and Benoit, J.P. (1994) Fate and biocompatibility of three types of microspheres implanted into the brain. *Journal of Biomedical Materials Research*, **28**, 1079–1085.

Narkis, M., Sibony-Chaouat, S., Shkolnik, S. and Bell, J.P. (1985) Irradiation effects on polycaprolactone. *Polymer*, **26**, 50–54.

Olabisi, O. Robeson, L.M. and Shaw, M.T. (1972) *Polymer-Polymer Miscibility*, Academic Press, New York.

Ookuma, K. and Kurachi, K. (1993) Pseudomonas for degradation of polycaprolactone. Japanese Kokai Tokkyo Koho, patent #94319532, Sumitomo Metal Industries.

Ory, S.J., Hammond, C.B., Yancy, S.G., Hendren, R.W. and Pitt, C.G. (1983) The effect of a biodegradable contraceptive capsule (Capronor) containing levonorgestrel on gonadotropin, estrogen, and progesterone levels. *American Journal of Obstetrics & Gynecology*, **145**, 600–605.

Pandey, S., Singh, U.V. and Udupa, N. (1994) Implantable flurbiprofen for treating inflammation associated with arthritis. *Indian Drugs*, **31**, 254–257.

Pelerin, J. (1993) Method for making a custom impression tray. US Patent #5,213,498, Advantage Dental Products.

Perego, G., Cella, G.D., Aldini, N.N., Fini, M. and Giardino, R. (1994) Preparation of a new nerve guide from a poly(L-lactide-co-6–caprolactone). *Biomaterials*, **15**, 189–93.

Perrin, D.E. (1989) unpublished results.

Phillips, B., Pollart, D.F. and Koleske, J.V. (1972) Formable orthopaedic cast materials, resultant casts and method. US Patent #3,692,023, Union Carbide Corporation.

Pitt, C.G., Chasalow, Y.M., Hibionada, Y.M., Klimas, D.M. and Schindler, A. (1981b) Aliphatic polysters I. The degradation of poly(ε-caprolactone) *in-vivo*. *Journal of Applied Polymer Science*, **26**, 3779–3787.

Pitt, C.G., Gratzl, M.M., Jeffcoat, A.R., Zweidinger, R. and Schindler, A. (1979a) Sustained drug delivery systems II. factors affecting release rates from poly(ε-caprolactone) and related biodegradable polyesters. *Journal of Pharmaceutical Science*, **68**, 1534–1538.

Pitt, C.G., Gratzl, M.M., Kimmel, G.L., Surles, J. and Schindler, A. (1981a) Aliphatic polyesters II. The degradation of poly (D,L,-lactide, poly (ε-caprolactone), and their copolymers *in-vivo*. *Biomaterials*, **2**, 215–219.

Pitt, C.G., Jeffcoat, A.R., Zweidinger, R.A., Schindler, A. (1979b) Sustained drug delivery systems. I. the permeability of poly(ε-caprolactone) poly(D,L-lactic acid) and their copolymers. *Journal of Biomedical Materials Research*, **13**, 497–507.

Pitt, C.G., Marks, T.A., Schindler, A. (1980) Biodegradable drug delivery systems based upon aliphatic polyesters: application to contraceptives and narcotic antagonists. In R. Baker, (ed.), *Controlled Release of Bioactive Materials*, Academic Press, New York, pp. 232–253.

Pitt, C.G. and Schindler, A. (1980a) The design of controlled drug delivery systems based upon biodegradable polymers. In H. Ese and V.O. Waa, (eds.), *Biodegradables & Delivery Systems for Contraception volume 1, Progress in Contraceptive Delivery Systems*, MTP Press, Lancaster, England, pp. 17–46.

Pitt, C.G. and Schindler, A. (1980b) Capronor – a biodegradable delivery system for levonorgestrel. In G. Zatuchni, (ed.), *Long Acting Contraceptive Delivery Systems*, Proceedings of an International Workshop, pp. 48–63, 1980.

Pitt, C.G., Schindler, A. (1983a) Biodegradable polymers of lactones. US Patent #4,379,138, Research Triangle Institute.

Pitt, C.G. and Schindler, A. (1983b) Biodegradation of polymers. In S.D. Bruck, (ed.), *Controlled Drug Delivery*, CRC Press, Boca Raton, FL, pp. 55–80.

Radiology and Imaging Letter, vol. 2, pp. 164–165, (1992).

Rajasubramanianm G., Meidell, R.S., Landau, C., Dollar, M.L., Holt, D.B., Willard, J.E., Prager, M.D. and Eberhart, R.C. (1994) Fabrication of resorbable microporous intravascular stents for gene therapy. *Journal of the American Society for Artificial Internal Organs*, **40**, 584–589.

Schindler, A., Hibionada, Y.M. and Pitt, C.G. (1982) Aliphatic polyesters III. Molecular weight and molecular weight distributions in alcohol-initiated polymerizations of ε-caprolactone. *Journal of Polymer Science, Polymer Chemistry Edition*, **20**, 319–326.

Schindler, A., Jeffcoat, R., Kimmel, G.L., Pitt, C.G., Wall, M.E. and Zweidinger, R. (1977) Biodegradable polymers for sustained drug delivery. In E.M. Pearce and J.R. Schaefgen, (eds.), *Contemporary Topics in Polymer Science volume 2*, Plenum Press, New York, pp. 251–289.

Shalaby, S.W. (1980) Isomorphic copolymers of ε-caprolactone and 1,5-dioxepan-2-one. US Patent #4,190,720, Ethicon.

Shalaby, S.W., Jamiolkowski, D.D. (1985) Synthesis and intrinsic properties of crystalline copolymers of ε-caprolactone and glycolide. *Polymer Preprints, American Chemical Society Division of Polymer Chemistry*, **26**, 190.

Shalaby, S.W. and Jamiolkowski, D.D. (1986) Surgical articles of copolymers of glycolide and ε-caprolactone and methods of producing the same. US Patent #4,605,730, Ethicon.

TONE P-700 Polymer. (1983) Brochure #181D700-1a, 77796 00 from Union Carbide Corporation.

TONE Polymers. (1990) UCAR Coatings Brochure #F-60745, 5/90-5M from Union Carbide Corporation.

Wada, R., Hyon, S.H., Nakamura, T. and Ikada, Y. (1991) *In-vitro* evaluation of sustained drug release from a biodegradable elastomer. *Pharmaceutical Research*, **8**, 1292–1296.

Williams, D.F. (1994) Molecular biointeractions of biomedical polymers with extracellular exudate and inflammatory cells and their effect on biocompatibility, *in-vivo*. *Biomaterials*, **15**, 779–785.

Woodward, S.C., Brewer, P.S., Moatmed, F., Schindler, A, and Pitt, C.G. (1985) The intracellular degradation of poly (ε-caprolactone). *Journal of Biomedical Materials Research*, **19**, 437–444.

Yasin, M, Tighe, B.J. (1992) Polymers for biodegradable medical devices. viii. hydroxybutyrate-hydroxyvalerate copolymers: physical and degradative properties of blends with polycaprolactone. *Biomaterials*, **13**, 9–16.

APPENDIX

Polymer Name	Structure	Precursor/polymer	Source
Poly(ϵ-caprolactone)	$((CH_2)_5\text{-}COO)n$	ϵ-caprolactone	Union Carbide Corp.
		Stannous Octoate (catalyst)	Sigma Chemical Company
		1,6-Hexanediol	Aldrich Chemical Company (& other chain control agents)
		Poly(ϵ-caprolactone) (various Mw)	Birmingham Polymers, Inc. Union Carbide Corp.
Poly(ϵ-caprolactone)-co-(glycolide)	$((CH_2)_5\text{-}COO)n\text{-}(CH_2\text{-}COO)y$	ϵ-caprolactone	Union Carbide Corp.
		Glycolide	Boehringer Ingelheim KG B.I. Chemicals, Inc. PURAC biochem bv PURAC America
		Poly(ϵ-caprolactone)-co-(glycolide) (various Mw)	Birmingham Polymers, Inc.
Poly(ϵ-caprolactone)-co-(L-lactide)	$((CH_2)_5\text{-}COO)n\ (CH(CH_3)\text{-}COO)y$	ϵ-caprolactone	Union Carbide Corp.
		L-lactide	Boehringer Ingelheim KG B.I. Chemicals, Inc. PURAC biochem bv PURAC America
		Poly(ϵ-caprolactone)-co-(L-lactide) (various Mw)	Birmingham Polymers, Inc.

4. POLYHYDROXYALKANOATES

YOSHIHARU DOI

Polymer Chemistry Laboratory, The Institute of Physical and Chemical Research (RIKEN), Hirosawa, Wako-shi, Saitama 351–01, Japan

INTRODUCTION

A wide variety of microorganisms synthesize an optically active polymer of (R)-3-hydroxybutyric acid and accumulate it as a reserve energy source (Anderson and Dawes 1990; Doi 1990a; Steinbüchel 1991). Poly((R)-3-hydroxybutyrate) (P(3HB)) isolated from such microorganisms is a biodegradable and biocompatible thermoplastic with a melting temperature around 180°C. Many prokaryotic organisms, such as bacteria and cyanobacteria, have been found to accumulate P(3HB) up to 80% of their cellular dry weight when growth is limited by the depletion of an essential nutrient such as nitrogen, oxygen, phosphorus, or magnesium. Recently, many bacteria have been found to produce copolymers of (R)-3-hydroxyalkanotic acids with a chain length ranging from 4 to 14 carbon atoms (Steinbüchel 1991). In addition, 3-hydroxypropionic acid and 4-hydroxybutyric acid have been found as new constituents of microbial polyhydroxyalkanoates (PHA) (Doi 1990a). The general class of microbial polyesters is called PHA. The thermal and mechanical properties of PHA can be regulated by varying the compositions of the copolymers (Holms 1988; Inoue and Yoshie 1993).

A remarkable characteristic of PHA is their biodegradability in various environments (Lenz 1993). A number of microorganisms such as bacteria and fungi in soil, sludge, and sea water excrete PHA depolymerases to hydrolyze the solid PHA into water-soluble oligomers and monomer, and they utilize the resulting products as nutrients within cells. Aerobic and anaerobic PHA-degrading microorganisms have been isolated from various ecosystems, and the properties of their extracellular PHA depolymerases have been studied (Shirakura *et al.*, 1986; Schirmer *et al.*, 1993; Mukai *et al.*, 1993). The purified PHA depolymerases consist of a single polypeptide chain and their molecular weights are in the range of 37,000–59,000, having a serine residue at the active site.

These biodegradable PHA polymers have attracted attention as environmentally degradable thermoplastics to be used for a wide range of agricultural, marine, and medical applications. This paper surveys the biosynthesis and properties of bacterial PHA.

Biosynthesis of Polyhydroxyalkanoates

Copolymers containing hydroxyalkanoic acids with a chain length ranging from 3 to 14 carbon atoms have been produced from various carbon substrates (such as sugars, alkanoic acids, alcohols, and alkanes) by a variety of bacteria over 70 strains (Steinbüchel 1991). The PHA compositions produced by a bacterium are dependent of the substrate specificities of enzymes in the PHA biosynthetic pathway.

Y. DOI

Table 1 Microbial synthesis of PHA copolymers containing (R)-3HB as a constituent

Bacterial strain	Carbon substrate	Random copolymer
Alcaligenes eutrophus	propionic acid pentanoic acid	 (R)-3HB (R)-3HV
Alcaligenes eutrophus *Alcaligenes latus*	3-hydroxypropionoc acid 1,5-pentanediol	 (R)-3HB 3HP
Aeromonas cavie	olive oil	 (R)-3HB (R)-3HHx
Alcaligenes eutrophus *Alcaligenes latus* *Comamonas acidovorans*	4-hydroxybutyric acid γ-butyrolactone 1,4-butanediol 1,6-hexanediol	 (R)-3HB 4HB

Table 1 shows the PHA copolymers containing (*R*)-3-hydroxybutyrate as a constituent. A random copolymer of 3-hydroxybutyrate and 3-hydroxyvalerate, P(3HB-*co*-3HV), has been produced commercially by Zeneca in the United Kingdom in a fed-batch culture of *Alcaligenes eutrophus* by feeding propionic acid and glucose (Byrom 1987). The copolymer composition varies from 0 to 47 mol % 3HV,depending on the compoition of carbon substrates in feed. The Zeneca process is based on a large-scale, two-stage fermentation in a batch reactor. In the first stage, *A. eutrophus* cells grow and multiply in a glucose-salts medium under conditions of carbon and nutrient excess. Then the phosphate supply becomes depleted and the feeding of propionic acid is started. The P(3HB-*co*-3HV) copolymers are produced in the second stage of phosphate limitation. At present, Zeneca produces 500–1000t of P(3HB- *co*-3HV) per year.

P(3HB-*co*-3HV) copolymers with a wide range of compositions (up to 95 mol%) were produced in *A. eutrophus* by using pentanoic and butyric acids as the carbon sources (Doi *et al.*, 1988). The P(3HB) homopolymer was produced from butyric acid, while a P(3HB- *co*-3HV) copolymer with 95 mol% 3HV was produced from pentanoic acid. P(3HB-*co*-3HV) copolymers with a wide composition range were produced by varying the ratio of pentanoic acid to butyric acid in the feed.

A random copolymer of 3HB and 3-hydroxypropionic acid, P(3HB-*co*-3HP), was produced by *A.eutrophus* from carbon substrates such as 3-hydroxypropionic acid and 1,5-pentanediol (Nakamura *et al.*, 1991). The mole fraction of 3HP unit in the copolymer was lower than 7 mol%. The copolymer was synthesized in *Alcaligenes latus* from mixed carbon substrates of sucrose and 3-hydroxypropionic acid (Shimamura *et al.*, 1994a). *A. latus* accumulated P(3HB) homopolymer in the cells up to 60% of the dry weight during the course of growth, when sucrose was used as the sole carbon source. In contrast, *A. latus* did not grow in the medium containing 3-hydroxypropionic acid as the sole carbon source. When 3-hydroxypropionic acid was fed with sucrose, P(3HB-*co*-3HP) copolyesters were accumulated in the cells. The composition of copolymers varied from 0 to 26 mol% 3HP, depending on the fraction of carbon substrates. In addition, the P(3HB-*co*-3HP) copolymers with a wide range of compositions varying from 0 to 88 mol% 3HP were produced by *A. latus* from the mixed carbon substates of 3-hydroxybutyric and 3-hydroxypropionic acids. The copolymers were shown to have a random sequence distribution of 3HB and 3HP monomeric units by analysis of ^{13}C NMR spectra. The Mn of copolymers were in the range $1.1–3.5 \times 10^5$.

A random copolymer of 3HB and 3-hydroxyhexanoic acid, P(3HB-*co*-3HH), was produced by *Aeromonas caviae* from olive oil (Shimamura *et al.*, 1994b). The mole fraction of 3HH unit in copolymers was in the range of 11–17 mol%, depending on the fermentation conditions.

A random copolyester of 3HB and 4-hydroxybutyric acid, P(3HB-*co*-4HB), was produced by *A. eutrophus* (Kunioka *et al.*, 1988, 1989; Nakamura *et al.*, 1992), *A. latus* (Hiramitsu *et al.*, 1993) or *Comamonas acidovorans* (Saito and Doi, 1994), when 4-hydroxybutyric acid, 1,4-butanediol or γ-butyrolactone was used as the carbon source. The P(3HB-*co*-4HB) copolymers with a wide range of compositions from 0 to 100 mol% 4HB was produced by *A. eutrophus* from the mixed carbon substrates of 3-hydroxybutyric and 4-hydroxybutyric acids in the presence of some additives (Nakamura *et al.*, 1992). When 4-hydroxybutyric acid, citrate, and ammonium sulfate were fed to *A. eutrophus*, P(3HB-*co*-4HB) copolymers with compositions of 70–100 mol% 4HB were produced. In contrast, *C. acidovorans* produced a P(4HB) homopolymer in the presence of 1,4–butanediol or 4-hydroxybutyric acid without additives (Saito and Doi, 1994).

Physical Properties of Polyhydroxyalkanoates

The physical and thermal properties of bacterial PHA copolymers can be regulated by varying their molecular structure and copolymer compositions. The P(3HB) homopolymer is a relatively stiff and brittle material. The introduction of hydroxyalkanoate comonomers into a P(3HB) chain greatly improves its mechanical properties. The PHA family of polyesters offers a wide variety of polymeric materials, from hard crystalline plastic to elastic rubber. The PHA materials behave as thermoplastics with melting temperatures of 50–180°C and can be processed by conventional extrusion and molding equipment (Holm 1988).

The optically active P(3HB) homopolymer is a highly crystalline thermoplastic with a highly crystalline thermoplastic with a melting temperature around 180°C. The unit cell is orthorhombic [a = 0.576 nm, b = 1.320 nm, c(fiber axis) = 0.596 nm, space group $P2_12_12_1$], and two molecules of P(3HB) pass through the unit cell. The conformation of the P(3HB) molecule has a compact, right-handed 2_1 helix with a twofold screw axis and a fiber repeat of 0.596 nm (Cornibert and Marchessault 1972). The mechanical properties of Young's modulus (3.5 GPa) and the tensile strength (43 MPa) of P(3HB) material are close to those of isotactic polypropylene. The extension to break (5%) for P(3HB) is, however, markedly lower than that of polypropylene (400%).

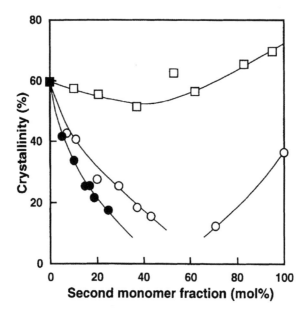

Figure 1 Degrees of X-ray crystallinities of various PHA copolymers containing 3HB unit. (●): P(3HB-co-3HH), (□): P(3HB-co-3HV), (○): P(3HB-co-3HP).

Figure 1 shows the degrees of X-ray crystallinities of various PHA copolymers containing 3HB unit as a constituent. The P(3HB-co-3HV) copolymers have approximately the same high degree of crystallinity (50–70%) throughout a wide range of compositions (Kunioka et al., 1989). A structural characteristic of P(3HB-co-3HV) is isodimophism, i.e., P(3HB-co-3HV) copolymers crystallizing either in a P(3HB) or P(3HV) crystal lattice for the 3HV fraction's lower or higher than 40 mol%, respectively (Bluhm et al., 1986). On the other hand, the crystallinities of P(3HB-co-3HP) and P(3HB-co-3HH) copolymers decrease with an increase in the content of second momer unit. The crystallographic parameters of P(3HB-co-3HP) with compositions up to 43 mol% 3HP were little influenced by the presence of 3HP unit, and the crystallinities decreased from 60 to 7% as the 3HP content was increased from 0 to 67 mol% (Shimamura et al., 1994a). The crystallinities of P(3HB-co-3HH) decrease with the 3HH fraction (Shimamura et al., 1994b). These results indicate that 3HP and 3HH units cannot crystallize in the sequence of 3HB units and act as defects in the P(3HB) crystal lattice.

The physical and thermal properties of P(3HB-co-4HB) copolymers are listed in Table 2. The crystallinity of P(3HB-co-4HB) decreased from 60 to 14 % as the 4HB content increased from 0 to 49 mol% (Kunioka et al., 1989). Only one crystalline form of the P(3HB) lattice was observed for the X-ray diffractions of P(3HB-co-4HB) copolymers with compositions of 0–29 mol% 4HB. In contrast, only the P(4HB) lattice was observed for the P(3HB-co-4HB) copolymers with compositions of 78–100 mol% 4HB (Saito and Doi 1994). The tensile strength of P(3HB-co-4HB) films with compositions of 0–16 mol% 4HB decreased from 43 to 26 MPa with an increase in the 4HB fraction, while the elongation to break increased from 5 to 444%.

Table 2 Physical and thermal properties of P(3HB-co-4HB) copolymers

	4HB fraction (mol%)										
	0	3	7	10	16	27	64	78	82	90	100
Melting temp. (°C)	178		172		130		50	49	52	50	53
Glass transition temp. (°C)	4	-2			-7		-35	-37	-39	-42	-48
Crystallinity (%)	60	55	50	45	45	40	15	17	18	28	34
Density (g·cm^{-3}	1.250		1.232	1.234	1.234						
Water uptake (wt%)	0.32		0.20	0.14	0.45						
Stress at yield (MPa)[a]		34		28	19						14
Elongation at yield (%)[a]		4		5	7						17
Tensile strength (MPa)[a]	43	28		24	26		17	42	58	65	104
Elongation to break (%)[a]	5	45		242	444		591	1120	1320	1080	1000

[a]at 23°C.

Table 3 Changes in the thickness of bacterial P(3HB) films in various environments

Environment	Change in the thickness (μm/week)
Sea water (22°C)	5
Activated sludge (25°C)	7
Soil (25°C)	5
Sterile sea water (25°C)[a]	0

[a]Autoclaved at 121°C for 15 min.

The tensile strength of the films with compositions of 64–100 mol% 4HB increased from 17 to 104 MPa with increasing the 4HB fraction. The true tensile strength of P(4HB) homopolymer is calculated to be as large as 1 GPa if the cross-section is corrected. Thus, P(3HB-co-4HB) copolymers exhibit a wide range of material properties.

Biodegradation of Polyhydroxyalkanoates

One of unique properties of bacterial PHA is their biodegradability in various environments. P(3HB) is the best studied material of PHA and probably most abundant in the environment. Table 3 shows the rates of biodegradation of P(3HB) films in various environments such as soil, sludge, and sea water. Table 3 also shows that no simple hydrolysis of P(3HB) takes place in the water without microorganisms. Since PHA is a solid polymer of high molecular weights and incapable of being transported through the cell wall, PHA-degrading microorganisms excrete extra-cellular PHA depolymerases to degrada environmental PHA. Such PHA-degrading microorganisms were isolated from various environments such as soil (*Pseudomonas lemoignei*), activated sludge (*Alcaligenes faecalis*), laboratory atmosphere (*Pseudomonas pickettii*), sea water (*Comamonas testosteroni*), and lake water (*Pseudomonas stutzeri*).

The extracellular PHA depolymerases have been purified from different isolates, and their properties have been studied. (Mukai *et al.*, 1993, 1994). Almost all

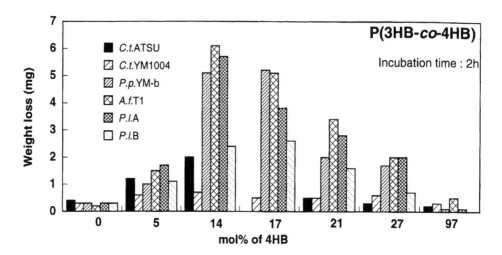

Figure 2 Weight loss (erosion) of P(3HB-*co*-4HB) films with various compositions after the enzymatic degradation of 2h at 37°C with different PHA depolymerases in a 0.1 M phosphate buffer (pH 7.4).

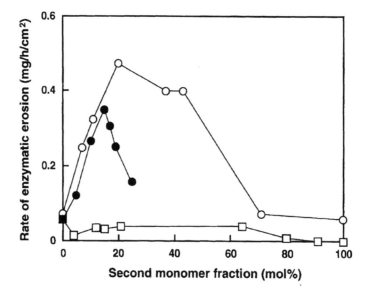

Figure 3 Rates of enzymatic erosion of various PHA copolymer films in the aqueous solution of PHA depolymerase (from *Alcaligenes faecalis*) at 37°C and pH 7.4. (●): P(3HB-*co*-3HH), (□): P(3HB-*co*-3HV), (○) : P(3HB-*co*-3HP).

extracellular PHA depolymerases can be prepared by hydrophobic interaction column chromatography. The purified PHA depolymerases consisted of a single polypeptide chain and their molecular weights were in the range of 37,000–59,000. All enzymes have a serine residue at the active site.

Enzymatic degradation of PHA films were performed in an aqueous solution of purified PHA depolymerases at 37°C. The enzymatic degradation occurred on the surface of PHA films, and the thickness of the films decreased with time (Doi *et al.*, 1990). Figures 2 and 3 show the weight loss of different PHA films by PHA depolymerases from various microorganisms. The rate of film erosion by PHA depolymerases was strongly dependent of the copolymer composition. The rate of enzymatic degradation remarkably increased with an increase in the 4HB fraction, and the highest rate was observed at 14 mol% 4HB (see Figure 2). Thus, the presence of 4HB units in the P(3HB) sequence accelerated the enzymatic degradation. We showed that the rate of surface erosion of P(3HB) film by PHA depolymerase was markedly decreased with an increase in the crystallinity, and that the rate of enzymatic degradation on the amorphous phase of the P(3HB) film was about twenty times higher than the rate on the crystalline phase (Kumagai *et al.*, 1992). As Figure 3 shows, the acceleration of enzymatic degradation was observed on the films of P(3HB-*co*-3HP) and P(3HB-*co*-3HH) as well as on the P(3HB-*co*-4HB) films, which may be caused by the decrease in crystallinity.

REFERENCES

Anderson, A.J. and Dawes, E.A. (1990) Occurrence, metabolism, metabolic role, and industrial use of bacterial polyhydroxyalkanoates, *Microbiol. Rev.* **54**, 450–472.

Bluhm, T.L., Hamer, G.K., Marchessault, R.H., Fyfe, C.A. and Veregin, R.P. (1986) Isodimorphism in bacterial poly(b-hydroxybutyrate-*co*-b-hydroxyvalerate), *Macromolecules* **19**, 2871–2876.

Byrom, D. (1987) Polymer synthesis by microorganisms:technology and ecomomics, *Trends Biotechnol.* **5**, 246–250.

Cornibert, J. and Marchessault, R.H. (1972) Physical propeties of poly-b-hydroxybutyrate: conformational analysis and crystalline structure, *J. Mol. Biol.* **72**, 735–756.

Doi, Y., Tamaki, A., Kunioka, M. and Soga, K. (1988) Production of copolyester of 3-hydroxybutyrate and 3-hydroxyvalerate by *Alcaligenes eutrophus* from butyric and pentanoic acids, *Appl. Microbiol. Biotechnol.*, **28**, 330–334.

Doi, Y. (1990a) Microbial polyesters, VCH Publishers, New York.

Doi, Y., Kanesawa, Y., Kunioka, M. and Saito, T. (1990b) Biodegradation of microbial copolyesters: poly(3-hydroxybutyrate-*co*-3-hydroxyvalerate) and poly(3-hydroxybutyrate-*co*-4-hydroxybutyrate), *Macromolecules* **23**, 26–31.

Hiramitsu, M., Koyama, N. and Doi, Y. (1993) Production of poly(3–hydroxybutyrate-co-4-hydroxybutyrate) by *Alcaligenes latus*, *Biotechnol. Lett.*, **15**, 461–464.

Holm, P.A. (1988) Biologically produced (R)-3-hydroxyalkanoate polymers and copolymers, In Development in Crystalline Polymer (Ed. by D. C. Bassett), Elsevier, London, p1–65.

Inoue, Y. and Yoshie, N. (1992) Structure and physical properties of bacterially synthesized polyesters, *Prog. Polym. Sci.*, **17**, 571–610.

Kumagai, Y., Kanesawa, Y. and Doi, Y. (1992) Enzymatic degradation of microbial poly(3-hydroxybutyrate) films, *Makromol. Chem.*, **193**, 53–57.

Kunioka, M., Nakamura, Y. and Doi, Y. (1988) New bacterial copolyesters produced in *Alcaligenes eutrophus* from organic acids, *Polym. Commun.*, **29**, 174–176.

Kunioka, M., Kawaguchi, Y. and Doi, Y. (1989a) Production of biodegradable copolyesters of 3-hydroxybutyrate and 4-hydroxybutyrate by *Alcaligenes eutrophus*, *Appl. Microbiol. Biotechnol.*, **30**, 569–573.

Kunioka, M., Tamaki, A. and Doi, Y. (1989b) Crystalline and thermal properties of bacterial copolyesters: poly(3-hydroxybutyrate-*co*-3-hydroxyvalerate) and poly(3-hydroxybutyrate-*co*-4-hydroxybutyrate), *Macromolecules*, **22**, 694–697.

Lenz, R.W. (1993) Biodegradable polymers, *Adv. Polym. Sci.* **107**, 1–40.

Mukai, K., Yamada, K. and Doi, Y. (1993a) Kinetics and mechanism of heterogeneous hydrolysis of poly[(R)-3-hydroxybutyrate] film by PHA depolymerases, *Int. J. Biol. Macromol.*, **15**, 361–366.

Mukai, K., Doi, Y., Sema, Y. and Tomita, K. (1993b) Substrate specificities in hydrolysis of polyhydrox-yalkanoates by microbial esterases, *Biotechnol. Lett.*, **15**, 601–604.

Mukai, K., Yamada, K. and Doi, Y. (1994) Efficient hydrolysis of polyhydroxyalkanoates by *Pseudomonas stutzeri* YM1414 isolated from lake water, *Polym. Deg. Stab.*, **43**, 319–327.

Nakamura, S., Kunioka, M. and Doi, Y. (1991) Biosynthesis and characterization of bacterial poly(3-hydroxybutyrate-*co*-3- hydroxypropionate), *Macromol. Rep.*, **A28**, 15–24.

Nakamura, S., Doi, Y. and Scandola, M. (1992) Microbial synthesis and characterization of poly(3-hydroxybutyrate-*co*-4-hydroxybutyrate), *Macromolecules*, **25**, 4237–4241.

Saito, Y. and Doi, Y. (1994) Microbial synthesis and properties of poly(3-hydroxybutyrate-*co*-4-hydroxy-butyrate) in *Comamonas acidovorans*, *Int. J. Biol. Macromol.* **16**, 99–104.

Schirmer, A., Jendrossek, D. and Schlegel, H.G. (1993) Degradation of poly(3-hydroxyocdtanoic acid) by bacteria: purification and properties of a P(3HO) depolymerase from *Pseudomonas fluorescens* GK13, *Appl. Environ. Microbiol.*, **59**, 1220–1227.

Shimamura, E., Scandola, M. and Doi, Y. (1994a) Microbial synthesis and characterization of poly(3-hydroxybutyrate-*co*-3-hydroxypropionate), *Macromolecules*, **27**, 4429–4435.

Shimamura, E., Kasuya, K., Kobayashi, G., Shiotani, T., Shima, Y. and Doi, Y. (1994b) Physical properties and biodegradability of microbial poly(3- hydroxybutyrate-*co*-3-hydroxyhexanoate), *Macromolecules*, **27**, 878–880.

Steinbüchel, A. (1991) Polyhydroxyalkanoic acids, In Biomaterials (Ed. by Byrom, D.), Macmillan Pub., Basingstoke, pp. 123–213.

Shirakura, Y., Fukui, T., Saito, T., Okamoto, Y., Narikawa, T., Koide, K., Tomita, K., Tamemura, T. and Masamune, S. (1986) Degradation of poly(3- hydroxybutyrate) by poly(3-hydroxybutyrate) depolymerase from A *lcaligenes faecalis* T1, *Biochim. Biophys. Acta.* **880**, 46–53.

5. POLY(PROPYLENE FUMARATE)

SUSAN J. PETER[1], MICHAEL J. MILLER[2], MICHAEL J. YASZEMSKI[3]
and ANTONIOS G. MIKOS[1,*]

[1] *Cox Laboratory for Biomedical Engineering, Institute of Biosciences and Bioengineering
and Department of Chemical Engineering, Rice University,
P.O. Box 1892, Houston, Texas 77251–1892*
[2] *Department of Plastic and Reconstructive Surgery, M.D. Anderson Cancer Center,
The University of Texas, 1515 Holcombe Blvd.,
Box 62, Houston, Texas 77030*
[3] *Department of Orthopaedic Surgery, Wilford Hall Medical Center,
Lackland Air Force Base, Texas 78236*

INTRODUCTION

Synthetic biodegradable polymers have become an invaluable asset in the medical field. Polymers have an advantage over other synthetic materials because their properties, such as degradation rate and mechanical strength, can be tailored to specific applications. They are also free of the immunological response that biological materials can cause. For example, poly(α-hydroxy esters), such as poly(lactic acid) and poly(glycolic acid), are used as sutures for surgery. Poly(anhydrides) are used for drug delivery. Specifically for orthopaedic applications, a synthetic, biodegradable, injectable polymer that crosslinks *in situ* would be useful for filling osseous defects and inducing bone regeneration. One such polymer, poly(propylene fumarate) (PPF), is being studied in an attempt to create a biodegradable bone cement.

PPF is an unsaturated linear polyester (Figure 1) that degrades into fumaric acid and propylene glycol, both biocompatible degradation products. Many formulation methods have been proposed for orthopaedic applications. Once PPF is produced, it can be crosslinked through the fumarate double bond with a vinyl monomer. Various filler particles can be incorporated to create porosity, increase mechanical properties, and enhance osteoconductivity. This mixture creates a moldable paste suitable for filling irregularly shaped bone defects that hardens in a short period of time. This review describes the many methods used to synthesize and characterize PPF, and explores possible applications.

SYNTHESIS

Sanderson (1988) produced PPF by a transesterification of diethyl fumarate and propylene glycol with a para-toluene sulfonic acid catalyst (Figure 2). The two components were combined and slowly heated to a temperature of 250°C over a five hour period. This mixture was cooled to 100°C and placed under a vacuum of 1 mm Hg to remove any remaining volatile components. Still under vacuum, the reaction

---

*To whom correspondence should be addressed.

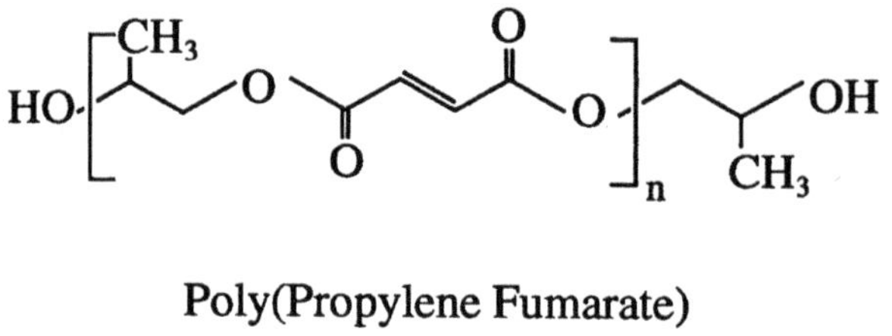

Poly(Propylene Fumarate)

Figure 1 Poly(Propylene Fumarate).

Diethyl Fumarate   Propylene Glycol

Poly(Propylene Fumarate)

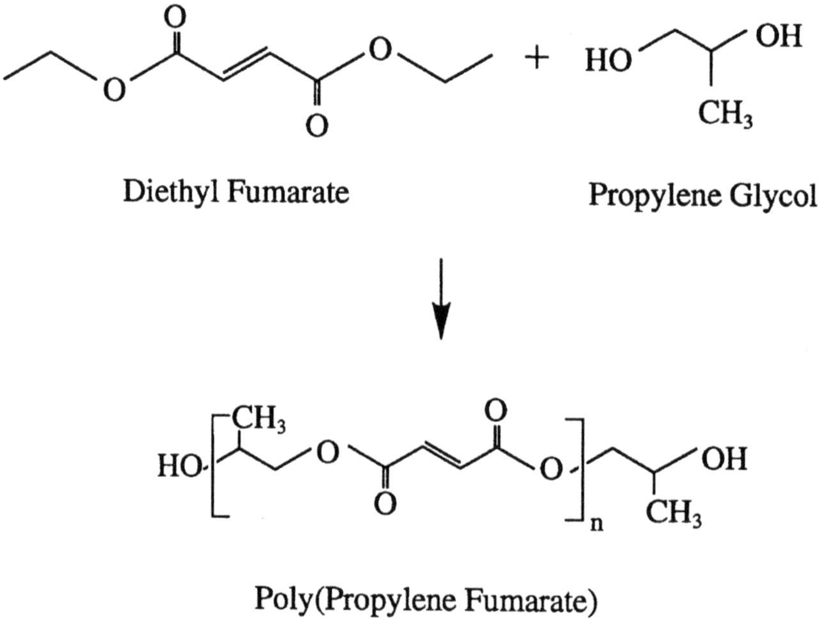

Figure 2 Poly(Propylene Fumarate) Synthesis Method of Sanderson (1988).

vessel was heated to 220°C and held there for four hours. The resulting polymer was a clear, thick, viscous mixture separated by fractionation from methylene chloride with ether. When vacuum dried, the PPF was a white to yellow powder with a 35% yield by weight of the reactants.

Gerhart and Hayes (1989) prepared PPF through a condensation reaction of propylene glycol and fumaric acid (Figure 3). The mixture was initially heated to 145°C for five hours, with a subsequent increase in temperature to 180°C. Samples were removed during the course of the reaction to monitor the viscosity of the product. Initially the viscosity was at a value of 2 poise, and rose steadily with increasing reaction time. The reaction was terminated when the viscosity reached 10–15 poise. The product was consistently within a number average molecular weight (M_n) range of 500 to 1200, with a polydispersity index (PI) ranging from 3 to 4.

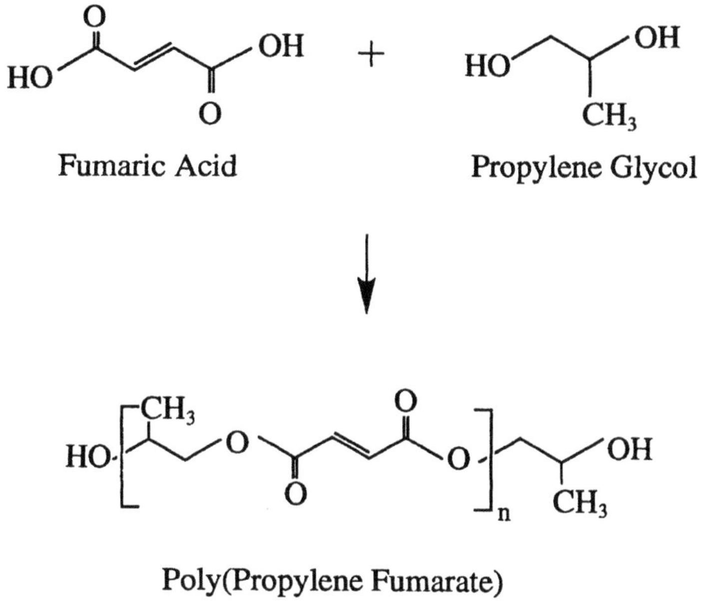

Figure 3 Poly(Propylene Fumarate) Synthesis Method of Gerhart and Hayes (1989).

Domb *et al.* (1989) prepared PPF through several reaction methods. The first method involved preparing bis-propylene glycol fumarate (PFP trimer) by reacting fumaric acid with propylene glycol for 20 hours, using pyridine as a catalyst (Figure 4). The viscous liquid trimer was shaken with Na_2HPO_4, and then the phases were allowed to separate. The lower water phase was discarded, while the upper phase containing the trimer was washed with KCl, filtered, and dried. Propylene glycol dibutenoic ester (MPM trimer) was prepared through the reaction of maleic anhydride in toluene with propylene glycol at 100°C for 24 hours. The mixture was cooled and allowed to separate into two phases. The ether phase was washed with hexane and rotary evaporated to remove the solvents. PPF was formulated by the melt polymerization of PFP with MPM at 180°C for two hours, giving a liquid product. Higher molecular weights were achieved by using pentamer or higher molecular weight forms of PFP and MPM. PPF was also prepared by the step polymerization reaction of the PFP trimer with maleic anhydride at 110°C for 24 hours. The final formulation method to prepare PPF was through the controlled reaction of propylene glycol and fumaric acid at 130°C for 10 hours and then at 180°C for 2 hours. The reaction yielded a clear viscous liquid. The PPF samples prepared by this final method had an M_n of 300 to 2000, with a PI range of 1.5 to 1.7.

Domb *et al.* (1996) also investigated variation of the terminal groups of PPF in an attempt to improve the mechanical properties of PPF composites without the use of a crosslinking monomer (Figure 5). Trimeric bis(2–hydroxypropyl fumarate) (HPF) was prepared through the dropwise addition of propylene glycol to a mixture of fumaric acid and dry pyridine in acetone (Domb *et al.*, 1990). The solvents were removed by drying, yielding a slightly viscous yellow liquid. Propylene bis(hydrogen

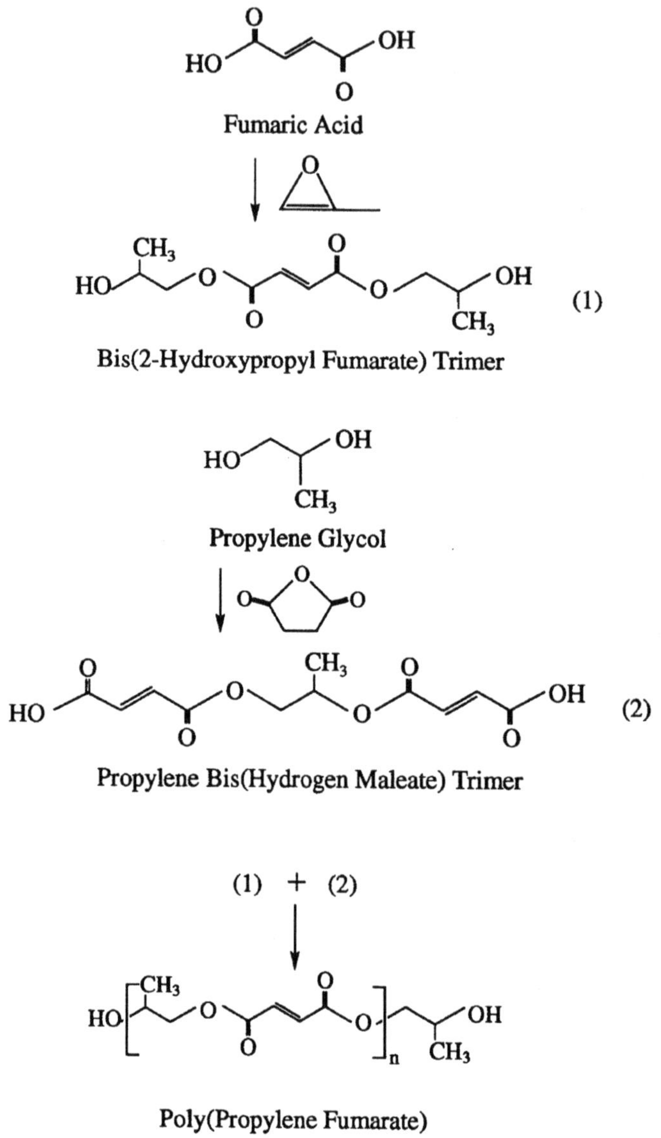

Figure 4 Poly(Propylene Fumarate) Synthesis Method of Domb *et al.* (1989).

maleate) (PHM) was formed through the dropwise addition of propylene glycol to a solution of maleic anhydride in toluene. The addition took place at $100°C$ over a 24 hour period. The product was cooled and any remaining solvents were evaporated. The intermediate product, PPF, was prepared by a melt condensation reaction between HPF and PHM at $140°C$ for 24 hours. The system was then placed under a 5 mmHg vacuum to remove any volatile components. The molecular weight was found to be M_n 750 with a PI of 2.4. This polymeric form of PPF was reacted

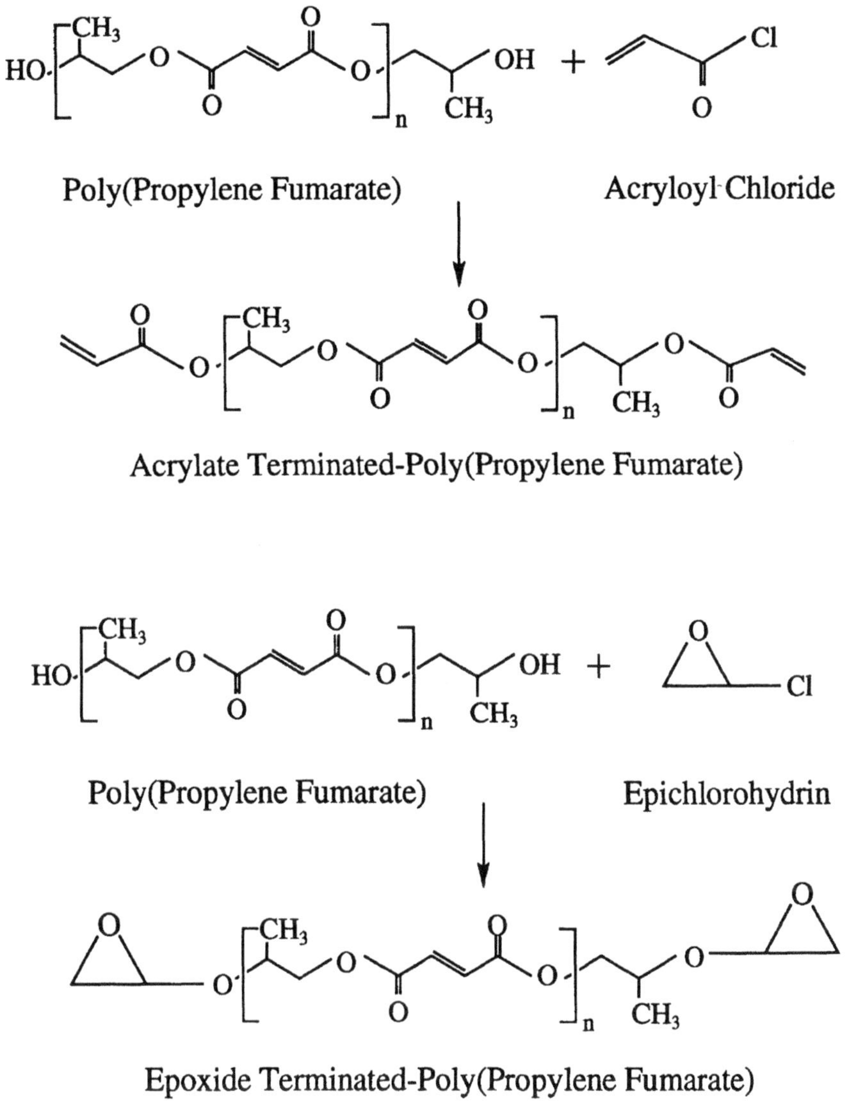

Figure 5 Two Methods of Modifying the Terminal Groups of Poly(Propylene Fumarate) devised by Domb *et al.* (1989).

with acryloyl chloride in a chloroform solution over a three day period of time. The solvent was then evaporated, yielding a viscous oligomer of acrylate terminated PPF. A second variation of PPF was formed through a reaction with epichlorohydrin. PPF was dissolved in dry toluene and held at 90°C. Tin chloride catalyst was added, followed by epichlorohydrin, under nitrogen. After 3 hours at 90°C the reaction was brought to room temperature for a period of 3 days. Upon evaporation of the solvent, the diepoxide-PPF oligomer was obtained.

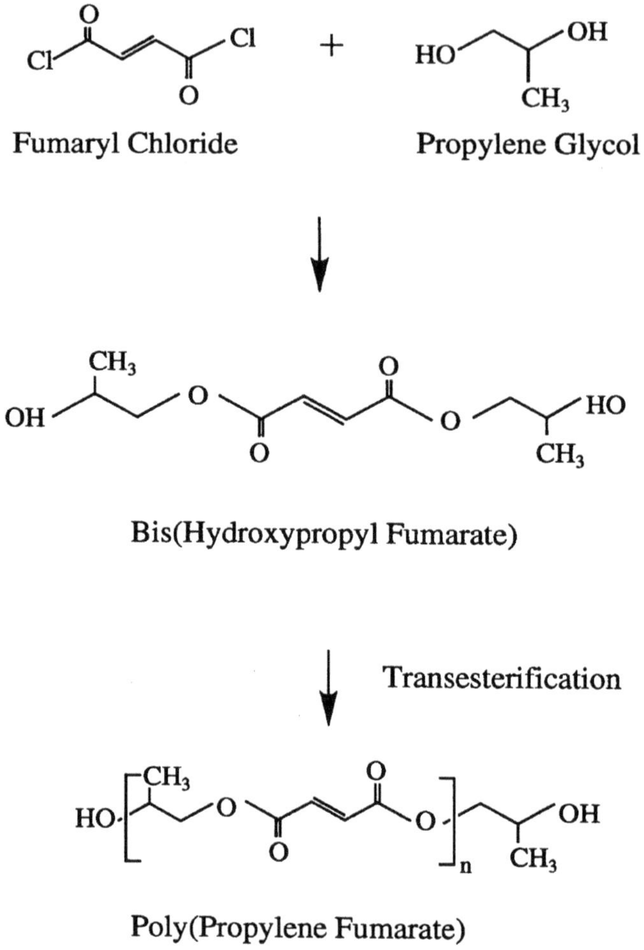

Figure 6 Poly(Propylene Fumarate) Synthesis Method of Yaszemski *et al.* (1994).

Yaszemski *et al.* (1994) prepared PPF through an initial reaction of fumaryl chloride and propylene glycol at room temperature (Figure 6). The product, bis(2-hydroxypropyl fumarate), was purified through solution-precipitation in THF and petroleum ether, respectively. This was followed by a transesterification at 160°C under vacuum with antimony trioxide as a catalyst in a nitrogen atmosphere. Excess propylene glycol was boiled off and collected during the reaction. The reaction time ranged from 4 hours to 24 hours. Purification was completed through a solution-precipitation as mentioned above, followed by drying in a rotary evaporator. The resulting polymer was a yellow viscous liquid of M_n 750 and PI of 1.7 when reacted for 7 hours, and of M_n 850 and PI of 2 when reacted for 8 hours (Yaszemski et al., 1994). Without the use of the catalyst, a reaction time of 30 hours produced PPF with M_n 1225 and PI of 1.6. The highest molecular weights were found with the use of the catalyst for longer reaction times, reaching an M_n of 1500 with PI of 3.

Gresser *et al.* (1995) prepared PPF through a direct esterification of fumaric acid and propylene glycol, using *p*-toluene sulfonic acid monohydrate as a catalyst, a modification of the method by Sanderson (Figure 2). Addition of t-butyl hydroquinone prevented crosslinking at high temperatures. The polymer was dissolved in methylene chloride and filtered to remove residual fumaric acid. Subsequent washing with 20% aqueous methanol was used to remove remaining propylene glycol. The product was precipitated from the methylene chloride with diethyl ether. Further purification was carried out by dissolution in acetone, filtration, drying, and removal of the solvent under vacuum. The M_n was found to be 2600, with a PI of 2.6.

CHARACTERIZATION

Sanderson (1988) combined PPF with a calcium sulfate filler, benzoyl peroxide, and *N*-vinyl pyrrolidinone (VP). The mixture hardened in 15 to 30 minutes. Mechanical studies gave a compressive strength of 50 MPa for the crosslinked composites. This value was alterable by varying the molecular weight of the polymer and the ratio of the polymer to the filler. Beads 1.5 mm in diameter were created with the crosslinking formula and cured at 37°C. These beads were placed in distilled water at 37°C and agitated for 16 days. A 50% weight loss was found without a macroscopic change in structural integrity. The weight loss was most likely due to dissolution of the filler, creating a porous structure. Further crosslinking was performed in saline at 37°C, demonstrating that the crosslinking can be carried out in a wet surgical environment. *In vivo* studies were performed using a rat model. A composite formulation including powdered human bone as a filler component was prepared and used to fill one-eighth inch holes in the rat tibia, cranium, tooth or jaw. This study showed that the composite mixture hardened *in situ* and bone grew into the crosslinked areas within four to six weeks, leaving little residual putty.

Gerhart and Hayes (1989) crosslinked their PPF with methyl methacrylate (MMA) in a ratio of 85:15 respectively. A free radical initiator (benzoyl peroxide) and an accelerator [dimethyl-*p*-toluidine (DMT)] were used to crosslink the PPF-MMA mixture. Calcium phosphate ceramic and a resorbable calcium salt were added to the PPF-MMA matrix to promote osteoconduction and create porosity, respectively. The compressive strength and modulus of the composite were found to be 19 MPa and 200 MPa, respectively. In *in vitro* experiments, crosslinked samples were immersed in water buffered at neutral pH. Slight swelling was observed, causing a decrease in mechanical strength and stiffness. Secondary calcium ions from the ceramic were believed to have complexed with PPF carboxylic end groups, causing the samples to regain their initial mechanical strength after several days. When the environment was made alkaline, the samples swelled further, causing a significant decrease in mechanical strength, as well as an increase in degradation rate. However, an increased ratio of MMA to PPF allowed for a longer degradation time. *In vivo* studies were performed in rabbits using a subcutaneous implantation site. Samples crosslinked with varying amounts of MMA were mechanically tested at regular intervals post-implantation. At one day following implantation the specimens showed a decrease in strength, believed to be due to swelling. After four days, the mechanical properties increased, again believed to be due to secondary calcium ion reactions. Later samples showed varying strengths dependent on MMA content: the larger the MMA ratio, the higher the maintained level of mechanical strength.

Additional studies were performed using a paraspinal, subcutaneous rat model, with 6×12 mm cylinders being implanted (Frazier *et al.*, 1995). Implant composition and length of degradation time were investigated. Increasing the amount of methyl methacrylate and benzoyl peroxide was found to significantly increase the compressive strength over time, but had no effect on the modulus over time. With a formulation consisting of PPF, 10% MMA, 15% benzoyl peroxide, 33% tricalcium phosphate and 33% calcium carbonate, the compressive strength was found to increase linearly through day 21 to 5 MPa, and then decreased linearly to 1 MPa by the 84th day of the study.

Proximal femur fractures were studied as well through the use of a PPF composite consisting of PPF, MMA, tricalcium phosphate and calcium phosphate (Witschger *et al.*, 1991). Dynamic hip screws were placed in cadaveric proximal femora and fixed in place with the PPF composite mixture. The yield load when placed under mechanical testing was 1130 N, just slighly under 1750 N corresponding to the use of MMA alone.

Domb *et al.* (1990) mixed their PPF with calcium particulates and crosslinked this mixture with MMA in order to assess the mechanical properties of the crosslinked polymer. Benzoyl peroxide and DMT were used to effect the crosslinking process. The compressive strength of the PPF bone cement was found to increase with increasing polymer chain length and narrowing polydispersity, reaching a maximum of 7 MPa. *In vitro* degradation studies showed an initial decrease in mechanical properties due to the swelling of the matrix (Domb, 1989). After several days, the samples regained their mechanical strength. Under alkaline conditions, increased degradation was evident. In an acidic environment, the phosphate was leached out, causing a decrease in stiffness, though the samples retained their shape. Increased MMA content delayed degradation. Due to the stepwise addition during polymerization, the terminal group can represent either the carboxylic acid of the fumaric acid or the hydroxyl group of the propylene glycol. The balance of hydroxyl end groups to carboxylic acids end groups was found to effect the mechanical properties of the composites, showing an increase in compressive strength with higher carboxylic acid end group content.

PPF with acrylate and epoxide terminal groups was produced by Domb *et al.* (1996). The polymers were crosslinked in a ratio of 30 wt% PPF and 70 wt% calcium carbonate-tricalcium phosphate mixture. The strongest samples required incorporation of both VP and MMA. The PPF-diacryl alone gave a stress level of 43.4 MPa, but had values as high as 130 MPa when mixed with the vinyl monomers. The PPF diol had a compressive strength of 56 MPa and 38 MPa when mixed with VP and MMA respectively, but had a value of less than 10 MPa without incorporation of either vinyl monomer. The epoxide terminated PPF gave strong composites, but was difficult to crosslink through a free radical reaction. Again, the best mechanical properties were found when one of the vinyl monomers was added to the crosslinking mixture, giving a maximum stress of 37 MPa with both monomers, and a value of 24 MPa without.

Yaszemski *et al.* (1995) incorporated sodium chloride as a pore forming component into a composite formulation also including β-tricalcium phosphate (β-TCP) or a calcium phosphate matrix. The mixture was crosslinked with VP to form a biodegradable bone cement. The mechanical properties of the composite were studied over several varying factors in a fractional factorial design. These mixtures were crosslinked using benzoyl peroxide and DMT. The results were as follows: increasing the sodium chloride content, using a lower molecular weight PPF, and decreasing the amount of VP all increased the compressive strength and compressive modulus; increasing the amount of β-TCP effected a higher compressive modulus (Yaszemski *et al.*, 1995). By changing the levels of the above components, the compressive strength was varied from 2 to 14 MPa, and the compressive modulus was varied from 13 to 138 MPa. These values encompass the target strength and modulus for bone replacement (Goldstein *et al.*, 1991).

In vitro mechanical studies were performed over time with varying amounts of crosslinker in the formulation (Yaszemski *et al.*, 1996). A composite material formulation consisting of PPF of molecular weight 1780, β-TCP (at a ratio of β-TCP/PPF of 0.75/1), and VP (at the same ratio of VP/PPF 0.75/1) showed an initial compressive strength of 18 MPa which rose to a final value of 21 MPa after 12 weeks in phosphate-buffered saline (PBS) at $37°C$. The compressive modulus for the same formulation also increased during the 12 week testing period from 113 MPa to 696 MPa. The increase in mechanical properties with degradation time was related to the degradation of the composite material and may prove to be useful for orthopaedic applications.

In vivo studies were performed with the same formulation through a 2-mm proximal tibia defect in rats (Yaszemski *et al.*, 1995). Histologically it was seen that increasing amounts of bone grew into the polymer over time, beginning at the peripheral surface and progressing towards the center of the defect. The polymeric material was seen to degrade over time and be replaced by new woven bone, without eliciting an immune response. The composite was found to degrade faster at the surface of the implant, allowing the interior of the implant to maintain its mechanical integrity for a longer period of time. This will be useful in repairing bone defects located in load bearing areas of the skeleton, lending stability until the bone has fully replaced the polymer's function in filling the defect as well as in support.

Gresser *et al.* (1995) combined their PPF with VP and hydroxyapatite, crosslinking this mixture with benzoyl peroxide and DMT. Samples were allowed to crosslink at $37°C$ for 24 hours. These hardened samples were ground to analyze the amount of PPF and VP incorporated. The ground power was run in acetonitrile on reverse-phase thin layer chromatography plates, and developed with iodine vapor or ultraviolet light. Over 90% of the PPF was found to have been crosslinked in mixtures containing between 0.33 to 2 PPF/VP ratios. The fraction of VP incorporated was dependent on the ratio of the two components. The fraction of PPF or VP incorporated was independent of the amount of hydroxyapatite, benzoyl peroxide, or DMT used. The VP monomer was found to add to poly(vinyl pyrrolidinone) twice as often as it added to the fumarate double bond.

ORTHOPAEDIC APPLICATIONS

Orthopaedic surgeons must fill defects created by trauma, removal of cancerous tumors, or abnormal development. Bone replacement and fixation are also issues for plastic surgeons in craniofacial procedures, hand and foot deformities, and extremity injuries. Though poly(methyl methacrylate) (PMMA) bone cement is currently in use to address these problems, it in nonbiodegradable, remaining in the patient's body forever, often leading to the problems of stress shielding and particulate wear. If a biodegradable, injectable bone cement were available, many of these problems would be avoided. For orthopaedic applications, a bone cement that polymerizes *in situ* with minimal heat emission and maintains its mechanical integrity as it degrades and is replaced by new bone is preferred. The mechanical properties of a biodegradable bone cement must approximate those of the material it is replacing: the compressive strength and modulus of human trabecular bone are 5 MPa and 50 MPa, respectively (Goldstein *et al.*, 1991). It is the goal of researchers working on PPF to create crosslinked composites of PPF with all of these required properties.

PPF can further be used as a drug delivery agent to deliver bioactive molecules to bone. Gerhart *et al.* (1994) used as their design goals a carrier that would release a drug within four to twelve weeks, and degrade in a period of sixteen to twenty weeks. The study involved the use of gentamicin and vancomycin for treating *Staphylococcus aureus* osteomyelitis. The antibiotics were impregnated into PPF-MMA and PMMA cylinders (Gerhart *et al.*, 1988). These were implanted subcutaneously in rats, with the blood and wound fluid levels monitored over a two week time period. The PPF-MMA was found to have higher wound antibiotic levels than the PMMA bone cement. However, these levels remained below maximum safety levels. Mechanical testing showed no significant loss of mechanical integrity due to antibiotic incorporation. PPF-MMA was found to be a successful antibiotic carrier, with the advantage over PMMA of being biodegradable. Other drugs, such as thioridazine hydrochloride or thioridazine base were also incorporated into PPF with much success (Sanderson, 1988).

FUTURE DIRECTIONS

PPF is a versatile polymeric material with great potential for use in bone healing. Composite materials of crosslinked PPF have sufficient mechanical stability to replace the structural role of the missing bone, and retain this strength during bone ingrowth and remodeling. With the incorporation of a leachable porogen, pores for cells to migrate into the composite are formed. Moreover, drugs can be incorporated into the matrix to fight infection and enhance healing. Further work can be in the avenue of functionalization of the polymeric backbone with cell adhesion specific molecules to allow for modulation of receptor-mediated processes which contribute to tissue development and growth. There is also potential for use as a carrier for cells, genes, and bioactive molecules to induce wound repair and facilitate tissue formation and remodeling. Further applications are being investigated through the development of a PPF copolymer with poly(ethylene glycol) for use as a biodegradable, injectable vascular stent (Suggs *et al.*, 1995). Poly(propylene fumarate) has the potential to have an enormous impact in the engineering of hard and soft tissues.

REFERENCES

Domb, A.J. (1989) "Poly(Propylene Glycol Fumarate) Compositions for Biomedical Applications," United States Patent 4,888,413, 1–32.

Domb, A.J., Laurencin, C.T., Israell, O., Gerhart, T.N. and Langer, R. (1990) "The Formation of Propylene Fumarate Oligomers for Use in Bioerodible Bone Cement Composites," *J. Polym. Sci., Polym. Chem.*, **A28**, 973–985.

Domb, A.J., Manor, N. and Elmalak, O. (1996) "Biodegradable Bone Cement Compositions Based on Acrylate and Epoxide Terminated Poly(Propylene Fumarate) Oligomers and Calcium Salt Compositions," *Biomaterials*, **17**, 411–417.

Frazier, D.D., Lathi, V.K., Gerhart, T.N., Altobelli, D.E. and Hayes, W.C. (1995) "*In-Vivo* Degradation of a Poly(Propylene-Fumarate) Biodegradable Particulate Composite Bone Cement. In: *Polymers in Medicine and Pharmacy*," Mikos A.G., Leong K.W., Yaszemski M.J., Tamada J.A., Radomsky M.L., eds., **394**, Pittsburgh: Materials Research Society, 15–19.

Gerhart, T.N. and Hayes, W.C. (1989) "Bioerodible Implant Composition," United States Patent 4,843,112, 1–16.

Gerhart, T.N., Laurencin, C.T., Domb, A.J., Langer, R.S. and Hayes, W.C. (1994) "Bioerodible Polymers for Drug Delivery in Bone," United States Patent 5,286,763, 1–10.

Gerhart, T.N., Roux, R.D., Horowitz, G., Miller, R.L., Hanff, P. and Hayes, W.C. (1988) "Antibiotic Release From an Experimental Biodegradable Bone Cement," *J. Orthop. Res.*, **6**, 585–592.

Goldstein, S., Matthews, L., Kuhn, J. and Hollister, S. (1991) "Trabecular Bone Remodeling: An Experimental Model," *J. Biomech.*, **24**, 135–150.

Gresser, J.D., Hsu, S.H., Nagaoka, H., Lyons, C.M., Nieratko, D.P., Wise, D.L., Barabino, G.A. and Trantolo, D.J. (1995) "Analysis of a Vinyl Pyrrolidinone/Poly(Propylene Fumarate) Resorbable Bone Cement," *J. Biomed. Mater. Res.*, **29**, 1241–1247.

Sanderson, J.E. (1988) "Bone Replacement and Repair Putty Material from Unsaturated Polyester Resin and Vinyl Pyrrolidone," United States Patent 4,722,948, 1–14.

Suggs, L.J., Payne, R.G., Kao, E.Y., Alemany, L.B., Yaszemski, M.J., Wu, K.K. and Mikos, A.G. (1995) "The Synthesis and Characterization of a Novel Block Copolymer Consisting of Poly(Propylene Fumarate) and Poly(Ethylene Oxide)," In: *Polymers in Medicine and Pharmacy*, Mikos A.G., Leong K.W., Yaszemski M.J., Tamada J.A., Radomsky M.L., eds., **394**, Pittsburgh: Materials Research Society, 167–173.

Witschger, P., Gerhart, T., Goldman, J., Edsberg, L. and Hayes, W. (1991) "Biomechanical Evaluation of a Biodegradable Composite as an Adjunct to Internal Fixation of Proximal Femur Fractures," *J. Orthop. Res.*, **9**, 48–53.

Yaszemski, M.J., Mikos, A.G., Payne, R.G. and Hayes, W.C. (1994) "Biodegradable Polymer Composites for Temporary Replacement of Trabecular Bone: The Effect of Polymer Molecular Weight on Composite Strength and Modulus," In: *Biomaterials for Drug and Cell Delivery*, Mikos A.G., Murphy R., Bernstein H., Peppas N.A., eds., **331**, Pittsburgh: Materials Research Society, 251–256.

Yaszemski, M.J., Payne, R.G., Hayes, W.C., Langer, R.S., Aufdemorte, T.B. and Mikos, A.G. (1995) "The Ingrowth of New Bone Tissue and Initial Mechanical Properties of a Degrading Polymeric Composite Scaffold," *Tissue Eng.*, **1**, 41–52.

Yaszemski, M.J., Payne, R.G., Hayes, W.C., Langer, R. and Mikos, A.G. (1996) "*In Vitro* Degradation of a Poly(Propylene Fumarate)-Based Composite Material," *Biomaterials*, **17**, 2127–2130.

6. POLY(ORTHO ESTERS)

JORGE HELLER

APS Research Institute, Redwood City, CA 94063, USA

DEVELOPMENT OF POLY(ORTHO ESTERS)

Development of poly(ortho ester) started in 1970 and four distinct families of such polymers have been now prepared. These will be designated as poly(ortho ester) I, poly(ortho ester) II, poly(ortho ester) III and poly(ortho ester) IV.

Poly(Ortho Ester) I

The first example of a poly (ortho ester) was described in a series of patents assigned to Alza in 1978–79 (Choi *et al.*, 1978a, 1978b, 1978c, 1979a, 1979b). The polymer is prepared as shown in Scheme 1.

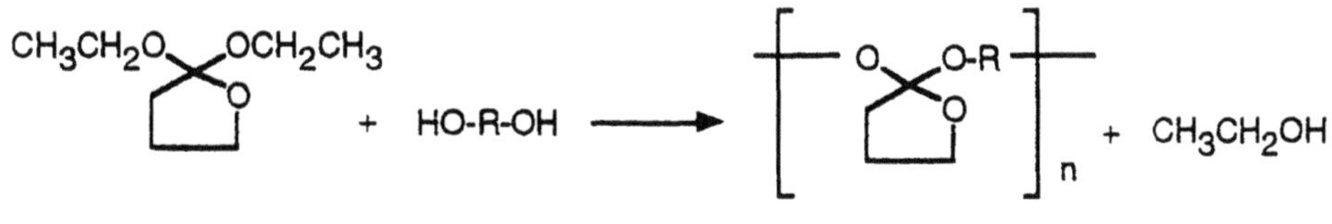

Scheme 1

Typical Experimental Procedure

To 45 g (0.312 mole) of anhydrous *trans*-1,4–cyclohexanedimethanol and 0.05 g polyphosphoric acid in a commercially available polymerization reactor is added with constant stirring under an inert nitrogen environment and normal atmospheric pressure 50 g (0.312 mole) of anhydrous 2,2-diethoxytetrahydrofuran. Next, the mixture is heated to 110–115°C and held at that temperature for 1.5 to 2 hr with slow distillation of ethanol. Then, while maintaining the temperature, the pressure is gradually reduced to 0.01 mm of mercury and at this reduced pressure the temperature is slowly increased to 180°C. The reaction is continued at this temperature for 24 hr. The polymer is isolated by extrusion from the reactor.

Polymer Hydrolysis

When this poly (ortho ester) is placed in an aqueous environment, an initial hydrolysis to a diol and γ-butyrolactone takes place. The γ-butyrolactone then rapidly hydrolyzes to γ-hydroxybutyric acid. This reaction path is shown in Scheme 2.

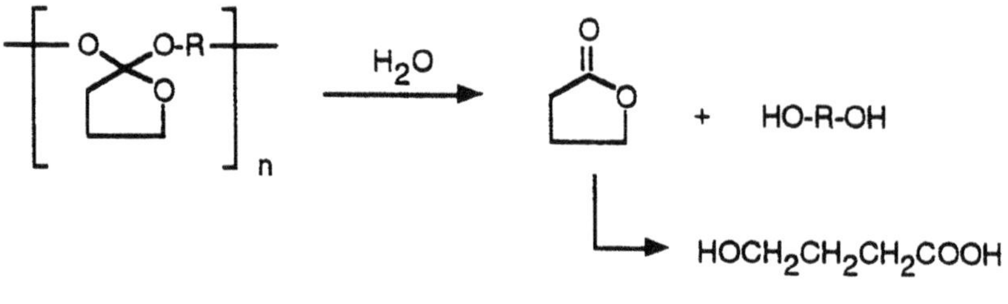

Scheme 2

The hydrolysis is an autocatalytic process because the γ-hydroxybutyric acid hydrolysis product accelerates hydrolysis of the acid-sensitive ortho ester linkages. Therefore, in order to prevent autoacceleration of the hydrolysis, a base such as $CaCO_3$ is incorporated into the polymer to neutralize the γ-hydroxybutyric acid.

Applications

Despite the fact that the polymer was developed in the early 1970's, very little information related to its actual use in controlled drug release has been published and the polymer has not been identified beyond the code names of C111 and C101ct. However, it can be inferred that C101ct refers to a polymer prepared from diethoxytetrahydrofuran and *cis/trans*-cyclohexane dimethanol and C111 refers to a polymer prepared from diethoxytetrahydrofuran and 1,6-hexanediol. The polymer has been used for the release of naltrexone (Capoza *et al.*, 1978) the release of contraceptive steroids (Gabelnick, 1983) and in the treatment of burns (Vistness *et al.*, 1976). Very recently (Solheim *et al.*, 1995) described the use of C101ct for the release of indomethacin in the prevention of reossification of experimental bone defects.

Commercial Availability

The polymer is covered by US Patents 4,079,038; 4,093,709; 4,131,648; 4,138,344 and 4,180,646 assigned to the Alza corporation, was originally designated as Chronomer™ and is currently known as Alzamer®. It is not commercially available nor has had FDA approval.

Poly(Ortho Ester) II

Ortho esters can also be prepared by the reaction between a ketene acetal and an alcohol as shown in Scheme 3 (Heller *et al.*, 1980).

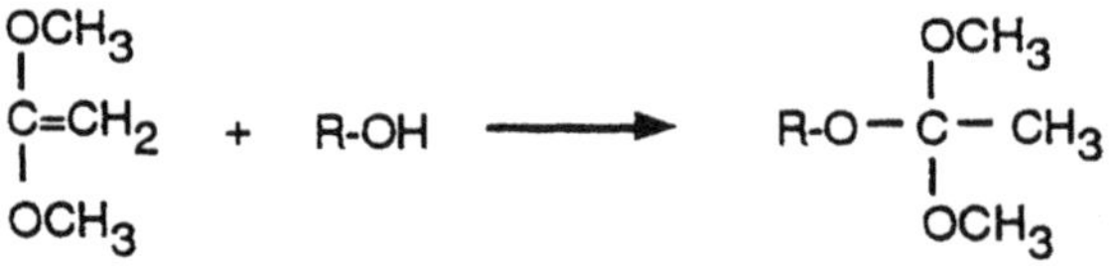

Scheme 3

Clearly, formation of polymers requires the use of a diketene acetal. Early work with this polymer system was based on the reaction of the diketene acetal 3,9-bis(methylene) 2,4,8,10-tetraoxaspiro[5,5]undecane and 1,6-hexanediol as shown in Scheme 4 (Heller *et al.*, 1980).

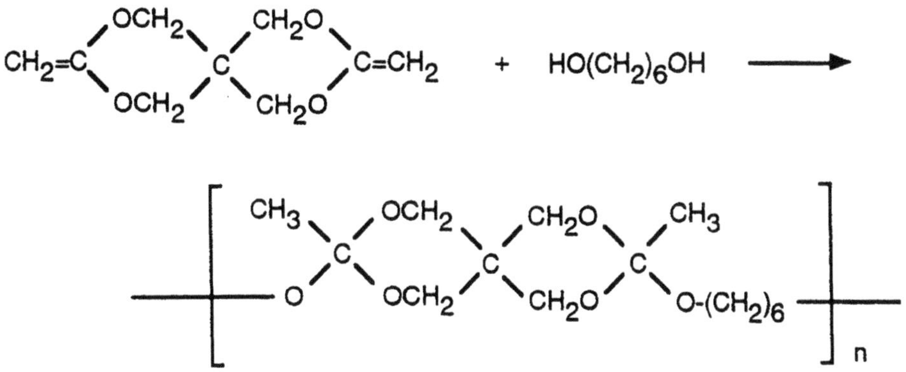

Scheme 4

However, the diketene acetal 3,9-bis(methylene) 2,4,8,10-tetraoxaspiro [5,5] undecane contains a double bond connected to two electron donor groups and is thus extremely susceptible to a cationic polymerization. For this reason, the synthesis of a more useful diketene acetal was evolved as shown in Scheme 5. In this diketene acetal, the cationic polymerization has been inhibited by the introduction of a methylene group which sterically hinders the facile cationic polymerization. This diketene acetal can be readily prepared by a rearrangement of the commercially available diallylpentaerythritol as shown in scheme 5 (Ng *et al.*, 1992).

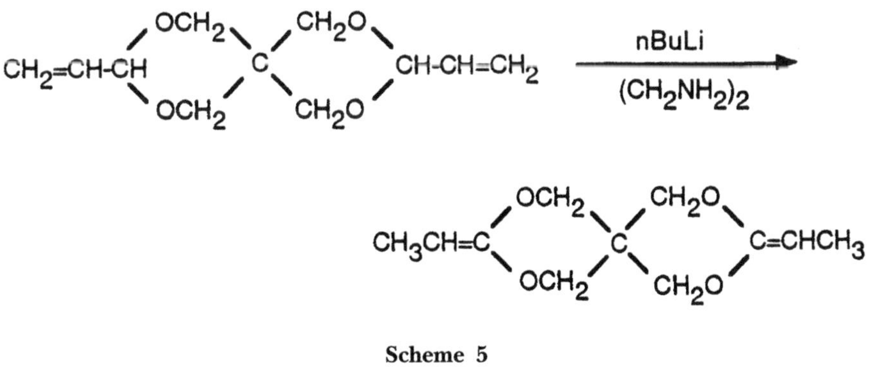

Scheme 5

Polymers are easily prepared by the addition of a diol to the diketene acetal as shown in Scheme 6 (Ng *et al.*, 1992).

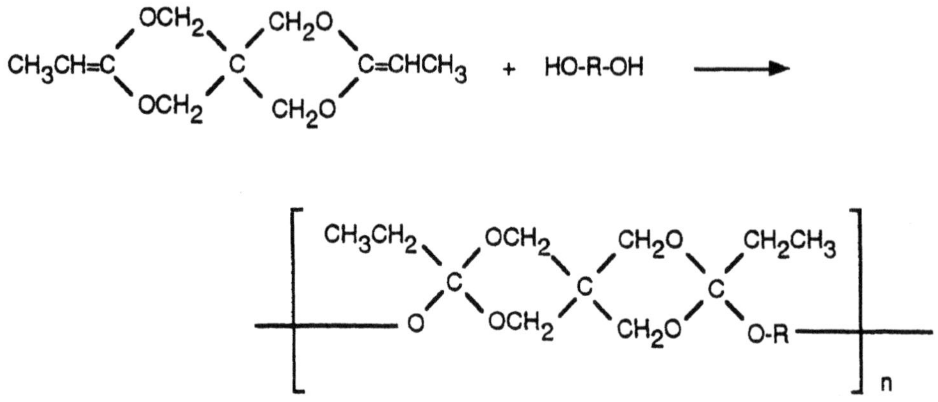

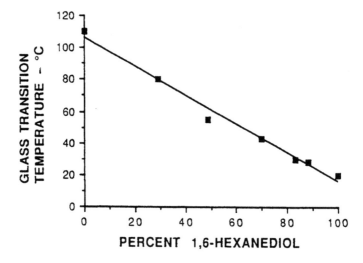

Scheme 6

The reaction proceeds readily at room temperature and to prepare polymers it is merely necessary to dissolve the monomers in a polar solvent such a tetrahydrofuran and add a trace of an acid catalyst. Polymerization is exothermic and high molecular weight polymers are formed virtually instantaneously.

Mechanical properties of the polymer can be controlled by an appropriate choice of the diols used in the condensation reaction (Heller *et al.*, 1983). Use of the rigid diol *trans*-cyclohexanedimethanol produces a rigid polymer having a glass transition temperature of 120°C while use of the flexible diol 1,6-hexanediol produces a soft material having a glass transition temperature of 20°C. Mixtures of the two diols produce polymers that have glass transition temperatures between these two values.

Figure 1 Glass transition temperature of polymer prepared from 3,9-bis (ethylidene)-2,4,8,10-tetraoxaspiro [5,5] undecane, 1,6-hexanediol and *trans*-cyclohexanedimethanol as a function of mol % 1,6-hexanediol. Reprinted with permission from Heller *et al.*, 1983.

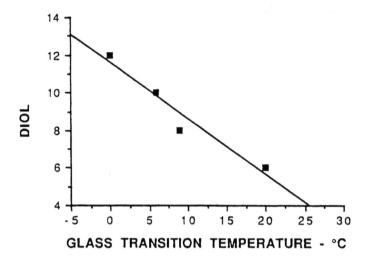

Figure 2 Effect of diol chain length on the glass transition temperature of polymers prepared from 3,9-bis (ethylidene)-2,4,8,10-tetraoxaspiro [5,5] undecane and α-ω diols. Reprinted with permission from Heller *et al.*, 1995.

Variation of the glass transition temperature with composition of the diol mixture is shown in Figure 1 (Heller *et al.*, 1983). It is also possible to prepare materials having lower glass transition temperatures by using diols that have more than six methylene groups (Heller *et al.*, 1995). Glass transition temperature of the polymer as a function of the number of methylene groups is shown in Figure 2.

The preparation of poly(ortho esters) by the addition of diols to diketene acetals is similar to the preparation of polyurethanes by the addition of diols to diisocyanates. Because the condensation between a diketene acetal and a diol, just like that between a diisocyanate and a diol, proceeds without the evolution of volatile by-products, dense, crosslinked materials can be produced by using reagents having a functionality greater than two (Heller *et al.*, 1985a).

To prepare crosslinked materials, a molar excess of the diketene acetal is used and the resulting prepolymer with ketene acetal end-groups is reacted with a triol or a mixture of diols and triols. This synthesis is shown in Scheme 7.

Typical Experimental Procedures

Preparation of 3,9-bis (ethylidene)-2,4,8,10-tetraoxaspiro[5,5]undecane

In a 3-L three necked flask fitted with a mechanical stirrer, argon inlet tube, thermometer and rubber septum is placed 1.2-L of ethylene diamine. The flask is cooled with ice water and the contents kept at about 8°C under an argon atmosphere. A hexane solution of 130 g (2 moles) of n-butyllithium is added via a stainless steel hypodermic U-tube pushed through the rubber septum using carefully controlled argon pressure over a period of 1 hr.

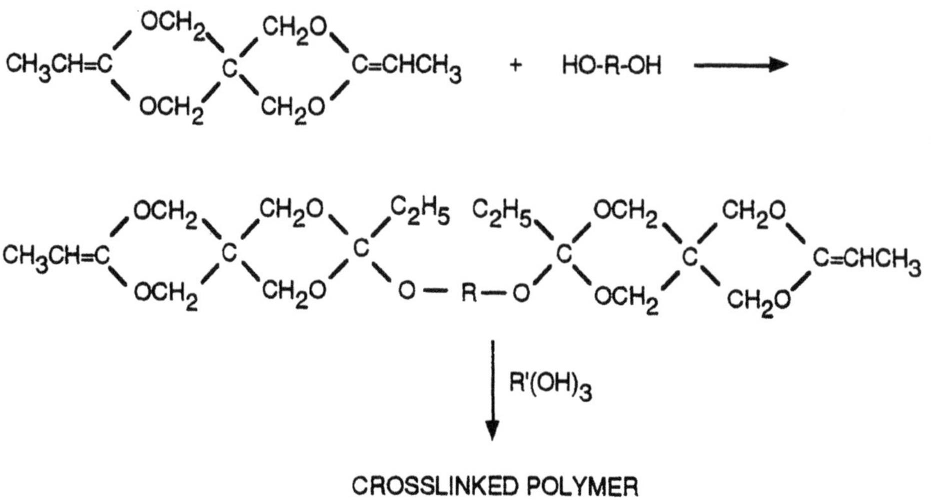

$$CH_3CH=C \begin{matrix} OCH_2 \\ \\ OCH_2 \end{matrix} C \begin{matrix} CH_2O \\ \\ CH_2O \end{matrix} C=CHCH_3 \quad + \quad HO\text{-}R\text{-}OH \longrightarrow$$

$$CH_3CH=C \begin{matrix} OCH_2 \\ \\ OCH_2 \end{matrix} C \begin{matrix} CH_2O \\ \\ CH_2O \end{matrix} C \begin{matrix} C_2H_5 \\ \\ O-R-O \end{matrix} \begin{matrix} C_2H_5 \\ \\ \end{matrix} C \begin{matrix} OCH_2 \\ \\ OCH_2 \end{matrix} C \begin{matrix} CH_2O \\ \\ CH_2O \end{matrix} C=CHCH_3$$

$$\Big\downarrow R'(OH)_3$$

CROSSLINKED POLYMER

Scheme 7

Next, a mixture of 530 g (2.5 moles) of 3,9-bis (vinyl)-2,4,8,10-tetraoxaspiro [5,5] undecane and 0.5-L of ethylenediamine is cooled to 8°C and added to the three necked flask. After stirring at 8°C for 3 hr, the reaction mixture is poured into 3-L of ice-water with vigorous stirring. The aqueous mixture is extracted twice with 1-L portions of hexane. The combined hexane extracts are washed three times with 1-L portions of water, dried over anhydrous magnesium sulfate and filtered under suction. The filtrate is evaporated to dryness on a rotary evaporator to give 413 g (78%) of crude material containing 90% of 3,9-bis (ethylidene)-2,4,8,10-tetraoxaspiro [5,5] undecane.

The crude product is dissolved in 2-L of hexane containing 10 ml of triethylamine and the solution placed in a 4-L filter flask, sealed and stored in a freezer at −20°C for 2 days. The crystals thus formed are collected by basket centrifugation at −5°C under an argon atmosphere. Distillation of the brownish product through a 12 in. vigreaux column at reduced pressure gives 313 g (61%) of 3,9-bis (ethylidene)-2,4,8,10-tetraoxaspiro [5,5] undecane as a colorless liquid, b.p. 82°C (0.1 torr) which crystallizes at room temperature, m.p. 30°C; characteristic IR band at 1700 cm^{-1}.

Preparation of Linear Polymers

Into a 5-L, three necked flask equipped with an overhead stirrer, an argon inlet tube and a condenser are placed 89.57g (0.621 mole) of *trans*-cyclohexanedimethanol, 39.52 g (0.334 mole) of 1,6-hexanediol, and 1.8 L of distilled tetrahydrofuran. The mixture is stirred until all solids have dissolved; then, 200 g (0.942 mole) of 3,9-bis (ethylidene)-2,4,8,10-tetraoxaspiro [5,5] undecane is added. The polymerization is initiated by the addition of 2 mL of a solution of *p*-toluenesulfonic acid (20 mg/mL) in tetrahydrofuran.

The polymerization temperature rapidly rises to the boiling point of tetrahydrofuran and then gradually decreases. Stirring is continued for about 2 hr, 10 mL of triethylamine stabilizer added, and the reaction mixture then very slowly poured with vigorous stirring into about 15 gallons of methanol containing 100 mL of triethylamine. The precipitated polymer is collected by vacuum filtration and dried in a vacuum oven at 60°C for 24 hr. The weight of the dried polymer is 325 g (98.8% yield)

Preparation of Crosslinked Polymers

To a solution of 31.84 g (0.159 mole) of 3,9-bis (ethylidene)-2,4,8,10-tetraoxaspiro [5,5] undecane in 200 mL of distilled tetrahydrofuran is added 10.42 g (0.100 mole) of 2-methyl-1,4-butanediol. The solution is stirred under argon and 0.5 mL of p-toluenesulfonic acid solution in tetrahydrofuran (20 mg/ml) is added to initiate the reaction. After the heat of reaction has subsided, the solution is stirred until the temperature returns to ambient and then concentrated on a rotary evaporator followed by heating in a vacuum oven at 40°C to remove residual solvent.

Devices are then prepared by mixing into the prepolymer an excess of 1,2,6-hexanetriol and the desired excipients and curing the mixture in a mold at 75°C for 5 hr. Best results are obtained when the mole ratio hydroxyl to ketene acetal is about 1.3.

Characterization

Figure 3 shows a 25.2 MHz ^{13}C-NMR spectrum of a polymer prepared from 3,9-bis(methylene) 2,4,8,10-tetraoxaspiro[5,5]undecane and 1,6-hexanediol. Band assignments are shown in Figure 3 (Heller *et al.*, 1980). The spectrum verifies that the expected structure was obtained and that there are no extraneous bands indicating abnormal linkages.

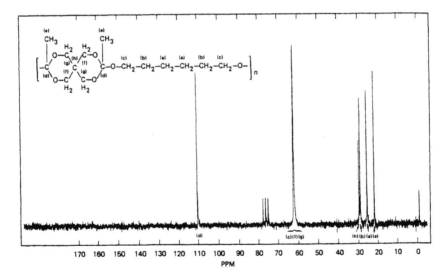

Figure 3 25.2 MHz ^{13}C NMR spectrum of a polymer prepared from 3,9-bis (methylene)-2,4,8,10-tetraoxaspiro [5,5] undecane and 1,6-hexanediol in CDCl$_3$ at room temperature. Reprinted with permission from Heller *et al.*, 1981.

Polymer Hydrolysis

When these poly (ortho esters) are placed in an aqueous environment they hydrolyze as shown in Scheme 8 (Heller *et al.*, 1983). A mechanistic study establishing the nature of all degradation products of the hydrolysis has been carried out (Heller *et al.*, 1987a).

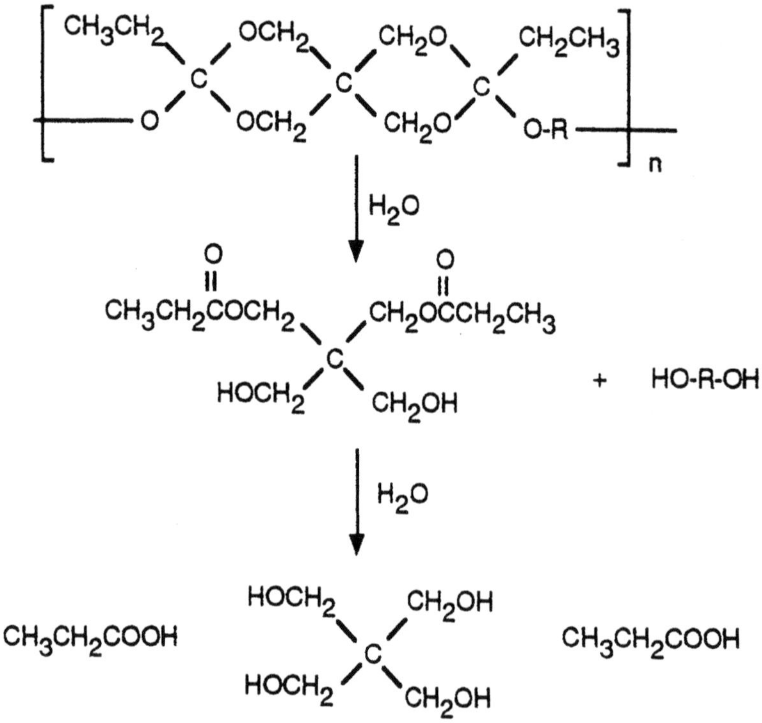

Scheme 8

Even though the hydrolysis eventually produces an acid, polymer erosion rate is controlled by hydrolysis of the ortho ester bonds. The subsequent hydrolysis of the ester bonds takes place at a much slower rate so that the neutral, low molecular weight reaction products can diffuse away from the bulk polymer before hydrolysis to an acid takes place. Thus, unlike the poly (ortho ester) system I, no autocatalysis is observed and it is not necessary to use basic excipients to neutralize the acidic hydrolysis product.

Control of Polymer Hydrolysis Rate

Although ortho ester linkages are very labile in solution, when they are incorporated into a highly hydrophobic polymer matrix, rate of hydrolysis is very slow. This is illustrated in Figure 4 which shows weight loss of a polymer prepared from

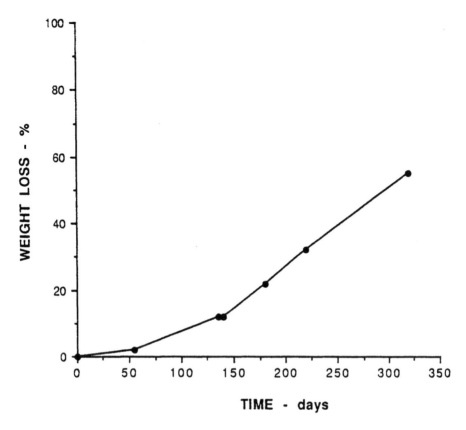

Figure 4 Weight loss of a polymer prepared from 3,9-bis (methylene)-2,4,8,10-tetraoxaspiro [5,5] undecane and 1,6-hexanediol as a function of time. Disks, 37°C, pH 7.4 phosphate buffer.

3,9-bis (methylene)-2,4,8,10-tetraoxaspiro [5,5] undecane and 1,6-hexanediol. As shown, weight loss of the polymer disk is extremely slow and in one year, only about 50% weight loss is observed. To accelerate polymer hydrolysis and the concomitant release of incorporated therapeutic agents, the hydrolysis needs to be accelerated and such acceleration can be achieved by the incorporation of acidic excipients into the polymer matrix. Rate of hydrolysis can also be accelerated by increasing matrix hydrophilicity which accelerates water penetration into the matrix (Heller *et al.*, 1987b).

When a poly(ortho ester) with a physically dispersed acidic excipient is placed into an aqueous environment, water will diffuse into the polymer, dissolve the acidic excipient in the surface layers and the lowered pH will accelerate hydrolysis of the ortho ester bonds. This process is schematically shown in Figure 5A where it has been analyzed in terms of the movement of two fronts, V_1, the movement of a hydrating front and V_2, the movement of an erosion front (Heller, 1985b). Clearly, the ultimate behavior of a device will be determined by the relative movement of these two fronts. If $V_1 > V_2$, the thickness of the reaction zone will gradually increase

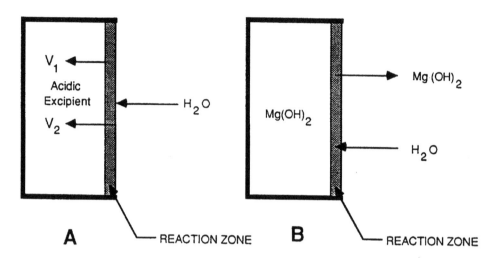

Figure 5 Schematic representation of water intrusion and erosion for one side of a bioerodible device. (A) device containins dispersed acidic excipient and (B) device contains dispersed $Mg(OH)_2$.

and at some time, the matrix will be completely permeated by water. At that point, all ortho ester linkages will hydrolyze at comparable rates and bulk hydrolysis will take place. However, if $V_1 = V_2$, then hydrolysis is confined to the surface layers and only surface hydrolysis will take place. In this latter case, rate of polymer erosion will be completely determined by the rate at which water intrudes into the polymer.

The sorption of water by poly (ortho esters) has been found to be relatively small, about 0.30 to 0.75% with a diffusion coefficient ranging from a high of 4.07×10^{-8} $cm^2.s^{-1}$ for a polymer based on 1,6-hexanediol (T_g 22°C) to a low of 2.11×10^{-8} $cm^2.s^{-1}$ for a polymer based on *trans*-cyclohexanedimethanol (T_g 122°C) (Nguyen *et al.*, 1985). Thus, assuming a disk thickness of about 2 mm and using the lowest diffusion coefficient, the disk would be completely permeated by water in about 10 days. Thus, the use of acidic excipients limits the design of surface eroding devices to lifetimes that do not exceed two to four weeks, depending on the actual thickness of the device. However, if bulk erosion is acceptable, then longer delivery times are possible.

If delivery times of many months are desired and if bulk erosion needs to be prevented, a basic excipient must be incorporated into the matrix. Because ortho ester linkages are stable in base, polymer hydrolysis in the interior of the device is prevented even though the matrix is completely permeated by water. During early work, bases such as Na_2CO_3 were used (Heller *et al.*, 1981). However, use of a water-soluble base leads to osmotic imbibing of water with consequent matrix swelling (Fedors, 1980). For this reason, $Mg(OH)_2$ which has a water solubility of only 0.8 mg/100 ml was selected. No swelling was noted when $Mg(OH)_2$ was used and for this reason, this basic salt was used in all subsequent studies. A plausible mechanism for erosion of devices that contain $Mg(OH)_2$ is shown in Figure 5B (Heller, 1985c).

According to this mechanism, $Mg(OH)_2$ stabilizes the interior of the device and erosion can only occur in the surface layers where the base has been eluted or neutralized. This is believed to occur by water intrusion into the matrix and diffusion of the slightly water-soluble $Mg(OH)_2$ out of the device where it is neutralized by the external buffer. Polymer erosion then occurs in the $Mg(OH)_2$-depleted layer.

Applications

This polymer system has been extensively investigated in a number of applications which can be divided into (a) use in an insulin self-regulated delivery system, (b) in short term delivery and in (c) long term delivery.

(a) *Insulin Self-Regulated Delivery System*
A self-regulated insulin delivery system, contains insulin dispersed in a bioerodible polymer and variable insulin delivery is achieved by changes in polymer erosion rate which occur in response to small pH changes caused by the interaction of glucose and glucose oxidase. To be useful in such an application, the poly(ortho ester) must be able to significantly change rate of hydrolysis in response to small changes of external pH. Since the poly(ortho ester)s II thus far discussed are unable to achieve the desired pH sensitivity, the structure was modified by the incorporation of a tertiary amine function as shown in Scheme 9 (Heller *et al.*, 1990).

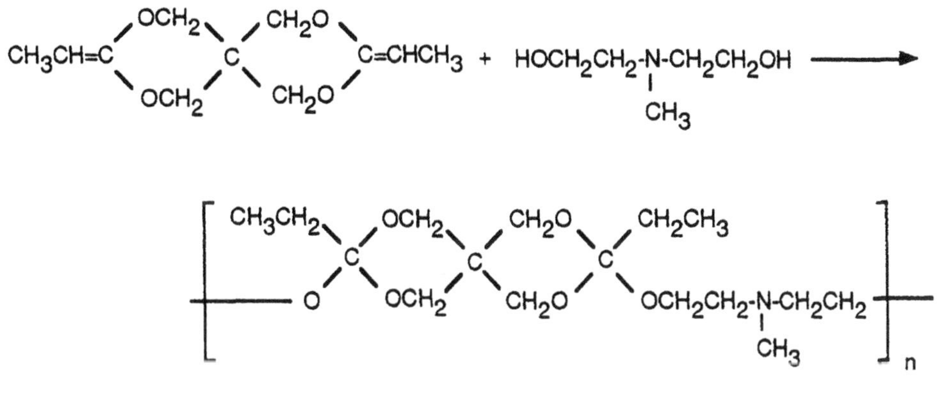

Scheme 9

The significantly enhanced rate of hydrolysis, as measured by cumulative insulin release, in response to small pH changes is shown in Figure 6. Unfortunately, despite the greatly enhanced rate of hydrolysis in response to small pH changes noted *in vitro*, subsequent work has shown that this behavior was due to a general acid catalysis specific to a citrate buffer and that pH changes induced by a glucose-glucose oxidase reaction had no significant effect on polymer hydrolysis.

(b) *Short Term Delivery*
Two applications have been extensively investigated. One application deals with the development of a delivery system for 5-fluorouracil for the treatment of cancer and

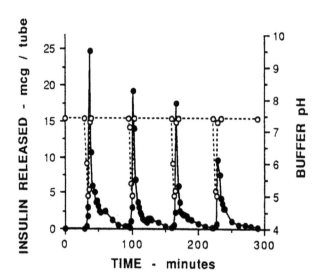

Figure 6 Release of insulin from a polymer prepared from 3,9-bis (ethylidene)-2,4,8,10-tetraoxaspiro [5,5] undecane and N-methyldiethanolamine as a function of external pH variation between pH 7.4 and 5.0 at 37°C. Buffer was continuosly perifused at a flow rate of 2 ml/min and a total effluent collected at 1–10 min intervals. (○) buffer pH (•) insulin release. Reprinted with permission from Heller *et al.*, 1990a.

the other application deals with the delivery of naltrexone for the treatment of narcotic addiction. In the 5-fluorouracil delivery, 0.15 wt% suberic acid incorporated into a polymer prepared from 3,9-bis (ethylidene) 2,4,8,10-tetraoxaspiro [5,5] undecane and 1,6-hexanediol resulted in a system that completely eroded in about 2 weeks and was able to release 10 wt% incorporated 5-fluorouracil in a linear manner. Devices implanted intraperitoneally in DBA$_2$ mice inoculated with L1210 tumor cells increased survival times from 16 days to 25 days (Seymour *et al.*, 1994).

In the other application naltrexone pamoate was incorporated into a polymer prepared from 3,9-bis (ethylidene) 2,4,8,10-tetraoxaspiro [5,5] undecane and 1,6-hexanediol. In this particular application, it was necessary to use naltrexone pamoate since naltrexone free base with a pKa of 8.13 will produce a saturated aqueous solution having a pH of about 10. Because this would stabilize the polymer, the slightly acidic naltrexone pamoate was used. In this particular case, naltrexone pamoate without the incorporation of suberic acid was able to catalyze polymer erosion and excellent, linear release was achieved over about 50 days. Incorporation of 1 and 3 wt% suberic acid decreased the time to total release to 30 and 10 days, respectively (Maa *et al.*, 1990).

(c) *Long Term Delivery*

The objective of this study was to develop an implant capable of delivering the contraceptive steroid levonorgestrel for one year. As already discussed, long erosion times requires the use of a base which will stabilize the interior of a device and only allow erosion to take place in the outer layers from which the base has been depleted by diffusion. In the development of these devices a crosslinked polymer was used. Devices were fabricated by first preparing a ketene acetal terminated prepolymer derived from two equivalents of the diketene acetal 3,9-bis(ethylidene) 2,4,8,10 tetraoxaspiro [5,5] undecane and one equivalent of the diol 3–methyl-1,5-pentanediol. Then, 30 wt% levonorgestrel, 7 wt% $Mg(OH)_2$ and a 30 mole% excess of 1,2,6-hexanetriol were mixed into the prepolymer and this mixture then extruded into rods and cured. Erosion and drug release from these devices was studied by implanting the rod shaped devices subcutaneously into rabbits, explanting at various time intervals and measuring weight loss and residual drug (Heller *et al.*, 1985).

Levonorgestrel blood plasma levels determined by radioimmunoassay showed a reasonably constant level for one year, once the initial burst subsided. However, the steady state plasma level was too low and thus a more rapidly eroding polymer was needed. To achieve a more rapid erosion, a material containing 7 wt% $Mg(OH)_2$ and 1 mole% copolymerized 9,10-dihydroxystearic acid was prepared. The devices were again implanted into rabbits and levonorgestrel blood plasma level determined. Results of these studies have shown a much higher, satisfactory drug plasma level (Heller, 1993).

The explanted devices were also examined by scanning electron microscopy which clearly showed surface erosion as a progressive diminution of a central uneroded zone and the development of voids around the periphery of the rod-shaped device (Heller, 1985). The presence of voids suggest that once erosion starts, generation of hydrophilic degradation products at that location accelerate further polymer hydrolysis.

Commercial Availability

A composition of matter patent, US Patent No. 4,304,767, issued in December 8, 1981 is held by SRI International. The polymer is not commercially available. The rearrangement of diallylpentaerythritol is covered by US Patent No. 4,513,143, issued in April 23, 1985, also held by SRI International.

Although the polymer has been used extensively in a number of *in vivo* studies and no evidence of adverse toxicological reaction have ever been observed, no GMP toxicological studies have been submitted to the FDA and the polymer does not have FDA approval.

Poly(Ortho Ester) III

This family of poly (ortho esters) can be prepared as shown in Scheme 10 by reacting a triol with two vicinal hydroxyl groups and one removed by at least three methylene groups with an alkyl orthoacetate (Heller *et al.*, 1990).

The intermediate does not have to be isolated and continuous reaction produces a polymer. The use of flexible triols such as 1,2,6-hexanetriol produces highly flexible polymers that have ointment-like properties even at relatively high molecular weights. Properties such as viscosity and hydrophobicity can be readily varied by controlling molecular weight and the size of the alkyl group R′. Use of a rigid triol such as 1,1,4-cyclohexanetrimethanol produces a solid polymer (Heller *et al.*, 1992). This synthesis is shown in Scheme 11. The solid polymer is designated as poly(ortho ester) IV and is covered later in this chapter.

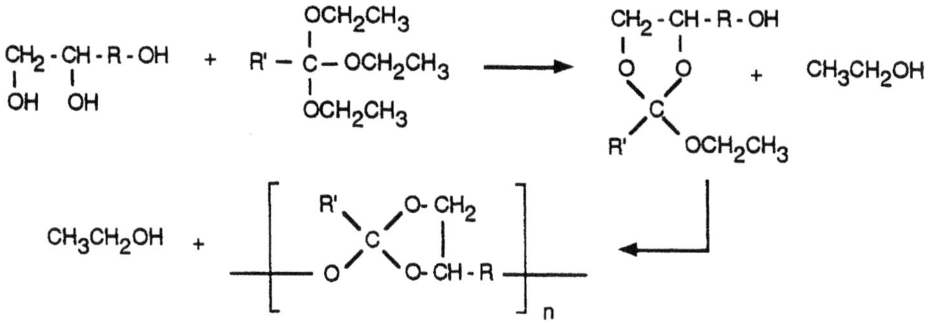

Scheme 10

Typical Experimental Procedure

Under anhydrous conditions 48.67 g (0.30 mole) of triethyl orthoacetate, 40.25 g (0.30 mole) of 1,2,6-hexanetriol and 20 mg of p-toluenesulfonic acid were weighed into a 500 mL round bottom flask equipped with a magnetic stirring bar. Next, 300 mL of cyclohexane was added and the flask adapted to a 60 cm spinning band column. The reaction flask was heated to 100°C with vigorous stirring and the ethanol-cyclohexane azeotrope was rapidly removed at 55°C. Throughout the procedure, strictly anhydrous conditions were maintained. When the boiling point began to climb above 55°C, the take-off ratio of the column was reduced to 1/20 distillation/reflux ratio until the boiling point reached 81°C where the take-off was set for total reflux. After heating for an additional 4 hr, the solution was cooled to room temperature. Five drops of triethylamine were then added to stabilize the product and the solvent removed by distillation. The product was a viscous, ointment-like material with an average molecular weight of 29,000 as determined by gel permeation chromatography using a Waters 150-C instrument with Waters ultrastyrogel 10^3 and 10^4 columns, with tetrahydrofuran solvent at 30°C with a small amount of triethylamine stabilizer. Polystyrene was used for calibration.

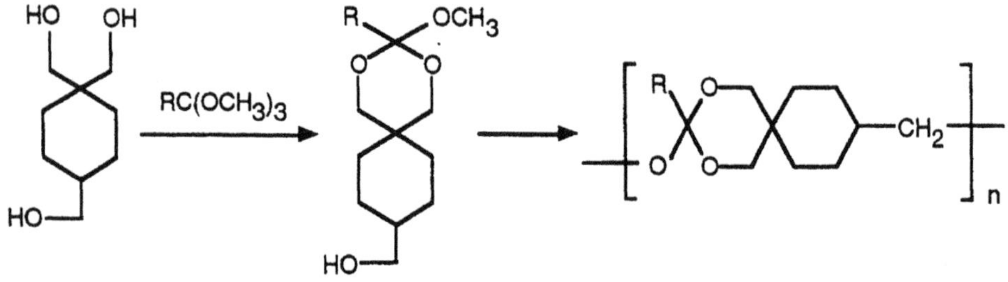

Scheme 11

Polymer Hydrolysis

Polymer hydrolysis occurs as shown in Scheme 12 for a polymer prepared from 1,2,6-hexanetriol. Initial hydrolysis occurs at the labile ortho ester bonds to generate one or more isomeric monoesters of the triol (Wuthrich *et al.*, 1992). This initial hydrolysis is followed by a much slower hydrolysis of the monoesters to produce a carboxylic acid and a triol. Thus, as with the poly (ortho ester) II, no autocatalysis is observed.

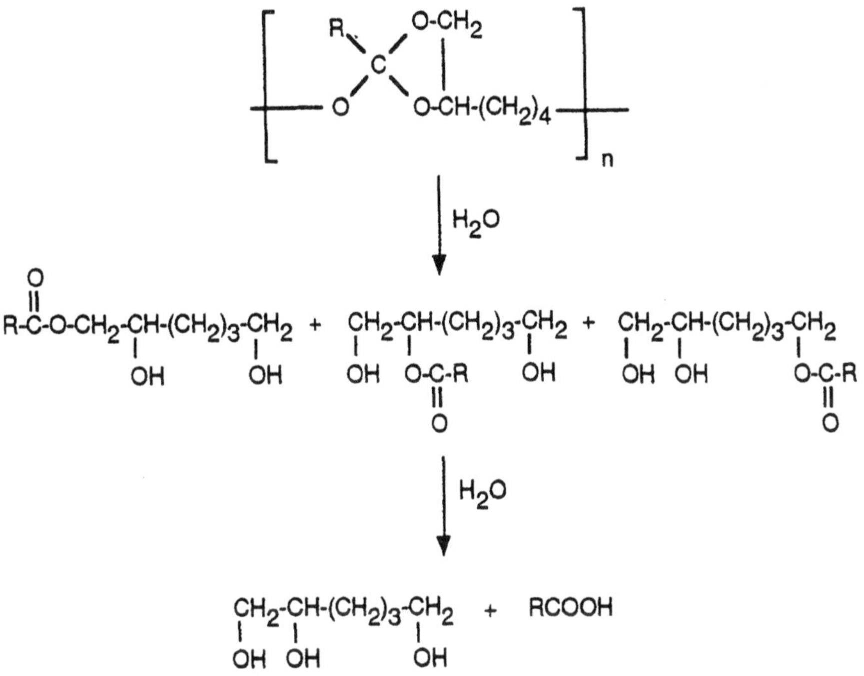

Scheme 12

Applications

Because this polymer has an ointment-like consistency at room temperature, a number of unique applications are possible. Of particular interest are applications where sensitive therapeutic agents, such as proteins, are incorporated into the polymer at room temperature and without the use of solvents. Initial results carried out at room temperature with lysozyme incorporated into polymers having different molecular weights and R-groups suggested that pulsatile release could be achieved with pulses spaced more than one month (Wuthrich *et al.*, 1992). However, when these experiments were repeated at the physiological temperature of 37°C, pulse times were shortened to only a few days (Heller *et al.*, 1995). Thus, despite initial encouraging results, prospects of developing a useful multipulse vaccine did not materialize.

Another potential application investigated in some detail, was treatment of periodontal disease. *In vitro* studies have shown that good control over release of tetracycline could be achieved and very good *in vitro* adhesion to bovine teeth was demonstrated (Roskos *et al.*, 1995). However, studies in beagle dogs with naturally occurring periodontitis were not successful because ointment-like polymers with a relatively low viscosity are squeezed out of the pocket within about one day, despite good adhesiveness. Additional studies are currently underway in an attempt to develop materials that have the right combination of adhesiveness and mechanical consistency.

Commercial Availability

The polymer is covered by a composition of matter patent, US 4,066,747 issued on January 3, 1978 and assigned to the Alza Corporation. The use of acidic and basic excipients to control hydrolysis rate is covered by US Patents 5,030,457, issued in July 9, 1991 and is held by Pharmaceutical Delivery Systems and by US Patent 5,336,505, issued in August 9, 1994 also held by Pharmaceutical Delivery Systems. It is not commercially available, nor has had FDA approval.

POLY (ORTHO ESTER) IV

The general synthetic procedure used to prepare the ointment-like poly(ortho esters) III can also be used in the preparation of solid polymers. To do so, it is only necessary to replace the flexible triol with a rigid one, such as 1,1,4-cyclohexanetrimethanol, as already shown in Scheme 11 (Heller *et al.*, 1992). As before, the intermediate does not have to be isolated and continuing reaction produces a polymer. The triol, 1,1,4-cyclohexanetrimethanol can be prepared as shown in Scheme 13 (Heller *et al.*, 1992).

Typical Experimental Procedure

1,4-Cyclohexanedimethanol monoacetate. A mixture of *cis* and *trans* 1,4-cyclohexane dimethanol, (930 g, 6.448 moles) was dissolved in 3 L of tetrahydrofuran and 550 mL of pyridine, (7.69 moles) added. The solution was cooled in an ice bath and stirred under argon. An acetyl chloride solution, (506.4 g, 6.45 moles) in 500 mL of tetrahydrofuran was added dropwise over a 2 hour period. The ice bath was removed and the reaction mixture stirred for 2 hours at room temperature. It was then filtered to remove the pyridine HCl salt and evaporated to remove the tetrahydrofuran. The residue was dissolved in 2 L of ethyl acetate and the solution extracted with diluted aqueous HCl (2×300 mL), warm water (2×300 mL) and aqueous NaHCO$_3$ solution (2×300 mL). The ethyl acetate solution was dried over anhydrous MgSO$_4$ and the ethyl acetate removed on a rotoevaporator. Vacuum distillation of the residue yielded 335 g of product. GC analysis showed that the product contained 59% cyclohexanedimethanol monoacetate and 41% cyclohexanedimethanol diacetate. The overall yield of the monoacetate was 16.5%.

4-Acetoxymethyl-1-cyclohexanecarboxaldehyde. Under anhydrous conditions, oxalyl chloride (358 g, 2.82 moles) was dissolved in 2.5 L of methylene chloride and the solution cooled to $-40°C$. Dimethylsulfoxide (407 g, 5.2 moles) dissolved in 200 mL of methylene chloride was then added via a dropping funnel while the reaction mixture was vigorously stirred and the temperature maintained between $-40°C$ and $-20°C$. Next, a solution of cyclohexanedimethanol monoacetate (600 g, 59% pure, 1.9 moles) in 200 mL of methylene chloride was added dropwise while the reaction temperature was kept below $-20°C$. After the addition of cyclohexanedimethanol monoacetate solution was completed, the reaction mixture was stirred for an additional 15 min. and triethylamine (658 g, 6.5 moles) added. The cooling bath was removed and the reaction mixture stirred for 2 hrs. It was then extracted successively with diluted aqueous HCl, aqueous $NaHCO_3$ and aqueous NaCl. After drying over anhydrous $MgSO_4$, the methylene chloride solution was distilled under argon to remove the solvent. The residue was distilled at $80°C$ at 0.4 mm to give the aldehyde (248 g, 70.8% yield).

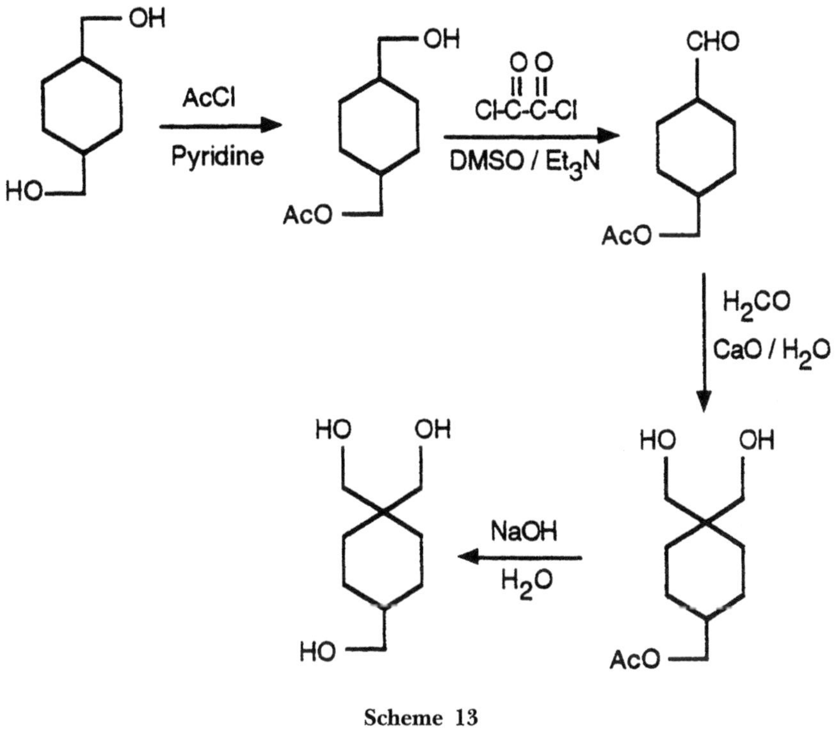

Scheme 13

4-Acetoxymethyl-1,1-cyclohexanedimethanol. A mixture of 4-acetoxymethyl-1-cyclohexa-necarboxaldehyde, (248 g, 1.46 moles), a 37 wt% formaldehyde solution (700 mL, 8.6 moles) and tetrahydrofuran (200 mL) was cooled in an ice water bath. Calcium oxide was then added in small portions while the mixture was vigorously stirred with an overhead mechanical stirrer. After the addition of CaO was completed, the ice bath was removed and the mixture stirred for 2 hrs. It was then evaporated to dryness and the product extracted into acetone. Evaporation of the acetone solution produced a viscous oil.

1,1,4-cyclohexanetrimethanol. The crude 4-acetoxymethyl-1,1-cyclohexanedimethanol was added to 1L of an aqueous 2N NaOH solution and the mixture heated at 100°C for 2 hrs. After cooling to room temperature, the reaction mixture was neutralized with aqueous HCl and extracted with methylene chloride. The aqueous solution was evaporated to dryness and the residue extracted with acetone. After drying over anhydrous $MgSO_4$, the acetone solution was evaporated to dryness. Distillation of the crude product from the acetone solution at 175°C and 0.1 mm yielded a viscous liquid. Repeated trituration with methylene chloride produced a solid product (150 g, 59% yield, 98.8% purity by GC).

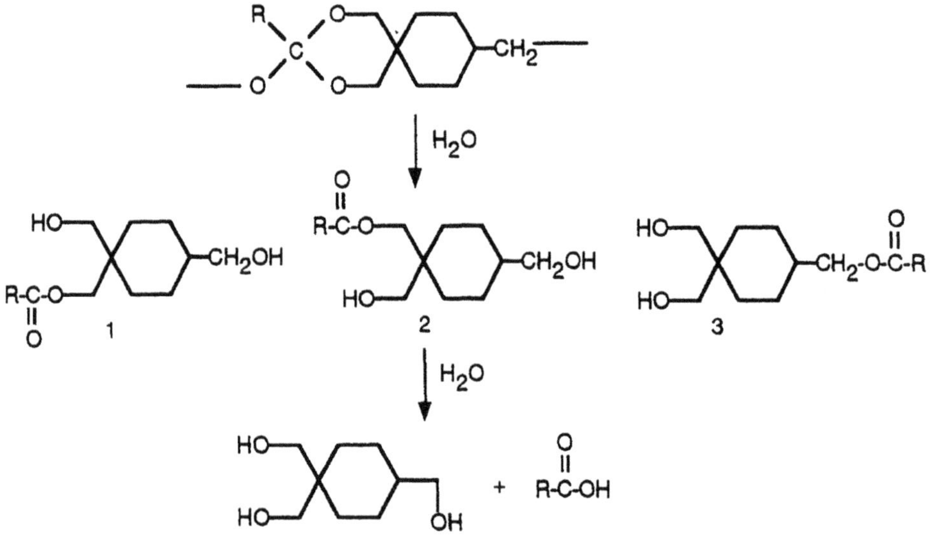

Scheme 14

Preparation of Polymer

Under anhydrous conditions, 1,1,4-cyclohexanetrimethanol (3.524 g, 20 mmoles), trimethyl orthoacetate (2.403 g, 20 mmoles), *p*-toluenesulfonic acid (3 mg) and distilled cyclohexane (80 mL) were added to a pre-dried flask. The flask was fitted with a spinning band column and heated at 100°C under argon. Methanol was removed azeotropically at 56°C at a fast rate and as the boiling point began to rise, the distillation rate was reduced to 4 drops/min. and heating continued for 15 hrs. The polymer thus prepared precipitated out of cyclohexane. The powdery polymer is crystalline with a melting point (DSC) of 212°C. It is insoluble in the usual organic solvents such as methylene chloride, chloroform, ether, tetrahydrofuran, ethyl acetate, acetone, dimethylformamide, and dimethylsulfoxide.

In a similar manner, 1,1,4-cyclohexanetrimethanol (3.524 g, 20 mmoles) was allowed to react with triethyl orthopropionate (3.634 g, 20 mmoles). This reaction produced a polymer which remained in the cyclohexane solution. Precipitation into methanol yielded a polymer having a MW of 51,000 (GPC using polystyrene standards) and a T_g of 67.8°C. The polymer was soluble in organic solvents with low or medium polarities such as methylene chloride, chloroform, ether, tetrahydrofuran and ethyl acetate.

Polymer Hydrolysis

When these poly (ortho esters) are placed in an aqueous environment they hydrolyze as shown in Scheme 14 (Heller *et al.*, 1992).

Polymer Physical Properties

Recent work has shown that when $R=CH_3$, the polymer is crystalline, but that the crystallinity disappears when $R=CH_3CH_2$ (Heller *et al.*, 1992). This is the only example of a crystalline poly(ortho ester).

Commercial Availability

The polymer is not commercially available and is covered by US Patent 5,336,505, issued in August 9, 1994 and held by Pharmaceutical Delivery Systems. The monomer, 1,1,4-cyclohexanetrimethanol can be readily prepared in small quantities, but scale-up has not been successful.

REFERENCES

Capozza, R.C., Sendelbeck, S.L. and Balkenhol, W.J. (1978) Preparation and evaluation of a bioerodible naltrexone delivery system, In: *Polymeric Delivery Systems*, Kostelnik, R.J. (Ed), Gordon and Breach, New York, p. 59.
Choi, N.S. and Heller, J. (1978a) US Patent 4,079,038.
Choi, N.S. and Heller, J. (1978b) US Patent 4,093,709.
Choi, N.S. and Heller, J. (1978c) US Patent 4,131,648.
Choi, N.S. and Heller, J. (1979a) US Patent 4,138,344.
Choi, N.S. and Heller, J. (1979b) US Patent 4,180,646
Fedors, R.F. (1980) Osmotic effects in water absorption by polymers, *Polymer*, **21**, 207.
Gabelnick, H.L. (1983) Biodegradable implants. alternate approaches, In. *Long-Acting Steroid Contraception*, Mishel, D.R. Jr. (Ed), Raven Press, New York, p. 149.
Heller, J., Penhale, D.W.H and Helwing R.F. (1980) Preparation of poly(ortho esters) by the reaction of diketene acetals and polyols, *J. Polymer Sci.*, Polymer Letters Ed., **18**, 619.
Heller, J., Penhale, D.W.H., Helwing, R.F. and Fritzinger, B.K. (1981) Release of norethindrone from poly(ortho esters), *Polymer Eng. Sci.*, **21**, 727.
Heller, J., Penhale, D.W.H., Fritzinger, B.K., Rose, J.E. and Helwing, R.F. (1983) Controlled release of contraceptive steroids from biodegradable poly(ortho esters), *Contracept. Deliv. Syst.*, **4**, 43.
Heller, J., Fritzinger, B.K., Ng, S.Y. and Penhale, D.W.H. (1985a) In vitro and in vivo release of levonorgestrel from poly(ortho esters), II. Crosslinked polymers, *J. Controlled Release*, **1**, 233.
Heller, J. (1985b) Controlled drug release from poly(ortho esters) - A surface eroding polymer, *J. Controlled Release*, **2**, 167.
Heller, J. (1985c) Control of polymer surface erosion by the use of excipients, In: *Polymers in Medicine II*, Chielini, E., Giusti, P., Migliaresi, C. and Nicolais, L. (Eds.), Plenum Press, New York, 357.
Heller, J., Ng, S.Y., Penhale, D.W.H., Sanders, L.M., Burns, R.A., Gaynon, M.S. and Bhosale, S.S. (1987a) Use of poly(ortho esters) for the controlled release of 5-fluorouracil and a LHRH analogue, *J. Controllesd Release*, **6**, 217.

Heller, J., Penhale, D.W.H., Fritzinger, B.K. and Ng, S.Y. (1987b) The effect of copolymerized 9,10-dihydroxystearic acid on erosion rates of poly(ortho esters) and its use in the delivery of levonorgestrel, *J. Controlled Release*, **5**, 173.

Heller, J., Chang, A.C., Rodd, G. and Grodsky, G.M. (1990a) Release of insulin from pH-sensitive poly(ortho esters), *J. Controlled Release*, **13**, 295.

Heller, J., Ng, S.Y., Fritzinger, B.K. and Roskos, K.V. (1990b) Controlled drug release from bioerodible hydrophobic ointments, *Biomaterials*, **11**, 235.

Heller, J., Ng, S.Y. and Fritzinger, B.K. (1992) Synthesis and characterization of a new family of poly(ortho esters), *Macromolecules*, **25**, 3362.

Heller, J. (1993) Poly(ortho esters), *Adv. in Polymer Sci.*, **107**, 43.

Heller, J., Rime, A-F., Rao, S.S., Fritzinger, B.K. and Ng, S.Y (1995) Poly(ortho esters) for the pulsed and continuous delivery of peptides and proteins, In: *Trends and Future Perspectives in Peptide and Protein Drug Delivery*, Lee, V.H.L., Hashida, M. and Mizushima Y. (Ed.), Harwood Academic Publishers, GmbH, 39.

Maa, Y.F. and Heller, J. (1990) Controlled release of naltrexone pamoate from linear poly(ortho esters) *J. Controlled Release*, **14**, 21.

Ng, S.Y., Penhale, D.W.H. and Heller, J. (1992) Poly(ortho esters) by the addition of diols to a diketene acetal, *Macromolecular Synthesis*, **11**, 23.

Nguyen, T.H., Himmelstein, K.J. and Higuchi, T. (1985) Some equilibrium and kinetics aspects of water sorption in poly(ortho esters), *Int. J. Pharmaceut.*, **25**, 1.

Seymour, L.W., Duncan, R., Duffy, J., Ng, S.Y. and Heller, J. (1994) Poly(ortho ester) matrices for controlled release of the antitumour agent 5-fluorouracil, *J. Controlled Release*, **31**, 201.

Roskos, K.V., Fritzinger, B.K., Rao, S.S., Armitage, G.C. and Heller, J. (1995) Development of a drug delivery system for the treatment of periodontal disease based on bioerodible poly(ortho ester), *Biomaterials*, **16**, 313.

Solheim, E., Pinholt, E.M., Andersen, R., Bang, G. and Sudmann, E. (1995) Local delivery of indomethacin by a polyorthoester inhibits reossification of experimental bone defects, *J. Biomed Mater. Sci.*, **29**, 1141.

Vistnes, L.M., Schmitt, E.E., Ksander, G.A., Rose, E.H., Balkenhol, W.J. and Coleman, C.L. (1976) Evaluation of a prototype therapeutic system for prolongued, continuous topical delivery of homosulfanilamide in the management of Pseudomonas burn wound sepsis, *Surgery*, **79**, 690.

Wuthrich, P., Ng, S.Y., Fritzinger, B.K., Roskos, K.V. and Heller, J. (1992) Pulsatile and delayed release of lysozyme from ointment-like poly(ortho esters), *J. Controlled Release*, **21**, 191.

7. OTHER POLYESTERS

VENKATRAM R. SHASTRI[1,*], PATRICE P. HILDGEN[1,3], ROBERT LANGER[1],
YOUSEF NAJAJRAH[2], ISRAEL RINGEL[2] and ABRAHAM J. DOMB[2]

[1]*Department of Chemical Engineering, Massachusetts Institute of Technology,
Cambridge, MA 02139, USA*
[2]*Department of Pharmaceutical Chemistry, School of Pharmacy,
The Hebrew University, Jerusalem, 91120, Israel*
[3]*University of Montreal, Faculty of Pharmacy, Montreal, H3C-3J7, Canada*

INTRODUCTION

In the past 3 decades the use of synthetic polymers in medicine has increased tremendously. Dexon© (American Cyanamid), a homopolymer of glycolic acid has been in use in surgical medicine as biodegradable and bioabsorbable suture material since the early 1960's. Furthermore, in the past decade or so, the environmental damage caused due to inertness of several polymers in land fills has lead to a concerted effort in developing biodegradable polymers for daily consumer use. The ester bond unlike the amide linkage, which is quite stable under various conditions and undergoes only enzymatic cleavage *in vivo* by amidases, is more susceptible to hydrolysis under both mildly alkaline and acidic conditions and also by esterases. The hydrolytic susceptibility of the ester linkage has thus, made it a mainstay in developing biodegradable polymers. Among the many applications where degradation of the polymer is desirable, the biomedical applications such as polymer based drug delivery systems and synthetic polymer scaffolds for tissue engineering have been areas of main focus in the recent years. The most widely used polyesters are the ones derived from lactic acid, glycolic acid, hydroxybutyric acid and caprolactone and are described in separate chapters in this book. This chapter has been divided into two parts. The first part will cover the synthesis and characterization of some polyesters namely, poly(ether-esters), unfunctionalized and functionalized poly(ester amides), poly(ester-urethanes), and their block copolymers. In the second part we have included a brief description of poly(phosphate ester) as it is not described elsewhere in this book.

PART I

Poly(ether-esters)

Among the various poly(ether-esters) (see chapter by Bezwada) the aliphatic poly(ether-esters) obtained by the ring opening homo and co-polymerization of

---

*Correspondence: Venkatram Shastri, 45 Carleton st., Bldg. E-25, Rm. 342, Massachusetts Institute of Technology, Cambridge, MA 02139.

1,4-dioxan-2-one (A), 1,5-dioxepan-2-one (DXO) (B) and its cyclic dimer 1,4,6-trioxaspiro[4.4]nonane (C) (Scheme 1) with lactides, glycolides and ε-caprolactones have been of immense interest due their *in vivo* hydrolytic degradation and bioabsorption.

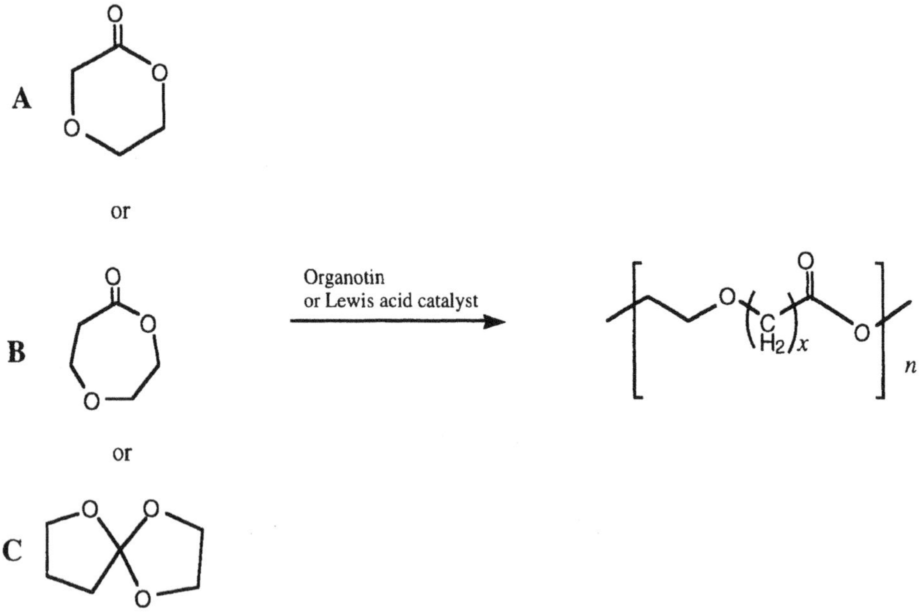

A

or

B

Organotin
or Lewis acid catalyst

C

A = 1,4-dioxan-2-one, $x = 1$
B = 1,5-dioxepan-2-one, $x = 2$
C = 1,4,6-trioxaspiro[4.4]nonane, $x = 3$

Scheme 1 (Adapted from Mathisen, 1989).

While, the homopolymer of 1,4-dioxan-2-one (poly(p-dioxanone) can be viewed as an alternating co-polymer of α-hydroxy acetic acid and ethylene glycol, the homopolymer of 1,4,6-trioxaspiro[4.4]nonane can be viewed as an alternating co-polymer of γ-hydroxy butanoic acid and ethylene glycol and the homopolymer of 1,5-dioxepan-2-one in essence is an alternating co-polymer of 3-hydroxy propanoic acid and ethylene glycol. Poly(p-dioxanone), which is a partly crystalline polymer (melting temperature (Tm) of 110°C and a glass transition temperature (Tg) of −16°C), and currently in use as a biodegradable suture has been discussed in detail elsewhere in this book. This section will focus on the synthesis, characterization and polymerization of 1,5-dioxepan-2-one. Recently, Albertsson *et al.*, (Mathisen, 1989) have developed efficient routes to synthesize DXO (Mathisen, 1989; Lofgren, 1994).

Synthesis of 1,5-dioxepan-2-one (DXO) polymers (Mathisen, 1989; Lofgren, 1994)

In the first step, 3-chloropropionyl chloride is converted to 1,5 dichloropentane-3-one via Friedel-Crafts acylation in presence of ethylene at 20°C as described by

Arentzen *et al.*, (Arentzen 1975). In the subsequent step, 1,5-dichloropentane-3-one is converted to the cyclic intermediate tetrahydro-4H-pyran-4-one(THP-one) via cyclization under basic conditions. In the next step, THP-one is converted to DXO by Baeyer-Villiger oxidation ring expansion using m-chloro peroxybenzoic acid (mCPBA). In brief, 0.55 moles of THP-one is added to a slurry of 0.75 moles of 82 % mCPBA and 1 mole of sodium bicarbonate in 800 ml of dry methylene chloride. The slurry is initially maintained at 0°C under constant stirring for a period of 1–2 hours following which stirring is continued for an additional 16 hours at 20°C. At the end of the reaction the excess per-acid is neutralized by washing the organic (methylene chloride) phase first with sodium bisulfite followed by sodium bicarbonate. The pure DXO is then obtained by distillation of the resulting yellow oil under reduced pressure in 50% overall yield. DXO is purified by re-crystallization from anhydrous diethyl ether and stored under dry conditions until use.

The homopolymerization of DXO was reported by Albertsson and co-workers (Mathisen, 1989). The homopolymers were synthesized via the ring opening of the cyclic monomers using organo-metallic catalysts. In brief, monomer and initiator in the desired ratio are transferred into a 20 ml serum bottle equipped with a magnetic stir bar, purged with nitrogen and sealed. The typical polymerization time for DXO is 6 h at 100°C. At the end of the polymerization, the polymer is dissolved in methylene chloride and then washed with 2 M HCl to remove traces of catalyst and residual monomer. The homopolymer of DXO is completely amorphous with a Tg around −40°C. Albertsson and co-workers have synthesized homo and di and tri block co-polymers of DXO with ε-caprolactone (ε-CL) using aluminum alkoxides as initiators (Lofgren, 1994). The "living" characteristic of the polymerization was demonstrated by the formation of block co-polymers by sequential addition of monomers. The narrow polydispersity of these polymers is typically in the range of 1.03–1.16, a further indication of the "living" nature of this polymerization. These di and tri block co-polymers of DXO with ε-CL exhibit two distinct Tg's one at −55°C due to the ε-CL block and one at −40°C due to the DXO block. These block co-polymers, unlike the homopolymers of DXO, exhibit a melting transition (Tm) around 59°C. In contrast the homopolymer of ε-CL has a Tg around −65°C and Tm around 57°C. The reaction conditions, conversions and molecular weights of some block co-polymers of DXO with ε-CL are listed in Table 1.

Table 1 Reaction conditions, conversions and molecular weights of some block copolymers of DXO with ε-CL (Data from Lofgren 1994)

block copolyesters	polym. time* (h)	conv. (%)	$\overline{M_n}$ GPC	$\overline{M_W}/\overline{M_n}$
PCL block	2.5	>99	15500	1.18
PCL/PDXO	2.5	99	26500	1.15
PCL/PDXO/PCL	2.0	37	31500	1.18
PCL block	1.0	>99	10500	1.15
PCL/PDXO	1.0	98	16500	1.20
PCL/PDXO/PCL	3.3	90	26000	1.21

*All copolymerizations were initiated by Al(O *i*Pr)₃ in toluene at 0°C.

Degradation of homo and co-polymers of DXO with L- and D,L-lactide have been carried out for over 20 months under physiologically relevant conditions of pH 7.4 phosphate buffered saline (PBS) at 37° C. The analysis of degradation products using headspace gas chromatography (Karlsson 1994) has shown that these polymers do undergo biodegradation and 2-ethoxy hydroxy propionic acid is the degradation product of p(DXO).

Poly(ester-ether) networks

ABA type block co-polymer of lactide, glycolide (Kricheldorf, 1993; Xiong, 1995; Du, 1995) or ε-caprolactone ((Sawhney 1990) containing a hydrophilic 'B' block is typically synthesized by using polyethylene glycol (PEG) diols as ring opening polymerization initiation sites using stannous octoate as initiator. In a recent study, Pluronic F-68 (BASF) a hydroxy terminated di-block co-polymer of polyethylene oxide (PEO) and polypropylene oxide (PPO) in 20:80 wt%.) was utilized as the hydrophilic block. These ter polymers exhibit a wide variety of physical properties ranging from materials that are highly crystalline to totally amorphous and those that are brittle to ones that are syrupy in consistency. Multiblock copolymers of polyethylene glycol and poly(hydroxy acids) have been synthesized (Gref 1995). First, several methoxy PEG amine (M-PEG-NH$_2$) chains were attached together at one chain end by reaction with citric, mucic, tartaric acid or other functional molecules. The remaining hydroxyl groups were further used to initiate the ring-opening polymerization of the α-hydroxy acids. The structure of the copolymers of lactide is described in Scheme 2. These copolymers were used to prepare drug loaded PEG coated nanospheres for prolonged blood circulation. Lidocaine was used as a model hydrophobic drug to study the factors that determine encapsulation efficiency, nanosphere mean diameter and release patterns (Gref, 1994, 1995). Regardless of the polymer used, high entrapment yields were achieved. The nanosphere diameter and the entrapment efficiency depend upon the polymer physico-chemical characteristics. In the case of diblock PEG-PLA polymers, the entrapment efficiency is independent of the loaded drug and on the PEG molecular weight. Conversely, in the case of PEG3-PLA brush polymer, a slight decrease of the entrapment yield with the decrease in MW was observed. Nanospheres injected in mice show about 20% of the injected dose in the blood circulation, 5 hours post injection.

Okada *et al.*, have synthesized homo and copolyesters containing 2,5-linked and 2,6-linked tetrahydropyran rings and pendant carboxylic acids by cationic ring opening polymerization of their respective bicyclic lactones (Okada, 1994a, 1995). These polymers were found to undergo degradation both in soil and activated sludge into low molecular weight compounds and finally to hydroxy-tetrahydropyran carboxylic acid. In general the polymers without pendant groups and with 2,6-linked tetrahydropyran in the backbone underwent degradation much readily then their 2,5 counterparts (Okada, 1994b). However, among polymers containing pendant groups a decrease in hydrolysis rates with increasing hydrophobicity of pendant groups was observed. Even small concentrations of free carboxylic acids on the polymer backbone significantly enhanced polymer hydrolysis. Recently, Okada et. al., synthesized polyesters from 1,4:3,6-dianhydrohexitols and aliphatic dicarboxylic acids such as, adipic acid (Okada, 1995) which undergo spontaneous hydrolysis in phosphate buffer at elevated temperatures.

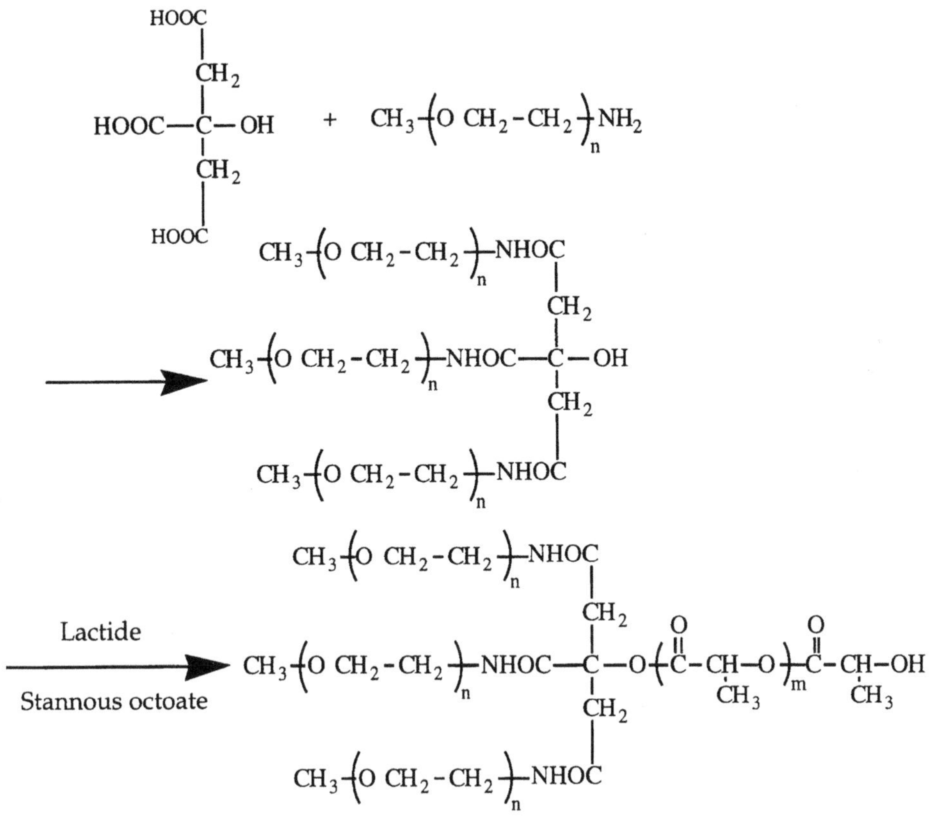

$$\text{HOOC}$$
$$\text{HOOC}-\overset{\displaystyle CH_2}{\underset{\displaystyle CH_2}{C}}-\text{OH} \quad + \quad CH_3\text{--}\!\!\left(O\ CH_2\text{-}CH_2\right)_{\!n}\!\!\text{--}NH_2$$
$$\text{HOOC}$$

$$CH_3\text{--}\!\!\left(O\ CH_2\text{-}CH_2\right)_{\!n}\!\!\text{--}NHOC$$
$$|$$
$$CH_2$$
$$\xrightarrow{\hspace{2cm}} \quad CH_3\text{--}\!\!\left(O\ CH_2\text{-}CH_2\right)_{\!n}\!\!\text{--}NHOC-\overset{\displaystyle CH_2}{C}-\text{OH}$$
$$CH_2$$
$$|$$
$$CH_3\text{--}\!\!\left(O\ CH_2\text{-}CH_2\right)_{\!n}\!\!\text{--}NHOC$$

$$CH_3\text{--}\!\!\left(O\ CH_2\text{-}CH_2\right)_{\!n}\!\!\text{--}NHOC$$
$$|$$
$$CH_2$$

Lactide
$$\xrightarrow[\textbf{Stannous octoate}]{\hspace{2cm}} \quad CH_3\text{--}\!\!\left(O\ CH_2\text{-}CH_2\right)_{\!n}\!\!\text{--}NHOC-\overset{\displaystyle CH_2}{\underset{\displaystyle CH_2}{C}}-O\text{--}\!\!\left(\overset{\displaystyle O}{\overset{\|}{C}}\text{-}\underset{\displaystyle CH_3}{CH}\text{-}O\right)_{\!m}\!\!\overset{\displaystyle O}{\overset{\|}{C}}\text{-}\underset{\displaystyle CH_3}{CH}\text{-}OH$$
$$|$$
$$CH_3\text{--}\!\!\left(O\ CH_2\text{-}CH_2\right)_{\!n}\!\!\text{--}NHOC$$

PEG_3-PLA Brush copolymer

Scheme 2 (Adapted from Gref, 1994).

Bengs *et al.*, have synthesized polytartaric acids with alternating ethyl ester and ketal side groups (Bengs, 1995). By varying the chemical composition of the side chain, polymers with wide range of properties could be obtained. Buserelin-an LHRH antagonist, has been encapsulated in microsphere formulations and released in a reproducible manner in rats and dogs. Polypeptide release upto twelve weeks without any initial burst was obtained. Single and multiple dose toxicity studies in rats and mice reveal no adverse side effects or skin irritation. Also, an Ames test on the degradation products of the polymer was negative for mutagenecity. These polymers hold considerable promise as controlled release matrices.

In view of synthesizing polymers with non-toxic degradation products, in recent years increased attention has been paid to synthesis of polymers derived from intermediates of the Krebs cycle. Poly(β-malic acid), which is an aliphatic polyester with pendant carboxyl groups derived from malic acid, a Krebs cycle intermediate has immense potential in controlled drug delivery and tissue engineering. Guerin *et al.*, have synthesized optically active poly(benzyl β-malate) by anionic ring opening polymerization of benzyl malolactonates derived from L-Aspartic acid, using triethyl an amine-aluminum porphyrin as initiator (Guerin, 1986). Polymers with molecular weight as high as 41,000 were obtained. In an other approach Guerin *et al.*, have synthesized poly(β-malic acid alkyl esters) by the ring opening of β-substituted β-lactones which was synthesized by the cycloaddition of a ketene to an alkyl glyoxylate (Ramiandrasoa 1993). The final step in the synthesis of poly(β-malic acid) involves a palladium catalyzed hydrogenolysis of the benzyl protecting group. By comparing the carbonyl NMR resonance's of copolymers of β-malic acid and benzyl β-malate with different comonomer sequences using a specialized one-dimensional NMR technique called INEPT it has been shown that palladium catalyzed hydrogenolysis of benzyl β-malate leads to the formation of block copolymer of β-malic and benzyl β-malate (Guerin, 1992).

Poly(ester amides) (PEA) and Functionalized Polyesters

The utility of polyamides in biomedical applications has been limited due to the chemical stability of the amide linkage *in vivo*. Unlike the ester bond, which can undergo hydrolysis under mildly basic conditions such as the *in vivo* environment, the amide linkage is not easily hydrolyzed even under strong acidic or basic conditions. *In vivo*, the only available route for cleavage of an amide bond is enzymatic. The enzymes that specifically cleave amide linkages, known as amidases however are very site and amino acid specific in their activity and action. Hence, in the recent years attempts have been made at improving the utility of polyamides in *in vivo* biomedical applications by incorporating easily hydrolyzable bonds such as aliphatic ester linkages in the polymer backbone.

In the late eighties Barrows and his team (Barrows, 1994, 1982, 1984) at 3M Corp. (St. Paul, Minnesota) designed and synthesized poly(ester-amides) which could exploit the inductive effect of the amide linkage in enhancing polymer degradation. Their efforts lead to the development of the 3M poly(ester-amides). However, recent attempts to synthesize poly(ester-amides) have focused on the incorporation of an aliphatic ester linkage in the poly(ester-amide) derived from α-hydroxy acids such as glycolic and lactic acid and their derivatives. In an other approach (Andini, 1988; Simone, 1992) degradable blocks of polyesters derived from lactides or glycolides were incorporated in the polyamide backbone via an interfacial polycondensation reaction. A second approach is based on the ring opening polymerization of depsipeptides (morpholine-2,5-diones) (Veld, 1990, 1992). Morpholine-2,5-diones and their derivatives are formed by the condensation of α-hydroxy acid with an α-amino acid. The resulting polymer i.e., the polydepsipeptide is an alternating poly(ester-amide). Functionality can be introduced into the polymer backbone by utilizing morpholine-2,5-dione derivatives containing lysine, aspartic acid or cysteine as the α-amino acid component.

In yet another approach Katsarava *et al.*, synthesized regular PEAs based on natural amino acids (Arabuli, 1994). Their choice of an α-amino acid to introduce hydrolytic susceptibility was based on the fact that the esters of N-acyl-L-α-amino acids are easily cleaved by α-chymotrypsin and at a rate five orders in magnitude greater than the corresponding amides. In their synthetic approach, in the first step N-BOC-protected L-phenyl alanine is coupled with dibromo alkane under basic conditions to yield the diamine-diester. In the subsequent step the amine group is deprotected and then converted to the salt of p-toluenesulfonic acid by refluxing in a mixture of benzene and p-nitro benzene (1:1) with removal of water. Condensation of the salt with bis(p-nitrophenyl adipate) in presence of triethylamine in N-methyl-2-pyrrolidone (NMP) or chloroform yielded the polymer in almost quantitative yield. The intrinsic viscosity's of the polymers ranged from 0.3–0.6 dL/g depending on the alkyl chain length coupling the L-phenyl alanine and the solvent used during polymerization. These polymers exhibit very good film forming properties and are fusible at low temperatures. Biodegradation studies carried out on polymer films at $37°C$ under nitrogen atmosphere in presence of α-chymotrypsin showed that these films do undergo enzymatic hydrolysis with 50% of the ester groups cleaved at 4 h. Furthermore, an increase in polymer degradation rate with increasing alkyl chain length is observed. It appears that the degradation process occurs from the surface inwards and is controlled by the adsorption of the enzyme onto the polymer surface.

3M poly(ester-amides)

In order to exploit the inductive effect of the amide bond on the hydrolysis of the ester linkage in the polymer backbone the amide and ester linkage should be ideally separated by one methylene unit. It was apparent that a hydroxyacetamide structure would yield such an linkage. Hydroxyacetamide amidediols were obtained in high yields as the thermodynamically favorable product upon heating glycolic acid with alkane diamine neat with removal of water (Barrows, 1982, 1988, 1994). The polymer is then obtained either by trans esterification polymerization of the diol (bis-hydroxyacetamides) with the desired diester of the diacid or by the reaction of bis-hydroxyacetamides with diacid chlorides (Scheme 3) (Barrows 1982, 1988, 1994). Poly(ester-amides) can be prepared from a wide variety of different bis-hydroxyacetamide amidediols and diacids.

Trans esterification polymerization of the diol monomer with the respective dimethyl ester of a diacid yielded low molecular weight polymers. However, condensation of the diol monomer with adipoyl chloride yielded high MW polymers with an inherent viscosity of 0.95. Linear aliphatic diamines and diacids were chosen for biodegradable polymers. Fibers made of poly(ester-amides) based on hexamethylene diamine-glycolic acid-succinic acid (Scheme 3, x=2, y=6) had comparable mechanical properties to the commercial erodible sutures Maxon® and PDS® with an estimated 8 month for bioabsorption which is similar for the commercial sutures. Extensive *in vivo* studies were carried out to ascertain the metabolic fate of the degradation products. The metabolism of PEA primarily resulted in urinary excretion of mostly unchanged amidediol as the major degradation product, as determined by carbon-14 labeled fibers implanted in rats. It was found that the elimination or clearance time for the diols ranged from three days for the water soluble C_6-diol to sixteen days for the water insoluble C_{12}-diol (Barrows 1988). Furthermore, the acute toxicity of the C_6-amidediol was determined by the intraperitoneal administration at a dose of 5000 mg/kg in rats. No evidence of any acute toxicity was observed upto two weeks post-administration. (Barrows 1994).

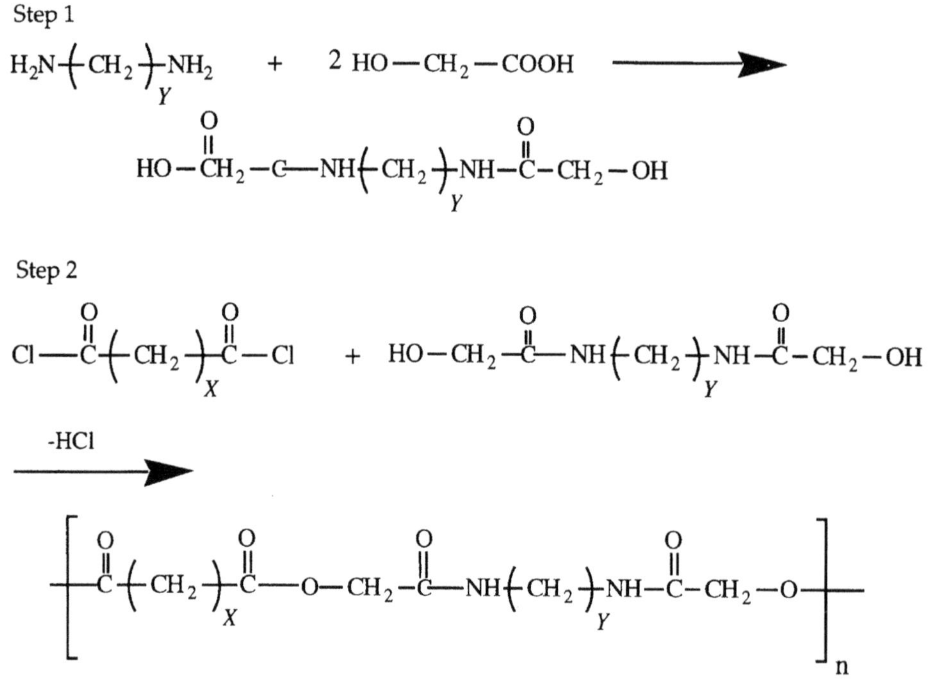

Step 1

$$H_2N\text{---}(CH_2)_Y\text{---}NH_2 \quad + \quad 2\ HO\text{---}CH_2\text{---}COOH \longrightarrow$$

$$HO\text{---}CH_2\overset{\overset{\displaystyle O}{\parallel}}{C}\text{---}NH\text{---}(CH_2)_Y\text{---}NH\text{---}\overset{\overset{\displaystyle O}{\parallel}}{C}\text{---}CH_2\text{---}OH$$

Step 2

$$Cl\text{---}\overset{\overset{\displaystyle O}{\parallel}}{C}\text{---}(CH_2)_X\text{---}\overset{\overset{\displaystyle O}{\parallel}}{C}\text{---}Cl \quad + \quad HO\text{---}CH_2\text{---}\overset{\overset{\displaystyle O}{\parallel}}{C}\text{---}NH\text{---}(CH_2)_Y\text{---}NH\text{---}\overset{\overset{\displaystyle O}{\parallel}}{C}\text{---}CH_2\text{---}OH$$

$$\xrightarrow{\text{-HCl}}$$

$$\left[\overset{\overset{\displaystyle O}{\parallel}}{C}\text{---}(CH_2)_X\text{---}\overset{\overset{\displaystyle O}{\parallel}}{C}\text{---}O\text{---}CH_2\text{---}\overset{\overset{\displaystyle O}{\parallel}}{C}\text{---}NH\text{---}(CH_2)_Y\text{---}NH\text{---}\overset{\overset{\displaystyle O}{\parallel}}{C}\text{---}CH_2\text{---}O\right]_n$$

PEA-X,Y

Scheme 3 (Adapted from Barrows, 1988).

Synthesis of Poly(ester-amides) Containing Degradable Blocks

Palumbo and co-workers (Andini, 1988; Simone, 1992) have synthesized poly(ester-amides) containing hydrolyzable segments of L-lactide via a two step approach. In the first step, a telechelic oligomer of L-lactide bearing hydroxyl end groups is synthesized by standard ring opening polymerization of the lactide monomer. In the second step, the oligomer is converted to the diacid chloride derivative by reacting it with an excess of alkyl diacid chloride such as sebacoyl dichloride, followed by interfacial condensation with an alkyl diamine to yield the poly(ester-amide). L-lactide telechelic oligomer (PLA-OH) was synthesized by reacting 85 mmole of L-lactide, 8.5 mmole of 1,4-butanediol and 13 mg of 2-ethylhexanoate in a sealed glass ampule heated at 100°C for 3 h. By changing the monomer/initiator ratio (M/I), PLA-OH of different molecular weights can be obtained. Telechelic hydroxy functionalized oligomers ranging from 600–1500 in molecular weight have been synthesized using this approach with a yield of 85–95%. The structure of PLA-OH oligomer is shown in Scheme 4.

Synthesis of functionalized biodegradable polymers is key in the development of the next generation of bio-interactive polymers. The ability to attach bio-active ligands such as GRGDY, which contains the cell adhesion sequence RGD, and cell

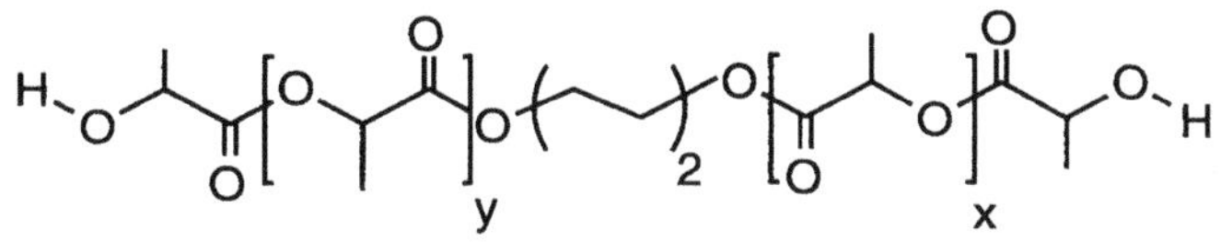

Scheme 4 (Adapted from Andini, 1988; Simone, 1992).

homing molecules in a biodegradable polymer backbone will have a profound impact on both drug delivery as well as tissue engineering. Recognition between the cellular matrix and the polymer scaffold at a molecular level is key in attaining maximum tissue-polymer matrix interaction and tissue regeneration.

Synthesis of Morpholine 2,5-dione and Derivatives

In one synthetic approach (Veld 1990, 1992), alanine is first converted to the bromo derivative via the reaction of the diazonium salt of the amine with hydrogen bromide. In the next step the acid functionality is activated by converting it to the acid chloride using thionyl chloride. The activated acid is condensed with a protected α-amino acid to yield the dipeptide intermediate which is then cyclized by heating in presence of Celite (ion exchange resin) to yield the final product. However, the overall yield of this reaction is fairly low. A more elegant approach is shown in Scheme 5. In this approach, an α-amino acid with a protected side chain (e.g., ε-Z-lysine) is reacted with 2-bromo-propionyl bromide under Schotten-Bauman conditions (Fischer 1908) to yield the intermediate 5a, which is then cyclized under basic conditions to the depsipeptide 5b.

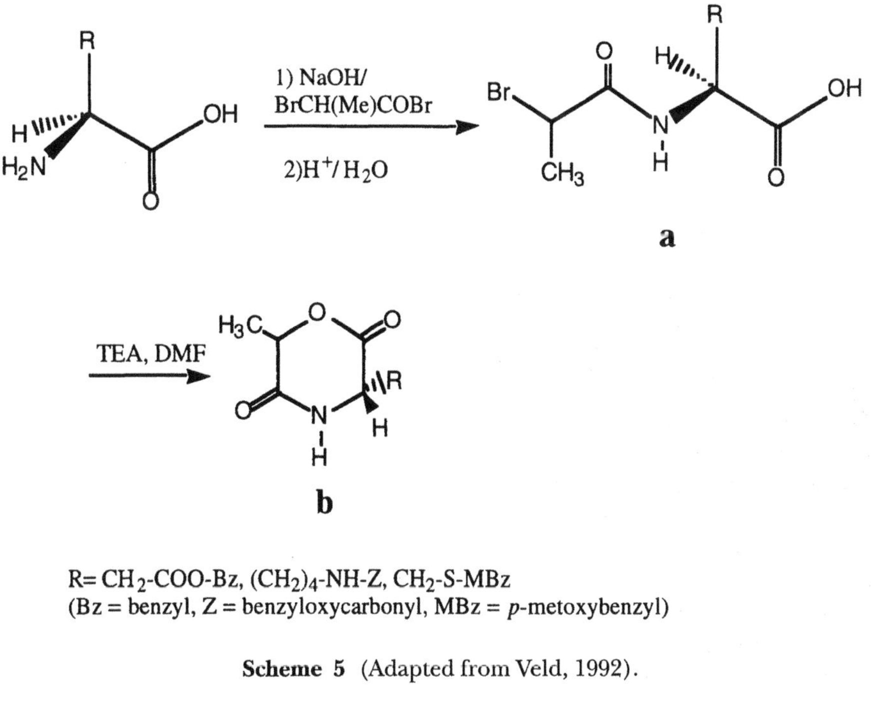

R= CH$_2$-COO-Bz, (CH$_2$)$_4$-NH-Z, CH$_2$-S-MBz
(Bz = benzyl, Z = benzyloxycarbonyl, MBz = p-metoxybenzyl)

Scheme 5 (Adapted from Veld, 1992).

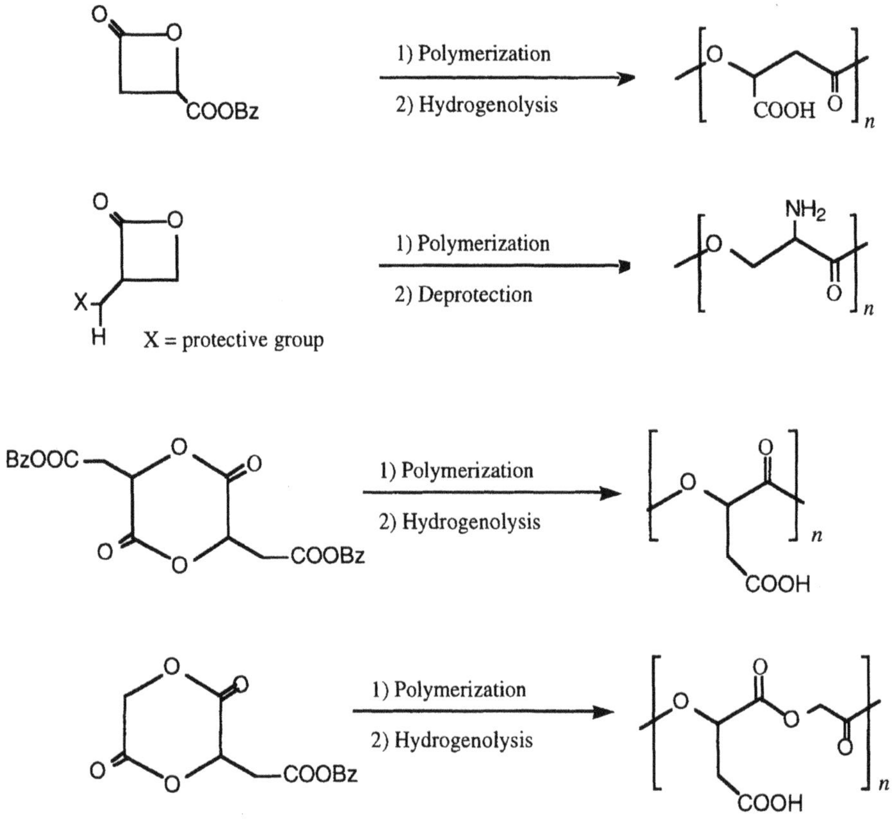

Scheme 6

The chemical structures of some representative monomers and functionalized biodegradable polymers obtained by the ring opening polymerization are shown in Scheme 6. Copolymers of 5b with other lactones were synthesized by ring opening polymerization of the desired derivatized morpholine-2,5-dione (5b) and the co-monomer such as lactide or ε-CL initiated by stannous octoate at 130°C for 48 h (Scheme 7). Copolymers of aspartic acid, lysine and cysteine derivative of morpholine-2,5-dione with lactide and ε-CL had a molecular weight ranging from 20,000 to 60,000 with Tg's in the range of 40°C and −30°C for the lactide and caprolactone, respectively (Veld 1990, 1992).

Recently, Langer and co-workers have attached cell adhesion peptide sequence such as RGD onto co-polymers of lactides containing lysine residues (Barrera, 1993). These materials exhibit improved cell interactions in comparison with the homopolymer of lactic acid. They have also synthesized comb polymers containing poly(L-lysine) pendant groups (Hrkach, 1995) by the nucleophilic ring opening polymerization of the N-carboxy anhydride of ε-Z-L-lysine, which was initiated by the ε-amino terminus of the lysine residue in the co-polymer backbone.

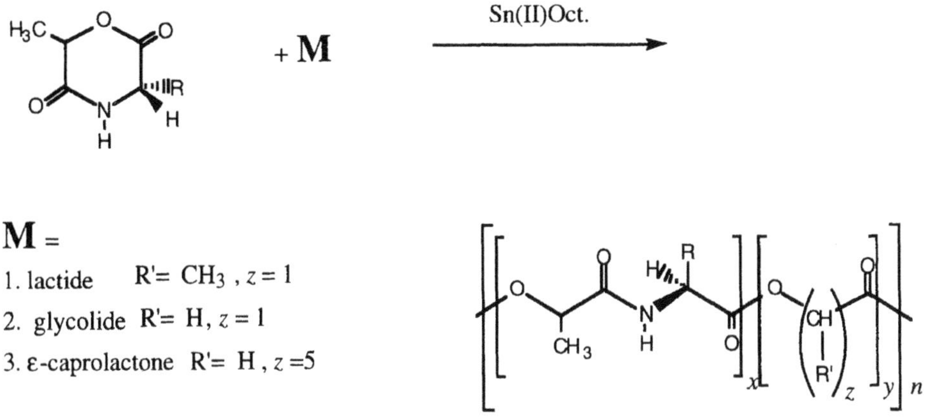

Scheme 7

M =
1. lactide   R'= CH$_3$, z = 1
2. glycolide R'= H, z = 1
3. ε-caprolactone R'= H , z =5

Biodegradable Polyester Networks

Poly(ester-urethane) Elastomeric Networks

Polyurethanes have found extensive use in several in vivo biomedical applications such as blood catheters and artificial heart valves, due to their excellent blood contacting and mechanical properties. Biomer®, which is the most widely used medical grade polyurethane, is a poly(ether-urethane), synthesized using aromatic diisocyanate. These aromatic diisocyanate units are hydrolyzed *in vivo* into their respective diamines, which are toxic and potential carcinogens and teratogens. In order to overcome this deficiency diisocyanates derived from di-amine containing amino acids such as lysine have been explored by several researchers (Bruin, 1988; Wiggins, 1992). Furthermore, for certain biomedical applications such as, vascular grafts or cell-scaffolds for tissue engineering shorter and more predictable degradation behavior might be desirable. Towards this goal biodegradable poly(ester-urethane) networks derived from lysine diisocyanate and degradable polyester blocks of lactide and glycolide have been synthesized (Wiggins, 1992). In all the approaches a core molecule, typically a multifunctional alcohol, such as pentaerythyritol, meso-inositol, sorbitol or glycerol is utilized to form the highly branched (star) polyester block. This is then crosslinked using lysine diisocyanate to establish the ester-urethane network. Ethyl 2,6-diisocyanatohexanoate (L-lysine diisocyanate) is synthesized by reacting L-lysine monohydrochloride ethyl ester with phosgene in dichlorobenzene at 100–110°C for 8–12 h. The branched biodegradable polyester core is synthesized by reacting the monomer(s) (lactide, caprolactone etc.), the multifunctional hydroxy core molecule, and the initiator in the desired ratio in dry DMF at 100°C for 20 h under dry nitrogen. The poly(ester-urethane) elastomeric network is obtained by cross-linking the polyester core with ethyl 2,6-diisocyanatohexanoate ([OH]/[isocyanate = 1) in a mixed solvent system of toluene and methylene chloride.

PART II

Polyphosphate Esters

The earliest report on the nonbiological synthesis of polyphosphonates and polyphosphate ester was by Arvin. (Arvin, 1936). Much of the earlier work in the synthesis of polyphosphate ester was carried out by Millich *et al.*, (Millich, 1969, 1974). One of the earliest reviews in the interfacial synthesis of polyphosphonates, polyphosphates and polyphosphites by Millich *et al.*, (Millich, 1977) is highly recommended.

Millich *et al.*, have synthesized several poly(phosphonates) derived from both aromatic diols such as, hydroquinone and resorcinol and, aliphatic diols such as, 1,3-dihydroxy acetone, ethylene glycol, 1,4-butane diol, tartaric acid, 1,3-propane diol and derivatives via an interfacial synthesis involving the diol and phenyl phosphonic dichloride under basic conditions (Millich, 1969, 1974). They found that the polymer yield was significantly higher at higher pH's with a maximum yield around pH 12. Polycondensation of diphosphoric monomers with aromatic diols such as, bisphenol A (BPA) has also been explored by Corallo *et al.*, (Corallo 1978). However, the molecular weights (M_n) of the polymers so obtained were rather low ranging from 1800–4300. Ring opening polymerization of cyclic phosphates under elevated conditions in presence of acid or base catalyst do yield linear polymers but with a rather low DP of about 10–20 monomer units (Millich 1977 and references therein). Recently, Nishikobu *et al.*, (Nishikobu, 1994) reported the synthesis of polyphosphonates by the addition of a diepoxide, bisphenol A diglycidyl ether (BPGE) with phenyl phosphonic dichloride in the presence of quaternary ammonium salts. Polymers were obtained in 20–95% yield with M_n ranging from 4000–17,000.

In the late sixties and early seventies several organic polyphosphonates and polyalkyl polyphosphonates derived from low molecular weight polyethylene were evaluated as possible synthetic materials for bone repair and dental restoration (Anbar, 1971, 1974) with quite promising results. Recently, polyphosphates and polyphosphonates have been explored as a biodegradable carrier for drug delivery and fracture fixation by Leong *et al.*, (Richards, 1991a, 1991b; Leong 1994). The polymers were synthesized by interfacial condensation of either ethyl or phenylphosphorodichloridates and various dialcohols including BPA and polyethylene glycol under phase transfer conditions (Scheme 8). The interfacial polycondensation was found to be dependent on the catalyst concentration with an optimum around 5–10 mole% of phase transfer catalyst such as tetra alkylammonium halide. Typical molecular weights for the BPA polymers ranged from 20,000 to 40,000 with a Tg of around 110°C.

These polymers are expected to degrade in the presence of a nucleophile, such as a hydroxyl ion, to release a diol and a phosphate molecule. The rate of degradation is influenced by the chemical characteristic of the side chain wherein, polymers with aliphatic pendant groups are more susceptible to cleavage and degradation than aromatic pendant groups. The polymers based on BPA degraded over several months, with a weight loss of 5% and 20% for the ethyl and phenyl phosphonate derivative respectively, after 8 months in buffer pH 7.4 at 37°C. The in vitro swelling results corresponds well to those for degradation. After 10 days, the ethyl derivative had swollen more than 60% while values for the phenyl

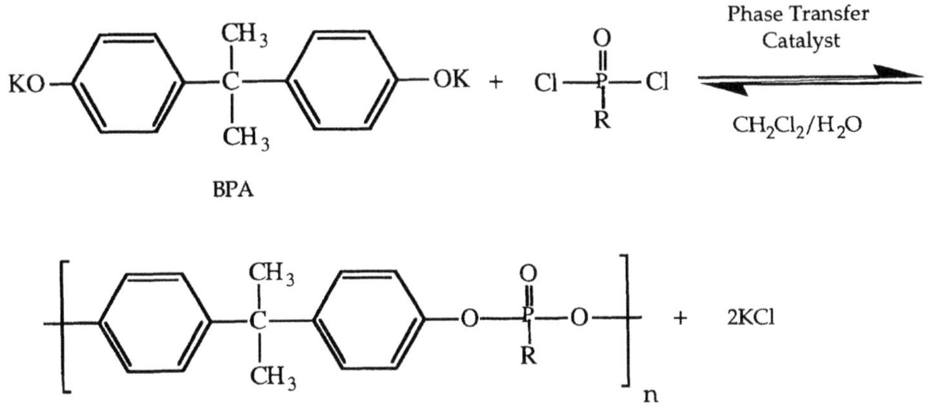

Scheme 8 (Adapted from Richards. 1991).

phosphate were less than 20%. However, when implanted in rabbits these polymers exhibited an initial weight loss of about 10% presumably due to the extraction of low molecular weight oligomers which are more soluble in body fluids. A weight loss of around 80% was observed at 15 months post implantation for the ethyl phosphate derivative. Histological evaluation of the tissues at the implant site displayed an initial inflammatory response at 3 to 7 weeks with lymphocytes, giant cells, and macrophages present in varying amounts.

This chapter summarizes several co-polyesters that may have potential biomedical applications.

REFERENCES

Anbar, M., Feldman, C. and Wolf, P. (1971) Effect of polyalkyl polyphosphonates on bone development in young rats. *Experientia*, **27**(6), 664–665.

Anbar, M. and Farley, E.P. (1974) Potential use of organic polyphosphonates as adhesives in the restoration of teeth. *J. Dent. Res.*, **53**(4), 879–888.

Andini, S., Ferrara, L., Maglio, G. and Palumbo, R. (1988) Synthesis of block polyesteramides containing biodegradable poly(L,L-lactide) segments. *Makromol. Chem.: Rapid Commun.*, **9**, 119–124.

Arabuli, N., Tsitlanadze, G., Edilashvili, L., Kharadze, D., Goguadze, T., Beridze, V., Gomurashvili, Z. and Katsarava, R. (1994) Heterochain polymers based on natural amino acids. Synthesis and enzymatic hydrolysis of regular poly(ester-amide)s based on bis(L-phenylalanine) α,ω-alkylene diesters and adipic acid. *Macromol. Chem. Phys.*, **195**, 2279–2289.

Arentzen, R., YanKui, Y.T. and Reese, C.B. (1975) Improved procedures for the preparation of tetrahydro-4H-pyran-4-oneand 5,6-dihydro-4-methoxy-2H-pyran. *Synthesis*, 509–510.

Arvin, J. (1936) United States patent 2,058,394.

Barrera, D.A., Zylstra, E., Peter T. Lansbury, J. and Langer, R. (1993) Synthesis and RGD Peptide Modification of a New Biodegradable Copolymer: Poly(lactic acid-co-lysine) *J. Am. Chem. Soc.*, **115**, 11010–11011.

Barrows, T.H. (1982) United States patent 4,343,931.

Barrows, T.H. (1988) Comparison of bioabsorbable poly(ester-amide) monomers and polymers In Vitro using radiolabeled homologs. *Polym. Mat. Sci. Eng.*, **58**, 376–377.

Barrows, T.H. (1994) Bioabsorbable poly(ester-amides) In S.W. Shalaby (Ed.), Biomedical polymers: Designed-to-degrade systems. Cincinnati: Hanser/Gardner Publications, pp. 97–116.

Bengs, H., Bayer, U., Ditzinger, G., Krone, V., Lill, N., Sandow, J. and Walch, A. (1996) Polytartarate-A new biodegradable polymer. *Proceedings of the International Symposium on Biodegradable Materials*, Hamburg, Germany, 79–80.

Bruin, P., Veenstra, G.J., Nijenhuis, A.J. and A.J.P. (1988) Design and synthesis of biodegradable poly(ester-urethane) elastomer networks composed of non-toxic building blocks. *Makromol. Chem.: Rapid Commun.*, **9**, 589–594.

Corallo, M. and Pietrasanta, Y. (1978) Synthese de Polyphosphonates et Polyphosphonamides a partir de monomers diphosphoniques polycondensation en solution. *Polym. J.*, **14**, 265–272.

Du, Y.J., Lemstra, P.J., Nijenhuis, A.J., Aert, H.A.M.v. and Bastiaansen, C. (1995) ABA Type Copolymers of Lactide with Poly(ethylene glycol) Kinetic, Mechanistic and Model Study. *Macromolecules*, **28**, 2124–2132.

Fischer, E. and Scheibler, H. (1908) III. Derivate der activen valine. *Justus Liebig's Ann. Chem.*, **363**, 136–167.

Gref, R., Minamitake, Y., Peracchia, M. T., Trubetskoy, V., Torchilin, V. and Langer, R. (1994) Biodegradable long-circulating nanospheres. *Science*, **263**, 1600–1603.

Gref, R., Domb, A.J., Quellec, P., Blunk, T., Müller, R.H., Verbavatz, J.M. and Langer, R. (1995) The controlled intravenous delivery of drugs using PEG-coated sterically stabilized nanospheres. *Adv. Drug. Del. Rev.*, **16**, 215–233.

Guerin, P., Francillette, J., Braud, C. and Vert, M. (1986) Benzyl esters of optically active malic acid stereocopolymers as obtained by ring-opening polymerization of (R)-(+) and (S)-(−)-benzyl malolactonates. *Makromol. Chem.: Macromol. Symp.*, **6**, 305–314.

Guerin, P., Girault, J.P., Caron, A., Francillette, J. and Vert, M. (1992) Selective INEPT as an NMR tool for studying repeat unit distribution and stereosequences in poly(β-malic acid) copolymers. *Macromolecules*, **25**, 143–148.

Hrkach, J. S., Ou, J., Lotan, N. and Langer, R. (1995) Synthesis of Poly(L-Lactic acid-co-lysine) Graft Copolymer. *Macromolecules*, **28**, 4736–4739.

Karlsson, S., Hakkarainen, M. and Albertsson, A.-C. (1994) Identification by headspace gas chromatography-mass spectrometry of in vitro degradation products of homo- and copolymers of L- and D,L-lactide and 1,5-dioxepan-2-one. *J. Chromatography A*, **688**, 251–259.

Kricheldorf, H.R. and Meier-Haack, J. (1993) ABA Triblock Copolymers of L-lactide and Poly(ethylene glycol) *Macromol. Chem.*, **194**, 715–725.

Leong, K.W. (1995) Alternative materials for fracture fixation. *Conn. Tissue Res.*, **31**(4), S69–S75.

Lofgren, A., Alberstsson, A.C., Dubois, P., Jerome, R. and Teyssie, P. (1994) Synthesis and characterization of biodegradable Homopolymers and Block-Copolymers Based on 1,5-dioxepan-2-one. *Macromolecules*, **27**, 5556–5562.

Mathisen, T., Masus, K. and Albertsson, A.-C. (1989) Polymerization of 1,5-dioxepan-2-one. 2. Polymerization of 1,5-dioxepan-2-one and Its Cyclic Dimer, Including a New Procedure for the synthesis of 1,5-dioxepan-2-one. *Macromolecules*, **22**, 3842–3846.

Millich, F. and C.E. Carraher, J. (1969) Interfacial syntheses of Polyphosphonate and Polyphosphate esters. I. Effects of alkaline medium. *J. Polym. Sci.: Part A-1*, **7**, 2669–2678.

Millich, F., Lambing, L. and Teague, J. (1977) Polyphosphonates, Polyphosphates, and Polyphosphities. In F. Millich and J. Charles Carraher (Ed.), Interfacial Synthesis. New York and Basel: Marcel Dekker, pp. 309–350.

Nishikubo, T., Kameyama, A. and Minegishi, S. (1994) A novel synthesis of Poly(phosphonate) by the addition reaction of diepoxide with phenylphosphonic dichloride. *Macromolecules*, **27**, 2641–2642.

Okada, M., Ito, S., Aoi, K. and Atsumi, M. (1994(a)) Spontaneous hydrolytic degradability of copolyesters having tetrahydropyran rings in their backbones. *J. Appl. Polym. Sci.*, **51**, 1035–1043.

Okada, M., Ito, S., Aoi, K. and Atsumi, M. (1994(b)) Biodegradability of polyesters having tetrahydropyran rings in their backbones. *J. Appl. Polym. Sci.*, **51**, 1045–1051.

Okada, M., Okada, Y. and Aoi, K. (1995) Synthesis and degradabilities of polyesters from 1,4:3,6–dianhydrohexitols and aliphatic dicarboxylic acids. *J. Polym. Sci.: Part A: Polym. Chem.*, **33**, 2813–2820.

Ramiandrasoa, P., Guerin, P., Girault, J.P., Bascou, P., Hammouda, A., Cammas, S. and Vert, M. (1993) Poly(β-maic acid alkyl esters) derived from 4-alkyloxycarbonyl-2-oxetanones obtained via the ketene route. *Polym. Bull.*, **30**, 501–508.

Richards, M., Dahiyat, B.I., Arm, D.M., Lin, S. and Leong, K.W. (1991(a)) Interfacial polycondensation and characterization of polyphosphates and polyphosphonates. *J. Polym. Sci.: Part A: Polym. Chem.*, **29**, 1157–1165.

Richards, M., Dahiyat, B.I., Arm, D.M., Brown, P.R. and Leong, K.W. (1991(b)) Evaluation of polyphosphates and polyphosphonates as degradable biomaterials. *J. Biomed. Mat. Res.*, **25**, 1151–1167.

Sawhney, A.S. and Hubbell, J.A. (1990) Rapidly Degraded Terpolymers of DL-lactide, Glycolide, and ε-caprolactone with Increased Hydrophilicity by Copolymerization with Polyethers. *J. Biomed. Mat. Res.*, **24**(10), 1397–1411.

Simone, V.D., Maglio, G., Palumbo, R. and Scardi, V. (1992) Synthesis, characterization, and degradation of block Polyesteramides Containing Poly(L-Lactide) Segments. *J. Appl. Polym. Sci.*, **46**, 1813–1820.

Veld, P.J.A., Dijkstra, P.J., Lochem, J.H. and Feijen, J. (1990) Synthesis of alternating polydepsipeptides by ring-opening polymerization of morpholine-2,5-dione derivatives. *Makromol. Chem.*, **191**, 1813–1825.

Veld, P.J.A., Dijkstra, P.J. and Feijen, J. (1992) Synthesis of biodegradable polyesteramides with pendant functional groups. *Makromol. Chem.*, **193**, 2713–2730.

Wiggins, J.S. and Storey, R. (1992) Synthesis and characterization of L-lysine based poly(ester-urethane) networks. *Polym. Preprints.*, **32**, 516–517.

Xiong, C.D., Cheng, L.M., Xu, R.P. and Deng, X.M. (1995) Synthesis and Characterization of Block Copolymers from D,L-Lactide and Poly(tetramethylene ether glycol) *J. Appl. Polym. Sci.*, **55**, 865–869.

ABBREVIATIONS

BPA = Bisphenol-A

BPGE = Bisphenol A diglycidyl ether

BOC = Butyloxycarbonyl

ε-CL = ε-Caprolactone

mCPBA = meta-Chloro peroxybenzoic acid

DMF = N,N-dimethylformamide

DP = Degree of polymerization

DXO = 1,5-dioxepan-2-one

GRGDY = Glysine-Arginine-Glysine-Aspartic acid-Tyrosine

I = Initiator

INEPT = Intensive Nuclear Enhancement by Polarization Transfer

LHRH = Lutenizing Hormone Releasing Hormone

M = Monomer

M-PEG-NH$_2$ = Methoxy-PEG-Amine

NCA = N-carboxy anhydride

NMP = N-methyl-2–pyrrolidone

NMR = Nuclear Magnetic Resonance

PEA = Poly(ester-amides)

PEG = Polyethylene glycol

PEO = Polyethylene oxide

PLA = Poly(lactic acid)

PLA-OH = L-lactide telechelic oligomer

PGA = Poly(glycolic acid)

PCL = Poly(ε-Caprolactone)

PPO = Polypropylene oxide

RGD = Arginine-Glysine-Aspartic acid

Tg = Glass transition temperature

Tm = Melting temperature

THP-one = Tetrahydro-4H-pyran-4-one
M_n = Number average molecular weight
M_w = Weight average molecular weight
MW = Molecular weight
PDI = Polydispersity Index
Z = Benzyloxycarbonyl

8. POLYANHYDRIDES

ABRAHAM J. DOMB[1], OMAR ELMALAK[1], VENKATRAM R. SHASTRI[2],
ZEEV TA-SHMA[3], DAVID M. MASTERS[4], ISRAEL RINGEL[1],
DORON TEOMIM[1] and ROBERT LANGER[2]

[1] *The Hebrew University of Jerusalem, School of Pharmacy-Faculty of Medicine,
Jerusalem 91120 Israel*
[2] *MIT, Department of Chemical Engineering, Cambridge, MA 02139, USA*
[3] *Israel Defense Forces (IDF)*
[4] *Mayo Clinic and foundation, Anethesia Research, Rochester MN 55905, USA*

INTRODUCTION

Polyanhydrides are useful bioabsorbable materials for controlled drug delivery. They hydrolyze to dicarboxylic acid monomers when placed in aqueous medium. Since their introduction to the field of controlled drug delivery, about 15 years ago, extensive research has been conducted to study their chemistry as well as their toxicity and medical applications. Several review articles have been published on polyanhydrides for controlled drug delivery applications (Domb *et al.*, 1992; Leong *et al.*, 1989; Laurencin 1995).

The earliest report on the synthesis of poly(anhydrides) was by Bucher and Slade (1909). Years later, Hill and Carothers (1930 and 1932) had synthesized polymers based on aliphatic diacid monomers for textile applications. During the 1950s and 1960s, Conix (1958) and Yoda (1963, 1962) synthesized over a 100 new polyanhydrides based on aromatic and heterocyclic diacid monomers. In 1980 Langer proposed the use of polyanhydrides as biodegradable carriers for controlled drug delivery systems (Rosen, 1983), resulting so far into two implantable devices for human use (e.g. Gliadel[TM] implant for the treatment of brain tumors and Septacin[TM] implant for treating chronic bone infections (Brem *et al.*, 1995; Domb *et al.*, 1994). This chapter reviews the chemistry, degradation, biocompatibility and applications of polyanhydrides.

SYNTHESIS

Polyanhydrides have been synthesized by melt condensation of activated diacids (Domb *et al.*, 1987), ring opening polymerization, dehydrochlorination, and dehydrative coupling agents (Leong *et al.*, 1987; Domb *et al.*, 1988). Solution polymerization yielded in general low molecular weight polymers. The most widely used method is the melt condensation of dicarboxylic acids treated with acetic anhydride:

$$\text{HOOC-R-COOH} + (\text{CH}_3\text{-CO})_2\text{O} \xrightarrow{\text{reflux}} \text{CH}_3\text{-CO-(O-CO-R-CO-)}_m\text{O-CO-CH}_3$$
$$\text{(I)}$$

$$\text{(I)} \xrightarrow{180°\text{C}} \text{CH}_3\text{-CO-(O-CO-R-CO-)}_n\text{O-CO-CH}_3$$

$$m = 1\text{--}20; \quad n = 100\text{--}1000$$

The polycondensation takes place in two steps. In the first step the dicarboxylic acid monomers are reacted with excess acetic anhydride to form acetyl terminated anhydride prepolymers with a degree of polymerization ranging from 1 to 20, which are then polymerized at elevated temperature under vacuum to yield polymers with Dp ranging from 100 to over 1000. Acetic acid mixed anhydride prepolymers were also prepared from the reaction of the diacid monomers with ketene or acetyl chloride. A typical procedure for polyanhydride synthesis is as follows: To a 1L flask containing 500 ml refluxing acetic anhydride, 100 g of recrystalized sebacic acid is added. The sebacic acid is dissolved in acetic anhydride (approx 5 minutes) and the reaction is stopped 15 minutes after the complete solubilization of the diacid. The solution is filtered and concentrated at 60°C under reduced pressure. The liquid residue is then mixed with 70 ml dichloromethane and the solution is added to a 1:1 mixture of petroleum ether: ether (500 ml) to precipitate the prepolymer. The white precipitate is isolated and dried under vacuum before further use. The polymerization is carried-out in a 1L polymerization kettle equipped with a vacuum sealed overhead stirrer and a vacuum inlet. The kettle is immersed in a 180°C silicone oil bath to melt the prepolymer after which a high vacuum (<0.5 mm Hg) is applied to the melt to initiate the condensation polymerization. The polymerization is continued for 90 minute to yield a viscous yellowish melt which in cooling solidifies into an off-white polymer (>90% yield) with molecular weight in excess of 50,000. The polymer is then purified by precipitation into excess of petroleum ether from a solution in methylene chloride. The polymer is then transferred to a glass container flushed with dry argon and stored at 0°C or below.

The condensation reaction of diacetyl mixed anhydrides of aromatic or aliphatic diacids is carried out in the temperature range of 150 to 200°C (Domb and Langer 1987). A variety of catalysts have been used in the synthesis of a range of polyanhydrides. Significantly higher molecular weights in shorter reaction times were achieved by utilizing cadmium acetate, earth metal oxides, and $\text{ZnEt}_2\text{-H}_2\text{O}$. Except for calcium carbonate which is a safe natural material, the use of these catalysts for the production of medical grade polymers is limited because of their potential toxicity.

Polyanhydrides can be synthesized by melt condensation of trimethylsilyl dicarboxylates and diacid chlorides to yield polymers with an intrinsic viscosity 0.43 dl/g (Gupta 1988). Direct polycondensation of sebacic acid and adipic acid at a high temperature under vacuum resulted in low molecular weight oligomers (Knobloch *et al.*, 1975). The preparation of adipic acid polyanhydride from cyclic adipic anhydride (oxepane-2,7–dione) was investigated (Albertsson *et al.*, 1990).

A variety of solution polymerizations at ambient temperature have been reported (Leong *et al.*, 1987, Domb at al. 1988). Partial hydrolysis of terephthalic acid chloride in the presence of pyridine as an acid acceptor yielded a polymer of MW = 2100. The use of N,N-bis(2-oxo-3-oxazolidinyl)phospharamido chloride, dicyclohexylcarbodiimide, chlorosulfonyl isocyanate, and phosgene as coupling agents produced low molecular weight polymers. Furthermore, homo- and copolyanhydrides have also been synthesized via aqueous and non aqueous interfacial reaction conditions with limited success. Various aromatic polymers were prepared from the reaction of equimolar amounts of the acid dissolved in aqueous base and the corresponding diacid chloride dissolved in an organic solvent (Leong *et al.*, 1987; Domb *et al.*, 1988; Subramanyam *et al.*, 1985). Because the reaction is between a dibasic acid in one phase and an acid chloride in the other phase, the copolymers may present a regularly alternating structure. In the reaction between sebacoyl chloride in chloroform and isophthalic acid sodium salt in water a copolymer that contain mostly sebacic acid units, was obtained. This result can be explained on the bases of a side reaction occurring between sebacic acid which is formed on the two imisible phases with the acid chloride in the organic phase resulting in the formation of a polymer rich in sebacic acid.

POLYANHYDRIDE STRUCTURES

Since the discovery of polyanhydrides in 1909, hundreds of polymer structures have been reported (Encyclopedia, 1969). A representative list of polymers developed till the 70's is shown in Table 1. Polyanhydrides intended for use in medicine that have been developed since 1980 are described below:

Unsaturated Polymers

A series of unsaturated polyanhydrides were prepared by melt or solution polymerization of fumaric acid (FA), acetylenedicarboxylic acid (ACDA), and 4,4'-stilbendicarboxylic acid (STDA) (Domb *et al.*, 1991).

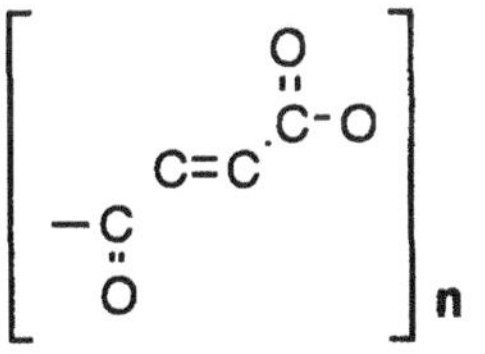

P(FA) 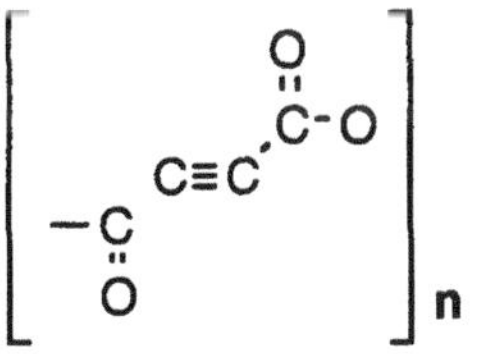   P(ACDA)

The double bonds remain intact throughout the polymerization process and were available for a secondary reaction to form a crosslinked matrix. The unsaturated homopolymers were crystalline and insoluble in common organic solvents whereas copolymers with aliphatic diacids were less crystalline and were soluble in chlorinated hydrocarbons (Domb *et al.*, 1991).

Table 1 Representative polyanhydrides synthesized during the years 1909–1980

Polymer structure	Melting point (°C)
	400
	256
	91
	250
	330
	151
	91
	160
	67
	78
	188
	230
	>300

Table 1 Continued

Polymer structure	Melting point (°C)

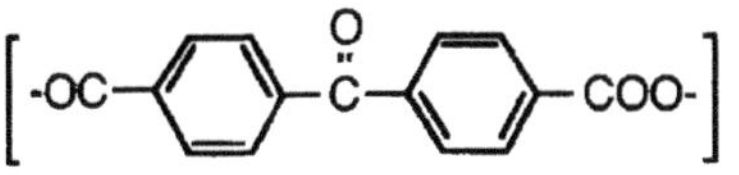

338

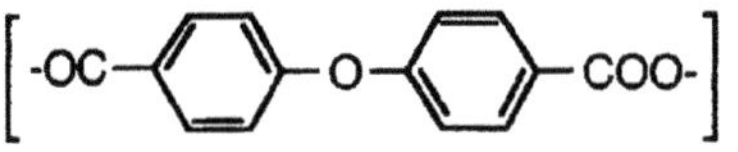

295

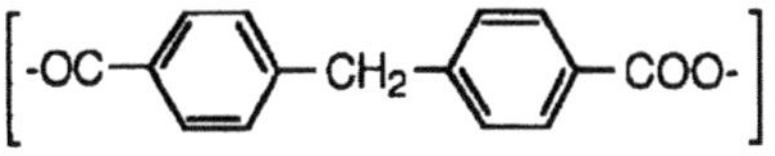

332

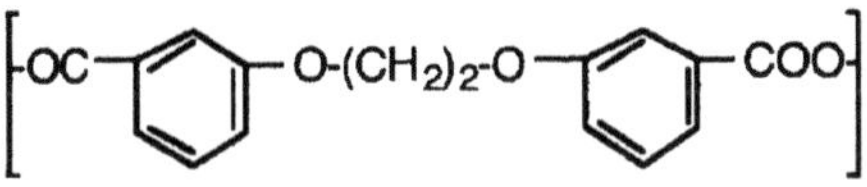

237

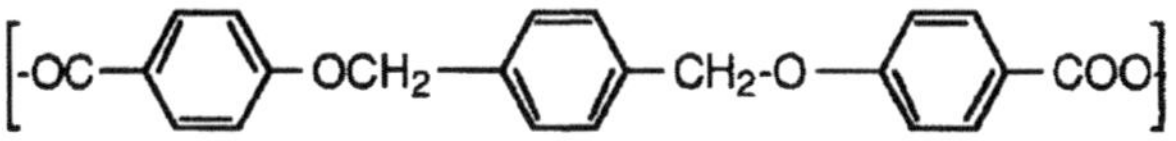

140

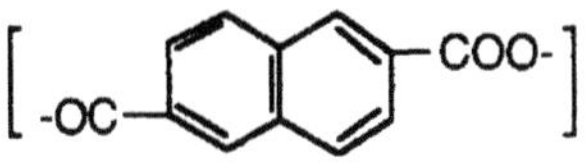

450

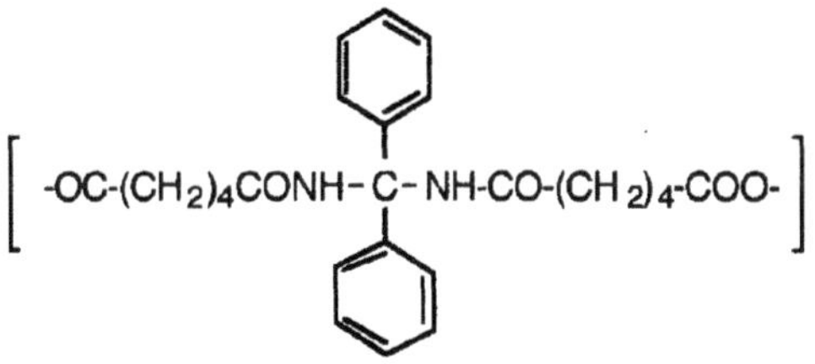

>300

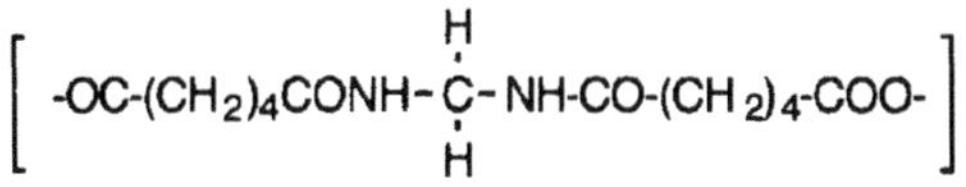

285

[OC-(CH2)2-SO2-(CH2)2-SO2-(CH2)2-COO]   185

[OC-(CH2)2-S-(CH2)2-S-(CH2)2-COO]   81

[OC-P-COO]

[OC-(CH2)x-COO] x=4-16   60-100

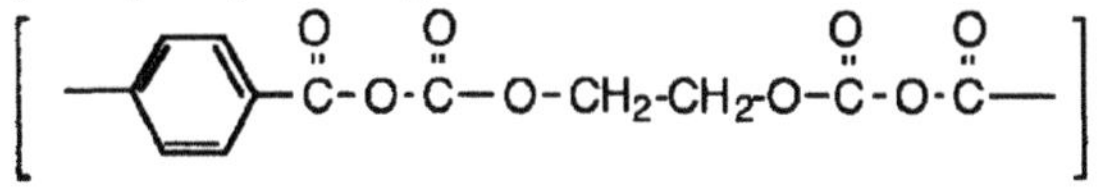

155

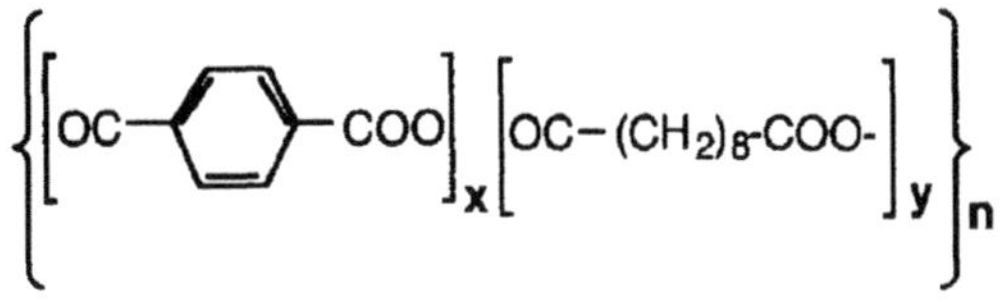

220

taken from Domb *et al.*, 1992.

Amino Acid Based Polymers

General methods for the synthesis of poly(amide-anhydrides) and poly(amide-esters) based on naturally occurring amino acids were described (Domb *et al.*, 1990). The polymers were synthesized from dicarboxylic acids prepared by amidation of the amino group of an amino acid with a cyclic anhydride, or by the amide coupling of two amino acids with a diacid chloride. Low molecular weight polymers from methylene bis(p-carboxybenzamide) were synthesized by melt condensation (Hartmann *et al.*, 1989). A series of amido containing polyanhydrides based on p-aminobenzoic acid were synthesized by melt condensation. The polymers melted at 58 to 177°C and had a molecular weight of 2500 to 12400.

Poly[methylenebis(p-carboxybenzamide)]

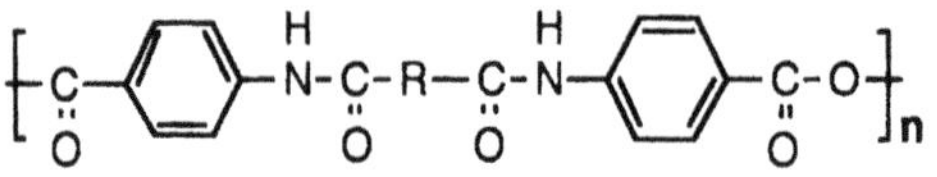

R= (CH₂)₃; (CH₂)₂-O-(CH₂)₈-O-(CH₂)₂

Trimelitic-and pyromelitic- mino acid based polymers

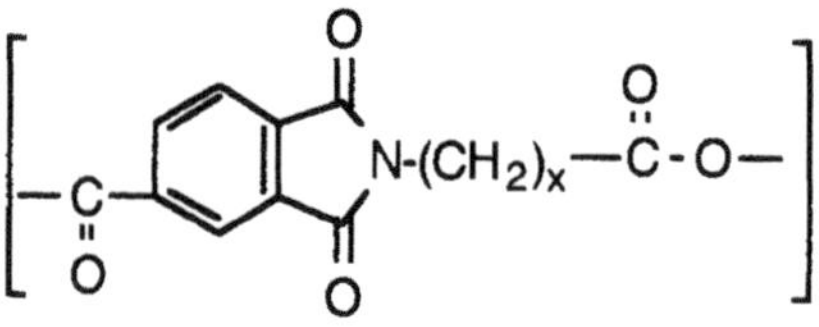

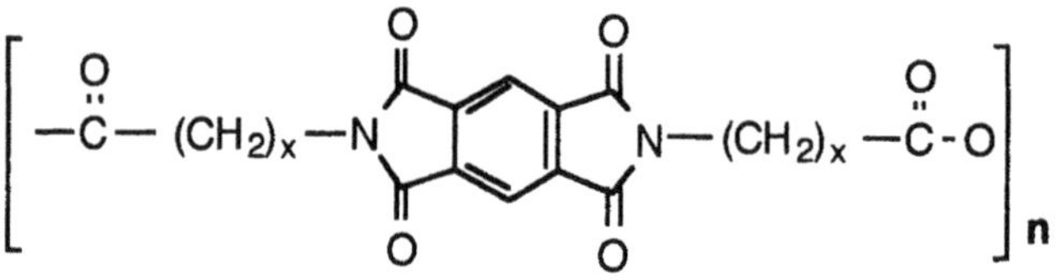

x=1-5

The trimellitic-amino acid polymers and its copolymers were extensively studied for use as drug carriers (Staubli *et al.*, 1990, 1991). The following amino acids were incorporated in a cyclic imide structure to form a diacid monomer: glycine, β-alanine, γ-aminobutyric acid, L-leucine, L-tyrosine, 11-aminoundecanoic acid and 12-aminododecanoic acid. The homopolymers of all N-trimellitylimido acids containing amino acids were rigid and brittle with MW below 10,000 (Staubli *et al.*, 1991). Higher molecular weight polymers were obtained by incorporation of flexible segments, i.e. copolymers with aliphatic diacids, in the polymer backbone. Copolymers of N-trimellitylimido-glycine or aminodecanoic acid with either sebacic acid (SA) or 1,6-bis(p-carboxyphenoxy)hexane (CPH) were prepared in defined ratios. High molecular weight copolymers (>100,000) were generally obtained with an increasing content of the SA or CPH comonomer.

Poly(imide-anhydride) based on trimellitic-imide diacids

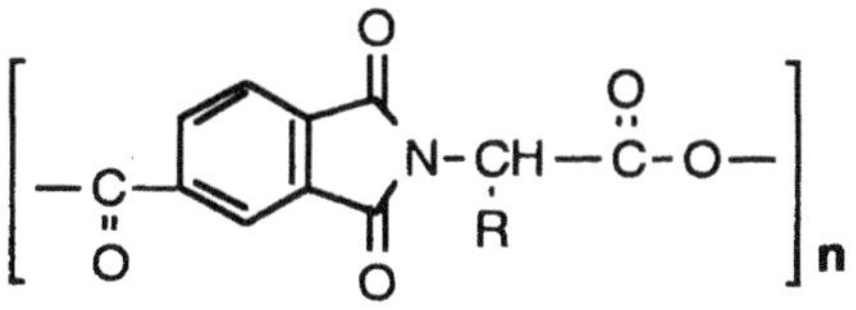

$R = -CH_2-CH-(CH_3)_2$

$R = -CH_2-C_6H_5-OH$

$R = -CH_2-S-CH_2-C_6H_5$

Poly(imide-anhydride) copolymers with sebacic acid

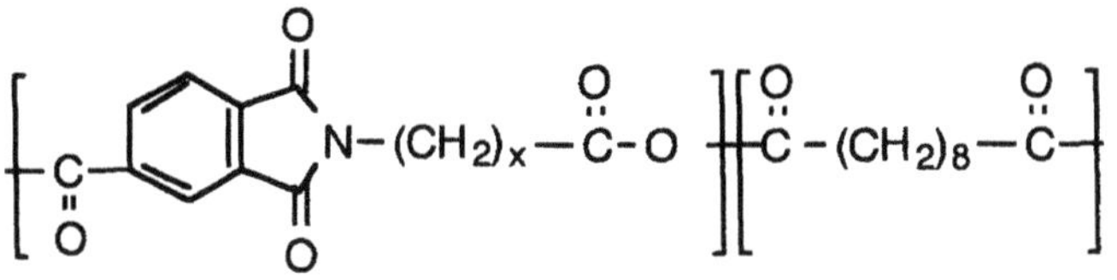

Aliphatic-aromatic Homopolymers

Polyanhydrides of diacid monomers containing aliphatic and aromatic moieties, poly[p-carboxyphenoxy)alkanoic anhydride], were synthesized by either melt or solution polymerization with molecular weights of up to 44,600 (Domb *et al.*, 1989).

Aliphatic-Aromatic homopolyanhydrides

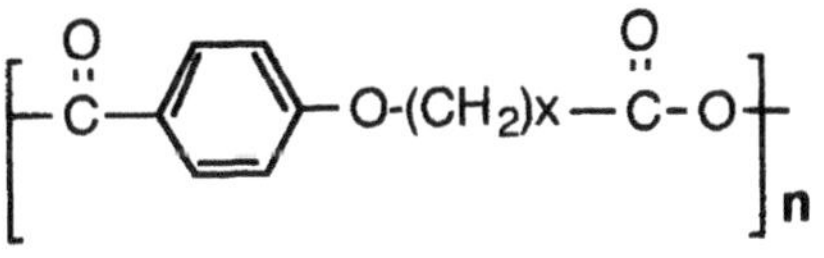

x=1-10

The polymers of carboxyphenoxy alkanoic acid of n = 3,5, and 7 methylenes were soluble in chlorinated hydrocarbons and melted at temperatures below 100°C. These polymers displayed zero-order hydrolytic degradation profile for 2 to 10 weeks. The degradation time was dictated by the length of the alkanoic chain wherein, an increasing degradation time was observed with an increasing chain length.

Soluble Aromatic Copolymers

Aromatic homopolyanhydrides are insoluble in common organic solvents and melt at temperatures above 200°C (Domb *et al.*, 1992b). These properties limit the use

Table 2 Soluble and Low Melting Aromatic Copolymer Compositions[a]

Copolymer of:	Compositions, % monomer	Copolymer of:	Compositions, % monomer
TA-CPP	20 to 30 TA	TA-SA	0 to 30 TA
CPP-IPA	10 to 60 CPP	CPP-SA	0 to 65 CPP
TA-IPA	10 to 40 TA	IPA-SA	0 to 70 IPA

[a]Polymers in this range have a solubility of >1% in dichloromethane and melting points below 150°C. TA-terephthalic acid; IPA- isophthalic acid; CPP-bis(p-carboxyphenoxy propane); SA-sebacic acid. The estimated range is in mole % with an error of ±5% (taken from Domb *et al.*, 1992b).

of purely aromatic polyanhydrides, since they can not be fabricated into films or microspheres using solvent or melt techniques. Fully aromatic polymers that are soluble in chlorinated hydrocarbons and melt at temperatures below 100°C were obtained by copolymerization of aromatic diacids such as isophthalic acid (IPA), terephthalic acid (TA), 1,3-bis(carboxyphenoxy)-propane (CPP) or hexane (CPH) (Table 2).

Copolyanhydrides of terephthalic acid and isophthalic acid

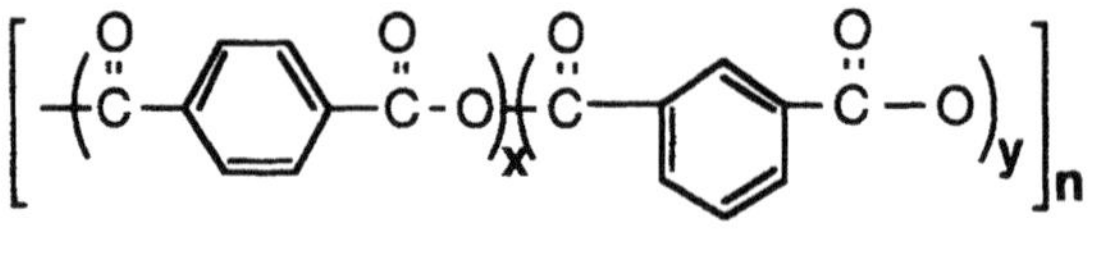

TA   IPA

Poly(ester-anhydrides)

4,4′-alkane- and oxa-alkanedioxydibenzoic acids were used for the synthesis of polyanhydrides (McIntyre 1964). The polymers melted at a temperature range of 98 to 176°C and had a molecular weight of up to 12,900. Di- and tri-block copolymers of poly(caprolactone), poly(lactic acid) and poly(hydroxybutyrate) have been prepared from carboxylic acid terminated low molecular weight polymers copolymerized with sebacic acid prepolymers by melt condensation (Abuganima 1996). Similarly, di-tri- and brush copolymers of poly(ethylene glycol) (PEG) with poly(sebacic anhydride) have been prepared by melt copolymerization of carboxylic acid terminated PEG (Gref 1995).

Ester Containing Polyanhydrides

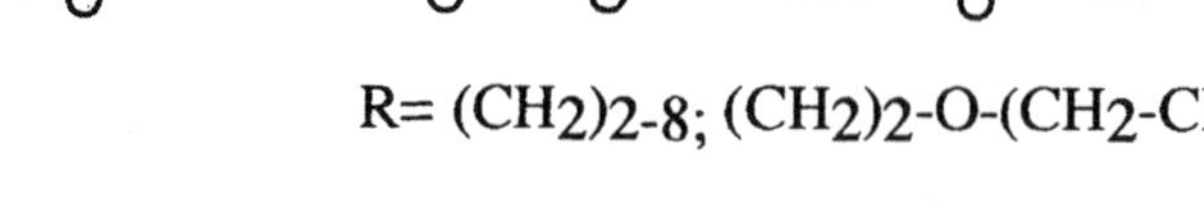

R= $(CH_2)_{2-8}$; $(CH_2)_2$-O-$(CH_2$-CH_2-O$)_2$-$(CH_2)_2$

Fatty Acid Based Polyanhydrides

Polyanhydrides were synthesized from dimer and trimer unsaturated fatty acids (Domb *et al.*, 1993). The dimers of oleic acid and erucic acid, are liquid oils containing two carboxylic acids available for anhydride polymerization. The homopolymers are viscous liquids, copolymerization with increasing amounts of sebacic acid forms solid polymers with increasing melting points as a function of SA content. The polymers are soluble in chlorinated hydrocarbons, tetrahydrofuran, 2-butanone, and acetone. Polyanhydrides synthesized from nonlinear hydrophobic fatty acid esters, based on ricinoleic, maleic acid and sebacic acid, possessed desired physico-chemical properties such as low melting point, hydrophobicity and flexibility, in addition to biocompatibility and biodegradability. The polymers were synthesized by melt condensation to yield film-forming polymers with molecular weights exceeding 100,000 (Domb *et al.*, 1995).

Incorporation of long chain fatty acid terminals such as stearic acid, in the polymer composition alters its hydrophobicity and decreases its degradation rate (Domb and Maniar, 1993b). Since natural fatty acids are monofuctional they would act as polymerization chain terminators and control the molecular weight. A detailed analysis of the polymerization reaction show that up to about 10 mole% content of stearic acid, the final product is essentially a stearic acid terminated polymer. Whereas, at higher amounts of acetyl stearate in the reaction mixture resulted in the formation of increasing amounts of stearic anhydride by-product with minimal effect on the polymer molecular weight which remains in the range of 5,000. Physical mixtures of polyanhydrides with triglycerides and fatty acids or alcohols did not form uniform blends.

poly(erucic acid dimer-co-sebacic acid)

poly(ricinoleic acid maleate-co-sebacic acid)

Stearic acid terminated poly(sebacic acid)

MODIFIED POLYANHYDRIDES AND BLENDS

The physical and mechanical properties of polyanhydrides can be altered by modification of the polymer structure with a minor change in the polymer composition. Several such modifications include the formation of polymer blends, branched and crosslinked polymers, partial hydrogenation and reaction with epoxides.

Biodegradable polymer blends of polyanhydrides and polyesters have been investigated as drug carriers (Abuganima 1996, Domb 1993c). In general, polyanhydrides of different structures form uniform blends with a single melting temperature. Low molecular weight poly(lactic acid) (PLA), poly(hydroxybutyrate) (PHB), and poly(caprolactone) (PCL) are miscible with polyanhydrides while high molecular weight polyesters (MW>10,000) are not compatible with polyanhydrides. Uniform blends of PCL with 10 to 90% by weight of poly(dodecanedioic anhydride), PDD were prepared by melt mixing and exhibited good mechanical strength. DSC thermograms showed two separate peaks (at $55°C$ and $75–90°C$ for PCL and PDD, respectively) for all compositions. IR and weight loss measurements of the blends during hydrolysis indicated a rapid degradation of the anhydride component. After 20 days, the blends contain only PCL with some diacid degradation products and no anhydride polymer. This study indicate that the anhydride component degraded and released from the blend composition without affecting the PCL degradation.

Branched and crosslinked polyanhydrides were synthesized in the reaction of diacid monomers with tri- or polycarboxylic acid branching monomers (Maniar *et al.*, 1990). Sebacic acid was polymerized with 1,3,5 benzenetricarboxylic acid (BTC) and poly(acrylic acid) (PAA) to yield random and graft-type branched polyanhydrides. The molecular weights of the branched polymers were significantly higher (mol. wt. 250,000) than the molecular weight of the respective linear polymer (mol. wt. 80,000). The specific viscosities of the branched polymers were lower than linear polyanhydrides with similar molecular weights. Except for the difference in molecular weights, there were no noticeable changes in the physico-chemical or thermal properties of the branched polymers and the linear polymer. Release of drug was faster from the branched polymers as compared to the respective linear polymer of a comparable molecular weight.

CHARACTERIZATION

The characterization of polyanhydrides and data obtained about their chemical composition and structure, crystallinity and thermal properties, mechanical properties, and thermodynamic and hydrolytic stability is summarized in this section.

Composition by ^{1}H NMR

The following copolymer characteristics have been studied by ^{1}H NMR (Ron *et al.*, 1991), the degree of randomness that suggests whether the polyanhydride is either a random or block copolymer, the average length of sequence (Ln), and the frequency of occurrence of specific comonomer sequences. Copolymers of 1,3-bis(carboxyphenoxypropane) (CPP) and sebacic acid (SA) were used as

Table 3 Comonomer sequence distribution of the poly(CPP-SA) series

mole ratio 0f SA-CPP in the polymer, p(SA)	probability of finding the diad SA-SA, p(SA-SA)	probability of finding the diad SA-CPP, p(SA-CPP)	average block length L(SA)	degree of randomness
0.96	0.86	0.14	12.3	0.3
0.87	0.76	0.22	7.8	0.4
0.82	0.67	0.30	5.5	0.6
0.63	0.45	0.36	3.5	0.7
0.59	0.36	0.47	2.5	0.9
0.49	0.24	0.49	2.0	1.0

Data taken from (Ron *et al.*, 1991).

model polymer. The protons on the aromatic ring close to the anhydride groups experience a lower density of shielding electrons and absorb at lower frequency. On the other hand, the protons next to aliphatic comonomers, absorb at higher frequency. Accordingly, the CPP-CPP and CPP-SA dyads were represented by peaks at 8.1 and 8.0 respectively, and the triplets at 2.6 and 2.4 represent the SA-CPP and SA-SA diads, respectively. By integration of the 'H NMR spectra of poly(CPP-SA) of various compositions the degree of randomness, average block length, and the probability of finding the diacid SA-SA or SA-CPP were calculated (Table 3). Similar data analysis was applied also to aromatic-aliphatic homopolymers (Domb *et al.*, 1989) and copolymers of sebacic acid with fumaric acid (Domb *et al.*, 1991), 1,3-bis(p-carboxyphenoxy)hexane (CPH) (Uhrich *et al.*, 1995), and trimellitic-imide derivatives (Staubli *et al.*, 1991b).

Molecular Weight

The molecular weight of polyanhydrides were determined by viscosity measurements and gel permeation chromatography (GPC) (Ron *et al.*, 1991). The weight average molecular weight (Mw) of polyanhydrides ranges from 5,000 to 300,000 with a polydispersity of 2 to 15 which increases with the increase in Mw. The intrinsic viscosity [ζ] increases with the increase in Mw. The Mark-Houwink relationship for poly(CPP-SA) was calculated from the viscosity data and the Mw values as determined by universal calibration of the GPC data using polystyrene standards.

$$[n]_{CHCl_3}^{23°C} = 3.88 \times 10^{-7}Mw^{0.658}$$

The acetic acid end group determination for molecular weight estimation was not used because the polymer may contain cyclic macromolecules with no acetate end groups (Domb and Langer 1989b).

Crystallinity

Since crystallinity is an important factor in controlling polymer erosion, analysis of the effect of polymer composition on crystallinity was studied (Ron *et al.*, 1991; Uhrich *et al.*, 1995; Staubli *et al.*, 1991b). Polymers based on sebacic acid (SA), CPP, CPH and fumaric acid (FA) were investigated. The crystallinity was determined by

Table 4 Heat of fusion and crystallinity of poly(CPP-SA).

Polymer	Tm °C	Tg °C	Heat of fusion cal/g	Crystallinity Xc %	Crystallinity Wc %
Poly(SA), 100%	86.0	60.1	36.6	66.0	
poly(CPP-SA)4:96	76.0	41.7	24.9	46.5	58.7
poly(CPP-SA)13:87	75.0	47.0	20.7	39.5	40.5
poly(CPP-SA)22:78	66.0	47.0	15.3	30.0	35.0
poly(CPP-SA)31:69	66.0	40.0	5.1	10.6	14.5
poly(CPP-SA)41:59	178.0	4.2	2.0	4.0	16.2
poly(CPP-SA)46:54	185.0	1.8	3.1	6.1	14.2
poly(CPP-SA)60:40	200.0	0.2	6.0	13.9	15.0
poly(CPP-SA)80:20	205.0	15.0	8.2	17.6	19.5
poly(CPP),100%	240.0	96.0	26.5		61.4

Tm, Tg and heat of fusion were determined by DSC. The crystallinity was determined from the X-ray diffraction and the heat of fusion. Data taken from Ron *et al.* (1991).

X-ray diffraction, a combination of X-ray and DSC, and data generated from ^{1}H NMR spectroscopy and Flory's equilibrium theory. Homopolyanhydrides of aromatic and aliphatic diacids were crystalline (>50% crystallinity). The copolymers possess high degree of crystallinity at high mole ratios of either aliphatic or aromatic diacids. The heat of fusion values for the polymers demonstrated a sharp decrease as CPP is added to SA or vice versa (Table 4). The trend of decreasing crystallinity, as one monomer is added, appeared using the X-ray or DSC methods. The decrease in crystallinity is a direct result of the random presence of other units in the polymer chain. A detailed analysis of the copolymers of sebacic acid with the aromatic and unsaturated monomers, CPP, CPH, FA, and trimellitic-amino acid derivative was reported (Staubli *et al.*, 1991b). Copolymers with high ratios of SA and CPP, TMA-gly, or CPH were crystalline while copolymers of equal ratios of SA and CPP or CPH were amorphous. The poly(FA-SA) series displayed high crystallinity regardless of comonomer ratio.

Infra Red and Raman Analysis

Anhydrides present characteristic peaks in the IR and Raman spectra. In general, aliphatic polymers absorb at 1740 and 1810 cm^{-1} and aromatic polymers at 1720 and 1780 cm^{-1}. A typical IR spectra of aliphatic and aromatic polymers that contain aliphatic and aromatic anhydride bonds may present 3 distinct peaks, where the aliphatic peak is shown at 1810 cm^{-1}, the aromatic peak is shown at 1780 cm^{-1} and the peaks at 1720–1740 cm^{-1} in general overlap. The presence of carboxylic acid groups in the polymer can be determined from the presence of a peak at 1700 cm^{-1}. The degradation of polyanhydrides can be followed by IR from the ratio between the anhydride peak at 1810 and 1700 cm^{-1}. The significance of this analysis is that it measures the degradation of the anhydride bonds and not the dissolution of the degradation products which is dependent on the solubility of the degradation products.

Fourier-Transform Raman Spectroscopy (FTRS) was used to characterize an homologous series of aliphatic poly(anhydrides), poly(carboxyphenoxy)alkanes, and copolymers of carboxyphenoxy propane (CPP) and sebacic acid. All anhydrides show two diagnostic carbonyl bands, the aliphatic polymers has the carbonyl pairing at 1803/1739 cm^{-1}, and the aromatic polymers have the band pair at 1764 and 1712 cm^{-1} (Tudor *et al.*, 1991; Davies *et al.*, 1991). All the homo- and copolymers showed methylene bands due to deformation, stretching, rocking and twisting; the spectra for the aromatic poly(anhydrides) such as PCPP also showed diagnostic benzene para-substitution bands. It was possible to differentiate between aromatic and aliphatic anhydrides bonding and in conjunction with other diagnostic bands to monitor the change in individual monomer composition within a copolymer mixture.

FTRS was used to study the hydrolytic degradation of polyanhydrides (Davies *et al.*, 1991). PSA rods exposed to water for 15 days were analyzed daily by FTRS. The carbonyl anhydride band pair (1803/ 1739 cm^{-1}) diminished in intensity from day zero to 15, with the emergence of the complimentary acid carbonyl band (1640 cm^{-1}) which increased in intensity over the same period. Similarly, the increase in the intensity of the C-C deformation at 907 cm^{-1} with hydrolysis reflects the increased freedom of the methylene chain in the low molecular weight oligomers.

Surface and Bulk Analysis

The morphology of polyanhydrides was studied by Scanning Electron Microscope (SEM) to elucidate the mechanism of polymer degradation and drug release from polyanhydrides (Mathiowitz *et al.*, 1990). The surface chemical structure of aliphatic polyanhydride films has been examined using time-of-flight secondary ion mass spectroscopy (ToF-SIMS) and X-ray photoelectron spectroscopy (XPS) (Davies *et al.*, 1996). The main peak at 285 eV corresponds to the C-H. The peak at 289.5 eV arises from O-C=O . The XPS data confirmed the purity of the surface, and the experimental surface elemental ratios were in good agreement with the known stoichiometry of the examined polyanhydrides. The ToF-SIMS spectra of the polyanhydrides are shown to reflect the polymer structure. The SIMS data confirms a systematic fragmentation, in both negative- and positive-ion SIMS spectra, occurring throughout the entire series of the polyanhydrides examined. Radical cations were observed in the positive-ion spectra. The ion at m/z 71 which may arise from the fragmentation of the anhydride unit, CH2=CHCOO- was seen for all polyanhydrides. The combined use of ToF-SIMS and XPS is shown to provide a detailed insight into the interfacial chemical structure of polyanhydrides. Atomic force microscopy (AFM) was used to follow the degradation of poly(sebacic acid) and its blends with poly(lactide) (Shakesheff *et al.*, 1994, 1995). These AFM studies reveal the surface polymer morphology to a resolution comparable to vacuum based scanning electron microscopy and demonstrate the influence of a variety of factors including polymer crystallinity and the pH of the aqueous environment on the kinetics of the degradation of biodegradable polymers and their blends. Typical AFM images of the degradation of PSA/PLA blends are shown in Figure 1.

STABILITY

The stability of polyanhydrides in solid state and dry chloroform solution was studied (Domb and Langer 1989b). Aromatic polymers such as poly(CPP) and poly(CPM) maintained their original molecular weight for at least one year in solid state.

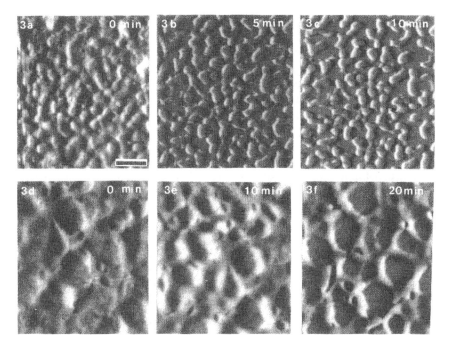

Figure 1 Atomic force microscopy (AFM) images of the degradation of poly(sebacic anhydride)/poly(lactide) blends spun cast on glass from chloroform solution (1% w/w). Parts a-c show the exposure of isolated islatd of PLA on a glass substrated as PSA is preferentially lost from a 70% PSA: 30% PLA films. Degradation of a 1:1 PSA:PLA film resulted in the exposure of network of PLA (taken from Shakesheff 1994).

In contrast, aliphatic polyanhydrides such as PSA decreased in molecular weight over time. The decrease in molecular weight shows first-order kinetics, with activation energies of 7.5 Kcal/mole-$^\circ$K. The decrease in molecular weight was explained by an internal anhydride interchange mechanism, as revealed from elemental and spectral analysis. This mechanism was supported by the fact that the decrease in molecular weight was reversible and heating of the depolymerized polymer at 180°C for 20 min. yielded the original high molecular weight polymers. However under similar conditions the hydrolyzed polymer did not increase in molecular weight. It was also found that in many cases, the stability of polymers in the solid state or in organic solutions did not correlate with its hydrolytic stability. A similar decrease in molecular weight as function of time was also observed among the aliphatic-aromatic co-polyanhydrides and imide containing co-polyanhydrides. (Domb *et al.*, 1989; Staubli *et al.*, 1991, 1991b).

The effect of γ-irradiation on polyanhydrides for sterilization purposes has been studied (Mader *et al.*, 1996). Aliphatic and aromatic homo- and copolymers were γ-irradiated at a 2.5 Mrad dose under dry ice and the properties of the polymer before and after radiation were monitored. All polymers did not change in physical or mechanical properties and their ^{1}H-NMR and IR spectra remain the same. A slight increase in molecular weight was found for aliphatic polyanhydrides.

Electron paramagnetic resonance (EPR) spectroscopy (1.2 GHz and 9.25 GHz, room temperature) was used to characterize free radicals in gamma sterilized biodegradable polymers. Free radicals were detected in all irradiated polymer samples. The temperature of irradiation (room temperature vs. dry ice) is only of minor influence of the radical yield and the shape of the EPR spectra. In contrast, polymer composition and incorporated drugs strongly influence the radiation induced free radical formation and reactivity. In general, polymers with high melting points and crystallinity give the highest yields of room temperature observable radicals. These endogenous free radicals were used to study processes of water penetration and polymer degradation *in vivo* (Mader *et al.*, 1996). The detection of gamma sterilization induced free radicals *in vivo* using EPR, could be of significance in that, the changes in the mobility of the radicals can be used as the tool to the study of drug release kinetics in an non-invasive and continuos fashion without the need of the introduction of paramagnetic species.

FABRICATION OF DELIVERY SYSTEMS

The low melting point to the solubility of the polymers in common organic solvents such as methylene chloride, allows for the easy dispersion of drug into the polymer matrix. Drugs can also be incorporated via compression or melt molding processes. For example, drugs can be incorporated into a slab either by melt mixing the drug into the melted polymer or by solvent casting. Polymer slabs loaded with a drug can also be prepared by compression molding a powder containing the drug. Similarly, one can injection mold the drug-polymer formulation into beads or rods. Polymer films can be prepared by solvent evaporation by casting the polymer solution containing the drug onto a Teflon coated dish. Microspheres based delivery systems can be formulated by the common techniques including solvent removal, hot-melt encapsulation and spray drying (Mathiowitz *et al.*, 1987; Pekarek *et al.*, 1994; Bindschaedler *et al.*, 1988; Tabata *et al.*, 1994). However, it is essential that all processes be performed under anhydrous conditions to avoid hydrolysis of the polymer.

IN VITRO DEGRADATION AND DRUG RELEASE

The degradation of polyanhydrides, in general, varies with a number of factors. These factors include, the chemical nature and the hydrophobicity of the monomers used to produce the polymer, the level of drug loading in the polymeric matrix, the pH of the surrounding medium (the higher the pH, the more rapidly the polymers degrade), the shape and geometry of the implant (the degradation is a function of the surface area) and the accessibility of the implant to water (porous materials will degrade more rapidly than non-porous). The porosity in an implant is dependent on the method of fabrication. For example, a compression-molded device will degrade at a much more rapid rate than an injection molded device due to the presence of a higher porosity in the polymer as compared to the latter.

The degradation rates for a number of polyanhydrides are available in the literature (Leong *et al.*, 1986; Shieh *et al.*, 1994; Domb *et al.*, 1995b; Tamada and Langer, 1993; Dang *et al.*, 1996). Most studies focused on the degradation of the clinically tested plyanhydrides namely, poly(CPP-SA) and poly(FAD-SA). In general, during the initial 10 to 24 hours of incubation in aqueous medium, the molecular weight dropped rapidly with no water mass loss. This period was followed by a fast decrease in wafer mass accompained by a very small change in polymer molecular weight. The period of extensive mass loss starts when the polymer molecular weight reaches a number average molecular weight (Mn) of about 2,000 regardless of the initial molecular weight of the polymer. During this period which lasts for about one week, sebacic acid, the relatively water soluble comonomer, is released from the wafer leaving the less soluble comonomer, CPP or FAD, which is slow to solubolize (Dang et al., 1996) increasing the content of sebacic acid in the copolymer increases the hydrophilicity of the copolymer, which results in a higher erosion rate and hence higher drug release rates. This could be explained by the fact that the anhydride linkages in the polymer are hydrolyzed subsequent to penetration of water into the polymer. The penetration of water or water uptake depends on the hydrophobicity of the polymer and therefore, the hydrophobic polymers which prevent water uptake, have slower erosion rates and lower drug release rates. This is valuable information since one can alter the hydrophobicity of the polymer by altering the structure and/or the content of the copolymer, thereby being able to alter the drug release rate. Since in the P(CPP-SA) and P(FAD-SA) series of copolymers, a 10 fold increase in drug release rate was achieved by alteration of the ratio of the monomers, both polymers can be used to deliver drugs over a wide range of release rates.

As mentioned above, there is no correlation between the rate of drug release and polymer degradation expressed as % decrease in the molecular weight, which at first glance might appear to be contradictory (D'Emanuele et al., 1994). However, on closer examination, it appears that drug dispersed in the polymer matrix is released when the eroding polymer brings the drug with it into solution. Thus, the release rate would depend on the rate of erosion expressed as volume of the matrix dissolved per unit time, times the drug load rather than the rate of polymer degradation. The implication being that drug release should correlate with weight loss, which is a more appropriate indicator of erosion rate than the decrease in molecular weight. Another feature of surface erosion is that while the molecular weight of the polymer at the surface may decrease, the interior of the device may still retain the same molecular weight. Furthermore, the lower molecular weight fragments so formed may not diffuse out or dissolve into the release medium. Therefore, it is not the decrease in molecular weight but the subsequent weight loss to the diffusion and erosion of molecular weight fragments, which should correlate with drug release. This also explains why drug release from these polymer devices, was independent of the initial molecular weight of the copolymer (D'Emanuele et al., 1994).

As mentioned earlier, factors that affect drug release from polyanhydrides include polymer composition, fabrication method, size and geometry, particle size of incorporated drug, drug solubility and drug loading. In general, hydrophilic drugs are released much more rapidly than the hydrophobic drugs and also suffer from a significant initial burst effect, which is also a function of drug particle size. Also the correlation between drug release and polymer degradation is better for injection or melt molded devices compared to compression molded devices (Domb et al., 1994b). In a recent report, the release of indomethacin from poly(CPP-SA) and poly(FAD-SA) was studied and was found to be independent of drug loading

(Gopferich and Langer 1993). A simple model that takes into account the following kinetic steps namely, the spontaneous degradation of polymer to crystallized monomer, the creation of pores, the dissolution of monomers inside the pores and the final release of monomer via diffusion through the pore network was putforth by Gopferich & Langer to explain this unusual release behavior (Gopferich and Langer, 1993, 1995, 1995b, 1995c). Erosion was then simulated using a Monte Carlo method that describes these morphological changes during erosion (Gopferich and Langer 1995).

A noninvasive *in vivo* and *in vitro* monitoring of drug release and polymer degradation was done using EPR spectroscopy (Mader *et al.*, 1996b, 1996c). By the incorporation of a nitroxide radical probe such as 2,2,5,5-tetramethyl-3-carboxyl-pyrrollidine-1-oxyl (= PCA) at a 5 mmole/kg, the microviscosity and drug mobility as well as the pH within a device can be monitored both *in vivo* and *in vitro* using EPR methods. Low frequency EPR spectrometer was used to study the *in vivo* and *in vitro* degradation of polyanhydrides. Tablets loaded with PCA were exposed to 0.1 M phosphate buffer pH 7.4 at $37°C$ or implanted subcutaneously in the back of rats or mice. Measurements were carried out using a standard 9.4 Ghz EPR spectrometer or a 1.1 GHz spectrometer equipped with a surface coil. EPR measurements of pH are based on the effect of protonation / deprotonation of groups located in close proximitly to the radical moiety which induces changes in the hyperfine splitting constant α_N and the γ-value. The measurements of pH values in the range between 0 and 9 are possible by using probes with different pKa values.

g = 2.0054; a_N = 1.557 mT   g = 2.0057; a_N = 1.43 mT

In vitro and *in vivo* studies have demonstrated the environment within the polyanhydride tablets is acidic pH (around 4 and less) when the degradation is carried out in a 0.1 M pH 7.4 phosphate buffered solution. One approach that has been taken to counteract the acidity is the incorporation of buffering substances into during the device fabrication. It has been shown that the incorporation of buffering substances within the polymer does result in an increase of the pH inside the delivery system (device). The microviscosity and drug mobility as well as the formation of radicals within a device were monitored both *in vivo* and *in vitro* using EPR methods. Spectral spatial EPR imaging (EPRI) which introduces a spatial dimension by means of additional gradients was applied to characterize the degradation front and the microenvironment of polyanhydride disks loaded with a pH sensitive nitroxide (Mader *et al.*, 1996c). Exposure to buffer (pH 7.4) resulted in the formation of a front of a degraded polymer from outside to inside. A pH gradient was found to exist within the polymer matrix which raises with time from 4.7 to 7.4. The issue of possible chemical reaction between amine and hydroxyl group containing drug moieties with the anhydride bonds in the polymer during drug incorporation and release has also been investigated (Domb *et al.*, 1994c).

BIOCOMPATIBILITY AND ELIMINATION

The biocompatibility and safety of polyanhydrides were established following the 1986 guidelines by the Food and Drug Administration (FDA) for testing and evaluating new biomaterials. Several accepted criteria and tests to evaluate new biomedical materials were used to assess the safety of polyanhydrides (Leong *et al.*, 1986b; Braun *et al.*, 1982; Laurencin *et al.*, 1990). In this study, poly[bis(p-carboxy-phenoxy)propane anhydride (PCPP) and its copolymers with sebacic acid were tested. Neither mutagenicity nor cytotoxicity or teratogenicity was associated with the polymers or their degradation products, as evaluated by mutation assays [66,68]. The tissue response of these polyanhydrides was studied by subcutaneous implantation in rats and in the cornea of rabbits. The polymers did not provoke inflammatory responses in the tissues over a six week implantation period. Histological evaluation indicated relatively minimal tissue irritation with no evidence of local or systemic toxicity (Laurencin *et al.*, 1990). Systemic response to the polymer was evaluated by monitoring of blood chemistry and hematological values, and by comprehensive examination of organ tissues. Both methods revealed no significant response to the polymer.

Since the CPP-SA copolymer was designed to be used clinically to deliver an anticancer agent directly into the brain for the treatment of brain neoplasms, *in vivo* safety evaluations and brain biocompatibility were assessed in rats (Tamargo *et al.*, 1989), rabbits (Brem *et al.*, 1989), and monkeys (Brem *et al.*, 1988). In the rat brain study, the tissue reaction of the polymer (PCPP-SA 20:80) was compared to the reaction observed with two standard materials used in surgery, which have been extensively studied namely, Gelfoam® (absorbable gelatin sponge), and Surgicel® (oxidized cellulose absorbable hemostat commonly used in brain surgery). Histological evaluation of the tissue demonstrated a small rim of necrosis around the implant, and a mild to marked cellular inflammatory reaction limited to the area immediately adjacent to the implantation site. The pathological response associated with poly(CPP-SA) copolymer was slightly more pronounced than Surgicel® at the earlier time points, but noticeably less marked than Surgicel® at the later times. The reaction to Gelfoam® was essentially equivalent to that observed in control rats. In a similar brain biocompatibility study carried out in monkeys, no tissue abnormalities were noted either in CT and MRI's. Furthermore, no abnormalities were observed either in the blood chemistry or hematology evaluations (Brem *et al.*, 1988). Their appeared to be no adverse systemic effects due to the implants as assessed by the histological evaluation of tissue tested. Overall, no unexpected or untoward reaction to the treatment was observed. Copolymers of sebacic acid with several aliphatic comonomers such as dimer of erucic acid (FAD), fumaric acid and isophthalic acid were also tested subcutaneously and in the rat brain were found to be biocompatible as well (Rock *et al.*, 1991). The hydrolysis and elimination processes of polyanhydrides has been studied using a series of polyanhydrides derived from different linear aliphatic diacids (Domb *et al.*, 1995b). These polymers degrade into their monomer or oligomers units at about the same rate but differ in the water solubility of their degradation products. Polymers based on natural diacids of the general structure $-[OOC-(CH_2)x-CO]-$ where x is between 4 and 12, were implanted subcutaneously in rats and the elimination of the polymers from the implantation site studied. The *in vitro* hydrolysis of this polymer series was studied by monitoring the weight loss, release of monomer degradation products and the changes in the content of anhydride bonds in the polymer as a function of time. It was observed that, both *in vitro* and *in vivo* the rate of polymer elimination was a function of monomer solubility. The elimination time for polymers based on soluble monomers (x = 4–8) was 7–14 days, while the polymers based on monomers with lower solubility (x = 10–12) were eliminated only after 8 weeks. All polymers were found to be biocompatible and useful as carriers for drug delivery.

The elimination of the biodegradable polymer poly(CPP-SA) based implant (Gliadel™), which is currently in clinical use for the treatment of brain cancer, was studied in rabbit and rat brains using radioactive polymer and drug (Domb *et al.*, 1994d, 1995c). The implant is composed of N,N-bis(2-chloroethyl)-N-nitrosourea (BCNU) dispersed in a copolyanhydride matrix of CPP and sebacic acid (SA). Four groups of rabbits were implanted with wafers loaded with BCNU, one in a ^{14}C-SA-labeled polymer, another in a ^{14}C-CPP- labeled polymer, and two groups with ^{14}C-BCNU in a nonlabeled polymer, one for BCNU disposition study and one for residual drug study. In the rabbits implanted with the ^{14}C-SA-labeled polymer, approximately 10% of the radioactivity was found in the urine and 2% in the feces, and about 10% remained in the device seven days after implantation. In contrast, only 4% of the radioactivity associated with the ^{14}C-CPP labeled polymer was found in urine and feces during this period. However, a drastic increase in the CPP excretion was found after 9 days, and at 21 days during which 64% of the implanted ^{14}C-CPP was recovered in the urine and feces, and 29% was still in the recovered wafers. Studies with radiolabeled BCNU in rabbit brain revealed that approximately 50% of the BCNU in the wafers was released in 3 days, and over 95% was released after 6 days in the rabbit brain. Excretion of this polymer after implantation in the rat brain using radiolabeled polymers showed that over 70% of the sebacic acid comonomer was excreted in seven days with about 40% of the sebacic acid metabolized to CO_2 (Domb *et al.*, 1995c). The elimination of poly(FAD-SA) rods loaded with 0, 10 and 20 weight% of gentamicin sulfate after implantation in the femoral muscle and bone of dogs was studied as part of the preclinical studies for Septacin™-bone implant (Domb and Amselem, 1994). Most of the polymer implant was gradually eliminated from bone and muscle within 4 to 8 weeks post implantation with the elimination from bone being faster leading to new bone formation in the implant site without any polymer entrapment. The elimination rate was dependent mainly on the amount of polymer implanted. Gentamicin was released for a period of about 3 weeks with no residual drug detected in the polymer remnants 8 weeks post implantation. In all experiments, no local or systemic toxicity was observed.

Mader (Mader *et al.*, 1996b) utilized the non-invasive technique Magnetic Resonance Imaging (MRI), to visualize both the polymer erosion (*in vivo* and *in vitro*) and the physiological response (edema, encapsulation) to the implant. MRI enables one to monitor *in vivo*, in non-invasive manner the water content, implant shape, and the response of the biological system to an implant in real time without stopping the experiment. MRI images taken during the course of degradation of slabs of PSA- a fast degrading polyanhydride, placed in physiologic buffer solution and subcutaneously implanted in rats are shown in Figure 2. The bright contrast indicates high water content in and around the device. The PSA slab absorbed water within 3 days *in vitro* and completely eroded in 9 days *in vivo*. In contrast, in the case of the control polymer PLGA no changes were observed

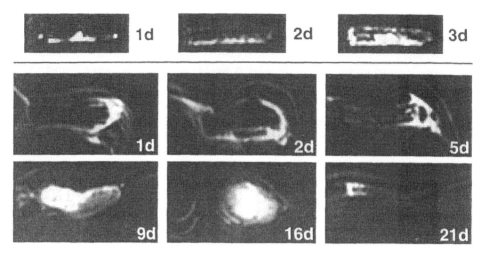

Figure 2 T2 magnetic resonance images taken during the course of degradation of 2 mm thick poly(sebacic anhydride) (PSA) implants placed in physiologic buffer solution (upper row) or subcutaneously implanted in rats (lower row). The bright contrast indicates high water content in and around the device (taken from Mader 1996c).

during the first week. However, a water penetration front is clearly visible at day 21 *in vitro* and at day 32 the bright image shows that the entire polymer matrix is filled with water. This brightness was never noticed *in vivo*. Instead, a deformation of the implant is observed starting with the rounding of the corners (day 9) which progressively increases (days 16 and 20) and completed at day 28. MRI makes it possible to monitor water content, implant shape, and the response of the biological system in real time without the need to stop the experiment at certain time points.

APPLICATIONS

Applications of Polyanhydrides have been recently reviewed by Domb (Domb *et al.*, 1992) and Laurencin (Laurencin, 1995). Several drug delivery applications have been realized using polyanhydrides. For example, local anesthetics were successfully delivered from polyanhydride cylinders in close proximity to the sciatic nerve to produce a neural block for several days (Masters *et al.*, 1993a, 1993b). Anticancer agents were incorporated in polyanhydride wafers are currently approved by the FDA for the site-specific chemotherapy for the treatment of brain tumors (Domb and Ringel, 1994e). Phase III human clinical trials have demonstrated that site specific delivery of BCNU (carmustin) from a poly(CPP-SA)20:80 wafer (Gliadel[R]) in patients with recurring brain cancer (glioblastoma multiforme) significantly prolongs patient survival (Brem *et al.*, 1995). Gladel[R] has recently won approval from the FDA adjuct therapy for the treatment of brain tumors. In the past 5 years investigations has expanded to newer polymers and other drugs such as 4-hydroperoxy cyclophosphamide (4HC), cisplatin, carboplatin, Taxol and several alkaloid drugs in an effort to develop a better system for treating brain tumors (Brem *et al.*, 1994; Olivi *et al.*, 1996; Judy *et al.*, 1995). Carboplatin incorporated in poly(FAD-SA), prepared by mixing the drug in the melted polymer has been evaluated for the treatment of brain tumors in laboratory animals with promising results (Olivi *et al.*, 1996). Poly(FAD-SA) has also been used to develop a delivery system for gentamicin sulfate for the treatment of osteomyelitis (Domb and Amselem, 1994; Laurencin *et al.*, 1993). A sustained release of gentamicin sulfate over a period of few weeks was obtained both *in vivo* and *in vitro* using this system. The delivery device which is in the form of a chain of beads is currently undergoing Phase I human clinical trials in the USA. The effect of long term glutamic acid stimulation of trigeminal motoneurons, using poly(FAD-SA) microspheres has also been explored. This study was undertaken to determine the role of glutamate in possible growth disorders of the craniofacial skeleton. Pronounced skeletal changes in the snout region were observed in rats that received glutamate showing that sustained release of glutamic acid *in vivo* can effect the development of skeletal tissue in growing rats (Hamilton-Byrd *et al.*, 1992).

ACKNOWLEDGMENTS

This work was supported in part by NCDDG grant CA52857–06 and the German-Israel Foundation (GIF).

REFERENCES

Abuganima, E. (1996) Synthesis and characterization of copolymers and blends of polyanhydrides and polyesters, M.Sc. Thesis, The Hebrew Univ. of Jerusalem.

Albertsson, A. and Lundmark, S. (1990) Synthesis of poly(adipic) anhydride by use of ketene. *J. Macromol. Sci. Chem.*, **A22**, 23.

Bindschaedler, C., Leong, K., Mathiowitz, E. and Langer, R. (1988) Poly(anhydride) microspheres formulation by solvent extraction. *J. Pharm. Sci.*, **77**, 699.

Braun, A.G., Buckner, C.A., Emerson, D.J. and Nichinson, B.B. (1982) Quantitative correspondence between the *in vivo* and *in vitro* activity of teratogenic agents. *Proc. Natl. Acad. Sci. USA*, **79**, 2056.

Brem, H., Tamargo, R.J., Pinn, M. and Chasin, M. (1988) Biocompatibility of a BCNU-loaded biodegradable polymer: a toxicity study in primates. *Amer. Assoc. Neuml. Surg.*, **24**, 381.

Brem, H., Kadei, A., Epstein, J.I., Tamargo, R.J., Domb, A.J., Langer, R. and Leong, K.W. (1989) Biocompatibility of bioerodible controlled release polymer in the rabbit brain, *Sel. Cancer Ther.*, **5**, 55.

Brem, H., Piantadosi, S., Burger, P.C., Walker, M., Selker, R. Vick, N.A., Black, K., Sisti, M., Brem, S., Mohr, G. *et al.* (1995) Placebo-controlled trial of safety and efficacy of intraoperative controlled delivery by biodegradable polymers of chemotherapy for recurrent gliomas. The Polymer-brain Tumor Treatment Group, *Lancet*, **345: 8956**, 1008–12.

Brem, H., Walter, K. A., Tamargo, R.J., Olivi, A. and Langer, R. (1994) Drug delivery to the brain, In: *Polymeric site specific pharmacotherapy*, A. Domb (Ad.) Wiley, Chichester, pp. 117–140.

Bucher, J.E. and Slade, W.C. (1909) The anhydrides of isophthalic and terephthalic acids, *J. Am. Chem. Soc.*, **31**, 1319.

Conix, A. (1958) Aromatic poly(anhydrides): a new class of high melting fiber-forming polymers, *J. Polym. Sci.*, **29**, 343.

Dang, W., Daviau, T. and Nowotnik, D. (1996) *In vitro* erosion kinetics of implantable polyanhydride Gliadel™. *Proceed. Intern. Symp. Control. Rel. Bioact. Mater.*, **23**, 731–2.

D'Emanuele, A., Hill, J., Tamada, J.A., Domb, A.J. and Langer, R. (1994) Molecular weight changes in polymer erosion. *Pharmaceut. Res.*, **9**, 1279–1283.

Davies, M.C., Khan, M.A., Domb, A.J., Langer, R., Watts, J.E. and Paul, A. (1991) The analysis of the surface chemical structure of biomedical aliphatic poly(anhydrides) using XPS and ToF-SIMS. *J. Appl. Polym. Sci.*, **42**, 1597.

Davies, M.C., Shakesheff, K.M., Shard, K.M., Domb, A.J., Roberts, C.J. (1996) Tendler, S.J.B. and Williams, P.M., Surface analysis of biodegradable polymer blends of poly(sebacic anhydride) and poly(DL-lactic acid). *Macromolecules*, **29**, 2205–2212.

Domb, A.J. and Langer, R. (1987) Poly(anhydrides). I. Preparation of high molecular weight polyanhydrides. *J. Polym. Sci. Polym. Chem.*, **25**, 3373.

Domb, A.J., Ron, E. and Langer, R. (1988) Poly(anhydrides). II. One step polymerization using phosgene or diphosgene as coupling agents. *Macromolecules*, **21**, 1925.

Domb, A.J., Gallardo, C.E. and Langer, R. (1989a) Poly(anhydrides). 3. Poly(anhydrides) based on aliphatic-aromatic diacids. *Macromolecules*, **22**, 3200.

Domb, A.J. and Langer, R. (1989b) Solid-state and solution stability of poly(anhydrides) and poly(esters). *Macromolecules*, **22**, 2117.

Domb, A.J. (1990) Biodegradable polymers derived from amino acids. *Biomaterials*, **11**, 680.

Domb, A.J., Mathiowitz, E., Ron E., Giannos, S. and Langer, R. (1991) Polyanhydrides IV. Unsaturated and cross-linked poly(anhydrides). *J. Polym. Sci. Part A: Polymer Chemistry*, **29**, 571.

Domb, A.J., Amselem, S., Shah, J. and Maniar, M. (1992a) Poly(anhydrides): synthesis and characterization, in: *Advances in Polymer Sciences*, Peppas N.A. and Langer R. (Eds.), Springer-Verlag, pp.93

Domb, A.J. (1992b) Synthesis and characterization of bioerodible aromatic anhydride copolymers. *Macromolecules*, **25**, 12.

Domb, A.J. and Maniar, M. (1993a) Absorbable biopolymers derived from dimer fatty acids, *J. Poly. Sci.: Polymer Chem.*, **31**, 1275.

Domb, A.J. and Maniar, M. (1993b) Fatty acid terminated polyanhydrides. US Patent 5,179,189.

Domb, A.J. (1993c)Biodegradable polymer blends: Screening for miscible polymers, *J. Polym. Sci.: Polym. Chem.*, **31**, 1973–1981.

Domb, A.J. and Amselem, S. (1994a) Antibiotic Delivery Systems for the Treatment of Bone Infections. In: *Polymeric Site-Specific Pharmacotherapy*, Domb A.J. (Ed.) pp. 242–265.

Domb, A. J., Amselem, S., Langer, R. and Maniar, M. (1994b) Polyanhydrides as carriers of drugs, In: *Designed to Degrade Biomedical Polymers*, S. Shalaby, Ed., Carl Hauser Verlag, pp. 69–96.

Domb, A.J., Turovsky, L. and Nudelman, R. (1994c) Chemical interactions between drugs containing reactive amines with hydrolyzable biopolymers in aqueous solutions. *Pharm. Research*, **11**, 865–868.

Domb, A.J., Rock, M., Perkin, C., Proxap, B., Villemure, J.G. (1994d) Metabolic disposition and elimination studies of a radiolabelled biodegradable polymeric implant in the rat brain. *Biomaterials*, **15**, 681–688.

Domb, A.J. and Ringel, I. (1994e) Polymeric Drug Carrier Systems in the Brain, In: *Providing Pharmaceutical Access to the Brain, Methods in Neuroscience*, Flanaga, T.R., Emerich, D.F., Winn, S.R. (Eds.) CRC Press, **21**, pp. 169–183.

Domb, A.J. and Nudelman, R. (1995a) Biodegradable polymers derived from natural fatty acids. *J. Polym. Sci.*, **33**, 717–725.

Domb, A.J. and Nudelman, R. (1995b) *In vivo* and *in vitro* elimination of aliphatic polyanhydrides. *Biomaterials*, **16**, 319–323.

Domb, A.J., Perkin, C., Proxap, B. and Villemure J.G. (1995c) Excresion of a radiolabelled biodegradable polymeric implant in the rabbit brain. *Biomaterials*, **16**, 319–323.

Gopferich, A. and Langer, R. (1993) The influence of microstructure and monomer properties on the erosion mechanism of a class of polyanhydrides. *J. Polym. Sci.*, **31**, 1445–2458.

Gopferich, A. and Langer, R. (1995a) Modeling monomer release from bioerodible polymers. *J. Controll. Rel.*, **33**, 55–69.

Gopferich, A. and Langer, R. (1995b) Modeling of polymer erosion in three dimentions: Rotationally symmetric devices. *AIChE J.*, **41**, 2292–2299.

Gopferich, A., Karydas, D. and Langer, R. (1995c) Predicting drug release from cylindrical polyanhydride matrix discs. *Eur. J. Pharm. Biopharm.*, **41**, 81–87.

Gref, R., Minamitake, Y., Peracchia, M.T., Domb, A.J., Trubetskoy, V., Torchilin, V. and Langer, R. (1995) Poly(ethylene glycol) coated nanospheres, *Advanced Drug Delivery Reviews*, **16**, 215–233.

Gupta, B. (1988) US patent 4,868,265 .

Hamilton-Byrd, E.L., Sokoloff, A.J., Domb, A.J., Terr, L. and Byrd, K.E. (1992) L-Glutamate microsphere stimulation of the trigeminal motor nucleus in growing rats. *Polym. Adv. Techn.*, **3**, 337–344.

Hartmann, M. and Schultz, V. (1989) Synthesis of poly(anhydride) containing amido groups, *Macromol. Chem.*, **190**, 2133.

Hill, J.W. (1930) Studies on polymerization and ring formation. VI. Adipic anhydride. *J. Am. Chem. Sec.*, **52**, 4110.

Hill, J.W. and Carothers, H.W. (1932) Studies of polymerization and ring formation. XIV. A linear super poly(anhydride) and a cyclic dimeric anhydride from sebacic acid. *J. Am. Chem. Sec.*, **54**, 5169.

Judy, K.D., Olivi, A., Buahin, K.G., Domb, A.J., Epstein, J.I., Colvin, O.M. and Brem, H. (1995) Effectiveness of controlled release of a cyclophosphamide derivative with polymers against rat gliomas, *J. Neurosurg.*, **82**, 481–486.

Knobloch, J.O. and Ramirez, F. (1975) *J. Org. Chem.*, **40**, 1101.

Laurencin, C., Gerhart, T., Witschger, P., Satcher, R., Domb, A.J., Hanff, P., Edsberg, L., Hayes, W. and Langer, R. (1993) Biodegradable polyanhydrides for antibiotic drug delivery, *J. Orthop. Res.*, **11**, 256–262.

Leong K.W., Kost J., Mathiowitz E. and Langer R. (1986a) Poly(anhydrides) for controlled release of bioactive agents. *Biomaterials*, **7**, 364, .

Leong, K., D'Amore, P. and Langer, R. (1986b) Bioerodible poly(anhydrides) as drug carrier matrices. II. Biocompatibility and chemical reactivity. *J. Biomed. Mater, Res.*, **20**, 51.

Leong, K.W., Simonte, V. and Langer, R. (1987) Synthesis of poly(anhydrides): Melt-polycondensation, dehydrochlorination, and dehydrative coupling. *Macromolecules*, **20**, 705.

Leong, K.W., Domb, A., and Langer, R. (1989) Poly(anhydrides), in *Encyclopedia of Polymer Science and Engineering*, 2nd ed., Wiley and Sons.

Laurencin, C.T., Domb, A.J., Morris, C., Brown, V., Chasin, M., Mc-connell, R., Lange, N. and Langer, R. (1990) Poly(anhydrides) administration in high doses *in vivo*: studies of biocompatibility and toxicology. *J. Biomed. Mater. Res.*, **24**, 1463.

Laurencin, C.T., Ibim, S.E.M. and Langer, R. (1995) Poly(anhydrides). In *Biomedical applications of synthetic biodegradable polymers*, Hollinger J.O. (Ed.) CRC Press, Boca Raton, pp. 59–102.

Mader, K., Domb, A.J., Swartz, H.M. (1996) Gamma sterilization induced radicals in biodegradable drug delivery systems, *J. Appl. Rad. Isotop.*

Mader, K., Cremmilleleux, Y., Domb, A.J., Dunn, J.F. and Swartz, H.M. (1996b) *in vitro/ in vivo* comparison of drug release and polymer erosion from biodegradable P(FAD-SA) polyanhydrides-a noninvasive approach by the combined use of Electron Paramagnetic Resonance Spectroscopy and Nuclear Magnetic Resonance Imaging.

Mader, K., Bacic, G., Domb, A.J., Elmalak, O., Langer, R. and Swartz, H.M. (1996c) Noninvasive *in vivo* monitoring of drug release and polymer erosion from biodegradable polymers by EPR spectroscopy and NMR imaging. *J. Pharm. Sci.*, in press.

Mader, K., Nitschke, S., Stosser, R., Hans-Hubert, and Domb, A.J. (1996d) Nondestractive and localized assessment of acidic microenvironments inside biodegradable polyanhydrides by spectral spatial Electron Paramagnetic Resonance Imaging. *Polymer*, in press.

Maniar, M., Xie, X. and Domb, A.J. (1990) Poly(anhydrides). V. Branched poly(anhydrides). *Biomarerials*, **11**, 690.

Masters, D.B., C.B., Dutta, S., Turek, T. and Langer, R. (1993a) Sustained local anesthetic release from bioerodible polymer matrices: a potential method for prolonged regional anesthesia. Pharm. Res. 10(10): 1527–1532.

Masters, D.B., Berde, C.B., Dutta, S.K., Griggs, C.T., Hu, D., Kupsky, W. and Langer, R. (1993b) Prolonged regional nerve blockade by controlled release of local anesthetic from a biodegradable polymer matrix. Anesthesiol. 79(2): 340–346.

Mathiowitz, E. and Langer, R. (1987) Poly(anhydride) microspheres as drug carriers. I. Hot-melt microencapsulation. *J. Control Rel.*, **5**, 13.

Mathiowitz, E., Kline, D. and Langer, R. (1990) Morphology of poly(anhydride) microspheres delivery systems. *J. Scanning Microscopy*, **4**, 329.

McIntyre, J.E. (1964) British Patent No. 978,669.

Olivi, A., Awend, M.G., Utsuki, T., Tyler, B., Domb, A.J., Brat, D.J. and Brem, H. (1996) Interstitial delivery of carboplatin via biodegradable polymers is effective against experimental glioma in the rat. *Cancer, Chemotherap. Pharmacol.*

Pekarek, K.J., Jacob, J.S. and Mathiowitz, E. (1994) *Nature* **357**, 258–260.

Poly(anhydrides) (1969) in *Encyclopedia of Polymer Science and Technology*, Mark H.E. *et al.*, (Eds.) John Wiley & Sons, New York, **10**, 630.

Rock, M., Green, M., Fait, C., Gell, R., Myer, J., Maniar, M. and Domb, A.J. (1991) Evaluation and comparison of biocompatibility of various classes of polyanhydrides. *Palym. Preprints.*, **32**, 221.

Ron, E., Mathiowitz, E., Mathiowitz, G., Domb, A.J. and Langer, R. (1991) NMR characterization of erodible copolymers, *Macromolecules*, **24**, 2278.

Rosen, H.B., Chang, J., Wnek, G.E., Lindhardt, R.J. and Langer, R. (1983) Biodegradable poly(anhydrides) for controlled drug delivery. *Biomaterials*, **4**, 131.

Shakesheff, K.M., Davies, M.C., Roberts, C.J., Tendler, S.J.B., Shard, K.M. and Domb, A.J. (1994) In situ AFM imaging of polymer degradation in an aqueous environment, *Langmuir*, **10**, 4417–4419.

Shakesheff, K.M., Chen, X., Davies, M.C., Domb, A.J., Roberts, M.C., Tendler, S.J.B. and Williams, P.M. (1995) Relating the phase morphology of a biodegradable polymer blend to erosion kinetics using simultaneous in situ atomic force microscopy and surface plasmon resonance analysis, *Langmuir*, **11**, 3921–3927.

Shieh, L., Tamada, J., Chen, I., Pang, J., Domb, A.J., Langer, R. (1994) Erosin of a new family of biodegradable polyanhydrides. *J. Biomat. Mater. Res.*, **28**, 1465–1975.

Staubli, A., Ron, E. and Langer, R. (1990) Hydrolytically degradable amino acid containing polymers. *J. Am. Chem Sec.*, **112**, 4419.

Staubli, A., Mathiowitz, E., Lucarelli, M. and Langer, R. (1991) Characterization of hydrolytically degradable amino acid containing poly(anhydride-co-imides). *Macromolecules*, **24**, 2283.

Staubli, A., Mathiowitz, E. and Langer, R. (1991b) Sequence distribution and its effects on glass transition temperatures of poly(anhydrides-co-imides) containing asymmetric monomers, *Macromolecules*, **24**, 2291.

Subramanyam, R. and Pinkus, A.G. (1985) Synthesis of poly(terephthalic anhydride) by hydrolysis of terephthaloyl chloride triethylamine intermediate adduct: characterization of intermediate adduct. *J. Macromol. Sci. Chem.*, **A22**, 23.

Tabata, Y., Domb, A.J. and Langer, R. (1994) Polyanhydride granules provide controlled release of water-soluble drugs with reduced initial burst. *J. Pharm. Sci.*, **83**, 5–11.

Tamada, J.A. and Langer R. (1993) Erosion kinetics of hydrolytically degradable polymers. *Proc. Natl. Acad. Sci. UAS*, **90**, 552.

Tamargo, R.J., Epstein, J.I., Reinhard, C.S., Chasin, M. and Brem, H. (1989) Brain biocompatibility of a biodegradable controlled-release polymer in rats. *J. Biomed. Mater. Res.*, **23**, 253.

Tudor, A.M., Church, S., Domb, A.J., Hendra, P.J., Langer, R., Celia, C.D. and Davies, M.C. (1991) The application of the Fourier-Transform Raman spectroscopy to the analysis of poly(anhydride) homo and copolymers. *Spectrochimica Act.*, **47A**, 1335.

Uhrich, K.E., Gupta, A., Thomas, T.T., Laurencin, C.T. and Langer, R. (1995) Synthesis and characterization of degradable poly(anhydride -co-imides). *Macromolecules*, **28**, 2184–2193.

Yoda, N. (1963) Synthesis of poly(anhydrides). Crystalline and high melting poly(amide) poly(anhydrides) of methylene bis(p-carboxyphenyl) amide. *J. Polym. Sci., Part A*, **1**, 1323.

Yoda, N. (1962) Synthesis of poly(anhydrides). Poly(anhydrides) of five-membered heterocyclic dibasic acids. *Makromol. Chem.*, **55**, 174.

ABBREVIATIONS

ACDA = acetylenedicarboxylic acid
BTC = 1,3,5–benzenetricarboxylic acid
Co = drug loading in gm/cc
CPH = 1,6-bis(p-carboxyphenoxy)hexane
CPP = 1,3-bis(p-carboxyphenoxy)propane
CPV = carboxphenoxy valerate
DMF = N,N- dimethylformamide
DSC = differential scanning calorimetry
EPR = electron paramagnetic resonance
EPRI = electron paramagnetic resonance imaging
FA = fumaric acid

FAD = fatty acid dimer
Gelfoam© = absorbable gelatin sponge
Gliadel© = polyanhydride brain tumor implant containing BCNU
GPC = gel permeation chromatography
4HC = 4-hydroperoxycyclophosphamide
Ln = average length of sequence
MIT = Massachusetts Institute of Technology
Mw = weight average molecular weight
Mn = number average molecular weight
MRI = magnetic resonance imaging
NMR = nuclear magnetic resonance
PLA = poly(lactic acid)
PCL = poly(caprolactone)
PHB = poly(hydroxybutyrate)
PSA = poly(sebacic acid)
SA = sebacic acid
SEM = scenning electron microscope
Septicin™ = polyanhydride antibacterial bone implant
STDA = 4,4′stilbendicarboxylic acid
Surgicel© = oxidized cellulose absorbable hemostat
TA = terephthalic acid
Tg = glass transition temperature
TMA-gly = trimellitimide-glycine
ToF-SIMS = time-of-flight secondary ion mass spectroscopy
Vycryl© = synthetic absorbable suture
Xc = degree of crystallinity
XPS = X-ray photoelectron spectroscopy

9. BIODEGRADABLE POLYPHOSPHAZENES FOR BIOMEDICAL APPLICATIONS

J. VANDORPE, E. SCHACHT, S. DEJARDIN and Y. LEMMOUCHI

Polymer Materials Research Group, Department of Organic Chemistry,
University of Gent, Krijgslaan 281 S4bis, Gent, Belgium

GENERAL INTRODUCTION

Poly[(organo)phosphazenes] (Figure 1) are polymers with an inorganic backbone, consisting of alternating nitrogen and phosphorous atoms linked by alternating single and double bonds.

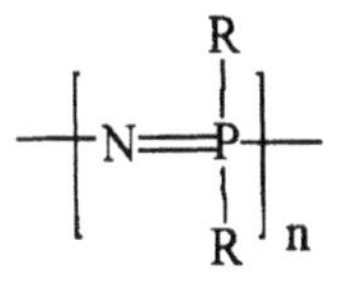

Figure 1 Structure of poly[(organo)phosphazene].

The R group can be an alkoxy, aryloxy, amino, alkyl, heterocyclic ring or an inorganic or organometallic unit. Allcock and co-workers were the first who did study extensively the synthesis of poly[(organo)phosphazene] derivatives. Most polyphosphazene derivatives are prepared starting from a precursor polymer, poly[(dichloro)phosphazene]. A variety of polymers with variable properties can be prepared by nucleophilic displacement reactions (Allcock, 1972, 1979, 1986) (Figure 2).

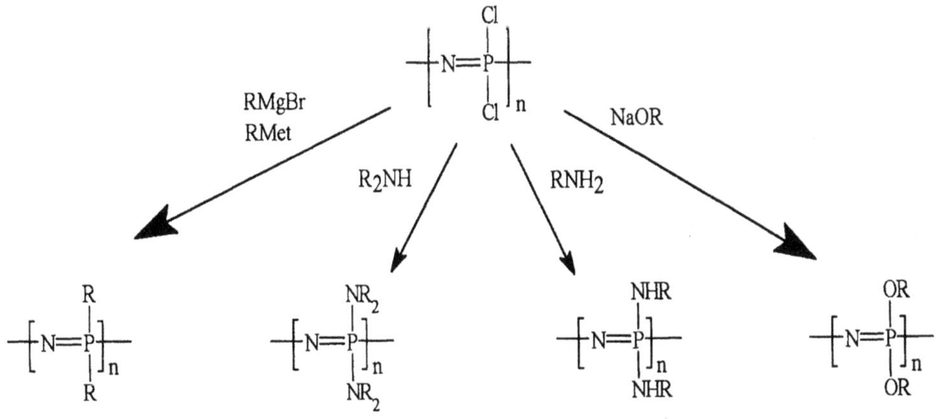

Figure 2 Overall synthesis procedure for phosphazene polymers.

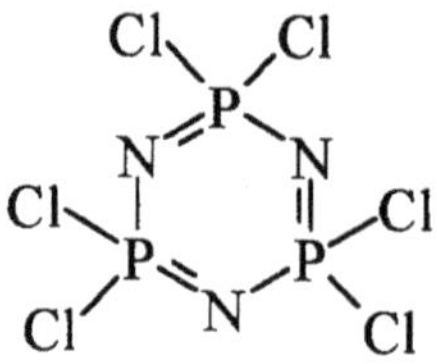

Figure 3 Structure of hexachlorocyclotriphosphazene.

METHODS OF PREPARATION ON POLY[(DICHLORO)HOSPHAZENE]

Several methods are described for the preparation of poly[(dichloro)phosphazene].

Ring-opening Polymerization

The chemistry of small molecule phosphazene can be traced back to the discovery that phosphorous pentachloride and ammonium chloride react to yield a mixture of compounds of which the main product is a volatile white solid, known to have the cyclic trimer structure, shown in Figure 3 (Ross, 1832; Liebich, 1834; Stokes, 1897).

It was later noticed that the hexachlorocyclotriphosphazene can be polymerized by heating in the melt to yield an uncrosslinked linear high polymer, poly[(dichloro)-phosphazene] (Allcock *et al.*, 1965–1966; Rose, 1968). Further heating results in the formation of an insoluble crosslinked material. This leads in both cases to a transparent, rubbery elastomer which hydrolyzes slowly, when exposed to moisture, forming phosphate, ammonia and hydrochloric acid. At temperatures above 350°C, the polymer depolymerizes to cyclic oligomers. The uncrosslinked species serve as highly reactive polymeric precursor.

This ring-opening polymerization has been studied extensively and several mechanisms have been proposed (Allcock, 1972–1980; Emsley *et al.*, 1984; Gleria *et al.*, 1984; MacCallum *et al.* 1967; Singler *et al.* 1975). The polymerization process is currently thought to occur via a cationic ring-opening mechanism, initiated by the cleavage at the polar P+-Cl- (Figure 4).

It has been mentioned that traces of impurities can function as powerful catalysts for the polymerization reaction (Allcock *et al.*, 1975; Ganapathianppan *et al.*, 1987; Gimblett, 1960; MacCallum *et al.*, 1969). During the extended polymerization of hexachlorocyclotriphosphazene there is a tendency for crosslinking but these side reactions are generally avoided by the use of low trimer conversion degrees (30–60%). A typical uncatalyzed bulk polymerization at 250°C yields very high molecular weight polymers (e.g. M_w $2-10 \times 10^6$) with a broad distribution (10).

More recently, the solution polymerization in chloroform, hydrocarbons, or CS_2 has also been investigated (Kajiwara *et al.*, 1984; Konecny *et al.*, 1960; Mujumdar *et al.*, 1989–1990; Patat *et al.*, 1951; Scopelianos *et al.*, 1987). In these cases, polymers having medium molecular weights (10^5) and narrower molecular weight distributions are generally formed.

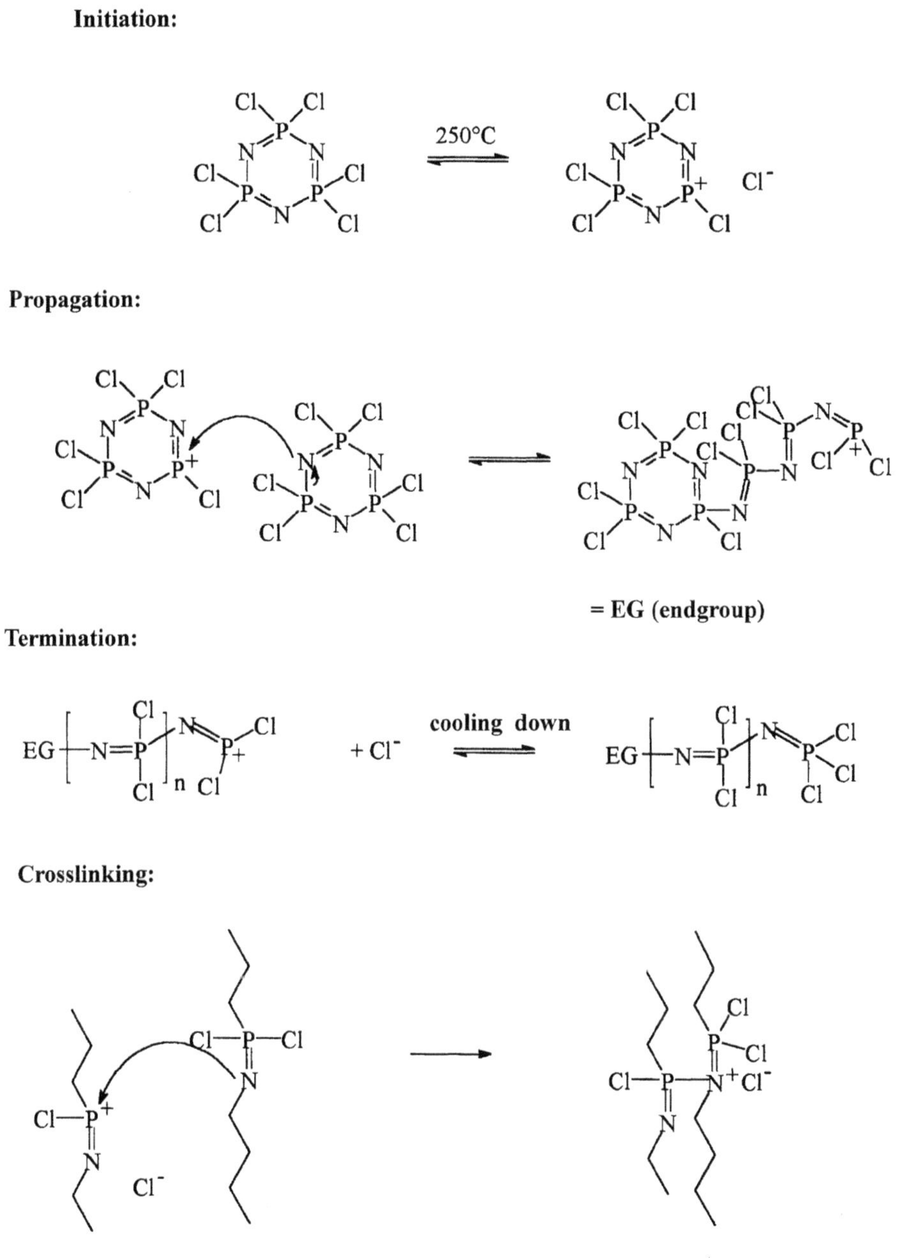
Initiation:
250°C
Propagation:
= EG (endgroup)
Termination:
cooling down
Crosslinking:

Figure 4 Polymerization mechanism of $[NPCl_2]_3$.

PCl$_5$ + (NH$_4$)$_2$SO$_4$

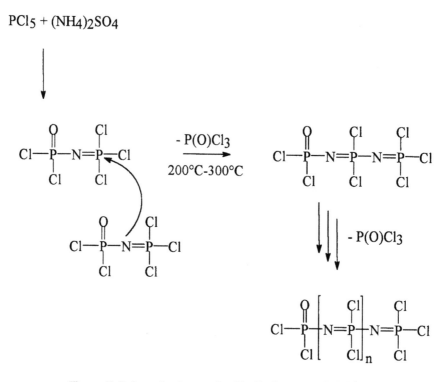

Figure 5 Polymerization method by De Jaeger and Haluin.

Polymerization Mechanism by De Jaeger and Haluin

A new polymerization method for the formation of poly[(dichloro)phosphazene] was introduced by De Jaeger and Haluin (D'Halluin *et al.*, 1989–1992; Helioui *et al.*, 1982). This polymerization involves a high conversion condensation of the monomeric phosphoranimine, N-diclorophosphoryl-P-trichloromonophosphazene Cl$_2$P(O)N = PCl$_3$ with the elimination of phosphoryl trichloride, POCl$_3$ (Figure 5). The polymers prepared according to this method have a molecular weight in the order of 5×10^5 with a very broad molecular weight distribution (15).

Polymerization Method by Hornbaker and Li

Another polymerization method was mentioned by Hornbaker and Li (1978). This method describes the direct formation of poly[(dichloro)phosphazene] starting from the basic compounds PCl$_3$, Cl$_2$ and NH$_4$Cl and without intermediate isolation of a precursor compound (Figure 6).

PCl$_5$ + NH$_4$Cl $\longrightarrow$ Cl$_3$P$-$[N$=$PCl$_2$]$_n$Cl

Figure 6 Polymerization mechanism by Hornbaker and Li

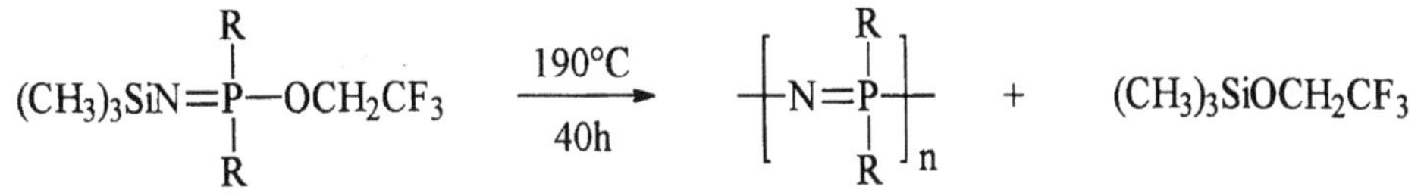

$$(CH_3)_3SiN=\overset{\overset{R}{|}}{\underset{\underset{R}{|}}{P}}-OCH_2CF_3 \quad \xrightarrow[40h]{190°C} \quad \left[N=\overset{\overset{R}{|}}{\underset{\underset{R}{|}}{P}}\right]_n \quad + \quad (CH_3)_3SiOCH_2CF_3$$

R = Me, Phe, Et,...

Figure 7 Polycondensation Reaction of N-silyl-phosphoranimines

Thermal Polymerization of Phosphoranimines

Poly[(organo)phosphazenes] can also be synthesized by the condensation reaction of N-silyl-phosphoranimines (Neilson *et al.*, 1988; Wisian-Neilson *et al.*, 1980–1986). This method yields to polyphosphazenes directly substituted with alkyl- or aryl- side groups (Figure 7).

The molecular weights of the polymers prepared by this method are varying between 10.000 and 200.000. The mechanism of this new synthetic route has been intensively studied by Matyjaszewski *et al.* (1992), who also reported the synthesis of poly[(organo)phosphazenes] with alkoxygroups by the condensation of phosphoranimines (1992, 1993, 1994).

BIODEGRADABLE POLYPHOSPHAZENES

Amino Acid Derivatives

The first biodegradable polyphosphazenes that have been synthesized were substituted by amino acid derivatives. Allcock *et al.* (1977) prepared polyphosphazenes with glycine ethylester, leucine methylester, alanine methylester and phenylalanine methylester by reaction of poly[(dichloro)phosphazene] with the amino acid ester in the presence of triethylamine. Since only glycine ethylester gave total halogen replacement, complete substitution of the chlorine was achieved by subsequent introduction of methylamino as cosubstituent. A general synthesis procedure is given in Figure 8.

These materials are susceptible to a slow hydrolytical degradation leading to "harmless" products: amino acid(ester), phosphate, ammonia, ethanol or methanol and methylamine. *In vitro* degradation studies on polyphosphazenes with different amino acid derivatives show that it takes the materials several months to degrade (Allcock *et al.*, 1994; Crommen *et al.*, 1992–1993; Ruiz *et al.*, 1993) depending on the amino acids present. The rate of degradation could be varied by preparation of copolymers containing different amino acid esters in different ratios. Goedemoed *et al.* (1988) prepared copolymers containing ethylglycinate and glutamic acid diethyl ester (Figure 9) or phenylalanine ethylester (Figure 10) to achieve a different rate of polymer degradation and subsequent release of a drug.

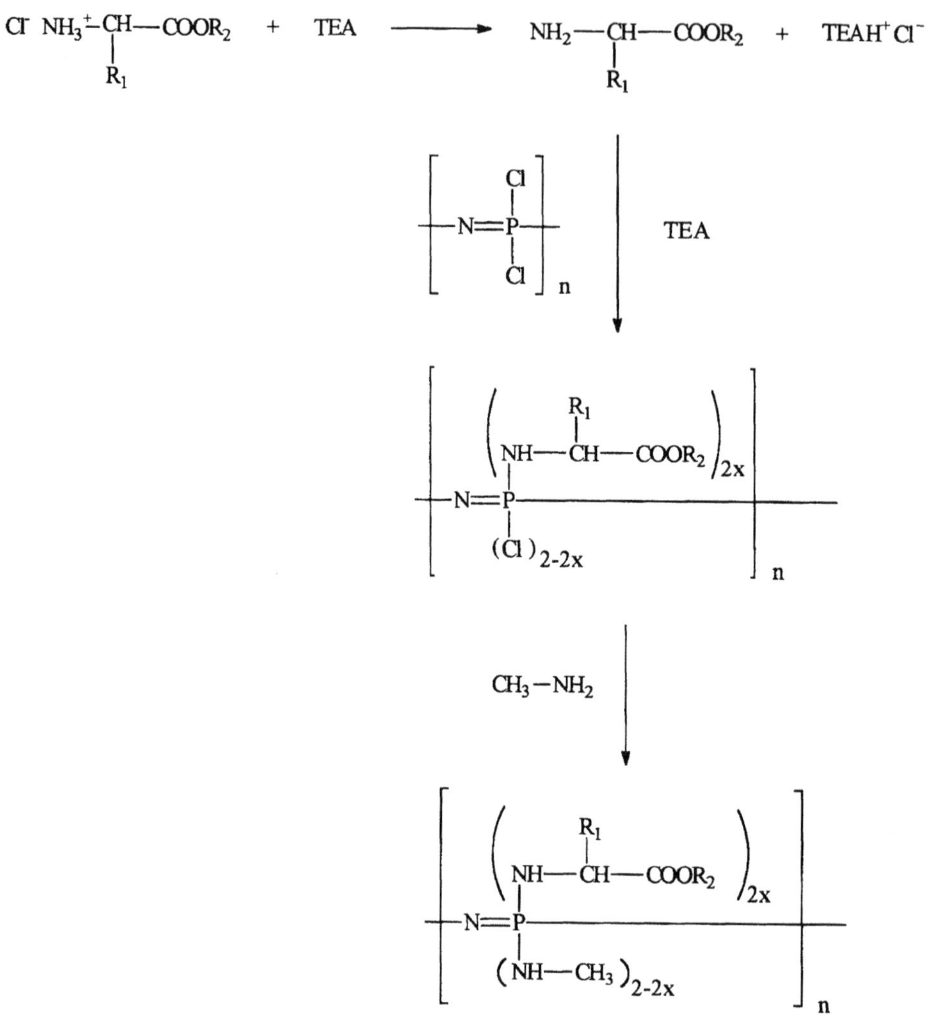

Figure 8 Synthesis method for poly[(amino acid ester)-co-(methylamine)phosphazenes].

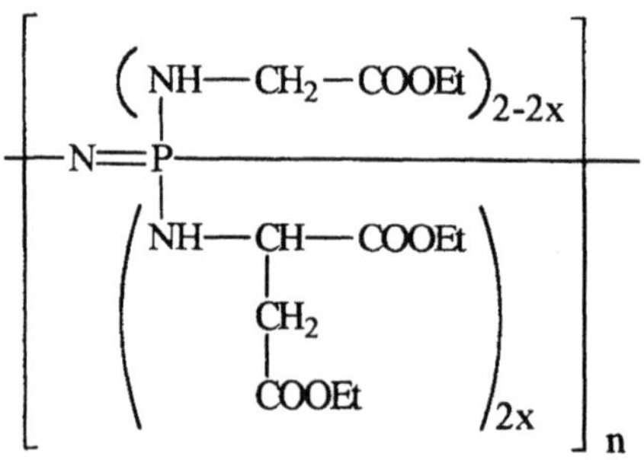

Figure 9 Poly[(glutamic acid diethylester)-co-(ethylglycinate)phosphazene].

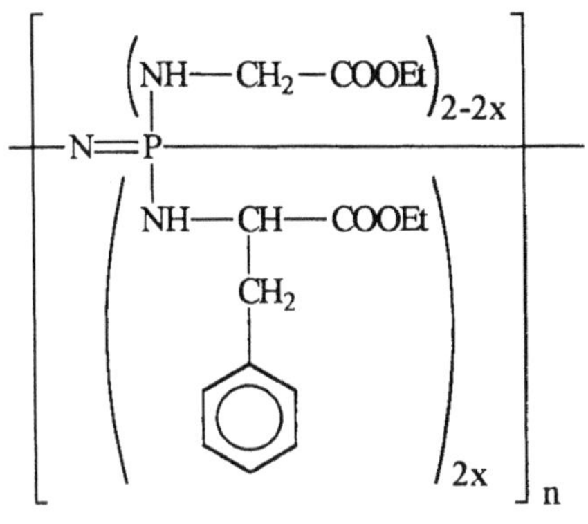

Figure 10 Poly[(phenyl alanine ethylester)-co-(ethylglycinate)phosphazene].

Allcock *et al.* (1977) synthesized copolymers containing amino acid esters and methylamino side groups (Figure 11). The *in vitro* degradation and water solubility of these polymers could be varied by using different amino acid esters and different ratios of methylamine.

The same group also prepared 3 different poly[(amino acid ester)phosphazenes]: poly[di(ethylglycinato)phosphazene], poly[di(ethylalanato)phosphazene] and poly[di(benzylalanato)phosphazene] (Allcock *et al.*, 1994). These polymers show, in the order given above, a decreasing molecular weight decline, a decreasing mass loss and a decreasing release of small molecules. The release occurred through diffusion of the small molecules through the solid and by decomposition of the polymer.

For the hydrolytic degradation of polyphosphazenes several mechanisms have been proposed (Allcock *et al.*, 1977–1994; Goedemoed *et al.*, 1988). However, analysis of the degradation products revealed that the main hydrolysis pathway involves release of the amino acid ester side group, followed by hydrolysis of the ester with formation of the amino acid and the alcohol (Crommen *et al.*, 1992) (Figure 12).

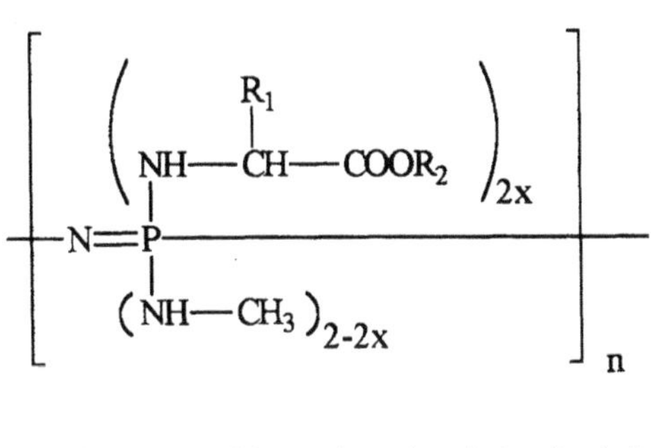

Figure 11 Poly[(amino acid ester)-co-(methylaminc)phosphazene].

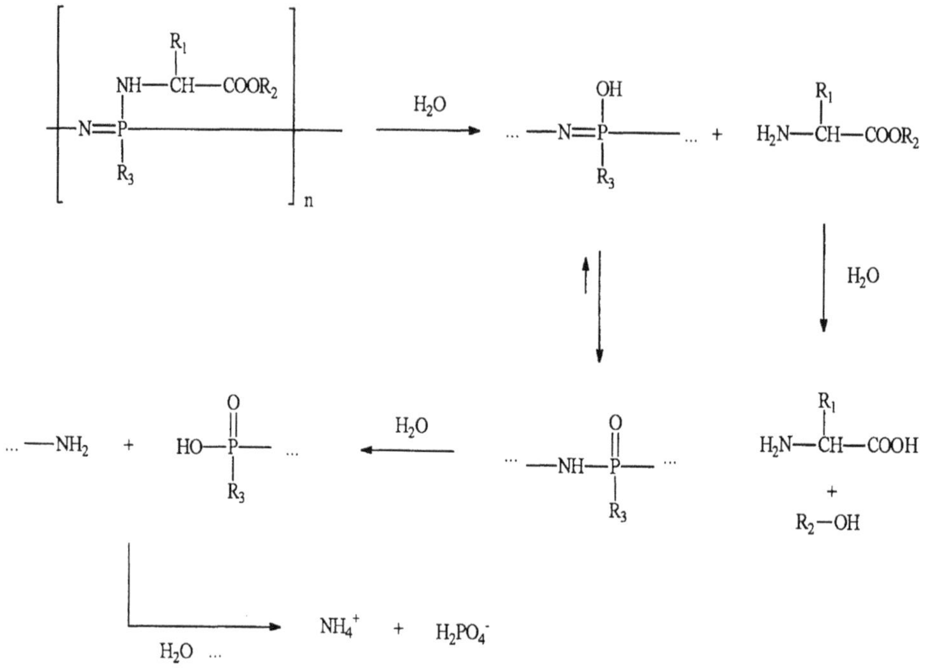

Figure 12 Degradation of poly[(amino acid ester)phosphazene].

Initial hydrolysis of the ester bond with subsequent release of the amino acid cannot be excluded but is probably predominant (Figure 13).

Another way to control the rate of polymer degradation is to introduce hydrolytical sensitive groups. Allcock *et al.* (1981–1982) showed that polyphosphazenes containing imidazol side groups were unstable in aqueous medium and that it was the imidazole side chain substituent which conferred hydrolytic instability to the polymer. Laurencin *et al.* (1987) prepared hydrophobic polyphosphazenes (Figure 14) containing p-methyl-phenoxy side groups and different amounts of imidazol to control polymer degradation and hereby control the drug release.

An increase in imidazole content, resulted in a faster initial mass loss. Release of p-nitro-aniline appeared to follow $t^{1/2}$ diffusional release kinetics as predicted by Higuchi (1963). Release of progesteron and BSA was possible over long periods of time.

Caliceti *et al.* (1994) on the other hand prepared hydrophobic polyphosphazenes containing L-alanine ethyl ester and different amounts of imidazole (Figure 15). These polymers were used as a matrix for the release of naproxen, acetyltryptophanamide and narciclasine. Again, higher amounts of imidazol led to increased rates of drug release.

It has been suggested by Allcock *et al.* (1982) that polyphosphazene degradaton is catalyzed by acids. This leads to the idea that the presence of a controlled amount of acid groups on the polymer could be a method for controlling polymer degradation.

Crommen *et al.* (1992) prepared poly[(organo)phosphazenes] bearing side groups with hydrolysis sensitive ester functions: the so called depsipeptide-esters (Figure 16). He proved (Crommen *et al.*, 1992) that the rate of biodegradation can be varied by changing the chemical composition of the polymer (Figures 17 and 18).

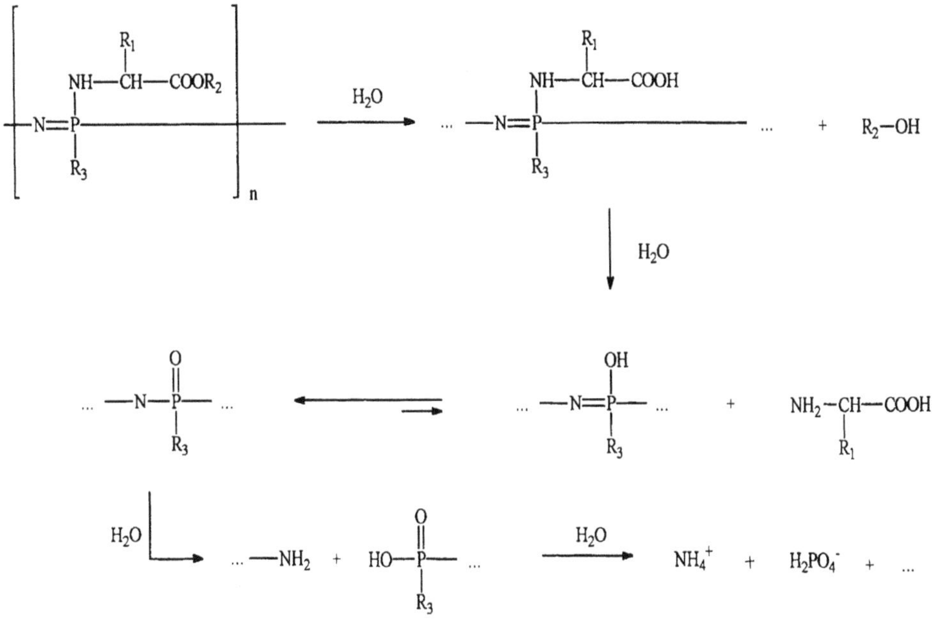

Figure 13 Alternative degradation mechanism for poly[(amino acid ester)phosphazene].

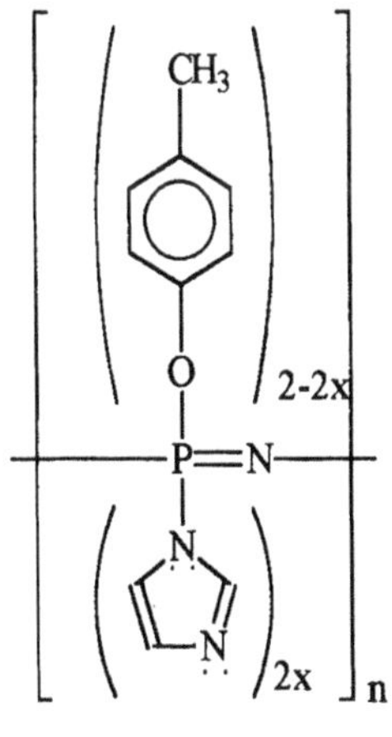

Figure 14 Poly [(imidazole)-co-(p-methylphenol)phosphazene].

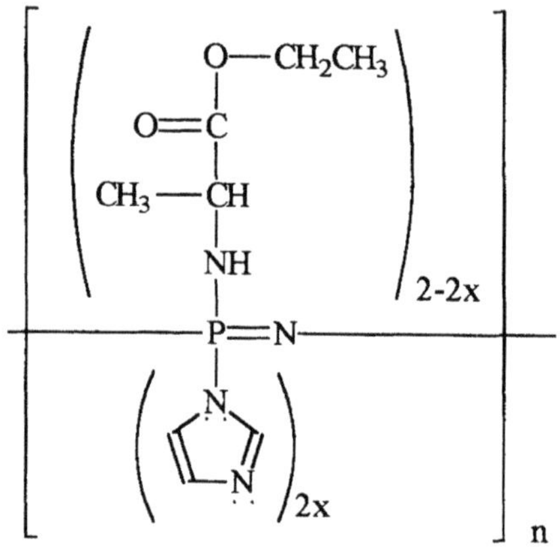

Figure 15 Poly [(imidazole)-co-(L-alanine ethyl ester)phosphazene].

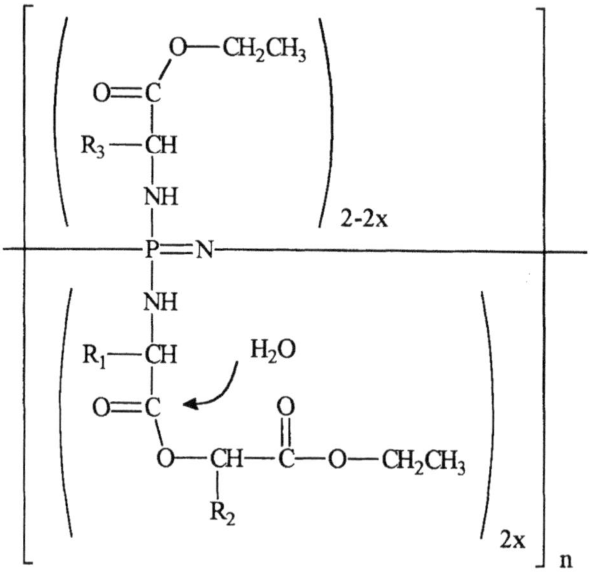

Figure 16 Poly [(depsipeptide ester)-co-(amino acid ester)phosphazene].

The rate of drug release from a polymer matrix cannot only be controlled by polymer degradation but also by the rate of hydrolysis of the covalent linkage between polymer and drug. Grolleman *et al.* (1986) prepared biodegradable amino acid ester substituted polyphosphazenes for the controlled release of a covalently linked drug naproxen. The drug was linked to the polymer chain by an amino acid residue as a spacer group (Figure 19). No tissue inflammation was detected after implantation in rats. The rate of naproxen release was larger than *in vitro* but was still very low.

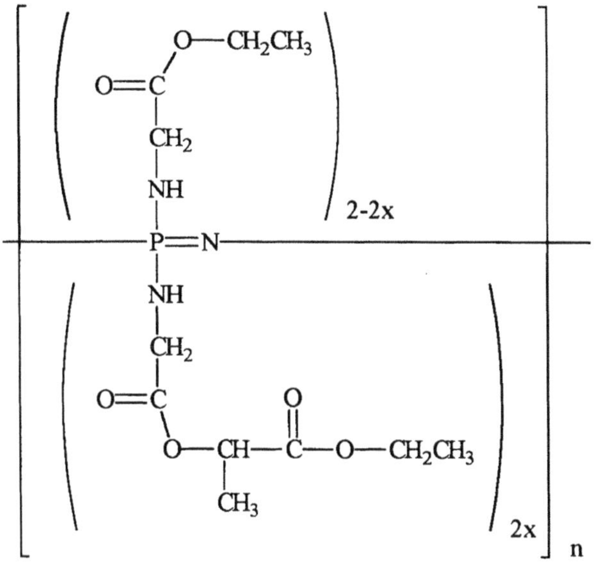

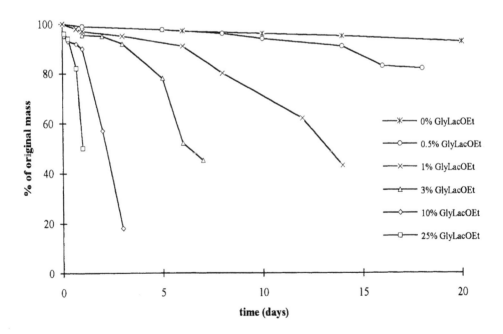

Figure 17 Poly [(ethyl 2-(O-glycyl)lactate)-co-(glycine ethyl ester)phosphazene].

Figure 18 Mass loss of poly[(ethylglycinate)phosphazene] containing different amounts of the depsipeptide glylacOEt.

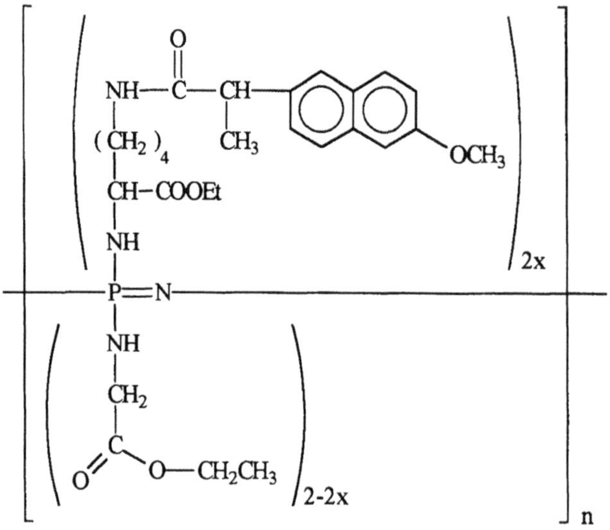

Figure 19 Poly [(naproxen)-co-(ethylglycinate)phosphazene].

Biodegradable polyphosphazenes containing steroidal and amino acid ester groups have been prepared by Allcock *et al.* (1980) (Figure 20).

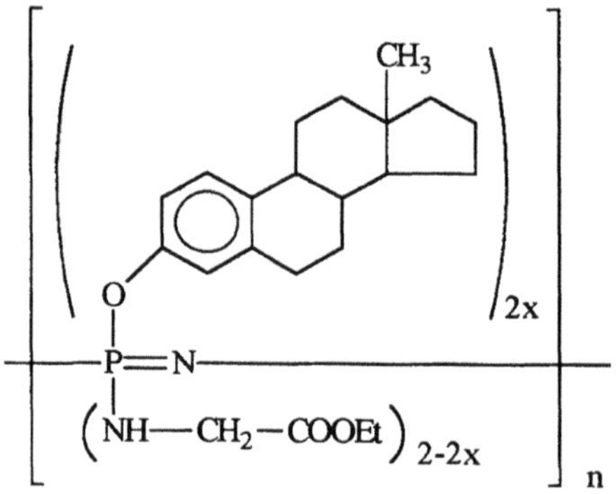

Figure 20 Poly [(ethylglycinate)-co-(steroid)phosphazene].

Hydrophilic Polyphosphazenes

Water-soluble polyphosphazenes have been prepared by Allcock *et al.*, (1986) by substitution of poly[(dichloro)phosphazene] with alkylether side groups ($-O(CH_2 CH_2O)_nCH_3$ with n = 1, 2, 7, 12, 17). Cosubstitution with hydrophobic groups ($-OCH_2CF_3$) gave a range of polyphosphazenes with, depending on the ratio of these substitutes, different water-solubility properties and different applications. Another possibility is the substitution of glyceryl side groups, as done by Allcock *et al.* (1988) (Figure 21).

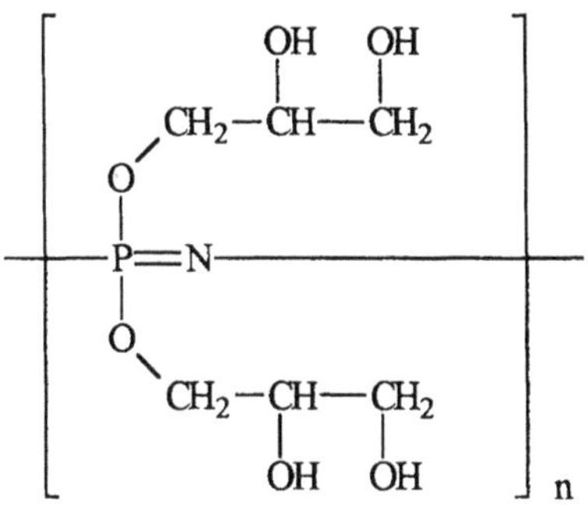

Figure 21 Poly[(diglyceryl)phosphazene].

To avoid crosslinking during the synthesis of the polymer, glycerol had to be protected before substitution to the polymer and deprotected after complete polymer substitution. The resulting polymers are water soluble and hydrolyze slowly in aqueous media to yield glycerol, phosphoric acid and ammonia. The hydroxyl units provide sites for the covalent linkage of drug molecules. The protected and deprotected polymers could be crosslinked and produced hydrogels.

Polyphosphazenes that bear -D-glucosyl side groups cosubstituted with methylamine, alkoxy or aryloxy side groups have been synthesized (Allcock *et al.*, 1991) (Figure 22) resulting in a range of polymers with varying water solubility, hydrophilicity or hydrophobicity.

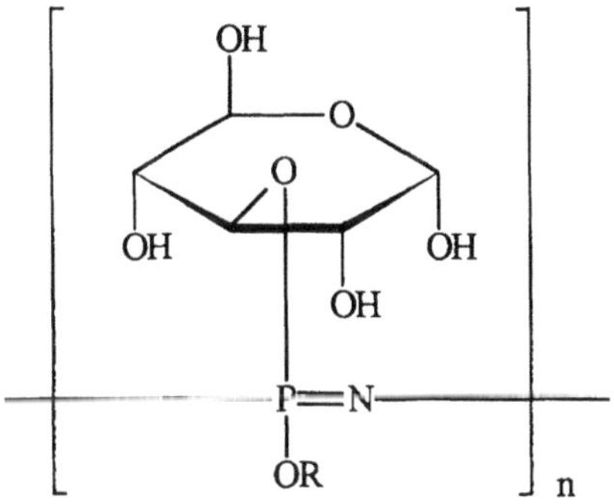

Figure 22 Polyphosphazenes bearing -D-glucosyl side groups.

Again subsequent crosslinking of the resulting polymers offered a new series of polymers that could have applications as membranes, hydrogels, biomaterials and bioerodible polymers.

Other Biodegradable Polyphosphazenes

Biodegradable polyphosphazenes containing esters of glycolic or lactic acid as side groups (Allcock *et al.*, 1994) have been described (Figure 23). The degradation

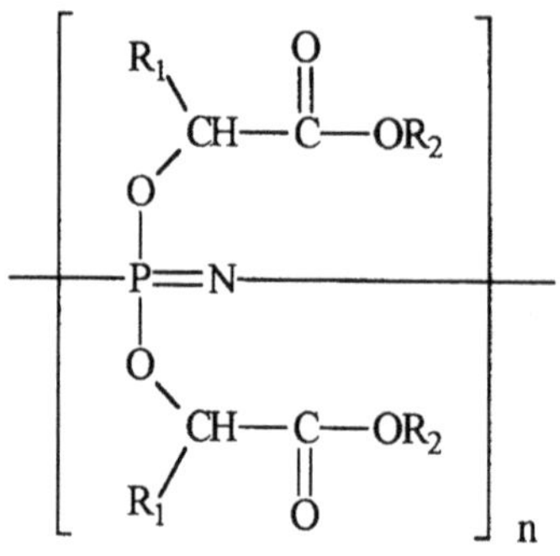

Figure 23 Poly[(glycolic acid ester) phosphazene] (R1 = H) or poly [(lactic acid ester)phosphazene] (R1 = CH3).

products of subsequent hydrolysis, ethanol, benzylalcohol, glycolic or lactic acid, phosphate and ammonia are considered nontoxic and easily metabolized. These materials are non crystalline and are hydrolized relatively fast.

Poly[(amino)phosphazenes] have been synthetised by the replacement of chlorine atoms in poly[(dichloro)phosphazene] by amines (Allcock *et al.*, 1966). Drug molecules bearing an amino-group are substituted in the same way, such as the anesthetic molecules procaine, benzocaine, chloroprocaine, butyl-p-aminobenzoyl and 2-amino-4-picoline (Allcock *et al.*, 1982). The poly[(diamino)phosphazene] is not water soluble but this can be achieved by cosubstitution with methylamine, procaine or 2-amino-4-picolino.

Molecules, bearing a carboxylic group, can be attached to a polyphosphazene chain through linkage to a spacer group. For this, p-(aminomethyl)phenoxy was used by Allcock *et al.* (1982) to substitute bioactive molecules such as N-acetyl-DL-penicillamine, p-(dipropylsulfamoyl)benzoic acid and 2,4-dichlorophenoxyacetic acid.

BIOMEDICAL APPLICATIONS OF POLY[(ORGANO)PHOSPHAZENES]

A wide range of polyphosphazenes have been used for a number of biomedical applications. Examples are inert biomaterials for cardiovascular and dental uses, bioerodible and water soluble polymers for controlled drug delivery applications (Allcock *et al.*, 1990).

Some poly[(organo)phosphazene] materials are of interest as biomedical polymers as non-interacting tissue replacement materials. Allcock examined the possibility of poly[(organo)phosphazenes] to be used as coating materials for artificial implants (Allcock *et al.*, 1992). This coating materials should be able to enhance the antibacterial activity of the surface of implanted materials.

Studies were made by Laurencin *et al.* (1993) to examine the possible use of polyphosphazenes for skeletal tissue regeneration. Polyphosphazenes with imidazole or ethylglycinate side groups next to methylphenoxy groups were evaluated. The *in vitro* evaluation on several cell cultures of these materials suggested that these polyphosphazene materials could be suitable candidate biomaterials for the reconstruction of a cell-polymer matrix for tissue regeneration.

The applicability of poly[(ethylalanate)-co-(imidazole)phosphazene]derivatives as bioresorbable tubular nerve guides prosthesis was verified by Langone *et al.* (1995). In this case the implant has only a temporary function and needs to be present only during the time necessary for growth and maturation of the nerve within the tube.

Applications of poly[(fluoroalkoxy)phosphazenes] in dentistry have also been reported (Dootz *et al.*, 1992; Kawano *et al.*, 1994). The use of a polyphosphazene fluoroelastomer as denture liner for the covering of dental prosthesis has been described. This material is known under the name of "Novus" (Hygenic Corp. Akron, USA).

Another field of biomedical applications in which polyphosphazenes can be used, is the field of drug delivery systems. One can distinguish two types of drug delivery systems. In a first type of systems, the drug to be released, is covalently attached to the polymer backbone. In a second concept, the polymer material is used as matrix system in which the drug is physically dispersed.

A large variety of drugs and other bioactive molecules have been covalently linked to the polyphosphazene chain. Steroids were covalently linked by reaction of the sodium salt of steroidal alcohol functional groups with poly[(dichloro)phosphazene] (Allcock *et al.*, 1980). Remaining chlorine side groups were replaced with methylamine, a side group which promotes the water solubility of the resulting polymer.

The linkage of local anesthetic drugs, as procaine, benzocaine, chloroprocaine, butyl-p-aminobenzoate and 2-amino-4-picoline, via their amino group to the phosphazene chain was also studied in more detail (Allcock *et al.*, 1982–1986) (Figure 24). Homopolymers and polymers with methylamino groups as co-substituents have been described.

Biologically active amines can also be linked indirectly to aryloxyphosphazenes by Schiff's base formation through a pendant aldehyde group. Drugs such as sulfadiazine have been coupled in this way (Allcock *et al.*, 1981). A similar procedure was used for the immobilization of the enzymic glucosephosphate dehydrogenase and trypsin (Allcock *et al.*, 1986) (Figure 25).

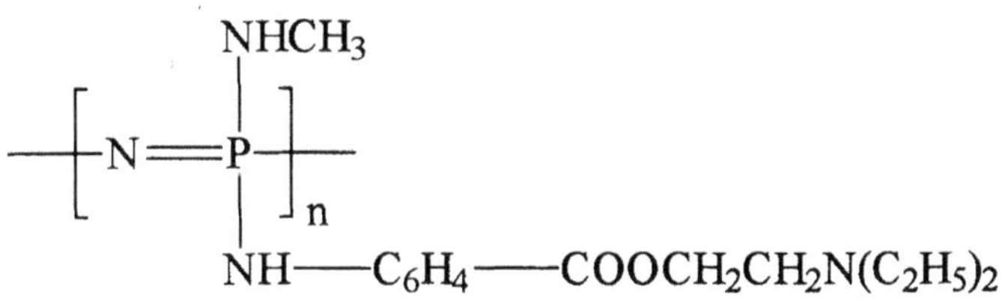

Figure 24 Procaine-substituted poly[(organo)phosphazene].

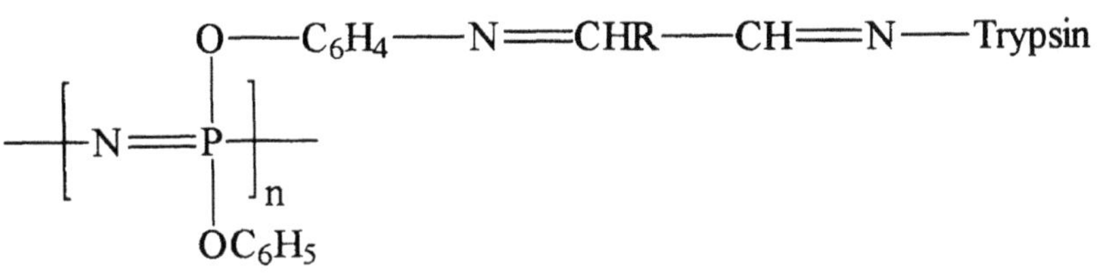

Figure 25 Immobilization of enzymes by covalent linkage with poly[(organo)phosphazenes].

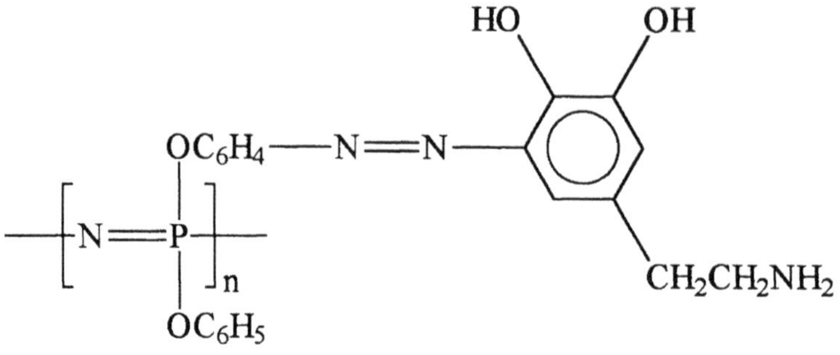

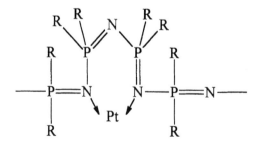

Figure 26 Diazo coupling of dopamine with poly[(organo)phosphazene].

An active carboxylic acid, such as nicotinic acid or N-acetylpenicilamine, can be linked to an aminoethylenearyloxy phosphazene substituent by DCC-induced peptide coupling techniques (Allcock *et al.*, 1982).

Another example of a non degradable pendant chain system involves the chemical linkage of catecholamines, such as dopamine, to amino-aryloxy side groups by diazo-coupling techniques (Allcock *et al.*, 1983) (Figure 26).

The coupling of the anti-inflammatory agent naproxen to the phosphazene precursor polymer via a lysine ester spacer and subsequent replacement of residual chlorine by glycine ethyl ester has been performed by Grolleman *et al.* (1986). Both *in vitro* and *in vivo* release tests were performed.

Furthermore, the anticoagulant heparin has been coupled ionically to a quaternized ammonium-aryloxyphosphazene resulting in an anti-thrombogenic activity of the material surface (Neenan *et al.*, 1982).

The use of poly[bis(methylamino)phosphazene] as a coordinative ligand for platinum anticancer drugs $(NH_3)_2PtCl_2$, has been described (Allcock *et al.*, 1977). It was found that $[NP(NHCH_3)_2]n$ is able to substitute the NH_3-ligands of the $(NH_3)_2PtCl_2$ and to form a coordination complex with $PtCl_2$ through the N-atoms of the polyphosphazene chain (Figure 27). Tests did show that part of the activity was preserved in the polymer-drug-conjugate.

Physical incorporation of a bioactive compound into an implantable or injectable matrix system could provide controlled release at a certain site in the body. Physical incorporation of drugs in a matrix of a given shape and size, offers distinct advantages over pendant chain devices. Fabrication is relatively simple and no chemical modification of the incorporated drug need to take place.

Figure 27 Complex of platina with poly[(organo)phosphazene].

As mentioned before, Laurencin *et al.* (1987) used polyphosphazene derivatives with imidazole and methylphenoxy side groups as monolithic controlled release system. *In vitro* and *in vivo* release studies were performed, using progesterone as a low molecular weight drug example and bovine serum albumine (BSA) as a polypeptide drug example. It was found that both, low molecular weight molecules as well as polypeptides, could be released over prolonged periods.

Goedemoed *et al.* (1988–1991) did use different polyphosphazene polymers based on amino acid esters as substitutes on the polyphosphazene chain, to develop locally acting antitumor devices. A model anticancer agent, melphalan, was physically incorporated. The drug was incorporated into polymer microspheres by means of a solvent evaporation process. *In vitro* tumor models were used for the drug release studies.

Ethidium has been succesfully incorporated in poly[(amino acid ester)phosphazene]devices by Crommen *et al.* (1990). This drug is used for the cure of bovine trypanosomiases (sleeping disease) which is a chronic disease very widely spread by the cattle in Africa. Successful release studies are performed both *in vitro* and *in vivo*.

Allcock also used different types of poly[(organo)phosphazenes] based on ethyl and benzyl esters of amino acids as side groups (Allcock *et al.*, 1994). Small molecules like ethacrynic acid, a diuretic, and Biebrich Scarlet, an azo dye, were incorporated as model release components.

Polyphosphazenes with amino acid esters and imidazole as cosubstituent were also used by Caliceti *et al.* (1994). Matrices were prepared in the form of films by solvent evaporation and spheres by solvent extraction. The release of model drugs naproxen, narciclasine and acetyl trytophanamide was evaluated. The films were found more suitable for drug release.

BIOCOMPATIBILITY OF POLY[(ORGANO)PHOSPHAZENES]

Biocompatibility of polyphosphazenes is often mentioned, but few detailed studies about the biocompatibility of polyphosphazenes *in vivo* have been made. Biocompatibility itself is a very complex topic. The biocompatibility of a material is very much dependent upon the situation of testing. Animal species, place of implantation, shape of material, etc., are all very important factors that will determine the biocompatibility of the material. For that reason it is very important that specifications about the test conditions are given. This, however, is often lacking.

The first type of poly[(organo)phosphazenes] that can be considered are those with fluoroalkoxy side groups. Subcutaneous *in vivo* testing of these polymers have shown minimal tissue response (Wade *et al.*, 1978). Poly[(aryloxy)phosphazene]-derivatives also show great promise as inert biomaterials on the basis of preliminary *in vivo* tissue biocompatibility tests (Wade *et al.*, 1978).

Lora *et al.*, did try to enhance the biocompatibility of poly[bis(trifluoroethoxy)-phosphazenes] (PTFP) and poly[bis(phenoxy)phosphazenes] (PPP) by grafting different side groups on the polymer surface (Figure 28). Graft copolymerization with dimethylaminoethylmethacrylate (DMAEM) onto the polyphosphazene surfaces highly enhances their biocompatibility. Subsequent heparinization has a negative effect, which is more appreciable with the PPP-based samples (Lora *et al.*, 1991). Surface modification of poly[bis(trifluoroethoxy)phosphazene] with

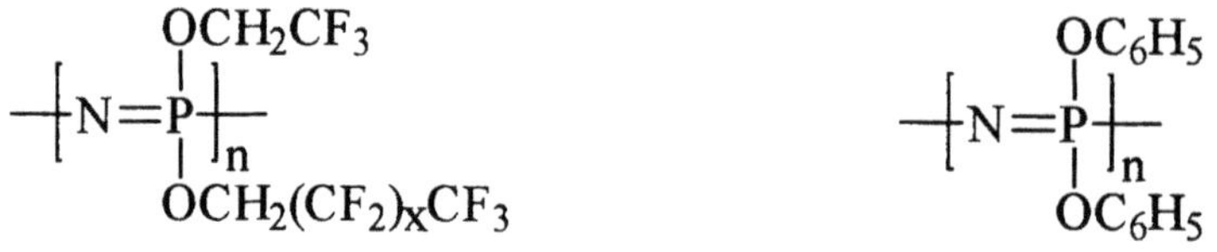

$$\begin{array}{ccc} & \overset{\displaystyle OCH_2CF_3}{\underset{|}{|}} & \\ \text{---}[N=P]\text{---} & & \\ & \underset{|}{\overset{|}{OCH_2(CF_2)_xCF_3}} & \end{array}$$

$$\begin{array}{ccc} & \overset{\displaystyle OC_6H_5}{\underset{|}{|}} & \\ \text{---}[N=P]\text{---} & & \\ & \underset{|}{\overset{|}{OC_6H_5}} & \end{array}$$

Figure 28 Structure of polyphosphazenes with fluoroalkoxy or aryloxy side groups.

polyethylene glycols was heading to materials with enhanced biocompatibility in comparison to non modified polymers (Lora *et al.*, 1993). In a third study the PTFP was grafted with hydrophilic monomers like dimethylacrylamide (DMAA) and acrylamide (AAm) (Lora *et al.*, 1994). Large differences in the biocompatibility of these materials were found. For the AAm-grafted samples the biocompatibility was greatly enhanced, while for the DMAA-grafted samples an opposite behavior was found, and a severe reaction against the samples was observed. All studies with these samples were done by intraperitoneal implantation of thin films in adult male wistar rats. Tissue around the implantation site was evaluated by histological examination after 25 days.

Laurencin *et al.*, (1987) studied the biocompatiblity of poly[(organo)phosphazene] derivatives containing imidazole groups and methylphenoxy side units. (Figure 14) Polymer matrix systems were implanted subcutaneously on the dorsum of female Sprague-Dawley rats. After one month, polymer and surrounding tissue were removed and histologically examined. At time of explantation no gross areas of surrounding inflammation was found. Polymer was found to be intact and surrounded by a fibrous capsule.

The same result has been found by Lagone *et al.*, (1995) who used poly[(ethylalanate)-co-(imidazole)phosphazene] derivatives for *in vivo* evaluation. Films of this polymer type were subcutaneously implanted in rats. The animals were killed after 30 days, or 60 days. Biopsy samples of the implant zone were histologically examined. In both cases the animals were healthy. The biological material surrounding the polymer was found to correspond to fibroblast collagen with only a few monocytes in the internal site.

Some *in vivo* experiments about the biocompatibility of poly[(organo)phosphazenes] were also performed by Grolleman *et al.* (1986). They used polyphosphazenes with aminoacid ester side groups. Biocompatibility of some naproxen-containing poly[(aminoacid ester)phosphazene]derivatives was also performed. Implants were placed subcutaneously in rats and removed after 4 weeks, and 4 and 6 months. Sample tissue containing the implants were excised and histologically evaluated. In none of the experiments inflammation was observed at the implantation site. A slight encapsulation but no "foreign-body reaction" was observed. There was no significant difference in interaction with the material, whether the implant contained naproxen or not.

In vivo experiments of poly[bis(ethylglycinate)phosphazene] with and without ethidium were performed by Crommen *et al.* (1990). Polyphosphazene sample rods were implanted subcutaneously by injection in the thorax of rabbits, just behind the shoulder. The implantation sites were checked after 4 days for possible tissue

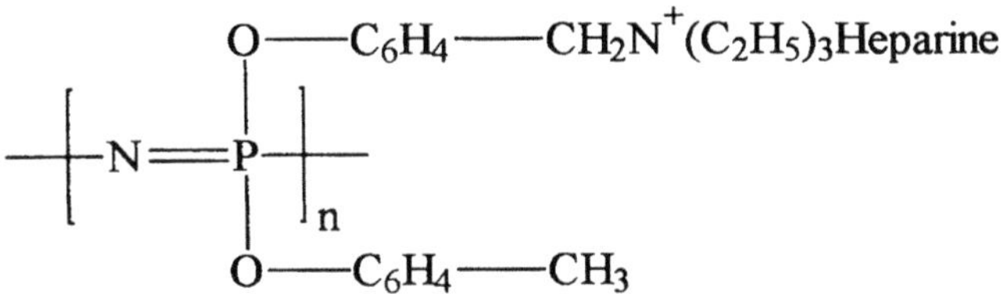

Figure 29 Heparinized poly[(aryloxy)phosphazene].

reaction. No or very little reaction was observed around the implants devoid of ethidium. At the most, a very small membrane was formed around the implant. There was more tissue reaction observed around the implants which contained ethidium.

In vitro biocompatibility tests were also performed by Neenan *et al.* (1982). They used heparinized poly[(organo)phosphazenes] to enhance the anti-thrombogenic properties of polyphosphazenes. Films of these materials containing heparine were subjected to bovine blood clotting tests (Figure 29).

An appreciable non-thrombogenic response was observed for these materials. They also examined the antibacterial activity and mutagenicity of 8 different poly[(organo)phosphazenes] (Allcock *et al.*, 1992). They reported that several of those polyphosphazenes showed a strong antibacterial activity. None of the phosphazene compounds was found to be mutagenic under the circumstances.

GENERAL CONCLUSION

Starting from poly[(dichloro)phosphazene] a whole range of polyphosphazenes with different side groups can be synthesized. A versatility of polyphosphazene derivatives with different composition and different properties can be made available. Depending upon the nature of side groups, polymers that are stable or hydrolytically degradable can be obtained. Both, biodegradable and non-degradable polyphosphazene derivatives, can be used for a variety of biomedical applications. Biodegradable poly[(organo)phosphazenes] can be used as biomaterials or as drug delivery systems. The rate of degradation can be varied by a proper choice of the chemical composition.

REFERENCES

Allcock, H.R. (1972) Recent advances in phosphazene (phosphonitrilic) chemistry. *Chem. Rev.*, **72**, 315–356.

Allcock, H.R. (1972) *Phosphorous-nitrogen compounds*, Academic Press, New York and London.

Allcock, H.R. (1979) Small-molecule phosphazene rings as models for high polymeric chains. *Acc. Chem. Res.*, **12**(10), 351–358.

Allcock, H.R. (1980) Polymerization of cyclic phosphazenes. *Polymer*, **21**, 674–683.

Allcock, H.R. (1986) Poly(organophosphazenes) : synthesis, unique properties and applications . *Makromol. Chem., Macromol. Symp.*, **6**, 101–108.

Allcock, H.R. (1990) Polyphosphazenes as new biomedical and bioactive materials. In: R. Langer and M. Chasin, (eds.), *Biodegradable polymers as drug delivery systems*, M. Dekker, New York, pp. 163–193.

Allcock, H.R., Allen, R.W. and O'Brien, J.P. (1977) Synthesis of platinum derivatives of polymeric and cyclic phosphazenes. *J. Am. Chem. Soc.*, **99**, 3984–3991.

Allcock, H.R. and Austin, P.E. (1981) Schiff base coupling of cyclic and high-polymeric phos-phazenes to aldehydes and amines : chemotherapeutic models. *Macromolecules*, **14**, 1616–1622.

Allcock, H.R., Austin, P.E. and Neenan, T.X. (1982) Phosphazene high polymers with bioactive substituent groups: prospective anesthetic aminophosphazenes. *Macromolecules*, **15**(3), 689–693.

Allcock, H.R., Austin, P.E. and Neenan, T.X. (1985) Anesthetic polyorganophosphazenes. US Patent, no. 4,495,174, Jan 22.

Allcock, H.R., Austin, P.E., Neenan, T.X., Sisko, J.T., Blonsky, P.M. and Shriver, D.F. (1986) Polyphos-phazenes with etheric side groups : prospective biomedical and solid electrolyte polymers. *Macro-molecules*, **19**, 1508–1512.

Allcock, H.R. and Fuller, T.J. (1980) Phosphazene high polymers with steroidal side groups. *Macromolecules*, **13**, 1338–1345.

Allcock, H.R. and Fuller, T.J. (1981) Synthesis and hydrolysis of hexakis(imidazolyl)cyclotri-phosphazene *J. Am. Chem. Soc.*, **103**, 2250–2256.

Allcock, H.R., Fuller, T.J., Mack, D.P., Matsumura, K. and Smeltz, K.M. (1977) Synthesis of poly[(amino acid alkyl ester)phosphazenes. *Macromolecules*, **10**(4), 824–830.

Allcock, H.R., Fuller, T.J. and Matsumura K. (1982) Hydrolysis pathways for aminophosphazenes. *Inorganic Chemistry*, **21**, 515–521.

Allcock, H.R., Gardner, L.E. and Smeltz, K.M. (1975) Polymerization of hexachlorocyclotriphosphazene. The role of phosphorus pentachloride, water and hydrogen chloride. *Macromolecules*, **8**(1), 36–42.

Allcock, H.R., Hymer, W.C. and Austin, P.E. (1983) Diazo coupling of catecholamines with poly(organo-phosphazenes). *Macromolecules*, **16**, 1401–1406.

Allcock, H.R. and Kugel, R.L. (1965) Synthesis of high polymeric alkoxy- and aryloxy-phosphonitriles. *J.Am.Chem.Soc.*, **87**, 4216–4217.

Allcock, H.R. and Kugel, R.L. (1966) Phosphonitrilic compounds V; Cyclized products from the reactions of hexachlorocyclotriphosphazene (phosphonitrilic chloride trimer) with aromatic dihydroxy, dithiol and diamino compounds. *Inorg. Chem.*, **5**(10), 1709–1715.

Allcock, H.R. and Kugel, R.L. (1966) Phosphonitrilic compounds VII. High molecular weight poly(di-aminophosphazenes). *Inorg. Chem.*, **5**(10), 1716–1718.

Allcock, H.R. and Kwon, S. (1986) Covalent linkage of proteins to surface-modified poly(organo-phosphazenes): immobolization of glucose-6-phosphate dehydrogenase and trypsin. *Macromolecules*, **19**, 1502–1508.

Allcock, H.R. and Kwon, S. (1988) Glyceryl polyphosphazenes : synthesis, properties and hydrolysis. *Macromolecules*, **21**(7), 1980–1985.

Allcock, H.R., Neenan, T.X. and Kossa, W.C. (1982) Coupling of cyclic and high-polymeric (aminoaryl) oxyphosphazenes to carboxylic acids: prototypes for bioactive polymers. *Macromolecules*, **15**(3), 693–696.

Allcock, H.R. and Pucher, S.R. (1991) Polyphosphazenes with glucosyl and methylamino, tri-fluoroethoxy, phenoxy or (methoxyethoxy)ethoxy side groups. *Macromolecules*, **24**(1), 23–34.

Allcock, H.R., Pucher, S.R., Fitzpatrick, R.J. and Rashid, K. (1992) Antibacterial activity and mutagenicity studies of water-soluble phosphazene high polymers. *Biomaterials*, **13**(12), 857–862.

Allcock, H.R., Pucher, S.R. and Scopelianos, A.G. (1994) Synthesis of poly(organophosphazenes) with glycolic acid ester and lactic acid ester side groups : prototypes for new bioerodible polymers. *Macromolecules*, **27**(1), 1–4.

Allcock, H.R., Pucher, S.R. and Scopelianos, A.G. (1994) Poly[(amino acid ester)phosphazenes: synthesis, crystallinity and hydrolytic sensitivity in solution and the solid state. *Macromolecules*, **27**(5), 1071–1075.

Allcock, H.R., Pucher, S.R. and Scopelianos, A.G. (1994) Poly[(amino acid ester)phosphazenes] as substrates for the controlled release of small molecules. *Biomaterials*, **15**(8), 563–569.

Caliceti, P., Lora, S., Marsilio, F., Guiotto, A. and Veronese, F.M. (1994) Amino acid esters and imidazole derivatives of polyphosphazenes : influence on release of drugs, degradability and swelling. *Il Farmaco*, **49**(1), 69–74.

Crommen, J.H.L., Schacht, E.H. and Mense, E.H.G. (1992) Biodegradable polymers. I. Synthesis of hydrolysis-sensitive poly[(organophosphazenes]. *Biomaterials*, **13**(8), 511–520.

Crommen, J.H.L., Schacht, E.H. and Mense, E.H.G. (1992) Biodegradable polymers. II. Degradation characteristics of hydrolysis-sensitive poly[(organo)phosphazenes]. *Biomaterials*, **13**(9), 601–611.

Crommen, J.H.L., Vandorpe, J. and Schacht, E.H. (1993) Degradable polyphosphazenes for biomedical applications *J. Control. Rel.*, **24**, 167–180.

D'Halluin, G., De Jaeger, R. and Potin, Ph. (1989) Polydichlorophosphazenes : synthese a partir de $Cl_3PNP(O)Cl_2$. *Bull. Soc. Chim. Belge*, **98**(9–10), 653–665.

D'Halluin, G., De Jaeger, R., Chambrette, J.P. and Potin, Ph. (1992) Synthesis of poly(dichloro-phosphazenes) from $Cl_3P = NP(O)Cl_2$. *Macromolecules*, **25**, 1254–1258.

Dootz, E.R., Koran, A. and Craig, R.G. (1992) Comparison of the physical properties of 11 soft denture liners. *J. Prosthet. Dent.*, **67**, 707–712.

Emsley, J. and Udy, P.B. (1972) Polymerization of hexachlorotriphosphonitrile, $(NPCl_2)_3$. *Polymer*, **13**, 593–594.

Ganapathianppan, S., Dhathathreyan K.S. and Krishnamurthy S.S. (1987) New initiators for the ring-opening thermal polymerization of hexachlorocyclotriphosphazene: synthesis of linear poly(dichloro-phosphazene) in high yields. *Macromolecules*, **20**(7), 1501–1505.

Gimblet, F.G.R. (1960) Catalysts for the bulk polymerization of phosphonitrilic chloride trimer. *Inorganic Polymers*, **1**, 65–73.

Gleria, M., Audisio, G., Daolio, S., Tralde, P. and Vecchi, E. (1984) Mass spectrometric studies on cyclo- and polyphosphazenes. 1. Polymerization of hexachlorocyclophosphazene. *Macromolecules*, **17**, 1230–1233.

Goedemoed, J.H. and de Groot, K. (1988) Development of implantable antitumor devices based on polyphosphazenes. *Makromol. Chem., Macromol. Symp.*, **19**, 341–365.

Goedemoed, J.H., de Groot, K., Claessen, A.M.E. and Scheper, R.J. (1991) Development of implantable antitumor devices based on polyphosphazenes, II. *J. Control. Rel.*, **17**, 235–244.

Grolleman, C.W.J., de Visser, A.C., Wolke, J.G.C., van der Groot, H. and Timmerman, H. (1986) Studies on a bioerodible drug carrier system based on polyphosphazene part I. Synthesis. *J. Control. Rel.*, **3**, 143–154.

Grolleman, C.W.J., de Visser, A.C., Wolke, J.G.C., van der Groot, H. and Timmerman, H. (1986) Studies on a bioerodible drug carrier system based on a polyphosphazene part II. Experiments *in vitro. J. Control. Rel.*, **4**, 119–131.

Grolleman, C.W.J., de Visser, A.C., Wolke, J.G.C., Klein, C.P.A.T., van der Groot, H. and Timmerman, H. (1986) Studies on a bioerodible drug carrier system based on a poly-phosphazene part III. Experiments *in vivo. J. Control. Rel.*, **4**, 133–142.

Helioui, M., De Jaeger, R., Puskaric, E. and Heubel, J. (1982) New preparation of linear poly-(chlorophosphazene)s. *J. Macromol. Chem.*, **183**(5), 1137–1143.

Higuchi, T. (1963) Mechanism of sustained action medication. *J. Pharm. Sci.*, **52**, 1145–1149.

Hornbaker, E.D. and Li, H.M. (1978) Low-molecualr-weight linear phosphonitrilic chloride oligomers. US Patent, US 4 198 381.

Kajiwara, M. and Shiomoto, K. (1984) The solution polymerization reaction of hexachlorocyclo-triphosphazene by sulfur compounds. *Polymer*, **25**, 93–96.

Kawano, F., Dootz, E.R., Koran, A. and Craig, R.G. (1994) Sorption and solubility of 12 soft denture liners *J. of Prosthet. Dent.*, **72**(4), 393–398.

Konecny, J.O., Douglas, C.M. and Gray, M.Y. (1960) Polymerization of dichlorophosphinic nitrides *J. Pol. Sci.*, **42**, 383–390.

Langone, F., Lora, S., Veronese, F.M., Caliceti, P., Parnigotto, P.P., Valenti, F. and Palma, G. (1995) Peripheral nerve repair using a poly(organo)phosphazene tubular prosthesis. *Biomaterials*, **16**(5), 347–353.

Laurencin, C.T., Koh, H.J., Neenan, T.X., Allcock, H.R. and Langer, R. (1987) Controlled release using a new bioerodible polyphosphazene matrix system. *J. Biomed. Mater. Res.*, **21**, 1231–1246.

Laurencin, C.T., Norman, M.E., Elgendy, H.M., El-Amin, S.F., Allcock, H.R., Pucher, S.R. and Ambrosio, A.A. (1993) Use of polyphosphazenes for skeletal tissue regeneration. *Journ. Biomed. Mat. Res.*, **27**, 963–973.

Liebich, J. (1834). *Justus Liebichs Ann. Chem.*, **11**, 139.

Lora, S., Carenza, M., Palma, G., Pezzin, G., Caliceti, P., Battaglia, P. and Lora, A. (1991) Biocompatible polyphosphazenes by radiation-induced graft copolymerization and heparinization. *Biomaterials*, **12**(3), 275–280

Lora, S., Palma, G., Bozio, R., Caliceti, P. and Pezzin, G. (1993) Polyphosphazenes as biomaterials: surface modification of poly(bis(trifluoroethoxy)phosphazene) with polyethylene glycols. *Biomaterials*, **14**(6), 430–436.

Lora, S., Palma, G., Carenza, M., Caliceti, P. and Pezzin, G. (1994) Radiation grafting of hydrofilic monomers onto poly[bis(trifluoroethoxy)phosphazene]. *Biomaterials*, **15**(11), 937–943.

MacCallum, J.R. and Wernick, A. (1967) Kinetics of bulk polymerization of trimeric phosphonitrilic chloride. *J. Polym. Sci., Part A-1*, **5**, 3016.

MacCallum, J.R. and Tanner, J. (1969) The effect of sulfur on the polymerization of hexachlorotriphosphazene. *J. Pol. Sci. B, Polymer Letters*, **7**, 743–747.

Matyjaszewski, K., Cypryk, M., Dautch, J., Montague, R. and White, M. (1992) New synthetic routes towards polyphosphazenes. *Makromol. Chem., Makromol. Symp.*, **54/55**(13), 13–30.

Matyjaszewski, K., Moore, M.K., White, M.L. (1993) Synthesis of polyphosphazene block copolymers bearing alkoxyethoxy and trifluoroethoxy groups. *Macromolecules*, **26**, 6741–6748.

Matyjaszewski, K., Lindenberg, M.S., Moore, M.K. and White, M.L. (1994) Synthesis of poly-phosphazenes bearing alkoxyethoxy and trifluoroethoxy groups *J. Pol. Sci., Part A: Polym. Chem.*, **32**, 465–473.

Matyjaszewski, K., Franz, U., Montague, R.A., White, M.L. (1994) Synthesis of poly-phosphazenes from phosphoranimines and phosphine azides. *Polymer*, **35**(23), 5005–5011.

Mujumdar, N.M., Young, S.G., Merker, R.L., Magill, J.H. (1989) Solution polymerization of selected polyphosphazenes. *Makromol. Chem.*, **190**, 2293–2302.

Mujumdar, N.M., Young, S.G., Merker, R.L. and Magill, J.H. (1990) A study of solution polymerization of polyphosphazenes. *Macromolecules*, **23**(1), 14–21.

Neenan, T.X. and Allcock, H.R. (1982) Synthesis of a heparinized poly(organophosphazene). *Biomaterials*, **3**, 78–80.

Neilson, R.H. and Wisian-Neilson, P. (1988) Poly(alkyl/arylphosphazenes) and their precursors. *Chem. Rev.*, **88**, 541–562.

Patat, F. and Kollinsky, F. (1951) Polymerization of phosphonitrilic chloride. *Macromol. Chem.*, **6**, 292.

Rose, S.H. (1968) Synthesis of phosphonitrilic fluoroelastomers *J. Pol. Sci., Part B6*, 837–839.

Ross, H. (1832). *Am. Phys. Chem.*, **24**, 295.

Ruiz, E.M., Ramírez, C.A., Aponte, M.A. and Barbosa-Cánovas, G.V. (1993) Degradation of poly[(bis-(glycine ethyl ester)phosphazene] in aqueous media. *Biomaterials*, **14**(7), 491–496.

Schacht, E. and Crommen, J. (1990) Bioerodible sustained release implants. US Patent no. 4,975,280, Dec. 4.

Scopelianos, A.G. and Allcock, H.R. (1987) Polymerization of hexachlorocyclotriphosphazene in the presence of carbon disulfide. *Macromolecules*, **20**, 432–433.

Singler, R.E., Schneider, N.S. and Hagnauer, G.L. (1975) Polyphosphazenes : synthesis-properties-applications. *Polym. Engin. Sci.*, **5**, 321–338.

Stokes, H.N. (1897) *Am. Chem. J.*, **19**, 782.

Wade, C.W.R., Gourlay, S., Rice, R., Heggeli, A., Singler, R. and White, J. (1978) Bio-compatibility of eight poly(organophosphazenes). In C.E. Carraher, J.E. Sheats and C.U. Pittman, (eds.), *Organometallic polymers*, Academic Press, New York, pp. 283–288.

Wisian-Neilson, P. and Neilson, R.H. (1980) Poly(dimethylphosphazene), (Me$_2$PN)$_n$. *J. Am. Chem. Soc.*, **102**, 2848–2489.

Wisian-Neilson, P., Ford, R.R., Neilson, R.H. and Roy, A.K. (1986) Silylated derivatives of poly-[methylphenylphosphazene]. *Macromolecules*, **19**, 2089–2091.

10. POLY (ALKYLCYANOACRYLATES)

ELIAS FATTAL, M. TERESA PERACCHIA and PATRICK COUVREUR

Laboratoire de Physico-Chimie, Pharmacotechnie, Biopharmacie,
URA CNRS 1218, University of Paris XI, School of Pharmacy, France

INTRODUCTION

Cyanoacrylates polymers are considered of high interest because of the strong reactivity of the corresponding monomers, able to polymerize easily in various media including water. Their adhesive properties are also remarkable. For these reasons, cyanoacrylates have been started to be investigated and extensively used as surgical glue. At the moment, an important application consists in their use in pharmacy as drug particulate carriers: poly(alkylcyanoacrylate) (PACA) nanoparticles are now among the most promising colloidal drug delivery system. A phase I clinical trial of PACA nanoparticles loaded with doxorubicin has been conducted in 21 patients with refractory solid tumors (Kattan *et al.*, 1992).

This review provides information on the synthesis of the monomers, the preparation of the corresponding polymers, the mechanism of polymerization and the characterization of the polymer, as well as the technologies available for nanoparticle preparation. Finally, the review focuses on the application of PACA and PACA nanoparticles in medicine and pharmacy.

SYNTHESIS AND PURIFICATION OF CYANOACRYLATE MONOMERS

The chemical structure of cyanoacrylate esters is represented on Figure 1. Monomers present different lateral alkyl chain of length ranging from methyl to decyl. Methods of synthesis of cyanoacrylates were described in details by Leonard *et al.* (1966). Typical synthesis is as follows: paraformaldehyde (135g), 300 ml of methanol, 100 ml of diglyme (dimethyl ether of ethylene glycol), 2.0 ml of piperidine, are placed in a two-liter three-necked flask, fitted with a mechanical stirrer, water-cooled condenser, Dean and Stark trap, and dropping funnel. This mixture is stirred and heated until the methanol refluxed vigorously. Then a 5-mole portion of butyl cyanoacetate (705 g) is added through a dropping funnel at a rate sufficient to maintain reflux, after removal of the external heat source. Methanol is then distilled off until the vapor temperature reached 88°C. Benzene (250 ml) is added, and water is removed from the reaction mixture by azeotropic distillation. The apparatus is then converted to a conventional distillation set up, and a total of 3.5 mole water is removed. At this stage, 15 g of phosphorus pentoxide is added, benzene is removed under water aspirator, and residual benzene and diglyme are removed at 3 mm Hg. The vacuum distillation is continued until the pot temperature reaches 160°C to remove any residual unreacted butylcyanoacetate. At this point, cracking begin to occur and a receiver with small amounts of pyrogallol and phosphorus pentoxide is attached to the apparatus, and the monomer is collected. At this stage, some polymer

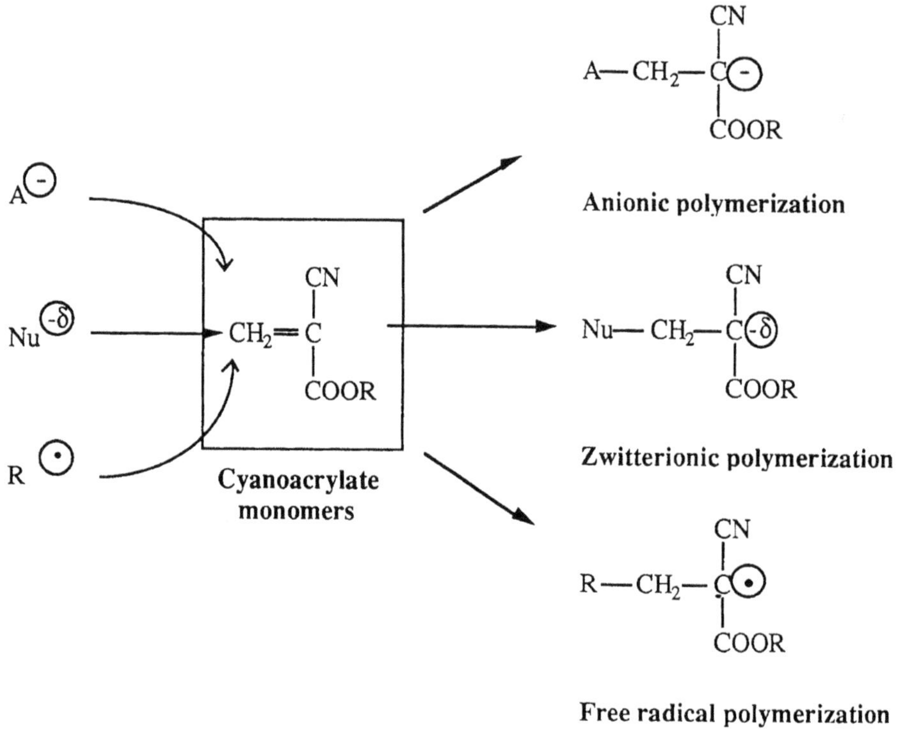

$$A—CH_2—\underset{\underset{\text{COOR}}{|}}{\overset{\overset{\text{CN}}{|}}{C}}\text{(−)}$$

Anionic polymerization

$$CH_2{=}\underset{\underset{\text{COOR}}{|}}{\overset{\overset{\text{CN}}{|}}{C}}$$

Cyanoacrylate monomers

$$Nu—CH_2—\underset{\underset{\text{COOR}}{|}}{\overset{\overset{\text{CN}}{|}}{C}}\text{(−δ)}$$

Zwitterionic polymerization

$$R—CH_2—\underset{\underset{\text{COOR}}{|}}{\overset{\overset{\text{CN}}{|}}{C}}\text{(•)}$$

Free radical polymerization

Figure 1 Structure of cyanoacrylates monomers and mechanism of polymerisation.

may tend to form on the sides of the distilling adaptors; however, the replacement of the adaptors and addition of another charge of 15 g of phosphorus pentoxide to the reaction mixture minimizes this condition. Over 500 g of crude monomer collected is redistilled through a 6 inches Vigreux column. A sulfur dioxide bleed is introduced through a capillary tube as an inhibitor to prevent anionic polymerization and also bumping during distillation.

The monomers obtained by this procedure were found to have average purities of 98.5% and more, based on the peak areas on the gas chromatograms.

It is very difficult to handle the monomer in pure form, i.e. when freed from the acid stabilizers (SO_2, sulfonic acid, etc) or free radical scavengers (hydroquinone) whose presence is normally necessary to prevent spontaneous polymerization.

SYNTHESIS OF THE POLYMER AND MECHANISM OF POLYMERIZATION

Polymerization of cyanoacrylates can be achieved by both free radical (Alvotidinov, 1982) and anionic/zwitterionic mechanisms (Donelly, 1977; Pepper, 1978, 1992) (Figure 1).

The free radical mechanism of polymerization is determined by a high activation energy (125 KJ/mol). This polymerization is very slow and the reaction rate depends strongly on the temperature and the quantity of radicals.

The anionic/zwitterionic polymerization mechanisms hold much greater interest because it is more rapid and easier to handle, thus suitable for biomedical applications. Furthermore, even very weak bases, such as OH-ions deriving from water dissociation, are capable to initiate the reaction. Classical initiators are clearly ionic compounds (e.g. I^-, CH_3COO^-, Br^-, OH^-, etc), but also nucleophilic compounds, such as tertiary bases (phosphine, pyridine) exhibit activity as initiators (Donelly et al., 1977; Pepper 1978, 1992). They presumably generate zwitterions, rather than singly charged anionic species, capable of anionic propagation.

Propagation of the polymerization occurs after formation of carbanions that are able to react with another monomer molecule leading to the formation of living polymer chains.

Termination is due to the presence of a cation that leads to the end of the polymerization. After testing different anionic chains terminators (O_2, CO_2, H_2O, HCl), Pepper (1978) found out that only strong acids were efficient to stop polymerization. However, water is also able to terminate the polymerization through the conjugate acid H_3O^+ (Eromoselle, 1989). Anionic polymerization leaves a polymer which consists in an "onium salt". In the case of zwitterionic polymerization, termination generally occurs with the formation of intramolecular bonds, leading to the cyclization of the polymer, or intermolecular bonding, leading the doubling of the polymer chain length (Pepper, 1978). The zwitterion appears to be stable and insensitive to other anionic chains terminators such as water and carbon dioxide, leading to the formation of high molecular weight polymers (MW) (Pepper, 1978).

CHARACTERIZATION AND METHOD OF ANALYSIS

The main experimental difficulty with cyanoacrylate characterization, as emphasized by Pepper (1978), is the extreme reactivity of the monomer and its sensitivity to apparently spontaneous polymerization, caused by traces of unknown initiators. To prevent this, commercial preparations of the monomer contain a few parts per million of undisclosed "stabilizers" (e.g. sulphur dioxide). If these stabilizers are removed, the monomers can be preserved only in the frozen state, and after liquifying, they polymerize to glassy solids of very high MW (10^6–10^7 Da).

Concerning the polymer, the main problem is the difficulty to find suitable solvents for PACA characterization, especially for the determination of their MW. The strongly polar solvents (nitromethane, acetonitrile and DMF) dissolve the polymers from methyl, ethyl and n-butyl esters (PMCA, PECA, PBCA), but they are suspected to cause degradation. Ether-type solvents are supposed to be inert, thus suitable. THF is indicated only for PBCA, since PMCA is insoluble, and PECA of high MW in THF shows a high viscosity and GPC anomalies.

Spectroscopic Analysis

The mechanism of anionic polymerization of alkylcyanoacrylates requires that the base initiating specie is present as an endgroup in the polymer, as reported by

Leonard *et al.* (1966). This can be demonstrated by the infrared (IR) spectra of the polymer. Polymer preparation, and in particular the composition of the polymerization medium, is very important and can be evidenced in the spectra obtained. For example, polymers prepared in water can show the OH peak at 3600 cm^{-1}, whereas the polymer prepared in undried methanol shows a suppression of this band.

This indicates an apparent competition between the methoxy and hydroxy nucleophiles for the initiation of the polymerization. Polymers prepared in aqueous solutions of pyridine, cysteine, alanine and glycine, show also a suppression of the OH band, suggesting that nucleophiles other than the OH may preferably be involved in the initiation of the polymerization. This fact has important consequences for *in vivo* applications of PACA as adhesives, since such initiation of the monomer could lead to primary chemical bonding of the adhesives to the tissue substrate.

Molecular Weight

Polymer MW can be determined by gel permeation chromatography (GPC) calibrated with polystyrenes (Pepper, 1978). Polymers number average MW can be determined in acetonitrile by using vapor phase osmometer (Leonard *et al.*, 1966).

PACA MW can vary from a few hundreds Da for oligomers up to more than one million Da, due to the nature of the polymerization medium, which, as explained previously, can affect the polymerization process. In general, free radical polymerization yields high molecular weight polymer (Coover *et al.*, 1990). As postulated by Pepper (1978), polymer produced by stronger bases should have MW determined by the "living polymer" relationship, with a high monomer conversion. Zwitterions formed from weaker bases, prone to a termination process increasing with temperature, would be expected to form polymers with MW decreasing with temperature increase. On the contrary, the complete conversion produced by these initiators is always of very high MW, irrespective of temperature or reagent concentrations. This can be explained by the fact that the elimination of base by combination of "onium" and carbanion ends can take place intermolecularly between two zwitterions as well as between the two ends of one zwitterion. Such intermolecular elimination of base is still a termination reaction, but it causes an effective doubling of the polymer chain length. Repetition of such intermolecular termination processes would determine indefinite growth. However, working at low temperatures and thus with termination reactions relatively slow, it was observed an increase of intrinsic viscosity with conversion in polymerization, that led to an increase in MW.

Some salts, such as sodium tetraphenyl borate, were found to cause reduction of MW, but increase in polymerization rate. This behavior is consistent with the salt's providing counter ions for both ends of propagation zwitterion, reducing the degree of intermolecular association and hence the rate of termination by the mutual chain-doubling process. Salts of alkyl ammonium cations determined strong reductions of both polymerization rate and MW, suggesting their role as transfer/termination agents, presumably by alkyl-group transfer to the growing carbanionic ends (Pepper, 1978).

Thermal Degradation and Stability

Low Molecular Weight Oligomers

As reported by Rooney (1981a), thermal degradation of methyl and ethyl cyanoacrylates oligomers of number-average MW 370–1350 is shown to proceed by depolymerization at temperatures in the range 140°–180°C, with the formation of monomeric products, as observable by isothermal pyrolysis as thermogravimetric analysis. Liquid chromatograms of low MW oligomers were depicted by Rooney (1981b). Spectroscopic analysis (NMR) of the degradation residues during pyrolysis gave no evidence of structural change in the oligomers, except a gradual MW decrease (Rooney, 1981a). Results indicated that weight loss kinetics appear to be governed by the diffusion of the monomer vapor through the liquid sample (Rooney, 1981a).

High Molecular Weight Polymers

The thermal degradation and stability of PACA with high chain regularity and of MW ranging from a few thousands to more than a million, was investigated by Birkinshaw and Pepper (1986) using thermogravimetry, DSC and pyrolis-gas-liquid chromatography. The polymer of *n*-butylcyanoacrylate (PBCA) was selected as a model, since it is soluble in THF. PBCA was found to degrade between 150°C and 320°C, leading to the formation of volatile products, detectable by an interfaced TG-IR technique. Almost quantitative conversion to monomer occurred, and MW characterization of polymer thermally degraded to half of its original mass showed that the MW of the residual material was substantially the same as that of the original one. The mechanism of thermal degradation is thus considered to be chain unzipping with a zip length greater than the degree of polymerization of the longest chain (Figure 2).

Birkinshaw and Pepper (1986) also showed that the degradation kinetics and the thermal stability of the polymer strongly depend on the nature of the polymerization initiator and on the chain length. This implies that the depolymerization process is end initiated at the residue of the polymerization initiator. Thus, for example pyridine initiated polymers, having pyridinium chloride salts ends, are more stable than those having phosphonium or quaternary ammonium salts, and exhibit a degradation temperature 50°C higher, and initial rates of mass loss an order of magnitude lower.

Adhesive Properties

PACA are hard glassy polymers with a high polarity, and this explains their adhesive properties (Coover et al., 1959). The adhesive action of alkyl 2-cyanoacrylates was discovered during an investigation of a series of polymers derived from 1,1-disubstituted ethylenes. It was found that certain compounds of general structure:

$$CH_2 = C \overset{\diagup \quad X}{\diagdown \quad Y}$$

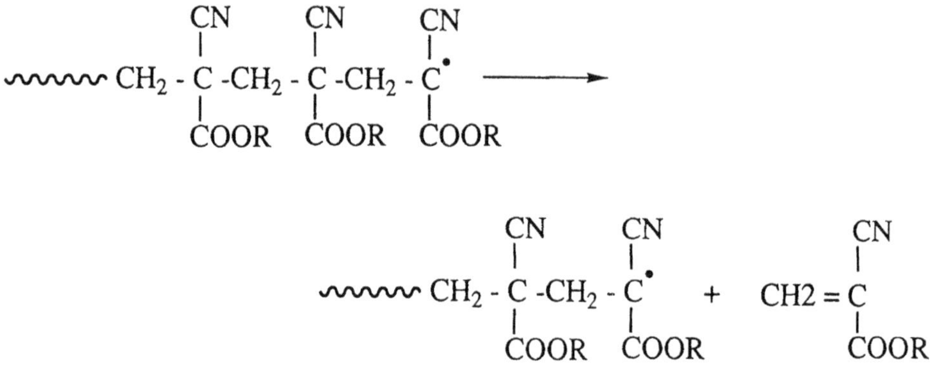

$$
\text{\raisebox{0ex}{$\sim\!\!\sim\!\!\sim$}} \ CH_2 - \underset{\underset{COOR}{|}}{\overset{\overset{CN}{|}}{C}} - CH_2 - \underset{\underset{COOR}{|}}{\overset{\overset{CN}{|}}{C}} - CH_2 - \underset{\underset{COOR}{|}}{\overset{\overset{CN}{|}}{\overset{\cdot}{C}}} \longrightarrow
$$

$$
\text{\raisebox{0ex}{$\sim\!\!\sim\!\!\sim$}} \ CH_2 - \underset{\underset{COOR}{|}}{\overset{\overset{CN}{|}}{C}} - CH_2 - \underset{\underset{COOR}{|}}{\overset{\overset{CN}{|}}{\overset{\cdot}{C}}} \ + \ CH2 = \underset{\underset{COOR}{|}}{\overset{\overset{CN}{|}}{C}}
$$

Figure 2 Thermal degradation mechanism of poly(alkylcyanoacrylate) by alkyl chain unzipping.

where X and Y are strongly electronegative groups (i.g. cyanoacrylates), exhibit adhesive properties. The mechanism of this bonding action is a consequence of the polymerization that occurs, catalyzed by minute amounts of water or other weak bases present on the adherent surface. Thus, polymerization occurs at room temperature and does not require the use of any solvents or catalysts. Tensile strength, or adhesiveness, decreases with increase in the length of the alkyl side chain, as reported by Coover *et al.* (1959).

Polymer Degradability

The mechanism of *in vivo/in vitro* degradation held for a long time as the main one was proposed by Leonard *et al.* (1966). It was observed that PACA degrades in the presence of water. In this aqueous degradation process, a random addition of water molecules to the polymer chains occurs, leading to a hydrolytic scission of a carbon-carbon bond on the polymer chain backbone. The fact that the slow degradation in water, at neutral pH, is highly accelerated in the alkaline medium, suggested to these authors that the degradation was started by the initial attack of the hydroxyl ion, leading to the reverse Knövenagel reaction with the formation of formaldehyde and polymeric degradation products (Figure 3a). The carbanions formed according to this mechanism can recombine with the formaldehyde product, making the whole process reversible.

Degradation can be followed by detection of formaldeyde, one of the main degradation products, by the method of Bricker and Johnson (1945), in which formaldehyde is allowed to react with chromotropic acid in presence of sulfuric acid, leading to a coloration that absorbs at 570 nm. Thus, as described by Leonard *et al.* (1966), formaldehyde formation can be evidenced by the preparation of its derivatives 2,4-dinitrophenylhydrazine and dimedone. The rate of aqueous degradation is considerably slower for the polymers of the higher alkyl esters. PMCA degrades much faster, and the diminution in this rate becomes smaller in the higher members of the homologous series. Concomitantly, the production of formaldeyde is accompanied by a decrease in number-average MW of the polymer.

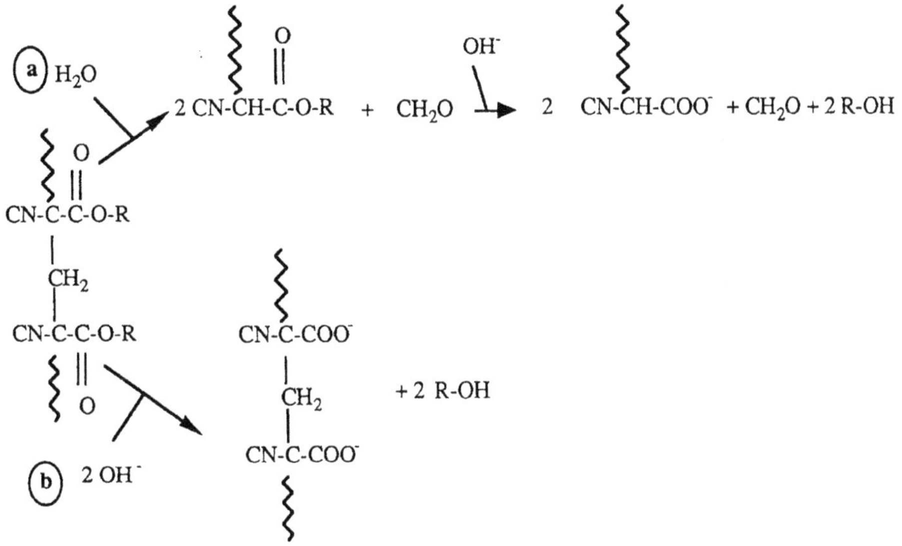

a) H$_2$O → 2 CN-CH-C-O-R (C=O) + CH$_2$O →[OH$^-$] 2 CN-CH-COO$^-$ + CH$_2$O + 2 R-OH

CN-C-C-O-R (C=O)
|
CH$_2$
|
CN-C-C-O-R (C=O)

b) 2 OH$^-$ → CN-C-COO$^-$ — CH$_2$ — CN-C-COO$^-$ + 2 R-OH

Figure 3 Degradation mechanisms of poly(alkylcyanoacrylate): a) hydrolysis of the Carbon-Carbon bond with formation of formaldehyde. b) ester hydrolysis with formation of acidic moieties and alcohol.

Vezin and Florence (1980) used monodisperse powdered polymer particles for studying the *in vitro* degradation of poly(*n*-alkyl α-cyanoacrylates), with a range of alkyl side chains and MW. These authors showed that degradation in aqueous buffer depends not only upon the pH and the length of the polymer alkyl side chain, but also critically on the polymer particle surface, particle size, polymer MW and MW distribution. They concluded that at low MW (below the characteristics of effective tissue adhesiveness), increased water solubility, plasticity and diffusivity may result in a bulk rather than surface polymer degradation.

Wade and Leonard (1972) postulated another mode by which *in vitro* degradation may occur, by ester hydrolysis. This reaction results in the formation of acidic moieties (Figure 3b). The combination of the two degradation pathways, random hydrolytic chain scission with the formation of formaldehyde, and ester hydrolysis with the formation of acidic moieties, results in the formation of low molecular weight and water soluble fraction *in vitro*. Wade and Leonard (1972) studied the degradation of PMCA implanted subcutaneously in dogs and analyzed their urine for formaldehyde and its metabolites, in order to determine whether hydrolytic scission of the implanted polymer plays a major role in the mechanism of *in vivo* degradation of PACA. It is known that *in vivo* formaldehyde is oxidized via formic acid to carbon dioxide and, if injected s.c. in dogs, causes an increase in urinary formate. Thus, it was expected that, if formaldehyde had been produced *in vivo* as *in vitro* from degradation of PACA, its metabolism should have led to urine-soluble products. Their data were consistent with the hypothesis that both hydrolytic chain scission and ester hydrolysis may be involved in PACA degradation *in vivo*. Finally, Leonard *et al.* (1966) showed that the higher homologues degrade at a slower rate, allowing the degradation products to be metabolized and limiting local inflammatory response.

APPLICATIONS OF CYANOACRYLATES IN BIOMEDICINE

As mentioned above, an important property of cyanoacrylates is their ability to form rapidly strong bonds as they undergo anionic/zwitterionic polymerization in the presence of hydroxyl or amino groups, such as with proteins or weak bases. For this reason, PACA have been extensively used in surgery because of their adhesive properties.

A main application has been in vascular surgery. There have been reports on the direct closure or repairs of arterial lesions by the means of cyanoacrylates (Carton *et al.*, 1961; Nathan *et al.*, 1960; Casanova *et al.*, 1987; Takenaka *et al.*, 1992). Also, cyanoacrylates adhesives were used as an auxiliary after placing several stay sutures (Hosbein and Blumenstock 1964; Hafner *et al.*, 1963; Tschopp 1975). Some complications were due to bleeding in the early stages (Carton *et al.*, 1961). This bleedings were attributed by Takenaka *et al.* (1992) to technical errors such as the insufficient application of adhesive or inadequate compression of the anastomotic part when pulling out the sutures.

Transcatheter occlusive therapy was even performed with isobutylcyanoacrylate (Freeny *et al.*, 1979; Giuliani *et al.*, 1979). This monomer was found to be a good occlusive agent after polymerization, achieving complete artery occlusion and thus reducing blood loss as shown for duodenal hemorrhage (Goldin, 1976) or bleeding in cases of urogenital carcinomas. Cyanoacrylates have also been used as embolic agent for cranial arterious malformations (Vinters *et al.*, 1980) or the treatment of cerebrospinal fluid leaks (Maxwell and Goldware, 1973).

Cyanoacrylates were shown to form a strong durable bond between bones *in vitro.* Tensile adhesive strength between smooth bovine cortical bone specimens bonded together with the isobutyl monomer and tested after one day storage in water was approximately 6.5 mPa (Brauer *et al.*, 1979). The monomer was used, without evidence of histotoxicity, to repair osteochondral fractures (Harper and Ralston, 1983) and recently to improve meniscal repairs (Koukabis *et al.*, 1995). Butylcyanoacrylate was also used in facial bone surgery for frontal bone reconstitution (Avery and Ord, 1982).

In dentistry, cyanoacrylates have been proposed as adhesives for hard tooth structure, for sealing of extraction wounds and as periodontal dressing (Beech 1972; King *et al.*, 1967; Lobene and Sharawy, 1968).

In skin surgery, cyanoacrylates were used with success to close skin wounds or to fix skin grafts (Reynolds *et al.*, 1966). An experiment conducted with methylcyanoacrylate showed that it was adsorbed readily and eliminated by the body. Removal from the site of skin incision was complete by 107 days after application.

Finally, cyanoacrylates were also applied to the eye to treat persistent epithelial defects and exposure keratitis (Donnenfeld *et al.*, 1991), and after glaucoma surgery for sealing leaking filtering blebs (Weber and Baker, 1989).

APPLICATIONS OF CYANOACRYLATES IN THE DESIGN OF DRUG PARTICULATE CARRIERS

The idea to use PACA for the design of colloidal drug carriers (nanoparticles) has emerged in the eighties (Couvreur *et al.*, 1982) due to the fact that:

(1) cyanoacrylate monomers were able to polymerize in aqueous medium which is of great interest for pharmaceutical purposes.
(2) PACA were biodegradable polymers, which is necessary for intravascular administration.
(3) the wide use of PACA as a surgical glue in humans was a guarantee for the low toxicity of these polymers.

Preparation of Nanoparticles from Alkylcyanoacrylate Monomers

Nanoparticles may be defined as submicronic (<1 μm) colloidal systems generally made of synthetic or natural polymers. According to the morphological structure, *nanospheres* or *nanocapsules* can be obtained. Nanospheres are matrix systems in which the drug is dispersed throughout the particle or adsorbed onto its surface; nanocapsules are vesicular systems in which the drug is confined to a cavity surrounded by a unique polymeric shell (Couvreur *et al.*, 1995).

PACA nanospheres are prepared by emulsion/polymerization in which droplets of water-insoluble monomers are emulsified in an aqueous phase (Couvreur *et al.*, 1979; Couvreur *et al.*, 1982). The mechanism of the polymerization is an anionic process initiated by covalent bases present in the polymerization medium (i.e. OH-ions deriving from water dissociation), as already previously discussed. The reaction takes place in micelles after diffusion of monomer molecules through the water phase. The pH of the medium determines both the polymerization rate and the adsorption of drugs when the latter is ionizable (Couvreur *et al.*, 1979). Drug can be combined with nanoparticles after dissolution in the polymerization medium either before the introduction of the monomer or after its polymerization. The size of the nanospheres obtained is approximately 200 nm, but it can be reduced to 30–40 nm using a non ionic surfactant in the polymerization medium (Seijo *et al.*, 1990) or by adding SO_2 to the monomer (Lenaerts *et al.*, 1989).

The method for preparing biodegradable poly(alkylcyanoacrylate) nanospheres has provided the basis for the development of polycyanoacrylate nanocapsules. Al Khouri-Fallouh *et al.* (1986) proposed an original method in which the monomer is solubilized in an alcohol phase containing an oil and is then dispersed in an aqueous phase containing surfactants. In contact with water, the alcohol phase diffuses and favours the formation of a very fine oil-in-water emulsion. The monomer, insoluble in water, polymerizes at the interface of the phases to form the wall of the nanocapsules.

Degradation of Poly(alkylcyanoacrylates) Nanoparticles

As previously described, degradation of alkyl cyanoacrylates polymers follows two pathways, hydrolytic polymer side alkyl chain scission (with formation of formaldehyde) and enzymatic ester hydrolysis (with formation of soluble acidic moieties) (Figure 3). Initially, it was believed that only the first degradation pathway existed (Leonard *et al.*, 1966). Then, Wade and Leonard (1972) observed *in vivo*, after the implantation of a PACA polymer block, a contribution in the degradation mechanism by enzymatic ester hydrolysis, and concluded that both pathways were involved in PACA degradation. Later Lenaerts *et al.* (1984), investigating the degradation of PACA nanoparticles, found that the major pathway for the degradation of polymer particles was the cleavage of the side chain ester bond. It was verified a poor contribution of formaldheyde products, at any pH, thus the previous hypothesis by Vezin and Florence (1980) was contested. The results of Lenaerts were later verified by Stein and Hamacher (1992), who observed a much faster degradation rate in dog serum than in phosphate buffer. Furthermore, it was not observed a particle destruction, that would have been a consequence of the breakdown of the polymer backbone (*via* inverse Knövenagel reaction). By the second mechanism (ester hydrolysis), the polymer chain remains intact, but it increases its solubility until it is completely soluble. Due to the low MW, the resulting water soluble polymer acid is then rapidly excreted from the body (Grislain *et al.*, 1983), as discussed later. This degradation pathway is consistent with the production of the corresponding alchohol during the bioerosion of the nanoparticles *in vitro* in presence of esterases (Lenaerts *et al.*, 1984a; Stein and Hamacher, 1992).

Morphologically, the degradation of PACA nanoparticles was found to be a surface erosion process (Müller *et al.*, 1990). Müller *et al.* (1990) measured the degradation of the nanoparticles in a different way than assaying the final degradation products as Lenaerts *et al.* (1984a), or Vezin and Florence (1980). Degradation was followed by spectrophotometric measurement at 400 nm, during incubation of nanoparticles in aqueous solutions at diffrent pH. Absorbance measurement provided informations about the time required for the complete degradation of nanoparticles. Indeed, PCS could be used to follow particle degradation as a function of size only for the initial phase of the degradation process, since then a plateau is reached and larger particles or aggregates are formed. Also Tuncel *et al.* (1995), who observed particle degradation in phosphate buffer solution by electron microscopy, concluded that particle degradation occur mainly by surface erosion.

Therapeutic Applications of PACA Nanoparticles

Experimental Cancers

Most of the research for improving therapeutic efficiency with nanoparticles concerns experimental cancer chemotherapy. The most spectacular results were often obtained when the carrier was administered in the compartment where the neoplasic cells were localized (i.e., murine leukemia L 1210) (Couvreur *et al.*, 1986). However, it is still unclear whether the enhanced anticancer activity of cytostatics associated to nanospheres is related to a better targeting to tumor cells or to a slow drug-release. A first answer to this question was obtained by Chiannikulchai *et al.* (1989, 1990). These authors have clearly shown that the important increase of targeted doxorubicin in the tumoral tissue (hepatic metastases of sarcoma M 5076) was due to an important capture of doxorubicin-loaded nanospheres by the Kupffer cells. This effect should lead to the formation of an effective gradient of concentration favourable for a massive and prolonged diffusion of the drug towards the neighbouring malignant cells. As a consequence, and irrespective to the dose and the administration regime, the reduction of the number of metastases was far much larger with doxorubicin-loaded nanospheres than with free doxorubicin (Chiannikulchai *et al.*, 1989). The superiority of the targeted drug was clearly evidenced by histological examinations, showing that both the number and the size of the tumoral cores were lower when doxorubicin was associated with nanospheres.

More recently, the problem of the cancer multidrug resistance has been approached using PIHCA nanospheres (Cuvier *et al.*, 1992). In fact, multidrug resistance is the main cause of chemotherapy failure in cancer. It is often associated with the overexpression of a cell membrane glycoprotein of 170 kD molecular weight (Kartner and Ling, 1989; Robinson, 1987). This glycoprotein, designated P-glycoprotein, could act as an efflux pump and reject numerous drugs from the cells as shown for bacterial transport proteins (Gros *et al.*, 1986). Indeed, multidrug resistance is associated with a low intracellular accumulation of some drugs. To solve this problem, the efficacy of nanospheres loaded with doxorubicin has been evaluated by Cuvier *et al.* (1992) on five different multidrug resistant cell lines, whose mechanism of pleiotropic resistance is known to be related to the presence of P-glycoprotein. A complete reversion of drug resistance was obtained *in vitro*, i.e. a cell growth inhibition comparable to that obtained with sensitive cells exposed to free doxorubicin. Similar results were obtained by Kubiak *et al.* (1989) on the resistant DC3F-AD/AZA subline. Thus, it has been demonstrated that *in vitro* resistance of tumor cells to doxorubicin can be fully circumvented using biodegradable PIHCA nanospheres. The mechanism behind this important observation is not fully understood although it was found that PIBCA nanospheres allowed higher intracellular concentrations of doxorubicin to be reached, which was correlated with a higher cytotoxicity compared to the free drug (Nemati *et al.*, 1994). However, the fact that drug incorporation by the cells was not influenced by cytochalasin B suggested that endocytosis was not the main mechanism of nanoparticle-cell association, although this was the most likely mechanism of interaction between such colloidal carriers and cells (Colin de Verdière *et al.*, 1994). It has been hypothesized that the rapid drug release of doxorubicin from nanospheres adhering to the cell membrane should lead to an overflow and saturation of the Pgp, resulting in the observed increase of drug diffusion in the intracellular compartment (Colin de Verdière *et al.*, 1994). Finally, doxorubicin nanospheres have been tested in a model of doxorubicin-resistant C6 rat glioblastomia lines differing by their degree and mechanism of resistance (Bennis *et al.*, 1993). The key finding of this study was that the reversal of doxorubicin resistance by nanospheres was closely dependent on the nature of the resistance: nanospheres were only efficient on pure MDR phenotype cells and not on the additional mechanism of resistance to doxorubicin (Bennis *et al.*, 1993).

A phase I clinical trial has been conducted in 21 patients with refractory solid tumors using doxorubicin loaded onto PIHCA nanospheres. In this study (Kattan *et al.*, 1992), granulocytopenia appeared to be the dose-limiting toxicity. However, the comparison of this toxicity to historical data relative to free doxorubicin suggests that nanoparticle doxorubicin is, in some respects, less myelotoxic. Using Doppler-Echocardiography, no cardiotoxicity was observed among 18 evaluable patients. However, this study needs to be confirmed with a higher number of patients receiving a bigger cumulative dose of doxorubicin (maximum cumulative dose was 180 mg/m^2). SGPT and SGOT serum levels remained normal throughout the treatment in all patients (Kattan *et al.*, 1992). The lack of hepatic toxicity could eliminate the possible risk of liver damage due to the accumulation of doxorubicin as a consequence of nanoparticle capture by hepatic Kupffer cells. The appearance of an allergy shortly after the beginning of the infusion in 2 patients at 15 mg/m^2, without a history of allergic reactions, was an unexpected and a disturbing event

(Kattan *et al.*, 1992). This complication was almost circumvented by diluting the drug in 250 ml Dextrose 5 % and increasing the duration of perfusion to 60 min.

Intracellular Infections

Intracellular infections were found to be another field of interest for drug delivery by means of PACA nanospheres. Indeed, infected cells may constitute a "reservoir" for microorganisms which are protected from antibiotics inside lysosomes. The resistance of intracellular infections to chemotherapy is often related to the low uptake of commonly used antibiotics or to their reduced activity at the acidic pH of lysosomes. To overcome these effects, the use of antibiotics-loaded nanospheres was proposed as endocytozable formulation (Fattal *et al.*, 1989). The preparation of ampicillin-loaded biodegradable nanospheres (PIBCA, PIHCA) was achieved and the antimicrobial activity of that drug was shown to remain unaltered upon linkage to the carrier (Henry-Michelland *et al.*, 1987; Fattal *et al.*, 1991). The effectiveness of PIHCA nanospheres was tested in the treatment of two experimental intracellular infections. Firstly, ampicillin-loaded nanospheres were tested in the treatment of experimental *Listeria monocytogenes* infection in congenitally athymic nude mice, a model involving a chronic infection of both liver and spleen macrophages (Youssef *et al.*, 1988). After adsorption of ampicillin onto nanospheres, the therapeutic activity of ampicillin was found to increase dramatically over that of the free drug. Bacterial counts in the liver were at least 20-fold diminished after linkage of ampicillin to PIHCA nanospheres. In addition, nanoparticulate ampicillin was capable of ensuring liver sterilization after two injections of 0.8 mg of nanospheres-bound drug whereas no such sterilization was ever observed with any of other regimens tested. Reappearance of living bacteria in the liver after the end of the treatment was probably due to a secondary infection derived from other organs such as the spleen which was not completly sterilized by the treatment (Youssef *et al.*, 1988).

Secondly, nanospheres-bound ampicillin was tested in the treatment of experimental salmonellosis in C57/BL6 mice, a model involving an acute fatal infection (Fattal *et al.*, 1989). All mice treated with a single injection of nanoparticle-bound ampicillin survived whereas all control mice and all those treated with unloaded nanospheres died within 10 days postinfection. With free ampicillin, an effective-curative effect required 3 doses of 32 mg each. Lower doses (3×0.8 mg and 3×16 mg) delayed but did not reduce mortality. Thus, the therapeutic index of ampicillin, calculated on the basis of mice mortality, was increased by 120–fold when the drug was bound to nanospheres.

Controlled Delivery of Peptides and Proteins

Oral delivery of peptides is severely hampered by their degradation in the gastro-intestinal tract. Spectacular results have been obtained with insulin encapsulated into nanocapsules of PIBCA (Damgé *et al.*, 1988). In diabetic rats, insulin nanocapsules given by a single intragastric administration after an overnight fast, reduced glycemia by 50–60%. This effect appeared 2 days after administration and was maintained for a period of 20 days. On the contrary, free insulin did not affect glycemia when administered orally under the same experimental conditions (Damgé *et al.*, 1988). It has been hypothesized that the long-term hypoglycemic effect could be due in part to the progressive arrival of nanocapsules from the stomach to the gut, leading to a delayed absorption. Grislain *et al.* (1983) have found intense radioactivity in the gastrointestinal tract of rats 4 h after intubation of radiolabelled PACA nanospheres; radioactivity was still present in the mucosa 24 h later. After arrival into the gut, insulin nanocapsules may be protected against proteolytic enzymes and absorbed by the intestinal mucosa. Then, nanocapsules are probably transported to the liver by the portal route. Thus, a slow process of redistribution from that organ and/or a slow release of insulin from nanocapsules could occur. Finally, because insulin is encapsulated in nanocapsules formed by an interfacial anionic polymerization process, it is not out of question that insulin is at least partly covalently linked to the polymer. This could explain the observed long-term hypoglycemic effect.

Even if the main limitation to oral adminstration of PACA nanoparticles is that their passage through the intestinal barrier is probably restricted and sometimes erratic, they represent an interesting tool for oral delivery of antigens. Indeed, M-cells appear to be the main site for the uptake of PACA nanoparticles after oral administration (Damgé *et al.*, 1990) and, furthermore, it is generally accepted that limited doses of antigen are sufficient for a mucous immunization. In fact, oral delivery of antigens may be considered as the most convenient means of producing an IgA antibody response. However, it is importantly limited by enzymatic degradation of antigens in the GI tract and, additionally by their poor absorption. Thus, it has been postulated that the use of micro- or nanoparticles for the oral delivery of antigens should be efficient if those systems are able to achieve the protection of the antigenic molecule. PACA nanoparticles have been shown to enhance the secretory immune response after their oral administration in association with ovalbumin (O'Hagan *et al.*, 1989). This result was not fully reproduced in the case of polyacrylamide nanoparticles loaded with the same antigen. It was postulated that in the case of polyacrylamide nanospheres, much of the antigen was located at the surface of the polymer and could have been degraded during its passage through the gut. The relatively high surface concentration of ovalbumin adsorbed onto PBCA particles may have reduced the ability of the proteolytic enzymes in the gut to gain access to and to degrade the antigen, resulting in a greater antigen availability.

Controlled Delivery of Antisense Oligonucleotides

Among the different strategies developed in order to fight against viral infections and cancer, the development of antisense oligonucleotides have received special attention. Indeed, these compounds are very specific because they only recognize and bind the complementary sequence of a gene or of its messenger ribonucleic acid. Both effects result in the alteration of the expression of the gene. However, the therapeutic potential of oligonucleotides *in vivo* remains limited because of their extremely high instability and the difficulty of those molecules to penetrate into cells. Thus, Chavany *et al.* (1992) have proposed to associate oligonucleotides with PACA nanoparticles. Oligo(thymidilate) was used as a model and was efficiently adsorbed (5 μmol/g polymer) at the surface of PIBCA and PIHCA nanospheres precoated with hydrophobic cations. The adsorption of the oligonucleotide was attributed to the formation of ion-pairs between the negatively charged phosphate groups of the nucleic acid chain and the hydrophobic cations. Binding was also dependent on the salt concentration, and on the length of the oligonucleotide. The most important point of this study was that the oligonucleotide associated with nanoparticle remained stable. Cell uptake studies have shown that the intracellular penetration of oligonucleotide nanoparticle was 8 times greater than the intracellular capture of the free oligonucleotide (Chavany, 1994). Recently, Nakada et al. (1995) have shown that in vivo, PIBCA nanospheres were able to increase the liver concentration and to reduce kidney elimination of an oligonucleotide. In addition, it was found that PIBCA nanospheres could partially protect the oligonucleotide against degradation in the plasma and in the liver, whereas free oligonucleotide was totally degraded. Finally, Schwab et al. (1994) reported that when applied to anti-ras oligonucleotides, this delivery system markedly inhibited Ha-ras-dependent tumor growth in nude mice after intratumoral injection.

Opthalmic Diseases

The first report on a nanoparticulate system for ocular delivery was achieved in 1980 by Gurny and Taylor (1980). Since then, various types of nanospheres were used in order to take advantage of the prolonged residence time of this system since the short elimination half-life of eye-drop solutions remains a major problem in ocular therapy. PACA nanospheres were found able to prolong the intraocular pressure-reducing effect of pilocarpine in rabbits for more than 9 hours (Diepold et al., 1989a). Similar results were obtained by Marchal-Heussler et al. (1990, 1992) using betaxolol as a drug model. It was found by Diepold et al. (1989b) that the concentration of ^{14}C-labelled PACA nanospheres in the cornea, conjunctiva, nictitating membrane and aqueous humor was 3 to 5-times higher in eyes in which a chronic inflammation was induced. This is an interesting observation which suggests that this type of nanospheres has an enhanced bioadhesiveness on inflamed tissues. In consequence, it should represent an interesting alternative to conventional eye-drops for the administration of antiinflammatory compounds.

CONCLUSION

Among the biodegradable materials in medicine and pharmacy, polyalkylyanoacrylates are very interesting compounds because they are easily obtained from the corresponding monomer in aqueous medium or when in contact with biological substrates. This property allows their application as surgical and dental adhesives, since the monomer can polymerize in situ.

Drug particulate injectable carriers (nanospheres, nanocapsules) can also be easily prepared by emulsion/polymerization of alkylcyanoacrylates, without the use of aggressive solvents or high energy application. The most significant advantage of PACA compared to other materials is that PACA nanoparticles allow to modify the intracellular traffic of drugs. Therefore, ways are now open for targeting intracellular bacteria responsible of opportunistic diseases or for administering anticancer compounds to cells in a manner that is able to bypass the cell detoxification process of P-glycoprotein. They could also represent, potentially, an attractive alternative to viral carriers for the administration of genetic material into cells.

The oral administration of peptides and proteins by means of PACA nanoparticles is another interesting perspective. Although the oral absorption of nanoparticles remains a controversial field of research, this approach achieved interesting and promising results. Thus, nanoparticles could also open very interesting perspectives for the oral delivery of antigens.

PACA nanoparticles represent an original drug delivery system which has emerged from the academic pharmaceutical research in Europe. In fifteen years, their technology has been improved and their potentials in therapeutics well documented. However, their future as drug carriers will now heavily depend on the will of the pharmaceutical industry to develop them and to release the financial means for determining their safety.

REFERENCES

Al Khouri-Fallouh, N., Roblot-Treupel, L., Fessi, H., Devissaguet, J.P. and Puisieux, F. (1986) Development of a new process for the manufacture of polyisobutylcyanoacrylate nanocapsules. *Int. J. Pharm.*, **28**, 125–132.

Alvotidinov,A. (1982) Study of the kinetics of polymerisation of Ethyl-a-cyanoacrylate. *Dokl. Akad. Nauk. Uzb. SSR*, **7**, 41–43.

Avery, B.S. and Ord, R.A. (1982) The use of butylcyanoacrylate as a tissue adhesive in maxillo-facial and cranio-facial surgery. *British J. Oral Surg.*, **20**, 84–95.

Beech, D.R. (1972) Bonding of Alkyl 2–cyanoacrylates to human dentin and enamel. *J. Dent. Res.*, **51**, 1438–1442.

Bennis, S., Chapey, C., Couvreur, P. and Robert, J. (1994) Enhanced cytotoxicity of doxorubicin encapsulated in poly-isohexylcyanoacrylate nanospheres against multidrug-resistant tumor cells in culture. *Eur. J. Cancer*, **30A**, 89–93.

Birkinshaw, C. and Pepper, D.C. (1986) The thermal degradation of polymer of n-butylcyanoacrylates prepared using phosphine and amine initiators, *Polym. Degradation and Stability*, **16**, 241–259.

Brauer, G.M., Kumpula, J.W., Termini, D.J. and Davidson, K.M. (1979) Durability of the bond between bone and various 2-cyanoacrylates in an aqueous environment. *J. Biomed. Mat. Res.*, **13**, 593–606.

Bricker, C.E. and Johnson, H.R. (1945) Spectrophotometric method of determining CH_2O. *Ind. Eng. Chem.*, **17**, 400–406.

Carton, C.A., Kessler, L.A., Seidenberg, B., Hurwitt, E.S. (1961) Experimental studies in surgery of small blood vessels. *J. Neurosurg.*, **18**, 188–194.

Casanova, R., Herrera, G.A., Engels, B.V., Velasquez, C. and Grotting, J.C. (1987) Micorarterial sutureless sleeve anastomosis using a polymeric adhesive: an experimental study. *J. Reconsrtuct. Micros.*, **3**, 201–210.

Chavany, C., Le Doan, T., Couvreur, P., Puisieux, F. and Helene, C. (1992) Polyalkylcyanoacrylate nanoparticles as polymeric carriers for antisense oligonucleotides. *Pharm. Res.*, **9**, 441–449.

Chavany, C., Saison-Behmoaras, T., Doan, T.L., Puisieux, F., Couvreur, P. and Helene, C. (1994) Adsorption of oligonucleotides onto polyisohexylcyanoacrylate nanoparticles protects them against nucleases and increases their cellular uptake. *Pharm. Res.*, **11**, 1370–1378.

Chiannilkulchai, N., Driouich, Z., Benoît, J.P., Parodi, A.L. and Couvreur, P. (1989) Doxorubicin-loaded nanoparticles: increased efficiency in murine hepatic metastases. *Select. Cancer Ther.*, **5**, 1–11.

Chiannilkulchai, N., Ammoury, N., Caillou, B., Devissaguet, J.Ph. and Couvreur, P. (1990) Hepatic tissue distribution of doxorubicin-loaded nanoparticles after, I.V. administration in reticulosarcoma M 5076 metastases-bearing mice. *Cancer Chemother. Pharmacol.*, **26**, 122–126.

Colin de Verdiere, A., Dubernet, C., Nemati, F., Poupon, M.F., Puisieux, F. and Couvreur, P. (1994) Uptake of doxorubicin from loaded nanoparticles in multidrug resistant leukemic murine cells. *Cancer Chemother. Pharmacol.*, **33**, 504–508.

Coover, H.W., Joyner, F.B., Shearer, N.H. and Wicker, T.H. (1959) Chemistry and performance of cyanoacrylate adhesives, *SPE J.*, 413–417.

Coover, H.W., Dreifus, D.W. and Connor, J.T. (1990) Cyanoacrylates adhesives. In Handbook of Adhesives, 3rd Edition, Skeist I. Ed., Van Nostrand Reinhold press, New York, pp. 463–477.

Couvreur, P., Kante, B., Roland, M., Guiot, P., Baudhuin, P. and Speiser, P. (1979) Polycyanoacrylate nanocapsules as potential lysosomotropic carriers: preparation, morphological and sorptive properties. *J. Pharm. Pharmacol.*, **31**, 331–332.

Couvreur, P., Kante, B., Lenaerts, V., Scailteur, V., Roland, M. and Speiser, P. (1980) Tissue distribution of antitumor drugs associated with polyalkylcyanoacrylate nanoparticles. *J. Pharm. Sci.*, **69**, 199–202.

Couvreur, P., Roland, M. and Speiser, P. (1982) Biodegradable submicroscopic particles containing a biologically active substance and composition containing them. US Patent no. 4,329,332.

Couvreur, P., Grislain, L., Lenaerts, V., Brasseur, P., Guiot, P. and Biernacki, A. (1986) Biodegradable polymeric nanoparticles as drug carrier for antitumor agents. In *Polymeric nanoparticles and microspheres* (Guiot P. and Couvreur P., Eds.). *CRC Press*, Boca Raton, Florida, pp. 27–93.

Couvreur, P., Dubernet, C. and Puisieux, F. (1995) Controlled drug delivery with nanoparticles: current possibilities and future trends, *Eur. J. Pharm. Biopharm.*, **41**, 2–13.

Cuvier, C., Roblot-Treupel, L., Millot, J.M., Lizard, G., Chevillard, S., Manfait, M., Couvreur, P. and Poupon, M.F. (1992) Doxorubicin-loaded nanospheres bypass tumor cell multidrug resistance. *Biochem. Pharmacol.*, **44**, 509–517.

Damgé, C., Michel, C., Aprahamiam, M., Pinget, M. and Couvreur, P. (1988) A new approach for oral administration of insulin using polyalkylcyanoacrylate nanocapsules as a drug carrier. *Diabetes*, **37**, 246–251.

Damgé, C., Michel, C., Aprahamiam, M., Couvreur, P. and Devissaguet, J.P. (1990) Nanocapsules as carriers for oral peptide delivery. *J. Control. Release*, **13**, 233–239.

Diepold, R., Keuter, J., Himber, J., Gurny, R., Lee, V.H.L., Robinson, J.R., Seattone, M.F. and Schnaudingel, O.E. (1989a) Comparison of different models for the testing of pilocarpine eyedrops using conventional eyedrop and a novel depot formulation (nanoparticles), *Graef Arch. Clin. Exp. Ophthal.*, **227**, 188–193.

Diepold, R., Keuter, J., Guggenbuhl, P. and Robinson, J.R. (1989b) Distribution of poly-hexyl-2-cyano-[3-^{14}C]acrylate nanoparticles in healthy and chronically inflamed rabbit eyes. *Int. J. Pharm.*, **54**, 149–153.

Donelly, E.F., Johnston, D.S. and Pepper, D.C. (1977) Ionic and zwitterionic polymerisation of n-alkyl 2-cyanoacrylates. *Polymer lett.*, **15**, 399–405.

Donnenfeld, E.D., Perry, H.D. and Nelson, D.B. (1991) Cyanoacrylate temporary tarsorrhapy in the management of corneal epithelial defects. *Opht. Surg.*, **22**, 591–593.

Douglas, S.J., Davis, S.S. and Illum, L. (1986) Biodistribution of polybutyl 2-cyanoacrylate nanoparticles in rabbits. *Int. J. Pharm.*, **34**, 145–152.

Eromoselle, I.C., Pepper, D.C. and Ryan, B. (1989) Water effects on the zwitterionic polymerization of cyanoacrylates. *Makromol. Chem.*, **190**, 1613–1622.

Fattal, E., Youssef, M., Couvreur, P. and Andremont, A. (1989) Treatment of experimental salmonellosis in mice with ampicillin-bound nanoparticles. *Antimicrob. Agents Chemother.*, **33**, 1540–1543.

Freeny, P.C., Bush, W.H. and Kidd Reiley (1979) Trancatheter occlusive therapy of genitourinary abnormalities using Isobutyl 2–cyanoacrylate (Bucrylate). *A.J.R.*, **133**, 647–656.

Giuliani, L., Carmignani, G., Belgrano, E. and Puppo, P. (1979) Gelatin foam and Isobutyl 2-cyanoacrylate in the treatment of life threatening bladder haemorraghe by selective transcatheter embolisation of the internal iliac arteries. *Brit. J. Urol.*, **51**, 125–128.

Goldin, A.R. (1976) Control of duodenal haemorrhage with cyanoacrylate., *British J. Radiol.*, **49**, 583–588.

Grislain, L., Couvreur, P., Lenaerts, V., Roland, M., Deprez-Decampeneere, D. and Speiser, P. (1983) Pharmacokinetics and distribution of a biodegradable drug-carrier. *Int. J. Pharm.*, **15**, 335–345.

Gros, P., Croop, J. and Houssman, D. (1986) Mammalian multidrug resistance gene: Complete cDNA sequence indicates strong homology to bacterial transport proteins. *Cell.*, **47**, 371–380.

Gurny, R. and Taylor, D. (1980) Development and evaluation of a prolonged acting drug delivery system for the treatment of glaucoma. In Proceedings of the International Symposium of the British Pharmaceutical Technology Conference, Rubinstein, M.H. Ed., Solid Dosage Research Unit, Liverpool, England.

Hafner, C.D., Fogarty, T.J. and Cranley, J.J. (1963) Nonsuture anastomosis of small arteries using a tissue adhesive. *Surg. Gyn. Obst.*, **116**, 417–421.

Harper, M.C. and Ralston, M. (1983) Isobutyl 2-cyanoacrylate as an osseous adhesive in the repair of osteochondral fractures. *J. Biomed. Mat. Res.*, **17**, 167–177.

Henry-Michelland, S., Alonso, M.J., Andremont, A., Maincent, P., Sauzières, J. and Couvreur, P. (1987) Attachment of antibiotics to nanoparticles: preparation, drug-release and antimicrobial activity *in vitro*. *Int. J. Pharm.*, **35**, 121–127.

Hosbein, D.J. and Blumenstock, D.A. (1964) Anastomosis of small arteries using a tissue adhesive. *Surg. Gyn. Obst.*, **118**, 112–114.

Hubert, B., Atkinson, J., Guerret, M., Hoffman, M., Devissaguet, J.P. and Maincent, P. (1991) The preparation and acute hypertensive effects of a nanocapsular form of darodipine, a dihydropiridine calcium entry blocker. *Pharm. Res.*, **8**, 734–738.

Kartner, N. and Ling, V. (1989) Multidrug resistance in cancer. *Sci. Am.*, 44–51.

Kattan, J., Droz, J.P., Couvreur, P., Marino, J.P., Boutan-Laroze, A., Rougier, P., Brault, P., Vranckx, H., Grognet, J.M., Morge, X. and Sancho-Garnier, H. (1992) Phase I clinical trial and pharmacokinetic evaluation of doxorubicin carried by polyisohexylcyanoacrylate nanoparticles. *Invest. New Drugs*, **10**, 191–199.

King, D.R., Reynolds, D.C. and Kruger, G.O. (1967) A plastic adhesive for nonsuture sealing of extraction wound in heparinized dogs. *Oral Surg. Oral Med. Oral Pathol.*, **24**, 307–312.

Koukabis, T.D., Glisson, R.R., Feagin, J.A., Seaber, Jr. A.V. and Parker Vail, T. (1995) Augmentation of meniscal repairs with cyanoacrylate glue, *J. Biomed. Mat. Res.*, **29**, 715–720.

Kubiak, C., Manil, L., Clausse, B. and Couvreur, P. (1989) Increased cytotoxicity of nanoparticle-carried adriamycin *in vitro* and potentiation by verapamil and amiodarone. *Biomaterials*, **10**, 553–556.

Lenaerts, V., Couvreur, P., Christiaens-Leyh, D., Joirris, E., Roland, M., Rollman, B. and Speiser, P. (1984) Identification and study of a degradation way for polyisobylytcyanoacrylate nanoparticles. *Biomaterials*, **5**, 65–68.

Lenaerts, V., Raymond, J., Juhaz, J., Simard, M.A. and Jolicoeur, C. (1989) New method for the preparation of cyanoacrylic nanoparticles with improved colloidal properties. *J. Pharm. Sci.*, **70**, 1051–1052.

Leonard, F., Kulkarni, R.A., Brandes, G., Nelson, J. and Cameron, J.J. (1966) Synthesis and degradation of poly(alkyl α-cyanoacrylates). *J. Appl. Polym. Sci.*, **10**, 259–272.

Lobene, R.R. and Sharawy, A.M. (1968) The response of alveolar bond to cyanoacrylate tissue adhesive. *J. Periondontol.*, **39**, 150–156.

Maxwell, J.A. and Goldware, S.I. (1973) Use of tissue adhesive in the surgical treatment of cerebrospinal fluid leaks. *J. Neurosurg.*, **39**, 332–336.

Müller, R.H., Lherm, C., Herbort, J. and Couvreur, P. (1990) *In vitro* model for the degradation of alkylcyanoacrylate nanoparticles. *Biomaterials*, **11**, 590–595.

Nakada, Y., Fattal, E., Foulquier, M. and Couvreur, P. (1995) Pharmacokinetics and biodistribution of oligonucleotide adsorbed onto polyalkylcyanoacrylate nanoparticles after iv administration in mice. *Pharm. Res.*, In press.

Nathan, H.S., Nachlas, M.M., Solomon, R.D. and Halpern, B.D. and Seligman, A.M. (1960) Nonsuture closure of arterial incisions using a rapidly polymerizing adhesive. *Ann. Surg.*, **152**, 648–665.

Nemati, F., Dubernet, C., Colin de Verdiere, A., Poupon, M.F., Treupel-Acar, L., Puisieux, F. and Couvreur, P. (1994) Some parameters influencing cytotoxicity of free doxorubicin and doxorubicin-loaded nanoparticles in sensitive and multidrug resistant leukemic murine cells: incubation time, number of nanoparticles per cell. *Int. J. Pharm.*, **102**, 55–62.

O'Hagan, D.T., Palin, K. and Davis, S.S. (1989) Poly(butyl-2-cyanoacrylate) particles as adjuvants for oral immunization. *Vaccine*, **7**, 213–216.

Pepper, D.C. (1978) Anionic and zwitterionic polymerisation of α-cyanocrylates. *J. Polym. Sci.*, **62**, 65–77.

Pepper, D.C. (1992) Zwitterionic chain polymerisation of cyanoacrylates. *Makromol. Chem. Macromol. Symp.*, **60**, 267–277.

Pinto-Alphandary, H., Balland, O., Laurent, M., Andremont, A., Puisieux, F. and Couvreur, P. (1994) Intracellular visualization of ampicillin-loaded nanoparticles in peritoneal manophages infected *in vitro* with Salmonella typhimurium. *Pharm. Res.*, **11**, 38–46.

Reynolds, R.C., Fasset, D.W, Astill, B.D. and Casarett, L.J. (1966) Absorption of Methyl-2-Cyanoacrylate-^{14}C from full-thickness skin incisions in the guinea pig and its fate *in vivo*. *J.S.R.*, **6**, 132–136.

Robinson, I.B. (1987) Molecular mechanism of multidrug resistance in tumor cells. *Clin. Physiol. Biochem.*, **5**, 140–151.

Rooney, J.M. (1981a) Thermal degradation of methyl and ethyl cyanoacrylates oligomers, *Brit. Polym. J.*, 160–163.

Rooney, J.M. (1981b) On the mechanism of oligomer formation in condensations of alkyl cyanoacetates with formaldehyde, *Polym. J.*, **13**(10), 975–978.

Schwab, G., Chavany, C., Duroux, I., Goubin, G., Lebeau, J., Helene, C. and Behmoaras, T.S. (1994) Antisense oligonucleotides adsorbed to polyalkylcyanoacrylate nanoparticles specifically inhibit mutated Ha-ras-mediated cell proliferation and tumorigenicity in nude mice. *Proc. Natl. Acad. Sci. USA.*, **91**, 10460–10464.

Seijo, B., Fattal, E., Roblot-Treupel, L. and Couvreur, P. (1990) Design of nanoparticles of less than 50 nm in diameter, preparation, characterization and drug loading. *Int. J. Pharm.*, **62**, 1–7.

Stein, M. and Hamacher, E. (1992) Degradation of polybutyl 2–cyanoacrylate microparticles. *Int. J. Pharm.*, **80**, R11–R13.

Takenaka, H., Esato, K., Ohara, M. and Zempo, N. (1992) Sutureless anastomosis of blood vessels using cyanoacrylate adhesives. *Surg. Today*, **22**, 46–54.

Tschopp, H.M. (1975) Small artery anastomosis using a cuff of dura mater and a tissue adhesive. *Plast. Reconstr. Surg.*, **55**, 606–611.

Tuncel, A., Çiçek, H., Hayran, M. and Piskin, E. (1995) Monosize poly(ethylcyanoacrylate) microspheres: preparation and degradation properties, *J. Biomed. Mat. Res.*, **29**, 721–728.

Vezin, W.R. and Florence, A.T. (1980) *In vitro* heterogeneous degradation of poly(n-alkyl-cyanoacrylates). *J. Biomed. Mater. Res.*, **14**, 93–106.

Vinters, H.V. and Ho, H.W. (1988) Effects of isobutyl 2-cyanoacrylate polymer on cultured cells derived from murine cerbral microvessels . *Toxic. in Vitro*, **2**, 37–41.

Vinters, H.V., Mark, M.D., Lundie, J. and Kaufmann, J.C.E. (1986) Long-term pathological follow-up of cerebral arteriovenous malformations treated by embolization with bucrylate, N. *Eng. J. Med.*, **314**, 477–483.

Weber, P.A. and Baker, N.D. (1989) The use of cyanoacrylate adhesives with a collagen shield in leaking filtering blebs, *Ophthalmic Surg.*, **20**, 284–285.

Wade, C.L. and Leonard, F. (1972) Degradation of poly(methyl 2-cyanoacrylates), *J. Biomed. Mat. Res.*, **6**, 215–220.

Youssef, M., Fattal, E., Alonso, M.J., Roblot-Treupel, L., Sauzières, J., Tancrède, C., Omnes, A., Couvreur, P. and Andremont, A. (1988) Effectiveness of nanoparticle-bound ampicillin in the treatment of *Listeria monocytogenes* infection in athymic nude mice. *Antimicrob. Agents Chemother.*, **32**, 1204–1207.

APPENDIX

CYANOACRYLATE

Monomer structure

R =	Sources	FDA Status	Price	Clinical trial or clinics
Methyl Ethyl	SIGMA CHEMICAL COMPANY, PO Box 14508 Saint Louis, MO 63178 USA	NO	100$/10g 100$/10g	NO
Butyl Hexyl	HENKEL, SICHEL-WERKE GMBH Postfach 911380 D-3000 Hanover GERMANY	NO	on request	NO
Methyl Ethyl	LOCTITE 4450 Cranwood Parkway Cleveland, OH 44128 USA	NO	on request	NO
Methyl Ethyl	POLYSCIENCES Inc. 400 Valley Road Warrington, PA 18976 USA	NO	260$/10g 210$/10g	NO

ABBREVIATIONS

DMF = dimethylformamide
DSC = differential scanning calorimetry
GPC = gel permeation chromatography
MW = molecular weight
HPLC = high performance liquid chromatography
NMR = nuclear magnetic resonance
PACA = polyalkylcyanoacrylate
PCS = photon correlation spectroscopy
SEM = scanning electron microscopy
TEM = transmission electron microscopy
THF = tetrahydrofuran
PEG = Polyethylene glycol
PMCA = polymethylcyanoacrylate
PECA = polyethylcyanoacrylate
PBCA = polybutylcyanoacrylate

PIBCA = polyisobutylcyanoacrylate
PHCA = polyhexylcyanoacrylate
PIHCA = polyisohexylcyanoacrylate
MPS = mononuclear phagocytic system

11. DEGRADABLE HYDROGELS

JUN CHEN, SEONGBONG JO and KINAM PARK*

Purdue University, School of Pharmacy, West Lafayette, IN 47907, USA

INTRODUCTION

Hydrogels are polymer networks which swell in water without dissolving. Hydrogels are usually made of hydrophilic polymers which are crosslinked by various interactions such as chemical bonds, hydrogen bonding, ionic interactions, and hydrophobic interactions. Hydrogels containing a significant portion of hydrophobic polymers can also be made. Hydrophobic polymers can be blended with hydrophilic polymers or form interpenetrating networks with hydrophilic polymer networks. Hydrophobic monomers can be copolymerized with hydrophilic monomers to form random copolymers or block copolymers in the presence of crosslinking agent. The properties of hydrogels can be controlled by careful choice of the type of monomers.

Even with a rather short history, hydrogels have been exploited extensively in various areas such as biomaterials (Pulapura and Kohn, 1992; Zhang *et al.*, 1993), agriculture (Rehab *et al.*, 1991), pharmaceutics (Kamath and Park, 1995; Park, *et al.*, 1993; Bae and Kim, 1993; Heller, 1993; Kamath and Park, 1993; Yoshida *et al.*, 1993), and biotechnology (Daubresse, 1994). Hydrogels have a high water holding capacity which is very important for applications in biological environments. Hydrogels possess good biocompatibility due to their unique properties such as low interfacial tension with surrounding biological fluids and the rubbery nature which can minimize mechanical and frictional irritation to the surrounding tissues. Hydrogels can be designed to respond (i.e., either swell or shrink) to changes in environmental conditions such as pH, temperature, electrical field, ionic strength, salt type, solvent, and light. It is these unique properties that have made hydrogels useful in various applications.

Hydrogels can be classified into matrix, film (or membrane), and microsphere according to their physical appearance; synthetic hydrogels and natural hydrogels according to their origin; homopolymer, copolymer, block copolymer, IPN (interpenetrating polymer networks), and semi-IPN hydrogels based on their structure; and physical gels and chemical gels based on the nature of crosslinking. All of these different forms of hydrogels can be made degradable, and this chapter deals those hydrogels which degrade under suitable conditions.

The concept of degradable materials, especially polymers, has a relatively short history. Recent development of degradable polymers is mainly due to environmental urgency caused by waste crisis and demand for sophisticated materials used in biomedical and pharmaceutical fields. Although many definitions exist for degradable polymers, there are no guidelines and classification methods generally accepted to define degradation of polymers. In the dictionary, degradation is defined as "the

---

*Correspondence: Kinam Park, Ph.D., Purdue University, School of Pharmacy, West Lafayette, IN 47907. Tel: (317) 494–7759; Fax: (317) 496–1903; E-mail: esp@omni.cc.purdue.edu

breakdown of organic compounds". All materials are degradable if sufficient time is allowed, but only some of them are degradable in our time scale. Therefore, kinetic classification of degradable materials is necessary. Albertsson has applied the kinetic concept (i.e., time required to degrade polymers) to classify degradable polymers (Albertsson, 1993). Inert polymers such as polyethylene is classified as a "polymer which degrade over a long period of time". Polyethylene/starch blend is an example of "polymers with medium long degradation time". Hydrolyzable polymers are assigned to the "polymers with short degradation time".

In addition to the kinetic concept, mechanisms of degradation, sources of degradable materials, and preparation methods can be used to classify degradable polymers (Hocking, 1992). Polymers can be degraded by photolysis, photo-oxidation, thermolysis, thermooxidation, and biodegradation. Biodegradation of polymers are of great importance for biomedical, pharmaceutical, and environmental applications. Biodegradation has been defined as conversion of materials into less complex intermediate or end-products by solubilization, simple hydrolysis, or the action of biologically formed entities (Park *et al.*, 1993; Kamath and Park, 1993). Bioabsorption, bioresorption, bioerosion, and biodeterioration are other expressions of *in vivo* degradation of polymeric materials. Various factors affecting biodegradability of synthetic polymeric materials such as water, temperature, pH, and oxygen have been discussed in the literature (Satyanaryana and Chatterji, 1993). The uniqueness of (bio)degradable hydrogels is that the degradation occurs in the presence of abundant water.

HYDROGEL PREPARATION

Various materials originated from natural resources or chemical synthesis can be used for the preparation of degradable hydrogels. Methods of hydrogel preparations have been reviewed extensively in many references (Bae and Kim, 1993; Heller, 1993; Kamath and Park, 1993; Yoshida *et al.*, 1993; Yokoyama, 1992). For the discussion of degradable hydrogels, we will classify hydrogels into chemical and physical gels.

Chemical Gels

Monomers with vinyl groups (e.g., acrylic acid, acrylamide, hydroxyethyl methacrylate, and vinylpyrrolidone) or macromolecules modified with vinyl groups (e.g., albumin, gelatin, dextran, and starch which are modified with glycidyl acrylate) are usually polymerized by chemical initiation, photoinitiation, or gamma irradiation in the presence of molecules with divinyl group as a crosslinker. Polymers with other functional groups can be crosslinked by chemical reactions. For example, albumin and gelatin can be crosslinked with dialdehyde (Goosen *et al.*, 1982; Lee *et al.*, 1981; Silvio *et al.*, 1994; Akin and Hasirci, 1995; Tabata and Ikada, 1989), and cystein-bearing polypeptides can be crosslinked through cystein bonds (Lizuka *et al.*, 1993). Degradable hydrogels prepared by chemical crosslinking have either degradable polymer backbone or degradable crosslinking agent. The hydrogels with degradable backbones can be prepared from crosslinking polymers which are degradable chemically or enzymatically in biological environments.

Table 1 Types of physical gels

1. Blend hydrogels
2. IPNs and semi-IPNs
3. Block copolymer hydrogels
4. Polyelectrolyte complex hydrogels
5. Counter ion induced hydrogels
6. Thermally induced hydrogels
7. Specific interaction induced hydrogels

Degradable polymer backbones can be prepared easily by a simple modification of degradable polymers with polymerizable substituents. Degradable polyesters containing double bond can be used as prepolymers. Hydrogels with degradable crosslinkers can be prepared from polymerization of monomers which form non-degradable polymer backbones with degradable crosslinking agents.

Physical Gels

Some hydrogels are formed by physical interactions between polymer chains. These interactions include hydrogen bonding, hydrophobic interactions, and ionic interactions. Several types of physical gels are listed in Table 1.

Blend hydrogels are usually prepared by solvent casting or precipitation of a solution in which two or more different polymers are dissolved. The blend hydrogels normally have properties different from hydrogels made of individual polymers. When hydrogels are prepared this way, at least one polymer must be hydrophilic to provide the water absorbing property. Poly(ethylene oxide)-poly(propylene oxide)-poly(ethylene oxide) (PEO-PPO-PEO) block copolymers were blended with poly(lactic acid) (PLA) to make partially degradable hydrogels (Park *et al.*, 1992). In this system, PEO-PPO-PEO block copolymers provided hydrophilicity while PLA provided degradability. Cautions must be taken when blend hydrophilic and hydrophobic polymers because phase separation is easy to occur in blend hydrogels.

Another type of physical gels is interpenetrating polymer networks (IPNs) or semi-IPNs. IPNs are any materials containing two different types of polymers, each in a network form (Sperling, 1981; Frisch, 1985). In semi-IPNs only one type of polymer exists as a network form. IPNs and semi-IPNs are used to enhance the compatibility and degradability of polymer components and prevent phase separation. IPNs and semi-IPNs of poly(hydroxyethyl methacrylate) (PHEMA) and polycaprolactone (PCL) were prepared (Eschbach and Huang, 1993; Davis *et al.*, 1988; Eschbach and Huang, 1991, 1994). The hydrophobic PCL was added to enhance the mechanical properties and to provide degradability.

Physical gels can also be prepared from block copolymers. A number of degradable, water-insoluble polymers can be copolymerized with hydrophilic polymers to form block copolymers with degradability. PEO-PLA block copolymers, PEO-poly(glycolic acid) (PGA) block copolymers, and PEO-PCL block copolymers are good examples (Sawhney *et al.*, 1993, 1994; Hubbell *et al.*, 1993). Some degradable hydrogels are made of polysalt or polyelectrolyte complexes. Ionic interactions between two oppositely charged polyelectrolytes lead to the formation of polyelectrolyte complexes which swell in water. One example is the formation of polyelectrolyte film of amino group-containing chitosan with sodium alginate, a polyanion (Hagino and Huang, 1995). Certain polyelectrolytes can form hydrogels in the presence of counter ions. A typical example is sodium alginate which gels in the presence of calcium ions.

Other types of physical gels include thermally induced networks and hydrogels based on specific interactions. Thermally induced networks are formed when thermal energy induces the structural change of polymer in solution or change in the balance between hydrogen bonding and hydrophobic interactions. One example is the gelation of gelatin solution as temperature is lowered (Djabourov *et al.*, 1985). Agarose gels can also be prepared by this method (Bellamkonda *et al.*, 1995). Polymer networks can also be formed by specific interactions such as interactions between glucose and concanavalin A (Lee and Park, 1994, in press). Vinyl group-modified glucose was copolymerized with vinylpyrrolidone, and the glucose-containing polymer chains were then crosslinked with concanavalin A, a lectin with specific binding sites for glucose, to form a network. The network (i.e., gel) degrades (i.e., becomes a sol) in the presence of free glucose molecules. In the absence of free glucose, the sol becomes a gel again, and the sol-gel process can be repeated.

TYPE OF DEGRADATION

There are several types of hydrogel degradation. Backbone degradation occurs when polymer backbone chains are cleaved into lower molecular weight, water-soluble fragments. Table 2 lists some examples of synthetic hydrogels with degradable backbone chains. Well known degradable synthetic polymers, are PLA, PGA, PCL, and polyhydroxybutyrate. Since these degradable polymers are generally too hydrophobic to make hydrogels, hydrophilic polymers need to be added to impart enough hydrophilicity. One method is to graft degradable hydrophobic polymers onto hydrophilic polymers and then crosslink the prepolymers into hydrogels. Sawhney *et al.* (1993, 1994) grafted PLA, PGA, and PCL onto PEO chain to make hydrogels. The hydrogel backbone was degradable because PLA, PGA, and PCL could be hydrolyzed. Another method is to physically blend hydrophilic polymers with degradable but hydrophobic polymers. Pitt *et al.* (1992) blended poly(vinyl alcohol) (PVA) with poly(glycolic acid-co-lactic acid) (PGLA) to make degradable hydrogel films. Some researchers made semi-IPN or IPN using those hydrophobic polymers as backbone chains. An example is semi-IPN or IPN prepared from PHEMA and PCL (Eschbach and Huang, 1991).

Since many of the synthetic hydrophilic polymers are not degradable, another approach to design degradable hydrogels is to use degradable crosslinkers. When the degradation of crosslinker is the main mode of degradation, it is called "crosslinker degradation". This degradation mode leads to degradation products of high molecular weight. Many types of degradable crosslinking agents have been developed. Table 3 shows some examples of commonly used degradable crosslinkers. The crosslinkers vary widely in size ranging from small molecules such as azoaromatic compounds (Brondsted and Kopecek, 1992; Kopecek *et al.*, 1992), to oligopeptides (Rejmanova *et al.*, 1981; Ulbrich *et al.*, 1982), and to macromolecules such as

Table 2 Synthetic hydrogels with degradable polymer backbone

Backbone polymers	Preparation methods	Degradation modes	Applications	Ref.
PEO/PLA, PGA, or PCL	PEO was central block with PLA, PGA, or PCL extended on both side and terminated with acrylate group. This precursor was photoinitiated by 2,2-dimethoxy-2-phenylacetophenone to form gel.	Degradation of backbone in pH 7.3 PBS buffer at 37°C within 1 day to several months.	Control postoperative wound healing and prevent postoperative adhesion.	Sawhney et al., 1993, 1994; Hubbell et al., 1993
PEO-PPO-PEO/PLA	PEO-PPO-PEO block copolymers were reacted with D,L-lactide to extend them with lactyl group on both ends. The resulting diols were further modified with acryloyl chloride to make diacrylates. These macromonomers were photoinitiated to form hydrogels.	Degradation of backbone in pH 7.4 PBS buffer solution.		Barman et al., 1995
	PLA was blended with PEO-PPO-PEO block copolymers. The block copolymers were used to increase the hydrophilicity of the system.	No effect of PEO-PPO-PEO block copolymers on PLA degradation in water.	Controlled release of protein. BSA was used as a model protein.	Park et al., 1992
Polyesters	Water-soluble polyesters containing double bonds were crosslinked with N-vinylpyrrolidone.	Chemical hydrolysis of polymer backbone.	Controlled release of macromolecules.	Heller et al., 1983
Poly(ester-urethane)	Crosslinking of polyester triols prepared from D,L-lactide, ε-caprolactone, or comonomer mixtures with ethyl 2,6-diisocyanatohexanoate.	Chemical hydrolysis of polymer backbone.		Storey et al., 1994

Table 2 *Continued*

Backbone polymers	Preparation methods	Degradation modes	Applications	Ref.
PHEMA/PCL	Semi-IPNs were prepared by crosslinking HEMA with EGDMA in the presence of PCL. PCL provided degradability.	Chemical hydrolysis, enzymatic degradation of PCL by *Cryptococcus laurentii*.	PCL was used to increase the mechanical strength of PHEMA gel.	Eschbach and Huang, 1993, 1994; Davis *et al.*, 1988
	IPNs were prepared by dissolving HEMA, EGDMA, and PCL which had itaconic anhydride group on both ends to form homogeneous solution. HEMA was first crosslinked by EGDMA using AIBN as initiator. Then functionized PCL was crosslinked using dicumyl peroxide as initiator.			Eschbach and Huang, 1991
Polyhydroxy-butyrate	Polyhydroxybutyrate and its copolymer with polyhydroxyvalerate were blended with dextran or amylose and compressed into disc shape. Dextran and amylose were to enhance the enzyme degradability.	Hydrolysis of polyhydroxybutyrate and its copolymer. Also degradation by amylase when blended with amylose.		Yasin and Tighe, 1993
Polyphos-phazene	Poly[di(carboxylatophenoxy)-phosphazene] and Poly[(carboxylatophenoxy) (glycinato)-phosphazene] were crosslinked by Ca²⁺ to make microspheres. Some were further coated by polylysine.	Degradation of backbone in pH 7.4 buffer solution at 37°C.		Andrianov *et al.*, 1994
PVA/PGLA	PVA was blended with PGLA to make film which was used to coat drug spheres.	Hydrolysis of PGLA at 37°C.	Control the permeability of drugs from the membrane.	Pitt *et al.*, 1992
Synthetic copolymers	Free radical copolymerization of 2-*p*-nitrobenzoxyethyl methacrylate and methoxydiethoxyethyl methacrylate.	Chemical hydrolysis of polymer backbone.	Drug delivery system.	Pitt and Shah, 1995

AIBN = 2, 2'-azobisisobutyronitrile; BSA = bovine serum albumin; EGDMA = ethylene glycol dimethacrylate; HEMA = hydroxyethyl methacrylate; IPN = interpenetrating polymer network; PBS = phosphate buffered solution; PCL = polycaprolactone; PEO = poly (ethylene oxide); PGA = poly(glycolic acid); PGLA = poly (glycolic acid-co-lactic acid); PHEMA = poly(hydroxyethyl methacrylate); PLA = poly(lactic acid); PPO = poly (propylene oxide); PVA = poly (vinyl alcohol).

Table 3 Degradable crosslinking agents used in the preparation of hydrogels

Degradable crosslinking agent	Degradation	Ref.
1. Small molecules		
N,N'-methylenebisacrylamide	Hydrolysis to formaldehyde.	Torchilin et al., 1977
N,O-dimethacryloyl hydroxylamine	Chemical hydrolysis.	Ulbrich et al., 1993, 1995
Azoaromatic compounds	Degradation by microbial azoreductase in colon.	Brøndsted and Kopecek, 1992; Kopecek et al., 1992
– p,p'-Divinyl azobenzene		
– N,N'-Bis(p-styrylsulfonyl)-4,4'-diamino azobenzene		
– N,N'-Dimethacryloyl-4,4'-diamino azobenzene		
– 4,4'-Di(N-methacryloyl-ω-aminocapryloyl amino) azobenzene		
– N,N'-(ω-Aminocapryloyl)-4,4' diamino azobenzene		
– 3,3',5,5'-Tetra bromo-4,4',4'-tetra (methacryloylamino) azobenzene		
2. Oligopeptides		
N-methacryloylated oligopeptide + diamine	Degradation by enzymes.	Rajmanova et al., 1981; Ulbrich et al., 1982
3. Macromolecules		
Intact proteins	Degradation by enzymes.	Kopecek and Rejmanova, 1979; D'Urso and Fortier, 1994
– insulin, albumin		
Functionalized protein		Park, 1988; Shalaby et al., 1990, 1992
– Glycidyl acrylate modified albumin		
Modified polysaccharides		
– Allyl carboxymethylcellulose		Buyanov et al., 1989
– Allyl hydroxyethylcellulose		
– Allyl dextran		Buyanov et al., 1992

modified proteins (Park, 1988; Shalaby *et al.*, 1990; Shalaby *et al.*, 1992) and polysaccharides (Buyanov *et al.*, 1992). Table 4 lists examples of synthetic hydrogels with degradable crosslinkers. A degradable azoaromatic compound, N,N'-(ω-aminocaproyl)-4,4'-diaminoazobenzene, was used to crosslink copolymers of N,N-dimethylacrylamide, N-tert-butylacrylamide, acrylic acid, and N-methacryloyl-glycylglycine *p*-nitrophenyl ester. This hydrogel system was used for colon-specific drug delivery because the crosslinker could be degraded by azoreductases, the bacterial enzymes, present predominantly in the colon (Brondsted and Kopecek, 1992; Kopecek *et al.*, 1992; Yeh *et al.*, 1994; Kopecek *et al.*, 1993). Degradable hydrogels of PHEMA were made by using modified synthetic polymers, such as α, ω-dimethacryloyl-poly(D,L)-lactide (Barakat *et al.*, 1994) and PCL endcapped with itaconic anhydride (Eschbach and S.J. Huang, 1994a, 1994b), as degradable crosslinkers. Glycidyl acrylate (GA)-modified albumin was used to crosslink poly-acrylamide, poly(acrylic acid) (Park, 1988), and poly(vinylpyrrolidone) (Shalaby *et al.*, 1990; Shalaby, Chen and Park, 1992; Shalaby, Blevins and Park, 1992; Shalaby, Peck and Park, 1991; Shalaby and K. Park, 1990). These hydrogels were degraded by trypsin and pepsin which acted on the crosslinker albumin.

Sometimes both backbone degradation and crosslinker degradation occur simultaneously. This degradation mode mostly happens to hydrogels from natural polymers such as proteins and polysaccharides. Another type of hydrogel degradation is pendent group degradation. Degradable pendent groups such as oligopeptides are used as spacer to attach drugs to the polymer backbone (Subr *et al.*, 1988).

MECHANISMS OF DEGRADATION IN BIODEGRADABLE HYDROGELS

The most common mechanisms of biodegradation are solubilization, hydrolysis, and enzyme catalyzed degradation. In solubilization, polymer matrices composed of hydrophilic polymers absorb water and swell to form hydrogels in aqueous environment. The diffusion of water molecules into hydrogels would be facilitated by the swelling of polymer matrixes. More water absorption by swollen hydrogel matrix leads to the breakdown of polymer networks and individual polymer molecules are dissolved in water. This phenomenon is generally known as erosion. A special type of solubilization is associated with charge formation. Polymer molecules which have ionizable functional groups such as amine group and carboxyl group are ionized dependent upon pH of the aqueous environment. The increase of ionized groups in a polymer chain increases the hydrophilicity of the polymer and induces the dissolution of the polymer chain. The degradation of pendent groups also can impart ionized groups into polymer chain. Hydrolysis of pendent anhydrides or esters results in ionized acids which can induce polymer dissolution.

Chemical hydrolysis is another important degradation mechanism. Hydrogels based on biodegradable polyesters such as poly(glycolic acid), poly(lactic acid), poly(ε-caprolactone), poly(β-hydroxybutyric acid), polyphosphazene, and poly(ester urethane) are degraded by simple hydrolysis (Table 2). Hydrogels made of natural resources can also undergo hydrolysis. Examples are hydrogels made of carrageenan (Singh, 1994), chitosan (Chandy and Sharma, 1991), α-methylgalactoside (Chen *et al.*, 1995), and sucrose (Chen and Park, unpublished data).

In addition to solubilization and hydrolysis, enzymatic degradation is also a very common degradation mechanism. The enzymatic degradation can act on

Table 4 Synthetic hydrogels with degradable crosslinkers

Backbone polymers	Preparation methods	Degradation modes	Applications	Ref.
PEG	Crosslinking the poly(ethylene glycol) activated with *p*-nitrophenyl chloroformate using albumin.	Degradation of crosslinker.	Potential biomaterial.	D'Urso and Fortier, 1994
Polyacrylamide	Acrylamide was crosslinked by degradable polypeptide, methacryloyl chloride modified poly[N-(2-hydroxyethyl)-L-glutamide].	Degradation of crosslinker by papain in PBS buffer.	Controlled release of biologically active macromolecules. Trypsin was used as a model protein.	Skorda *et al.*, 1993
	Acrylamide was crosslinked by GA modified albumin.	Degradation of albumin crosslinker by trypsin and pepsin.	Drug delivery.	Park, 1988
	Acrylamide was crosslinked with functionalized poly[N⁵-(2-hydroxyethyl)-L-glutamide].	Enzymatic degradation of the crosslinker by papain.	Controlled release of drug.	Skarda *et al.*, 1993
Poly(acrylic acid)	Acrylic acid was crosslinked by GA modified albumin.	Degradation of albumin crosslinker by trypsin and pepsin.	Drug delivery.	Park, 1988
PHEMA	PHEMA was crosslinked by degradable crosslinker, α, ω-dimethacryloyl-P(D,L)-lactide.	Hydrolysis of crosslinker.		Barakat *et al.*, 1994
	PHEMA was crosslinked by difunctional PCL endcapped with itaconic anhydride on both ends.	Degradation of crosslinker by hydrolysis, enzyme, and chemicals. PCL provided degradability while PHEMA brought the system with hydrophilicity which further accelerated the degradation.		Eschbach and Huang, 1994a, 1994b

Table 4 *Continued.*

Backbone polymers	Preparation methods	Degradation modes	Applications	Ref.
Poly[N-(2-hydroxypropyl)-methacrylamide]	Polymerization of N-(2-hydroxypropyl)-methacrylamide using N,O-dimethacryloyl hydroxylamine as crosslinker.	Chemical hydrolysis of N,O-dimethacryloyl hydroxylamine.	Drug delivery system with free or conjugated doxorubicin.	Ulbrick et al., 1993, 1995
	Copolymers of N-(2-hydroxypropyl) methacrylamide and *p*-nitrophenyl esters of N-methacrylated oligopeptides were crosslinked with diamine.	Enzymatic degradation of oligopeptides by chymotrypsin.	Drug carriers.	Rejmanova et al., 1981; Ulbrich et al., 1982
	Copolymers of N-(2-hydroxypropyl) methacrylamide and *p*-nitrophenyl esters of N-methacryloylated ε-aminocaproyl L-leucine were crosslinked with insulin.	Enzymatic degradation of insulin by chymotrypsin.		Kopecek and Rejmanova, 1979
PVP	VP was crosslinked by GA modified albumin.	Degradation of crosslinker by pepsin.	Gastric retention device targeted for long-term oral drug delivery.	Shalaby et al., 1990, 1991, 1992, 1993
	PVP hydrogel crosslinked by GA modified albumin was swelled in acryloxyethyltrimethylammonium chloride and GA modified albumin solution. This system was then subjected to γ-irradiation to form IPN.	Degradation of modified albumin by pepsin.		Shalaby et al., 1993
PVP/polyacrylamide	VP and acrylamide were crosslinked by BIS.	Chemical hydrolysis. Degradation depended upon crosslinking density.	Immobilization of enzyme.	Torchilin et al., 1977
Synthetic copolymers	Copolymers of N,N-dimethylacrylamide, N-tert-butylacrylamide, acrylic acid, and N-methacryloylglycylglycine *p*-nitrophenyl ester was crosslinked by degradable crosslinker N,N'-(ω-aminocaproyl)-4,4'-diaminoazobenzene.	Degradation of crosslinker by bacterial enzyme, azoreductases, present predominantly in the colon.	pH-sensitive, colon-specific drug delivery system.	Brondsted and Kopecak, 1992; Kopecak et al., 1992; Yeh et al., 1994; Kopecak et al., 1993

BIS = N,N'-methylenebisacrylamide; GA = glycidyl acrylate; PEG = poly (ethylene glycol); PVP = poly(vinylpyrrolidone); VP = vinylpyrrolidone.

both hydrogel backbone and crosslinker. Most enzyme-degradable hydrogels are prepared from natural polymers, such as polypeptides and polysaccharides, since most of them are inherently degradable by specific enzymes. Various enzymes which participate in the metabolism of bioorganisms can be exploited for the breakdown of hydrogel networks. Table 5 lists degradable hydrogels based on proteins and polypeptides. Albumin and gelatin are two proteins most widely used for preparation of enzyme-degradable hydrogels. Most of the hydrogels based on proteins or polypeptides can be degraded by proteases such as chymotrypsin, trypsin, pepsin, and papain. The most common method of preparing protein hydrogels is to crosslink the protein solution with dialdehyde such as glutaraldehyde. Many protein hydrogels have been made by this method (Goosen *et al.*, 1982; Lee *et al.*, 1981; Silvio *et al.*, 1994; Cremers *et al.*, 1990; Rathna *et al.*, 1994; Kaur and Chatterji, 1990; Ma *et al.*, 1995; Hiroyuki *et al.*, 1995). Other methods of preparing protein hydrogels, such as cooling (Djabourov *et al.*, 1985; Livingston *et al.*, 1994) and heat denaturation (Yapel, 1985), are also listed in Table 5.

Table 6 lists degradable hydrogels prepared from polysaccharides. Alginate, chitosan, dextran, and starch have been exploited extensively. Degradation of polysaccharide hydrogels occurs mostly through hydrolysis and enzymatic reaction. Unlike the protein hydrogels, the enzymatic degradation of polysaccharide hydrogels requires specific enzymes, such as κ-carrageenase for κ-carrageenan (Knutsen and Grasdalen, 1992), chitosanase for chitosan (Okajima *et al.*, 1994), dextranase for dextran (Kamath and K. Park, 1995), and α-amylase for starch (Heller *et al.*, 1990). The methods of preparation of polysaccharide hydrogels are more versatile. Many polysaccharide hydrogels are prepared through ionic interaction. For example, alginate (Bowersock *et al.*, 1994) or pectin (El-Nawawi and Heikal, 1995) molecules were crosslinked by Ca^{2+}; chitin by Fe^{3+} (Tokura *et al.*, 1990); and chitosan by sodium tripolyphosphate (Zihnioglu and Telefoncu, 1995). Thermally induced gelation is also a common way of preparing polysaccharide hydrogels. Agarose (Bellamkonda *et al.*, 1995), amylopectin (Durrani and Donald, 1995), and carrageenan (Singh, 1994) gels were made by cooling. Epichlorohydrin, which reacts with free hydroxy groups, has been used to crosslink many polysaccharides. Cellulose (Rehab *et al.*, 1991) and dextran (Pereswetoff-Morath and Edman, 1995) were two examples. Polysaccharides containing amine groups, such as chitin (Pangburn *et al.*, 1982) and chitosan (Beena *et al.*, 1995), have been crosslinked by glutaraldehyde. Some polysaccharides have been modified with vinyl groups and then polymerized to form hydrogels. Glycidyl acrylate and glycidyl methacrylate are two common alkylating agents used to modify polysaccharides. Alginate (Paparella and Park, 1994), dextran (Daubresse *et al.*, 1994; Edman *et al.*, 1980; Dijk-Wolthuis *et al.*, 1995), gellan (Paparella and Park, 1994), hyaluronic acid (Paparella and Park, 1994), and starch (Heller *et al.*, 1990; Artursson *et al.*, 1984; Stjarnkvist *et al.*, 1989) hydrogels have been prepared by this method. Table 6 also lists other methods of preparing polysaccharide hydrogels.

APPLICATIONS OF DEGRADABLE HYDROGELS

Hydrogels have been extensively used as drug delivery systems. Since hydrogels are hydrophilic in nature, they are generally very biocompatible. This is especially important when a hydrogel is to be implanted into the body. Unlike conventional

Table 5 Degradable hydrogels made of proteins and polypeptides

Hydrogel components	Preparation methods	Degradation modes	Applications	Ref.
Albumin	Albumin microspheres were prepared by heat denaturation.			Yapel, 1985
	Albumin was crosslinked with glutaraldehyde to make microbeads.	Chymotrypsin digestion on subcutaneous implantation in diabetic rats.	Controlled release of insulin.	Goosen *et al.*, 1982
	Albumin was crosslinked with glutaraldehyde to make microspheres.	Degradation of albumin by chymotrypsin.	Sustained drug delivery. Progesterone was the model drug.	Lee *et al.*, 1981
Albumin/ heparin	Albumin-heparin microspheres were prepared using double crosslinking technique, i.e., first by 1-ethyl-3-(3-dimethylaminopropyl)carbodiimide and subsequently by glutaraldehyde.	No complete degradation in the mice liver after 6 weeks.	Drug delivery system. Adriamycin was the model drug.	Cremers *et al.*, 1990, 1995; Kwon *et al.*, 1992
Albumin/PEG	Albumin was crosslinked with activated PEG, [di(*p*-nitrophenyl)-PEG-carbonate].		Enzyme immobilization.	D'Urso *et al.*, 1994; Gayet and Fortier, 1994
Collagen	Hydrogels of collagen and collagen with added laminin or fibronectin were prepared by cooling.		Incorporate nerve growth factor to control the nervous tissue regeneration.	Livingston *et al.*, 1994
	Lyophilized mixture of collagen solution with interferon was shaped to minipellets.	Degradation of minipellets administered subcutaneously to mice.	Interferon delivery.	Fujioka *et al.*, 1995a, 1995b
Collagen/ PHEMA	HEMA mixed with collagen and anticancer drugs was polymerized by chemical initiation to make hydrogel.		Controlled release of anticancer drugs, such as 5-fluorouracil, bleomycin, and mitomycin.	Jeyanthi and Rao, 1990

Table 5 *Continued*

Hydrogel components	Preparation methods	Degradation modes	Applications	Ref.
Fibrin	Fibrin beads were prepared in two different size distributions by varying the ratio of oleic acid to mineral oil in the oil phase of the emulsion system.		Delivery of macromolecules.	Ho et al., 1995
Gelatin	Gel formed by cooling.			Djabourov et al., 1985
	Microspheres were prepared by crosslinking gelatin with glutaraldehyde.	Degradation of gelatin by pepsin, collagenase, and digested by macrophages.	Delivery system for growth hormone and interferon.	Silvio et al., 1994; Akin and Hasirci, 1995; Tabata and Ikada, 1989
	Gelatin was crosslinked by hexamethylene diisocyanate via urea linkages in 2,2,2-trifluoroethanol.			Zhao et al., 1995
	Hydrogels were prepared by gamma irradiation of GA modified gelatin.	Degradation of gelatin by pepsin.		Kamath and Park, 1992
Gelatin/CMC	CMC was mixed with gelatin which was then crosslinked by glutaraldehyde to make semi-IPNs. CMC was used to control the swelling.			Rathna et al., 1994
Gelatin/Polyacrylamide	IPNs were prepared by crosslinking gelatin with glutaraldehyde and acrylamide with BIS respectively.			Kaur and Chatterji, 1990; Chatterji, 1991

Table 5 *Continued*

Hydrogel components	Preparation methods	Degradation modes	Applications	Ref.
Gelatin/ PHEMA	IPNs were prepared by crosslinking gelatin with glutaraldehyde and HEMA with BIS respectively.		Smart materials which bent in respond to the environment change such as pH, ion strength, and electric field.	Ma *et al.*, 1995
Lectin	Lectins are carbohydrate binding proteins. Hydrogen bonds and van der Waals forces were involved in stabilization of the interaction.		For self-regulated insulin delivery.	Lee and Park, 1994, in press; Heller, 1988
Polypeptides	Poly(L-ornithine) and copoly(ornithine tyrosine) were crosslinked by glutaraldehyde.	Degradation of backbone by chymotrypsin.	A polylysine homologue, they were used as polycations.	Hiroyuki *et al.*, 1995
	Poly(N-hydroxyethyl L-glutamine), poly(N-hydroxypropyl L-glutamine), poly(N-hydroxypentyl L-glutamine), and their copolymers were crosslinked by octamethylenediamine.	Degradation of backbone by papain.	Used as a skin substitutes.	Nakanishi *et al.*, 1991
	Poly(2-hydroxyethyl-L-glutamine) was crosslinked by diaminododecane to make hydrogel films.	Degradation of backbone *in vivo* through hydrolysis by proteolytic enzymes released during the normal inflammatory response. Degradation *in vitro* by pronase and papain.		Dickson *et al.*, 1981; Dickson and Hiltner, 1981
	Copolymer of L-cystein, L-alanine, L-glutamic acid was crosslinked through cystein residue by potassium ferricyanide oxidation.	Degradation of backbone by papain.	Used as a skin substitutes.	Lizuka *et al.*, 1993

CMC = carboxymethylcellulose.

Table 6 Degradable hydrogels made of polysaccharides

Hydrogel components	Preparation methods	Degradation modes	Applications	Ref.
Agarose	Physical gels were prepared by cooling agarose solution.		An extracellular matrix equivalent with a physicochemical structure capable of supporting neurite extension from primary neural cells.	Bellamkonda et al., 1995
Alginate	Alginate microspheres were prepared by crosslinking with Ca^{2+} followed by coating with polylysine.		Oral delivery of antigen and vaccine.	Bowersock et al., 1994 Kwok et al., 1991
	Alginate microbeads were prepared by crosslinking with Ca^{2+}.		Immobilization of endothelial cell growth factor. Controlled release of sulphamethoxazole. Controlled release of herbicides.	Pfister et al., 1986; Ko et al., 1995; Badwan et al., 1985
	A binary system. The core was viscous microorganism suspension while the coating was alginate gel crosslinked by Ca^{2+}.	Fast degradation ($\sim$1 hr) in $(NH_4)_2CO_3$ and Na_2CO_3 solution.	Microbial inoculants.	Digat, 1993
	A binary system. The core was erythroleukaemia cells. The shell was alginate crosslinked by Ca^{2+} and then coated with polylysine. The shell was further modified with PEO and PVA to increase the membrane strength.		Cell inoculants.	Kung et al., 1995
	Kaltostat, a commercial sodium calcium alginate wound dressing.		A absorbent haemostatic dressing for skin wounds.	Matthew et al., 1995
	Alginate from bacterial, P. syringae was crosslinked with Ca^{2+}.			Day and Ashby, 1995

Table 6 *Continued*

Hydrogel components	Preparation methods	Degradation modes	Applications	Ref.
	Gels were prepared by crosslinking alginate with Ca^{2+} ion.	Degradation by an endo-type alginate lyase from a bacteria called *Pseudomonus* SP.		Kinoshita *et al.*, 1991; Yotsuyanagi *et al.*, 1987
	GA modified alginate was polymerized by gamma irradiation.		Potential use as drug delivery system.	Paparella and Park, 1994
Alginate/ chitosan	Alginate microspheres were prepared by crosslinking with Ca^{2+} followed by coating with chitosan.		Affinity microspheres. Blue dextran was encapsulated as the ligand to isolate BSA from saline solution and plasma.	Achow and Goosan, 1994
Amylopectin	Solution of sufficiently high concentration formed physically crosslinked thermoreversible gels upon cooling to below room temperature.			Durrani and Donald, 1995
Carrageenan	κ-carrageenan gel was prepared by cooling its solution.	Degradation by acid hydrolysis in LiCl/HCl pH 2 buffer.		Singh, 1994
		Degradation by κ-carrageenase.		Knutsen and Gradalen, 1992
Cellulose/PVA	Gel films were prepared by precipitation of cellulose and PVA blends. Strong hydrogen bonding existed between the two different polymers.			Patel and Manley, 1995
Cellulose	Hydroxyethylcellulose mixed with PVP and poly(vinyl acetate) was crosslinked with epichlorohydrin to form beads containing herbicides.		Controlled release of herbicides.	Rehab *et al.*, 1991

Table 6 *Continued*

Hydrogel components	Preparation methods	Degradation modes	Applications	Ref.
Chitin	6-O-carboxylmethyl-chitin was crosslinked by Fe^{3+}.	Degradation by lysozyme.	Controlled drug delivery. Doxorubicin was the model drug.	Tokura *et al.*, 1990
	Partially deacetylated chitin was crosslinked by glutaraldehyde.	Degradation by lysozyme in pH 5.5 buffer solution.		Pangburn *et al.*, 1982
Chitosan	Chitosan membrane was prepared by crosslinking with glutaraldehyde.		Transdermal drug delivery system. Proparanolol hydrochloride was the model drug.	Thacharodi and Rao, 1995
	Chitosan microspheres were prepared from deacetylated chitin by the glutaraldehyde crosslinking.	No significant degradation in the skeletal muscle of rats.	Drug delivery vehicle. Mitoxantrone was used as model drug.	Jameela and Jayakrishnan, 1995
	Chitosan beads were prepared by cosslinking with sodium tripolyphosphate.		Immobilization of rabbit hepatic microsomal UDP-glucuronyl transferase to make biocatalyst for detoxification.	Zihnioglu and Telefoncu, 1995a,b
	Amino group containing chitosan was blended with polyanion, sodium alginate, to form polyelectrolyte film.		Super absorbing gel.	Hagino and Huang, 1995
		Degradation by chitosanase from *Amycolatopsis* SP. OsO-2.		Okajima *et al.*, 1994
	Chitosan films and beads were made by solvent evaporation and precipitation respectively.	Slow degradation in pH7.4 PBS buffer solution.	Used as drug delivery system. Steroid drugs, testosterone, progesterone, and β-oestradiol were used as model drugs.	Chandy and Sharma, 1991

Table 6 *Continued*

Hydrogel components	Preparation methods	Degradation modes	Applications	Ref.
Chitosan/ collagen	Membranes were prepared by solvent evaporation of chitosan and collagen solution.		Controlled release of propranolol hydrochloride.	Tacharodi and Rao, 1995
Chitosan/PEG	Chitosan was crosslinked by glutaraldehyde in the blends of PEG. Hydrogen bonding formed between the amino hydrogen in chitosan and polyether oxygen in PEG to form IPN.		Immobilization of heparin to prevent thrombus formation.	Beena *et al.*, 1995
Chitosan/PEO	Hydrogel was prepared by crosslinking chitosan with glyoxal. Semi-IPNs were made by crosslinking chitosan in a blend with PEO by glyoxal.		PH sensitive drug delivery system. Riboflavin was the model drug.	Amiji and Tatel, 1994
	Chitosan was blended with PEO to make membranes.		Improved permeability and reduced thrombogenicity for haemodialysis.	Amiji, 1995
Chitosan/PVA	Chitosan and PVA blend membrane were prepared by solvent casting while using glutaraldehyde as crosslinker.		pH sensitive membrane for controlled release of insulin and riboflavin.	Lee and Kim, 1994
Chondroitin	Chondroitin sulphate was crosslinked by 1,12-diaminododecane to give a series of hydrogels.	Specific degradation by colonic enzyme.	Served as a colon-specific drug carrier. Indomethacin was the model drug.	Sintov *et al.*, 1995; Rubinstein *et al.*, 1992a,b
Dextran	Epichlorohydrin crosslinked dextran to make microspheres (Sephadex®).		A potential nasal drug delivery vehicle for insulin.	Pereswetoff-Morath and Edman, 1995
	Dextran nanoparticles were prepared by polymerization of GA modified dextran.		Immobilization of proteins such as alkaline phosphatase.	Daubresse *et al.*, 1994; Edman *et al.*, 1980

Table 6 *Continued*

Hydrogel components	Preparation methods	Degradation modes	Applications	Ref.
	Dextran was crosslinked by diethylenetriaminepentaacetic acid dianhydride.		Potential drug delivery vehicle.	Metaggart and Halbert, 1993
	GA modified dextran was polymerized by gamma irradiation.	Degradation by dextanase.	Platform for protein drug delivery. Invertase was used as model drug.	Kamath and park, 1995; Paparella and Park, 1994
	Glycidyl methacrylate modified dextran was polymerized by chemical initiation.			Dijk-Wolthuis *et al.*, 1995
	Dextran mixed with PVP and poly(vinyl acetate) was crosslinked with epichlorohydrin to form beads containing herbicides.		Controlled release of herbicides.	Rehab *et al.*, 1991
Gellan	Crosslinked gellan gels were prepared using deacetylated gellan by means of an interchain partial esterification procedure.			Crescanzi *et al.*, 1995
	GA modified gellan was polymerized by gamma irradiation.		Potential use as drug delivery system.	Paparella and Park, 1994
Hyaluronic acid	Crosslinking hyaluronic acid with ethyleneglycol diglycidylether.	Degradation by hydroxyl radical (*in vitro*) or inflammation (*in vivo*).	Implantable drug delivery system which respond to inflammation.	Yui *et al.*, 1992
	GA modified hyaluronic acid and its benzyl ester derivatives were polymerized by gamma irradiation.		Potential use as drug delivery system.	Paparella and Park, 1994
	Sodium hyaluronate was crosslinked by diethylenetriaminepentaacetic acid dianhydride.		A potential drug delivery vehicle.	Mataggart and halbert, 1993

Table 6 *Continued*

Hydrogel components	Preparation methods	Degradation modes	Applications	Ref.
	Hyaluronic acid microspheres were prepared using solvent evaporation technique. Hydrocortisone, a model drug, was physically incorporated in or chemically bonded to the matrix.		Controlled drug delivery device. Hydrocortisone was the model drug.	Benesetti *et al.*, 1990
Hyaluronic acid/PAA or PVA	Hyaluronic acid was blended with PAA or PVA to make hydrogels.		A drug delivery system for growth hormone.	Cascone *et al.*, 1995
Hylan	Hyaluronan chains were crosslinked by vinyl sulphone.			Takigami *et al.*, 1995
Lignin	Naturally occurred.	Degradation of the crosslinking by chemical process during pulping.		Leclerc and Olsen, 1992
α-methyl galactoside	α-methyl galactoside 6-acrylate was chemoenzymatically synthesized and crosslinked with BIS.	Hydrolysis under high pH and high temperature.	Potential use as water absorbents.	Chen *et al.*, 1995
Pectin	Pectin was crosslinked in the presence of persulfate ion by oxidation.			Guillon and Thibault, 1990
	Low-ester pectin gels were prepared in the presence of sucrose, calcium ion, and proper pH.		Used in food industry.	El-Nawawi and Heikal, 1995
Starch	Microspheres were prepared by crosslinking of GA modified starch.	Degradation by lysosomal milieu.	Controlled release of dextranase.	Artursson *et al.*, 1984; Stjarnkvist *et al.*, 1989

Table 6 *Continued*

Hydrogel components	Preparation methods	Degradation modes	Applications	Ref.
	Gels were prepared by free radical polymerization of glycidyl methacrylate modified starch.	Degradation by α-amylase at 37°C in pH 7.2 buffer solution.	For triggered delivery of naltrexone.	Heller *et al.*, 1990
	Inactive form of starch hydrolytic enzyme, α-amylase, was incorporated in starch matrix.	Degradation of the starch matrix by α-amylase when the incorporated enzyme was activated in the presence of Ca^{2+}.	Intelligent biosensor for Ca^{2+}.	Kost, 1994
	Starch was crosslinked with phosphate or adipate.		A binding agent in conventional wet granulation process.	Visavarungroj *et al.*, 1990
Sucrose	Sucrose acrylate was chemoenzymatically synthesized and crosslinked by ethyleneglycol dimethacrylate or β-methylglucoside 2,6-diacrylate.	Degradation of the pendent sucrose by a alkaline protease, Proleather.	Potential use as water absorbents.	Dordick *et al.*, 1993, 1995
	Sucrose was modified by GA, glycidyl methacrylate, 1,2-epoxy hexene, methacryloyl chloride and acetyl chloride and then was polymerized by chemical initiator or gamma irradiation.	Hydrolysis in basic solution and strong acidic solution.	Potential use as superabsorbents.	Chen and Park, unpublished data
	Sucrose was modified by methacrylic anhydride, methacryloyl chloride, and methyl methacrylate and then was polymerized to form hydrogels.			Bruber, 1981

PAA = Poly(acrylic acid).

drug delivery systems, drug release from hydrogel dosage forms is achieved by diffusion of water molecules and degradation of polymer chains. One of the advantages of the degradable hydrogel systems is that the drug release can be easily controlled by adjusting the swelling and degradation kinetics. Invertase was incorporated into degradable hydrogels made of glycidyl acrylate-modified dextran (Kamath and Park, 1995). In the absence of dextranase, the total amount of invertase released from the dextran hydrogel was less than 30% even after 30 hrs. In contrast, in the presence of dextranase, all the incorporated invertase was released in less than 30 hrs. This clearly indicates that the release of drug can be adjusted by controlling the degradation kinetics of the hydrogel system. A hydrogel of hyaluronic acid crosslinked with glycidylether was found to be degraded by hydroxyl radicals produced by inflammation *in vivo* (Yui *et al.*, 1992). This hydrogel can be used as an implantable drug delivery device. Antiinflammatory drugs incorporated into the gel matrix can be released in response to inflammation. One of the unique advantages of degradable hydrogels is that the drug carriers will be removed from the administration site and subsequently from the body when they are no longer necessary. Progesterone was incorporated into albumin microbeads which were then injected intramuscularly or subcutaneously into rabbits. After the drug was released, the microbeads were degraded by the proteolytic enzymes at the injection site (Lee *et al.*, 1981). In addition, drug targeting can be realized if a hydrogel only swells or degrades in a target region of the body. Azobenzene derivatives were used as the crosslinkers to make hydrogels (Brondsted and Kopecek, 1992; Kopecek *et al.*, 1992; Yeh *et al.*, 1994; Kopecek *et al.*, 1993). The drug delivery systems based on these hydrogels were targeted to colon because the crosslinkers were degraded by an enzyme present predominantly in the colon. Another type of colon-specific drug delivery system was based on hydrogels made of chondroitin sulfate. The release of indomethacin from this system was dependent upon the biodegradation action of the caecal content (Sintov *et al.*, 1995; Rubinstein *et al.*, 1992; Rubinstein *et al.*, 1992). Artursson *et al.* prepared polyacryl starch microspheres as drug delivery vehicles (Artursson *et al.*, 1984). After i.v. injection in mice and rats, the microspheres were removed from the blood circulation by macrophages, accumulated in the lysosomes, and eventually degraded by the enzymes in lysosomes. This system allowed targeting of drugs to lysosomes. Recently, hydrogel foams were prepared as a gastric retention device for the long-term oral drug delivery (Park and Park, 1994). The biodegradable gastric retention device have an advantage in that it is removed from the stomach after all the drugs are released.

In addition to applications in drug delivery, degradable hydrogels have also been successfully used as biomaterials. Hydrogels made of polypeptides were used as temporary artificial skin substitutes (Nakanishi *et al.*, 1991). These skin substitutes can gradually be degraded by the proteases presented at the wound site. Degradable hydrogels were also used to prevent postoperation adhesion and promote healing. Prepolymers were made by using PEO as central block with PLA, PGA, or PCL extended on both sides and terminated with acrylate groups. Hydrogels were synthesized at the surgical site by photoinitiation of the prepolymers. The hydrogels adhered to the tissue and thus prevented the adhesion between organs. The hydrogels were degraded *in vivo* so that there was no need for surgical removal of the gels (Sawhney *et al.*, 1993, 1994; Hubbell *et al.*, 1993). Biosensors were made with degradable hydrogels. Inactive form of starch hydrolytic enzyme, α-amylase, was incorporated in starch matrix. When α-amylase sensed the presence of Ca^{2+}, it turned into active form and started to degrade starch (Kost, 1994). Microbial inoculants were made by entrapping microorganism suspension into degradable calcium alginate hydrogels. When the entrapped microorganisms needed to be released, the gel envelop was quickly degraded in ammonium carbonate or sodium carbonate solution (Digat, 1993). In addition to the above mentioned examples, degradable hydrogels have many other applications in biotechnology (Torchilin et al., 1977) and agriculture (Rehab et al., 1991; Pfister et al., 1986).

Progresses made in the area of degradable hydrogels during the last decade are impressive as shown by many examples in this chapter. Preparation of new degradable hydrogels provides more opportunities in various applications and the necessity of new applications prompts development of new degradable hydrogels. Continued collaboration among scientists in different disciplines will undoubtedly produce novel, degradable hydrogels with superb biocompatibility.

REFERENCES

Achaw, O.-W. and M.F.A. Goosen, Affinity microcapsules: chitosan-alginate membrane stability and isolation of bovine serum albumin, *Polymer Preprints*, **35** (1994) 75–76.

Akin, H. and N. Hasirci, Thermal properties of crosslinked gelatin microspheres, *Polymer Preprints*, **36** (1995) 384–385.

Albertsson, A.-C. Degradable polymers, *J.M.S.-Pure Appl. Chem.*, **A30** (1993) 757–765.

Amiji, M.M. and V.R. Patel, Chitosan-poly(ethylene oxide) semi-IPNs as a pH-sensitive drug delivery system, *Polymer Preprints*, **35** (1994) 403–404.

Amiji, M.M. Permeability and blood compatibility properties of chitosan-poly(ethylene oxide) blend membranes for haemodialysis, *Biomaterials*, **16** (1995) 593–599.

Andrianov, A.K., L.G. Payne, K.B. Visscher, H.R. Allcock and R. Langer, Hydrolytic degradation of ionically cross-linked polyphosphazene microspheres, *Journal of Applied Polymer Science*, **53** (1994) 1573–1578.

Artursson, P., P. Edman and I. Sjoholm, Biodegradable Microspheres. 1. Duration of action of dextranase entrapped in polyacrylstarch microparticles *in vivo*, *The Journal of Pharmacology and Experimental Therapeutics*, **231** (1984) 705–712.

Badwan, A.A., A. Abumalooh, E. Sallam, A. Abukalaf and O. Jawan, A sustained release drug delivery system using calcium alginate beads, *Drug Development and Industrial Pharmacy*, **11** (1985) 239–256.

Bae, Y.H. and S.W. Kim, Hydrogel delivery systems based on polymer blends, block copolymers or interpenetrating networks, *Advanced Drug Delivery Reviews*, **11** (1993) 109–135.

Barakat, I., P. Dubois, R. Jerome, P. Teyssie and E. Goethals, Macromolecular engineering of polylactones and polylactides. XV. Poly(D,L)-lactide macromonomers as precursors of biocompatible graft copolymers and bioerodible gels, *Journal of Polymer Science: Part A: Polymer Chemistry*, **32** (1994) 2099–2110.

Barman, S.P., M. Man, F. Yao, A.L. Coury and C.P. Pathak, Amphiphilic biodegradable *in situ* photopolymerizable macromonomers: modulation of hydrogel properties, *Polymer Preprints*, **36** (1995)

Beena, M.S., T. Chandy and C.P. Sharma, Heparin immobilized chitosan-poly ethylene glycol interpenetrating network: antithrombogenicity, *Art. Cells, Blood Subs., and Immob. Biotech.*, **23** (1995) 175–1923.

Bellamkonda, R., J.P. Ranieri, N. Bouche and P. Aebischer, Hydrogel-based three-dimentional matrix for neural cells, *Journal of Biomaterials Research*, **29** (1995) 663–671.

Benedetti, L.M., E.M. Topp and V.J. Stella, Microspheres of hyaluronic acid esters-fabrication methods and *in vitro* hydrocortisone release, *Journal of Controlled Release*, **13** (1990) 33–41.

Bowersock, T.L., H. HogenEsch, M. Suckow, E. Davis-Snyder, D. Borie, H. Park and K. Park, Oral administration of mice with ovalbumin encapsulated in alginate microspheres, *Polymer Preprints*, **35** (1994) 405–406.

Brondsted, H. and J. Kopecek, Hydrogels for site-specific drug delivery to the colon: *in vitro* and *in vivo* degradation, *Pharmaceutical Research*, **9** (1992) 1540–1545.

Buyanov, A.L., L.G. Revel'skaya and G.A. Petravlovskii, Formation and swelling behavior of polyacrylate superabsorbent hydrogels crosslinked by allyl ethers of polysaccharides, *Proc. ACS Div. Polym. Mat. Sci. Eng.*, **66** (1992) 87–88.

Buyanov, A.L., L.G. Revel'skaya and G.A. Petrovlovskii, Mechanism of formation and structural features of highly swollen acrylate hydrogels cross-linked with cellulose allyl ethers, *J. Appl. Chem. USSR*, **62** (1989) 1723–1728.

Cascone, M.G., B. Sim and S. Downes, Blends of synthetic and natural polymers as drug delivery systems for growth hormone, *Biomaterials*, **16** (1995) 569–574.

Chandy, T. and C.P. Sharma, Biodegradable chitosan matrix for the controlled release of steroids, *Biomat., Art. Cells & Immob. Biotech.*, **19** (1991) 745–760.

Chatterji, P.R., Cross-link dimensions in gelatin-poly(acrylamide) interpenetrating hydrogel networks, *Macromolecules*, **24** (1991) 4214–4215.

Chen, J. and K. Park, Unpublished data.

Chen, X., J.S. Dordick and D.G. Rethwisch, Chemoenzymatic synthesis and characterization of poly(α-methyl galactoside 6–acrylate) hydrogels, *Macromolecules*, **28** (1995) 6014–6019.

Cremers, H.F.M., J. Feijen, G. Kwon, Y.H. Bae, S.W. Kim, H.P.J.M. Noteborn and J.G. McVie, Albumin-heparin microspheres as carriers for cytostatic agents, *Journal of Controlled Release*, **11** (1990) 167–179.

Cremers, H.F.M., R. Verrijk, L.G. Bayon, M.M. Wesseling, J. Wondergem, Heuff, S. Meijer, G.S. Kwon, Y.H. Bae, S.W. Kim and J. Feijen, Improved distribution and reduced toxicity of adriamycin bound to albumin-heparin microspheres, *International Journal of Pharmaceutics*, **120** (1995) 51–61.

Crescenzi, V., M. Dentini and F. Siivi, Case studies of physical and chemical gels based on microbial polysaccharides, *Journal of Bioactive and Compatible Polymers*, **10** (1995) 235–248.

D'Urso, E.M. and G. Fortier, New bioartificial polymeric material: poly(ethylene glycol) cross-linked with albumin. I. Synthesis and swelling properties, *Journal of Bioactive and Compatible Polymers*, **9** (1994) 367–387.

D'Urso, E.M., J. Jean-Francois and G. Fortier, New bioartificial hydrogel: biomedical matrix for enzyme immobilization, *Polymer Preprints*, **35** (1994) 444–445.

Daubresse, C., C. Grandfils, R. Jerome and P. Teyssie, Enzyme immobilization in nanoparticles produced by inverse microemulsion polymerization, *Journal of Colloid and Interface Science*, **168** (1994) 222–229.

Davis, P.A., L. Niconais, L. Ambrosio and S.J. Huang, Poly(2–hydroxyethyl methacrylate)/poly(caprolactone) semi-interpenetrating polymer networks, *Journal of Bioactive and Compatible Polymers*, **5** (1988) 194–211.

Day, D.F. and R.D. Ashby, Bacterial alginates — a potentially useful class of biopolymers, *Proceedings of ACS Division of Polymeric Materials: Science and Engineering*, **72** (1995) 141–142.

Dickinson, H.R. and A. Hiltner, Biodegradation of a poly(α-amino acid) hydrogel. II. *in vitro*, *Journal of Biomedical Materials Research*, **1981** (1981) 591–603.

Dickinson, H.R., A. Hiltner, D.F. Gibbons and J.M. Anderson, Biodegradation of a poly(α-amino acid) hydrogel. I. *in vivo*, *Journal of Biomedical Materials Research*, **15** (1981) 577–589.

Digat, B. A new bioencapsulation technology for microbial inoculants, *Biomat., Art. Cells & Immob. Biotech.*, **21** (1993) 299–306.

Dijk-Wolthuis, W.N.E.v., O. Franssen, H. Talsma, M.J.v. Steenbergen, J.J.K.-v.d. Bosch and W.E. Hennink, Synthesis, characterization, and polymerization of glycidyl methacrylate derivatized dextran, *Macromolecules*, **28** (1995) 6317–6322.

Djabourov, M., J. Maquet, H. Theveneau, J. Leblond and P. Papon, Kinetics of gelation of aqueous gelatin solutions, *Br. Polym. J.*, **17** (1985) 169–174.

Dordick, J.S., B.D. Martin, R.J. Linhardt, X. Chen and D.G. Rethwisch, Chemoenzymatic routes to the synthesis of biodegradable, sugar-based water absorbent, *Proceedings of the ACS Division of Polymeric Materials: Science and Engineering*, **69** (1993) 562–563.

Dordick, J.S., B.D. Martin, X. Chen, J.O. Rich, N. Patil, R.J. Linhardt and D.G. Rethwisch, Enzymatic synthesis of polymeric materials, *Proceedings of ACS Division of Polymeric Materials: Science and Engineering*, **72** (1995) 90–91.

Durrani, C.M. and A.M. Donald, Physical characterization of amylopectin gels, *Polymer Gels and Networks*, **3** (1995) 1–27.

Edman, P., B. Ekman and I. Sjoholm, Immobilization of proteins in microspheres of biodegradable polyacryldextran, *Journal of Pharmaceutical Sciences*, **69** (1980) 838–842.

El-Nawawi, S.A. and Y.A. Heikal, Factors affecting the production of low-ester pectin gels, *Carbohydrate Polymers*, **26** (1995) 189–193.

Eschbach, F.O. and S.J. Huang, Hydrophilic-hydrophobic binary systems of poly(2-hydroxyethyl methacrylate) and polycaprolactone. Part II: degradation, *Journal of Bioactive and Compatible Polymers*, **9** (1994) 210–221.

Eschbach, F.O. and S.J. Huang, Hydrophilic-hydrophobic binary systems of poly(2–hydroxyethyl methacrylate) and polycaprolactone. Part I: synthesis and characterization, *Journal of Bioactive and Compatible Polymers*, **9** (1994) 29–54.

Eschbach, F.O. and S.J. Huang, Hydrophilic-hydrophobic binary systems: poly(2-hydroxyethyl methacrylate) and polycaprolactone, *Polymer Preprints*, **34** (1993) 848–849.

Eschbach, F.O. and S.J. Huang, Hydrophobic-hydrophilic IPN and SIPN, *Proc. ACS Div. Polym. Mater. Sci. Eng.*, **65** (1991) 9–10.

Frisch, H.L. Interpenetrating polymer networks, *Br. Polym.*, **17** (1985) 149–153.

Fujioka, K., Y. Takada, S. Sato and T. Miyata, Long-acting delivery system of interferon: IFN minipellet, *Journal of Controlled Release*, **33** (1995) 317–323.

Fujioka, K., Y. Takada, S. Sato and T. Miyata, Novel delivery system for proteins using collagen as a carrier material: the minipellet, *Journal of Controlled Release*, **33** (1995) 307–315.

Gayet, J.-C. and G. Fortier, New bioartificial hydrogels: synthesis, characterization and physical properties, *Polymer Preprints*, **35** (1994) 440–441.

Goosen, M.F.A., Y.F. Leung, S. Chou and A.M. Sun, Insulin-albumin microbeads: an implantable, biodegradable system, *Biomater. Med. Dev. Artif. Organs*, **10** (1982) 205–218.

Gruber, H., Hydrophile polymergele mit reaktiven gruppen, 1. mitt.: herstellung und polymerisation von glucose- und saccharosemethacrylaten, *Monatshefte fur Chemie*, **112** (1981) 273–285.

Guillon, F. and J.-F. Thibault, Oxidative cross-linking of chemically and enzymatically modified sugar-beet pectin, *Carbohydrate polymers*, **12** (1990) 353–374.

Hagino, Y. and S.J. Huang, Super absorbing gels derived from chitosan and sodium alginate, *Proceedings of ACS Division of Polymeric Materials: Science and Engineering*, **72** (1995) 249–250.

Heller, J. Chemically self-regulated drug delivery systems, *Journal of Controlled Release*, **8** (1988) 111–125.

Heller, J. Polymers for controlled parenteral delivery of peptides and proteins, *Advanced Drug Delivery Reviews*, **11** (1993) 163–204.

Heller, J., R.F. Helwing, R.W. Baker and M.E. Tuttle, Controlled release of water-soluble macromolecules from bioerodible hydrogels, *Biomaterials*, **4** (1983) 262–266.

Heller, J., S.H. Pangburn and K.V. Roskos, Development of enzymatically degradable protective coatings for use in triggered drug delivery systems. II. Derivatized starch hydrogels, *Biomaterials*, **11** (1990) 345–350.

Hiroyuki, Yamamoto and Y. Hirata, Gel properties as a hydrogel of crosslinked poly(L-ornithine) using organic crosslinking agents, *Polymer Gels and Networks*, **3** (1995) 71–84.

Ho, H.-O., C.-C. Hsiao, T.D. Sokoloski and C.-Y. Chen, Fibrin-based drug delivery systems III: The evaluation of the release of macromolecules from microbeads, *Journal of Controlled Release*, **34** (1995) 65–70.

Hocking, P. The classification, preparation, and utility of degradable polymers, *J.M.S.-Rev. Macromol. Chem. Phys.*, **C32** (1992) 35–54.

Hubbell, J.A., C.P. Pathak and A.S. Sawhney, *In vivo* photopolymerization of PEG-based biodegradable hydrogels for the control of wound healing, *Polymer Preprints*, **34** (1993) 846–847.

Jameela, S.R. and A. Jayakrishnan, Glutaraldehyde cross-linked chitosan microspheres as a long acting biodegradable drug delivery vehicle: studies on the *in vitro* release of mitoxantrone and *in vivo* degradation of microspheres in rat muscle, *Biomaterials*, **16** (1995) 769–775.

Jeyanthi, R. and K.P. Rao, Controlled release of anticancer drugs from collagen-poly(HEMA) hydrogel matrices, *Journal of Controlled Release*, **13** (1990) 91–98.

Kamath, K.R. and K. Park, Biodegradable hydrogels in drug delivery, *Advanced Drug Delivery Reviews*, **11** (1993) 59–84.

Kamath, K.R. and K. Park, Study on the release of invertase from enzymatically degradable dextran hydrogels, *Polymer Gels and Networks*, **3** (1995) 243–254.

Kamath, K.R. and K. Park, Use of gamma irradiation for the preparation of hydrogels from natural polymers, *Proc. Int. Symp. Control. Release Bioact. Mater.*, **19** (1992) 42–43.

Kaur, H. and P.R. Chatterji, Interpenetrating hydrogel networks. 2. swelling and mechanical properties of the gelatin-polyacrylamide interpenetrating networks, *Macromolecules*, **23** (1990) 4868–4871.

Kinoshita, S., Y. Kumoi, A. Ohshima, T. Yoshida and N. Kasai, Isolation of a alginate degrading organism and purification of its alginate lyase, *Journal of fermentation and bioengineering*, **72** (1991) 74–78.

Knutsen, S.H. and H. Grasdalen, Analysis of carrageenans by enzymic degradation, gel filtration and ^{1}H NMR spectroscopy, *Carbohydrate Polymers*, **19** (1992) 199–210.

Ko, C., V. Dixit, W. Shaw and G. Gitnick, *in vitro* slow release profile of endothelial cell growth factor immobilized within calcium alginate microbeads, *Art. Cells, Blood Subs., and Immob. Biotech.*, **23** (1995) 143–151.

Kopecek, J. and P. Rejmanova, Reactive copolymers of N-(2–hydroxypropyl) methacrylamide with N-methacryloylated derivatives of L-leucine and L-phenylalanine. II. Reaction with the polymeric amine and stability of cross-links towards chymotrypsin *in vitro, Journal of Polymer Science: Polymer Sympo.*, **66** (1979) 15–32.

Kopecek, J., P. Kopeckova, H. Brondsted, R. Rathi, B. Rihova, P.-Y. Yeh and K. Ikesue, Polymers for colon-specific drug delivery, *Journal of Controlled Release*, **19** (1992) 121–130.

Kopecek, J., P.Y. Yeh, P. Kopeckova and K. Ulbrich, Tailor-made synthesis of hydrogels, *Polymer Preprints*, **34** (1993) 833–834.

Kost, J., Biosensors based on intelligent polymeric systems, *Artificial Cells, Blood Substitute and Immobilization Biotechnology*, **22** (1994) A25.

Kung, I.M., F.F. Wang, Y.C. Chang and Y.J. Wang, Surface Modifications of alginate/poly(L-lysine) microcapsular membranes with poly(ethylene glycol) and poly(vinyl alcohol), *Biomaterials*, **16** (1995) 649–655.

Kwok, K.K., M.J. Groves and D.J. Burgess, Production of 5–15 μm diameter alginate-polylysine microcapsules by an air-atomization technique, *Pharmaceutical Research*, **8** (1991) 341–344.

Kwon, G.S., Y.H. Bae, H. Cremers, J. Feijen and S.W. Kim, Release of macromolecules from albumin-heparin microspheres, *Journal of Controlled Release*, **79** (1992) 191–198.

Leclerc, D.F. and J.A. Olson, A percolation — theory model of lignin degradation, *Macromolecules*, **25** (1992) 1667–1675.

Lee, S.J. and K. Park, Synthesis and characterization of sol-gel phase-reversible hydrogels sensitive to glucose, *Journal of Molecular Recognition*, (in press)

Lee, S.J. and K. Park, Synthesis of sol-gel phase-reversible hydrogels sensitive to glucose, *Proceed. Intern. Symp. Control. Rel. Bioact. Mater.*, **21** (1994) 93–94.

Lee, T.K., T.D. Sokoloski and G.P. Royer, Serum albumin beads: an injectable, biodegradable system for the sustained release of drugs, *Science*, **213** (1981) 233–235.

Lee, Y.M. and J.H. Kim, Controlled release of insulin through crosslinked poly(vinyl alcohol)/chitosan blend membrane, *Proceedings of ACS Division of Polymeric Materials: Science and Engineering*, **70** (1994) 322–323.

Livingston, T.L., C.E. Krewson, W. Dai and W.M. Salzman, Cell migration and aggregation in protein hydrogels, *Polymer Preprints*, **35** (1994) 399–400.

Lizuka, Y., M. Oya, M. Iwatsuki and T. Hayashi, Enzymatic hydrolysis of multi-component random copolypeptides *in vitro, Polymer Journal*, **25** (1993) 285–290.

Ma, J.T., L.R. Liu, X.J. Yang and K.D. Yao, Bending behavior of gelatin/poly(hydroxyethyl methacrylate) IPN hydrogel under electric stimulus, *Journal of Applied Polymer Science*, **56** (1995) 73–77.

Matthew, I.R., G.M. Browne, J.W. Frame and B.G. Millar, Subperiosteal behaviour of alginate and cellulose wound dressing materials, *Biomaterials*, **16** (1995) 265–274.

Mctaggart, L.E. and G.W. Halbert, Assessment of polysaccharide gels as drug delivery vehicles, *International Journal of Pharmaceutics*, **100** (1993) 199–206.

Nakanishi, E., Y. Shimiru, K. Ogura, S. Hibi and T. Hayashi, Effects of side chain length on membrane properties of copoly(N-hydroxyalkyl L-glutamine) hydrogels, *Polymer Journal*, **23** (1991) 1061–1068.

Okajima, S., A. Ando, H. Shinoyama and T. Fujii, Purification and characterization of an extracellular chitosanase produced by *Amycolatopsis* sp. CsO-2, *Journal of Fermentation and Bioengineering*, **77** (1994) 617–620.

Pangburn, S.H., P.V. Trescony and J. Heller, Lysozyme degradation of partially deacetylated chitin, its films and hydrogels, *Biomaterials*, **3** (1982) 105–108.

Paparella, A. and K. Park, Synthesis of polysaccharide chemical gels by gamma-irradiation, *Polymer Preprints*, **35** (1994) 884–885.

Park, H. and K. Park, Honey, I blew up the hydrogels!, *Proceed. Intern. Symp. Control. Rel. Bioact. Mater.*, **21** (1994) 21–22.

Park, K., Enzyme-digestible swelling hydrogels as platforms for long-term oral drug delivery: synthesis and characterization, *Biomaterials*, **9** (1988) 435–441.

Park, K., W.S.W. Shalaby and H. Park, *Biodegradable hydrogels for drug delivery*, Technomic, (1993).

Park, T.G., S. Cohen and R. Langer, Poly(L-lactic acid)/Pluronic blends: characterization of phase separation behavior, degradation and morphology and use as protein-releasing matrices, *Macromolecules*, **25** (1992) 116–122.

Patel, K. and R.S.J. Manley, Carbon dioxide sorption and transport in miscible cellulose/poly(vinyl alcohol) blends, *Macromolecules*, **28** (1995) 5793–5798.

Pereswetoff-Morath, L. and P. Edman, Dextran microspheres as a potential nasal drug delivery system for insulin — *in vitro* and *in vivo* properties, *International Journal of Pharmaceutics*, **124** (1995) 37–44.

Pfister, G., M. Bahadir and F. Korte, Release characteristics of herbicides from Ca alginate gel formulations, *Journal of Controlled Release*, **3** (1986) 229–233.

Pitt, C.G. and S.S. Shah, Manipulation of the rate of hydrolysis of polymer-drug conjugates: the degree of hydration, *Journal of Controlled Release*, **33** (1995) 397–403.

Pitt, C.G., Y. Cha, S.S. Shah and K.J. Zhu, Blends of PVA and PGLA: control of the permeability and degradability of hydrogels by blending, *Journal of Controlled Release*, **19** (1992) 189–200.

Pulapura, S. and J. Kohn, Trends in the development of bioresorbable polymers for medical applications, *Journal of Biomaterial Applications*, **6** (1992) 216–250.

Rathna, G.V.N., D.V.M. Rao and P.R. Chatterji, Water-induced plastisization of solution cross-linked hydrogel networks: energetics and mechanism, *Macromolecules*, **27** (1994) 7920–7922.

Rehab, A., A. Akelah, R. Issa, S. D'Antone, R. Solaro and E. Chiellini, Controlled release of herbicides supported on polysaccharide based hydrogels, *Journal of Bioactive and Compatible Polymers*, **6** (1991) 52–63.

Rejmanova, P., B. Obereigner and J. Kopecek, Polymers containing enzymatically degradable bonds, 2. Poly[N-(2–hydroxypropyl) methacrylamide] chains connected by oligopeptide sequences cleavable by chymotrypsin, *Makromol. Chem.*, **182** (1981) 1899–1915.

Rubinstein, A., D. Nakar and A. Sintov, Chondroitin sulfate: a potential biodegradable carrier for colon-specific drug delivery, *International Journal of Pharmaceutics*, **84** (1992) 141–150.

Rubinstein, A., D. Nakar and A. Sintov, Colonic drug delivery: enhanced release of indomethacin from cross-linked chondroitin matrix in rat cecal content, *Pharmaceutical Research*, **9** (1992) 276–278.

Satyanaryana, D. and P.R. Chatterji, Biodegradable polymers: challenges and strategies, *J.M.S-Rev. Macromol. Chem. Phys.*, **C33** (1993) 349–368.

Sawhney, A.S., C.P. Pathak and J.A. Hubbell, Bioerodible hydrogels based on photopolymerized poly(ethylene glycol)- co-poly(α-hydroxy acid) diacrylate macromers, *Macromolecules*, **26** (1993) 581–587.

Sawhney, A.S., C.P. Pathak, J.J.v. Rensburg, R.C. Dunn and J.A. Hubbell, Optimization of photopolymerized bioerodible hydrogel properties for adhesion prevention, *Journal of Biomedical Materials Research*, **28** (1994) 831–838.

Shalaby, W.S.W. and K. Park, Biochemical and mechanical characterization of enzyme-digestible hydrogels, *Pharmaceutical Research*, **7** (1990) 816–823.

Shalaby, W.S.W., G.E. Peck and K. Park, Release of dextromethorphan hydrobromide from freeze-dried enzyme-degradable hydrogels, *Journal of Controlled Release*, **16** (1991) 355–364.

Shalaby, W.S.W., M. Chen and K. Park, A mechanistic assessment of enzyme-induced degradation of albumin-crosslinked hydrogels, *Journal of Bioactive and Compatible Polymers*, **7** (1992) 257–274.

Shalaby, W.S.W., R. Jackson, W.E. Blevins and K. Park, Synthesis of enzyme-digestible, interpenetrating hydrogel networks by gamma-irradiation, *Journal of Bioactive and Compatible Polymers*, **8** (1993) 3–23.

Shalaby, W.S.W., W.E. Blevins and K. Park, Enzyme-induced degradation behavior of albumin crosslinked hydrogels, *Polymer preprints*, **31** (1990) 169–170.

Shalaby, W.S.W., W.E. Blevins and K. Park, *In vitro* and *in vivo* studies of enzyme-digestible hydrogels for oral drug delivery, *Journal of Controlled Release*, **19** (1992) 131–144.

Silvio, L.D., N. Gurav, M.V. Kayser, M. Braden and S. Downes, Biodegradable microspheres: a new delivery system for growth hormone, *Biomaterials*, **15** (1994) 931–936.

Singh, S.K. Kinetics of acid hydrolysis of κ-carrageenan as determined by molecular weight (SEC-MALLS-RI), gel breaking strength, and viscosity measurements, *Carbohydrate Polymers*, **23** (1994) 89–103.

Sintov, A., N. Di-Capua and A. Rubinstein, Cross-linked chondroitin sulphate: characterization for drug delivery purposes, *Biomaterials*, **16** (1995) 473–478.

Skarda, V., F. Rypacek and M. Ilavsky, Biodegradable hydrogel for controlled release of biologically active macromolecules, *Journal of Bioactive and Compatible Polymers*, **8** (1993) 24–40.

Sperling, L.H. *Interpenetrating polymer networks and related materials*, Plenum Press, New York, (1981).

Stjarnkvist, P., T. Laakso and I. Sjoholm, Biodegradable microspheres XII: properties of the crosslinking chains in polyacryl starch microparticles, *Journal of Pharmaceutical Sciences*, **78** (1989) 52–56.

Storey, R.F., J.S. Wiggins and A.D. Puckett, Hydrolyzable poly(ester-urethane) networks from L-lysine diisocyanate and D,L-lactide/ε-caprolactone homo- and copolyester triols, *Journal of Polymer Science: Part A: Polymer Chemistry*, **32** (1994) 2345–2363.

Subr, V., J. Kopecek, J. Pohl, M. Baudys and V. Kostka, Cleavage of oligopeptide side-chains in N-(2–hydroxypropyl) methacrylamide co-polymers by mixtures of lysosomal enzymes, *Journal of Controlled Release*, **8** (1988) 133–140.

Tabata, Y. and Y. Ikada, Synthesis of gelatin microspheres containing interferon, *Pharmaceutical Research*, **6** (1989) 422–427.

Takigami, S., M. Takigami and G.O. Phillips, Effect of preparation method on the hydration characteristics of hylan and comparison with another highly cross-linked polysaccharide, gum arabic, *Carbohydrate Polymers*, **26** (1995) 11–18.

Thacharodi, D. and K.P. Rao, Collagen-chitosan composite membranes for controlled release of propranolol hydrochloride, *International Journal of Pharmaceutics*, **120** (1995) 115–118.

Thacharodi, D. and K.P. Rao, Development and *in vitro* evaluation of chitosan-based transdermal drug delivery systems for the controlled delivery of propranolol hydrochloride, *Biomaterials*, **16** (1995) 145–148.

Tokura, S., Y. Miura and Y. Uraki, Biodegradable chitin derivative as various types of drug carriers, *Polymer Preprints*, **31** (1990) 627.

Torchilin, V.P., E.G. Tischenko, V.N. Smirnov and E.I. Chazov, Immobilization of enzymes on slowly soluble carriers, *J. Biomed. Mater. Res.*, **11** (1977) 223–235.

Ulbrich, K., J. Strohalm and J. Kopecek, Polymers containing enzymatically degradable bonds. VI. Hydrophilic gels cleavable by chymotrypsin, *Biomaterials*, **3** (1982) 150–154.

Ulbrich, K., V. Subr, L.W. Seymour and R. Duncan, Novel biodegradable hydrogels prepared using the divinylic crosslinking agent N,O-dimethacryloylhydroxylamine. 1. Synthesis and characterization of rate of gel degradation, and rate of release of model drugs, *in vitro* and *in vivo*, *Journal of Controlled Release*, **24** (1993) 181–190.

Ulbrich, K., V. Subr, P. Podperova and M. Buresova, Synthesis of novel hydrolytically degradable hydrogels for controlled drug release, *Journal of Controlled Release*, **34** (1995) 155–165.

Visavarungroj, N., J. Herman and J.P. Remon, Crosslinked starch as binding agent. I. Conventional wet granulation, *International Journal of Pharmaceutics*, **59** (1990).

Yapel, A.F.J., Albumin microspheres: heat and chemical stabilization, *Methods in Enzymology*, K.J. Widder and R. Green, Academic Press, Inc., **112** (1985) 3–18.

Yasin, M. and B.J. Tighe, Strategies for the design of biodegradable polymer systems: Manipulation of polyhydroxybutyrate-based materials, *Plastics, Rubber and Composites Processing and Applications*, **19** (1993) 15–27.

Yeh, P.Y., P. Kopeckova and J. Kopecek, Biodegradable and pH-sensitive hydrogels: synthesis by crosslinking of N,N-dimethylacrylamide copolymer precursors, *Journal of Polymer Science: Part A: Polymer Chemistry*, **32** (1994) 1627–1637.

Yokoyama, M. Block copolymers as drug carriers, *Critical Reviews in Therapeutic Drug Carrier Systems*, **9** (1992) 213–248.

Yoshida, R., K. Sakai, T. Okano and Y. Sakurai, Pulsatile drug delivery systems using hydrogels, *Advanced Drug Delivery Reviews*, **11** (1993) 85–108.

Yotsuyanagi, T., T. Ohkubo, T. Ohhashi and K. Ikeda, Calcium-induced gelation of alginic acid and pH-sensitive reswelling of dried gels, *Chem. Pharm. Bull.*, **35** (1987) 1555–1563.

Yui, N., T. Okano and Y. Sakurai, Inflammation responsive degradation of crosslinked hyaluronic acid gels, *Journal of Controlled Release*, **22** (1992) 105–116.

Zhang, X., Mattheus, F.A. Goosen, U.P. Wyss and D. Pichora, Biodegradable polymers for orthopedic applications, *J.M.S.-Rev. Macromol. Chem. Phys.*, **C33** (1993) 81–102.

Zhao, W., A. Kloczkowski and J.E. Mark, Preparation of high-performance materials from biodegradable gelatin using a novel orientation technique, *Proceedings of ACS Division of Polymeric Materials: Science and Engineering*, **72** (1995) 86–87.

Zihnioglu, F. and A. Telefoncu, Preparation and characterization of chitosan-entrapped microsomal UDP-glucuronyl transferase, *Art. Cells, Blood Subs., and Immob. Biotech.*, **23** (1995) 545–552.

Zihnioglu, F. and A. Telefoncu, Substrate specificity and the use of chitosan-entrapped rabbit hepatic microsomal UDP-glucuronyl transferase for detoxification, *Art. Cells, Blood Subs., and Immob. Biotech.*, **23** (1995) 533–543.

12. THE POLOXAMERS: THEIR CHEMISTRY AND MEDICAL APPLICATIONS

LORRAINE E. REEVE

*Alliance Pharmaceutical Corp., 3040 Science Park Road,
San Diego, California 92121, USA*

SYNTHESIS AND CHEMISTRY OF THE POLOXAMERS

The poloxamers are a series of copolymers composed of two polyoxyethylene blocks separated by a polyoxypropylene block (Lundsted, 1954; Lundsted, 1976; Schmolka, 1991). Although all poloxamers are composed of the same two monomers, they vary in total molecular weight, polyoxypropylene to polyoxyethylene ratio, and surfactant properties. The poloxamers all have the general structure:

$$HO(C_2H_4O)_a(C_3H_6O)_b \ (C_2H_4O)_cH$$

in which a and c are approximately equal, and b is at least 15.

Synthesis

The hydrophobic center block is synthesized in a base catalyzed ether condensation reaction using propylene glycol as the initiator and sequentially adding propylene oxide under an inert, anhydrous atmosphere, at elevated temperature and pressure. Upon reaching the desired molecular weight of the center block, the propylene oxide reactant is replaced by ethylene oxide. The polymer is then extended by the addition of ethylene oxide to form a hydrophilic segment of polyoxyethylene at each end of the polyoxypropylene block. Finally, the reaction is terminated by neutralization of the catalyst with acid. More than thirty different poloxamers have been synthesized which range in molecular weight from 1,000 to 15,000 (Table 1). The polyoxyethylene content of the molecule may vary from 10% to 90% of the weight of the molecule. The poloxamers are freely soluble in water, and are soluble in polar solvents. They are liquids, pastes or solids, depending largely on their polyoxyethylene to polyoxypropylene ratio and, secondarily, on their total molecular weight. In general, poloxamers having an average molecular weight of 3,000 or less are liquids if their polyoxyethylene content is not more than 50%. Poloxamers with molecular weights between 3,000 and 5,000 are liquids only if their polyoxyethylene content is 20% or less. Pastes range in molecular weight from 3,300 to 6,600 with a polyoxyethylene content between 30% and 50%. Solid poloxamers are the largest, having molecular weights of 5,000 to 15,000 and a polyoxyethylene content of 70% or greater.

Table 1 Average molecular weights and compositions of the poloxamers

Poloxamer Number	Average Molecular Weight	Average Number Of Sub-Units In Each Block		
		Ethylene Oxide	Propylene Oxide	Ethylene Oxide
101	1100	2	16	2
105	1900	11	16	11
108	5000	46	16	46
122	1630	5	21	5
123	1850	7	21	7
124	2200	11	21	11
181	2000	3	30	3
182	2500	8	30	8
183	2650	10	30	10
184	2900	13	30	13
185	3400	19	30	19
188	8350	75	30	75
212	2750	8	35	8
215	4150	24	35	24
217	6600	52	35	52
231	2750	6	39	6
234	4200	22	39	22
235	4600	27	39	27
237	7700	62	39	62
238	10800	97	39	97
282	3650	10	47	10
284	4600	21	47	21
288	13500	122	47	122
331	3800	7	54	7
333	4950	20	54	20
334	5850	31	54	31
335	6500	38	54	38
338	14000	128	54	128
401	4400	6	67	6
402	5000	13	67	13
403	5750	21	67	21
407	11500	98	67	98

Nomenclature

A nomenclature system has been developed in which each poloxamer has been assigned a number composed of three digits (Schmolka, 1991). This number indicates the molecular weight of the hydrophobe and the polyoxyethylene content of the respective poloxamer. The average molecular weight of the hydrophobic polyoxypropylene block is obtained by multiplying the first two digits by 100. The approximate weight percent of polyoxyethylene is obtained by multiplying the third digit by 10. For example, poloxamer 188 is composed of a center block of polyoxypropylene having an approximate average molecular weight of 1,800, and an ethylene oxide content of approximately 80% of the total molecule. Since the number of ethylene oxide substituents in each polyoxyethylene block

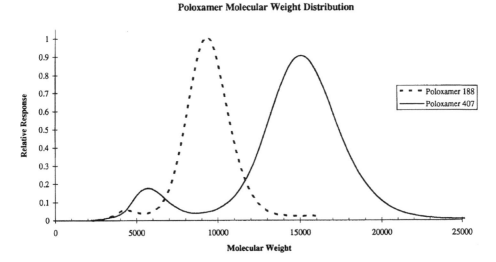

Figure 1 Molecular weight distribution of poloxamers 407 and 188, analyzed by gel permeation chromatography and refractive index detection.

is statistically equal to that in the other block, poloxamer 188 should consist of a polyoxypropylene block having a molecular weight of about 1,800, flanked on either end by polyoxyethylene segments with molecular weights of about 3,600. In fact, nominal molecular weights and polyoxyethylene to polyoxypropylene ratios vary among manufacturing lots and among suppliers.

Chemical and Physical Properties

Because the poloxamers are the products of a sequential series of reactions, the molecular weights of individual poloxamer molecules are statistical distributions about the average molecular weight. When poloxamers 188 and 407 were analyzed by gel permeation chromatography, a bimodal molecular weight distribution was observed for each (Figure 1). In the case of poloxamer 407, the major peak accounted for approximately 85% of the mass and was composed of molecules having molecular weights between 9,000 and 19,000. A second peak ranging in molecular weight from approximately 4,000 to 8,000 accounted for the remaining 15% of the poloxamer mass. A similar bimodal distribution and broad molecular weight range were observed for poloxamer 188, with a major peak having a molecular weight between 6,000 and 13,000. It has been reported, however, that poloxamer 407 obtained from various suppliers has significantly different molecular weight distributions (Porter, 1992a). This is presumably due to slightly different conditions during synthesis.

The poloxamers are unsaturated to the extent of about 0.02 to 0.07 mEq/g as determined by titration with mercuric acetate. Analysis of poloxamers 407 and 188 indicate that the unsaturated poloxamer molecules separate overwhelmingly into the lower molecular weight fraction during gel permeation chromatography.

This suggests that the broad molecular weight range observed for the commercially available poloxamers is at least in part due to elimination of a proton from either the proplyene oxide or ethylene oxide reactants, which results in formation of an allylic double bond during polyether synthesis. Such a molecule would function as a nucleophile, but would be able to add subunits at only one site. It has been claimed that the molecular weight distribution of the polymer can be minimized by using a cesium hydroxide catalyst in place of sodium or potassium hydroxide during polyether synthesis (Ott, 1988).

Aggregation of the Poloxamers in Solution

Because they contain both hydrophilic and hydrophobic regions, poloxamer molecules form micelles in aqueous solution if their polyoxypropylene region is large enough. Their aggregation behavior, however, is quite complex, and cannot be described by simply determining the critical micelle concentration for each polymer. Studies measuring changes in surface-tension with poloxamer concentration indicated two inflection points (Prasad, 1979). The first inflection point may be due to conformational changes of the individual poloxamer molecules or intermolecular interactions involving a only small number of molecules. The second inflection point, at a higher poloxamer concentration, is thought to result from multimolecular aggregation. Zhou and Chu (Zhou, 1988) observed that, as the concentration of poloxamer 188 in an aqueous solution increased, a plot of temperature versus hydrodynamic radius was shifted to the left, indicating an inverse relationship between temperature and critical micelle concentration. Between 40 and 80°C, the aggregation number of the micelles rose sharply. Rassing and Attwood (Rassing, 1983) studied the aggregation behavior of poloxamer 407 over a temperature range of 10 to 40°C. They concluded from measurements of the radius of gyration, that below 10°C the polymer exists largely as loosely associated aggregates of extended molecules, and, at least in this temperature range, questioned the presence of coiled single molecules as had been previously proposed (Prasad, 1979). Above 10°C light scattering measurements indicate that the aggregates become larger and more highly organized, and above approximately 25°C, they are spherical entities whose aggregation number increases with increasing temperature. Similar behavior has been reported for poloxamers 184 and 237 (Al-Saden, 1982). It is thus apparent that there are no single values for critical micelle concentration, critical micelle temperature, or aggregation number for the poloxamers.

Gellation of the Poloxamers in Solution

Concentrated aqueous solutions of many of the poloxamers form gels in a fully reversible, temperature dependent process (Schmolka, 1972; Schmolka, 1994). A 20% solution of poloxamer 407 forms a solid gel at 25°C (Table 2), but other poloxamers act similarly, at higher concentrations. As poloxamer concentration increases, gelation occurs at a lower temperature, and the viscosities of the resultant gels are higher at any given temperature (Figure 2). The gelation behavior of the poloxamers is explained by their complex aggregation behavior (Wanka, 1990; Attwood, 1985; Attwood, 1987; Nakashima, 1994). Although the polyoxypropylene

Table 2 Viscosities in millipascal of aqueous solutions of poloxamer 407. The abrupt increase in viscosity indicates gelation at a characteristic temperature for each concentration of poloxamer

Temperature °C	Concentration of poloxamer 407 in water (w/w %)				
	18	20	25	28	30
5	<1000	<1000	<1000	<1000	<1000
10	<1000	<1000	<1000	<1000	<1000
15	<1000	<1000	<1000	158000	254000
20	<1000	<1000	214000	248000	306000
25	76000	118000	250000	268000	320000
30	107000	138000	254000	272000	334000
35	120000	140000	258000	276000	338000

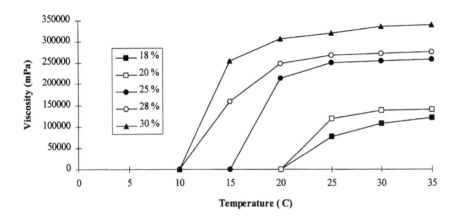

Figure 2 Changes of viscosity with temperature for varying concentrations of poloxamer 407 (% w/w) in aqueous solution.

segment of the poloxamer molecule is hydrophobic, its ether oxygens form hydrogen bonds with water molecules. For this reason, the transfer energy of a propyleneoxide unit from the aqueous to the micellar phase is a fairly low 0.3 kT at 25°C (Wanka, 1990), which results in a loose association of molecules. However, at higher temperatures, the polyoxypropylene segments are dehydrated. This has been shown to occur over a fairly broad temperature range, below the temperature at which gelation occurs (Wanka, 1990). In addition, the aggregation number of the micelles increases with increasing temperature. These two phenomena, loss of water and aggregation of more poloxamer molecules per micelle compensate for each other and the hydrodynamic radius remains approximately constant over a broad temperature range (Zhou, 1988; Attwood, 1985; Wanka, 1990; Attwood, 1987). At higher temperatures, the polyoxyethylene chains begin to dehydrate as well. Upon the loss of water, the polyoxyethylene chains of different micelles interact with each other if the poloxamer concentration is high enough, and gelation occurs. This process is fully reversible, and upon reduction of the temperature of a poloxamer gel, rehydration occurs and the solution returns to the liquid phase.

Commercial Uses of the Poloxamers

The poloxamers are widely used in topical and oral consumer products as solubilizers, emulsifiers, stabilizers, dispersing, suspending and coating agents (Schmolka, 1991; Schmolka, 1976). An extensive amount of research has been done on application of the micellar and gellation properties of the poloxamers for drug delivery (Kataoka, 1993; Miyazaki, 1984). A monograph setting specifications and analytical methods for poloxamers 124, 188, 237, 338 and 407 has been included in the United States National Formulary since 1990. Poloxamer 188 is included in the British Pharmacopea.

TOXICOLOGY OF THE POLOXAMERS

Because they contain ether linkages, the poloxamers are not readily metabolized in the body, and are therefore not truly biodegradable. They are generally nonirritating and do not cause sensitization when applied topically. Their potential for irritancy when administered intramuscularly, corresponds to their lipophilicity (Johnston, 1985). The oral LD_{50} values for the liquid poloxamers are generally greater than 1 gm/km of body weight, and for solids and pastes greater than 10 gm/km (BASF Corp., 1976). Other routes of administration, especially intravenous or intraperitoneal may indicate lower LD_{50} values, since many of the poloxamers are not well absorbed from the digestive tract. Intravenous administration of 10 doses of 4 g/kg of poloxamer 108 indicates that it is rapidly phagocytized, but otherwise well tolerated (Port, 1978). Two of the poloxamers, 188 and 407, have been studied extensively for internal medical use.

Toxicology of Poloxamer 188

Early studies on the toxicology of intravenously administered poloxamer 188 indicated that its systemic toxicity is low. The intravenous LD_{50} was reported to be greater than 3 gm/kg of body weight in both rats and mice (BASF Corp., 1976). Dogs tolerated continuous infusions of 500 mg/kg/hr for three days (Cytrx Corp., unpublished; Justicz, 1991). However, it has also been reported that infusions of poloxamer 188 resulted in elevated liver enzymes, primarily serum glutamic oxaloacetic transaminase, and a dose related cytoplasmic vacuolation of the proximal tubular epithelial cells of the kidneys (Culbreth, 1992). This study was also performed in dogs, but the dose range over which these observations were made was not specified.

Patients given an artificial blood substitute which contained 2.7% poloxamer 188 as an emulsifier showed signs of complement activation (Vercellotti, 1982). That their reaction was due to the poloxamer component was confirmed by infusing poloxamer 188 alone into rabbits and producing similar symptoms. Incubation of human plasma with poloxamer 188 caused C3 conversion and decreased CH_{50}, indicators of complement activation. However, the significance of this observation is in some doubt. Complement activation as determined from CH_{50} was not detected *in vivo* in the rat (Williams, 1988). Furthermore, poloxamer 188 has since been administered to a large number of patients, and no indication of complement activation has been observed.

Poloxamer 188 inhibited the growth of both cultured human Hela cells and B16 mouse melanoma cells and inhibited chemotaxis by neutrophils (Lane, 1988). The inhibition of cell growth was concentration dependent, although its mechanism was unclear. Whether this *in vitro* inhibition of cell growth is an indication of *in vivo* toxicity is also unknown. Furthermore, growth inhibition by poloxamer 188 may be specific to these cells, as poloxamer 188 has since been used in cell culture medium with no reported inhibition of cell growth. An effect on the normal migration of granulocytes, including neutrophils, has been observed *in vivo* in the rat (Williams, 1988). Lane and Lamkin (Lane, 1984) observed that poloxamer 188 inhibited human neutrophil chemotaxis *in vitro*. The potentially deleterious activities, inhibition of cell growth and neutrophil chemotaxis, were both partitioned into a fraction of poloxamer 188 by supercritical fluid fractionation (Lane, 1988). This suggests that a subpopulation of the poloxamer molecules or a contaminant may be responsible for these activities.

Toxicology of Poloxamer 407

Poloxamer 407 is nonirritating if applied topically or intracutaneously, and it exhibits little potential for irritation following intramuscular or intraperitoneal administration (Johnston, 1985; MDV Technologies, unpublished). Unlike some of the other poloxamers, it does not activate complement in human serum, nor does it cause a delayed-type hypersensitivity response in mice. Reproductive studies indicate that poloxamer 407 is nonteratogenic in both rats and rabbits.

Four doses of 1 gm/kg of poloxamer 407 given to rats caused hematological changes including decreased percent lymphocytes, red blood cell count, hemoglobin, and percent hematocrit. Enlargement of the spleen was noted 24 hours after the last dose, due to sequestration of macrophages containing lipid, and possibly the poloxamer itself (Johnston, 1993). Similar effects have been observed following equivalently high, repeated doses in the dog (MDV Technologies, unpublished). A no-adverse-effect dose of approximately 400 mg/kg body weight has been established in dogs and rats (Reeve, 1995). Doses above 120 mg/kg cause dose dependent increases in plasma cholesterol and triglyceride concentrations (Wout, 1992; Johnston, 1993, MDV Technologies, unpublished). After moderate doses of poloxamer 407, the plasma cholesterol and triglyceride concentrations return to the normal range within hours or days, but following very high doses, lipid concentrations may remain elevated for several weeks. Although the mechanism for this response to poloxamer 407 is unknown, a similar phenomenon has been observed following administration of other surfactants, most notably Triton WR-1339, a polyoxyethylene ether of an alkyl phenol (Ishikawa, 1979). It has been suggested that Triton WR-1339 associates with the circulating lipoproteins and forms a coating around the lipids, which prevents them from being hydrolized by lipoprotein lipase (Scanu, 1961). There is also evidence that the vascular lipoprotein lipase activity is reduced in the presence of either Triton WR-1339 or poloxamer 407 (Hayashi, 1981; Johnston, 1993). In any case, the lipoproteins are temporarily trapped in the plasma, and their concentrations rise. As the plasma surfactant concentration decreases, normal lipid metabolism is restored.

MEDICAL USES OF POLOXAMER 188

Proteins and cells are stable in the aqueous environment of the body because they maintain hydrophilic surfaces while being held together by strong hydrophobic interactions. Cellular damage exposes the hydrophobic surfaces, which in turn disrupt the hydrogen bonding of the surrounding water molecules. Conversely, if two molecules or particles with exposed hydrophobic regions approach each other, they adhere tightly, thus reducing the exposed hydrophobic surface area as well as the free energy due to disrupted hydrogen bonding with the surrounding water molecules. Such hydrophobic interactions are the underlying cause of platelet aggregation to damaged cells, reduced flexibility of red blood cells, and increased blood viscosity, all of which can sometimes be undesirable, especially following tissue injury. All of the observed therapeutic properties of poloxamer 188 are thought to be due to binding of its polyoxypropylene region to exposed hydrophobic surfaces, which removes them from the aqueous environment. The flexible hydrophilic segments of poloxamer 188 interact with the surrounding water molecules and restore hydrogen bonding (Culbreth, 1992; Raymond, 1989).

Treatment of Tissue Ischemia and Reperfusion Injury

Poloxamer 188 has been under investigation for treatment of circulatory pathologies including inhibition of thrombosis following myocardial infarction, decreasing blood viscosity, and improving the perfusion of damaged tissue. Arterial occlusion during myocardial infarction causes abrupt, diminished blood flow to the myocardium. The resulting ischemia, if prolonged, causes cell damage and finally cell death and necrosis. Myocardial cells surrounding the necrotic area may be damaged but still viable, due to collateral blood flow in the ischemic tissue. This flow may, however, be only 10% to 20% of normal. The arterial occlusion may be reversible, in which case reperfusion of the damaged area occurs, or it may be irreversible. In either case, the damaged cells elicit an inflammatory response and leukocytes concentrate in the damaged tissue (Entman, 1991). The leukocytes phagocytize the necrotic myocytes, a necessary step in the healing process. After about four days, the leukocyte population in the damaged area of the myocardium returns to the preocclusion level. Poloxamer 188 appears to affect this complex process at several steps.

Justicz et al. (Justicz, 1991) occluded the left anterior descending coronary artery of dogs for 75 minutes to simulate myocardial infarction. Poloxamer 188 was administered intravenously at a dose of 48 mg/kg during an additional 15 minutes of occlusion and a subsequent 45 minutes of reperfusion. Tissue damage due to ischemia was reduced by approximately 50% in animals treated with poloxamer 188. The reduction apparently resulted from stabilization of cellular membranes in the ischemic tissue. A similar beneficial effect was observed when poloxamer 188 was infused during the first 48 hours of a 72 hour reperfusion period (Schaer, 1994). These authors proposed that a possible mechanism for the beneficial effect of poloxamer 188 may be inhibition of neutrophil chemotaxis into the damaged area.

It has been observed that sometimes blood flow to the injured area is not readily re-established even when the arterial occlusion is removed. This no-reflow phenomenon is thought to be due, at least in part, to the presence of leukocytes in the injured area (Engler, 1986). Leukocytes are fairly large, inflexible cells, which may accumulate and obstruct the microvasculature due to reduced flow rate and pressure during arterial occlusion. As reperfusion is established, there is a rapid influx of neutrophils into the damaged area (Dreyer, 1991; Go, 1988), in response to the release of chemotactic factors as part of the inflammatory response. This influx may cause further obstruction of the capillaries and prevent adequate oxygenation from taking place, even after the initiation of normal flow and, in effect, prolong the occlusion. Thus a brief reduction of the influx of neutrophils during occlusion and early reperfusion may be beneficial.

Once neutrophils arrive at the site of injured tissue, they adhere to the vascular endothelium at the inflammatory sites. Expression of glycoproteins on the neutrophil surface is necessary for both their adherence and transendothelial migration into the damaged tissue (Entman, 1991). Poloxamer 188 has been shown to inhibit adherence of neutrophils *in vitro* to nylon wool by as much as 78% in a standardized assay (Lane, 1984). Whether the effect of poloxamer 188 on neutrophil adherence is due to alterations in glycoprotein expression or binding is unknown. However, this may be a second mechanism for the reduced accumulation of neutrophils in damaged tissue following ischemia (Schaer, 1994).

Reduction of Blood Viscosity

Under physiological conditions, blood viscosity is determined primarily by hematocrit and fibrinogen concentration, and is inversely related to shear rate. Following tissue injury due to trauma, surgery or myocardial infarction, soluble fibrin is present at elevated levels in the circulation. This frequently causes abnormally high blood viscosity, especially under the low shear conditions normally present in the capillaries. Poloxamer 188 has been shown to reduce blood viscosity without lowering the hematocrit (Grover, 1973) in dogs during cardiopulmonary bypass, a procedure known to cause increased blood viscosity. Blood flow was improved by infusion of poloxamer 188 under conditions of hemorrhagic shock (Mayer, 1994) and ischemia due to arterial occlusion (Colbassani, 1989) in rabbits. Both of these pathological conditions result in reduced blood flow rates, and therefore, low shear conditions, in which elevated viscosity due to soluble fibrin may be highly significant. These studies suggest that the viscosity lowering activity of poloxamer 188 may be due primarily to an effect on soluble fibrin.

Similar observations have been made in patients undergoing coronary bypass surgery (Hunter, 1990). Six hours following surgery, hematocrit and fibrinogen concentration, the determinants of viscosity under normal physiological conditions, had fallen, but both blood viscosity measured at low shear rate and the concentration of soluble fibrin had risen significantly. For six days postoperatively, soluble fibrin concentration paralleled the changes in blood viscosity while fibrinogen concentration varied inversely. Hematocrit dropped during surgery but returned to preoperative levels within 24 hours. Addition of high molecular weight oligomers of fibrin further increased viscosity. Thus it appears that *in vivo*, the rise in blood viscosity at low shear rates following surgery or other trauma is due to circulating soluble fibrin polymers. Addition of poloxamer 188 reversed the effect of the fibrin polymers and reduced the viscosity of the patient blood samples to the normal range, leading these authors to postulate that poloxamer 188 normalizes viscosity by effectively disrupting the hydrophobic adhesion between red cells and soluble fibrin polymers.

Interactions with Fibrin

Interactions between fibrin and poloxamer 188 also occur during clot formation (Carr, 1991; van Gelder, 1993). Addition of poloxamer 188 to either plasma or a purified system composed of fibrin, thrombin, and calcium in a buffer having controlled ionic strength, reduced the time of clot formation (Carr, 1991). The fibrin fibers which resulted from this reaction were larger as determined by mass to length ratio, and were more susceptible to lysis mediated by tissue plasminogen activator. In a second study (van Gelder, 1993), the effects of poloxamer 188 on clot formation were studied in plasma and whole blood. A small increase in fiber mass to length ratio, consistent with the previous study, was observed. However the permeability of the fibrin network was increased considerably more than could be accounted for by the change in the fiber radius, suggesting that poloxamer 188 affects the arrangement of the fibers within the network. Poloxamer 188 also enhanced platelet-induced clot retraction.

Nonspecific Cellular Effects

Two reported cellular effects of poloxamer 188 are reduction of hemolysis during cardiopulmonary bypass (Danielson, 1970) and reduction of leakage of fluorescent calcein from thermally or electrically damaged cells (Padanilam, 1994). Both of these effects are probably due to an interaction of poloxamer 188 with the cellular membranes, but the mechanism(s) of action are unknown.

USE OF POLOXAMER 407 FOR PREVENTION OF POSTSURGICAL ADHESIONS

Adhesions are an undesirable consequence of tissue repair, and can occur at any tissue surface following infection, injury or surgery. Peritoneal adhesions frequently occur following endometriosis or pelvic inflammatory disease in women, or following abdominal surgery in both men and women. Although generally asymptomatic, peritoneal adhesions sometimes cause infertility, pain, or intestinal obstruction. After orthopedic or neural surgery, adhesions may cause severe pain and limited range of motion.

Following tissue damage, blood or serous fluid is exuded from the injured site. This exudate contains fibrinogen, which quickly forms a clot composed of solid fibrin strands and seals the surface of the wound. Most such fibrin strands are removed at a later stage of the healing process. However, any fibrin strands remaining in place form a matrix for collagen deposition by fibroblasts. The resulting collagenous structures, called adhesions, may be quite large, and sometimes become vascularized. If adhesions form between the surfaces of two different tissues, they may restrict normal organ movement or function. Surgery is sometimes indicated to lyse symptomatic adhesions. However, this may provide only temporary improvement since adhesions frequently reform following surgery.

Various animal models have been developed to examine the sequelae of tissue injury inflicted during surgical procedures. These models require forming a lesion of a defined size and severity at a specified site on the surface of an organ, closing the animal and allowing healing to take place. After a defined period of time, usually from one to three weeks, the animal is reopened, and adhesion formation at the site of the original injury is evaluated. Models have been developed to test the efficacy of various drugs and biomaterials for prevention of adhesion formation involving specific tissues including intestines, uterus, neurons and lymph nodes.

Because aqueous solutions of greater than 20% poloxamer 407 are liquids at temperatures between 0 and 10°C, but form gels at body temperature, they have been examined for their ability to reduce or prevent adhesion formation. Such solutions are easily applied as liquids to injured tissues during laparoscopic or open surgery. Immediately upon being placed in contact with tissues at body temperature, they form firm gels and remain at the site of application for up to 24 hours. Formulations of poloxamer 407 have been studied for their ability to reduce adhesion formation in rats, rabbits, hamsters and swine.

Effect of Poloxamer 407 on Peritoneal Adhesions

Defined lesions were formed in rats by abrading the surface of the uterine horn and the adjacent sidewall. The injured tissues were kept in close proximity to each other by suturing the uterine horn to the side wall (Leach, 1990; Rice, 1993). Application of formulations containing poloxamer 407 to the injured tissues reduced adhesion formation by 30% to 55%.

Thermo- and electrocautery are used during surgery either to lyse tissue or to coagulate blood and stop bleeding from small vessels. Tissue damage resulting from this procedure includes cell death at the site of cautery, and devascularization and ischemia of the surrounding area. Animal models for injury due to cautery have been developed in rats (West, 1995), hamsters and rabbits (Steinleitner, 1991; Steinleitner, 1992). In these models, the vascular arcade to the uterine horn is ligated using cautery, and a defined area on the antimesenteric surface to the horn is cauterized. Application of 35% solutions of poloxamer 407 to the injured area reduced the extent of adhesion formation by approximately 49% in both the rat and hamster.

In order to study the effect of poloxamer 407 on reformation of adhesions, a more complex procedure, requiring two consecutive surgeries, was performed in rabbits. In the first, the standardized injury was formed using cautery. In a second surgery, the adhesions which had formed were lysed, also using cautery, and a solution of 28% poloxamer 407 was applied to the injured area. Reduction of adhesion reformation in this model ranged from about 47% (Steinleitner, 1991) to nearly complete elimination in some experiments (Steinleitner, 1992).

In a porcine model for pelvic and para-aortic lymphadenectomy, the lymph nodes were removed from four anatomically defined locations on each side of the animal (Deprest, 1995). Solutions of 28% poloxamer 407 were applied to the four sites on either the left or the right side of the animal, and all sites on the opposite side were left untreated. After healing had taken place, adhesion formation at all surgical sites was evaluated, and each treated site was compared to the contralateral, untreated site. Adhesion formation was reduced at the sites treated with poloxamer 407 by approximately 40%. In a subsequent experiment, the temperature of the poloxamer 407 solution was maintained at 0°C during application to the injured area. Adhesion reduction improved to about 55% (Deprest, unpublished).

Effect of Poloxamer 407 on Spinal Adhesions

Adhesions of the leptomeninges following spinal surgery often cause pain and physical limitations for the patient. In a rabbit model of human spinal surgery, the dura over the spinal cord was opened at two sites separated by one or two lumbar segments (Reigel, 1993). A sterile, buffered, hypo-osmotic solution of approximately 23% poloxamer 407 was inserted under the dura at one site. The other site was closed with no additional treatment, and served as a control. The rabbits were allowed to heal and then evaluated for adhesion formation. There was approximately a 50% reduction in leptomeningeal adhesion formation at the sites which received poloxamer 407.

Mechanisms of Adhesion Reduction

Although the precise mechanism of adhesion reduction by poloxamer 407 is unknown, at least two processes are thought to contribute to it. Upon reaching body temperature, poloxamer 407 solutions form gels which remain in place and are capable of coating and separating tissue surfaces for up to 24 hours, and thus preventing adhesion formation between them. Secondly, *in vitro* observations indicate that fibrin clot formation is altered by the presence of poloxamer 407 (Carr, 1995). As was previously observed with poloxamer 188, the fiber size was increased as determined from optical density measurements. Unlike poloxamer 188 however, fibrinolysis in the presence of tissue plasminogen activator was inhibited. Nevertheless, the clot may be less stable under physiological conditions if the permeability of the fibrin network is increased by the presence of poloxamer 407, as has been suggested for poloxamer 188 (van Gelder, 1993).

USES OF THE POLOXAMERS IN DRUG DELIVERY

Aqueous solutions of the poloxamers form micelles, which provide a lipophilic compartment surrounded by an aqueous environment. Drugs partition between the two compartments, leaving only that fraction of the drug which is present in the aqueous solution readily available for release. Generally, lipophilic drugs are sequestered in the micelles, and therefore released more slowly (Kataoka, 1993; Yokoyama, 1992). To confirm the effect of lipophilicity on drug release rates, Gilbert *et al.* used poloxamer 407 gels, and studied the release of the methyl-, ethyl-, propyl- and butyl-esters of p-hydroxybenzoate, a series of structurally related compounds with increasing lipophilic character (Gilbert, 1986). As predicted, as the lipophilicity of the test compound increased, its rate of release decreased and its energy of diffusion increased.

Site Specific Drug Delivery

Solutions of greater than 20% poloxamer 407 form gels at body temperature, and remain where they are placed, making them particularly attractive as site specific drug delivery vehicles, and for topical applications. Release rates of adriamycin and 5-fluorouracil from poloxamer 407 gels were shown to be proportional to drug concentration and temperature, and inversely proportional to poloxamer concentration (Miyazaki, 1984). Although they may be an effective release vehicle for topical drug delivery, the poloxamers do not increase skin permeability, and therefore do not enhance transdermal delivery (Cappel, 1991).

Poloxamer 407 solutions have been proposed for nasal delivery of peptides (Juhasz, 1989). Such formulations could be administered as solutions, but upon gelling, would provide slow release, and prolonged residence time in the nasal cavity. The release rate of atrial natriuretic factor, a peptide having a molecular weight of approximately 3,060, increased with temperature over a range from 0 to 40°C, despite the temperature dependent increase in poloxamer viscosity. Because the peptide is too hydrophilic to associate with the micelles, it is released by diffusion through the external aqueous compartment. Therefore, although the presence of the poloxamer micelles lengthen the diffusion pathway and delay release, the release rate of the peptide depends primarily on thermal energy.

Poloxamers in Complex Drug Delivery Systems

The poloxamers have been found to be capable of serving various functions as components of complex drug delivery systems. Poloxamer 188, for example, was shown to stabilize a positively-charged submicron emulsion (Klang, 1994). Poloxamers 331, 334, and 338 have been blended with poly(L-lactic acid) to form films (Park, 1992). When used as a release matrix, the films extended protein release, and minimized the initial burst of rapid release as compared to pure poly(L-lactic acid). Poloxamer 407 has also been blended with cellulose acetate phthalate to form a drug release matrix (Xu, 1993). At neutral pH, the rate of drug release was controlled by the rate of surface erosion. Laminates, composed of layers of matrix containing different amounts of drug, were found to erode at a constant rate and release the drug in a pulsatile, predictable manner over a period of hours or days.

Although microspheres are interesting potential drug carriers, they have the disadvantage of being rapidly taken up from the blood by the mononuclear phagocytes of the liver and spleen. However, coating the microspheres with poloxamer 338 reduced the rate of uptake by the liver, and the microspheres remained in the blood for a longer period of time (Illum, 1983; Davis, 1986). In another study, small (150 nm in diameter or less) polystyrene microspheres were coated with poloxamer 407 and administered intravenously to rabbits (Porter, 1992b).

The coated microspheres remained in the circulation for up to eight hours, but were then sequestered in the sinusoidal endothelial cells of the bone marrow. From these observations, it appears that poloxamer coated microspheres may provide a means of improving the efficacy of some drugs by increasing residency time in the circulation or specifically directing the drug to the bone marrow.

Drug-Like Properties of the Poloxamers

In addition to their drug delivery capabilities, the poloxamers exhibit some drug-like properties of their own. Poloxamer 331 was found to have a bacteriostatic effect on *Mycobacterium avium* complex (Hunter, 1995). Furthermore, the *in vitro* minimum effective concentration of rifampin against *M. avium* complex was reduced by a factor of 10 in the presence of 1.0 mg/ml poloxamer 331. Whether poloxamer 331 also has a synergistic effect *in vivo* with rifampin or other antibiotics remains to be determined.

Two investigational poloxamers have been synthesized and evaluated for their ability to treat acute infection with *Toxoplasma gondii* in mice (Krahenbuhl, 1993). Poloxamers CRL-8131 (molecular weight, 4,000; 10% ethylene oxide) and CRL-8142 (molecular weight, 5,100; 20% ethylene oxide) were both found to significantly reduce mortality in mice infected with between 10 and 1,000 times the lethal dose of T. *gondii*. During acute T. *gondii* infection, the pathogen is taken up by cellular macrophages. *In vitro* data suggested that the macrophages are, in fact, the site of the poloxamers' activity against T. *gondii*. Although CRL-8131 was consistently more efficacious than CRL-8142, the precise mechanism of the antitoxoplasma activity of these poloxamers is unknown.

USES OF THE POLOXAMERS IN ADJUVANT FORMULATIONS

Recombinant DNA technology is producing a wide variety of synthetic peptide antigens for immunization of humans and animals. In order to elicit cell-mediated and humoral immune responses, these antigens must be administered in effective adjuvant formulations. In an effort to develop an adjuvant formulation which would be effective with many antigens, a series of block copolymer surfactants were studied (Hunter, 1986). These copolymers included poloxamers, reversed triblock polymers, having a center block of polyoxyethylene flanked on either end by polyoxypropylene blocks, and octablock copolymers. All of these copolymers exhibit low solubility in water. At a concentration of 25 mg/ml, each of the copolymers produced a characteristic aggregate in the form of fibers, fine precipitates, or spherical droplets. Those copolymers which formed spheres, induced granulomatous responses, an undesirable reaction. Copolymers which formed fibers, however, also activated complement and thus elicited an immune response. Poloxamers 401 and 331 formed fibrous structures, activated complement, and also bound protein on the surface of the structure. Thus antigens, which are generally proteins, can readily be transferred from the surface of the poloxamer to the surface of antigen-presenting cells. Emulsions of poloxamer 401 with squalane and Tween 80 increase surface area and, therefore, maximize antigen availability (Allison, 1986). An adjuvant formulation prepared by adding to the emulsion the threonyl analog of muramyl dipeptide, which stimulated cell-mediated immunity, increased the efficacy of influenza, hepatitis B virus, herpes simplex virus, lentivirus and tumor vaccines in experimental animals (Allison, 1991).

REFERENCES

Al-Saden, A.A., Whateley, T.L. and Florence, A.T. (1982) Poloxamer association in aqueous solution. *Journal of Colloid and Interface Science*, **90**(2), 303–309.

Allison, A.C. and Byars, N.E. (1986) An adjuvant formulation that selectively elicits the formation of antibodies of protective isotypes and of cell-mediated immunity. *Journal of Immunological Methods*, **95**, 157–168.

Allison, A.C. and Byars, N.E. (1991) Immunological adjuvants: desirable properties and side-effects. *Molecular Immunology*, **28**(3), 279–284.

Attwood, D., Collett, J. and Tait, C. (1985) The micellar properties of the poly(oxyethylene)-poly(oxypropylene) copolymer Pluronic® F127 in water and electrolyte solution. *International Journal of Pharmaceutics*, **26**, 25–33.

Attwood, D., Tait, C.J. and Collett, J.H. (1987) "Thermally reversible gelation characteristics, poly(oxyethylene)-poly(oxypropylene) block copolymer in aqueous solution after exposure to high-energy irradiation" (Chapter 10), in *Controlled-Release Technology: Pharmaceutical Applications*, Lee P.I., and Good, W. (eds), American Chemical Society, Washington D.C.

BASF Wyandotte Corporation, Industrial Chemicals Group, (1976) Pluronic® polyols, toxicity and irritation data, Wyandotte, MI, 1976.

Cappel, M.J. and Kreuter, J. (1991) Effect of nonionic surfactants on transdermal drug delivery: II poloxamer and poloxamine surfactants. *International Journal of Pharmaceutics*, **69**, 155–167.

Carr, M.E., Jr., Carr, S.L. and High, A.A. (1995) Effects of Poloxamer 407 on the assembly, structure, and dissolution of fibrin clots. *Blood*, **86**, 886A.

Carr, M.E., Jr., Powers, P.L. and Jones, M.R. (1991) Effects of poloxamer 188 on the assembly, structure and dissolution of fibrin clots. *Thrombosis and Haemostasis*, **66 (5)**, 565–568.

Colbassani, H.J., Barrow, D.L., Sweeney, K.M., Bakay, R.A.E., Check, I.J. and Hunter, R.H. (1989) Modification of acute focal ischemia in rabbits by poloxamer 188. *Stroke*, **20**, 1241–1246.

Culbreth, P.H., Emanuele, R.M., Hunter, R.L. (1992) New polyoxypropylene-polyoxyethylene block copolymers for treating tissue damaged by ischaemia or reperfusion, stroke, myocardial damage and adult respiratory distress syndrome; Microbial Infection. European Patent WO 9216484.

Danielson, G.K., Dubilier, L.D. and Bryant, L.R. (1970) Use of Pluronic® F-68 to diminish fat emboli and hemolysis during cardiopulmonary bypass. *Journal of Thoracic and Cardiovascular Surgery*, **59**(2), 178–184.

Davis, S.S. and Illum, L. (1986) Colloidal delivery systems: opportunities and challenges, in *Site-Specific Drug Delivery*, Tomlinson, E. and Davis, S.S. (eds), John Wiley and Sons, New York.

Deprest, J.,Vergote, I., Evrard, V. and Brosens, I. (1995) Prevention of adhesions with the use of flogel after endoscopic pelvic and para-aortic lymphadenectomy in the porcine model. *American Society for Reproductive Medicine*, Program Supplement, S4.

Dreyer, W.J., Michael, L.H., West, M.S., Smith, C.W., Rothlein, R., Rossen, R.D., Anderson, D.C. and Entman, M.L. (1991) Neutrophil accumulation in ischemic canine myocardium: insights into time course, distribution, and mechanism of localization during early reperfusion. *Circulation*, **84**, 400–411.

Engler, R.L., Dahlgren, M.D., Morris, D.D., Peterson, M.A. and Schmid-Schonbein, G.W. (1986) Role of leukocytes in response to acute myocardial ischemia and reflow in dogs. *American Journal of Physiology*, **251**, H314–H322.

Entman, M.L., Michael, L., Rossen, R.D., Dreyer, W.J., Anderson, D.C., Taylor, A.A. and Smith, C.W. (1991) Inflammation in the course of early myocardial ischemia. *Federation of American Science and Experimental Biology Journal*, **5**, 2529–2537.

Gilbert, J.C., Hadgraft, J., Bye, A. and Brookes, L.G. (1986) Drug release from Pluronic F-127 gels. *International Journal of Pharmaceutics*, **32**, 223–228.

Go, L.O., Murry, C.E., Richard, V.J., Weischedel, G.R., Jennings, R.B. and Reimer, K.A. (1988) Myocardial neutrophil accumulation during reperfusion after reversible or irreversible ischemic injury. *American Journal of Physiology*, **255**, H1188–H1198.

Grover, F.L., Kahn, R.S., Heron, M.W. and Paton, B.C. (1973) A nonionic surfactant and blood viscosity. *Archives of Surgery*, **106**, 307–310.

Hayashi, H., Ninobe, S., Matsumoto, Y. and Suga, T. (1981) Effects of Triton WR-1339 on lipoprotein lipolytic activity and lipid content of rat liver lysosomes. *Journal of Biochemistry (Tokyo)*, **89**, 573–579.

Hunter, R.L. and Bennet, B. (1986) The adjuvant activity of nonionic block polymer surfacants. III. Characterization of selected biologically active surfaces. *Scandinavian Journal of Immunology*, **23**, 287–300.

Hunter, R.L., Papadea, C., Gallagher, C.J., Finlayson, D.C. and Check, I.J. (1990) Increased whole blood viscosity during coronary artery bypass surgery. *Thrombosis and Haemostasis*, **63**(1), 6–12.

Hunter, R.L., Chinnaswamy, J., Tinkley, A., Behling, C.A. and Nolte, F. (1995) Enhancement of antibiotic susceptibility and suppression of *mycobacterium avium* complex growth by poloxamer 331. *Antimicrobial Agents and Chemotherapy*, **39**(2), 435–439.

Illum, L. and Davis, S.S. (1983) Effect of the nonionic surfactant poloxamer 338 on the fate and deposition of polystyrene microspheres following intravenous administration. *Journal of Pharmaceutical Sciences*, **72**, 1086–1089.

Ishikawa, T. and Fidge, N. (1979) Changes in the concentration of plasma lipoproteins and apoproteins following the administration of Triton WR-1339 to rats. *Journal of Lipid Research*, **20**, 254–264.

Johnston, T.P. and Palmer, W.K. (1993) Mechanism of poloxamer 407 induced hypertriglyceridemia in the rat. *Biochemical Pharmacology*, **46**(6), 1037–1042.

Johnston, T.P. and Miller, S.C. (1985) Toxicological evaluation of poloxamer vehicles for intramuscular use. *Journal of Parental Science and Technology*, **39**(2), 83–88.

Johnston, T.P., Beris, H., Wout, Z.G. and Kennedy, J.L. (1993) Effects on splenic, hepatic,hematological, and growth parameters following high-dose poloxamer 407 administration to rats. *International Journal of Pharmaceutics*, **100**, 279–284.

Juhasz, J., Lenaerts, V., Raymond, P. and Ong, H. (1989) Diffusion of rat atrial natriuretic factor in thermoreversible poloxamer gels. *Biomaterials*, **10**, 265–268.

Justicz, A., Farnsworth, W., Soberman M., Tuvlin M., Bonner G., Hunter, R., Martino-Saltzman, D., Sink, J. and Austin, G. (1991) Reduction of myocardial infarct size by poloxamer 188 and mannitol in a canine model. *American Heart Journal*, **122**, 671–680.

Kataoka, K., Kwon, G., Yokoyama, M., Okano, T. and Sakurai, Y. (1993) Block copolymer micelles as vehicles for drug delivery. *Journal of Controlled Release*, **24**, 119–132.

Klang, S.H., Frucht-Pery, J., Hoffman, A. and Benita, S. (1994) Physicochemical characterization and acute toxicity evaluation of a positively-charged submicron emulsion vehicle. *Journal of Pharmaceutics and Pharmacology*, **46**, 986–993.

Krahenbuhl, J.L., Fukutomi, Y. and Gu, L. (1993) Treatment of acute experimental toxoplasmosis with investigational poloxamers. *Antimicrobial Agents and Chemotherapy*, **37**(11), 2265–2269.

Lane, T.A. and Lamkin, G.E. (1984) Paralysis of phagocyte migration due to an artificial blood substitute. *Blood*, **64**(2), 400–405.

Lane, T.A. and Krukonis, V. (1988) Reduction in the toxicity of a component of an artificial blood substitute by supercritical fluid fractionation. *Transfusion*, **28**, 375–378.

Leach, R.E. and Henry, R.L. (1990) Reduction of postoperative adhesions in the rat uterine horn model with poloxamer 407. *American Journal of Obstetrics and Gynecology*, **162**(5), 1317–1319.

Lundsted, L.G. and Schmolka, I.R. (1976) "The synthesis and properties of block copolymer polyol surfactants," in *Block and Graft Copolymerization*, vol. 2, Ceresa, R.J. (ed), John Wiley and Sons, New York, NY.

Lundsted, L.G. (1954) Polyoxyalkylene compounds. U.S. Patent 2,674,619.

Mayer, D.C., Strada, S.J., Hoff, C., Hunter, R.L. and Artman, M. (1994) Effects of poloxamer 188 in a rabbit model of hemorrhagic shock. *Annals of Clinical and Laboratory Science*, **24**(4), 302–311.

Miyazaki, S., Takeuchi, S., Yokouchi, C. and Takada, M. (1984) Pluronic® F-127 as a vehicle for topical administration of anticancer agents. *Chemical Pharmacy Bulletin*, **32**(10), 4205–4208.

Nakashima, K., Anzai, T. and Fujimoto, Y. (1994) Fluorescence studies on the properties of a Pluronic® F68 micelle. *Langmuir*, **10**, 658–661.

Ott, R.A. (1988) Process for the preparation of polyoxyalkylene block polyethers having enhanced properties. US Patent 4,764,567.

Padanilam, J.T., Bischof, J.C., Lee, R.C., Cravalho, E.G., Tompkins, R.G., Yarmush, M.L. and Toner, M. (1994) Effectiveness of poloxamer 188 in arresting calcein leakage from thermally damaged isolated skeletal muscle cells. *Annals New York Academy of Sciences*, **720**, 111–123.

Park, T.G., Cohen, S. and Langer, R. (1992) Poly(L-lactic acid)/Pluonic blends: characterization of phase separation behavior, degradation, and morphology and use as protein-releasing matrices. *Macromolecules*, **25**(1), 116–122.

Port, C.D., Garvin, P.J. and Ganote, C.E. (1978) The effect of Pluronicÿ20 F-38 (poloxamer 108) administered intravenously to rats. *Toxicology and Applied Pharmacology*, **44**, 401–411.

Porter, C.J.H., Moghimi, S.M., Davies, M.C., Davis, S.S. and Illum, L. (1992) Differences in the molecular weight profile of poloxamer 407 affect its ability to redirect intravenously administered colloids to the bone marrow. *International Journal of Pharmaceutics*, **83**, 273–276.

Porter, C.J.H., Moghimi, S.M., Illum, L. and Davis, S. (1992) The polyoxyethylene/polyoxypropylene block co-polymer Poloxamer-407 selectively redirects intravenously injected microspheres to sinusoidal endothelial cells of rabbit bone marrow. *Federation of European Biochemical Societies*, **305**(1), 62–66.

Prasad, K.N., Luong, T.T., Florence, A.T., Paris, J., Vaution, C., Seiller, M. and Puisieux, F. (1979) Surface activity and association of ABA polyoxyethylene-polyoxypropylene block copolymers in aqueous solution. *Journal of Colloid Interface Science*, **69**(2), 225–232.

Rassing J. and Attwood, D. (1983) Ultrasonic velicity and light-scattering studies on the polyoxyethylene-polyoxypropylene copolymer Pluronic® F127 in aqueous solution. *International Journal of Pharmaceutics*, **13**, 47–55.

Raymond, C. (1989) Copolymer, undergoing trials, could improve fibrinolytics' effectiveness. *Journal of the American Medical Association*, **261**(17), 2475, 2479–2480.

Reeve, L., Brown, L. and Foley, F. (1995) Treatment of peritoneal adhesions with flogel antiadhesion 28. *American Society for Reproductive Medicine*, Program Supplement, S229.

Reigel, D.H., Bazmi, B., Shih, S.R. and Marquardt, M.D. (1993) A pilot investigation of poloxamer 407 for the prevention of leptomeningeal adhesions in the rabbit. *Pediatric Neurosurgery*, **19**, 250–255.

Rice, V.M., Shanti, A., Moghissi, K.S. and Leach, R.E. (1993) A comparative evaluation of poloxamer 407 and oxidized regenerated cellulose (Interceed® [TC7]) to reduce postoperative adhesion formation in the rat uterine horn model. *Fertility and Sterility*, **59**(4), 901–906.

Scanu, A. and Orient, P. (1961) Triton hyperlipidemia in dogs. I. *In vitro* effects of the detergent on serum lipoproteins and chylomicrons. *Journal of Experimental Medicine*, **113**, 735–757.

Schaer, G.L., Hursey, T.L., Abrahams, S.L., Buddemeier, K., Ennis, B., Rodriguez, E.R.,Hubbell, J.P., Moy, J. and Parrillo, J.E. (1994) Reduction of reperfusion-induced myocardial necrosis in dogs by RheothRx injection (poloxamer 188 N.F.), a hemorheological agent that alters neutrophil function. *Circulation*, **90**, 2964–2975.

Schmolka, I.R. (1994) Physical basis for poloxamer interactions. *Annals of New York Academy of Science*, **720**, 92–97.

Schmolka, I.R. (1976) A review of block polymer surfactants. *Journal of American Oil Chemistry Society*, **54**, 110–116.

Schmolka, I.R. (1972) Silver ion gel compositions and method of using the same. U.S. Patent 3,639,575.

Schmolka, I.R. (1991) "Poloxamers in the pharmaceutical industry," in *Polymers for Controlled Drug Delivery*, Tarcha, P.J. (ed), CRC Press, Boca Ratan, FL.

Steinleitner, A., Lambert, H., Kazensky, C. and Cantor, B. (1991) Poloxamer 407 as an intraperitoneal barrier material for the prevention of postsurgical adhesion formation and reformation in rodent models for reproductive surgery. *Obstetrics and Gynecology*, **77**(48), 48–52.

Steinleitner, A., Lopez, G., Suarez, M. and Lambert, H. (1992) An evaluation of flowgel® as an intraperitineal barrier for prevention of postsurgical adhesion reformation. *Fertility and Sterility*, **57**(2), 305–308.

van Gelder J.M., Nair C.H. and Dhall, D.P. (1993) Effects of poloxamer 188 on fibrin network structure, whole blood clot permeability and fibrinolysis. *Thrombosis Research*, **71**, 361–376.

Vercellotti, G.M., Hammerschmidt, D.E., Craddock, P.R. and Jacob, H.S. (1982) Activation of plasma complement by perfluorocarbon artificial blood: probable mechanism of adverse pulmonary reactions in treated patients and rationale for corticosteroid prophylaxis. *Blood*, **59**(6), 1299–1304.

Wanka, G., Hoffmann, H. and Ulbricht, W. (1990) Polymer science: the aggregation behavior `of poly-(oxyethylene)-poly-(oxypropylene)-poly-(oxyethylene) block copolymers in aqueous solution. *Colloid Polymer Sciences*, **268**, 101–117.

West, J.L. and Hubbell, J.A. (1995) Comparison of covalently and physically cross-linked polyethylene glycol-based hydrogels for the prevention of postoperative adhesions in a rat model. *Biomaterials*, **16**(15), 1153–1156.

Williams, J.H., Chen, M., Drew, J., Panigan, E. and Hosseini, S. (1988) Modulation of rat granulocyte traffic by a surface active agent *in vitro* and bleomycin injury. *Proceedings of the Society for Experimental Biology and Medicine*, **188**, 461–470.

Wout, Z.G.M., Pec, E.A., Maggiore, J.A., Williams, R.H., Palicharla, P. and Johnston, T.P. (1992) Poloxamer 407 mediated changes in plasma cholesterol and triglycerides following intraperitoneal injection to rats. *Journal of Parenteral Science and Technology*, **46**(6), 192–200.

Xu, X. and Lee, P.I. (1993) Programmable drug delivery from an erodible association polymer system. *Pharmaceutical Research*, **10**(8), 1144–1152.

Yokoyama, M. (1992) Block copolymers as drug carriers. *Critical Reviews in Therapeutic Drug Carrier Systems*, **9**(3,4), 213–248.

Zhou, Z. and Chu, B. (1988) Light-scattering study on the association behavior of triblock polymers of ethylene oxide and propylene oxide in aqueous solution. *Journal of Colloid Interface Science*, **126** (1), 171–180.

APPENDIX
Commercial Sources of the Poloxamers

Chemical Name	Structure	Trade Names					Comments
		BASF (USA)[1]	BASF (Germany)[2]	ICI (UK)[3]	Sanyo (Japan)[4]	Rhone-Poulenc (USA)[5]	
Poloxamer 101	$HO(C_2H_3O)_2(C_3H_5O)_{16}(C_2H_3O)_2H$	Pluronic L31	Pluronic PE 3100	Synperonic PE L31			
Poloxamer 105	$HO(C_2H_3O)_{11}(C_3H_5O)_{16}(C_2H_3O)_{11}H$	Pluronic L35		Synperonic PE L35			
Poloxamer 108	$HO(C_2H_3O)_{46}(C_3H_5O)_{16}(C_2H_3O)_{46}H$	Pluronic F38		Synperonic PE F38			
Poloxamer 122	$HO(C_2H_3O)_5(C_3H_5O)_{21}(C_2H_3O)_5H$	Pluronic L42		Synperonic PE L42			
Poloxamer 123	$HO(C_2H_3O)_7(C_3H_5O)_{21}(C_2H_3O)_7H$	Pluronic L43	Pluronic PE4300	Synperonic PE L43			NF Grade available.
Poloxamer 124	$HO(C_2H_3O)_{11}(C_3H_5O)_{21}(C_2H_3O)_{11}H$	Pluronic L44		Synperonic PE L44			
Poloxamer 181	$HO(C_2H_3O)_3(C_3H_5O)_{30}(C_2H_3O)_3H$	Pluronic L61	Pluronic PE 6100	Synperonic PE L61	Newpol PE-61	Antarox L61, PGP 18-1	
Poloxamer 182	$HO(C_2H_3O)_8(C_3H_5O)_{30}(C_2H_3O)_8H$	Pluronic L62	Pluronic PE 6200	Synperonic PE L62	Newpol PE-62	Antarox L62, PGP 18-2	
Poloxamer 182 LF	$HO(C_2H_3O)_8(C_3H_5O)_{30}(C_2H_3O)_8H$	Pluronic L62 LF		Synperonic PE L62 LF		Antarox L62LF, PGP 18-2LF	
Poloxamer 183	$HO(C_2H_3O)_{10}(C_3H_5O)_{30}(C_2H_3O)_{10}H$	Pluronic L63					
Poloxamer 184	$HO(C_2H_3O)_{13}(C_3H_5O)_{30}(C_2H_3O)_{13}H$	Pluronic L64	Pluronic PE 6400	Synperonic PE L64	Newpol PE-64	Antarox L64, PGP 18-4	
Poloxamer 185	$HO(C_2H_3O)_{19}(C_3H_5O)_{30}(C_2H_3O)_{19}H$	Pluronic P65					
Poloxamer 188	$HO(C_2H_3O)_{75}(C_3H_5O)_{30}(C_2H_3O)_{75}H$	Pluronic F68	Pluronic PE 6800	Synperonic PE F68	Newpol PE-68	Antarox PGP 18-8	NF Grade available; clinical trial for treatment of myocardial infarction ongoing; clinical trial for sickle cell treatment ongoing. Fluosol (2.8% poloxamer 188) approved for angioplasty.
Poloxamer 212	$HO(C_2H_3O)_8(C_3H_5O)_{35}(C_2H_3O)_8H$	Pluronic L72		Newpol PE74			
Poloxamer 215	$HO(C_2H_3O)_{24}(C_3H_5O)_{35}(C_2H_3O)_{24}H$	Pluronic P75		Synperonic PE P75	Newpol PE-75		
Poloxamer 217	$HO(C_2H_3O)_{52}(C_3H_5O)_{35}(C_2H_3O)_{52}H$	Pluronic F77		Synperonic PE F77	Newpol PE-78		

Chemical Name	Structure	Trade Names					Comments
		BASF (USA)[1]	BASF (Germany)[2]	ICI (UK)[3]	Sanyo (Japan)[4]	Rhone-Poulenc (USA)[5]	
Poloxamer 231	$HO(C_2H_3O)_6(C_3H_5O)_{39}(C_2H_3O)_6H$	Pluronic L81	Pluronic PE 8100	Synperonic PE L81			
Poloxamer 234	$HO(C_2H_3O)_{22}(C_3H_5O)_{39}(C_2H_3O)_{22}H$	Pluronic P84		Synperonic PE P84		Antarox P84	
Poloxamer 235	$HO(C_2H_3O)_{27}(C_3H_5O)_{39}(C_2H_3O)_{27}H$	Pluronic P85		Synperonic PE P85			
Poloxamer 237	$HO(C_2H_3O)_{62}(C_3H_5O)_{39}(C_2H_3O)_{62}H$	Pluronic F87		Synperonic PE F87		Antarox PGP 23-7	NF Grade available.
Poloxamer 238	$HO(C_2H_3O)_{97}(C_3H_5O)_{39}(C_2H_3O)_{97}H$	Pluronic F88		Synperonic PE F88	Newpol PE-88	Antarox F88	
Poloxamer 282	$HO(C_2H_3O)_{10}(C_3H_5O)_{47}(C_2H_3O)_{10}H$	Pluronic L92	Pluronic PE 9200	Synperonic PE L92			
Poloxamer 284	$HO(C_2H_3O)_{21}(C_3H_5O)_{47}(C_2H_3O)_{21}H$	Pluronic P94	Pluronic PE 9400	Synperonic PE P94			
Poloxamer 288	$HO(C_2H_3O)_{122}(C_3H_5O)_{47}(C_2H_3O)_{122}H$	Pluronic F98					
Poloxamer 331	$HO(C_2H_3O)_7(C_3H_5O)_{54}(C_2H_3O)_7H$	Pluronic L101	Pluronic PE 10100	Synperonic PE L101			
Poloxamer 333	$HO(C_2H_3O)_{20}(C_3H_5O)_{54}(C_2H_3O)_{20}H$	Pluronic P103		Synperonic PE P103			
Poloxamer 334	$HO(C_2H_3O)_{31}(C_3H_5O)_{54}(C_2H_3O)_{31}H$	Pluronic P104				Antarox P104	
Poloxamer 335	$HO(C_2H_3O)_{38}(C_3H_5O)_{54}(C_2H_3O)_{38}H$	Pluronic P105	Pluronic PE 10500	Synperonic PE P105			
Poloxamer 338	$HO(C_2H_3O)_{128}(C_3H_5O)_{54}(C_2H_3O)_{128}H$	Pluronic F108		Synperonic PE F108			NF Grade available.
Poloxamer 401	$HO(C_2H_3O)_6(C_3H_5O)_{67}(C_2H_3O)_6H$	Pluronic L121		Synperonic PE L121			Used in adjuvants; clinical trials performed.
Poloxamer 402	$HO(C_2H_3O)_{13}(C_3H_5O)_{67}(C_2H_3O)_{13}H$	Pluronic L122					
Poloxamer 403	$HO(C_2H_3O)_{21}(C_3H_5O)_{67}(C_2H_3O)_{21}H$	Pluronic P123					
Poloxamer 407	$HO(C_2H_3O)_{98}(C_3H_5O)_{67}(C_2H_3O)_{98}H$	Pluronic F127		Synperonic PE F127			NF Grade available; clinical trial for postsurgical adhesion reduction ongoing; 510k approved for topical use.

1
BASF Corporation
3000 Continental Drive-North
Mount Olive, NJ 07828-1234
(201) 426-2600

2
BASF Aktiengesellschaft
Carl-Bosch-Straße 38
67056 Ludwigshafen
(49) 06 21 60-0

3
ICI (UK)
PO Box 90, Wilton Centre
Middlesbrough, Cleveland TS90 8JE
44 1642 454 144

4
Sanyo (Japan)
111 Ikkyo Nomoto Cho
Higashiyamaku, Kyoto 605
81 75 541 4311

5
Rhone-Poulenc (USA)
CN 7500
Cranberry, NJ 08512-7500
(609) 860-4000

13. DEGRADABLE POLYMERS DERIVED FROM THE AMINO ACID L-TYROSINE

JOHN KEMNITZER and JOACHIM KOHN*

Department of Chemistry, Rutgers University, Piscataway, NJ 08855–0939, USA

INTRODUCTION

Historic Overview

Over the last 20 years significant efforts have been devoted to the development of polymeric biomaterials. The vast majority of these efforts were focused on the identification of biologically inert materials that were stable under physiological conditions. These materials were used in long-term applications, such as artificial organ parts, bone and joint replacements, dental devices and cosmetic implants. In recent years, the emphasis has shifted to the development of bioresorbable polymers for short-term applications such as drug delivery, resorbable sutures, temporary vascular grafts and temporary bone fixation devices. One of the advantages of using degradable implants in short-term applications is that the problems associated with the long-term safety of permanently implanted devices can be circumvented. Unlike a biostable prosthesis, a bioresorbable implant may be replaced by fully functional tissue as part of the natural healing process.

Since poly(amino acids) are structurally related to natural proteins, these polymers were recognized as potential biomaterials. Starting from about 1970, the use of both homo and copolymers of amino acids was studied for a variety of biomedical applications (Hench, 1982; Lyman, 1983; Anderson, 1985; Marchant, 1985; Lescure, 1989). Several excellent, comprehensive reviews are available for developments prior to 1987 (Katchalski, 1958; Lotan, 1972; Katchalski, 1974; Anderson, 1985; Fasman, 1987). The early studies revealed that most poly(amino acids) could not be considered as potential biomaterials due to their immunogenicity and unfavorable mechanical properties. So far, only a small number of poly(γ-substituted glutamates) and copolymers thereof (Sidman, 1980; Sidman, 1983; Anderson, 1985; Bhaskar, 1985) have been identified as promising candidate materials for biomedical applications.

Amino Acid Derived Polymers with Modified Backbones

Attempts were made to utilize amino acids as monomeric building blocks for biomaterials, while avoiding the unfavorable physico-mechanical properties of poly(amino acids). These attempts resulted in the development of a wide range of amino acid derived polymers that do not have the conventional backbone structure

---
*Correspondence: Department of Chemistry, Rutgers University, Taylor Rd./Busch Campus, Piscataway, NJ 08855–0939, Tel: 908-445-3888; Fax: 908-445-5312; e-mail: Kohn@rutchem.rutgers.edu

found in peptides. Collectively, these materials are referred to as "non-peptide amino acid based polymers" or as "amino acid derived polymers with modified backbones". In spite of their large structural variability, it is possible to identify four main types of non-peptide amino acid based polymers.

Synthetic Polymers with Amino Acid Side Chains

These polymers consist of a synthetic polymer backbone onto which amino acids or peptides are grafted as side chains. Examples of materials with amino acids as side chains have been found to exhibit polyelectrolytic and metal complexation behaviour. Such systems include polymethacrylamides with glycylglycine and phenylalanine (Methenitis, 1994), as well as alanine, aspartic acid, asparagine, glutamic acid, and lysine (Morcellet-Sauvage, 1981; Morcellet, 1982; Lekchiri, 1987).

Copolymers of α-L-Amino Acids and Non-Amino Acid Monomers

A wide range of copolymers are known in which α-L-amino acids are copolymerized with appropriate non-amino acid monomers. Copolymers of α-amino acids and non-amino acid monomers have been prepared through a variety of functionalities and mechanistic features. For instance, the ring-opening co-polymerization of amino acid 2,5-morpholinedione derivatives and D,L-lactide gave copolymers of poly(α-amino acid-co-D,L-lactic acid) (ie. poly(amide esters)) (Samyn, 1988). Poly(tripeptide-co-α-hydroxy acids) (Yoshida, 1991) are excellent examples of how degradation can be modified with specificity control between protease-type and esterase-type enzymes. Additional relevant examples are polymers consisting of 1,2–ethanediol, adipic acid and either glycine, L-and D-leucine, and L-, D-, and D,L-phenylalanine as the amino acid components. These materials were found to exhibit a wide range of enzymatic susceptibility to varying degrees, with the interesting feature that these poly(amide esters) degraded primarily through the ester linkage. A new class of poly(anhydride-co-imides) based on naturally occurring α-amino acids or ω-amino acids linked via anhydride bonds represents the structure-property relationship control in materials for potential medical applications (Staubli, 1991). A novel copolymer of poly(lactic acid-co-lysine) has recently been prepared, with RGD modification on the ε-amine of lysine. This strategy appears to be a very promising approach to combining the advantages of synthetic and natural materials that can degrade into natural metabolites already present within the body (Barrera, 1993; Barrera, 1995). An additional example of copolymers based on amino acids and non-amino acid monomers are poly((amino acid ester) phosphazenes) (Allcock, 1994). Release of the amino acid from the phosphazene backbone followed by backbone hydrolysis giving phosphates and ammonia is another novel approach to controlling all aspects of biocompatibility, degradation, and structure-property relationships.

Pseudo-Poly(amino acids)

The development of pseudo-poly(amino acids) represents an attempt to circumvent some of the unfavorable material properties of conventional poly(amino acids), and to increase the range of amino acid derived polymers that can be considered for industrial and/or medical applications. In pseudo-poly(amino acids), naturally occurring amino acids are linked by non-amide bonds, e.g., ester, iminocarbonate, and carbonate bonds. The resulting polymers contain the same monomeric building blocks as conventional poly(amino acids), but do not have a peptide-like backbone structure.

Pseudo-poly(amino acids) were first described in 1984 (Kohn, 1984) and have since been evaluated for use in several medical applications (Kohn, 1987; Yu-Kwon, 1989; Zhou, 1990; Kohn, 1993; Mao, 1993). Although a range of different pseudo-poly(amino acids) has been prepared, detailed studies of the physical properties, biological properties, and possible applications of these polymers have so far been conducted only for a select group of new tyrosine- derived polycarbonates, polyiminocarbonates, and polyarylates. This review will encompass the work to date on these specific materials.

Block-Copolymers Containing Peptide or Poly (amino acid) Blocks

Most commonly, poly(ethylene glycol) is used as the non-peptide block. A large number A-B or A-B-A block copolymers are known in which the B block is a conventional poly(amino acid). Such block copolymers, however, will not be considered here, except to say that such systems represent promising materials for the delivery of therapeutic agents by control of supramolecular solution structures (Jeon, 1989; Yokoyama, 1989a; Yokoyama, 1989b; Yokoyama, 1990a; Yokoyama, 1990b; Yokoyama, 1990c; Cammas, 1995).

In this chapter, special emphasis has been placed on pseudo-poly(amino acids) which are the most recent addition to the family of non-peptide amino acid derived polymers. To provide a detailed description of the wide range of properties that can be achieved by modifications of the linking bond between individual amino acids, a series of tyrosine-derived polymers were selected as illustrative examples. For more information about other types of amino-acid derived polymers, the reader is referred to several recent publications (Wise, 1984; Anderson, 1985; Kumaki, 1985; Bennett, 1988; Li, 1991; Staubli, 1991; Tiffell, 1991; Caliceti, 1993) and to the many publications cited in the preeceding paragraphs.

DESIGN AND SYNTHESIS OF TYROSINE DERIVED POLYMERS

Monomer Design and Synthesis

Tyrosine is the only major, natural nutrient containing an aromatic hydroxyl group. Derivatives of tyrosine dipeptide can be regarded as diphenols and may be employed as replacements for the industrially used diphenols such as Bisphenol A in the design of medical implant materials (Figure 1). The observation that aromatic backbone structures can significantly increase the stiffness and mechanical strength of polymers provided the rationale for the use of tyrosine dipeptides as monomers.

In view of the nonprocessibility of conventional poly(L-tyrosine), which cannot be used as an engineering plastic, variational derivatives were envisioned. The development of tyrosine-based polycarbonates, polyarylates and polyiminocarbonates represents the first time tyrosine-derived polymers with favorable engineering properties have been identified.

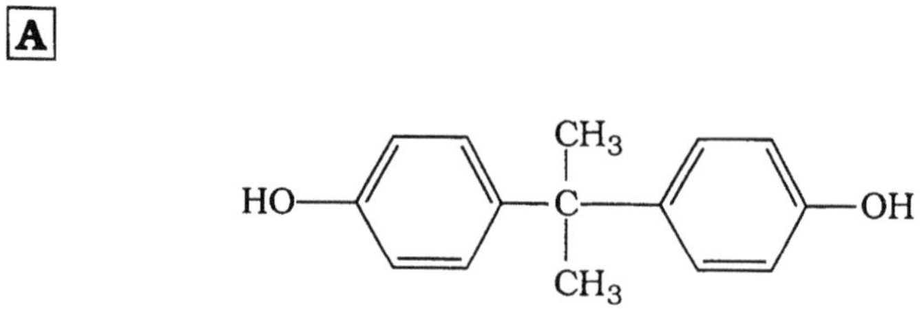

A

Bisphenol A (BPA)

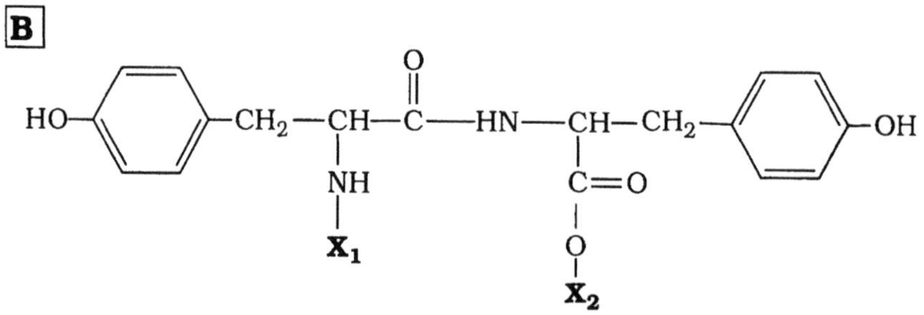

B

protected tyrosine dipeptide

Figure 1 Structures of (A) Bisphenol A, a widely used diphenol in the manufacture of commercial polycarbonate resins; (B) tyrosine dipeptide with specific chemical protecting groups X_1 and alkyl substituents X_2 attached to the N and C termini, respectively.

As shown in Figure 1, tyrosine dipeptide contains a free amino group and a free carboxylic acid group which have to be protected during polymer synthesis. The chemical structure of the protecting groups has a significant impact on the properties of the resulting polymers. The challenge of the early studies was to identify suitable protecting groups that will lead to nontoxic, fully degradable and processible polymers. The combination of these different properties within one single design proved to be a difficult task and early investigations did not lead to readily processible materials (Kohn, 1987; Kohn, 1988). Later, it was recognized that the number of interchain hydrogen bonding sites per monomer unit had to be minimized (Pulapura, 1990b). These studies led to the replacement of one tyrosine molecule by desaminotyrosine (3-(4'-hydroxyphenyl)propionic acid) and the identification of desaminotyrosyl-tyrosine alkyl esters (Figure 2) as fully biocompatible diphenolic monomers (Kohn, 1991a; Kohn, 1991b; Kohn, 1991c).

Monomer synthesis from 3-(4'-hydroxyphenyl)propionic acid and tyrosine alkyl esters was accomplished by carbodiimide mediated coupling reactions, following known procedures of peptide synthesis (Pulapura, 1992; Ertel, 1994) in typical yields of 70%. Monomers carrying an ethyl, butyl, hexyl, or octyl ester pendent chain were investigated extensively (Ertel, 1994; Hooper, 1995). These peptide-like diphenolic monomers were used as starting materials in the synthesis of polycarbonates, polyiminocarbonates, and polyarylates.

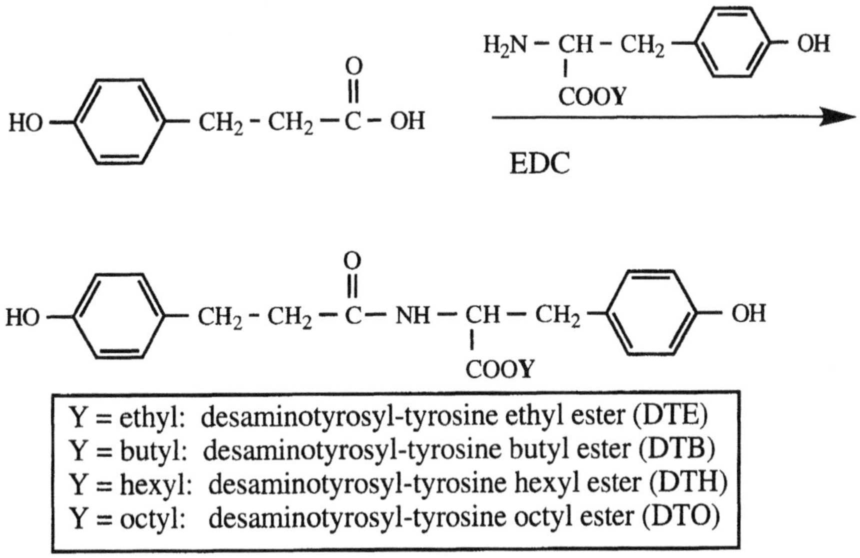

Figure 2 Reaction scheme for the coupling of desaminotyrosine and tyrosine alkyl esters to obtain a diphenolic monomer which carries an alkyl ester pendent chain (Y). Four specific monomers having an ethyl, butyl, hexyl and octyl ester pendent chain were investigated in detail. EDC = ethyl-3-(3'-dimethylamino)propyl carbodiimide hydrochloride salt (Hooper, 1995).

Polycarbonate Design and Synthesis

A particularly promising series of degradable polymers was obtained when the desaminotyrosyl tyrosine alkyl esters shown in Figure 2 were used in the synthesis of polycarbonates (Pulapura, 1992; Ertel, 1994). The exact chemical structures of the resulting polycarbonates are shown in Figure 3. These materials form a series of "homologous" carbonate-amide copolymers differing only in the length of their respective alkyl ester pendent chains. The ability to maintain the polymer backbone and independently alter the pendent chain structure is a powerful tool for the investigation of polymer structure-property relationships. For example, by varying the length of the alkyl ester pendent chain, polymer properties such as the glass transition temperature (Tg), surface free energy, strength, stiffness and the degradation rate can be readily controlled (Table 1).

The diphenolic monomers were polymerized using either phosgene or the more easily handled bis(chloromethyl) carbonate (triphosgene) (Pulapura, 1992; Ertel, 1994). It is noteworthy that the amide functionality present in the monomers did not interfere in the polymerization with phosgene, as long as mild reaction conditions and a low excess of phosgene were employed. It is vital to point out that, in correspondence with the mechanism of condensation polymerization reactions, the final polymer molecular weight was strongly dependent on monomer purity. Commercial 3-(4'-hydroxyphenyl)propionic acid, for example, may contain 3-(4'-methoxyphenyl) propionic acid which acts as a chain terminating agent, limiting the

Table 1 Some Physical Properties of Tyrosine-Derived Polycarbonates[a]

Polymer[a]	M_w[a] (Mw/Mn)	Tg[b] (°C)	Td[c] (°C)	Contact Angle (°)	Young's modulus[d] (GPa)	Tensile strength[d] (MPa)	Elongation at break (%)	Time constant[e] (weeks)	Surface Energy Parameters[f] γ_c	γ^d	γ^P
poly(DTE carbonate)	176,000 (1.8)	93	290	73	1.5	67	4	12	46.4	42.5	3.5
poly(DTB carbonate)	120,000 (1.4)	77	290	77	1.6	60	3	16	43.7	40.1	2.4
poly(DTH carbonate)	350,000 (1.7)	63	320	86	1.4	62	>400	21	40.6	37.5	1.1
poly(DTO carbonate)	450,000 (1.7)	52	300	90	1.2	51	>400	21	38.5	36.1	0.6

[a]Data from Ertel and Kohn (Ertel, 1994) Weight average molecular weights as determined by GPC.
[b]Glass transition temperature (midpoint) as determined by DSC.
[c]Decomposition temperature as determined by TGA. Measured at 2% weight loss.
[d]Unoriented samples.
[e]Degradation time constant for thin, solvent cast films under simulated physiological conditions (37°C, pH 7.4 PBS).
[f]Data obtained from Perez-Luna et al. (Perez-Luna, 1995) The critical surface tension (γ_c) was estimated using Zismann's method, and the dispersive and polar components (γ^d and γ^P) were calculated using the geometric mean approximation to the work of adhesion.

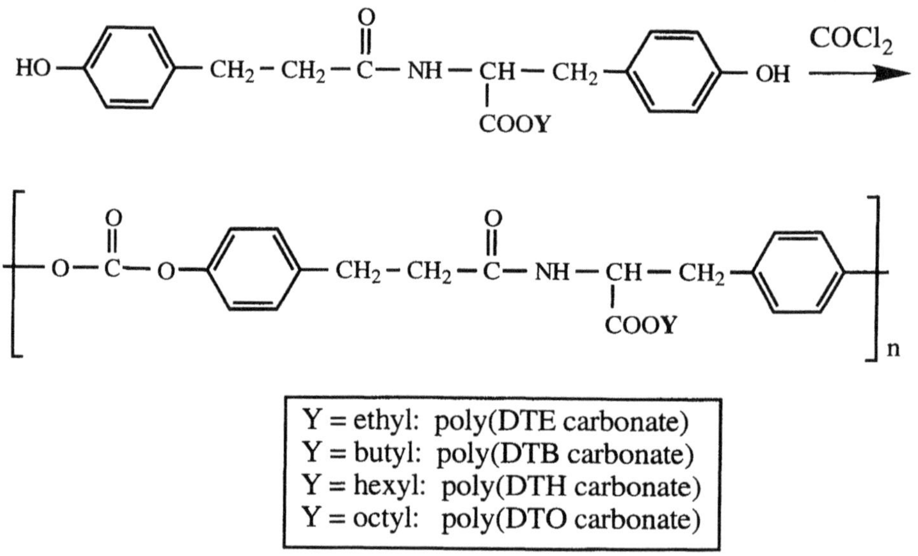

Y = ethyl: poly(DTE carbonate)
Y = butyl: poly(DTB carbonate)
Y = hexyl: poly(DTH carbonate)
Y = octyl: poly(DTO carbonate)

Figure 3 Reaction scheme for the preparation of tyrosine-derived polycarbonates. The monomers are polymerized by reaction with phosgene.

polymer molecular weight. With carefully purified monomers, polymers with weight average molecular weights of up to 400,000 g/mol (by GPC, relative to polystyrene standards) were obtained (Pulapura, 1992).

Polyarylate Design and Synthesis

Several polyarylates are liquid crystalline materials and are widely used as industrial plastics. The synthesis of polyarylates from industrially used diphenols and various diacids has been investigated by Moore and Stupp (Moore, 1990) who published a procedure based on a carbodiimide-mediated esterification in the presence of dimethylaminopyridinium p-toluene sulfonate (DPTS).

After suitable modifications, this procedure was adapted for the synthesis of tyrosine-derived polyarylates. Briefly, equimolar amounts of the tyrosine-derived diphenolic monomers and the alkyl diacids were polymerized by 1,3-diisopropyl-carbodiimide coupling, providing tyrosine-derived polyarylates with weight average molecular weights of about 100,000 g/mol (Figure 4).

The diphenol components selected were the desaminotyrosyl-tyrosine alkyl esters described above. The diacids included succinic, adipic, suberic and sebacic acid which contain, respectively, 2, 4, 6, and 8 methylene groups between two carboxylic acid functionalities. With this family of polymers it is possible to alter independently both the pendent chain lengths as well as the number of flexible methylene spacers in the backbone, creating a family of sixteen structural variants. Hence, these materials serve as a framework upon which to further investigate polymer structure-property relationships.

Polyiminocarbonate Design and Synthesis

Polyiminocarbonates are the "imine analogs" of the industrially used polycarbonates. In the late 1960s, the first synthesis of low molecular weight polyiminocarbonates by reaction of aqueous solutions of various chlorinated diphenolate sodium salts with cyanogen bromide was reported (Hedayatullah, 1967). Later, the reaction of a diphenol and a dicyanate to give a polyiminocarbonate, using both solution and bulk polymerization techniques, was suggested by Schminke et al. (Schminke, 1970). Tyrosine-derived polyiminocarbonates were first prepared in 1984 (Kohn, 1984) and were among the first pseudo-poly(amino acids) reported in the literature (Kohn, 1987). The first comprehensive and systematic study of polyiminocarbonate synthesis in general was published in 1989 (Li, 1989), followed in 1990 by a detailed exploration of the synthesis and characterization of tyrosine-derived polyiminocarbonates (Pulapura, 1990b). A comprehensive review of interfacial and solution polymerization procedures for the preparation of polyiminocarbonates of high molecular weight is available (Kohn, 1990).

The synthesis of tyrosine-derived polyiminocarbonates is based on the copolymerization of a diphenol and a dicyanate (Figure 5), requiring a two-step process. First, a diphenolic monomer is cyanylated to yield the corresponding dicyanate. Next, the diphenol and the dicyanate are mixed in stoichiometrically equivalent amounts. Upon addition of a strong base (as catalyst), the polymerization is spontaneous and rapid. Tyrosine-derived polyiminocarbonates with weight average

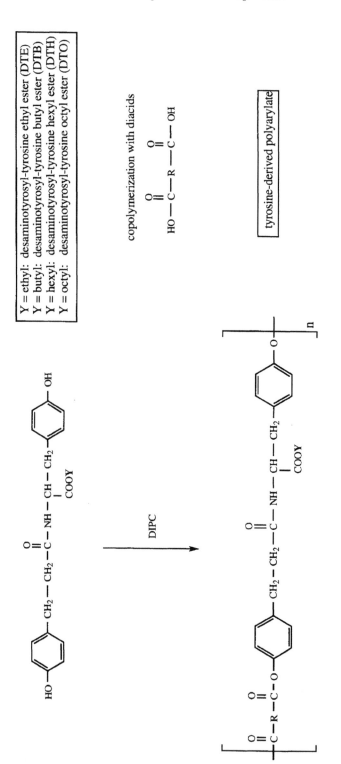

Figure 4 Reaction scheme for the preparation of tyrosine-derived polyarylates. This reaction scheme is a copolymerization of a diphenol component and a diacid component. The diphenol components are the same desaminotyrosyl-tyrosine alkyl esters used in the synthesis of polycarbonates (Figure 3). The pendent chain Y is ethyl, butyl, hexyl, or octyl. The diacid components are succinic acid, adipic acid, suberic acid, and sebacic acid providing a flexible backbone spacer (R) having 2, 4, 6, and 8 methylene groups respectively. DIPC = diisopropylcarbodiimide.

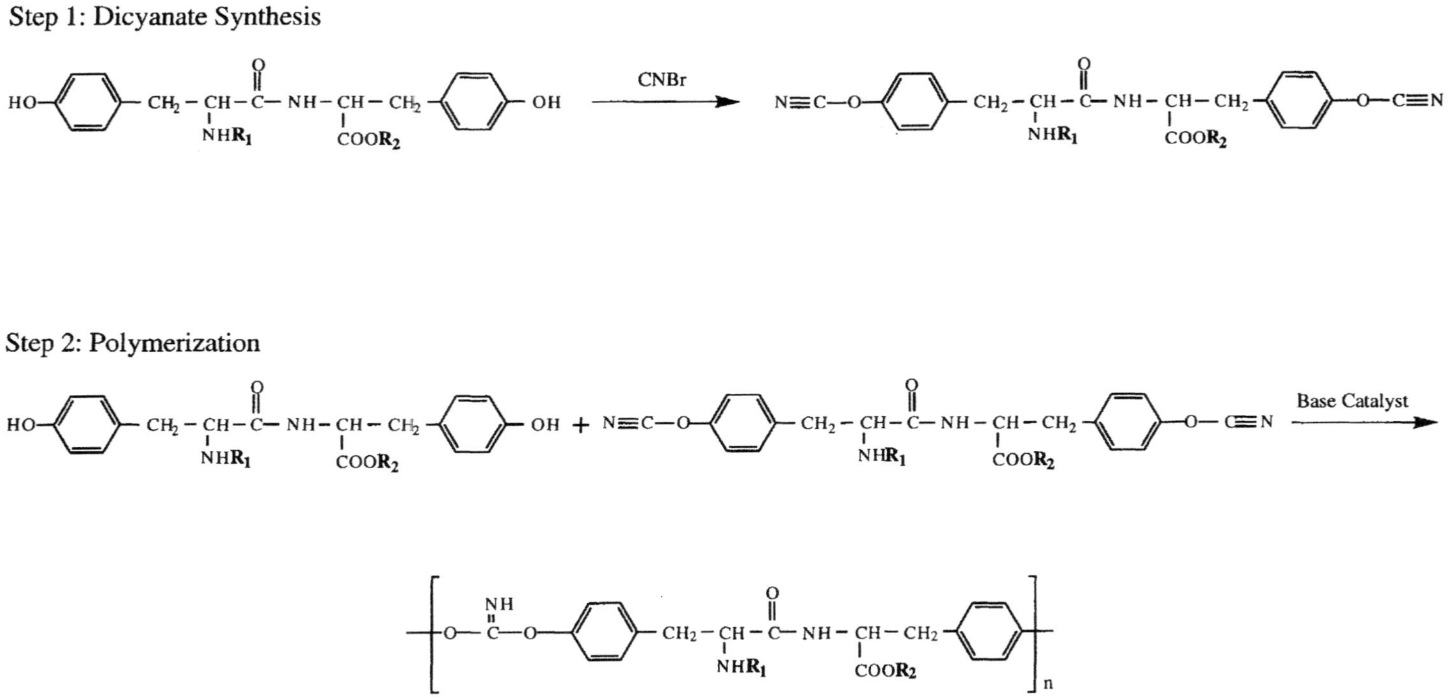

Figure 5 Reaction scheme for the preparation of tyrosine-derived polyiminocarbonates. The basic monomeric repeat unit is tyrosyl-tyrosine dipeptide. To optimize the polymer properties, the chemical structures of the N- and C-terminal protecting groups (R_1 and R_2) have to be designed carefully. In the first reaction step, protected tyrosyl-tyrosine dipeptide is cyanylated with cyanogen bromide (CNBr). In the next step, polymerization occurs when equimolar quantities of the dipeptide and the cyanylated dipeptide are mixed in the presence of a base catalyst.

Table 2 Some Physical Properties of Selected Tyrosine-Derived Polyarylates[a]

Polymer[a]	M_w	Tg[b] (°C)	Td[c] (°C)	Contact Angle (°)	Young's modulus[d] (GPa)	Tensile Strength[d] (MPa)	Elongation at break[d] (%)
poly(DTE adipate)	209,000	56	340	70	1.52	34	157
poly(DTH adipate)	232,000	34	356	82	0.43	30	418
poly(DTO adipate)	220,000	28	357	86	0.01	28	424

[a]Data from Kohn *et al.* (Fiordeliso, 1994; Kohn, 1994). Weight average molecular weights as determined by GPC.
[b]Glass transition temperature as determined by DSC.
[c]Decomposition temperature as determined by TGA. Measured at 10% weight loss.
[d]Unoriented samples; properties measured at room temperature.

Table 3 Some Physical Properties of Selected Tyrosine-Derived Polyiminocarbonates[a]

Diphenol Component	Dicyanate Component	Polymerization Method	Mw[a] (Mw/Mn)	Tg[b] (°C)	Td[c] (°C)
Dat-Tym	Dat-Tym	Interfacial	87,000 (1.33)	n/a	169
Z-Tyr-Tym	Z-Tyr-Tym	Interfacial	175,000 (1.50)	n/a	179
Dat-Tyr-Hex	Dat-Tyr-Hex	Solution	103,200 (1.32)	55	175
Z-Tyr-Tyr-Hex	Z-Tyr-Tyr-Hex	Solution	58,000 (1.49)	62	160
Z-Tyr-Tym	Dat-Tym	Interfacial	86,100 (1.16)	91	145
Dat-Tym	Dat-Tyr-Hex	Solution	60,200 (1.42)	55	157
Z-Tyr-Tyr-Hex	Dat-Tym	Solution	82,200 (1.14)	75	163
Dat-Tyr-Hex	Z-Tyr-Tyr-Hex	Solution	68,800 (1.49)	57	157

[a]Data from Kohn *et al.* (Kohn, 1990; Kohn, 1991a; Pulapura, 1990b). Weight average molecular weights as determined by GPC, relative to polystyrene standards in DMF containing 0.1% LiBr.
[b]Glass transition temperature (midpoint) as determined by DSC.
[c]Decomposition temperature as determined by TGA.

molecular weights in excess of 100,000 g/mol have been reported (Pulapura, 1990b) (Table 3). It is noteworthy that the same desaminotyrosyl-tyrosine alkyl esters were used as monomers in the synthesis of both polyiminocarbonates and polycarbonates. This fact was exploited in a study exploring miscible blends of polyiminocarbonates and polycarbonates (Pulapura, 1990a).

PROPERTIES OF TYROSINE DERIVED POLYMERS

Tyrosine-Derived Polycarbonates

Tyrosine-derived polycarbonates have important advantages when used in the design of implantable, degradable controlled release systems. First, all members of this series of polymers are amorphous materials with relatively low glass transition temperatures which are a function of the pendent chain length (Table 1). X-ray diffraction exhibited only an amorphous halo, indicative of the lack of crystalline domains. DSC analysis showed a glass transition and decomposition exotherm, but no melting endotherm, confirming the lack of crystallinity. Contrary to poly(L-tyrosine) which is a non-processible and insoluble polymer, tyrosine-derived polycarbonates are freely soluble in a variety of organic solvents, and are readily processible by conventional solvent casting and thermal processing techniques (extrusion, injection and compression molding) at relatively low temperatures. Typically, they form strong, transparent films. For extrusion and injection molding, processing temperatures were about 70 to 100°C above the glass transition temperature. Since the thermal decomposition temperature of all four polymers was about 300°C when measured by thermogravimetric analysis (TGA), there is a relatively large gap between the processing temperatures and the thermal decomposition temperatures.

The mechanical properties of these polycarbonates depend strongly on the length of the alkyl ester pendent chain. In general, increasing the length of the alkyl ester leads to a decrease in stiffness and an increase in ductility. Unoriented thin film specimens range in tensile modulus from 1.2 GPa for poly(DTO carbonate) to 1.6 GPa for poly(DTE carbonate). Tensile strength and elongation appear to be strongly affected by the temperature at which the tests are being conducted. At room temperature, the polymers with shorter pendent chains (ethyl and butyl) behave as brittle materials which failed without yielding after about 4% elongation. The polymers with longer pendent chains (hexyl and octyl) oriented under stress and could be elongated to over 400%. In their oriented state, the ultimate tensile strength at break surpassed 200 MPa. At slightly higher temperatures (40–60°C), all four polycarbonates could be elongated under stress. Under carefully controlled conditions, elongations of up to 1000% could be achieved, resulting in highly oriented specimens for which an ultimate tensile strength of up to 400 MPa has been observed (Kohn, unpublished results).

Tyrosine-derived polycarbonates are stiffer and stronger than many other degradable polymers of comparable molecular weights, such as polycaprolactone and polyorthoesters, but are not as stiff as poly(L-lactic acid) or poly(glycolic acid) (Daniels, 1990). Considering the strength and stiffness of these tyrosine-derived polycarbonates, it is conceivable to fabricate load-bearing devices (such as pins for small bone fixation) or load-bearing drug delivery systems which may find application in orthopedics.

Tyrosine-derived polycarbonates provided a convenient model system to study the effect of pendent chain length on the thermal properties and the enthalpy relaxation (physical aging). It is noteworthy that enthalpy relaxation kinetics are not usually reported in the biomedical literature and that a recent study by Tangpasuthadol (Tangpasuthadol, 1995) represents one of the first attempts to evaluate physical aging in a degradable biomedical polymer.

For the tyrosine-derived polycarbonates tested, the enthalpy relaxation process was not sensitive to the length of the pendent chain. This observation suggests that structural relaxation in these polymers is limited by backbone flexibility, and that the fraction of free volume in these polymers is not the limiting factor for polymer mobility. Furthermore, since the enthalpy relaxation time is short at aging temperatures of $T_g - 15°C$, a few hours of storage at that temperature will be sufficient to bring the physical aging process to completion. The results obtained by dynamic mechanical analysis support the general notion that an increase in the length of the pendent chain results in a more flexible material (Tangpasuthadol, 1995). This observation is in agreement with the results obtained in a previous study of these polymers (Ertel, 1994). These structure-property correlations can assist in the selection of suitable polymers for specific applications.

Another significant advantage of tyrosine-derived polycarbonates is their high hydrophobicity. Air-water contact angles increased with a corresponding decrease in the surface energy parameters as a function of increasing pendent chain length (Table 1). The increasing hydrophobicity imparted by longer alkyl ester pendent chains is also demonstrated by the equilibrium water content which is 4% at 37°C for poly(DTE carbonate) and less than 1% for poly(DTO carbonate).

It is a general observation that, consistent with the fairly stable polymer backbone, all tyrosine-derived polycarbonates degrade relatively slowly under physiological conditions. The mechanism of degradation has been carefully evaluated (Ertel, 1994). Based on evidence obtained from ESCA (XPS), ATR-FTIR and GPC, an *in vitro* degradation mechanism has been postulated. According to this mechanism, the ester bonds at the alkyl ester pendent chains are cleaved first, followed by the carbonate bonds in the polymer backbone. *In vitro*, the amide bonds are not hydrolyzed. Thus, desaminotyrosyl-tyrosine is the final degradation product. For thin solvent cast films of high initial molecular weight, the *in vitro* degradation time constants are listed in Table 1. The kinetic time constants calculated from Ertel's model (Ertel, 1994) indicate that poly(DTE carbonate) degrades hydrolytically almost twice as fast as the more hydrophobic poly(DTO carbonate). *In vivo*, the cleavage of the amide bond by enzymatic or cellular mechanisms may lead to additional degradation products. No detailed investigations as to the *in vivo* degradation products have been carried out.

To put these degradation properties into perspective, poly(DTH carbonate) and high molecular weight poly(L-lactic acid) exhibit comparable reductions in molecular weight when incubated in physiological buffer solution at 37°C. The similarity in the degradation profile between tyrosine-derived polycarbonates and high molecular weight poly(L-lactic acid) was also observed *in vivo* (Choueka, 1996). Extrapolating from a 26 week implantation study in the femur and tibia of rabbits, a resorption time of 2 to 3 years can be expected for high molecular weight poly(DTH carbonate) (Ertel, 1995).

Tyrosine-Derived Polyarylates

In this family of tyrosine-derived polyarylates, glass transition temperatures decrease as a function of the total number of carbons contained in the flexible spacers present in the pendent chain (Y) and backbone (R) (Figure 4). The changes in glass transition temperature follow a predictable pattern (Figure 6a) ranging from a low of 13°C to a high of 78°C (Brocchini, 1996, Tangpasuthadol, 1996). Evidence from X-ray scattering and DSC indicates that these tyrosine-derived polyarylates are amorphous. The polyarylates are thermally stable polymers with thermal decomposition temperatures in the range of 300°C. Therefore, a wide range of processing techniques can be used including solvent casting, compression molding, injection molding, and extrusion.

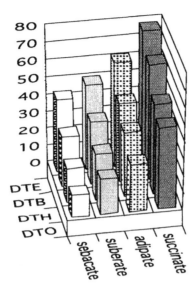

Figure 6a Glass transition temperatures of tyrosine-derived polyarylates. In this three dimensional presentation, the pendent chain length is plotted on the y axis, the length of the diacid component in the polymer backbone is plotted on the x axis, and the measured glass transition temperatures are plotted on the z axis.

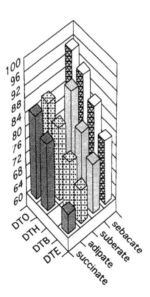

Figure 6b Air-water contact angles of tyrosine-derived polyarylates. In this three-dimensional presentation, the pendent chain length is plotted on the y axis, the length of the diacid component in the polymer backbone is plotted on the x axis, and the measured air-water contact angles are plotted on the z axis. The values for poly(DTB succinate) and poly(DTOsuberate) have not yet been determined.

As the number of carbons contained in the structural elements "Y" and "R" increases, the materials become more hydrophobic (Table 2). This is clearly demonstrated by the systematic and predictable increase in the air-water contact angle (Figure 6b) which increases from about 66° to over 96°. Figure 6b illustrates that relatively small changes in the polymer structure produce significant changes in surface hydrophobicity. This property has been used to study the effect of surface hydrophobicity on the attachment and growth of different cell lines (see below).

The mechanical properties of tyrosine-derived polyarylates vary dramatically, depending on the number and distribution of carbon atoms in the alkyl ester pendent chain (Y) and the flexible backbone spacer (R). Table 2 summarizes selected mechanical properties of poly(DTE adipate), poly(DTH adipate), and poly(DTO adipate). It is apparent that increasing the length of the pendent chain (Y) decreases both mechanical strength and stiffness of the polymer. While poly(DTE adipate) has approximately the same stiffness as poly(DTE carbonate), the tensile modulus of poly(DTO adipate) is reduced by over 2 orders of magnitude and is far below the value measured for the corresponding poly(DTO carbonate). Preliminary evidence suggests that changes in the length of the flexible backbone spacer (R) influence the mechanical properties in a similar fashion. Overall, the available mechanical properties range from soft, elastomeric materials (poly(DTO sebacate)) to fairly tough and strong materials (poly(DTE succinate)). No detailed correlations between polymer structure and the mechanical properties have so far been published.

Only very preliminary degradation studies of thin films of poly(DTE adipate), poly(DTH adipate) and poly(DTO adipate) have been carried out (Fiordeliso, 1993; Fiordeliso, 1994). Under simulated physiological conditions (37°C in pH 7.4 phosphate buffered saline) over a 26 week period, sample weights remained essentially unchanged while the polymer molecular weights decreased by about 50%. The inherent hydrophobicity of these polymers prevented significant water imbibition. For the tyrosine-derived polyarylates mentioned, currently available data indicate that hydrolysis of the arylate bonds in the polymer backbone is slow. The actual time frame of degradation of these polymers and the effect of polymer structure on the degradation rate have yet to be established.

Tyrosine-Derived Polyiminocarbonates

In comparison to the other polymers discussed above, polyiminocarbonates have very low decomposition temperatures (approximately 150°C) (Li, 1989; Pulapura, 1990b). This precludes the use of most thermal fabrication methods. However, solvent casting and wet spinning can be used to obtain films and fibers. Also, it is notable that tyrosine-derived polyiminocarbonates have relatively low glass transition temperatures. For example, poly(DTH iminocarbonate) has a T_g of 55°C, making it possible to compression mold devices of this material at approximately 70°C with only minor decomposition. This low processing temperature is advantageous when fabricating drug delivery systems for heat-sensitive drugs.

The polyiminocarbonates have been found to be slightly stiffer and stronger than the corresponding tyrosine-derived polycarbonates (Pulapura, 1990b; Engelberg, 1991). The most significant difference in the mechanical properties between polyiminocarbonates and polycarbonates is in their ductility. Polyiminocarbonates are brittle and break without plastic deformation after less than 5% of elongation, while polycarbonates can, under appropriate conditions, be elongated beyond 400%. A defining feature of tyrosine-derived polyiminocarbonates is the significant hydrolytic instability of the iminocarbonate bond which is responsible for the rapid degradation of high molecular weight polymers to low molecular weight (1,000–6,000 g/mol) oligomers within one week under simulated physiological conditions (Pulapura, 1990a; Pulapura, 1990b; Pulapura, 1991). The chemical mechanism of degradation has been shown to lead to the formation of ammonia, carbon dioxide and the partial regeneration of the diphenolic monomer used in the synthesis of the polymer (Kohn, 1986a). Since the low-molecular weight oligomers formed are highly insoluble, the rapid hydrolysis of the backbone does not lead to a concomitant mass loss or to rapid resorption of the polymeric device. This property has limited the application of polyiminocarbonates as biomaterials.

MEDICAL APPLICATIONS AND BIOCOMPATIBILITY OF TYROSINE DERIVED POLYMERS

Tyrosine-Derived Polycarbonates

Because of their strength and toughness, tyrosine-derived polycarbonates were recognized as possible candidates for the development of orthopedic implants (Lin, 1991). Particularly promising is the development of small bone fixation devices such as pins, screws, and plates that can be used to stabilize fractures in small, non-weight bearing bones. Currently, self-reinforced devices made of poly(glycolic acid) are in clinical use in the USA, but these devices have significant clinical side reactions due in part to the release of acidic degradation products (Böstman, 1991). Implants made of polydioxanone (Orthosorb®) are also in use, but these implants tend to degrade too fast, losing their mechanical strength within 1 to 2 months (Ertel, 1995).

In vitro attachment and proliferation of fibroblasts on tyrosine-derived polycarbonates was a function of the pendent chain length (Ertel, 1994). Consistent with the hypothesis that cells favor more hydrophilic and molecularly rigid surfaces, poly(DTE carbonate), the most hydrophilic and rigid of the surfaces tested (Table 1), supported cell growth and proliferation better than the more hydrophobic poly(DTO carbonate). Poly(DTB carbonate) and poly(DTH carbonate) were intermediate in their ability to support cell attachment and growth.

In an *in vivo* pilot study (Ertel, 1995), poly(DTH carbonate) pins were fabricated and compared to commercially available Orthosorb® pins made of polydioxanone. The pins were implanted transcortically in the distal femur and proximal tibia of New Zealand White rabbits for up to 26 weeks. In addition to routine histological evaluation of the implant sites, bone activity at the implant/tissue interface was visualized by UV illumination of sections labeled with fluorescent markers and the degree of calcification around the implants was ascertained by backscattered electron microscopy.

The bone tissue response was characterized by active bone remodeling at the surface of the degrading implant, the lack of fibrous capsule formation, and an unusually low number of inflammatory cells at the bone-implant interface. Poly(DTH carbonate) exhibited very close bone apposition throughout the 26 week period of this initial study. A roughened interface was observed which was penetrated by new bone as early as 2 weeks post implantation. Bone growth into the periphery of the implant material was visible at the 26 week time point (Ertel, 1995).

Some of the observations made in the transcortical rabbit model were recently confirmed in an independent study using the canine chamber model (Spivak, 1990) to compare the bone response to poly(L-lactic acid) and two tyrosine-derived polycarbonates, poly(DTE carbonate) and poly(DTH carbonate). In this study (Choueka, 1996), thin polymeric coupons were used to create narrow channels within a polyethylene housing. Upon implantation of the assembled device into the femur of dogs, the ability of bone to grow into the narrow channels lined by coupons made of different materials was evaluated. In addition, the host bone response was histologically evaluated. Test chambers containing coupons of poly(DTE carbonate) and poly(DTH carbonate) were characterized by sustained bone ingrowth throughout the 48 week study period. Histological sections revealed intimate contact between bone and tyrosine-derived polycarbonates. In contrast, bone ingrowth into the PLA chambers peaked at 24 weeks and dropped by half at the 48 week time point. A fibrous tissue layer was found surrounding the PLA implants at all time points.

The presence of a fibrous layer surrounding the PLA coupons is characteristic of a mild foreign body response. That such a fibrous layer was not formed at the bone/material interface for poly(DTE carbonate) and poly(DTH carbonate) is an important characteristic of these polymers. Intimate contact between bone and implant, even at 48 weeks post-implantation, is a strong indicator of the biocompatibility of the tyrosine-derived polycarbonates.

The long-term (48 week) *in vivo* degradation kinetics was also investigated. As expected, based on *in vitro* degradation studies, the two tyrosine-derived polycarbonates exhibited degradation kinetics comparable to PLA. Each test material lost approximately 50% of its initial molecular weight (Mw) over the 48 week test period. More detailed studies of the *in vivo* degradation mechanism and rate, as well as studies of *in vivo* resorption (mass loss) are currently in progress.

Although limited by its small sample size, this study suggests that tyrosine-derived polycarbonates exhibit an *in vivo* bone response that is fundamentally different from the response elicited by degradable polyesters (such as PLA or polydioxanone) for which long-term, direct bone apposition is not observed. Based on the results of this study, tyrosine-derived polycarbonates appear promising as a new class of degradable materials for orthopedic applications and are currently being evaluated in larger, longer term animal studies.

Poly(DTH carbonate) has been used in the design of an investigational long-term controlled-release device for the intracranial administration of dopamine (Coffey, 1992; Dong, 1993). For such applications, poly(DTH carbonate) has several potential advantages over other degradable polymers, which include the ease with which dopamine can be physically incorporated into the polymer (due to its relatively low processing temperature and the structural similarity between the drug and the polymer), the apparent protective action of the polymeric matrix on dopamine, the prolonged release of only about 15% of the total load of dopamine over about 180 days, and the high degree of compatibility with brain tissue.

The preliminary results from the investigation of the release of dopamine from poly(DTH carbonate) indicate an average release of about 1 to 2 μg/day over prolonged periods of time. Although this release rate was within the therapeutically useful range, no *in vivo* release experiments have so far been reported in the literature.

Tyrosine-Derived Polyarylates

In a series of preliminary publications, the cellular response to tyrosine-derived polyarylates was explored using chick embryo dorsal root ganglia cells (Kohn, 1994) and rat lung fibroblasts (Zhou, 1994), with no indications of cytotoxicity. Preliminary data indicate that the ability of cells to attach and grow on polyarylate surfaces was strongly correlated with the surface hydrophobicity of the polymers (as measured by the air-water contact angle).

Model drug release studies have been performed using the adipic acid series of polyarylates consisting of poly(DTE adipate), poly(DTH adipate), and poly(DTO adipate) (Fiordeliso, 1993; Fiordeliso, 1994). In this series of polymers, the backbone structure was kept constant while the length of the pendent chain was varied from 2 to 8 carbons. In line with this structural modification, glass transition temperature, physicomechanical properties, and polymer hydrophobicity changed as well. Solvent cast devices incorporating 5% (w/w) p-nitroaniline dye were incubated in pH 7.4 phosphate buffered saline (37°C) to simulate the release of low molecular weight drugs from these polymers. Over a 40 day period, all dye was released from poly(DTH adipate) and poly(DTO adipate). In contrast, poly(DTE adipate) released only half of the incorporated dye during the same incubation period. In spite of the difference in release rate, all polymers exhibited a diffusion controlled release mechanism, as indicated by the linear correlation between the cumulative release and the square root of the release time. These results are somewhat counter-intuitive since poly(DTE adipate) is the most hydrophilic of the three test polymers and should have exhibited the most rapid release profile. However, considering the glass transition temperatures, poly(DTE carbonate) was in the glassy state at 37°C while poly(DTH adipate) and poly(DTO adipate) were rubbery.

In Europe, tyrosine-derived polyarylates (Fiordeliso, 1994) were tested as haemo-compatible coatings for blood-contacting devices (Stemberger, 1994). Techniques were developed to incorporate anticoagulants into coatings made of tyrosine-derived polyarylates or lactide/glycolide copolymers. These coatings were then applied to carbon fibers. Without coating, the fibers were covered within minutes by a coagulation plug rich in fibrin and platelets. Degradable coatings without anticoagulants reduced the thrombogenicity of the test materials, but coatings releasing hirudin and prostacyclin inhibitors prevented the formation of thrombin at the coated surfaces.

Tyrosine-derived polyarylates offer the ability to alter widely the polymeric properties by changes in either the backbone or the pendent chain structure. These polymers appear most adept at addressing medical implant needs where a slowly degrading, relatively flexible and soft polymer is required.

Tyrosine-Derived Polyiminocarbonates

In a comparative study (Silver, 1992), solvent cast films of poly(DTH iminocarbonate) and poly(DTH carbonate) were evaluated in a subcutaneous rat model. In this study, high density polyethylene (HDPE) and medical grade poly(D,L-lactic acid) served as controls. Considering the significantly faster degradation rate of poly(DTH iminocarbonate), one would expect a different response from this material. Indeed, at 7 days post-implantation a greater cell density and inflammatory response was noted for poly(DTH iminocarbonate). However, at later time points, the biological response to poly(DTH iminocarbonate) was not notably different from the response observed for poly(DTH carbonate). The tissue response was characterized by a thin tissue capsule, absence of giant cells and a low inflammatory cell count, and was statistically not different from the response observed for polyethylene and poly(lactic acid) (Silver, 1992).

Since L-tyrosine is known for its adjuvant properties (Wheeler, 1982), the use of degradable tyrosine-derived polyiminocarbonates as antigen delivery devices appeared promising (Kohn, 1986b). This concept was tested using a system that consisted of a tyrosine-derived polyiminocarbonate implant releasing bovine serum albumin (BSA) subcutaneously in mice. Upon degradation of the polymeric matrix, the release of BSA and the concomitant presence of the tyrosine-derived poly-iminocarbonate (or its degradation products) resulted in an anti-BSA antibody titer that was comparable to the titer observed when BSA was administered repeatedly in complete Freund's adjuvant. This system effectively took advantage of the rapid degradation characteristics of tyrosine-derived polyiminocarbonates.

SUMMARY

All currently available pseudo-poly(amino acids) have been intentionally designed to degrade under physiological conditions. Although these materials are expected to be environmentally friendly, the relatively high cost of pseudo-poly(amino acids) excludes their use as degradable packaging and consumer plastics. However, because of their apparent non-toxicity and high degree of tissue compatibility, these materials are promising candidates for use in degradable medical implants and degradable drug delivery systems.

Pseudo-poly(amino acids) offer certain advantages over conventional poly(amino acids) such as lower cost of polymer synthesis and processibility by industrially used fabrication techniques. Contrary to petroleum-based polymers, the final degradation products of pseudo-poly(amino acids) can be expected to be simple, naturally occurring amino acids. In particular, studies of tyrosine-derived polycarbonates, polyarylates, and polyiminocarbonates confirmed that these materials are generally biocompatible and biodegradable materials that can be readily processed by a variety of means to yield microspheres, fibers, films, pins, and screws (Kohn, 1993). These characteristics lend themselves to the formulation of a wide range of implant and/or drug delivery systems.

The relatively high mechanical strength of tyrosine-derived polycarbonates has led to their evaluation as orthopedic implants. Tyrosine-derived polyarylates appear to be particularly useful in soft tissue applications requiring more flexible and softer materials. Some of the available polyarylates may be used as "thrombo-resistant" coatings for blood-contacting devices. The tyrosine-derived polyiminocarbonates tend to be very brittle and are subject to fast hydrolytic degradation. Unfortunately, the fast degradation of the polymer backbone is not matched by an equally fast resorption of the polymeric device. This mismatch between fast degradation and slow resorption has so far limited the practical applications for tyrosine-derived polyiminocarbonates.

The design of tyrosine-derived pseudo-poly(amino acids) has circumvented many of the traditional limitations of conventional poly(amino acids), giving rise to novel and useful biomaterials covering a wide range of potential biomedical applications.

COMMERCIAL AVAILABILITY

Several desaminotyrosyl-tyrosine alkyl esters and the corresponding tyrosine-derived polycarbonates are available commercially through Sigma Chemical Company. Tyrosine-derived polycarbonates are currently in preclinical evaluations for possible use in orthopaedic implants. Clinical trials in humans have not yet been conducted. Tyrosine-derived polyiminocarbonates and polyarylates are not available commercially.

REFERENCES

Allcock, H.R., Pucher, S.R. and Scopelianos, A.G. (1994) Poly[(amino acid ester)phosphazenes]: Synthesis, crystallinity, and hydrolytic sensitivity in solution and the solid state. *Macromolecules*, **27**, 1071–1075.

Anderson, J.M., Spilizewski, K.L. and Hiltner, A. (1985) Poly-α amino acids as biomedical polymers. *Biocompatibility of Tissue Analogs*. Boca Raton, CRC Press Inc. 67–88.

Barrera, D.A., Zylstra, E., Lansbury, P.T. and Langer, R. (1993) Synthesis and RGD peptide modification of a new biodegradable copolymer: Poly(lactic acid-co-lysine). *J. Am. Chem. Soc.*, **115**, 11010–11011.

Bennett, D.B., Adams, N.W., Li, X., Feijen, J. and Kim, S.W. (1988) Drug-coupled poly(amino acids) as polymeric prodrugs. *J. Bioact. Compat. Polym.*, **3**, 44–52.

Bhaskar, R.K., Sparer, R.V. and Himmelstein, K.J. (1985) Effect of an applied electric field on liquid crystalline membranes: Control of permeability. *J. Membr. Sci.*, **24**(1), 83–96.

Böstman, O.M. (1991) Absorbable implants for the fixation of fractures. *J. Bone Joint Surg.*, **73**(1), 148–153.

Brocchini, S. and Kohn, J. (1996) Pseudo-poly(amino acid)s. *The Polymeric Materials Encyclopedia*. Boca Raton, FL, CRC Press: in press.

Caliceti, P., Monfardini, C., Sartore, L., Schiavon, O., Baccichetti, F., Carlassare, F. and Veronese, F.M. (1993) Preparation and properties of monomethoxy poly(ethylene glycol) doxorubicin conjugates linked by an amino acid or a peptide as spacer. *Il Farmaco*, **48**(7), 919–932.

Cammas, S. and Kataoka, K. (1995) Functional poly[(ethylene oxide)-co-(β-benzyl L-aspartate)] polymeric micelles: block copolymer synthesis and micelles formation. *Makromol. Chem. Phys.*, **196**, 1899–1905.

Choueka, J., Charvet, J.L., Koval, K.J., Alexander, H., James, K.S., Hooper, K.A. and Kohn, J. (1996) Canine bone response to tyrosine-derived polycarbonates and poly(L-lactic acid). *J. Biomed. Mater. Res.*, **31**, 35–41.

Coffey, D., Dong, Z., Goodman, R., Israni, A., Kohn, J. and Schwarz, K.O. (1992) *Evaluation of a tyrosine derived polycarbonate device for the intracranial release of dopamine.* Symposium on Polymer Delivery Systems presented at the 203rd Meeting of the American Chemical Society, San Fransisco, CA, CELL 0058.

Daniels, A.U., Chang, M.K.O., Andriano, K.P. and Heller, J. (1990) Mechanical properties of biodegradable polymers and composites proposed for internal fixation of bone. *J. Appl. Biomaterials*, **1**, 57–78.

Dong, Z. (1993) Synthesis of four structurally related tyrosine-derived polycarbonates and *in vitro* study of dopamine release from poly(desaminotyrosyl-tyrosine hexyl ester carbonate) MSc. Thesis, Rutgers University.

Engelberg, I. and Kohn, J. (1991) Physico-mechanical properties of degradable polymers used in medical applications: A comparative study. *Biomaterials*, **12**(3), 292–304.

Ertel, S.I. and Kohn, J. (1994) Evaluation of a series of tyrosine-derived polycarbonates for biomaterial applications. *J. Biomed. Mater. Res.*, **28**, 919–930.

Ertel, S.I., Kohn, J., Zimmerman, M.C. and Parsons, J.R. (1995) Evaluation of poly(DTH carbonate), a tyrosine-derived degradable polymer, for orthopaedic applications. *J. Biomed. Mater. Res.*, **29**(11), 1337–1348.

Fasman, G.D. (1987) The road from poly(α-amino acids) to the prediction of protein conformation. *Biopolymers*, **26**, S59–S79.

Fiordeliso, J. (1993) Aliphatic polyarylates derived from L-tyrosine: A new class of biomaterials for biomedical applications. MSc Thesis, Rutgers University.

Fiordeliso, J., Bron, S. and Kohn, J. (1994) Design, synthesis, and preliminary characterization of tyrosine-containing polyarylates: New biomaterials for medical applications. *J. Biomater. Sci. (Polym. Ed.)*, **5**(6), 497–510.

Hedayatullah, M. (1967) Cyanates et iminocarbonates d'aryle. I. Action des halogenures de cyanogene sur les derives sodes de mono et de diphenols polysubstitues. *Bull. Soc. Chim. (France)*, 416–421.

Hench, L.L. and Ethridge, E.C. (1982) *Biomaterials – An interfacial approach*. New York, NY, Academic Press.

Hooper, K.A. and Kohn, J. (1995) Diphenolic monomers derived from the natural amino acid α-L-tyrosine: Large scale synthesis of desaminotyrosyl-tyrosine alkyl esters. *J. Bioact. Compat. Polym.*, **10**(4), 327–340.

Jeon, S.H., Park, S.M. and Ree, T. (1989) Preparation and complexation of an A-B-A type triblock copolymer consisting of helical poly(L-proline) and random-coil poly(ethylene oxide). *J. Polym. Sci. Part A: Polym. Chem.*, **27**, 1721–1730.

Katchalski, E. (1974) Poly(amino acids): Achievements and prospects. *Peptides, Polypeptides, and Proteins – Proceedings of the Rehovot Symposium on Poly(Amino Acids), Polypeptides, and Proteins and their Biological Implications*. New York, NY, John Wiley. 1–13.

Katchalski, E. and Sela, M. (1958) Synthesis and chemical properties of poly-(α-amino acids. *Advances in Protein Chemistry*. New York, NY, Academic Press. 243–492.

Kohn, J. (1990) Pseudopoly(amino acids). *Biodegradable Polymers as Drug Delivery Systems*. New York, NY, Marcel Dekker. 195–229.

Kohn, J. (1991a) Desaminotyrosyl-tyrosine alkyl esters: New diphenolic monomers for the design of tyrosine-derived pseudopoly(amino acids). *Polymeric Drugs and Drug Delivery Systems*. Washington DC, American Chemical Society. 155–169.

Kohn, J. (1991b) Pseudo-poly(amino acids). *Drug News and Perspectives*, **4**(5), 289–294.

Kohn, J. (1991c) The use of natural metabolites in the design of non-toxic polymers for medical applications. *Polymer News*, **16**(11), 325–332.

Kohn, J. (1993) Design, synthesis, and possible applications of pseudo-poly(amino acids). *Trends Polym. Sci.*, **1**(7), 206–212.

Kohn, J. (1994) *Tyrosine-based polyarylates: Polymers designed for the systematic study of structure-property correlations*. 20th Annual Meeting of the Society for Biomaterials, Boston MA, Society for Biomaterials, 67.

Kohn, J. and Langer, R. (1984) A new approach to the development of bioerodible polymers for controlled release applications employing naturally occurring amino acids. *Polymeric Materials, Science and Engineering*. Washington, DC, American Chemical Society. 119–121.

Kohn, J. and Langer, R. (1986a) Poly(iminocarbonates) as potential biomaterials. *Biomaterials*, **7**, 176–181.

Kohn, J. and Langer, R. (1987) Polymerization reactions involving the side chains of α-L-amino acids. *J. Am. Chem. Soc.*, **109**, 817–820.

Kohn, J. and Langer, R. (1988) Backbone modifications of synthetic poly-α-L-amino acids. *Peptides – Chemistry and Biology: Proceedings of the 10th American Peptide Symposium*. Leiden (The Netherlands), Escom Publishing. 658–661.

Kohn, J., Niemi, S.M., Albert, E.C., Murphy, J.C., Langer, R. and Fox, J.G. (1986b) Single-step immunization using a controlled release, biodegradable polymer with sustained adjuvant activity. *J. Immunol. Meth.*, **95**, 31–38.

Kumaki, T., Sisido, M. and Imanishi, Y. (1985) Antithrombogenicity and oxygen permeability of block and graft copolymers of polydimethylsiloxane and poly(α-amino acid). *J. Biomed. Mater. Res.*, **19**, 785–811.

Lekchiri, A., Morcellet, J. and Morcellet, M. (1987) Complex formation between copper (II) and poly(N-methacryloyl-L-asparagine). *Macromolecules*, **20**, 49–53.

Lescure, F., Gurny, R., Doelker, E., Pelaprat, M.L., Bichon, D. and Anderson, J.M. (1989) Acute histopathological response to a new biodegradable, polypeptidic polymer for implantable drug delivery system. *J. Biomed. Mater. Res.*, **23**, 1299–1313.

Li, C. and Kohn, J. (1989) Synthesis of poly(iminocarbonates): Degradable polymers with potential applications as disposable plastics and as biomaterials. *Macromolecules*, **22**(5), 2029–2036.

Li, X., Bennett, D.B., Adams, N.W. and Kim, S.W. (1991) Poly(α-amino acid)-drug conjugates. *Polymeric Drug and Drug Delivery Systems*. Washington, DC, American Chemical Society. 101–116.

Lin, S., Krebs, S. and Kohn, J. (1991) *Characterization of a new, degradable polycarbonate*. The 17th Annual Meeting of the Society of Biomaterials, Scottsdale, AR, Society for Biomaterials, 187.

Lotan, N., Berger, A. and Katchalski, E. (1972) Conformation and conformational transitions of poly-α-amino acids in solution. *Annual Review of Biochemistry*. Palo Alto (California), Annual Reviews Inc. 869–901.

Lyman, D.J. (1983) Polymers in medicine – An overview. *Polymers in Medicine. Biomedical and Pharmacological Applications*. New York, NY, Plenum Press. 215-218.

Mao, H.Q., Zhuo, R.X. and Fan, C.L. (1993) Synthesis and biological properties of polymer immuno-adjuvants. *Polym. J.*, **25**(5), 499–505.

Marchant, R.E., Sugie, T., Hiltner, A. and Anderson, J.M. (1985) Biocompatibility and an enhanced acute inflammatory phase model. *ASTM Spec. Tech. Publ. (Corros. Degrad. Implant Mater.)*, **859**, 251–266.

Methenitis, C., Morcellet, J., Pneumatikakis, G. and Morcellet, M. (1994) Polymers with amino acids in their side chain: Conformation of polymers derived from glycylglycine and phenylalanine. *Macromolecules*, **27**, 1455–1460.

Moore, J.S. and Stupp, S.I. (1990) Room temperature polyesterification. *Macromolecules*, **23**(1), 65–70.

Morcellet, M., Loucheux, C. and Daoust, H. (1982) Poly(methacrylic acid) derivatives. 5. Microcalori-metric study of poly(N-methacryloyl-L-alanine) and poly(N-methacrylol-L- alanine-co-N-phenylmeth-acrylamide) in aqueous solutions. *Macromolecules*, **15**, 890–894.

Morcellet-Sauvage, J., Morcellet, M. and Loucheux, C. (1981) Polymethacrylic acid derivatives. 1. Prepara-tion, characterization, and potentiometric study of poly(N-methacryloyl-L-alanine-co-N-phenylmeth-acrylamide). *Makromol. Chem.*, **182**, 949–963.

Pulapura, S. (1991) Biodegradable polymers for medical applications: The tyrosine derived poly(imino-carbonate-carbonate) system. Ph.D. thesis, Rutgers University.

Pulapura, S. and Kohn, J. (1990a) *The iminocarbonate-carbonate polymer system: A new approach for the design of controlled release formulations*. 17th International Symposium for the Controlled Release of Bioactive Materials, Reno, Nevada, Controlled Release Society, 154–155.

Pulapura, S. and Kohn, J. (1992) Tyrosine derived polycarbonates: Backbone modified, "pseudo"-poly(amino acids) designed for biomedical applications. *Biopolymers*, **32**, 411–417.

Pulapura, S., Li, C. and Kohn, J. (1990b) Structure-property relationships for the design of polyiminocar-bonates. *Biomaterials*, **11**, 666–678.

Samyn, C. and Beylen, M.v. (1988) Polydepsipeptides: Ring-opening polymerization of 3-methyl-2,5-morpholinedione, 3,6-dimethyl-2,5-morpholinedione and copolymerization thereof with D,L-lactide. *Makromol. Chem., Macromol. Symp.*, **19**, 225–234.

Schminke, H.D., Grigat, E. and Putter, R. (1970) *Polyimidocarbonic esters and their preparation*, US Patent 3,491,060.

Sidman, K.R., Schwope, A.D., Steber, W.D., Rudolph, S.E. and Poulin, S.B. (1980) Biodegradable implantable sustained release systems based on glutamic acid copolymers. *J. Membr. Sci.*, **7**, 277–291.

Sidman, K.R., Steber, W.D., Schwope, A.D. and Schnaper, G.R. (1983) Controlled release of macro-molecules and pharmaceuticals from synthetic polypeptides based on glutamic acid. *Biopolymers*, **22**, 547–556.

Silver, F.H., Marks, M., Kato, Y.P., Li, C., Pulapura, S. and Kohn, J. (1992) Tissue compatibility of tyrosine derived polycarbonates and polyiminocarbonates: An initial evaluation. *J. Long-Term Effects Med. Implants*, **1**(4), 329–346.

Spivak, J.M., Blumenthal, N.C., Ricci, J.L. and Alexander, H. (1990) A new canine model to evaluate the biological effects of implant materials and surface coatings on intramedullary bone ingrowth. *Biomaterials*, **11**(1), 79–82.

Staubli, A., Mathiowitz, E., Lucarelli, M. and Langer, R. (1991) Characterization of hydrolytically degradable amino acid containing poly(anhydride-co-imides). *Macromolecules*, **24**, 2283–2290.

Stemberger, A., Alt, E., Schmidmaier, G., Kohn, J. and Blümel, G. (1994) Blood compatible biomaterials through resorbable anticoagulant drugs with coatings. *Ann. Hematol.*, **68**(supplement II), A48.

Tangpasuthadol, V., Shefer, A., Hooper, K.A. and Kohn, J. (1995) *Evaluation of thermal properties and physical aging as function of the pendent chain length in tyrosine-derived polycarbonates, a class of new biomaterials.*

Symposium Proceedings Vol. 394: Spring Meeting of the Materials Research Society, San Fransisco, CA, Materials Research Society, 143–148.

Tirrell, D.A., Fournier, M.J. and Mason, T.L. (1991) New polymers from artificial genes: a progress report. *Polym. Prepr.*, **32**(3), 704–705.

Wheeler, A.W., Moran, D.M., Robins, B.E. and Driscoll, A. (1982) L-tyrosine as an immunological adjuvant. *Int. Archs. Allergy Appl. Immun.*, **69**, 113–119.

Wise, D.L. and Midler, O. (1984) Poly(alkylamino acids) as sustained release vehicles. *Biopolymers in Controlled Release Systems*. Boca Raton, Florida, CRC. 219–229.

Yokoyama, M., Anazawa, H., Takahashi, A. and Inoue, S. (1990a) Synthesis and permeation behavior of membranes from segmented multiblock copolymers containing poly(ethylene oxide) and poly(β-benzyl L-aspartate) blocks. *Makromol. Chem.*, **191**, 301–311.

Yokoyama, M., Inoue, S., Kataoka, K., Yui, N., Okano, T. and Sakurai, Y. (1989a) Molecular design for missile drug: Synthesis of adriamycin conjugated with immunoglobulin G using poly(ethylene glycol)-*block*-poly(aspartic acid) as intermediate carrier. *Macromol. Chem.*, **190**, 2041–2054.

Yokoyama, M., Miyauchi, M., Yamada, N., Okano, T., Sakurai, Y., Kataoka, K. and Inoue, S. (1990b) Characterization and anticancer activity of the micelle-forming polymeric anticancer drug adriamycin-conjugated poly(ethylene glycol)-poly(aspartic acid) block copolymer. *Cancer Res.*, **50**, 1693–1700.

Yokoyama, M., Miyauchi, M., Yamada, N., Okano, T., Sakuri, Y., Kataoka, K. and Inoue, S. (1990c) Polymer micelles as novel drug carrier: Adriamycin-conjugated poly(ethylene glycol)-poly(aspartic acid) block copolymer. *J. Controlled Release*, **11**, 269–278.

Yokoyama, M., Okano, T., Sakurai, Y., Kataoka, K. and Inoue, S. (1989b) Stabilization of disulfide linkage in drug-polymer-immunoglobulin conjugate by microenvironmental control. *Biochem. Biophys. Res. Commun.*, **164**(3), 1234–1239.

Yoshida, M., Asano, M., Kumakura, M., Katakai, R., Mashimo, T., Yuasa, H. and Yamanaka, H. (1991) Sequential polydepsipeptides containing tripeptide sequences and α-hydroxy acids as biodegradable carriers. *Eur. Polym. J.*, **27**(3), 325–329.

Yu-Kwon, H. and Langer, R. (1989) Pseudopoly(amino acids): A study of the synthesis and characterization of poly(trans-4-hydroxy-N-acyl-L-proline esters). *Macromolecules*, **22**, 3250–3255.

Zhou, J., Ertel, S.I., Buettner, H.M. and Kohn, J. (1994) *Evaluation of tyrosine-derived pseudo-poly(amino acids): In vitro cell interactions.* 20th Annual Meeting of the Society for Biomaterials, Boston MA, Society for Biomaterials, 371.

Zhou, Q.X. and Kohn, J. (1990) Preparation of poly(L-serine ester): A structural analogue of conventional poly(L-serine). *Macromolecules*, **23**, 3399–3406.

SECTION 2:
NATURAL, SEMI-SYNTHETIC AND
BIOSYNTHETIC POLYMERS

14. NATURAL AND MODIFIED POLYSACCHARIDES

JOSEPH KOST and RIKI GOLDBART

Department of Chemical Engineering, Ben-Gurion University, Beer-Sheva, Israel

INTRODUCTION

Carbohydrates are the most abundant biomolecules on earth. Photosynthesis, converts each year more than 100 billion tons of CO_2 and H_2O into cellulose and other plant products. Carbohydrates are one of the main sources of the human diet in most parts of the world, and the oxidation of carbohydrates is the main energy-yielding pathway in most non-photosynthetic cells. There are three major size classes of carbohydrates: monosaccharides, oligosaccharides and polysaccharides (the word "saccharide" is derived from the Greek sakkharon, meaning "sugar"). Monosaccharides consists of a single polyhydroxy aldehyde or ketone unit. The most abundant monosaccharide in nature is the six carbon sugar D-glucose. Oligosaccharides consist of short chains of monosaccharides units joined together by characteristic glycosidic linkage. The most abundant are the disaccharides, with two monosaccharide units. Polysaccharides consist of long chains having hundreds or thousands of monosaccharide units. The most abundant polysaccharides made by plants, starch and cellulose, consist of recurring units of D-glucose, but they differ in the type of glycosidic linkage (Lehninger *et al.*, 1993).

Polysaccharides also called glycans, differ from each other in the identity of their recurring monosaccharide units, in the length of their chain, in the type of the O-glycosidic bonds linking the units, and in the degree of branching. The variety of saccharide monomers and the variety of possible O-glycosidic linkages result in a diversity of polysaccharide structures and conformations. Some polysaccharides serve as storage forms of monosaccharides, whereas others serve as structural elements in cell walls and connective tissues. On complete hydrolysis with acid or by the action of specific enzymes, polysaccharides yield monosaccharides or their derivatives. The polysaccharides can be divided to several groups based on their source: microbial, plants, algae and seaweeds, and animals (Table 1).

Table 1 Polysaccharides source and structure

Polymer	Source	Structure
Agar	red sea weed (rhodophyceae: gractaria and gelidium)	extended ribbon like polymers, backbone glycosidic links are always di-equatorial, but the repeating unit is a dimmer agarobiose: 3,6 anhydro-4-O-(β-D-galactopyranosyl)-L-galactose
Agarose	red sea weed (rhodophyceae: gractaria and gelidium)	natural copolymer with D-galactose in $\beta(1-3)$ linkage, alternating with 3,6 anhydro-α-L-galactose in a(1–4) linkage

Table 1 *Continued*

Polymer	Source	Structure
Alginate	sea weed extracts	D-mannopyranosyluronate (D-mannuronic acid) in $\beta(1\text{--}4)$ linkage, and L-gulopyranosyluronate (L-guluronic acid) in a$(1\text{--}4)$ linkage linear blockcopolymer, anionic
Carrageenan $(\chi, \lambda, \mu, \kappa)$	sea weed extract (red algae – rhodophylae)	galactose residues which are sulfated and alternately linked in $\alpha(1\text{--}3)$ and $\beta(1\text{--}4)$ linkage. The 4-linked residue can be 3,6 anhydro-D-galactose and 3-linked residues are, at least partly, 4-sulfated
Cellulose	plant cell wall	D-glucose in $\beta(1\text{--}4)$ linkage, linear
Chitin	exoskeleton of insects and crustacea, fungi wall membrane (filamentous fungi)	D-N-acetyl glucosamine in $\beta(1\text{--}4)$ linkage, neutral
Chitosan	fungi wall membrane (Mucor rouxil), commercially derived from chitin by chemical conversion with strong alkali	2-amino-2-deoxy-D-glucose (D-N-glucosamine) in $\beta(1\text{--}4)$ linkage, cationic
Condroitin sulfate	animal	D-glucopyranuronic acid and 2-acetamido-2-deoxy-D-galactopyranose 6, or 4 sulfate, alternately linked in $\beta(1\text{--}3)$ and $\beta(1\text{--}4)$ linkage
Curdlan	bacteria, extracellular (alcaligenes faecalis, agrobacter spp.)	D-glucose in $\beta(1\text{--}3)$ linkage, neutral
Dextran	bacteria, extracellular (leuconstoc sp.)	D-glucose in $\alpha(1\text{--}6)$ linkage, branch linkages are $\alpha(1\text{--}4)$, $\alpha(1\text{--}2)$, $\alpha(1\text{--}3)$, neutral
Elsinan	fungi, extracellular	D-glucose in $\alpha(1\text{--}4)$ and $\alpha(1\text{--}3)$ linkages, in molar ratios of 2:1 to 2.5:1 approximately one in 140 linkages is $\alpha(1\text{--}6)$
Furcellran	sea weed extract (furcelleria fastigiata)	composed of D-galactose and 3,6-anhydro-D-galactose, with sulfate ester groups on both sugar components
Galactomannan (guar gum or guaran)	plant seed (endosperm of the seed of cyamopsis tetragonolobus)	D-mannopyranosyl units in $\beta(1\text{--}4)$ linkage, to which varying amounts of α-D-galactopyranosyl groups are joined in $\alpha(1\text{--}6)$ linkage
Gellan	bacteria, extracellular (pseudonomas elodea)	D-glucose, D-glucuronic acid and rhamnose in $\beta(1\text{--}4)$ linkage
Glycogen	animal cells (liver, skeletal, muscle)	D-glucose in $\alpha(1\text{--}4)$ linkage, branch linkages are $\alpha(1\text{--}6)$ every 8 to 12 glucose residues

Table 1 *Continued*

Polymer	Source	Structure
Gum arabic (acacia gum)	plant (acacia senegal or other related african species of acacia)	D-galactose in β(1–3) or β(1–6) linkage, with L-arabinose, L-rhamnose, D-galactose and D-glucuronic acid as side groups in (1–3) or (1–4) linkage
Hemicellulose	plant cell wall	D-xylose in β(1–4) linkage, with side chains of arabinose and other sugars.
Hyaluronic acid	human and animal (intercellular material in the space between skin, cartilage and muscle cells)	D-glucuronic acid and N-acetylglucosamine in β(1–3) and β(1–4) linkage, linear
Inulin	plant (artichoke)	D-fructose in β(1–2) linkage
Karaya gum (sterculia gum, Indian tragacanth)	plant (sterculia urens)	mixture of D-galactose, L-rhamnose and D-galacturonic acid. The galacturonic acid units are the branching points of the molecule
Levan	bacteria, extracellular	anhydro-D-fructo-furanose in predominantly β(2–6) linkage and some β(1–2) linkage
Pectin	plant cell wall	methyl D-galacturonate in α(1–4) linkage
Pollulan	fungi, extracellular (aurebasidium pullalans)	maltotriose in α(1–4) linkage, connected by α(1–6) linkage
Prophyran	red sea weed	partial 6-O-methylagarobiose together with partial replacement of L-galactose by L-galactose 6-sulfate
Psyllium-flea seed (plantago seed)	plant seed (plantego psyllium or plantego indica)	mixture of L-arabinose, D-galactose, D-galacturonic acid, L-rhamnose and D-xylose
Quince seed	plant seed (cydonia vulgaris)	
Scleroglucan	fungi, extracellular (sclerdium glucaricum)	D-glucose in β(1–3) linkage, with (1–6) β-D-glucoes side groups linked to every third β-D-glucose residue in the main chain
Starch	plant seed	D-glucose in α(1–4) linkage, branch linkages are α(1–6)
Tragacanth gum	plant (astragalus)	consists of water-soluble fraction known as tragacanthin (mixture of D-galacturonic acid, D-galactose, L-fucose, D-xylose and L-arabinose) and a water-insoluble fraction known as bassorin (the carboxyl groups of the galacturonic acid are mainly esterified with methanol)

Table 1 *Continued*

Polymer	Source	Structure
Welan	bacteria (alcaligenes spp.)	single sugar side groups: L-rhamnose or L-mannose are regularly linked in $\beta(1-3)$ linkage to a gellan backbone
Xanthan	bacteria, extracellular (xanthonomas compestris)	to a cellulose backbone (D-glucose in $\beta(1-4)$ linkage) are attached (every other β-D-glucose residue) side groups built up by D-mamose, D-glucuronic acid and D-mannose, anionic
Xylan	plant tissues (sugar cane bagasse, corn cobs and straw)	anhydroxylose with substituents of 4-O-methyl glucuronic acid, acetyl groups and anhydroarabinose units in $\beta(1-4)$ linkage, partially branched
Xyloglucan (amyloid)	plant seed (tamarindus indica)	D-glucose in $\beta(1-4)$ linkage, with side groups of D-galactose and D-xylose

With the exemption of some linear polysaccharides such as cellulose, polysaccharides are hydrophilic polymers which in a suitable solvent system, form hydrocolloids with different physical properties. Polysaccharides are of interest because of their unusual and functional properties (Kaplan et al., 1994). Some of these properties are listed in Table 2.

CHEMICAL MODIFICATIONS

The chemical modification of polysaccharide chain can occur at the hydroxyl, carboxyl, amino or sulfate moieties or by oxidation to open up the pyranose rings. Modification of hydroxyl group can be performed by alkylation, acylation and phosphorylation. Modification can be also obtained by the introduction of actvating groups to conjugate enzymes and proteins. Reagents used for the preparation of activated polysaccharides include: cyanogen bromide, carbonyldiimidazole, chloroformates (4-nitrophenylchloroformate, N-hydroxysuccinimidylchloroformate, trichlorophenylchloroformate), divinylsulfone, organosulfonyl, triazine, periodate and carbodiimide.

Table 2 Useful Properties of Polysaccharides

Film and gel forming capabilities
Stability over broad range of temperatures and pHs
Biocompatibility
Biodegradability
Water solubility
possible genetic manipulation to control product expression, molecular weight distribution, stereospecificity and functional properties

Grafting of natural polysaccharides is of considerable interest in biomedical, pharmaceutical, agricultural and consumer products. Graft copolymers are generally prepared by generating free radicals on the polysaccharide chain. The free radicals then initiate polymerization of vinylic or acrylic monomers. Common chemical initiators used include: ceric salts, potassium permanganate, trivalent manganese, cupric ions, ammonium persulfate, potassium persulfate and azobisisobutyronitrile. Initiation of polysaccharide molecules can also be achieved by ionizing radiation using ^{60}Co or electron beam radiation (Shalaby and Park, 1994).

BIODEGRADATION

Living organisms cannot only synthesize biopolymers such as proteins, nucleic acids and polysaccharides, but are also capable of degrading them. The general mechanism of degradation of polymers into the small molecules employed by nature is a chemical one. Living organisms are capable of producing enzymes which can attack biopolymers. The attack is usually specific with respect to both the enzyme/polymer couple and the site of attack at the polymer. Frequently, enzymes are designed according to their mode of action. Hydrolases, for instance, are enzymes catalyzing the hydrolysis of ester-, ether- or amide-linkages. Proteolytic enzymes (hydrolyzing proteins) are called proteases and enzymes hydrolyzing polysaccharides are called glycosidases. There are three types of glycosidases that hydrolyze either O-glycosyl, N-glycosyl, or S-glycosyl bonds in a polysaccharide chain (Park *et al.*, 1993; Shalaby and Park, 1994). Several enzymes together with their potential cleavage sites are pesented in Table 3.

Amylases operate at two modes: endo-(α amylases) or exo-(β amylases). α-amylases hydrolyze only (1–4) linkages and attack the amylose chain at random points. The end products are α-maltose and glucose. β-amylases can also hydrolyze only (1–4) linkages, but attack amylose specifically only at the non-reducing end of the molecule removing successively β-maltose molecules. Since neither α nor β-amylases can attack the (1–6) linkages, the end products of the hydrolysis of the branched amylopectin contain branched dextrins with intact (1–6) linkages.

Phosphorylases catalyze the phosphorolytic cleavage of α-D-glucose-l-phosphate from the non-reducing end of the amylose molecule.

Cellulases play an important role in the natural decomposition of plant residues and are produced by cyllolitic microorganisms including aerobic saprophytes, anearobic rumen bacteria and anearobic thermophilis spore formers. Similar to amylases, the cellulases operate in two modes: endo or exo-cellulase.

Lytic enzymes attack cell walls which consist mainly of peptidoglycans. The mode of action during lysis of peptidoglycans is depicted in scheme 1 (Schnabel, 1981).

POLYSACCHARIDES ANALYSIS

Mixture of polysaccharides can be revolved into their individual components by many techniques: differential centrifugation, ion-exchange chromatography, and gel filtration. Hydrolysis in strong acid yields a mixture of monosaccharides, which, after conversion to suitable volatile derivatives, may be separated, identified, and

Table 3 Enzymes capable of rupturing main chains in polysaccharides

Enzymes	Preferential cleavage sites
Agarase	hydrolysis of 1,3-β-D-galactosidic linkage in agarose
α-Amylas (glycogenase)	hydrolysis of 1,4-α-D-glucosidic linkages containing three or more 1,4-α-linked D-glucose units
β-Amylase	hydrolysis of 1,4-α-linked D-glucosidic linkages from nonreducing ends of chains
κ-Carageenase	hydrolysis of more 1,4-β-linkeges between D-galactose-4-sulfate and 3,6-anhydro-D-galactose in carrageenans
Cellulase	endohydrolysis of 1-4-β-D-glucosidic linkages in cellulose
Chitinase	random hydrolysis of N-acetyl-β-D-glucosaminide 1,4-β-linkages in chitin and chitodextrins
Dextranase	endohydrolysis of 1,6-α-D-glucosidic linkags in dextran
α-L-Fucosidase	α-L-Fucoside + water yields an alcohol + L-fucose
α-Galactosidase	hydrolysis of terminal, nonreducing α-D-galactosidase residues in α-D-galactosides
β-Galactosidase	hydrolysis of terminal, nonreducing β-D-galactosidase residues in β-D-galactosides
α-Glucosidase	hydrolysis of terminal, nonreducing 1,4-linked α-D-glucose residues
β-Glucosidase	hydrolysis of terminal, nonreducing β-D-glucose residues
Hyaluronidase	random hydrolysis of 1,4-linkages between N-acetyl-β-D-glucosamine and D-glucuronate residues in hyaluronate and random hydrolysis of 1,3-linkages between N-acetyl-β-D-glucosamine and N-acetyl-D-glucosamine residues in hyaluronate
Inulinase	endohydrolysis of 2,1-β-D-fructosidic linkages in inulin
Isoamylase	hydrolysis of 1,6-α-glucosidic branch linkages in glycogen, amylopectin, and their β-limit dextrin
Lysosyme	hydrolysis of 1,4-β-linkages between N-acetylmuramic acid and N-acetyl-D-glucosamine residues in peptidoglycan and betwwe N-acetyl-D-glucosamine residues in chitodextrin
Phosphorylase	cleavage of α-D-glucose-L-phosphate from the nonreducing end of the amylose molecule

quantified by gas-liquid chromatography to yield the overall composition of the polymer. For simple linear polymer such as amylose, the position of the glycosidic bond between monosaccharides is determined by treating the intact polysaccharide with methyl iodide to convert all free hydroxyls to acid-stable methyl ethers. When the methylated polysaccharide is hydrolyzed in acid, the only free hydroxyls present in the monosaccharides produced are those that were involved in glycosidic bonds. To determine the stereochemistry at the anomeric carbon, the intact polymer is tested for sensitivity to purified glycosidases known to hydrolyze only α- or only β-glycosides. Total structure determination for complex polysaccharides is much more difficult. Stepwise degradation with highly specific glycosidases, followed by isolation and identification of the products, is often helpful. Mass spectral analysis and high resolution nuclear magnetic resonance (NMR) spectroscopy are extremely powerful analytic tools (Lehninger *et al.*, 1993). brief statements on the basic principles of the methods with detailed description of experimental protocols can be found in (Chaplin and Kennedy, 1994).

PROPERTIES AND APPLICATIONS

Following is a more detailed description of properties, processing and applications of several polysaccharides.

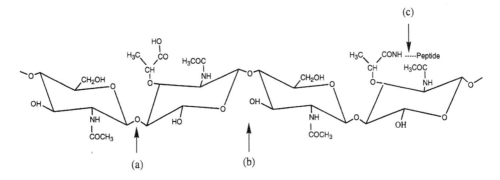

Scheme 1 Mode of action of lytic enzymes during the lysis of peptidoglycan. (a) Rupture of the N-acetylglucosaminyl-N-acetylmuramic acid bond by endoacetylglucosaminidases. (b) Endoacetylmuramidases split the N-acetylmuramyl-N-acetylglucosamine bond. (c) Rupture of the peptide bonds by peptidases.

Alginate

The affinity of alginates for cations and gel forming properties are mainly related to the content of contiguous guluronic residues. When two guluronic acid (G) residues are adjacent in the polymer, they form a binding site for polyvalent cations. Therefore the content of G-blocks is the main structural feature contributing to gel strength and stability of the gel. Alginates are widely used in pharmaceutical preparations, cell entrapment, and enzyme immobilization.

Entrapment of cells within spheres of calcium-alginate gels has in recent years become the most widely used technique for immobilizing living cells. The production of alginate beads containing cells is simple. A cell suspension is mixed with a aqueous solution of alginate (2–4%). The mixture is dripped into a solution containing a multivalent cation solution (CaCl$_2$, 0.05–0.1 M). The droplets instantaneously solidify forming ionotropic gel, entrapping the cells within. The advantages of alginates for cell entrapment are: the immobilization is performed under very mild condition, therefore little loss of viability for most cells. No chemical coupling is required, and the cells are not exposed to any harmful chemicals. The process can be performed in an isotonic buffered medium at neutral pH and a range of temperature from 0–100°C.

Alginate entrapment has now been applied for immobilization of different cells including bacteria, yeasts, molds, algae, plant and animal cells. Gel entrapment can be easily performed under sterile conditions, either by autoclaving or sterile filtration of the sodium alginate solution before mixing with cells. Since the gel is crosslinked by calcium ions, substances with high affinity for calcium will destabilize the gel. This preclude the use of buffers containing citrates or phosphates. It is therefore recommended to include 3 mM calcium in the medium, or to keep the sodium/calcium ratio less than 25:1.

Despite the extensive studies in the field since Lim and Sun (Lim and Sun, 1980) described the reversal of diabetes in rats by alginate-based encapsulated islets, there have been until recently, no reports of successful studies of the large

animal model nor any attempts in Type I diabetic patients. Recently Soon-Shiong
et al. (Soon-Shiong *et al.*, 1994) elucidated some of the factors affecting long term
viability of the implanted alginate based encapsulated islets. Based on their finding
the authors concluded that alginate material high in mannuronic acid content
should be avoided for in vivo microencapsulation. Also by controlling the kinetics
of gel formation and by the use of alginate high in G-content mechanical integrity
and chemical stability of the microcapsules were significantly improved. Phase I/II
human clinical trials demonstrated ongoing insulin secretion from the intraperitoneally implanted encapsulated islets for over 20 months. Following injection of
the initial dose of 10,000 islets per kg, the patient maintained normoglycemia.
Occasional hyperglycemic episodes were noted post-prandially and the patient was
placed on a minimal dose of insulin to maintain normoglycemia for a period of 6
months. In the seventh months additional 5000 islets per kg were injected which
resulted in insulin independence in the ninth month for the entire 30 days period.

Both Kelco from U.S and Protan from Norway produce medical grade alginates.
Kelco algin products have been affirmed GRASS by the U.S Food and Drug
Administration (FDA). Protan sells alginates for immobilization and encapsulation
with two guluronic content of 40 and 70% (PROTANAL LF 10/60, 10/40 RB, 20/60,
and 20/60RB)

Arabinogalactan

Arabinogalactan is a water soluble gum found in unusually high concentration (up to
35%) in the heartwood of all species of larch. It is a highly branched polysaccharide
of a 30,000 molecular weight, composed of L-arabinose and D-galactose units at
1:6 mole ratio. Arabinogalactan has many unique characteristics, such as complete
miscibility with water and low viscosity at high dissolved solids content (23.5 cps
at 40% concentration and 20°C). Temperature, electrolytes, and pH have little
effect on the viscosity of concentrated solutions of arabinogalactan. The refractive
index and specific gravity of arabinogalactan solutions are both linear functions of
concentration and can be used as analytical tools in measuring concentration of
the gum (Nazareth *et al.*, 1961). These unique characteristics have made the gum
useful in such industries as the printing, mining, carbon black, and food industries
(Adams and Ettling, 1973). In recent years arabinogalactan has been applied for
use in cell separation, cosmetic formulations, pharmaceutical dosage forms, and as
drug carrier for hepatic drug delivery (Groman *et al.*, 1994). It is FDA approved as
food ingredient, and commercially available since 1964, it is now available at various
grades from food grade to ultra pure (Larex International, St. Paul, USA).

Cellulosic Materials

These are discussed in Chapter 15 in this volume.

Chitin and Chitosan

Chitin, widely distributed in nature, is a substance that sustains and protects the body
of crustaceans and microorganisms. Chitin is the second most abundant natural

polymer after cellulose. Its annual production in nature totals thousands tons. Commercially chitin is mainly produced from shell wastes of crabs, shrimps and krills, which are generated from the seafood industries. Chitosan is the deacetylated product of chitin. Chitosan rarely exists naturally, except in a few species of fungi. Commercially, chitosan is produced from chitin by deacetylating the chitin with concentrated alkali solutions at elevated temperatures. The acetamide groups undergo a hydrolysis process and chitin is converted to chitosan as a result.

Pure chitin with 100% acetylation in the amine groups and pure chitosan with 100% deacetylation rarely exist. The natural chitin and chitosan often exist as a copolymer of glucosamine and acetylated glucosamines. The degree of acetylation is therefore one of the most important structural parameters in chitin and chitosan. Many methods have been developed for analyzing the degree of acetylation, ranging from elemental analysis, titration, circular dichroism, infrared spectroscopy, UV spectroscopy, pyrolysis-gas chromatography, thermal analysis, NMR and X-ray diffraction.

Chitin is insoluble in water and most ordinary solvents. This property has restricted its use to applications that do not require solubility of the polymer (Chandy and Sharma, 1990; Muzzarelli *et al.*, 1986). As chitosan contains the basic free amines, it can form salts with a variety of inorganic and organic salts. Most of these salts are water soluble and chitosan can be dissolved in aqueous solutions of almost all the organic and inorganic acids. The chitosan solution can be dried and then ground to form a powder which can be highly purified to provide a water soluble raw material for various uses.

Acylation of the amine groups of chitosan can be readily carried out using acyl anhydride as the reactant. During the acylation process of a chitosan solution, the chitosan slowly loses the solubility and a gel is formed. Muzzarelli *et al.* (Muzzarelli *et al.*, 1986) modified chitosan with aldehyde-acids and keto-acids, from which they obtained a series of polymers possessing such groups as carboxylic acids, primary and secondary amines and primary and secondary hydroxyl groups. The product showed excellent metal chelating properties (Muzzarelli, 1973).

Since Prudden *et al.* (Prudden *et al.*, 1970) reported chitin's ability to promote the healing of wounds, some bioactivating effects based on its immunological adjuvant activity have become apparent. Based on their basic properties, braided fibers, sheets and porous sponges made from chitin and chitosan have been applied clinically in the form of absorbable sutures thread and temporary dressing for the skin and soft tissues defects. Chitin and derivatives have been also applied as drug delivery matrices as filling for bone defects, as internal fixation devices for bone fracture, and as immunological activation adjuvants (Maeda *et al.*, 1995).

Chitin artificial skin, in the form of nonwoven fabric, is commercially available in Japan. A number of studies have already reported satisfactory results regarding its wound healing effects, histological reactions, and wound management abilities (Kifune, 1992).

Pangburn *et al.* (Pangburn *et al.*, 1984) suggested the application of partially deacetylated chitin in self regulated drug delivery systems. Suzuki *et al.* (Suzuki *et al.*, 1993) obtained the release of high concentration cisplatinum under an implantation to mouse muscle during more than 8 weeks. In recent years additional studies were aimed at the development of drug delivery systems using chitin and its derivatives (Watanabe *et al.*, 1992).

Following is a procedure described for the preparations of films and matrices made of chitosan (Golomb *et al.*, 1992). High and low molecular weight chitosan purchased from Fluka were used after being passed through a 10–mesh screen. The films were prepared by dissolving the polymer (91 gr) in 100 mL of 3% acetic acid. The viscous solution was poured into petri dishes, and the remaining acetic acid was removed from the film by lyophilization. For drug delivery matrices, the drug was dissolved in the polymer solution and casted as described, yielding matrices with 5–40% w/w drug in polymer. Crosslinking of the films or matrices in attempt to retard degradaton and drug release rates is possible by exposing them to glutaraldehyde vapors in closed chambers for 24 h at ambient temperatures.

Dextran

Many different types of esters and ethers of dextran provide macrmolecules with diverse properties and negative, positive or neutral charges. Properties depend upon the type and degree of substitution, and the molecular weight of the dextran. The most widely used dextran derivative is obtained by the reaction of an alkaline solution of dextran with epichlorohydrin to give cross-linked chains. The product is a gel that is used as a molecular sieve. With its commercial introduction in 1959 by Pharmacia Fine Chemicals, Ltd, Uppsala, Sweden, cross-linked dextran revolutionized the purification and bio-separation processes.

Commercial cross-linked dextran is know as Sephadex. It is produced in bead form by dissolving the dextran in sodium hydroxide solution, dispersing it in an immiscible organic solvent such as poly(vinyl acetate) in toluene, and adding epichlorohydrin. The reaction mixture is kept at $50°C$ until the beads gel. Several types of Sephadex have been developed with different degrees of cross-linking, giving different molecular exclusion limits. Cross-linked dextran matrix has also been used as a solid support for affinity chromatography in which the desired-affinity ligand is covalently coupled to the dextran by adding alkaline cyanogen bromide to the cross-linked dextran followed by the ligand.

Dextran of relatively low molecular weight can be used as therapeutic agent in restoring blood volume for mass casualties. The initial impetus for the commercial production was for this use. Natural dextran with a molecular weight of about 5×10^8 daltons is unsuitable as a blood plasma-substitute. The optimum size for blood-plasma dextran, so called clinical dextran, is 50,000–100,000 daltons. These relatively low molecular weight dextrans may be produced by controlled acid hydrolysis of natural dextran, followed by organic solvent fractionation (Gronwall, 1957).

Dextran have been also used as drug carrier. In these drug-linked dextrans the polysaccharides act as a shield, protecting the active moiety against chemical or biological degradation, and also as a carrier transferring the active drug via the blood to the organs, without decreasing its bioavailability. Hydroxyl groups of dextran can react directly with chlorides of organic acids in the presence of acceptors of chloride ions i.e. alkaline hydroxides or pyridine. Direct esterification of dextran is possible by diimides such as dicylohexylcarbodiimide. To synthesize derivatives that do not contain carboxylic groups capable of esterification, it is necessary to activate the hydroxyl groups of dextran and produce the desired compound by covalent bonds. The most frequently used agents for the activation are periodate, azide, cyanogen halides, organic cyanates and epoxyhalopropyl (Molteni, 1979; Schacht *et al.*, 1985) (Molteni, 1979; Schacht, Ruys *et al.*, 1985; Mocanu and Carpov, 1996).

Recently Kaplan and Park studied chemically crosslinked dextran hydrogels for application in the controlled delivery of bioactive proteins (Kaplan and Park, 1995). Dextran was functionalized by reacting it with glycidyl acrylate to introduce reactive double bonds. Upon exposure to γ-irradiation the functionalized dextran formed a crosslinked gel which could be degraded by dextranase.

Hyaluronic Acid

Hyaluronic acid (HA) is a natural component of connective tissue, and is also found in synovial fluid that cushions joints as well as the vitreous and aqueous humors of the eye. It is considered as a space-filling, structure-stabilizing, cell coating and cell-protective polysaccharide. Because of its properties, HA has found widespread use in applications such as ophthalmic surgery and viscosupplementation (the therapeutic process by which the normal rheological state of a tissue compartment with pathologically decreased viscoelasticity is restored by the introduction of viscosupplementary materials). HA has been proposed for use in a variety of medical treatments, including wound healing, adhesion prevention in postoperative orthopedic, abdominal gynecological surgery, artificial organs, and drug delivery. However, many clinical applications are precluded, because HA exists in the form of an aqueous gel that has a limited shelf life and is rapidly degraded on administration. Recently through chemical modification of HA a series of biopolymers with physiochemical properties that are different from those of HA, but which retain the desired biological properties of HA, have been produced (Benedetti, 1994).

The HA chain contain three types of functional groups that can be used for derivitizations, namely, hydroxy, carboxy, and acetamido groups. A large number of HA derivatives cross-linked via their hydroxy group were developed by Balazs and Leshchiner (Balazs *et al.*, 1995) named hylans. Cross-linking agents of various functionalities were used in the reactions, including formaldehyde, dimethylol urea, dimethylolethylene urea, polyisocyanate, and vinyl sulfone. Water-insoluble products were obtained in the shape of powder, a film, or a coating on various substrates. The products obtained with vinyl sulfone as the crosslinking agent were developed further to biomedical products with excellent biocompatibility (Balazs and Leshchinger,).

The first medical application of the noninflammatory fraction of Na-hyaluronan (NIF-NaHY) was in the ophthalmic surgery. The commercial product under the trademark Healon[R] was introduced by Kabi Pharmacia, Upsala, Sweden. Healon is primarily used in cataract surgery. It minimizes the surgical trauma during the removal of the cataractous lens and introduction of its replacement, the intraocular plastic lens.

In the late 1980s two additional hyalronan preparation for treatment of arthritis were marked; one in Japan (Arzt[R] Seikagaku and Kaken) and one in Italy (Hyalagan[R], Fidia). The preparations consist of a 1% solution of relatively low molecular hyloronan (700,000) which are injected for viscosupplementation of osteoarthritic joints (Balazs *et al.*, 1995). Addition applications of HA have been evaluated for management of postsurgical adhesion in: abdominal, cardiac, musculoskeketal and neurological procedures.

HA and hylans have been developed as topical, injectable, and implantable vehicles for the controlled and localized delivery of biologically active molecules (Larsen and Balazs, 1991). The most commonly studied HA delivery system is for the delivery of pilocarpine agent used in the management of the elevated intraocular pressure associated with glaucoma. Topical HA solutions (0.1–0.2%) have been shown to be effective therapy for dry eye syndrome.

Interferon is a protein with antiviral and antiproliferative properties that has been approved for treatment of hairy cell leukemia, genital warts, and Kaposi's sarcoma. In vitro studies performed using hylan vehicles loaded with alpha-interferon indicate that hylan gel-fluid mixtures have a substantial effect on the release period of interferon. In the presence of hylan vehicles, there was up to an eight fold reduction in the levels of interferon released over a 24-hour period, as compared with controls not containing hylan.

At Hyal Pharmaceutical Corp. a topical form of HA similar to drug patches is under development for delivery of analgesics and drugs for treatment of pre-cancerous skin. The topical products have not been approved yet but the company hopes for a 1997 launch of these products (Ziegler, 1995).

Starch

Starch forms the main source of carbohydrate in the human diet. Starch consists of two main polysaccharides, amylopectin and amylose: the former constitutes about 80% of the most common starches. Amylose is essentially a linear polymer having a MW of 100,000–500,000. Conversely amylopectin is a highly branched polymer with molecular weight in the millions. The branches of amylopectin contain about 20–25 glucose units. Amylases are capable of hydrolyzing starch completely to yield D-glucose. In humans amylases are produced in the salivary glands and the pancreas.

A wide range of modification mechanisms of starches are known (Yalpani, 1988). These include self-association (induced by changes of pH, ionic strength, or physical and thermal means) and complexation with salts and covalent cross-linking.

The ability of polysaccharides to form a network structure (gel), even at very low concentrations, constitutes one of their most important functional properties. The gelation process, the formation of a three-dimensional nonsoluble network structure, is different than the gelatinization process, which is a means of dissolving starch. Gelatinization is known to destroy the crystalline-like structure by opening starch tertiary and quaternary structures due to breakdown and rearrangement of hydrogen bonds. During this process the granular structure of starch is completely destroyed, but it is still in its macromolecular state.

Starch in its native or modified form, has been subjected to extensive study over the past 50 years. Early interest in starch was associated with the food and paper industry, textile manufacture, and pharmacology. With the increased interest of biomedical and pharmaceutical research in biodegradable polymers as matrices for controlled drug delivery systems, impressive activities on the modification of natural polymers to meet growing needs have been reported (Hag et al., 1990; Kost and Shefer, 1990; Shefer et al., 1992; Trimnell et al., 1982; Vandenbossche et al., 1992; Visavarungroj et al., 1990).

Kost and Shefer (Kost and Shefer, 1990) studied two methods of network formation (crosslinking by calcium or epichlorohydrin) to entrap drug molecules in the starch matrix for use in controlled drug delivery systems. Since the enzymatic degradation of starch is very sensitive to both its structure and morphology, the degradation rates, thus drug release, will be affected and depend upon the structure of the network formed. The first stage of network formation in both calcium and epichlorohydrin procedures involves the gelatinization of starch by addition of sodium hydroxide solution (6.6% (w/v), 50 mL) to a suspension of water (30 mL) and starch with the drug (12 g). The solution turns into a high viscosity mixture after 1 h of continuous stirring at 600 rpm. Solution of calcium chloride (12.2 g, 24.5 mL of 50% (w/v) solution in water) or epichlorohydrin (20.5 mL), depending on the type of network formed, were added upon continuos slow mixing ending in coagulating the mixture. The particles formed were washed with water and air dried at room temperature.

Recently Labopharm introduced an excipient for controlled release matrices under the trade name CONTRAMID. Contramid is a derivative of amylose, obtained by treating amylose with cross-linking agents. With Contramid it is possible to prepare tablets by direct compression of a drug-excipient mix. The release profiles are a function of the cross-linking degree, the hydrophilic character of the drug and the core-loading.

Xanthan Gum

In 1964 Kelco commercialized xanthan gum. Since then, xanthan gum has been used in numerous applications by a multitude of industries. Xanthan gum was approved in the United States by the U.S. Food and Drug Administration for general use in foods. (21 CFR 172,695). Xanthan gum appears as E-415 on the European Community Consolidated Directive for Food Additives. The EC's Scientific Committee for Foods assigned an "ADI unspecified" to xanthan gum in its 66th meeting.

Xanthan gum solutions are highly pseudoplastic. When shear stress is increased, viscosity is progressively reduced. Upon the reduction of shear, total viscosity is recovered almost instantaneously. This behavior results from the high-molecular weight molecule which forms complex molecular aggregates through hydrogen bonds and polymer entanglement. Also, this highly ordered network of entangled, stiff molecules accounts for high viscosity at low shear rates. Shear thinning results from disaggregation of this network and alignment of individual polymer molecules in the direction of shear force. However, when the shearing ceases, aggregation reform rapidly.

Xanthan gum is a valuable excipient in oral controlled release products, in gastric fluid it swells and therefore produces a viscous layer around the tablet through which the drug must diffuse. Matrix systems using xanthan gum can be easily produced by normal production processes such as direct compression or granulation (Dumoulin et al., 1993).

REFERENCES

Adams, M.F. and Ettling, B.V. (1973) Larch arabinogalactan. In *Industrial gums, polysaccharides and their derivatives*, R.L. Whistler, ed. (New York: Academic Press).

Balazs, E.A. and Leshchinger, E. US patents: 4,605,691, 4,852,865, 4,636,524.

Balazs, E.A., Leshchinger, E., Larsen, N.E. and Band, P. (1995) Hyaluronan biomaterials: medical applications. In *Encyclopedic handbook of biomaterials and bioengineering*, D.L. Wise, D.J. Trantolo, D. E. Altobelli, M.J. Yashemski, J.D. Gresser and E.R. Schwartz, eds. (New York: Marcel Dekker, Inc), pp. 1693–1715.

Benedetti, U. (1994) New biomaterials from hyaluronic acid. *Medical Device Technology*, 32–37.

Chandy, T. and Sharma, C.P. (1990) Chitosan – as a biomaterial. *Biomaterials Artificial Cells, Artificial Organs*, **18**, 1–24.

Chaplin, M.F. and Kennedy, J.F. (1994) *Carbohydrate analysis a practical approach* (Oxford: Oxford University Press).

Dumoulin, Y., Cartilier, L., Lenaerts, V. and Mateescu, M.A. (1993) Cross-linked amylose: enzymatically controlled drug release (ECDR) system. In *International symposiumon controlled release of bioactive materials*, T. Roseman, N. Peppas and H. Gabelnick, eds. (Washington D.C.: Controlled Release Society, Inc.), pp. 306–307.

Golomb, G., Levi, M. and Van Gelder, J. (1992) Controlled release of bisphosphonate from a biodegradable implant: evaluation of release kinetics and anticalcification efect. *Journal of Applied Biomaterials*, **3**, 3–28.

Groman, E.V., Enriquez, P.M., Jung, C. and Josephson, L. (1994) Arabinogalactan for hepatic drug delivery. *Bioconjugate Chem.*, **5**, 556–574.

Gronwall, A. (1957) Dextran and its use in colloidal infusion solutions (New York: Academic Press).

Hag, E., Teder, H., Roos, G., Christensson, P.I. and Stenram, U. (1990) Enhanced effect of adriamycin on rat liver adenocarcinoma after hepatic artery injection with defradable starch microspheres. *Selective Cancer Therapeutics*, **6**, 23–34.

Kaplan, D., Wiley, B., Mayer, J., Arcidiacono, S., Keith, J., Lambardi, S., Ball, D. and Allen, A. (1994) Biosynthetic polysaccharides. In *Biomedical polymers*, S. Shalaby, ed. (Munich: Hanser Publishres), pp. 188–212.

Kaplan, K. and Park, K. (1995) Study on the release of invertase from enzymatically degradable dextran hydrogels. *Polymer Gels and Networks*, **3**, 243–254.

Kifune, K. (1992) Clinical applications of chitin in artificial skin. In *Advances in chitin and chitosan*, C.J. Brine, P.A. Sandford and J. P. Zikakis, eds. (New York: Elsevier Applied Sciences), pp. 9–15.

Kost, J. and Shefer, S. (1990) Chemically-modified polysaccharides for enzymatically-controlled oral drug delivery. *Biomaterials*, **11**, 695–698.

Larsen, N.E. and Balazs, E.A. (1991) Drug delivery systems using hyaluronan and its derivatives. *Advanced Drug Delivery Reviews*, **7**, 279–293.

Lehninger, A., Nelson, D. and Cox, M. (1993) Carbohydrates. In *Principles of biochemistry* (New York: Worth Publisher), pp. 298–323.

Lim, F. and Sun, A. M. (1980) Microencapsulated islets as a bioartificial endocrine pancreas. *Science*, **210**, 908.

Maeda, M., Inoue, Y., Iwase, H. and Kifune, K. (1995) Biomedical properties and applications of chitin and its derivatives. In *Encyclopedic handbook of biomaterials and bioengineering*, D.L. Wise, D.J. Trantolo, D.E. Altobelli, M.J. Yashemski, J.D. Gresser and E.R. Schwartz, eds. (New York: Marcel Dekker, Inc), pp. 1585–1598.

Molteni, L. (1979) Dextran as drug carriers. In *Drug carriers in biology and medicine*, G. Gregoriadis, ed. (New York: Academic Press), pp. 107–125.

Muzzarelli, R. (1973) *Natural chelating polymer* (New York: Pergamon Press).

Muzzarelli, R., Jeniaux, C. and Gooday, G.W. (1986) *Chitin in nature and technology* (New York: Plenum Pess).

Nazareth, M.R., Kennedy, C.E. and Bhatia, V.N. (1961) Studies in larch arabinogalactan. *Journal Pharmaceutical Sciences*, **50**, 560–564.

Pangburn, S.H., Trescony, P.V. and Heller, J. (1984) Partially deacetylated chitin: its use in self-regulated drug delivery systems. In *Chitin, chitosan and related enzymes*, J.P. Zikakis, ed. (New York: Academic Press), pp. 3–19.

Park, K., Shalaby, W. and Park, H. (1993) *Biodegradable hydrogels for drug delivery* (Lancaster: Technomics).

Prudden, J.F., Migel, P., Hanson, P., Friedrich, L. and Balassa, L. (1970) The discovery of a potent pure chemical wound-healing accelerator. *American Journal of Surgery*, **119**, 560–564.

Schacht, E., Ruys, L., Vermeersch, J., Remon, J.P. and Duncan, R. (1985) Use of polysaccharides as drug carriers, dextran and inulin derivatives of procainamide. In *Macromolecules as drug and as carriers for biologically active materials*, D. Tirrell, G. Donaruma and A. Turek, eds. (New York: The New York Academy of Science), pp. 199–212.

Schnabel, W. (1981) Polymer degradation: principles and practical applications (Munchen: Hanser International).

Shalaby, W. and Park, K. (1994) Chemical modification of proteins and polysaccharides and its effect on enzyme-catalyzed degradation. In *Biomedical Polymers*, S. Shalaby, ed. (Munich: Hanser Publishers), pp. 213–258.

Shefer, A., Shefer, S., Kost, J. and Langer, R. (1992) Structural characterization of starch networks in the solid state by cross-polarization magic-angle spinning 13C NMR spectroscopy and wide angle X-ray diffraction. *Macromolecules*, **25**, 6756–6760.

Soon-Shiong, P., Heintz, R., Merideth, N., Yao Qiang, X., Yao, Z., Zheng, T., Murphy, M., Moloney, M., Schmehl, M., Harris, M., Mendez, R., Mendez, R. and Sandford, P. (1994) Insulin independence in a Type I diabetic patient after encapsulated islet transplantation. *Lancet*, **343**.

Suzuki, K., Yoshimura, H., Marsuura, H., Katoh, T., Nakamure, T., Tsurutani, R. and Kifune, K. (1993) A new slow-releasing drug delivery system for chemically combined cisplatin with chitin for intraoperative local applications: an experimental study. In *Recent advances in diseases of the esophagus* (Tokyo: Sp[ringer-Verlag), pp. 865–870.

Trimnell, D., Shasha, B.S., Wing, R.E. and Otey, F.H. (1982) Pesticide encapsulation using starch-borate complex as wall material. *Journal of Applied Polymer Science*, **27**, 3919–3928.

Vandenbossche, G., Leffebvre, R., De Wilde, G. and Remon, J.P. (1992) Performance of a modified starch hydrophilic matrix for the sustained release of thephylline in healthy volunteers. *Journal of Pharmaceutical Sciences*, **81**, 245–248.

Visavarungroj, N., Herman, J. and Remon, J. P. (1990) Crosslinked starch as sustained release agent. *Drug Development and Industrial Pharmacy*, **16**, 1091–1108.

Watanabe, K., Saiki, I., Matsumoto, Y., Azuma, I., Seo, H., Okuyama, H., Uraki, Iura, Y. and Tokura, S. (1992) Antimetastatic activity of neicarzinostatin incorporated into controlled release gel of CM-chitin. *Carbohydrate Polymers*, **17**, 29–37.

Yalpani, M. (1988) Polysaccharides synthesis, modification and structure/property relations (Amsterdam: Elasevier).

Ziegler, J. (1995) Hyaluronic acid: a biomaterial with promise. *Biomaterials Forum*, **17**, 13.

15. OXIDIZED CELLULOSE: CHEMISTRY, PROCESSING AND MEDICAL APPLICATIONS

REGINALD L. STILWELL, MICHAEL G. MARKS[*], LOWELL SAFERSTEIN[#]
and DAVID M. WISEMAN[†]

Johnson & Johnson Medical, Inc., 2500 Arbrook Blvd., Arlington, Texas 76014, USA
[†] *Currently: Synechion Inc., 6757 Arapaho, Suite 711, Dallas, TX 75248, USA*
[#] *Currently: Consultant; 14 Currey Lane, West Orange, NJ 07052, USA*

OVERVIEW

Oxidation of cellulose produces an absorbable biomaterial. Eastman Kodak pioneered an industrial scale oxidation process using nitrogen dioxide gas vapor. Johnson & Johnson subsequently modified this process using nitrogen dioxide in a solvent. Maintaining control over the oxidation reaction imparts desirable physical properties and degradation characteristics to the oxidized cellulose. Applications of this material are varied and each exploits the oxidation chemistry.

INTRODUCTION

Cellulose is the most abundant biomass on the surface of the earth. For millennia, it has provided mankind with a functional, low cost and renewable raw material. As a biomaterial, cellulose can be converted into a wide range of derivatives with desired properties for a variety of biomedical applications. The three hydroxyl groups of the cellulose molecule can undergo chemical reactions common to all primary and secondary alcohol groups, such as esterification, nitration, etherification and oxidation. From these reactions, a variety of useful polymers can be created (Serad & Sanders, 1979; Bogan *et al.*, 1979; Greminger, 1979). Oxidation, however, is the only process that renders cellulose bioabsorbable in man. Complete oxidation of cellulose yields carbon dioxide and water. Partial oxidation of cellulose results in a biomaterial with controlled degradation characteristics that are used to advantage by the biomedical industry. Since the oxidation process is complex, it often results in non-homogenous materials. The focus of this chapter is the oxidation of cellulose.

The action of numerous oxidants on cellulose has been studied under widely varying conditions of temperature, pH, time of reaction and concentration. The resulting polymer (see Figure 1) contains reactive groups (such as aldehyde, ketone or carboxylic acid) in fixed positions. A major problem with oxidation is the difficulty of producing materials that are homogeneous in chemical and physical properties. This complication arises from:

- the different reactivities of the three hydroxyl groups,
- the dissimilar availabilities of different parts of the cellulose molecule and
- the distinct behavior of different oxidants.

[*]Michael G. Marks, Johnson & Johnson Medical, Inc., 2500 Arbrook Blvd., Arlington, TX 76014. Tel: (817) 784-5165; Fax: (817) 784-5462.

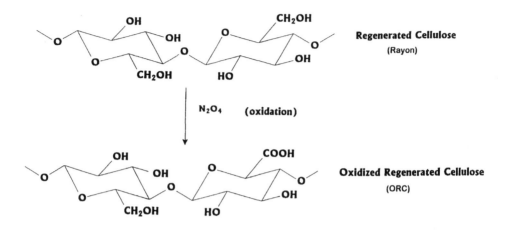

Figure 1 Chemical structure of oxidized cellulose.

The various types of oxidized cellulose therefore differ extensively in their properties. Several of the oxidants are not selective as to the particular hydroxyl groups they attack, so that distinct types of oxidation may occur simultaneously. Oxidation methods for bulk quantities of cellulose are typically topochemical; when the oxidation is mild, the products consist of an oxidized portion and an unchanged residue. More drastic conditions lead to increased degradation; some of the compounds no longer have an anhydroglucose ring. Therefore one of the major characteristics of the cellulose structure is lost and often results in materials that are friable and weak. By maintaining precise control over the oxidation reaction, bulk quantities of cellulose can be oxidized and create a biomaterial with specified properties for the intended application.

Today, oxidized cellulose materials are used in medical devices such as absorbable hemostatic agents and absorbable adhesion barriers. This chapter reviews the history of development, nitrogen dioxide oxidation, chemical characteristics, biodegradation, bioabsorption and applications of oxidized cellulose.

HISTORY OF DEVELOPMENT

Since the work of Witz describing the action of oxidizing agents on cellulose (1883), a vast amount of information has been reported (Nevell, 1983). Significant events in the development of oxidized cellulose are summarized below.

Cohen and Calvert (1897) prepared aldehydes from benzyl alcohol and its derivatives using nitrogen dioxide as an oxidizing agent. Shorygin and Khait (1937) prepared oxidized cellulose using nitrogen dioxide as an oxidizing agent, but the material was not well characterized. In the late 1930's at the Research Laboratories of Eastman Kodak, monocarboxycellulose was prepared to test whether the sodium salt was water soluble, similar to carboxymethyl cellulose. In the 1940's, fundamental research was conducted on the oxidation of cotton gauze with nitrogen dioxide

(Yackel & Kenyon, 1942; Unruh & Kenyon, 1942; Taylor *et al.*, 1947; McGee *et al.*, 1947a, b; Unruh *et al.*, 1947). Results indicated that the acidic oxidized cotton gauze material retained much of its original tensile strength and was insoluble in water but rapidly soluble in alkali to give clear, thin, low viscosity, light yellow solutions. The product was characterized as a copolymer of 35 mole percent anhydroglucose and 65 mole percent anhydroglucuronic acid displaying a 17 weight percent carboxylic acid content measured by titration.

Intrigued by the resulting gauze fabric, Kenyon quickly discovered that the oxidized cotton gauze also decomposed at the pH of blood. This raised the possibility that the fabric might be bioabsorbable. Implantation of the fabric in animals showed that it was bioabsorbable in 7 to 20 days with no anaphylactic reaction and minimal tissue irritation. The oxidized cotton gauze was then examined by Putnam (1943) as an absorbable carrier for thrombin, a blood clotting enzyme under development by Parke, Davis & Company. With slight modification, it was found to be efficacious as a carrier. Subsequent work by Frantz *et al.* (1944) showed that oxidized cellulose exhibited hemostatic properties.

Following clinical tests in 1945 that demonstrated oxidized cellulose was a worthwhile hemostatic agent, Eastman Kodak Company and Parke, Davis & Company began a cooperative arrangement in which the Eastman organization would manufacture oxidized cellulose and Parke, Davis & Company would process and market the product as a pharmaceutical material. Parke, Davis & Company began marketing Eastman's oxidized gauze and oxidized surgical cotton as Oxcel™ in 1946.

There is an inherent disadvantage when using cotton for pharmaceutical materials due to the lack of uniformity of chemical and physical properties of cotton. Since cotton fibers are tapered and thicker at one end than the other, a uniform and reproducible oxidation can not be carried out (Ashton & Moser, 1968). As an alternative, regenerated cellulose has been used as the starting material for oxidation. Regenerated cellulose is formed by subjecting cellulose to a dissolution and reconstitution process. A commonly used regeneration process is the viscose process. Cellulose, in the form of wood pulp, is mixed into a caustic solution of carbon disulfide. The slurry is then sprayed under pressure through nozzle jets. As the solution exits the jets, it is neutralized with acid. This produces a regenerated cellulose fiber, also called rayon, with a uniform and reproducible molecular size distribution.

Regenerated cellulose fibers are uniform in diameter. This permits a uniform oxidation and imparts uniform chemical and physical characteristics to the pharmaceutical material. In the early 1960's, Johnson & Johnson entered the market with an oxidized knitted rayon fabric, SURGICEL™ Absorbable Hemostat. Since then, Johnson & Johnson has developed a few other oxidized knitted rayon products. A list of currently available bioabsorbable oxidized cellulose and oxidized regenerated cellulose products, and the respective manufacturer is contained in Table 1. A list of relevant patents for oxidized cellulose technology is contained in Table 2.

NITROGEN DIOXIDE OXIDATION

Cellulose can be oxidized using metaperiodate, hypochlorite and dichromate. However, the only suitable method for preparing material with a high carboxyl

Table 1 Bioabsorbable Oxidized Cellulose Products

Product	Material	Manufacturer
SURGICEL™ Absorbable Hemostat	oxidized regenerated cellulose	Ethicon, Inc.
SURGICEL™ Nu-Knit Absorbable Hemostat	oxidized regenerated cellulose	Ethicon, Inc.
OXYCEL™ Absorbable Hemostat	oxidized cellulose	Becton-Dickinson, Inc.
INTERCEED™ Absorbable Adhesion Barrier	oxidized regenerated cellulose	Ethicon, Inc.

Table 2 Relevant Patents for Oxidized Cellulose Technology

Title	Assignee	Number	Year
Oxycellulose	Eastman Kodak	2,232,990	1941
Production of Acids by the Oxidation of Alcohols, Aldehydes or Ketones	Eastman Kodak	2,298,387	1943
Hemostatic Compositions	Ethicon	2,642,375	1953
Oxidized Cellulose Products and Methods for Preparing the Same	Johnson & Johnson	3,364,200	1968
Hemostatic Material	Johnson & Johnson	3,666,750	1972
Surgical Hemostat Comprising Oxidized Cellulose	Johnson & Johnson	4,626,253	1986
Heparin-Containing Adhesion Prevention Barrier and Process	Johnson & Johnson	4,840,626	1989
Method and Material for Prevention of Surgical Adhesions	Johnson & Johnson	5,002,551	1991
Method and Material for Prevention of Surgical Adhesions	Johnson & Johnson	5,007,916	1991
Process for Preparing a Neutralized Oxidized Cellulose Product and its Method of Use	Johnson & Johnson	5,134,229	1992
Cellulose Oxidation by a Perfluorinated Hydrocarbon Solution of Nitrogen Dioxide	Johnson & Johnson	5,180,398	1993
Calcium-Modified Oxidized Cellulose Hemostat	Johnson & Johnson	5,484,913	1996

content that retains good physical properties is oxidation with nitrogen dioxide. Both gas (Kenyon, 1949) and solution phase nitrogen dioxide oxidation processes (Ashton & Moser, 1968) have been commercialized in the United States.

Nitrogen dioxide dimerizes almost instantly to an equilibrium mixture with nitrogen tetroxide:

$$2NO_2 \rightleftarrows N_2O_4 \qquad \Delta H = -57.3 \text{ kJ/mole}$$

The mixture is a pungent brownish liquid that boils at $21°C$. Lower temperatures and increasing pressure shift the reaction to the production of nitrogen tetroxide. Nitrogen dioxide is a brown paramagnetic species while the dimer is colorless and diamagnetic; at $21°C$ the equilibrium mixture contains 0.08% NO_2.

Eastman Kodak Process – Gas Phase Oxidation

Investigators at Eastman Kodak produced equivalent oxidized cellulose fabrics using either nitrogen dioxide vapor (Yackel & Kenyon, 1942; Unruh & Kenyon, 1942) or a solution of nitrogen dioxide in carbon tetrachloride (McGee *et al.*, 1947b). They utilized the vapor process; assuming it was a lower cost method and allowed for easier removal of gaseous nitrogen oxide residues. Disadvantages of this procedure included the more complicated equipment needed to circulate the gas, to remove the heat of reaction and to wash the oxidized fabric with water after completion of the reaction.

The Eastman Kodak process (Kenyon, 1949) for making oxidized cellulose began with a 36 inch wide roll of surgical gauze plied into 300 layer pads and loaded onto a 20 foot long support frame. The frame was placed into a 40 inch diameter stainless steel tubular chamber. After the oxidizer was sealed, nitrogen dioxide was admitted to the reactor manifold from an evaporator. The gas was circulated through the layers of gauze by a high velocity stainless steel blower capable of moving 2500 cubic feet of nitrogen dioxide gas per minute against a 14 inch head of water pressure. A temperature of 25°C was maintained by a heat exchanger for 16–18 hours. During the reaction the percentage of nitric oxide was checked at 30 minute intervals and controlled by regulating the rate of return of condensate to the evaporator and then to the circulatory system. After completion of the oxidation, the reactor was evacuated and then vented for 30 minutes with filtered air. The gauze frame was removed from the reactor and placed in a washer where it soaked in several changes of deionized water to remove acidic by-products. The oxidized gauze was dried in a forced air dryer for 8 hours.

Johnson & Johnson Process – Liquid Phase Oxidation

In the early 1960's, Johnson & Johnson became the second US manufacturer of oxidized cellulose. Johnson & Johnson made several changes in the manufacturing process, replacing woven cotton gauze with knitted rayon fabric and employing a solvent based oxidation procedure (Ashton & Moser, 1968). In the Johnson & Johnson process, knitted rayon fabric was wrapped around a perforated mandrel and placed in a glass lined vertical reaction vessel. A solvent, inert to nitrogen dioxide, was pumped into the reactor and circulated from the bottom of the mandrel up and through the plies of rayon cloth. From an external gas cylinder, nitrogen dioxide was introduced into the circulating solvent to produce a 20% solution that was continuously pumped through the cloth during the reaction. The resulting fabric was uniformly oxidized to an 18–22 wt% carboxyl content.

An initial exotherm from the adsorption of the nitrogen dioxide onto the rayon was controlled by a water jacket that maintained the reaction temperature at 25°C. Chilled condensers at the top of the reactor returned solvent and nitrogen tetroxide to the vessel. When the oxidation was complete, the spent solution was pumped to a caustic tank to neutralize any unreacted nitrogen oxides. The oxidized cloth was then washed with fresh solvent and alcohol. Drying of the fabric was done in a separate forced air tunnel.

Mechanism of Oxidation

The main reaction path of cellulose with nitrogen dioxide gas or solution is the oxidation of the C-6 primary hydroxyl group of the anhydroglucose unit to carboxylic acid. Several schemes proposed to describe this reaction are based on the hypothesis that nitrites of cellulose form as intermediates and subsequently take part in the oxidation process.

Moisture on the cloth is necessary to achieve oxidation. Between 5.5 to 11% w/w moisture is needed for nitrogen tetroxide to adsorb on cellulose. With low moisture content, only nitration takes place while with high moisture content nitric acid is formed, but nitration predominates. The log of the amount of adsorbed nitrogen tetroxide is proportional to the reciprocal of the temperature (Yasnitskii *et al.*, 1972) and the oxidation of cellulose is proportional to the amount of previously adsorbed nitrogen tetroxide (Yasnitskii *et al.*, 1971).

At 22.5°C the adsorption of nitrogen tetroxide by cellulose rises rapidly to a maximum in 10 minutes, then declines quickly to a level where it remains constant (Yasnitski *et al.*, 1972). The formation of carboxylic acid groups steadily increases. Physical adsorption of nitrogen tetroxide is followed by chemisorption which is accompanied by heat evolution that reduces the adsorbent capacity of cellulose. Oxidation of the primary hydroxyl groups to carboxylic acid groups occurs as a homogeneous solid phase reaction, which takes place inside the adsorbed complex without participation of the gaseous phase. Yasnitskii *et al.* (1971) suggests that in the gas phase oxidation of cellulose with nitrogen dioxide, 1 to 1.2 moles of nitrogen tetroxide are adsorbed per mole of anhydroglucose and of this 0.8 moles are chemisorbed.

Chemisorption is the result of strong binding forces comparable with those leading to the formation of chemical compounds. In contrast to physical adsorption, chemisorption may be regarded as the formation of a type of surface compound that is seldom reversible. The energies of adsorption range from about 10^4 to 10^5 Joules per mole. The chemisorbed oxides of nitrogen dioxide are the precursors to the formation of nitrites; the principal intermediates of alcohols reacting with nitrogen dioxide (Fairlie *et al.*, 1953). The physically adsorbed nitrogen tetroxide converts the nitrite intermediates to carboxyl groups and forms oxidized cellulose. The concentration of nitrogen tetroxide in solution has the same effect as pressure in gas phase reactions; high concentrations of nitrogen tetroxide increase the amount of physically adsorbed nitrogen tetroxide but not chemically adsorbed. Therefore, the higher the concentration of nitrogen tetroxide in solution the more extensive the oxidation.

Kaputskii *et al.* (1973) have proposed that nitrogen oxides are both chemically and physically adsorbed on cellulose. The amount adsorbed increases with the moisture content of the cellulose until the moisture content reaches 18 mg/mmole or 1 mole water per cellulose repeat unit. At this point the total adsorption of nitrogen tetroxide reaches 1 mole per monomer of cellulose and oxidation proceeds as a heterogeneous reaction in which there is an exchange with the gaseous phase. One of the gaseous products is nitric oxide.

Kaputskii *et al.* (1975) demonstrated that the amount of bound nitrogen versus time it reaches a maximum and then declines, while the carboxyl group content in the treated cellulose continues to rise. This indicates the nitrite intermediates are not directly converted into carboxyl groups. Svetlov *et al.* (1974) also observed an increase in the nitrite concentration only at the initial moment of the reaction. The adsorption of nitrogen dioxide does not follow a simple kinetic law. Its disappearance is mirrored by the appearance of other gases such as nitrogen, nitric oxide, nitrous oxide, and carbon dioxide; nitric oxide and nitrogen are the principal reduced products of the reaction.

Investigations of the oxidation of cellulose by nitrogen tetroxide in various solvents show that the degree of dissociation of nitrogen tetroxide into nitrogen dioxide increases in non-polar solvents. This leads to an increase in the degree of oxidation (Pavyluechenko *et al.*, 1975). The hypothesis that nitrogen dioxide is the active oxidant is supported by other studies (Nevell, 1951a, b).

Spectroscopic analyses of the intensity of the carbonyl absorption at 1730 cm^{-1}, appearing during the nitrogen dioxide oxidation of cellulose paper and cellulose sulfate in carbon tetrachloride, show that this band increases with the reaction time. New bands appear at 1410 and 1610 cm^{-1} that pass through a maximum after four hours, as does the chemically bound nitrogen. The authors suggest that the bands are due to nitrites or hydroxyamic acids (Gusev *et al.*, 1976). Other investigators have shown that primary alcohols treated with nitrogen tetroxide in a mole ratio of one to one form only organic nitrites and nitric acid. Pure nitrites in the presence of one mole of nitrogen tetroxide do not oxidize further unless there are trace amounts of water or nitric acid present (Langenbeck & Richter, 1956). Langenbeck and Richter speculate that alkyl nitrites are converted with nitrous acid into nitrites of hydroxamic acid, which further react to form acyl nitrites and in the presence of moisture are converted to carboxylic acids.

When cellulose is oxidized with nitrogen dioxide in either a gas or solution phase process, nitrogen dioxide appears to be rapidly adsorbed both chemically and physically at certain preferred sites provided there is between 5–10% w/w moisture present on the cellulose. This adsorption is responsible for the initial exotherm during oxidation. The nitrogen content increases quickly as nitrites form, while the carboxylic acid content increases slowly. The primary hydroxyl nitrites of cellulose react with additional molecules of nitrogen oxides to form oximes or hydroxamic acid groups, which under the effect of concentrated acids present in the reaction mixture, hydrolyze to form monocarboxy cellulose. The presence of carbon dioxide as a by-product arises from decarboxylation side reactions. Nitrogen observed in the off-gases originates from the reaction of hydroxylamine with nitrous acid:

$$2NH_2OH + 2HNO_2 \rightleftarrows 4H_2O + 2NO + N_2$$

The hydroxylamine is produced during the hydrolysis of the hydroxamic acid. The overall stoichiometry of this proposed reaction is:

$$RCH_2OH + 2N_2O_4 \rightleftarrows RCOOH + 2HNO_3 + NO + \tfrac{1}{2}N_2$$

In a closed system the pressure increase follows first order kinetics and mirrors the formation of carboxylic acid content (see Figure 2). When the oxidation reaction is conducted under oxygen pressure in a closed system, the nitric oxide is oxidized back to nitrogen dioxide. Oxygen becomes the principal oxidant with the nitrogen dioxide acting only as a catalyst. However, some crosslinking of the oxidized cellulose

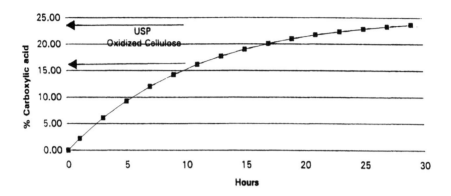

Figure 2 First order kinetics of weight % carboxylic acid versus time.

occurs. The resulting product is not completely soluble in caustic which predicts poor bioabsorption.

Cellulose cannot be oxidized in ether-like solvents such as dioxane, diphenyl ether or methyl tertiary butyl ether due to binding of nitrogen tetroxide by the solvent molecules and solvation of the reaction centers of cellulose by the solvent (Pavyluechenko *et al.*, 1975). The rate of the oxidation reaction in inert solvents such as Freon or fluorinated hydrocarbons is controlled in part by the partitioning of the oxidant between the solvent and the cellulose. In Freon 113, an excellent solvent for nitrogen dioxide (Ashton & Moser, 1968), the rate of oxidation is slow compared to fluorinated hexane, a poor solvent for the oxidant (Boardman *et al.*, 1993). Reaction rates are also controlled by the polarity of the solvent which affects the dissociation of the nitrogen tetroxide; dissociation and rate increase in non-polar solvents.

Cellulose is soluble or partially soluble in mixtures of nitrogen dioxide with certain organic solvents (Shorygin & Khait, 1937). High mole ratios of nitrogen dioxide to cellulose during an oxidation reaction can partially swell the cellulose fibrils. Upon removal of the oxidant, the fibrils fuse together producing stiff oxidized fabric.

Analysis of crystalline and amorphous cellulose samples treated with nitrogen dioxide by X-ray diffraction and deuteration indicates that oxidation initially proceedes rapidly in the amorphous regions and then more slowly in the crystalline regions (Takahashi & Takahashi, 1973). Rayon fabrics made by the viscose process are more readily oxidized with nitrogen dioxide than fabrics made from Tencel™, Courtaulds' new solvent spun cellulose fiber with a higher crystallinity than rayon. Under identical oxidation conditions rayon exhibits a carboxylic acid content of 18 wt% and is soluble in caustic; oxidized Tencel exhibits a 14 wt% carboxyl content and its caustic solutions are hazy with particulates. Microcrystalline cellulose on the other hand oxidizes readily due to the high surface area of the powder, but the oxidized product loses the unique aqueous suspensoid forming characteristics of microcrystalline cellulose.

CHEMICAL CHARACTERISTICS

Eastman Kodak Company and Parke, Davis & Company agreed to a collaborative effort in 1945 to produce and market oxidized cellulose as a pharmaceutical material. Specifications to standardize and describe the material were based on the physical, chemical and bioabsoprtion properties of the oxidized fabric. The minimum carboxylic acid content was established as no less than 16 wt% and the fabric could contain no more than 0.5% nitrogen. Specifications were based on the observation that all materials meeting this specification behaved similarly.

Today USP oxidized cellulose is characterized by no less than 16 and no more than 24 weight percent carboxylic acid corresponding to a 60 to 93 mole % conversion of the anhydroglucose units to anhydroglucuronic acid (United States Pharmacopeia, 1995). The polymer chain appears to be randomly composed primarily of uronic acid and anhydroglucose units, but also units containing other carbonyl groups. Oxidized cellulose fabrics with less than 5 weight percent carboxylic acid content or 18.3 mole percent conversion are not readily bioabsorbable (unpublished). The complete oxidation of cellulose with nitrogen dioxide leads to polyglucuronic acid with a 25.5 weight percent carboxylic acid content.

Structural Considerations

There is conjecture about the existence and identity of a moiety secondary to the C-6 carboxylic acid group due to the disparity between carboxyl titration and carbon dioxide generated from oxidized cellulose. While carboxyl titration measures the carboxylic acid content of the oxidized cellulose, the carbon dioxide generated reflects the total carbonyl moieties. The carbon dioxide method results in a higher carbonyl content than is reflected in the carboxyl titration method (Unruh *et al.*, 1947).

At Johnson & Johnson, the structure of oxidized cellulose was investigated by several techniques. Chemically, a differential titration was developed to measure the carboxylic acid content and any saponifiable secondary functions. The titration measures carboxylic protons by ion exchange. By treating the resulting ion exchanged solution with a measured excess of base, the saponifiable function is measured by acidimetric back-titration. Using the differential titration, a previously measured 20 wt% carboxylic acid content is revealed to be 18 wt% carboxylic acid and 2 wt% saponifiable moiety (unpublished).

Fourier Transform Infrared (FTIR) spectroscopy shows a strong absorption at 1735 cm^{-1}, a spectrum typical for cellulosic acid. When the sodium salt of oxidized cellulose is isolated and measured, a strong carboxylate band is seen at 1594 cm^{-1}. In addition to the carboxylate band a small band (approximately 2% intensity) is uncovered at 1737 cm^{-1}. This absorption may correspond to a strained ester or lactone moiety (unpublshed; Zhbankov RG, 1991).

When oxidized cellulose is studied in solid state with ^{13}C NMR (^{13}C CPMAS NMR), a typical carboxylic acid absorption is seen at 175ppm. When the sodium salt of oxidized cellulose is used, the carboxylate carbonyl shifts slightly and a small peak is revealed at 178 ppm. The secondary carbonyl peak is about 2% of the intensity of the carboxylate peak. The chemical shift of the smaller peak suggests an ester-like moiety (unpublished). It is noteworthy that in ^{13}C CPMAS NMR there are no absorptions seen in the ketone region of the spectrum for oxidized cellulose.

Thermal desorption of adsorbed water was studied using thermogravimetric analysis and FTIR (TGA/FTIR) to measure the loss of water from oxidized cellulose. When monitored over a temperature range, FTIR detected a gradual desorption of water. Abruptly around 95–110°C, a large decrease in oxidized cellulose weight was noted by TGA. FTIR revealed that carbon dioxide gas was given off as the heating continued. From TGA/FTIR it appeared that oxidized cellulose undergoes thermal decarboxylation at about 100°C (unpublished). Thermal decarboxylation of carbonyl compounds is a well known reaction but it is unusual to see decomposition of an organic acid to carbon dioxide at temperatures less than 100°C.

BIODEGRADATION

Biodegradation is defined as the ability of a material to degrade into nontoxic products. The carboxylic acid solubilizes the degraded fractions as water soluble salts of the acid. The degradation appears to be controlled by the carbonyl content arising from the oxidation of the hydroxyl groups at C-2 or C-3 to ketones, or other carbonyl moieties. The ketones introduce an alkali labile linkage as these groups on enolization yield enediols, rendering the adjacent glucosidic groups sensitive to alkali β-elimination (Kiss, 1974; Neukom & Deuel, 1958). Once degradation is initiated, it continues along the polymer chain. As degradation fragments are generated, they are further degraded due to the formation of new carbonyl groups. Oxidized cellulose dissolves within a few minutes in 0.1N sodium hydroxide to produce a clear, light yellow, low viscosity solution. The yellow color arises from the chromophoric vicinal dicarbonyl moiety formed from the β-elimination reaction of the segment containing the hydroxy carbonyl structure.

The sensitivity of oxidized cellulose to alkali was observed in the last century when cotton oxidized with acid dichromate solutions became tenderized and lost much of its original tensile strength (Jeanmaire, 1873). The importance of the carbonyl group in controlling the alkali stability of oxidized cellulose was convincingly demonstrated when periodate oxidized cellulose exhibiting high alkali sensitivity was converted to alkali stable cellulose by reducing the aldehyde content (Meller, 1955). There seems to be a relationship between carbonyl content and alkali lability of oxidized cellulose. Alginic acid is an example of a polysaccharide with carboxylic acid groups in the same C-6 position as oxidized cellulose but lacks carbonyl groups at C-2 or C-3 and is perfectly stable in alkali. Pectins are naturally occurring polysaccharides that also have a carboxylic acid function at the C-6 position and like oxidized cellulose are unstable in alkali medium; however, in the pectin molecule approximately 75% of the carboxylic acid groups are present as the methyl ester. Esters unlike free acids or their salts, have a strongly polarized carbonyl oxygen bond. The carbon atom of the carbonyl ester is slightly electropositive and contributes to making the alpha hydrogen labile that leads to β-elimination reactions and rapid decomposition in alkali.

In Vitro Degradation of Oxidized Regenerated Cellulose

When oxidized cellulose is suspended in neutral pH water, no appreciable hydrolysis occurs. However, a drop in pH is observed. If the pH of the water is increased, the structure breaks down. To simulate physiologic conditions of constant fluid replacement and maintenance of neutral to slightly alkaline pH, Dimitrijevich and coworkers (1990a) dialyzed oxidized regenerated cellulose (ORC) against phosphate buffered saline (pH 7.4). In this simulation, about 90% of the ORC material by weight was solubilized within 21 days and recovered as the white, powdery sodium salt of polyglucuronic acid. In the same series of studies, Dimitrijevich and coworkers also evaluated the breakdown products of ORC following solubilization in serum and found a higher proportion of degradation products with a low degree of polymerization compared with phosphate buffered saline incubation. These data suggest that biodegradation occurs via both chemical and enzymatic processes.

In Vivo Degradation of Oxidized Cellulose

Oxidized regenerated cellulose is one of the fastest resorbing polymers known. When implanted *in vivo*, the material imbibes surrounding fluid, loses much of its original tensile strength, turns a yellowish brown and becomes gelatinous within eight hours. The gelatinous material continues to erode as fragments dissolve into the surrounding fluids. Within a few days most of the fibrous structure is lost and by two weeks 80% of the original material has absorbed. *In vivo* studies by Dimitrijevich (1990b) showed that ORC implanted in the peritoneal cavity of rabbits produced oligomeric carbohydrate breakdown products present in the peritoneal fluid but absent in serum or urine. The size distribution and the quantity of the oligomers decreased after the first day and by day four were essentially gone. The *in vivo* degradation products were the same as previously identified from *in vitro* studies (Dimitrijevich, 1990a).

Among the many oligomeric carbohydrate breakdown products observed in the peritoneal fluid were glucose and glucuronic acid which are metabolized via well-established pathways. Since the degradation products did not accumulate in either the blood or urine, a local intraperitoneal clearance mechanism of oxidized cellulose was suggested. This proposed mechanism is consistent with an initial chemical degradation followed by macrophage digestion of the oligosaccharides carried out with hydrolytic enzymes including β-D-glucosidase and β-D- glucuronidase present in peritoneal macrophages. When the lead salt of oxidized cellulose was implanted in the peritoneal cavity of rabbits, macrophages were identified containing lead.

BIOABSORPTION

Bioabsoprtion is defined as the ability to metabolize and/or eliminate nontoxic products from the body. Oxidation of cellulose to a carboxylic acid content above approximately 10 wt% renders cellulose bioabsorbable. Above this level, the rate of bioabsoption seems to increase with the degree of oxidation, although the relationship is not linear. Within the range of 12–24 wt% carboxylic acid content, all oxidized cellulose materials appear to absorb within a period of 1 month.

Implantation of a bioabsorbable material alters the microenvironment of the cells surrounding the implant and the tissue in which it is implanted. The extent of the tissue response depends upon the chemical and physical nature of the material. A material is considered biocompatible if the tissue response it evokes does not compromise either its function or the function of the organ in which it is implanted.

The rate of absorption and tissue response are complicated by variables including the size of the implant, configuration of the material at the time of implantation, the degree of surgical trauma, amount of blood present, and presence of infection. Macroscopically, oxidized cellulose breaks down in 3 to 10 days. This early cellular reaction is characterized by the infiltration of histiocytes, fibroblasts and foamy macrophages. The reaction eventually resolves, and within about 1 month oxidized cellulose is completely resorbed with no traces of a foreign body reaction remaining (Pierce *et al.*, 1990; Diamond *et al.*, 1990). Tissue responses with oxidized cellulose are slightly more pronounced than with oxidized regenerated cellulose (Hurwitt *et al.*, 1960; unpublished).

APPLICATIONS

Early work by Frantz (1943) suggested clinical applications for oxidized cellulose. Frantz stated:

> *"A relatively nonirritating foreign material, oxidized cellulose . . . may be employed in hemostasis and possibly in protecting injured surfaces where a smooth membrane is desired in the final healing."*

Since that time, topical absorbable hemostatic agents (marketed as brand names of OXYCEL™, SURGICEL™, and SURGICEL Nu-Knit™) and an absorbable adhesion barrier (marketed as brand name INTERCEED™) have been approved for use in the United States by the Food & Drug Administration. Along with these two well-established clinical applications, oxidized cellulose has been used in other applications.

Topical Absorbable Hemostatic Agent

The pH of oxidized cellulose varies from about 2 to 4. When exposed to blood, oxidized cellulose turns to a dark brown or black color. This results from degradation of red blood cells with subsequent formation of acid hematin. The hemostatic action of oxidized cellulose is in part due to styptic action from the low pH of the material. However, the effect produced is slight, local, and transient since the carboxyl groups of the oxidized cellulose are rapidly neutralized by buffer systems in blood (Miller *et al.*, 1961). The hemostatic action of oxidized cellulose is primarily due to a combination of local mechanical action and chemical activity (Frantz *et al.*, 1944; Frantz & Lattes, 1945; Miller *et al.*, 1961) that results in the formation of an "artificial clot". Oxidized cellulose, along with the damage to the bleeding tissue surface, activates platelets. The material swells as it becomes hydrated with blood components and gels with increasing biodegradation; some of the damaged blood vessels are mechanically "sealed off" in the process, and an "artificial clot" is formed chemically using oxidized cellulose as a scaffold. SURGICEL™ and SURGICEL™ Nu-Knit Hemostats, for example, are manufactured by knitting rayon fibers into a mat. When exposed to fluids, the oxidized cellulose fibers swell and the pores narrow to form a gelatinous sheet (Diamond *et al.*, 1990).

Oxidized cellulose has been used as a topical absorbable hemostatic agent since the 1940s. During these early years of use, oxidized cellulose was used in a variety of surgical procedures including neurosurgery, orthopedic surgery, gynecologic surgery, anorectal surgery, abdominal surgery, cardiovascular surgery, thoracic surgery, head and neck surgery, pelvic surgery, and skin and subcutaneous tissue procedures (Frantz & Lattes, 1945; Frantz, 1946; Scarff *et al.*, 1949; Lebendiger *et al.*, 1959; Hanna *et al.*, 1963; Degenshein *et al.*, 1963; Venn, 1965). Today, oxidized cellulose is available in a variety of preparations including pads and pledgets (OXYCEL™) and knitted fabric strips (SURGICEL™ and SURGICEL™ Nu-Knit). Oxidized cellulose continues to be utilized topically in a variety of clinical situations with the greatest use being to control oozing from broad surfaces.

By utilizing the carboxyl groups available in the oxidized cellulose polymer chain, ion exchanges can be made with calcium. Since calcium is an important ion in the blood coagulation cascade, calcium modified oxidized cellulose is more efficient in promoting hemostasis (Stilwell *et al.*1996).

Absorbable Adhesion Barrier

Oxidized regenerated cellulose has been used clinically as an absorbable adhesion barrier since 1989. Available as a knitted fabric (INTERCEED™ Absorbable Adhesion Barrier), it is the first material in the United States indicated for the prevention of post-operative pelvic adhesions. Since its introduction, additional clinical studies have been completed (Azziz *et al.*, 1993; Sekiba *et al.*, 1992; Franklin *et al.*, 1995; Li & Cooke 1994; Nordic Study Group, 1995) that confirm the efficacy of oxidized regenerated cellulose as an adhesion barrier.

There are two main advantages to using oxidized regenerated cellulose (ORC) for this application. Firstly, since ORC is biocompatible, peritoneal cells beneath the adhesion barrier can repair damaged membranous surfaces while being separated from apposing tissues. Secondly, since ORC is bioabsorbable, the material serves its function as an adhesion barrier and is removed from the implantation site within 4 weeks so that no further foreign-body reactions can occur.

The efficacy of ORC as an adhesion barrier is compromised in the presence of excessive quantities of blood. To reduce the sensitivity of ORC to blood, the number of carboxyl groups available to react with the blood components can be reduced by creating a sodium salt of ORC. This also raises the pH of the material from 2–4 to 5–6. Animal studies have been conducted (Wiseman *et al.*, 1993) that illustrate improved efficacy in the presence of excessive blood.

Drug Delivery Matrix

Due to the availability of many carboxyl groups, oxidized cellulose has a low pH. This accounts for the antibacterial activity produced by oxidized regenerated cellulose

Table 3 Chemical Moieties Bound to Oxidized Cellulose for Evaluation of Drug Delivery

Chemical Moiety	Application	Reference
Heparin	Adhesion prevention	Reid *et al.*, 1993
Tissue Plasminogen Activator	Adhesion prevention	Wiseman *et al.*, 1992
Sodium	Adhesion prevention	US Patent 5,134,229 1992
Calcium	Hemostasis	US Patent 5,484,913 1996

against a broad range of pathogens (Dineen, 1976; Dineen, 1977; Scher & Coil, 1982). However, these carboxyl groups readily undergo ion exchange and remove cations from salt solutions passed over the polymer. The partial sodium salt of oxidized cellulose is less acidic and shows compatibility with acid sensitive biologics such as thrombin.When considering delivery of drugs, chemical moieties can easily be bound to the oxidized cellulose backbone. Binding can occur through absorption, adsorption, ionic bonds, or peptide bonds. Table 3 shows various moieties that have successfully been bound to oxidized cellulose.

SUMMARY

Cellulose is rendered bioabsorbable using an oxidation process. Oxidizing agents vary but the most widely used is nitrogen dioxide, in a gas or liquid phase. Nitrogen dioxide is the only suitable oxidizing agent for preparing material with a high carboxyl content that retains good physical properties. The high carboxyl content imparts desirable degradation characteristics to oxidized cellulose. By exploiting the chemistry, oxidized cellulose can be used as a hemostasis agent, an absorbable adhesion barrier and as a vehicle for drug delivery.

REFERENCES

Ashton, W.H. and Moser, C.E. (1968) Oxidized Cellulose Products and Methods for Preparing the Same. US Patent 3,364,200.

Azziz, R., and the INTERCEED(TC7) Adhesion Barrier Study Group II (1993) Microsurgery alone or with Interceed absorbable adhesion barrier for pelvic sidewall adhesion re-formation. *Surg Gynecol Obstet*, **177**, 135–9.

Boardman, F. and Saferstein, L. (1993) Cellulose oxidation by a perfluorinated hydrocarbon solution of nitrogen dioxide. US Patent 5,180,398.

Bogan, R.T., Kuo, C.M. and Brewer, R.J. (1979) Cellulose derivatives, ethers. In: *Kirk-Othmer, Encyclopedia of Chemical Technology*, M. Grayson (Ed.), Vol. 5, 3rd ed., 118–43, New York: Wiley-Interscience.

Chiang, L., Upasani, R., Swirczewski, J. and Soled, S. (1993) Evidence of hemiketals incorporated into the structure of fullerols derived from aqueous acid chemistry. *J. Am. Chem. Soc.*, **115**, 5453–57.

Cohen, J.C. and Calvert, H.J. (1897) *J. Chem. Soc.*, **71**, 1050.

Degenshein, G.A., Hurwitz, A. and Ribacoff, S. (1963) Experience with regenerated oxidized cellulose. *New York State J. Med.*, **Sept**, 2639–43.

Diamond, M.P., Cunningham, T., Linsky, C.B., Kamp, L., McConnell, R.F. and Gracy, R.W. (1990) Interceed™ (TC7) as an adjuvant for adhesion: animal studies. In: *Treatment of Post Surgical Adhesions*, G.S. diZerega, L.R. Malinak, M.P. Diamond, C.B. Linsky (Eds.) *Prog. Clin. Biol. Res.*, **358**, 131–43, New York: Wiley-Liss.

Dimitrijevich, S.D., Tatarko, M. and Gracy, R.W. (1990a) Biodegradation of oxidized regenerated cellulose. *Carbohydr. Res.*, **195**, 247–56.

Dimitrijevich, S.D., Tatarko, M., Gracy, R.W., Wise, G.E., Oakford, L.X. (1990b) *In vivo* degradation of oxidized, regenerated cellulose. *Carbohydr. Res.*, **198**, 331–41.

Dineen, P. (1976) Antibacterial activity of oxidized regenerated cellulose. *Surg. Gynecol. Obstet.*, **142**, 481–6.

Dineen, P. (1977) The effect of oxidized regenerated cellulose on experimental intravascular infection. *Surg.*, **82**, 576–9.

Franklin, R.R., and the Ovarian Adhesion Study Group (1995) Reduction of ovarian adhesions by the use of Interceed. *Obstet. Gynecol.*, **86**, 335–40.

Frantz, V.K. (1943) Absorbable cotton, paper and gauze (oxidized cellulose). *Ann. Surg.*, **118**, 116– 26.

Frantz, V.K., Clarke, H.J. and Lattes, R. (1944) Hemostasis with an absorbable gauze (oxidized cellulose). *Ann. Surg.*, **120**, 181–98.

Frantz, V.K. and Lattes, R. (1945) Oxidized cellulose-absorbable gauze (cellulosic acid). *J. Am. Med. Assoc.*, **129**, 798–801.

Frantz, V.K. (1946) New absorbable hemostatic agents. *Bull NY Acad. Med.*, **22**, 102–10.

Fairlie, A.M., Carberry, J.J. and Treacy, J.C. (1953) A study of the kinetics of the reaction between nitrogen dioxide and alcohols. *J. Am. Chem. Soc.*, **75**, 3786–89.

Greminger, G.K. Jr. (1979) Cellulose derivatives, ethers. In: *Kirk-Othmer, Encyclopedia of Chemical Technology*, M. Grayson (Ed.), Vol. 5, 3rd ed., 144–163, New York: Wiley- Interscience.

Gusev, S.S., Grinshpan, D.D. and Kaputskii, F.N. (1976) Infrared spectra of products of cellulose oxidation by nitrogen synthesis. *Zh. Prikl. Spektrosk.*, **24**, 716–19.

Hanna, C.B., Hastings, W.D. and Hague, E.E. Jr. (1963) Experience with the surgical use of a new absorbable hemostatic agent (Surgicel absorbable hemostat). *J. South Carolina Med. Assoc.*, **59**, 377–80.

Hurwitt, E.S., Henderson, J., Lord, G.H., Gitlitz, G.F. and Lebendiger, A. (1960) A new surgical absorbable hemostatic agent. Experimental and clinical evaluation. *Am. J. Surg.*, **100**, 439–446.

Jeanmaire, P. (1873) *Soc. Ind. Mulhouse Bull.*, **43**, 334.

Kaputskii, F.N. and Bashmakov, I.A. (1973) Role of the degree of oxidant sorption in the nitrogen oxide cellulose system. *Vestsi Akkad Navuk BSSR Ser Khim*, **4**, 29–32.

Kaputskii, F.N., Bashmakov, I.A., Balabaeva, M.D. and Tikavaya (1975) Nature of nitrogen in monocarboxylcellulose. *Vestsi Akkad Navuk BSSR Ser Khim Navuk*, **3**, 115–18.

Kenyon, R.L. (1949) Oxidation of cellulose. *Ind. Eng. Chem.*, **41**, 2–8.

Kiss, J. (1974) b-eliminative degredation of carbohydrates containing uronic acid residues. In: *Advances in Carbohydrate Chemistry and Biochemistry*, R. Tipson, R. Stuart, D. Horton (Eds.), Vol. 29, 229–303, New York: Academic Press.

Langenbeck, W. and Richter, M. (1956) The mechanism of the Maurer oxidation. *Chem. Berichte*, **89**, 202–8.

Lebendiger, A., Gitlitz, G.F., Hurwitt, E.S., Lord, G.H. and Henderson, J. (1959) Laboratory and clinical evaluation of a new absorbable hemostatic material prepared from oxidized regenerated cellulose. *Surg. Forum*, **10**, 440–3.

Li, T.C. and Cooke, I.D. (1994) The value of an absorbable adhesion barrier, Interceed™, in the prevention of adhesion reformation following microsurgical adhesiolysis *Br. J. Obstet. Gynecol.*, **101**, 335–9.

Mauer, K. and Grefahl, G. (1942) *Ber. Dtsch. Chem. Ges.*, **75**, 1489.

McGee, P.A., Fowler, W.F. Jr. and Kenyon, W.O. (1947a) Investigation of the properties of cellulose oxidized by nitrogen dioxide. III. The reaction of the carboxyl groups of polyuronides with calcium acetate. *J. Am. Chem. Soc.*, **69**, 347–49.

McGee, P.A., Fowler, W.F. Jr., Taylor, E.W., Unruh, C.C. and Kenyon, W.O. (1947b) Investigation of the properties of cellulose oxidized by nitrogen dioxide. V. Study of mechanism of oxidation in presence of carbon tetrachloride. *J. Am. Chem. Soc.*, **69**, 355–61.

Meller, A. (1955) *Tappi*, **38**, 682.

Miller, J.M., Jackson, D.A. and Collier, C.S. (1961) An investigation of the chemical reactions of oxidized regenerated cellulose. *Exp. Med. Surg.*, **19**, 196–202.

Neukom, H. and Deuel, H. (1958) Alkaline degradation of pectin. *Chemistry and Industry*, **June**, 683.

Nevell, T.P. (1951a) Oxidation of cotton cellulose by nitrogen dioxide. *J. Textile Inst.*, **42**, 91–129.

Nevell, T.P. (1951b) Qualitative x-ray study of the oxidation of cotton cellulose by nitrogen dioxide. *J. Textile Inst.*, **42**, 130–40.

Nevell, T.P. (1983) Oxidation of cellulose. In: *Cellulose Chemistry and Its Applications*, TP Nevell and JH Zeronian (Eds.), 243–65, New York: Ellis Horwood Limited.

Nordic Adhesion Prevention Study Group (1995) The efficacy of Interceed(TC7)* for prevention of reformation of postoperative adhesions on ovaries, fallopian tubes, and fimbriae in microsurgical operations for fertility: a multicenter study. *Fertil. Steril.*, **63**, 709–14.

Pavlyuchenko, M.M., Kaputskii, F.N. and Grinshpan, D.D. (1975) Effect of organic solvent nature on the interaction of cellulose with nitrogen tetroxide. *Zh. Prikl. Khim.*, **48**, 1822–5.

Pierce, A., Wilson, D. and Wiebkin (1987) Surgicel™: macrophage processing of the fibrous component. *Int. J. Oral. Maxillofac. Surg.*, **16**, 338–45.

Putnam, T.J. (1943) The use of thrombin on soluble cellulose in neurosurgery. *Ann. Surg.*, **118**, 119–29.

Reid, R.L., Lie, K., Spence, J.E., Tulandi, T. and Yuzpe, A. (1993) Clinical evaluation of the efficacy of heparin-saturated Interceed for prevention of adhesion reformation in the pelvic sidewall of the human. In: *Gynecologic Surgery and Adhesion Prevention*, M.P. Diamond, G.S. diZerega, C.B. Linsky, R.L. Reid (Eds.) *Prog. Clin. Biol. Res.*, **361**, 261–64, New York: Wiley-Liss.

Scarff, J.E., Stookey, B. and Garcia, F. (1949) The use of dry oxidized cellulose as a primary hemostatic agent in neurosurgery. *J. Neurosurg.*, **6**, 304–6.

Scher, K.S. and Coil, J.A. Jr (1982) Effects of oxidized cellulose and microfibrillar collagen on infection. *Surg.*, **91**, 301–4.

Sekiba, K., and the Obstetrics and Gynecology Adhesion Prevention Committee (1992) Use of Interceed (TC7) absorbable adhesion barrier to reduce postoperative adhesion reformation in infertility and endometriosis surgery. *Obstet. Gynecol.*, **79**, 518–22.

Serad, G.A. and Sanders, J.A. (1979) Cellulose acetate and triacetate fibers. In: *Kirk-Othmer, Encyclopedia of Chemical Technology*, M. Grayson (Ed.), Vol. 5, 3rd ed., 89–117, New York: Wiley-Interscience.

Shorygin, P.P. and Khait, E.V. (1937) Nitration of cellulose with nitric acid and nitrogen peroxide. *Zh. Obsch. Khim.*, **7**, 188–92.

Stilwell, R.L., Whitmore, E.J. and Saferstein, L.G. (1996) Calcium-modified oxidized cellulose hemostat. US Patent 5,484,913.

Svetlov, B.S., Lur'e. B.A. and Kornilova, G.E. (1974) Kinetics of cellulose oxidation by nitrogen dioxide. *Tr. Mosk. Khim. Teknol. Inst.*, **83**, 41–7.

Takahashi, A. and Takahashi, S. (1973) Oxycellulose. II. Oxidation reaction of cellulose with nitrogen dioxide. *Sen'i Gakkaishi*, **29**, T280–T284.

Taylor, E.W., Fowler, W.F. Jr., McGee, P.A. and Kenyon, W.O. (1947) Investigation of the properties of cellulose oxidized by nitrogen dioxide. II. The evolution of carbon dioxide from uronic acids and polyuronides. *J. Am. Chem. Soc.*, **69**, 342–47.

United States Pharmacopeia (1995), vol. 23. "Oxidized Cellulose" p. 318.

Unruh, C.C. and Kenyon, W.O. (1942) Investigation of the properties of cellulose oxidized by nitrogen dioxide. I. *J. Am. Chem. Soc.*, **64**, 127–31.

Unruh, C.C., McGee, P.A., Fowler, W.F. Jr. and Kenyon, W.O. (1947) Investigation of the properties of cellulose oxidized by nitrogen dioxide. IV. Potentiometric titration of polyuronides. *J. Am. Chem. Soc.*, **69**, 349–54.

Venn, R.D. (1965) Reduction of postsurgical blood-replacement needs with Surgicel™ hemostasis. *Med. Times*, **93**, 1113–16.

Wiseman, D.M., Kamp, L., Linsky, C.B., Jochen, R.F., Pang, R.H.L., Scholz, P.M. (1992) Fibrinolytic drugs prevent pericardial adhesions in the rabbit. *J. Surg. Res.*, **53**, 362–8.

Wiseman, D.M., Kamp, L.F., Saferstein, L., Linsky, C.B., Gottlick, L.E. and Diamond, M.P. (1993) Improving the efficacy of Interceed barrier in the presence of blood using thrombin, heparin or a blood insensitive barrier, modified Interceed (nTC7). In: *Gynecologic Surgery and Adhesion Prevention*, M.P. Diamond, G.S. diZerega, C.B. Linsky, R.L. Reid, (Eds.) *Prog. Clin. Biol. Res.*, **381**, 205–12, New York: Wiley-Liss.

Witz, G. (1883) *Bull. Soc. Ind. Mulhouse*, **43**, 334.

Yackel, E.C. and Kenyon, W.O. (1942) Oxidation of cellulose by nitrogen dioxide. *J. Am. Chem. Soc.*, **64**, 121–7.

Yasnitskii, B.G., Dolberg, E.B. and Oridoroga, V.A. (1971) Oxidation of cellulose and its derivatives. I. Basic principles of the reaction of cellulose with nitrogen oxides. *Zh. Prikl. Khim.*, **44**, 1615–19.

Yasnitskii, B.G., Dolberg, E.B. and Oridoroga, V.A. (1972) Mechanism of polyanhydroglucuronic acid formation during the interaction of cellulose with nitrogen oxides. *Ukr. Khim. Zh.*, **38**, 76–80.

Zhbankov, R.G. (1991) *Infrared Spectra of Cellulose and its Derivatives*. UMI Out of Print Books On Demand: Ann Arbor, MI. pp. 119–122.

16. GELATIN

Kind & Knox Gelatine, Inc.

Gelatin is a protein obtained by partial hydrolysis of collagen, the chief protein component in skin, bones, hides, and white connective tissues of the animal body. Type A gelatin is produced by acid processing of collagenous raw material; type B is produced by alkaline or lime processing. Because it is obtained from collagen by a controlled partial hydrolysis and does not exist in nature, gelatin is classified as a derived protein. Animal glue and gelatin hydrolysate, sometimes referred to as liquid protein, are products obtained by a more complete hydrolysis of collagen and thus can be considered as containing lower molecular-weight fractions of gelatin.

Use of animal glues was first recorded ca 4000 BC in ancient Egypt (Koepff, 1985). Throughout subsequent centuries, glue and crude gelatin extracts with poor organoleptic properties were prepared by boiling bone and hide pieces and allowing the solution to cool and gel. Late in the seventeenth century, the first commercial gelatin manufacturing began. At the beginning of the nineteenth century, commercial production methods gradually were improved to achieve the manufacture of high molecular weight collagen extracts with good quality that form characteristic gelatin gels (Koepff, 1985; Bogue, 1922; Smith, 1929).

Uses of gelatin are based on its combination of properties; reversible gel-to-sol transition of aqueous solution; viscosity of warm aqueous solutions; ability to act as a protective colloid; water permeability; and insolubility in cold water, but complete solubility in hot water. It is also nutritious. These properties are utilized in the food, pharmaceutical, and photographic industries. In addition, gelatin forms strong, uniform, clear, moderately flexible coatings which readily swell and absorb water and are ideal for the manufacture of photographic films and pharmaceutical capsules.

CHEMICAL COMPOSITION AND STRUCTURE

Gelatin is not a single chemical substance. The main constituents of gelatin are large and complex polypeptide molecules of the same amino acid composition as the parent collagen, covering a broad molecular weight distribution range. In the parent collagen, the 18 different amino acids are arranged in ordered, long chains, each having ~95,000 mol wt. These chains are arranged in a rod-like, triple-helix structure consisting of two identical chains, called α_1, and one slightly different chain called α_2 (Kang *et al.*, 1966; Kang *et al.*, 1969; Piez *et al.*, 1968). These chains are partially separated and broken, i.e., hydrolyzed, in the gelatin manufacturing process. Different grades of gelatin have average molecular weight ranging from 20,000 to 250,000 (Courts, 1954; Aoyagi, 1985; Larry *et al.*, 1985; Xiang-Fang *et al.*, 1985; Beutel, 1985; Koepff, 1984; Tomka, 1984; Bohonek *et al.*, 1976). Molecular weight distribution studies have been carried out by fractional precipitation with ethanol or 2-propanol and by complexing with anionic detergent molecules. The coacervates are isolated and recovered as gelatin fractions (Pouradier et al., 1950; Stainsby et al., 1954; Chen et al., 1992).

Analysis shows the presence of amino acids from 0.2% tyrosine to 30.5% glycine. The five most common amino acids are glycine, 26.4—30.5%; proline, 14.8–18%; hydroxyproline, 13.3–14.5%; glutamic acid, 11.1–11.7%; and alanine, 8.6–11.3%. The remaining amino acids in decreasing order are arginine, aspartic acid, lysine, serine, leucine, valine, phenylalanine, threonine, isoleucine, hydroxylysine, histidine, methionine, and tyrosine (Eastoe, 1967; Eastoe, 1955).

Warm gelatin solutions are more levorotatory than expected on the basis of the amino acid composition, indicating additional order in the molecule, which probably results from Gly-Pro-Pro and Gly-Pro-Hypro sequences (Josse et al., 1964). The α-chain form of gelatin behaves in solution like a random-coil polymer, whereas the gel form may contain as much as 70% helical conformation (Stainsby, 1977). The remaining molecules in nonhelical conformation link helical regions together to form the gel matrix. Helical regions are thought to contain both inter- and intramolecular associations of chain segments.

Gelatin structures have been studied with the aid of an electron microscope. The structure of the gel is a combination of fine and coarse interchain networks; the ratio depends on the temperature during the polymer-polymer and polymer-solvent interaction leading to hydrogen bond formation. The rigidity of the gel is approximately proportional to the square of the gelatin concentration. Crystallites, indicated by x-ray diffraction pattern, are believed to be at the junctions of the polypeptide chains.

Homogeneous α-chain gelatin has been prepared by pretreating collagen with pronase in the presence of 0.4 M $CaCl_2$ and extracting the gelatin with hot water at 80°C and pH 7.0 after inactivating the enzyme and removing the salts.

Stability

Dry gelatin stored in airtight containers at room temperature has a shelf life of many years. However, it decomposes above 100°C. For complete combustion, temperatures above 500°C are required. When dry gelatin is heated in air at relatively high humidity, 60% rh, and at moderate temperatures, i.e., above 45°C, it gradually loses its ability to swell and dissolve (Marks et al., 1968; Jones, 1987). Aqueous solutions or gels of gelatin are highly susceptible to microbial growth and breakdown by proteolytic enzymes. Stability is a function of pH and electrolytes and decreases with increasing temperature because of hydrolysis.

PHYSICAL AND CHEMICAL PROPERTIES

Commercial gelatin is produced in mesh sizes ranging from coarse granules to fine powder. In Europe, gelatin is also produced in thin sheets for use in cooking. It is a vitreous, brittle solid, faintly yellow in color. Dry commercial gelatin contains about 9–13% moisture and is essentially tasteless and odorless with specific gravity between 1.3 and 1.4. Most physical and chemical properties of gelatin are measured on aqueous solutions and are functions of the source of collagen, method of manufacture, conditions during extraction and concentration, thermal history, pH, and chemical nature of impurities or additives.

Gelation

Perhaps the most useful property of gelatin solution is its capability to form heat reversible gel-sols. When an aqueous solution of gelatin with a concentration greater than about 0.5% is cooled to about 35 to 40°C, it first increases in viscosity, then forms a gel. The gelation process is thought to proceed through three stages: (1) rearrangement of individual molecular chains into ordered, helical arrangement, or collagen fold (Gelatin Manufacturers Institute of America, Inc., 1986; von Hippel et al., 1959; von Hippel et al., 1960; Flory et al., 1960); (2) association of two or three ordered segments to create crystallites (von Hippel et al., 1960; Engel, 1962; Boedtker et al., 1954); and (3) stabilization of the structure by lateral interchain hydrogen bonding within the helical regions. The rigidity or jelly strength of the gel depends on the concentration, the intrinsic strength of the gelatin sample, pH, temperature, and additives.

Because the economic value of gelatin is commonly determined by jelly strength, the test procedure for its determination is of great importance. Commercially, gelatin jelly strength is determined by standard tests which measure the force required to depress the surface of a carefully prepared gel by a distance of 4 mm using a flat-bottomed plunger 12.7 mm in diameter. The force applied may be measured in the form of the quantity of fine lead shot required to depress the plunger and is recorded in grams. The measurement is termed the Bloom strength after the inventor of the lead shot device (Veis, 1964). In the early 1990s, sophisticated testing equipment utilizing sensitive load cells for the measurement are commonly used.

The conversion temperature for gelatin is determined as setting point, i.e., sol to gel, or melting point, i.e., gel to sol. Commercial gelatins melt between 23 and 30°C, with the setting point being lower by 2–5°C. One melting point determination method utilizes test tubes filled with gelatin solution that are gently chilled to form a gel. The tubes are tilted and colored carbon tetrachloride solution is placed on the gelatin surface. The tube is gradually warmed and the end point is determined when the descent of the colored solution is observed. Several methods have been used to determine the setting point of gelatin (Wainewright, 1977).

Solubility

In most commercial applications, gelatin is used as a solution. Gelatin is soluble in water and in aqueous solutions of polyhydric alcohols such as glycerol and propylene glycol. Examples of highly polar, hydrogen-bonding organic solvents in which gelatin dissolves are acetic acid, trifluoroethanol, and formamide (Kragh, 1977; Finch et al., 1977). Gelatin is practically insoluble in less polar organic solvents such as acetone, carbon tetrachloride, ethanol, ether, benzene, dimethylformamide, and most other nonpolar organic solvents. Many water-soluble organic solvents are compatible with gelatin, but interfere with gelling properties (Umberger, 1967). Dry gelatin absorbs water exothermally. The rate and degree of swelling is a characteristic of the particular gelatin. Swelled gelatin granules dissolve rapidly in water above 35°C. The cross-linking of gelatin matrix by chemical means is used extensively in photographic products, and this so-called hardening permanently reduces the solubility of gelatin (von Hippel *et al.*, 1959; Hornsby, 1956; Janus *et al.*, 1951; Itoh, 1992; Tabor *et al.*, 1992; Takahashi, 1992; Rottmann *et al.*, 1992).

Amphoteric Character

The amphoteric character of gelatin is due to the functional groups of the amino acids and the terminal amino and carboxyl groups created during hydrolysis. In strongly acidic solution the gelatin is positively charged and migrates as a cation in an electric field. In strongly alkaline solution, it is negatively charged and migrates as an anion. The intermediate point, where net charge is zero and no migration occurs, is known as the isoelectric point (IEP) and is designated in pH units (Weatherill *et al.*, 1992). A related property, the isoionic point, can be determined by utilizing a mixed-bed ion-exchange resin to remove all nongelatin cations and anions. The resulting pH of the gelatin solution is the isoionic point and is expressed in pH units. The isoionic point is reproducible, whereas the isoelectric point depends on the salts present. Type A gelatin has a broad isoionic region between pH 7 and pH 10; type B is in a lower, more reproducible region, reaching an isoionic point of 5.2 after 4 weeks of liming, which drops to 4.8 after prolonged or more vigorous liming processes (Maxey *et al.*, 1976; Li-juan, 1985; Toda, 1985; Toda, 1987). The isoelectric point can also be estimated by determining a pH value at which a gelatin solution exhibits maximum turbidity (Eastoe *et al.*, 1963). Many isoionic point references are recorded as isoelectric points even though the latter is defined as a pH at which gelatin has net charge of zero and thus shows no movement in the electric field (Veis, 1964).

Viscosity

The viscosity of gelatin solutions is affected by gelatin concentration, temperature, molecular weight of the gelatin sample, pH, additives, and impurities. In aqueous solution above 40°C, gelatin exhibits Newtonian behavior. Standard testing methods employ use of a capillary viscometer at 60°C and gelatin solutions at 6.67 or 12.5% solids (Gelatin Manufacturers Institute of America, Inc., 1986). The viscosity of gelatin solutions increases with increasing gelatin concentration and with decreasing temperature. For a given gelatin, viscosity is at a minimum at the isoionic point and reaches maxima at pH values near 3 and 10.5 (Stainsby, 1952). At temperatures between 30 and 40°C, non-Newtonian behavior is observed, probably due to linking together of gelatin molecules to form aggregates (Stainsby, 1962). Addition of salts decreases the viscosity of gelatin solutions. This effect is most evident for concentrated gelatin solutions (Cumper *et al.*, 1952; Stainsby, 1977).

Colloid and Emulsifying Properties

Gelatin is an effective protective colloid that can prevent crystal, or particle, aggregation, thereby stabilizing a heterogeneous suspension. It acts as an emulsifying agent in cosmetics and pharmaceuticals involving oil-in-water dispersions. The anionic or cationic behavior of gelatin is important when used in conjunction with other ionic materials. The protective colloid property is important in photographic applications where it stabilizes and protects silver halide crystals while still allowing for their normal growth and sensitization during physical and chemical ripening processes.

Coacervation

A phenomenon associated with colloids wherein dispersed particles separate from solution to form a second liquid phase is termed coacervation. Gelatin solutions form coacervates with the addition of salt such as sodium sulfate, especially at pH below the isoionic point. In addition, gelatin solutions coacervate with solutions of oppositely charged polymers or macromolecules such as acacia. This property is useful for microencapsulation (Zitko *et al.*, 1963; Wood, 1977; Upjohn Company, 1963; Croome, 1985; Kramer, 1992).

Swelling

The swelling property of gelatin is not only important in its solvation but also in the dissolution of pharmaceutical capsules. That pH and electrolyte content affect swelling has been explained by the simple Donnan equilibrium theory, treating gelatin as a semipermeable membrane (Procter *et al.*, 1916). This explains why gelatin exhibits the lowest swelling at its isoelectric pH. At pH below the isoelectric point, proper choice of anions can control swelling, whereas above the isoelectric point, cations primarily affect swelling. These effects probably involve breaking hydrogen bonds, resulting in increased swelling. The rate of swelling follows approximately a second-order equation (Libicky *et al.*, 1972). Conditioning at 90% rh and 20°C for 24 h greatly reduces swelling of hot dried film coatings (Jopling, 1956).

MANUFACTURE AND PROCESSING

Although new methods for processing gelatin, including ion exchange and cross-flow membrane filtration, have been introduced since 1960, the basic technology for modern gelatin manufacture was developed in the early 1920s. Acid and lime processes have separate facilities and are not interchangeable. In the past, bones and ossein, i.e., decalcified bone, have been supplied by India and South America. In the 1990s, slaughterhouses and meat-packing houses are an important source of bones. The supply of bones has been greatly increased since the meat-packing industry introduced packaged and fabricated meats, assisted by the growth of fast-food restaurants. Dried and rendered bones yield about 14–18% gelatin, whereas pork skins yield about 18–22%.

Most type A gelatin is made from pork skins, yielding grease as a marketable by-product. The process includes macerating of skins, washing to remove extraneous matter, and swelling for 10–30 h in 1–5% hydrochloric, phosphoric, or sulfuric acid. Then four to five extractions are made at temperatures increasing from 55–65°C for the first extract to 95–100°C for the last extract. Each extraction lasts about 4–5 h. Grease is then removed, the gelatin solution filtered, and, for most applications, deionized. Concentration to 20–40% solids is carried out in several stages by continuous vacuum evaporation. The viscous solution is chilled, extruded into thin noodles, and dried at 30–60°C on a continuous wire-mesh belt. Drying is completed by passing the noodles through zones of successive temperature changes wherein conditioned air blows across the surface and through the noodle mass. The dry gelatin is then ground and blended to specification.

Type B gelatin is made mostly from bones, but also from bovine hides and pork skins. The bones for type B gelatin are crushed and degreased at the rendering facilities, which are usually located at a meat-packing plant. Rendered bone pieces, 0.5–4 cm, with less than 3% fat, are treated with cool, 4–7% hydrochloric acid from 4 to 14 d to remove the mineral content. An important by-product, dibasic calcium phosphate, is precipitated and recovered from the spent liquor. The demineralized bones, i.e., ossein, are washed and transferred to large tanks where they are stored in a lime slurry with gentle daily agitation for 3–16 weeks. During the liming process, some deamination of the collagen occurs with evolution of ammonia. This is the primary process that results in low isoelectric ranges for type B gelatin. After washing for 15–30 h to remove the lime, the ossein is acidified to pH 5–7 with an appropriate acid. Then the extraction processing for type A gelatin is followed. Throughout the manufacturing process, cleanliness is important to avoid contamination by bacteria or proteolytic enzymes.

Bovine hides and skins are substantial sources of raw material for type B gelatin and are supplied in the form of splits, trimmings of dehaired hide, rawhide pieces, or salted hide pieces. Like pork skins, the hides are cut to smaller pieces before being processed. Sometimes the term calfskin gelatin is used to describe hide gelatin. The liming of hides usually takes a little longer than the liming of ossein from bone.

Most manufacturing equipment should be made of stainless steel. The liming tanks, however, can be either concrete or wood. Properly lined iron tanks are often used for the washing and acidification, i.e., souring, operations. Most gelatin plants achieve efficient processes by operating around the clock. The product is tested in batches and again as blends to confirm conformance to customer specifications.

ECONOMIC ASPECTS

World gelatin production in 1994 was believed to be nearly 220,000 t. The United States produced about 30,000 t, followed by France, Germany, Japan, Brazil, and Mexico. Of the gelatin produced in the United States, 55% is acid processed, i.e., type A. The U.S. food industry consumes about 20,000 t/yr, with an annual growth rate of 0.5%; the pharmaceutical industry consumes about 10,000 t/yr; and the photographic industry about 7,000 t/yr. In the United States, the pharmaceutical gelatin market is expected to grow on the average of 2.5% per year. The photographic gelatin market has been stable or growing slightly. Color paper and x-ray products use over 55% of the photographic gelatin in the United States, with graphic arts and instant films using an additional 30%.

ANALYTICAL TEST METHODS AND QUALITY STANDARDS

Gelatin is identified by a positive test for hydroxyproline, turbidity with tannic acid, or a yellow precipitate with acidic potassium dichromate or trinitrophenol. A 5% aqueous solution exhibits reversible gel-to-sol formation between 10 and 60°C. Gelatin gives a positive color test for aldehydes and sugars. Elemental analysis of commercial gelatin is reported as carbon, 50.5%; hydrogen, 6.8%; nitrogen, 17%; and oxygen 25.2% (Smith, 1921); a purer sample analyzed for 18.2-18.4% nitrogen (Eastoe, 1967; Eastoe, 1955). Regulations for quality standards vary from country to country, but generally include specifications for ash content, SO_2, heavy metals, chromium, lead, fluoride, arsenic, odor, and for the color or clarity of solutions (*United States Pharmacopeia* XXII, 1989). In addition, certain bacteriological standards, including *E. coli* and *Salmonella*, are specified. Restrictions on certain additives and preservatives are also listed. In the United States, the *Food Chemicals Codex* has been considering a new specification for food-grade gelatin; a final version should be issued soon (ca 1996). Standard testing procedures for viscosity, pH, ash, moisture, heavy metals, arsenic, bacteria, and jelly strength are described (*United States Pharmacopeia* XXII, 1989; Helrich, 1990; Vies, 1964). Additional test procedures have been published by the photographic and gelatin industries including the Japanese PAGI Method (PAGI Method, 1992).

USES

Food Products

Gelatin formulations in the food industry use almost exclusively water or aqueous polyhydric alcohols as solvents for candy, marshmallow, or dessert preparations. In dairy products and frozen foods, gelatin's protective colloid property prevents crystallization of ice and sugar. Gelatin products having a wide range of Bloom and viscosity values are utilized in the manufacture of food products, specific properties being selected depending on the needs of the application. For example, a 250-Bloom gelatin may be utilized at concentrations ranging from 0.25% in frozen pies to 0.5% in ice cream; the use of gelatin in ice cream has greatly diminished. In sour cream and cottage cheese, gelatin inhibits water separation, i.e., syneresis. Marshmallows contain as much as 1.5% gelatin to restrain the crystallization of sugar, thereby keeping the marshmallows soft and plastic; gelatin also increases viscosity and stabilizes the foam in the manufacturing process. Many lozenges, wafers, and candy coatings contain up to 1% gelatin. In these instances, gelatin decreases the dissolution rate. In meat products, such as canned hams, various luncheon meats, corned beef, chicken rolls, jellied beef, and other similar products, gelatin in 1–5% concentration helps to retain the natural juices and enhance texture and flavor. Use of gelatin to form soft, chewy candies, so-called gummie candies, has increased worldwide gelatin demand significantly (ca 1992). Gelatin has also found new uses as an emulsifier and extender in the production of reduced-fat margarine products. The largest use of edible gelatin in the United States, however, is in the preparation of gelatin desserts in 1.5–2.5% concentrations. For this use, gelatin is sold either premixed with sugar and flavorings or as unflavored gelatin packets. Most edible gelatin is type A, but type B is also used.

Pharmaceutical Products

Gelatin is used in the pharmaceutical industry for the manufacture of soft and hard capsules. The formulations are made with water or aqueous polyhydric alcohols. Capsules are usually preferred over tablets in administering medicine. Elastic or soft capsules are made with a rotary die from two plasticized gelatin sheets which form a sealed capsule around the material being encapsulated. Methods have been developed to encapsulate dry powders and water-soluble materials which may first be mixed with oil. The gelatin for soft capsules is low bloom type A, 170–180 g; type B, 150–175 g; or a mixture of type A and B. Hard capsules consisting of two parts are first formed and then filled. The manufacturing process is highly mechanized and sophisticated in order to produce capsules of uniform capacity and thickness. Medium-to-high bloom type A, 250–280 g; type B, 225–250 g; or the combination of type A and B gelatin are used for hard capsules. Usage of gelatin as a coating for tablets has increased dramatically. In a process similar to formation of gelatin capsules, tablets are coated by dipping in colored gelatin solutions, thereby giving the appearance and appeal of a capsule, but with some protection from adulteration of the medication. The use of glycerinated gelatin (United States Pharmacopeia XXII, 1989) as a base for suppositories offers advantages over carbowax or cocoa butter base (Tice *et al.*, 1953). Coated, cross-linked gelatin is used for enteric capsules. Gelatin is used as a carrier or binder in tablets, pastilles, and troches.

For arresting hemorrhage during surgery, a special sterile gelatin sponge known as absorbable gelatin sponge or Gelfoam is used. The gelatin is partially insolubilized by a cross-linking process. When moistened with a thrombin or sterile physiological salt splitting solution, the gelatin sponge, left in place after bleeding stops, is slowly dissolved by tissue enzymes. Special fractionated and prepared type B gelatin can be used as a plasma expander. Absorbable sterile gelatin powder is also available for use to control bleeding.

Gelatin can be a source of essential amino acids when used as a diet supplement and therapeutic agent. As such, it has been widely used in muscular disorders, peptic ulcers, and infant feeding, and to spur nail growth. Gelatin is not a complete protein for mammalian nutrition, however, since it is lacking in the essential amino acid tryptophan and is deficient in sulfur-containing amino acids.

Photographic Products

Gelatin has been used for over 100 years as a binder in light-sensitive products. The useful functions of gelatin in photographic film manufacture are a result of its protective colloidal properties during the precipitation and chemical ripening of silver halide crystals, setting and film-forming properties during coating, and swelling properties during processing of exposed film or paper.

Derivatized Gelatin

Chemically active groups in gelatin molecules are either the chain terminal groups or side-chain groups. In the process of modifying gelatin properties, some groups can be removed, e.g., deamination of amino groups by nitrous acid (Kenchington, 1958), or removal of guanidine groups from arginine by hypobromite oxidation (Davis, 1958); the latter destroys the protective colloid properties of gelatin. Commercially successful derivatized gelatins are made mostly for the photographic gelatin and microencapsulation markets. In both instances, the amino groups are acylated. Protein detergent is made by lauroylating gelatin. Phthalated gelatin is now widely used in the photographic industry (Heuer *et al.*, 1965). Arylsulfonylated gelatin has been patented for microencapsulation (Clark *et al.*, 1967). Gelatin which has been modified by reaction with succinic anhydride, succinylated gelatin, has been successfully used as a blood plasma volume expander (Tourtellotte *et al.*, 1963).

REFERENCES

Aoyagi, S. (1985) SDS gel electrophoresis of photographic gelatins. in H. Ammann-Brass, and J. Pouradier, eds., *Photographic Gelatin*, Proceedings of the Fourth IAG Conference, Internationale Arbeitsgemeinschaft für Photogelatine, Fribourg, Switzerland, 1983, pp. 79–94.

Bogue, R.H. (1922) *The Chemistry and Technology of Gelatine and Glue*, McGraw-Hill Book Co., Inc., New York.

Bohonek, J., Spühler, A., Ribeaud, M. and Tomka, I. (1976) Structure and formation of the gelatin gel. in R.J. Cox, ed., *Photographic Gelatin II*, Academic Press, Inc., New York, pp. 37–55.

Boedtker, H. and Doty, P. (1954) A study of gelatin molecules, aggregates and gels. *J. Phys. Chem., Ithica*, **58**, 968–983.

Beutel, J. (1985) Measurement and characterization of the molecular weight distribution of IAG gelatins. in H. Ammann-Brass, and J. Pouradier, eds., *Photographic Gelatin*, Proceedings of the Fourth IAG Conference, Internationale Arbeitsgemeinschaft für Photogelatine, Fribourg, Switzerland, 1983, pp. 65–78.

Chen, L.-J. and Shohei, A. (1992) Precipitation behaviour of photographic gelatin with alcohol. *J. Photogr. Sci.*, **40**, 159.

Clark, R.C. and co-workers (1967) Microscopic capsules and methods of making them. Brit. Pat. 1,075,952 (July 19, 1967) to Gelatin and Glue Research Assoc.

Courts, A. (1954) The N-terminal amino acid residues of gelatin. *Biochem. J.*, **58**, 70–79.

Croome, R.J. (1985) The flocculation of photographic silver halide emulsions precipitated in the presence of chemically modified gelatins. in H. Ammann-Brass, and J. Pouradier, eds., *Photographic Gelatin*, Proceedings of the Fourth IAG Conference, Internationale Arbeitsgemeinschaft für Photogelatine, Fribourg, Switzerland, 1983, pp. 267–282.

Cumper, C.W.N. and Alexander, A.E. (1952) The viscosity and rigidity of gelatin in concentrated aqueous systems. *Aust. J. Sci. Res.*, **A5**, 146–152,

Davis, P. (1958) The guanidino side chains and the protective colloid action of gelatin. in G. Stainsby, ed., *Recent Advances in Gelatin and Glue Research*, Pergamon press, London, pp. 225–230.

Eastoe, J.E. and Courts, A. (1963) Properties of the macromolecule. *Practical Analytical Methods for Connective Tissue Proteins*, Spon, London, Chap. 6.

Eastoe, J.E. (1955) The amino acid composition of mammalian collagen and gelatin. *Biochem. J.*, **61**, 589–600.

Eastoe, J.E. (1967) Composition of collagen and allied proteins. in G.N. Ramachandran, ed., *Treatise on Collagen*, Vol. 1, Academic Press, Inc., New York, pp. 1–72.

Engel, J. (1962) Investigation of the denaturation and renaturation of soluble collagen by light scattering. *J. Arch. Biochem.*, **97**, 150.

Finch, C.A. and Jobling, A. (1977) The physical properties of gelatin. in A.G. Ward and A. Courts, eds., *The Science and Technology of Gelatin*, Academic Press, Inc., New York, Chap. 8.

Flory, P.J. and Weaver, E.S. (1960) Helix-coil transitions in dilute aqueous collagen solutions. *J. Am. Chem. Soc.*, **82**, 4518–4525.

Heuer, R.P. and Fitzgerald, A.E. (1965) Refractory composition and method. U.S. Pat. 3,184,312 (May 18, 1965) (to General Refractories Co.)

Gelatin Manufacturers Institute of America, Inc. (1986), *Standard Methods for Sampling and Testing Gelatins*, New York.

Helrich, K. (1990) ed., *Official Methods of Analysis, 15th Ed.*, Association of Official Analytical Chemists, Arlington, Va.

Hornsby, K.M. (1956) Chemical reviews. *Brit. J. Photogr.*, **103**, 17, 28.

Itoh, N. (1992) Evaluation of hardening using sol fraction of gelatin films (review). *J. Photogr. Sci.*, **40**, 200.

Janus, J.W., Kenchington, A.W. and Ward, A.G. (1951) A rapid method for the determination of the isoelectric point of gelatin using mixed bed deionization. *Research*, **4**, 247.

Jones, R.T. (1987) 3. Gelatin: physical and chemical properties. in K. Ridgway, ed., *Hard Capsules Development and Technology*, The Pharmaceutical Press, London, pp. 41–42.

Jopling, D.W. (1956) The swelling of gelatin films. The effects of drying temperature and of conditioning the layers in atmospheres of high relative humidity. *J. Appl. Chem.*, **6**, 79.

Josse, J. and Harrington, W.F. (1964) Role of pyrrolidine residues in the structure and stabilization of collagen. *J. Mol. Biol.*, **9**, 269–287.

Kang, A.H. and co-workers (1969) Intramolecular cross-link of chick skin collagen. *Biochem. Biophys. Res. Commun.*, **36**, 345–349.

Kang, A.H. and co-workers (1966) Studies on the structure of collagen utilizing a collagenolytic enzyme from tadpole. *Biochemistry*, **5**, 509–515.

Kenchington, A.W. (1958) Chemical modification of the side chains of gelatin. *Biochem. J.*, **68**, 458–468.

Koepff, P. (1985) History of industrial gelatine production (with special reference to photographic gelatine). in H. Ammann-Brass and J. Pouradier, eds., *Photographic Gelatin*, Proceedings of the Fourth IAG Conference, Internationale Arbeitsgemeinschaft für Photogelatine, Fribourg, Switzerland, 1983, pp. 3–31.

Koepff, P. (1984) The use of electrophoresis in gelatin manufature. in H. Ammann-Brass and J. Pouradier, eds., *Photographic Gelatin Reports 1970–1982*, IAG, pp. 197–209.

Kragh, A.M. (1977) Swelling, adsorption and the photographic uses of gelatin. in A.G. Ward and A. Courts, eds., *The Science and Technology of Gelatin*, Academic Press, Inc., New York, Chapt. 14.

Kramer, D.L. (1992) Laboratory-scale test for studying the coagulation behavior of chemically modified gelatins. *J. Photogr. Sci*, **40**, 152.

Larry, D. and Vedrines, M. (1985) Determination of molecular weight distribution of gelatines by H.P.S.E.C. in H. Ammann-Brass and J. Pouradier, eds., *Photographic Gelatin*, Proceedings of the Fourth IAG Conference, Internationale Arbeitsgemeinschaft für Photogelatine, Fribourg, Switzerland, 1983, pp. 35–54.

Li-juan, C. (1985) Determination of iso-electric point and its distribution of IAG gelatins by iso-electric focusing method. in H. Ammann-Brass and J. Pouradier, eds., *Photographic Gelatin*, Proceedings of the Fourth IAG Conference, Internationale Arbeitsgemeinschaft für Photogelatine, Fribourg, Switzerland, 1983, pp. 95–106.

Libicky, A. and Bermane, D.I. (1972) Kinetics and equilibria of swelling of gelatin layers in water. in R.J. Cox, ed., *Photographic Gelatin*, Academic Press, Inc., New York, pp. 29–48.

Marks, E.M., Tourtellotte, D. and Andux, A. (1968) The phenomenon of gelatin insolubility. *Food Technol.*, **22**, 99–102.

Maxey, C.R. and Palmer, M.R. (1976) The isoelectric point distribution of gelatin. in R.J. Cox ed., *Photographic Gelatin II*, Academic Press, Inc., New York, pp. 27–36.

PAGI Method, 7th ed. (1992) Photographic and Gelatin Industries, Tokyo, Japan.

Piez, K.A. and co-workers (1968) Comparative studies on the chemistry of collagen utilizing cyanogen bromide cleavage. *Brookhaven Symp. Biol.*, **21**, 345–357.

Pouradier, J. and Venet, A.M. (1950) Contribution a l'etude de la structure des gelatines. *J. Chim. Phys.*, **47**, 391–398. Procter, H.R. and Wilson, J.A. (1916) The acid-gelatin equilibrium. *J. Chem. Soc.*, **109**, 307–319.

Procter, H.R. and Wilson, J.A. (1916) The acid-gelatin equilibrium. *J. Chem. Soc.*, **109**, 307–319.

Rottmann, J. and Pietsch, H. (1992) Crosslinking reactions of gelatin films. *J. Photogr. Sci.*, **40**, 217.

Smith, C.R. (1921) Osmosis and swelling of gelatin. *J. Am. Chem. Soc.*, **43**, 1350–1366.

Smith, P.I. (1929) *Glue and Gelatine*, Pitman Press, London.

Stainsby, G., Saunders, P.R. and Ward, A.G. (1954) The preparation and properties of some gelatin fractions. *J. Polym. Sci.*, **12**, 325–335.

Stainsby, G. (1977) The physical chemistry of gelatin in solution. in A. G. Ward and A. Courts, eds., *The Science and Technology of Gelatin*, Academic Press, Inc., New York, p. 127.

Stainsby, G. (1977) *ibid*, p. 109–136.

Stainsby, G. (1952) Viscosity of dilute gelatin solutions. *Nature (London)*, **169**, 662–663.

Stainsby, G. (1962) The gelatin-water system in the region of the melting point. in H. Sauverer, ed., *Scientific Photography*, Pergamon Press, London, p. 253.

Tabor, B.E., Owers, R. and Janus, J.F. (1992) The crosslinking of gelatin by a range of hardening agents. *J. Photogr. Sci.*, **40**, 205.

Takahashi, T. (1992) The mechanism of the crosslinking reaction of gelatin with hardener. *J. Photogr. Sci.*, **40**, 212.

Tice, L.F. and Abrams, R.E. (1953) Glycerinated gelatin as a suppository base. *J. Am. Pharm. Assoc.*, **14**, 24.

Toda, Y. (1985) The measurement of the isoelectric point distribution of gelatin. in H. Ammann-Brass and J. Pouradier, eds., *Photographic Gelatin*, Proceedings of the Fourth IAG Conference, Internationale Arbeitsgemeinschaft für Photogelatine, Fribourg, Switzerland, 1983, pp. 107–124.

Toda, Y. (1987) in S.J. Band, ed., "Photographic Gelatin," *Proceedings of the Fifth RPS Symposium*, Oxford, U.K. 1985, The Imaging Science and Technology Group of the Royal Photographic Society, pp. 28–37.

Tomka, I. (1984) Electrophoresis as a routine tool to investigate photographic gelatins. in H. Ammann-Brass and J. Pouradier, eds., *Photographic Gelatin Reports 1970–1982*, IAG, 210.

Tourtellotte, D. and Marks, E.M. (1963) Method of modifying type A gelatin and product thereof. U.S. Pat. 3,108,995 (Oct. 29, 1963) (to Charles B. Knox Gelatin Co., Inc.).

Umberger, J.Q. (1967) Solution and gelation of gelatin as related to solvent structure. *Photogr. Sci. Eng.*, **11**, 385–391.

United States Pharmacopeia XXII, (USP XXII-NFXVII), (1989) The United States Pharmacopeial Convention, Inc., Rockville, MD.

Upjohn Company (1963) Encapsulation of Particles by Liquid-liquid Phase Separation. Brit. Pat. 930,421 (July 3, 1963).

Veis, A. (1964) The gelatin-collagen transition. in B.L. Horecker, N.D. Kaplan and H.E. Sheraga, eds., *Molecular Biology*, Vol. V. Academic Press, Inc., New York, Chapt. 5.

Veis, A. (1964) The macromolecular chemistry of gelatin. in B.L. Horecker, N.D. Kaplan and H.E. Sheraga, eds., *Molecular Biology*, Vol. V. Academic Press, Inc., New York, p. 112.

von Hippel, P.H. and Harrington, W.F. (1959) Enzymic studies of the gelatin-collagen-fold transition. *Biochim. Biophys. Acta*, **36**, 427–447.

von Hippel, P.H. and Harrington, W.F. (1960) The structure and stabilization of the collagen macromolecule. *Brookhaven Symp. Biol.*, **13**, 213–231.

Wainewright, F.W. (1977) Physical tests for gelatin and gelatin products. in A.G. Ward and A. Courts eds., *The Science and Technology of Gelatin*, Academis Press, Inc., New York, pp. 531–532.

Weatherill, T.D., Henning, R.W. and Smith, K.A. (1992) Solid state structural characterization of gelatin crosslinked with a variety of hardeners. *J. Photogr. Sci.*, **40**, 220.

Wood, P.D. (1977) Technical and pharmaceutical uses of gelatine. in A. G. Ward and A. Courts, eds., *The Science and Technology of Gelatin*, Academic Press, Inc., New York, pp. 419–422.

Xiang-Fang, C. and Bi-Xian, P. (1985) Determination of molecular weight distribution of gelatin by high pressure gel permeation chromatography. in H. Ammann-Brass and J. Pouradier, eds., *Photographic Gelatin*, Proceedings of the Fourth IAG Conference, Internationale Arbeitsgemeinschaft für Photogelatine, Fribourg, Switzerland, 1983, pp. 55–64.

Zitko, V. and Rosik, J. (1963) The reaction of pectin with gelatin (IV) [and] the influence of tannin on the reaction of pectin with gelatin. *Chem. Zvesti*, **17**, 109–117.

17. COLLAGEN: CHARACTERIZATION, PROCESSING AND MEDICAL APPLICATIONS

FREDERICK H. SILVER[1,*] and ATUL K. GARG[2]

[1]*Division of Biomaterials, Department of Pathology, UMDNJ-Robert Wood Johnson Medical School, 675 Hoes Lane, Piscataway, New Jersey 08854, USA*
[2]*Integra LifeSciences, Co., P.O. Box 688, Plainsboro, New Jersey 08536, USA*

INTRODUCTION

Collagenous proteins, cells, elastic fibers, proteoglycans and cell attachments factors form the structural framework of all mammalian extracellular matrices. At least nineteen different molecular forms of collagen have been identified (Prockop, 1992; van der Rest and Bruckner, 1993; Myers *et al.*, 1994). These include the fibrillar collagens, types I, II, III, V, and XI which consist of a central uninterrupted triple-helical region of about 1000 amino acids in each of three polypeptide chains which are "capped" by propeptides that are usually removed once the molecules are incorporated into fibrils. The distribution of fibril forming collagens in vertebrates includes type I in bone and most other tissues together with some type III and V, except for cartilage in which types II and XI predominate.

A second class of non-fibrillar collagens includes type IV molecules that form antiparallel sheet-like structures that make-up basement membranes. The third major class is the fibril associated collagens (FACIT) with interrupted triple helices including types IX, XII, XIV and XVI. The FACIT collagens are separated into multiple domains of variable lengths.

Along with the nineteen types of collagen which maintain tissue shape and prevent premature mechanical failure (Silver *et al.*, 1992) extracellular matrix contains: elastic fibers that aid in recovery of tissue shape and act as highly efficient "springs" that store mechanical energy; proteoglycans that specifically attach to and surround collagen fibrils; and cell adhesion factors including fibronectin, vitronectin, entactin, laminin, heparan sulfate proteoglycans and integrins that connect the extracellular matrix to the cell cytoskeleton (Silver *et al.*, 1992).

Type I collagen, is the major fibrillar collagen and has been estimated to account for 25% of the dry protein found in mammals. It occurs in a variety of different structural hierarchies in tissue including parallel aligned collagen fibrillar networks (tendon, ligament, dura mater, and pericardium), alignable networks (skin, blood vessels and intestine walls) and mixtures of aligned and alignable collagen fibers (intervertebral disc) (Silver *et al.*, 1992). Although this collagen type is the one that is primarily used in the medical device industry, ultimately a variety of collagen types will be employed in devices.

The characteristic "finger prints" of type I collagen include:

*Correspondence: Frederick H. Silver, Biomaterials, V-14, UMDNJ-RWJMS, 675 Hoes Lane, Piscataway, New Jersey 08854–5635, USA.

- an amino acid composition consisting of about 33% glycine and 25% of proline and hydroxyproline;
- an elution profile based on chromatography of denatured α and β chains, γ components and higher molecular weight species;
- an intrinsic viscosity of about 1000 ml/gr;
- a molecular length of about 300 nm by rotary shadowing;
- a translational diffusion coefficient of about 0.85×10^{-7} cm^2/sec; and
- an ability to form fibrilar elements with a macroperiod of between 65 to 67 nm.

Type I collagen is used in a number of medical devices including hemostats, eye shields and cosmetic implants and has extensive potential in the growth of tissues in cell culture. Therefore, it is essential to understand the purification, characterization and usage of type I collagen .

Below we will describe methods for isolation of both insoluble and soluble collagen and typical characterization profiles of these materials as well as methods used to process collagen into a variety of shapes and end-products.

ISOLATION OF SOLUBLE AND INSOLUBLE TYPE I COLLAGEN

Type I collagen has been isolated from various tissues including bovine skin and tendon, human skin, bone and tendons and avian tissues. It is processed with and without enzymes that remove non-collagen proteins and/or the non-helical ends on the collagen molecule. In the preparation of type I collagen the appropriate procedures are chosen to optimize the yield and are dependent on whether the final product can be filtered and sterile filled into the final package or processed and packaged under class 10,000 conditions and then terminally sterilized. Terminal sterilization using ethylene oxide gas, gamma irradiation, or exposure to electron-beam are simple procedures that can be carried out by contract manufacturers. However, these processes tend to modify the physics and chemistry of the collagen molecule. Sterile filtration and filling operations are expensive to set-up and usually require enzymatic removal of the non-helical ends and therefore affect the ability of the collagen molecules to reassociate into materials with structural stability. However, little or no modification of the collagen structure occurs due to sterilization by filtration. Finally, processes where the raw materials are sterilized followed by aseptic processing of the sterile materials also yield a sterile final product. Therefore, a careful analysis of the design parameters are necessary prior to selection of the most appropriate procedure for type I collagen preparation. Table 1 lists a number of patents for production of collagen.

PREPARATION OF INSOLUBLE COLLAGEN

Insoluble type I collagen can be isolated from various animal tissues such as bovine corium and tendon. Fresh uncured hide can be obtained from a slaughterhouse and the lower layer, the corium, is frozen at -20°C prior to purification of collagen. In addition, limed corium is available as a by-product of leather manufacture. Typically,

Table 1 Patents Related To Preparation of Collagen

Title	Patent No.	Year	Inventor
Purification of Collagen	831,124 (Br.)	1960	Bloch and Oneson
Isolation of Collagen from Collagenous Sources	1,145,904 (Ger.)	1963	Nishihara
Method of Producing Soluble Collagen	3,131,130	1964	Oneson
Method for Colloidally Dispersing Collagen	3,121,049	1964	Nishihara
Method for Solubilizing Insoluble Collagen Fibers	3,314,861	1967	Fujii
Purified Collagen Fibrils	3,520,402	1970	Nichols and Oneson
Aqueous Collagen Composition	4,140,537	1979	Luck and Daniels
Process for Preparing Macromolecular Biologically Active Collagen	4,374,121	1981	Cioca

insoluble collagen is prepared from bovine corium that is exposed to lime and then precut, acidified and ground (Komanowsky *et al.*, 1974). The first step cuts the hide into strips 3/8" wide and then the strips are sequentially cut into thin pieces. Pieces are washed with water containing acid (e.g. formic, acetic, citric, propionic and benzoic) for about four hours at which time the pH is about 5.3. Acidified collagen is ground using equipment such as an Urschel grinder (Urschel Laboratories, Inc., Valparaiso, In) or a disc mill (Young Machinery Sales Company, Inc., Muncy, Pa). Microcut corium is then washed with distilled water followed by isopropanol until the product is free of fat (Chernomorsky, 1987). Some purification processes include enzymatic treatment to remove non-collagenous connective tissue components.

Other processing steps include precipitation of collagen from solution with sodium chloride, dialysis to remove excess salt, and filtration to remove large clumps of material.

Another source of raw material is bovine flexor tendon which can be processed by swelling and treatment with proteolytic enzymes. Although the treatment for producing insoluble collagen from tendon has been historically different than treatment of the corium, the final products in both cases are type I collagen. Methods for treatment with an alkali sulfate salt and an alkali metal hydroxide (Cioca, 1981;1983) as well as sodium borohydrate at alkaline pH (Oreson *et al.*, 1970) and sodium hydroxide (Courts, 1960) have been reported in the literature.

PREPARATION OF SOLUBLE COLLAGEN

Purification of soluble type I collagen dates back to the 1960s when Piez and coworkers (Piez *et al.*, 1961) reported the extraction of collagen from rat skin with either 1 M NaCl or 0.5 M acetic acid and to Oneson (1964) who received a patent for a method for producing soluble collagen. Both types of collagen were purified by repeated precipitation with 5% NaCl in 0.5 M acetic acid. The soluble supernatant was obtained after centrifugation at 30,000 g for 1 h. Drake *et al.* (1966) reported that enzyme treatment with noncollagenase proteases removed "extra-helix peptide appendages" (e.g. crosslinks and parts of the non-helical ends of the collagen molecule) from insoluble collagen releasing collagen molecules without changing the main structural feature of the molecule. They treated insoluble calf skin with a chymotrypsin, elastase, pepsin and trypsin which solubilized intact collagen molecules from insoluble skin. A process for enzymatic digestion of insoluble collagen was patented by Bloch and Oneson in 1960.

Miller and Rhodes (1982) have summarized the extraction of soluble collagen in three solvent systems: (1) a neutral salt solvent, (2) a dilute organic acid, and (3) 0.5 M acetic acid containing pepsin. In the neutral salt solvent (1.0 M NaCl, 0.05 M Tris, pH 7.5) the efficiency of extraction is increased at the higher salt concentrations. They advise to include a number of proteinase inhibitors in the solvent to minimize proteolytic degradation. Although the amount of neutral salt soluble collagen in most tissues is small, it can be increased when crosslinking is inhibited.

In dilute organic acid, such as 0.5 M acetic acid or 0.01 M HCl the solvent has increased ability to swell the tissue. Addition of proteinase inhibitors is suggested. Again as with neutral salt, the amount of soluble collagen obtained is small since the endogenous crosslinks are stable in this solvent.

Finally, 0.5 M acetic acid containing pepsin (1:10 W/W) in the pH range of 2.5 to 3.0 selectively degrades the crosslinks that allows solubilization of additional collagen molecules. The molecules so obtained are lacking part of, or all of the non-helical ends and therefore do not assemble into normal fibrillar elements.

Once dissolved in the above solvents, type I collagen is purified by salt precipitation using 2.2 to 2.5 M NaCl for neutral salt extraction or 0.7 M NaCl for extraction in 0.5 M acetic acid. Precipitations followed by resolubilization in the starting solvent (Miller and Rhodes, 1982). The precipitation step is then repeated until the product is free of non-collagenous material based on SDS PAGE, chromatography and amino acid analyses. When further purification is desired, chromatography further separates type I from other collagen types. However, this is only necessary when very high purity type I collagen is desired.

COLLAGEN CHARACTERIZATION

Soluble and insoluble type I collagen can be characterized using a variety of physical and chemical techniques (Miller and Rhodes, 1982; Silver, 1987). Typically an aliquot of the purified collagen is denatured and then characterized by ion-exchange or high pressure liquid chromatography (Miller and Rhodes, 1982) (see Table 2). Ion-exchange chromatography of denatured collagen in the past has been the "gold standard" by which collagen α-chains and β and γ components as well as high molecular weight components were evaluated. Specifically, chromatography on carboxymethyl (CM)-cellulose is used to evaluate chain cleavage that occurs during collagen purification. This approach has been replaced by high pressure liquid chromatography (HPLC).

To identify the level of non-type I collagen contamination in a preparation, fragmentation of the collagen α chains at methionyl residues with cyanogen bromide is used. The cyanogen bromide peptides thus obtained are separated using ion-exchange chromatography on CM-cellulose or by HPLC. Based on the elution pattern it is possible to identify cyanogen-bromide peptides not derived from type I collagen (Miller and Rhodes, 1982). Non-collagenous contaminants in insoluble

Table 2 Collagen Characterization

Method	Type of Information Obtained	Reference
Chromatography	Distribution of Collagen Components	Miller and Rhodes, 1982
Polyacrylamide Gel Electrophoresis	Collagen Type Determination Non-Collagenous Protein Determination	Miller and Rhodes, 1982
IR spectrum	Conformational changes in triple helix	Payne and Veis, 1988
UV absorbance spectrum	Conformational changes in triple helix, collagen concentration, number of tyrosure and phenylalamine residues	Na 1988; Silver and Trelstad, 1981
Polarimetry thermal denaturation curves	Fraction of nicked or fragmented molecules	Condell *et al.*, 1988
Thermal transition enthalpy	Transition temperature and enthalpy	McClain and Wiley, 1972
Optical Rotation	Collagen content	Nagelschmidt and Viell, 1987
Ultracentrifugation	Molecular weight	Obrink, 1972
Light scattering	Molecular weight, translational diffusion coefficient	Silver and Trelstad, 1980; Obrink, 1972; Fletcher, 1976; Birk and Silver, 1984; Silver 1987;Thomas and Fletcher, 1979; Silver *et al.*, 1979; Gelman and Piez, 1980
Intrinsic viscosity	Axial ratio	Davison and Drake, 1966; Obrink, 1972; Silver, 1987
Dynamic viscosity	Molecular stiffness	Henry *et al.*, 1983; Amis *et al.* 1985
Turbidity-Time Measurements	Self assembly, fibril diameter	Silver, 1987a; Birk and Silver, 1984
Electron Microscopy	D-period, banding pattern, molecular length	Schmitt *et al.*, 1942; Hall, 1956; Chapman and Hardcastle, 1974; Meek *et al.*, 1979

collagen can also be determined by measurement of the percentage of the sample not solubilized in 0.15 M trichloroacetic acid (Oneson *et al.*, 1970).

Polyacrylamide gel electrophoresis is also commonly used in the characterization of different collagen types. In this technique collagen is denatured in a sample buffer containing sodium dodecyl sulfate and is then layered onto a polyacrylamide gel. A current is applied to the gel and the peptides move at a rate that is proportional to the net charge. The migration pattern of the sample is then compared to that of a control.

Differences between triple-helical collagen and its denatured form gelatin, can be determined based on IR spectroscopy of collagen solutions. One study suggested that a baseline study of the fine structure of amide I band (which is assigned to the OH bonding motions of the carbonyl group of proline) indicated that a $-30\ \text{cm}^{-1}$ shift of this band occurs during denaturation (Payne and Veis, 1988). These authors concluded that temperature and conformationally dependent changes in the fine structure of amide I from dilute solutions of collagen can be monitored in a reproducible and quantitative fashion.

Na (1988) has demonstrated that the UV absorbance spectra of collagen in the range between 220 and 240 nm increases as the collagen unfolds to become gelatin. The absorbance increase on heating at 223 nm was found to parallel the % change collagen native fraction.

Quantitative analysis of nicked or shortened collagen molecules in pepsin treated bovine type I collagen preparations can be achieved using polarimetry thermal denaturation curves (Condell et al., 1988). Polarimetry denaturation curves of collagen are conducted at −400° and the specific rotation is monitored in the polarimeter while the temperature is raised. The areas of the nicked and complete molecular components of the derivative curve of the specific rotation are obtained by deconvolution. Since shortened or nicked collagen molecules melt at a lower temperature than intact molecules, the area of the Gaussian curve that fits the derivative curve is used to determine the amount of collagen fragments (Condell et al., 1988).

Determination of thermal transition enthalpy is another method for evaluation of both soluble and insoluble collagen (McClain and Wiley, 1972). Values of thermal transition enthalpy changes for tissues are virtually identical with those from solubilized collagen molecules. The experimentally determined value for the enthalpy change per residue is 1055 cal/mole for all collagens studied whereas the thermal denaturation or shrinkage temperature and total proline and hydroxyproline content varies from sample to sample. The transition temperature is determined from the temperature at which the onset of thermal denaturation occurs while the thermal transition enthalpy is obtained from the area under the thermogram. A direct relationship between the melting temperature of molecules (T_m) and the shrinkage temperature (T_s) of collagen fibrils exists (Danielsen, 1990).

In addition, the native collagen content can be calculated for soluble type I collagen by determination of the loss of optical rotation during the transition to gelatin. Nagelschmidt and Viell (1987) report that at 365 nm the specific optical rotation of type I collagen is between −1029° and −1334° while after denaturation it is between −382° and −509°.

A number of parameters for assessing collagen structure can be obtained from UV spectroscopic characterization of type I collagen (Silver and Trelstad, 1981; Na, 1988). The concentration of collagen solutions can be obtained from the absorbance measured at wavelengths between 190 nm and 230 nm. At 230 nm the collagen concentration in mg/ml in HCl pH 2.0 is obtained by multiplying the absorbance by 0.548 (Silver and Trelstad, unpublished observations). This applies up to an absorbance of 1.0. The relationship for other buffer systems is different; however, this approach is typically used to measure collagen concentration in fractions obtained by chromatography (Silver and Trelstad, 1981).

Na (1988) has measured the far UV absorption spectra of collagen in acetate buffer and guanidine hydrochloride. He concluded that in acetate buffer the absorption peak was found at its apex located between 199 and 203 nm and that the absorbance was linear with concentration. If the optical density was below one, the absorption coefficient was 49.0 cm^{-1} ml − mg^{-1}. In 6 M guanidine-hydrochloride the absorption peak was found around 217 to 221 nm. The absorption peak height was linear with concentration up to approximately 1.1 absorption units with an absorption coefficient of 9.43 cm^{-1} ml − mg^{-1}.

UV spectroscopy is also a tool that can be used to determine the number of tyrosine and phenylalanine residues per collagen molecule (Na, 1988). The number of tyrosine residues are determined from the peak at 276 nm. The absorption coefficient of tyrosine used is 1500 1 – mole $^{-1}$ – cm^{-1}. By fitting the spectrum in the range of 250 to 310 nm and knowing the number of tyrosine residues present, the number of phenylalanine residues per collagen molecule can be calculated (Na, 1988).

The molecular weight of vertebrate derived type I collagen molecules in solution have been measured by a number of techniques including light scattering (classical and low angle) and ultracentrifugation (Obrink, 1972; Silver and Trelstad, 1980; Birk and Silver, 1983; Silver and Birk, 1984 and Silver, 1987). From the average amount of light scattered at low angles from a solution of collagen molecules, the weight average molecular weight can be obtained. The accepted value for type I single molecules is 285,000. High values reflect the presence of crosslinks between molecules or physical associations. The state of collagen aggregation influences the rate of collagen self assembly and the final structure of any product. Weight average molecular weight can also be determined from equilibrium sedimentation studies (Silver, 1987).

Another means of characterizing the collagen molecule is based on intensity fluctuations of the intensity of scattered light. The time rate of change of intensity fluctuation can be autocorrelated and is related to the rate that rod-like molecules translate and rotate. Based on these principles the translational diffusion coefficient for type I collagen has been found to be between 0.78 and 0.86 × 10^{-7} cm^2/sec (Obrink, 1972; Fletcher, 1976; Thomas and Fletcher, 1979; Silver et al., 1979; Gelman and Piez, 1980; Silver and Trelstad, 1980; Bernego et al., 1983; Birk and Silver, 1983; Silver and Birk, 1984). The rotational diffusion coefficient for type I collagen has been found to be 1082 sec^{-1}. Intrinsic viscosity for type I collagen single molecules from vetebrate tissues are reported to be between 1000 and 1200 ml/g (Davison and Drake, 1966; Obrink, 1972) equivalent to an axial ratio of about 200 (Silver, 1987).

Dynamic viscoelastic measurements on collagen solutions yield information including the complex shear modulus, intrinsic viscosity and longest relaxation time for flexural motions. From the storage and loss moduli, which make up the complex shear modulus, the persistence length and Young's modulus can be calculated for collagen in solution (Henry et al., 1983). For pronase treated collagen at pH 4.0 a persistence length of 170 nm and a Young's modulus of 4×10^{10} dyne/cm^2 have been reported. These numbers were found to be consistent with the behavior of a semi-flexible rod (Henry et al., 1983). In a later study of a non-enzyme treated collagen the behavior at pH 4.0 was modelled by a rod-like molecule with partial flexibility along its entire length and a persistance length of 161 nm with no loose joints (Amis et al., 1985). At pH 7.4 the behavior was that expected for a semiflexible rod with two loose joints near the ends with a persistence length of 169 nm for the center segment. Measurements of the persistence length reflect the intactness of collagen single molecules.

The ability for collagen molecules to self assemble into fibrils and fibers with the native banding pattern is typically studied by measuring changes in turbidity as a function of time (Silver, 1987a; Garg et al., 1989). Turbidity measurements reflect changes in molecular weight per unit length and the kinetics and thermodynamics can be followed based on the change in lag time (time required for turbidity to become measurable) with increased concentration and the temperature dependence of the rate constant. The final turbidity per unit concentration is a measure of the final fibril diameter (Birk and Silver, 1984).

Collagen type I forms filamentous or fibrous structures when reconstituted under self-assembling conditions (Farber *et al.*, 1986). The molecules self-assemble into a quarter stagger array that can be characterized after staining with heavy metals such as phosphotungstic acid or uranyl ions (Silver, 1987). When heavy metal ions stain around the collagen molecule the result is termed negative staining. When the ions stain charged residues on the collagen molecule it is termed positive staining. Schmitt and coworkers (1942) reported that the banding pattern of positively stained collagen molecules repeated itself every 64 nm and this repeat could be broken down into 12 bands. The molecules can either be lined up in a side-to-side fashion an arrangement termed segment long-spacing crystallite or in a staggered fashion characteristic of the pattern found in tissues. These patterns can be correlated with the amino acid sequence.

Using the primary sequence for type I collagen, it has been shown that a correlation exists between the positively stained fibril banding pattern and the generated positions of the charged residus exists (Chapman and Hardcastle, 1974; Meek *et al.*, 1979). The same has been demonstrated for the SLS crystallite.

Direct observations of single type I collagen molecules with the electron microscope after shadowing with metals using freshly cleaved mica as substrate surface dates back to Schmitt et al. (1942) and Hall (1956). This approach is suited to facilitate the localization of specialized regions on the collagen triple helix (Furthmayr and Madri, 1982) and to evaluate intactness of the triple helix. Native type I collagen molecules are wound ropes with three regions of flexibility: one is located about 45 nm from the amino terminal end of the molecule, the second about 20 nm from the carboxyterminal end and the third around 60 nm from the carboxyterminal end. Using bead models, images from rotary shadowed collagen molecules can be converted into translational diffusion coefficients to be compared with the translational diffusion coefficient measured by light scattering studies (Silver, 1987).

COLLAGEN PROCESSING INTO DEVICES

Unlike many synthetic materials collagen molecules cannot be processed by injection molding or conventional extrusion techniques. Therefore the processing of collagen into films, sponges, beads, fibers and tubes involves modifications of three basic processes, e.g. casting, freeze drying and extrusion.

As with any other film forming polymer, collagen films are prepared by solvent evaporation. The clarity of the film is related to the size of the scattering units. Particles greater than about 29 nm in diameter cause the film to be translucent and not transparent (Silver, 1987). This is due to the fact that the collagen molecule has a length of about 300 nm. In the cornea, the geometric arrangement of the collagen fibrils results in transparency because the spaces between the fibrils are roughly equivalent to the fibril diameter.

Typically, dispersed or solubilized collagen is prepared at a concentration of about 1% (W/V) by grinding in a blender or homogenizer. About 6 ml of the material /cm^2 of surface is poured into a mold with the desired shape (Weadock et al., 1984; Weadock et al., 1986; Weadock et al., 1987). The material is allowed to air dry overnight at room temperature and the resulting film has a thickness of about 100 μm. Decreased film thicknesses are obtained by using less collagen dispersion.

Collagen sponge manufacture for hemostats, wound dressing materials, or substrates for cell growth in culture requires formation of a three dimensional structure that will either trap blood or allow for cellular ingrowth. The pore size required is dependent on the exact application.

Typically, insoluble collagen is used for sponge formation since the mechanical properties of sponges made out of soluble collagen are much inferior when compared to those made out of soluble collagen. In some cases small amounts of soluble collagen are added to the sponge to give the proper handling characteristics.

A typical procedure for preparing these sponges involves grinding the insoluble collagen into fine particles using a blender, homogenizer or bowl chopper at solid concentrations between one-half and several percent (typically 1.0% W/V is used) (Doillon et al., 1986). The pH of the dispersing medium depends on the exact purification treatment. Acid swollen collagen that has not been treated with alkali is dispersed at pHs between 2.0 and about 4.0. The exact pH is selected to minimize the dispersion viscosity. The dispersion viscosity is dependent in addition on extent of shear induced blending. Increased blending times tend to decrease the fiber size increasing the viscosity. High viscosity dispersions are difficult to pour into metal or plastic trays and then spread evenly. Trays of the collagen dispersion are frozen at temperatures between –20°C and –90°C to obtain the desired pore sizes. Average pore sizes as small as 14 μm are obtained at a freezing temperature of –80°C while at a freezing temperature of –30°C the average pore size is about 100 μm (Doillon et al., 1986).

Collagen microporous beads are produced by a modification of the process to produce collagen sponges. Insoluble collagen is again dispersed at a pH of between 2.0 and 4.0 by blending or homogenizing. Excess air is removed by vacuum degassing until no additional bubbles are detected. The dispersion is then passed through a vibrating tube or other device that is capable of producing droplets. Droplets are allowed to fall into a cryogenic bath at the desired temperature preferably between –20°C and –80°C where they freeze. Frozen beads are transferred to a lyophilizer where the frozen water is removed (Berg et al., 1989).

Finally, collagen fibers and tubes are manufactured using modified extrusion processes (Hughes et al., 1984; Hughes et al., 1985; Silver and Kato, 1992). Kato et al. (1989) described the extrusion of small diameter (50–100 μm) collagen fibers that can be formed into woven and non-woven products. The process involves use of a 1% (W/V) type I dispersion in dilute HCL at pH 2.0 that is deaerated under vacuum to remove any air bubbles. The dispersion is placed in 30 cc disposable syringes and then the collagen dispersion was extruded through polyethylene tubing with an inner diameter of 0.28 mm into a 37°C bath containing an aqueous fiber formation buffer at pH 7.5. Fibers are transferred after 45 min to a bath of isopropanol and then into distilled water prior to drying under tension.

Collagen tubes can be manufactured using a similar procedure except that the extrusion takes place through an annular die into a coagulation bath.

DEVICE STABILIZATION

To stabilize collagenous devices so that they do not lose their shapes on wetting or biodegrade rapidly on implantation, chemical modification is necessary. In the very simplest case the collagen is modified by chemical addition of an amine reactive acylating agent, a carboxylic acid reactive esterifying agent or a combination of these agents (Kelman and DeVore, 1994).

In cases where the implant must be able to withstand mechanical forces, chemical crosslinks are necessary to provide stability. Typically crosslinking involves bi-and multifunctional reagents containing groups that link two or more amino acid side chains. Such reagents include chemical agents such as aldehydes, alkyl and aryl halides, imidoesters, chromic acid, carbodiimides, aldehydes, N-hydroxysuccinimide active esters, N-substituted maleimides, acylating compounds, diphenylphosphorylazide and glucose (Bowes and Cater, 1968; Weadock et al., 1984; Kent et al., 1985; Petite et al., 1994). Physical crosslinking is achieved by exposure to ultraviolet radiation and other forms of radiation, heat, drying, severe dehydration (Weadock et al., 1984) and singlet oxygen (Hikichi et al., 1994).

The interaction of aldehydes with collagen has been the subject of extensive research. Glutaraldehyde has been shown to react with lysine residues forming a compound with more than one glutaraldehyde molecule (Bowes and Cater, 1968) and that in the presence of free glutaraldehyde the number of aldehyde molecules involved in the crosslink increases (Cheung and Nimni, 1982). Glutaraldehyde crosslinking has been shown to decrease the pitch of some of the turns of the collagen triple helix thereby somewhat altering the molecular structure (Jonak et al., 1979). However, the non-specificity and formation of polymeric crosslinks makes total removal of all free glutaraldehyde difficult (Bruck, 1980).

Crosslinking can be achieved by exposure to chemicals in either the wet or vapor states assuming that the vapor pressure of the crosslinking agent is high enough to penetrate the device. Physical crosslinking is usually performed in the dry state. Assuming there is no inherent cytotoxicity introduced as a result of crosslinking, the merits of each crosslinking procedure depend on the end-application. For instance, there has been concern over aldehyde crosslinking; however, glutaraldehyde treated porcine valves still remain the "gold standard" for bioprostheses. More critical than the exact agent used is the degree of crosslinking necessary for the proper performance of the final device and the extent that any unreacted chemical remains in the final device.

DEVICE PACKAGING AND STERILIZATION

Collagenous devices can be sterilized by filtration, exposure to cytotoxic chemicals, gamma irradiation or treatment with an electron beam. Injectable collagen is typically sterilized by sterile filtration which requires final packaging under sterile conditions. Zyplast™ and Zyderm™ injectable collagen implant materials are processed in this manner.

In contrast, collagen absorbable hemostatic sponges are sterilized either by exposure to ethylene oxide gas, γ-rays, dry heat or an electron beam. Collastat™ is an example of a material that is sterilized by exposure to ethylene oxide gas and Avitene™ is a microfibrillar collagen hemostat that is sterilized by exposure to dry heat. In some cases such as bioprosthetic heat valves, exposure to dilute glutaraldehyde solutions in the final packaged device is used for sterilization. Instat^{TM} and Bicol^{TM} are sterilized by gamma irradiation. Device sterilization influences the physical properties of the product. For instance, gamma irradiated materials tend to have a reduced shrinkage temperature and poor wetability compared to ethylene oxide treated materials. In the case of ethylene oxide, gamma irradiation as well as glutaraldehyde treatment the device is sterilized in the final package.

Sterilization is a necessary step in the processing of any medical device. Although ethylene oxide gas has been used to stabilize many implants, recent concern over hypersensitivity to devices sterilized in this fashion has been noted. In addition, exposure of collagenous devices to gamma rays leads to a modification of the triple helix by breakage of backbone peptide bonds (Bowes and Moss, 1962) thereby decreasing the stability of the product. Therefore, selection of any particular mode of sterilization is dependent on the end-use of the device and the structural stability desired.

FINAL DEVICE TESTING

Once a device is formulated, processed, stablized, packaged and sterilized, it is important to chemically and physically characterize the product to insure that the final device is safe and remains the same from lot to lot. These tests include biocompatibility assessment, mechanical stablity, biodegradation rate and in vivo function in an end-use application.

Biocompatibility testing for 510(k) and PMA filings have been reviewed elsewhere (Silver 1994) and will not be presented in detail in this chapter. As described in Table 3 for a 510(k) filing the typical tests and acceptable outcomes for collagenous devices are described. Uncrosslinked as well as crosslinked collagenous devices will pass all of these tests. In cases where the devices are crosslinked they will not fail these tests if all traces of crosslinking agents or other chemicals are removed.

One of the most important requirements for a medical device is biocompatibility. Regulatory agencies throughout the world have their own requirements regarding performance, quality and safety of devices. A device has to be safe both biologically and function appropriately in an end-use application. Safety of devices is evaluated through biological testing and clinical studies. There are several standards for evaluating biological safety of devices. These include but are not limited to the following tests:

1. Cytotoxicity Test
   This is an in-vitro toxicity test for materials and devices. Different methodologies have been used depending on cell lines, end points of cell responses and sample applications. Agar diffusion testing has been used as a standard for conducting cytotoxicity studies.

2. Sensitization Test
   Research has shown that allergic contact dermatitis is caused by various products, e.g. surgical gloves, bandages and dental alloys. The incident, population, and severity of allergic contact dermatitis are influenced by chemicals, products and individuals. The degree of sensitization is dependent on concentration of chemicals.

Table 3 Typical preclinical testing for a collagenous device (adapted from Silver, 1994)

Test	Acceptable outcome
Acute systemic toxicity	No signs of respiratory, motor, convulsive, ocular, salivary, piloerectal, gastrointestinal or skin reactions
Agar overlay cytotoxity	No detectable zone of cell reactivity
Guinea pig maximization test	No delayed skin irritation
Haemolysis	Less than 5% haemolysis
Inhibition of cell growth	No inhibition of cell growth at all extract concentrations
Limulus amebocyte lysate	No LAL activity present
MEM cytotoxicity	No cytotoxicity of test material
Primary skin irritation	No primary skin irritation
USP rabbit pyrogen	No rabbit shows individual temperature rise of $0.6°C$ or for three rabbits the temperature rise is below $1.4°C$

3. Mutagenicity (Gentotoxicity) Testing

There are three tests normally used as a standard. These are DNA effects, Salmonella/microsome or gene mutation test and chromosomal aberrations. The object is to find mutagens in the test material and to evaluate their potential in order to maintain the safety of the device. The tests are not only dependent on the mutagens but also on the process of extracting them.

4. Short-term implantation

The inflammatory reaction due to materials used in devices is usually very weak. To quantitatively and microscopically evaluate such weak reaction the thickness of the capsule or inflamed area around an implant can be correlated with the cytotoxicity index.

5. Pyrogenicity Test

There are two types of pyrogens, endotoxins and other pyrogenic chemicals. Cleanliness and sterility of device manufacture are important critera required to keep it from endotoxin contamination. Inactivation of endotoxins in aqueous solutions has been reported previously using heat treatment.

The ISO 10993 guidance document on testing of medical and dental materials and devices may become an international standard for device testing (Truscott, 1994). The recommended test procedures depend on whether the device is surface contacting (i.e. for contact with intact skin, mucosal membranes or breached or compromised surfaces), external communicating devices (i.e. blood path direct, tissue/bone/dentin communicating or circulating blood), or implant devices (i.e. tissue/bone or blood). The duration of contact with tissue is also an important aspect with the categories including limited exposure (less than 24 h), prolonged exposure (greater than 24 h but less than 30 days) and permanent contact (greater than 30 days). Typically, collagenous implants are designed to biodegrade in less than 30 days but more than 24 h and the testing simplifies to:

• Surface device – Irritation, sensitization and cytotoxicity
• Externally communicating devices:

Tissue/bone/dentin communicating – irritation, sensitization, cytotoxicity, pyrogenicity and genotoxicity

Blood path indirect – irritation, sensitization, cytotoxicity, systemic toxicity, haemocompatibility, pyrogenicity

Blood path direct – irritation, sensitization, cytotoxicity, systemic toxicity, hemocompatibility, pyrogenicity and genotoxicity

- Implant devices

Bone Tissue and Tissue Fluids – sensitization, cytotoxicity, implantation, genotoxicity

Blood – irritation, sensitization, cytotoxicity, systemic toxicity, hemocompatibility, implantation, pyrogenicity, genotoxicity

Both crosslinked and non-crosslinked collagen will pass all of these tests. The major concern with a biodegradable material like collagen is that the biodegradation is controlled such that the device maintains its intended purpose for a sufficient length of time depending on the end-use application.

Other tests for final device evaluation include analysis of complex collagen structures by light and electron microscopy as well as by measurement of the form birefringence. Light and electron microscopy of collagenous materials has been described in detail elsewhere (Doillon *et al.*, 1986; Doillon *et al.*, 1987; Doillon *et al.*, 1988; Wasserman *et al.*, 1988).

The ordered aggregation of collagen molecules into bundles can be assessed by measurement of the form birefringence (Pimentel, 1981). The measured birefringence of collagen is made of an intrinsic birefringence due to the anisotropy of the molecule and a form component due to alignment of molecules in fibrils. Total birefringence retardation when normalized by the thickness is constant for extensor tendon, irregardless of the diameter (McBride *et al.*, 1985). Direct measurement of forces between collagen triple helices in fibrillar structures has also been reported (Leiken *et al.*, 1994) making it possible to correlate fibrillar organization and the force required to separate individual molecules in collagenous materials.

Immunogenicity of Collagen

The immunogenicity of type I collagen has been a question of concern to the FDA and has resulted in warnings related to the use of injectable collagen products in the package insert. The insert for Contigen™, an injectable collagen used for correction of urinary incontinence, contains the following paragraphs:

"Some physicians have reported the occurrence of connective tissue diseases such as rheumatoid arthritis, systemic lupus erythematosus, dermatomyositis (DM), and polymyositis (PM) subsequent to collagen injections in patients with no previous history of these disorders. A comparison of the observed number of cases of PM/DM in the collagen-treated population with a estimate of the expected number of cases suggests an association between collagen injections and PM/DM: i.e. there appears to be a higher than expected incidence of PM/DM in the collagen-treated population. However, a causal relationship between collagen injection and the onset of autoimmune disease or systemic connective tissue disease has not been established. Contigen implant patients developing symptoms suggestive of autoimmune disease should be immediately referred to a physican trained in the diagnosis and treatment of rheumatological disease.

Also, an increased incidence of cell-mediated and humoral immunity to various collagens has been found in systemic connective tissue disease such as rheumatoid arthritis, juvenile rheumatoid arthritis, and progressive systemic sclerosis (scleroderma). Patients with these diseases may thus have an increased susceptibility to hypersensitivity responses and/or accelerated clearance of their implants when injected with bovine dermal collagen preparations. Therefore, caution should be used when treating these patients including consideration for multiple skin lesions.

Patients with a history of dietary beef allergy should be carefully evaluated before injectable bovine collagen therapy, since it is possible that the collagen component of the beef may be causing the allergy. More than one skin test is highly recommended prior to treating these patients".

Circulating antibodies to collagen have been found in serum from patients injected with fibrillar collagen (Cooperman and Michaeli, 1984; Ellingsworth *et al.*, 1986). Results of animal studies suggest that the level of circulating antibodies is highest in animals treated with microfibrillar collagen with pepsin-treated collagen exhibiting the lowest (DeLustro *et al.*, 1986). Antibody titers are increased to measurable levels in animals treated with booster injections and Freund's Complete Adjuvant. Crosslinking reduces the humoral response to collagen but may in some cases induce a cell-mediated response (Meade and Silver, 1990).

On a functional level Cooperman and Michael (1984a) reported that systemic complaints could not be correlated with skin reactions or antibody titers and that immune responses are typically localized reactions that manifest within the first two exposures to the implant material. A later study reported that all patients suffering adverse clinical reactions to bovine collagen implants were lacking the HLA-DR 4 antigen (Vanderveen *et al.*, 1986). All patients who received multiple bovine collagen injections without having adverse clinical reactions were lacking HLA-B5 and HL-DR5 and had a increased incidence of HLA-DR4. Antibodies to bovine collagen are always IgG and often IgA and multiple regions of the collagen molecule are recognized (McCoy *et al.*, 1985).

Bovine Spongiform Encephalopathy (BSE)

Collagen is typically derived from bovine tissues and converted into medical products. In a memorandum from Jane E. Henney, M.D. of the FDA to manufacturers of FDA-regulated products on December 17, 1993 it is stated that manufacturers are required to prevent use in regulated products of bovine-derived materials from cattle which have resided in or originated from countries where BSE has been documented.

Transmissible spongiform encephalopathies are degenerative central nervous system diseases of man and animals that are believed to be caused by abnormal neuronal membrane proteins which are resistant to sterilization. These proteins have been termed prions and can be transmitted by ingestion of slaughter byproducts including brain, spinal cord, spleen, thymus, tonsil, lymph nodes or intestines from affected animals. BSE is believed to be the bovine variant of scrappie, a slowly progressive disease of the central nervous system of sheep and goats that is manifested by sensory and motor malfunction and eventually leads to death. The agent is transmitted by direct or indirect contact and enters through the gastrointestinal tract. Processed tissues from sheep and goats are used as ingredients in animal feeds thereby explaining the occurrence of BSE in cattle.

On August 29, 1994 the FDA (Federal Register Vol 59, No. 166 page 44584) declared that the use of processed slaughter byproduct from adult sheep and goats in cattle feed not safe and is unapproved as a food additive when added to ruminant feed. Since there is no documented occurrence of BSE in cattle in the US, tissues derived from animals certified from herds in the US are acceptable as a source for the manufacture of collagen for medical devices.

MEDICAL APPLICATIONS

Collagen is biodegradable, biocompatible, hemostatic material that can be fabricated into different forms which makes it a good biomaterial candidate for medical device fabrication. There have been several excellent reviews written describing the medical applications of collagen (Chvapil *et al.*, 1962; Stenzel *et al.*, 1974; Chvapil, 1980; Pachence *et al.*, 1987). This chapter will present a summary of the medical applications as well as references after 1987 to avoid duplication of previous reviews. Table 4 lists some of the patents that have been issued on preparation of collagen products, while Tables 5 and 6 list the medical applications and U.S. companies involved in development of these products, respectively. Table 7 lists the PMAs issued on collagen products.

The use of collagen in the medical device industry is a consequence of its biological properties including controlled biodegradation rate, availability in commercial quantities, ability to trigger blood coagulation and platelet aggregation, stimulation of chemotaxis of connective tissue and inflammatory cells and capability to support cell attachment and growth (Pachence *et al.*, 1987). It has uses in the form of coatings, fibers, fabrics, films and membranes, solutions, matrices and sponges, suspensions, tapes, tubes and sealants.

Collagen coatings

Collagen coatings have been used for a number of applications including as a biological sealant and for improving the biological response to implants. Guidon *et al.* (1989) reported the use of collagen as a sealant for vascular grafts. Ksander (1988) coated silicone rubber implants with collagen and found that capsular contraction and contracture was absent. Other studies report the use of collagen coated Marlex mesh reinforced with continuous polypropylene spiral (Okumura *et al.*, 1994) and collagen coated Vicryl mesh for correction of full-thickness abdominal wall defects (Meddings *et al.*, 1993). These studies support the use of collagen coatings to modify the biological response to implant materials.

Drug Delivery

Collagen has been used as a biodegradable system for release of a variety of drugs including contraceptives, cyclosporine, heparin and fibroblast growth factor, 5-fluorouracil, dexamethasone, antibiotics, bone morphogenetic protein, insulin and growth hormone (Chvapil *et al.*, 1976; Moore *et al.*, 1981; Weiner *et al.*, 1985; Hwang *et al.*, 1989; Chen *et al.*, 1990; Hasty *et al.*, 1990; Takaoka *et al.*, 1991; Ipsen *et al.*, 1991; DeBlois *et al.*, 1994). For hydrophilic molecules, release occurs through

Table 4 Patents Related To Collagen Products

Title	Patent No.	Year	Inventor
Process of preparing surgical sutures	505,148	1893	Weaver
Suture and method for making same	1,254,031	1918	Davis
Method of spinning collagen solutions	2,475,129	1949	Cresswell
Surgical sponge and the preparation thereof	2,610,625,1952		Sifferd
Hemostatic sponges and method of preparing same	617,771 (Can. Pat.)	1961	Wiedentrimer and Valentine
Centrifugal casting of collagen to produce films and ribbons	3,036,341	1962	Taylor
Method of producing a collagen strand	3,114,593	1963	Griset et al.
Method for the manufacture of suture and ligatures from animal tendons	3,114,591	1963	Nichols and Reissmann
Preparation of a high density collagen sponge from collagen	3,157,524	1964	Artandi
Apparatus and method for producing collagen tubing	3,122,788	1964	Lieberman
Collagen-carbonic acid surgical sponge	3,368,911	1968	Kuntz and Nuwayser
Collagen matrix pharmaceuticals	3,435,100	1969	Nichols
Method and apparatus for obtaining electro deposited shaped articles from fibrous protein fibrils	3,556,939	1971	Mizuguchi et al.
Process for augmenting connective mammalian tissue with in situ polymerizible native collagen solution	3,949,072	1976	Daniels and Knapp
Collagen drug delivery device	4,164,559	1979	Miyata et al.
Chemically and enzymatically modified collagen hemostatic agent	4,215,200	1980	Miyata et al.
Collagen contact lens	4,264,155	1981	Miyata
Collagen gel contact lens and method of preparation	4,260,228	1981	Miyata
Regenerated fiber collagen condom and method of preparation	4,349,026	1982	Miyata
Collagen implant material and methods for augmenting soft tissue	4,424,208	1984	Wallace and Wade
Injectable cross-linked collagen implant material	4,582,640	1986	Smestad et al.

Table 4 *Continued*

Title	Patent No.	Year	Inventor
Collagen replacement prosthesis for the cornea	4,581,030	1986	Burns and Gross
Viscoelastic collagen solution for opthalmic use and method for preparation	4,713,446	1987	Devore et al.
Delivery system for implantation of fine particles in surgical procedures	4,657,548	1987	Nichols
Succinylated ateolocollagen solution for use in viscosurgery and as a vitreous substitute	4,748,152	1988	Miyata
Viscoelastic collagen solution for opthalmic use and method of preparation	4,851,513	1989	DeVore et al.
Weighted collagen microsponge for immobilizing bioactive materials	4,861,714	1989	Dean et al.
Collagen matrix beads for soft tissue repair	4,837,285	1989	Berg et al.
Non-biodegradable two phase corneal implant and method for preparing same	5,067,961	1991	Kelman and DeVore
Relates to a hollow conduit of a matrix of type I collagen and laminin for nerve regeneration	5,019,087	1991	Nichols
Synthetic collagen orthopaedics structures such as grafts and tendons and other structures	5,171,273	1992	Silver and Kato
Ophthalmological collagen coverings	5,094,856	1992	Fyodorov et al.
Collagen-based adhesives and methods of preparation thereof	5,219,895	1993	Kelman and DeVore

water filled channels and the release is therefore limited to short time periods (Weadock *et al.*, 1987). For hydrophobic molecules the release can be prolonged since diffusion occurs through hydrophobic portions of the collagen molecule.

Collagen Fiber, Fabrics and Tubes

Collagen in the form of fibers has been used in closing surgical wounds and incisions. In the case of suture materials composed of pure animal gut or collagen, it is treated to modify the rate of absorption and to allow for prolonged strength retention. Pure collagen sutures are used in ophthalmic surgery as well as for other applications (Brumback and McPherson, 1967; Adler *et al.*, 1967).

Collagen fibers can be processed into aligned filamentous structures for tendon and ligament devices (Kato *et al.*, 1989; Kato *et al.*, 1991) or made into meshes for hernia repair and other applications (Adler *et al.*, 1962; Cavallaro *et al.*, 1994).

Table 5 Medical Applications of Collagen

Specialty	Application
Cardiology	Heart Valves
Dermatology	Soft Tissue Augmentation
Dentistry	Oral Wounds
	Biocoating for dental implants
	Support for Hydroxyapatite
	Periodontal Attachment
General Surgery	Hemostasis
	Hernia Repair
	IV Cuffs
	Wound Repair
	Suture
Neurosurgery	Nerve Repair
	Nerve Conduits
Oncology	Embolization
Orthopaedic	Bone Repair
	Cartilage Reconstruction
	Tendon and Ligament Repair
Ophthalmology	Corneal Graft
	Tape or Retinal Reattachment
	Eye Shield
Plastic Surgery	Skin Replacement
Urology	Dialysis Membrane
	Sphincter Repair
Vascular	Vessel Replacement
	Angioplasty
Other	Biocoatings
	Drug Delivery
	Cell Culture
	Organ Replacement
	Skin Test

Collagen tubes have been used for replacement of structures including esophagus (Natsume *et al.*, 1993), peripheral nerve (Colin sand Donoff, 1984; Kljavin and Madison, 1991) and ureter (Tachibana *et al.*, 1985). It has also been used for culture of cells (Leighton *et al.*, 1985) and for alveolar ridge augmentation after being filled with hyrodxylapatite (Gongloff, 1988).

Collagen Films, Membranes and Matrices

Collagen films and membranes have been used for immobilization of biological materials such as factor XIII from blood (Blanchy *et al.*, 1986), for guided tissue regeneration (Yaffe *et al.*, 1982; Blumenthal *et al.*, 1986; Pitaru *et al.*, 1987; Tanner *et al.*, 1988; Blumenthal and Steinberg, 1990; Paul *et al.*, 1992; Quteish and Dolby, 1992; Blumenthal, 1993; Colangelo *et al.*, 1994; Black *et al.*, 1994), for filling of tooth extraction sites (Mannai *et al.*, 1986), as hemodialysis membranes (Rubin *et al.*, 1968), for retinal reattachment (L'esperance, 1965), as a dural substitute (Kline, 1965; Collins *et al.*, 1991; Pietrucha, 1991; Laquerriere a., 1993; Narotam *et al.*, 1993;1995), for stapesdectomy (Bellucci and Wolff, 1964), for prevention of

Table 6 U.S. Companies conducting research by application

Company	car	der	den	gen	neu	onc	ort	oph	pla	uro	vas	oth
Astra												x
Autogenesis									x			
Becton Dickinson												x
Biosurface								x				
B&L Pharma						x						
Bristol-Myers								x				
Chiron								x				
Collagen		x	x	x	x	x	x		x	x		x
C.R. BARD												x
Datascope					x							x
E.R Squibb												x
Hercules		x										
Integra LifeSci		x	x	x	x		x	x		x	x	x
J&J					x				x	x		x
Kensey Nash	x											
Med Chem			x	x	x		x				x	
Parke Davis				x								
Pfizer								x				x
Prodex												x
Organogenesis				x					x		x	x
Semex												x
Serono Labs		x							x			
St. Jude Med												x
Upjohn				x								
Verax												x

Codes for Speciality:

Cardiology	car
Dermatology	der
Dentistry	den
General Surgery	gen
Neurosurgery	neu
Oncology	onc
Orthopaedics	ort
Ophthalmology	oph
Plastic Surgery	plas
Urology	uro
Vascular	vas
Other	oth

anstomosis leakage (Marescaux *et al.*, 1991), for nerve regeneration (Yannas *et al.*, 1985), repair of the tympanic membrane (Goycoolea *et al.*, 1991), for cartilage, meniscus and bone repair (Riley and Leake, 1976; Chvapil, 1977; Stone *et al.*, 1990; Hogervorst *et al.*, 1992), for control of local bleeding (Coquin *et al.*, 1987; Ernst *et al.*, 1993; Foran *et al.*, 1993; Aker *et al.*, 1994;), repair of liver injuries (Peacock *et al.*, 1965) as a protective barrier during brain surgery (Kurze *et al.*, 1975) and for wound repair (Abbenhaus *et al.*, 1965; Poten and Nordgaard, 1976; Yannas *et al.*, 1982; Oliver *et al.*, 1982; Leipziger *et al.*, 1985; Doillon *et al.*, 1988).

Table 7 Approved PMAs by Manufacturer for Collagen Products

Specialty	Manufacturer	Product	Approval Date
Cardiovascular	Meadox Medicals	Microvel with Hemashield	04/26/89
Dental	Colla-Tec	CollaCote	11/08/85
Orthopaedic	Collagen	Alveoform Biograft	10/28/88
	Serono Labs	Fibrel	02/26/88
General and Plastic Surgery	Collagen	Zyderm CI Collagen Implant	07/22/81
	Colla-Tec	Helistat (Absorbable) Hemostatic (Sponge)	11/08/85
	Datascope	Hemopad (Absorbable) Hemostat	05/27/86
	Ethicon	Plain Chromic Absorbable Suture	09/26/63
	Johnson & Johnson	Instat (Absorbable Hemostat)	10/10/85
	Medchem Products	Avitene (Microfibrillar collagen) Hemostat Nonwoven Web Hemostat	08/26/76
	Sherwood Medical	Chronic Gut Suture	02/22/73
	Industries	Plain Gut Suture	01/01/76
	Viaphore	Collastat	10/10/81

Injectable Collagen

Collagen solutions and suspensions have been used in the form of an injectable augmentation system for arterial embolization (Cho *et al.*, 1989), cutaneous defects (Daniels and Knapp, 1976; Knapp *et al.*, 1977; Luck and Daniels, 1979; Klein , 1983; Kamer and Churukian, 1984), fracture healing (Benfer and Struck, 1973), spinal cord regeneration (De La Torre, 1982; Marchand and Woely, 1990), correction of urinary incontinence (Frey *et al.*, 1994; Appell, 1994), tendon gliding function restoration (Porat *et al.*, 1980) and lung gluing (Ennker *et al.*, 1994).

Cell Culture on Collagen (Tissue Engineering)

The growth of cells on collagen supports in tissue culture has been recently reviewed (Silver and Pins, 1992). A variety of types including cells from skin and cardiovascular system have found extensive applications (see Table 8).

Use of collagen substrates for growth of skin cells has been reported extensively (Yannas *et al.*, 1982, 1986; Alvarez and Biozes, 1984; Hansborough et al. 1989; Carter *et al.*, 1987; Bell *et al.*, 1981, 1981a; Hull, 1990 and Nanchahal *et al.*, 1989). Epidermal cell grafts produced on collagen sheets overcome the problem of implant fragility while at the same time inducing rapid dermal ingrowth. Cells inoculated on collagen sponges initiate rapid wound closure with no lag time prior to grafting. The limiting factor of these grafts is the availability of autogenous epithelial cells. Skin equivalents consisting of allogeneic, fibroblast-seeded collagen gels overlaid with epidermal cells prove effective in treating full thickness skin defects. Because these grafts require

Table 8 Growth of Skin Cells on Collagen

Cell Type(s)	Matrix	Tissue Produced	Ref.
Posembryonic skin epithelial cells	Collagen gel	Epidermis	Karasek and Charlton (1971)
Epidermal (perinatal mouse skin)	Rat tail tendon collagen	Epidermal cells cultured for 2 to 4 d in vitro	Worst *et al.* (1974)
Epidermal (rabbit skin)	Porcine skin	Expansion of epithelial cell surface by a factor of 50 within 7 to 21 d	Freeman et al. (1974)
Epidermal (human)	Collagen film (Helitrex Inc)	Single cell suspensions plated on collagen film	Eisinger *et al.* (1980)
Fibroblasts (human foreskin)	Rat tail tendon collagen	Fibroblasts condense a hydrated collagen lattice	Bell et al. (1979)
Fibroblasts + epidermal cells	Rat tail tendon collagen	Exposure of skin equivalent to UV light irradiation for 14 d stimulated pigment transfer from melaniocytes to keratinocytes	Bell et al. (1981; 1981a; 1983; 1984)
Smooth muscle cell	Types I and III collagen	Vascular media-cultured smooth muscle cells on collagen gels show suppression of cellular proliferation and enhanced differentiation in early stages of culture	Sakata *et al.* (1990)
Bovine aortic endothelial cells	Pepsin extracted bovine fetal skin types I and III collagen	Stress fibers more distinct in cells grown on type I collagen	Semich and Robenek (1990)
Rat cardiac myocytes	Collagen types I and III	Myofibrillogenesis was observed including formatiom of striated patterns after stress-fiber-like structures were observed	Hilenski *et al.* (1991)
Bovine aortic endothelial, smooth muscle and adventitial cells	Rat tail tendon collagen	Formed multilayered structure resembling artery with a lining of endothelial cells that produced Von Willebrand's factor and prostacyclin	Weinberg *et al.* (1986)

no autogenous tissue, they may be produced in large quantities and cryopreserved until the time they are required for treating wound defects.

Extensive research on growth of autogenous and allogeneic skin cells on collagen matrices has shown it is feasible to expect to grow a variety of tissues and organs in culture. Ultimately, the use of cryopreserved allogeneic cell cultured materials will have a significant impact on management of patients that experience organ and tissue failure.

Acknowledgements

The authors would like to thank Drs. Dale DeVore and Joseph Nichols for providing key references and Maria Ayash for help in preparing this manuscript.

REFERENCES

Abbenhaus, J.I., MacMahon, M.B., Rosenkrantz, J.G. and Paton, B.C. (1965) Collagen sheets as a dressing for large excised areas. *Surg. Forum*, **16**, 477–478.

Adler, R.H., Pelecanos, N.T., Geil, R.G., Rosenzeig, S.E. and Thorsell, H.G., (1962) Collagen mesh prosthesis for wound repair and hernia reinforcement. *Surgical Forum*, **13**, 29–31.

Adler, R.H., Montes, M., Dayer, R. and Harrod, D. (1967) A comparison of reconstituted collagen suture for colon anastomoses. *Surgery, Gynecology & Obstetrics*, **124**, 1245–1252.

Aker, U.T., Kensey, K.R., Heuser, R.R., Sandza, J.G. and Kussmaul, W.G. 3rd. (1994) Immediate arterial hemostasis after cardiac catherization: initial experience with a new puncture closure device. *Catheterization and Cardiovascular Diagnosis*, **31**, 228–232.

Alvarez, O.M. and Biozes, D.G. (1984) Cultured epidermal autografts, *Clin. Dermatol.*, **2**, 54–67.

Amis, E.J., Carriere, C.J., Ferry, J.D. and Veis, A. (1985) Effect of pH on collagen flexibility determined from dilute solution viscoelastic measurements. *Int. J. Biol. Macromol.*, **7**, 130–134.

Appell, R.A. (1994) Collagen injection therapy for urinary incontinence. *Urologic Clinics of North America*, **21**, 177–182.

Bell, E., Ivarsson, B. and Merrill, C. (1979) Production of a tissue-like structure by contraction of collagen lattices by human fibroblasts of different proliferative potential in vitro. *Proc. Natl. Acad. Sci. U.S.A.*, **76**, 1274–1278.

Bell, E., Ehrlich, P., Buttle, D.J. and Nakatsuji, T. (1981) Living tissue formed in vitro and accepted as skin equivalent tissue of full thickness. *Science*, **211**, 1052–1054.

Bell, E., Ehrlich, H.P., Sher, S., Merrill, C., Saber, R., Hull, B., Nakatsuji, T., Church, D. and Buttle, D. (1981a) Development and use of a living skin substitute. *Plast. Reconstr. Surg.*, **67**, 386–392.

Bell, E., Sher, S., Hull, B., Merrill, C., Rosen, S., Chamson, A., Asselineau, D., Dubertret, L., Coulomb, B., Lapiere, C., Nusgens, B. and Neveux, Y. (1983) The reconstitution of living skin. *J. Invest. Dermatol.*, **81**, 2S-10S.

Bell, E., Moore, H., Mitchie, C., Sher, S. and Coon, H. (1984) Reconstitution of a thyroid gland equivalent from cells and matrix materials. *J. Exp. Zool.*, 232–277–285.

Bellucci, R.J. and Wolff, D. (1964) Experimental stapedectomy with collagen sponge implant. *The Laryngoscope*, **LXIV**, 668–688.

Benfer, J. and Struck, H. (1973) Accelerated fracture healing through soluble heterologous collagen. *Arch. Surg.*, **106**, 838–842.

Berg, R.A., Silver, F.H. and Pachence, J.M. (1989) Collagen matrix beads for soft tissue repair. U.S. Patent 4,837,255.

Bernengo, J.C., Roziero, M.C. Bezot, P., Bezot, G., Hertage, D. and Veis, A. (1983) A hydrodynamic study of collagen fibrillogenesis by electric birefringence and quaselastic light scattering. *J. Biol. Chem.*, **258**, 1001–1006.

Birk, D.E. and Silver, F.H. (1983) Corneal and scleral type I collagens: analyses of physical properties and molecular flexibility. *Int. J. Biol. Macromol.*, **5**, 209–214.

Birk, D.E. and Silver, F.H. (1984) Collagen fibrillogenesis in vitro: Comparison of types I, II and III. *Archives Biochemistry and Biophysics*, **35**, 178–185.

Black, B.S., Gher, M.E., Sandifer, J.B., Fucini, S.E. and Richardson, A.C. (1994) Comparative study of collagen and expanded polytetrafluoroethylene membranes in the treatment of human class III furcation defects. *J. Periodontol.*, **65**, 598–604.

Blanchy, B.G., Coulet, P.R. and Gautheron, D.C. (1986) Immobilization of factor XIII on collagen membranes. *J. Biomed. Materials Res.*, **20**, 469–479.

Bloch, A. and Oneson, I.B. (1960) Purification of collagen by enzymatic digestion. Brit. Pat. #831, 124, 1960.

Blumenthal, N. Sabet, T. and Barrington, E. (1986) Healing responses to grafting of combined collagen decalcified bone in periodontal defects in dogs. *J. Periodontol.*, **57**, 84–90.

Blumenthal, N. and Steinberg, J. (1990) The use of collagen membrane barriers in conjunction with combined demineralized bone-collagen gel implants in human infrabony defects. *J. Periodontology*, **61**, 319–327.

Blumenthal, N.M. (1993) A clinical comparison of collagen membranes with ϵ-PTFE membranes in the treatment of human mandibular buccal class II furcation defects, *J. Periodontol.*, **64**, 925–933.

Bowes, J.H. and Cater, C.W. (1968) The interaction of aldehydes with collagen. *Biochim. Biophys. Acta*, **168**, 341–352.

Bowes, J.H. and Moss, J.A. (1962). The effect of gamma irradiation on collagen. *Radiation Research*, **16**, 211–223.

Bruck, S.D. (1980) *Properties of Biomaterials in the Physiological Environment*, CRC Press, Boca Raton FL., pp.112–115.

Brumback and McPherson, S.D. (1967) Reconstituted collagen sutures in corneal surgery: An experimental and clinical evaluation. *Am. J. Ophthalmology*, **64**, 222–227.

Carter, D.M., Lin, A.N., Varghese, M.C., Caldwell, D., Pratt, L.A. and Eisinger, M. (1987) Treatment of junctional epidermolysis bullosa with epidermal autografts. *J. Am. Acad. Dermatol.*, **17**, 246–250.

Cavallaro, J.F., Kemp, P.D. and Kraus, K.H. (1994) Collagen fabrics as biomaterials. *Biotechnology and Bioengineering*, **43**, 781–791.

Chapman, J.A. and Hardcastle, R.A. (1974) The staining pattern of collagen fibrils. II. A comparison with patterns computer-generated from amino acid sequence. *Conn. Tiss. Res.*, **2**, 151–159.

Chen, Y.F., Gebhardt, B.M., Reidy, J.J. and Kaufman, H.E. (1990) Cyclosporine-containing collagen shields suppress corneal allograft rejection. *Am. J. Ophthalmolgy*, **109**, 132–137.

Chernomorsky, A. (1987) Effect of purification procedure on the biocompatibility of collagen-based biomaterials. MS Thesis, Rutgers University.

Cheung, D.T. and Nimni, M.E. (1982). Mechanism of crosslinking of proteins by glutaraldehyde I: Reaction with model compounds. *Connective Tissue Research*, **10**, 187–199.

Cho, K.J., Fanders, B., Smid, A. and McLaughlin, P. (1989) Experimental hepatic artery embolization with a collagen embolic agent in rabbits: A microcirculatory study. *Investigative Radiology*, **24**, 371–374.

Chvapil, M., Kronenthal, R.L. and van Winkle, W., Jr. (1962) Medical and surgical applications of collagen. In: International review of *Connective Tissue Research*, **6**, 1–61.

Chvapil, M., Heine, M.W. and Horton, H. (1976) The aceptance of the collagen sponge diaphram as an intravaginal contraceptive in human volunteer. *Fertility and Sterility*, **27**, 1398–1406.

Chvapil, M., (1977) Collagen sponge: Theory and practice of medical applications. *J. Biomed. Mater. Res.*, **11**, 721–741.

Chvapil, M. (1980) Reconstituted collagen. *In: Biology of collagen*, edited by A. Vidiik and J. Vuust, Academic Press, London, pp 313–324.

Cioca, G. (1981) Process for preparing macromolecular biologically active collagen. US Patent #4,279,812.

Cioca, G. (1983) Macromolecular biologically active collagen articles. US Patent #4,374,121.

Colangelo, P., Piattelli, A., Barrucci, S., Trisi, P., Formisano, G., and Caiazza, S. (1994) Bone regeneration guided by resorbable collagen membranes in rabbits: A pilot study, *Implant Dentistry*, **2**, 101–105.

Colin, W. and Donoff, R.B. (1984) Nerve regeneration through collagen tubes. *J. Dent. Res.*, **63**, 987–993.

Collins RL. Christiansen D. Zazanis GA. Silver FH. (1991) Use of collagen film as a dural substitute: preliminary animal studies. *Journal of Biomedical Materials Research*, **25**, 267–276.

Condell, R.A., Sakai, N., Mercado, R.T. and Larenas, E. (1988) Quantitation of collagen fragments and gelatin by deconvolution of polarimetry denaturation curves. *Collagen Rel. Res.*, **8**, 407–418.

Cooperman, L. and Michael, D. (1984) The immunogenicity of injectable collagen., I. A 1-year prospective study, *J. Am. Acad. Dermatol.*, **10**, 638–646.

Cooperman, L. and Michaeli, D. (1984a) The immunogenicity of injectable collagen. II. A retrospective Review of seventy-two tested and treatment patients, *J. Am. Acad. Dermatol.*, **10**, 647–651.

Coquin, J.Y., Sebahoun, G., Fontaine, M., Debbas, N. and Carcassonne, Y. (1987) Use of collagen for control of local bleeding after jugular vein catherization in thrombopenic patients, *Current Therapeutic Research*, **42**, 1066–1072.

Courts, A. (1960) Structural changes in collagen. The action of alkalis and acids in the conversion of collagen into eucollagen. *Biochem. J.*, **74**, 238–247.

Daniels, J.R. and Knapp, R. (1976) Process for augmenting connective mammalian tissue within situ polymerizable native collagen solution. US Patent #3, 949, 073.

Danielsen, C.C. (1990) Age-related thermal stability and susceptibility to proteolysis of rat bone collagen. *Biochem. J.*, **272**, 697–701.

Davison, P.F. and Drake, M.P. (1966) The physical characterization of monomeric tropcollagen. *Biochemistry*, **5**, 313–321.

DeBlois, C., Cote, M.-F. and Doillon, C.J. (1994) Heparin-fibroblast growth factor-fibrin complex: In vitro and in vivo applications collagen-based materials. *Biomaterials*, **15**, 665–672.

De La Torre, J.C. (1982) Catecholamine fiber regeneration across a collagen bioimplant after spinal cord transection. *Brain Research Bulletin*, **9**, 545–552.

DeLustro, F., Condell, R.A., Nguyge, M.A. and McPherson, J.M. (1986) A comparative study of the biological and immunological response to medical devices derived from dermal collagen *J. Biomed. Mater. Res.*, **20**, 109–120.

Doillon, C.J., Whyne, C.E., Brandwein, S. and Silver, F.H. (1986) Collagen-based wound dressings: Control of the pore structure and morphology. *J. Biomedical Materials Research*, **20**, 1219–1228.

Doillon, C.J., Silver, F.H. and Berg, R.A. (1987) Fibroblast growth on a porous collagen sponge containing hyaluronic acid and fibronectin. *Biomaterials*, **8**, 195–200.

Doillon, C.J., Wasserman, A.J., Berg, R.A. and Silver, F.H. (1988) Behavior of fibroblasts and epidermal cells cultivated on analogues of extracellular matrix. *Biomaterials*, **9**, 91–96.

Drake, M.P., Davison, P.F., Bump, S. and Schmitt, F.O. (1966) Action of proteolytic enzymes on tropocollagen and insoluble collagen. *Biochemistry*, **5**, 301–312.

Eisinger, M., Monden, M., Raff, J.H. and Fortner, J.G. (1980) Wound coverage by a sheet of epidermal cells grown in vitro from dispersed single cell preparations. *Surgery*, **88**, 287–293.

Ellingsworth, L.R., DeLustro, F., Brennan, J.E., Sawamura, S. and McPherson, J. (1986) The human immune response to reconstituted bovine collagen. *J. Immunol.*, **136**, 877–882.

Ennker, I.C., Ennker, J., Schoon, D., Schoon, H.A., Rimpler, M., Hetzer, R. (1994) Formaldehyde-free collagen glue in experimental lung gluing. *Annals of Thoracic Surgery*, **57**, 1622–1627.

Ernst, S.M.P.G., Tjonjoegin, M., Schrader, R., Kaltenbach, M., Sigwart, U., Sanborn, T.A. And Plokker (1993) Immediate sealing of arterial puncture sites after cardiac catheterization and coronary angioplasty using a biodegradable collagen plug: Results of an international registry. *J. Am. Coll. Cardiol.*, **21**, 851–855.

Farber, S., Garg, A. and Silver, F.H. (1986) Collagen fibrillogenesis in vitro: Evidence of prenucleation and nucleation steps. *Int. J. Biol. Macromol.*, **8**, 37–42.

Fletcher, G.C. (1976) Dynamic light scattering from collagen solutions. I. Translational diffusion coefficient and aggregation effects. *Biopolymers*, **15**, 2201–2217.

Foran, J.P.M., Patel, D., Brookes, J. and Wainwright , R.J. (1993) early mobilization after percutaneous cardiac catheterization using collagen plug (VasoSeal) haemostasis. *Br. Heart J.*, **69**, 424–429.

Freeman, A.E., Igel, H.J., Waldman, N.L. and Losikoff, A.M. (1974) A new method for covering large surface area wounds with autografts. I. In vitro multiplication of rabbit-skin epithelial cells, *Arch. Surg.*, **108**, 721–723.

Frey P., Lutz N. Berger D. Herzog B. (1994) Histological behavior of glutaraldehyde cross-linked bovine collagen injected into the human bladder for the treatment of vesicoureteralreflux. *Journal of Urology*, **152**, 632–635.

Furthmayr, H. and Madri, J.A. (1982) Rotary shadowing of connective tissue macromolecules. *Collagen Rel. Res.*, **2**, 349–363.

Garg, A.K., Berg, R.A., Silver, F.H. and Garg, H.G. (1989) Effect of proteoglycans on type I collagen fiber formation. *Int. J. Biomaterials*, **10**, 413–419.

Gelman, R.A. and Piez, K.A. (1980) Collagen fibril formation in vitro. A quasielastic light-scattering study of early stages. *J. Biol. Chem.*, **255**, 8098–8102.

Gongloff, R.K. (1988) Use of collagen tube contained implants for particulate hydroxylapatite for ridge augmentation. *J. Oral Maxillofac. Surg.*, **46**, 641–647.

Goycoolea, M.V., Muchow, D.C., Scholz, M.T., Sirvio, L.M., Stypulkowski. P.H. (1991) In search of missing links in otology. I. Development of a collagen-based biomaterial. *Laryngoscope*, **101**, 717–726.

Guidon, R., Marceau, D., Couture, J., Rao, T.J., Merhi, Y., Roy, P.-E. and De la Faye, D. (1989) Collagen coatings as biological sealants for textile arterial prostheses. *Biomaterials*, **10**, 156–165.

Hall, C.E. (1956) Visualization of individual macromolecules with the electron microscope. *Proc. Natl. Acad. Sci., USA*, **42**, 801–807.

Hansborough, J.F., Boyce, S.T., Cooper, M.L. and Foreman, T.J. (1989) Burn wound closure with cultured autologous keratinocytes and fibroblasts attached to a collagen-glycosaminoglycan substrate. *JAMA*, **262**, 2125–2130.

Hasty, B., Heuer, D.K. and Minckler, D.S. (1990) Primate trabeculectomies with 5–fluorouracil collagen implants. *Amer. J. Ophthalmolgy*, **109**, 721–725.

Henry, F., Nestler, M., Hvidt, S., Ferry, J.D. and Veis, A. (1983) Flexibility of collagen determined from dilute solution viscoelastic measurements. *Biopolymers*, **22**, 1747–1758.

Hikichi, T., Uend, N., Trempe, C.L. and Chakrabarti, B. (1994) Cross-linking of dermal collagen induced by singlet oxygen. *Biochemistry and Molecular Biology International*, **33**, 497–504.

Hilenski, L.L., Terracio, L. and Borg, T.K. (1991) Myofibrillar and cytoskeletal assembly in neonatal rat cardiac myocytes cultured on collagen and laminin. *Cell Tissue Res.*, **264**, 577–587.

Hogervorst, T., Meijer, D.W. and Klopper, P.J. (1992) The effect of a TCP-collagen implant on the healing of articular cartilage defects in the rabbit knee joint. *J. Applied Biomaterials*, **3**, 251–258.

Hughes, K.E., Fink, D.J., Hutson, T.B. and Veis, A. (1984) Oriented fibrillar collagen and its application to biomedical devices. *J. American Leather Chemists Association*, **LXXIX**, 146–158.

Hughes, K.E., Hutson, T.B. and Fink, D.J. (1985) Collagen orientation. U.S. Patent # 4,544,516.

Hull, B.E., Finley, R.K. and Miller, S.F. (1990) Coverage of full-thickness burns with bilayered skin equivalents: A preliminary clinical trial. *Surgery*, **107**, 496–502.

Hwang, D.G., Stern, W.H., Hwang, P.H. and MacGowan-Smith, L.A. (1989) Collagen shield enhancement of topical dexamethasone penetration. *Archives Ophthalmology*, **107**, 1375–1380.

Ipsen, T., Jorgensen, P.S., Damholt, V. and Torholm, C. (1991) Gentamicin-collagen sponge for local applications. 10 cases of chronic asteomyelitis followed for 1 year. *Acta Orthop. Scand.*, **62**, 592–594.

Jonak, R., Nemetschek-Gansler, H., Nemetschek, Th., Riedl, H., Bordas, J. and Koch, M. (1979) Glutaraldehyde-induced states of stress of the collagen triple helix. *J. Mol. Biol.*, **130**, 511–512.

Kamer, F.M and Churukian, M.M. (1984) Clinical use of Injectable collagen: A three-year retrospective review. *Archives of Otolaryngology*, **110**, 93–98.

Karasek, M.A. and Charlton, M.E. (1971) Growth of post-embryonic skin epithelial cells on collagen gels. *J. Invest. Dermatol.*, **56**, 205–210.

Kato, Y.P., Christiansen, D.L., Hahn, R.A., Shieh, S.-J., Goldstein, J. and Silver, F.H. (1989) Mechanical properties of collagen fibers: A comparison of reconstituted and rat tail tendon fibers. *Biomaterials*, **10**, 38–42.

Kato, Y.P., Dunn, M.G., Zawadsky, J.P., Tria, A.J., and Silver, F.H. (1991) Regeneration of Achilles tendon using a collagen tendon prosthesis: Results of a year long implantation study. *Journal Bone and Joint Surgery*, **73–A**, 561–574.

Kelman, C.D. and DeVore, D.P. (1994) Human collagen processing and auto implant use. U.S. Patent # 5,332,802.

Kent, M.J.C., Light, N.D. and Bailey, A.J. (1985) Evidence for glucose-mediated covalent cross-linking of collagen after glycosylation in vitro. *Biochem. J.*, **225**, 745–752.

Kljavin, I.J. and Madison, R.D. (1991) Peripheral nerve regeneration within tubular prostheses: Effects of laminin and collagen matrices on cellular ingrowth. *Cells and Materials*, **1**, 17–28.

Klein, D.G. (1965) Dural replacement with resorbable collagen. *Arch. Surg.*, **91**, 924–929.

Klein, A.W. (1983) Implantation technics for injectable collagen: Two and one-half years of personal clinical experience. *J. Amer. Acad. of Dermatology*, **9**, 224–228.

Knapp, T.R., Kaplan and Daniels, J.R. (1977) Injectable collagen for soft tissue augmentation. *Plastic and Reconstructive Surgery*, **60**, 398–405.

Komanowsky, M., Sinnamon, H.I., Elias, S., Heiland, W.K. and Aceto, N.C. (1974) Comminuted collagen for novel applications. *J. of the American Leather Chemists Association*, **LXIX**, 9 410–422.

Ksander, G.A. (1988) Collagen coatings reduce the incidence of capsular contracture around soft silicone rubber implants in animals. *Annals of Plastic Surgery*, **20**, 215–224.

Kurze, T., Apuzzo, M.L.J., Weiss, M.H. and Heiden, J.S. (1975) Collagen sponge for surface brain protection. *J. Neurosurg.*, **43**, 637–638.

Laquerriere, A., Yun, J., Tiollier, J., Hemet, J. and Tadie, M. (1993) Experimental evaluation of bilayered human collagen as a dural substitute. *J. Neurosurg.*, **78**, 487–491.

Leiken, S., Raw, D.C. and Parsegian, V.A. (1994) Direct measurement of forces between self-assembled proteins: Temperature-dependent exponential forces between collagen triple helices. *Proc. Natl. Acad. Sci.*, **91**, 276–280.

Leipziger, L.S., Glushko, V., DiBernardo, B., Shafaie, F., Noble, J., Nichols and Alvarez, O.M. (1985) Dermal wound repair: Role of collagen matrix implants and synthetic polymer dressings. *J. Am. Acad. Dermatol.*, **12**, 409–419.

Leighton, J., Tchao, R. and Nichols, J. (1985) Radial gradient cell culture on the inner surface of collagen tubes. *In Vitro Cell Develop. Biol.*, **21**, 713–715.

L'esperance, F.A. (1965) Reconstituted collagen tape in retinal detachment surgery. *Archives of Ophthalmolgy*, **73**, 472–475.

Luck, E.E. and Daniels, J.F. 1979) Aqueous collagen composition. US Patent #4,140,537.

Marchand, R. and Woerly, S. (1990) Transected spinal cords grafted with in situ self-assembled collagen matrices. *Neuroscience*, **36**, 45–60.

Marescaux, J.F., Aprahamian, M., Mutter, D., Loza, E., Wilhelm, M., Sonzini, P. and Damge, C. (1991) Prevention of anastomosis leakage: An artificial connective tissue. *Br. J. Surg.*, **78** : 440–444.

McBride, D.J., Hahn, R.A. and Silver, F.H. (1985) Morphological characterization of tendon development during chick embryogenesis: Measurement of birefringence retardation. *Int. J. Biol. Macromol.*, **7**, 71–76.

McClain, P.E. and Wiley, E.R. (1972) Differential scanning colorimeter studies of the thermal transitions of collagen. Implications on structure and stability. *J. Biol. Chem.*, **247**, 692–697.

McCoy, J.P., Schade, W.J., Siegle, R.J., Waldinger, T.P., Vanderween, E.E. and Swanson, N.A. (1985) Characterization of humoral immuno response to bovine collagen implants. *Arch. Dermatol*, **121**, 990–994.

Meade, K.R. and Silver, F.H. (1990) Immunogenicity of collagenous implants. *Biomaterials*, **11**, 176–180.

Meddings RN. Carachi R. Gorham S. French DA. (1993) A new bioprosthesis in large abdominal wall defects. *Journal of Pediatric Surgery*, **28**, 660–663.

Meek, K.M., Chapman, J.A. and Hardcastle, R.A. (1979) Improved correlation with sequence data. *J. Biol. Chem.*, **254**, 10710–10714.

Miller, E.J. and Rhodes, R.K. (1982) Preparation and characterization of the different types of collagen. *Methods In Enzymology*, **82**, 33–64.

Moore, W.S., Chvapil, M., Seiffert, G. and Keown, K. (1981) Development of an infection-resistant vascular prosthesis. *Arch. Surg.*, **116**, 1403–1407.

Myers, J.C., Yang, H., D'Ippolito, Presente, A., Miller, M.K. and Dion, A.S. (1994) The triple-helical region of human type XIX collagen consists of multiple collagenous subdomains and exhibts limited sequence homology to $\alpha 1$(XVI). *J. Biol. Chem.*, **269**, 18549–18557.

Na, G.C. (1988) UV spectroscopic characterization of type I collagen. *Collagen Rel. Res.*, **8**, 315–330.

Nagelschmidt, M. and Viell, B. (1987) Polarimetric assay for the determination of the native collagen content of soluble collagen. *J. Biomedical Materials Research*, **21**, 201–209.

Nanchahal, J., Otto, W.R., Dover, R. and Dhital, S.K. (1989) Cultured composite skin grafts: Biological skin equivalents permitting massive expansion, *Lancet*, 191–193.

Nashihara, T. (1963) Isolation of collagen from collagenous sources. Ger. Pat. #1,145,904.

Natsume, T., Ike, O., Okada, T., Takimoto, N., Shimizu, Y. and Ikada, Y. (1993) Porous collagen sponge for esophageal replacement. *J. Biomedical Materials Research*, **27**, 867–875.

Obrink, B. (1972) Non-aggregated tropocollagen at physiological pH and ionic strength: A chemical and physio-chemical characterization of tropocolagen isolated from the skin of lathytic rats. *Eur. J. Biochem.*, **25**, 563–572.

Okumura N., Nakamura T., Natsume T., Tomihata K., Ikada Y. Shimizu Y. (1994) Experimental study on a new tracheal prosthesis made from collagen-conjugated mesh. *Journal of Thoracic & Cardiovascular Surgery*, **108**, 337–45.

Oliver, R.F., Barker, H., Cooke, A. and Grant, R.A. (1982) Dermal collagen implants. *Biomaterials*, **3**, 38–40.

Oneson, I.B. (1964) Method of producing soluble collagen. US Patent #3,131,130.

Oneson, I.B., Fletcher, D., Olivo, J., Nichols, J. and Kronenthal, R. (1970) The preparation of highly purified insoluble collagens. *J. American Leather Chemists Association*, **LXV**, 440–450.

Pachence, J.M., Berg, R.A. and Silver, F.H. (1987) Collagen: Its place in the medical device industry. Medical Device Diagnostic Industry January, 49–55.

Paul, P.F., Mellonig, J.T., Towle, H.J, 3rd, and Gray, J.L. (1992) Use of a collagen barrier to enhance healing in human periodontal furcation defects. *Intl. J. Periodontics and Restorative Dentistry*, **12**, 123–131.

Payne, K.J. and Veis, A. (1988) Fourier transform IR spectroscopy of collagen and gelatin solutions: Deconvolution of amide I band for conformational studies. *Biopolymers*, **27**, 1749–1760.

Peacock, E.E., Jr., Seigler, H.F. and Biggers, P.W. (1965) Use of tanned collagen sponges in the treatment of liver injuries. *Annals of Surgery*, **161**, 238–247.

Petite, H., Free, V., Huc, A. and Herbage, D. (1994) Use of diphenylphosphorylazide for crosslinking collagen-based biomaterials. *J. Biomedical Materials Research*, **28**, 159–165.

Pietrucha, K. (1991) New collagen dural implant as dural substitute. *Biomaterials*, **12**, 320–323..

Piez, K.A., Lewis, M.S., Martin, G.R. and Gross, J. (1961) Subunits of the collagen molecule. *Biochem. Biophys. Acta*, **53**, 596–598.

Pimentel, E.R. (1981) Form birefringence of collagen bundles. *Acta Histochem. Cytochem.*, **14**, 35–40.

Pitaru, S., Tal, H., Soldinger, M., Azar-Avidan, O. and Noff, M. (1987) Collagen membranes prevent the apical migration of epithelium during periodontal wound healing, *J. Periodontology Research*, **22**, 331–333.

Porat, S., Rousso, M. and Shoshan, S. (1980) Improvement of gliding fuction of flexor tendons by topically applied enriched collagen solution. *J. Bone and Joint Surgery*, **62–B**, 208–213.

Poten, B. and Nordgaard, J.A. (1976) The use of collagen film (Cutycol ^{R}) as a dressing for donor areas in split skin grafting. *Scand. J. Plast. Reconstr. Surg.*, **10**, 237–240.

Prockop, D.J. (1992) Mutations in collagen genes as a cause of connective tissue diseases. *N. Engl. J. Med.*, **326**, 540–546.

Quteish, D. and Dolby, A.E. (1992) The use of irradiated-crosslinked human collagen membrane in guided tissue regeneration. *J. Clin. Periodontol.*, **19**, 476–484.

Riley, R.W. and Leake, D.L. (1976) Cancellous bone grafting with collagen stents. *International Journal of Oral Surgery*, **5**, 29–32.

Rubin, A.L., Riggio, R.R., Nachman, R.L., Schwartz, G.H., Miyata, T. and Stenzel, K.H. (1968) Collagen materials in dialysis and implantation. *Trans. Amer. Soc. Artif. Int. Organs*, **14**, 169–174.

Sakata, N., Kawamura, K. and Takebayashi, E. (1990) Effects of collagen matrix on proliferation and differentiation of vascular smooth muscle cells in vitro, *Exp. Mol. Pathol.*, **52**, 179–191.

Schmitt, F.O., Hall, C.E. and Jakus, M.A. (1942) Electron Microscope Investigation of the Structure of Collagen. *J. Cell and Comp. Physiol.*, **20**, 11–33.

Semich, R. and Robenek, H. (1990) Organization of the cytoskeleton and focal contacts of bovine arotic endothelial cells cultured on type I and III collagen, *J. Histochem. Cytochem.*, **38**, 59–67.

Silver, F.H., Langely, K.H. and Trelstad, R.L. (1979) Type I collagen fibrillogenesis: Initiation via reversible linear and lateral growth steps. *Biopolymers*, **18**, 2523–2535.

Silver, F.H. and Trelstad, R.L. (1980) Type I collagen in solution. Structure properties of fibril fragments. *J. Biol. Chem.*, **255**, 9427–9433.

Silver, F.H. and Trelstad, R.L. (1981) Physical properties of type I collagen in solution: structure of α-chains and β and γ-components and two component mixtures. *Biopolymers*, **20**, 359–371.

Silver, F.H. and Birk, D.E. (1984) Molecular structure of collagen in solution. Comparison of types I, II, III and V. *Int. J. Biol. Macromol.*, **6**, 125–132.

Silver, F.H. (1987) Physical structure and modeling. *Biological Materials: Structure, Mechanical Properties and Modeling of Soft Tissues*, NYU Press, Chapter 4.

Silver, F.H. (1987a) Self-assembly of connective tissue macomolecules, *Biological Materials: Structure, Mechanical Properties and Modeling of Soft Tissue*, NYU Press, Chapter 5.

Silver, F.H. and Kato, Y.P. (1992) Synthetic collagen orthopaedic structures such as grafts, tendons and other structures. U.S. Patent 5,171,273.

Silver, F.H., Kato, Y.P., Ohno, M. and Wasserman, A.J. (1992) Analysis of mammalion connective tissue: Relationship between hierarchical structures and mechanical properties. *J. Long-Term Effects of Medical Implants*, **2**, 165–198.

Silver, F.H. and Pins, G. (1992) Cell growth on collagen: A review of tissue engineering using scaffolds containing extracellular matrix. *J. Long-Term Effects of Medical Implants*, **2**, 67–80.

Silver, F.H. (1994) 510(k) and PMA Regulatory Filings in the US, *Biomaterials Medical Devices and Tissue Engineering: An Integrated Approach*, Chapman & Hall, London, Chapter 9.

Stenzel, K.H., Miyata, T. and Rubin, A.L. (1974) Collagen as a biomaterial. *Annu. Rev. Biophys.*, **3**, 231–253.

Stone, K.R., Rodkey, W.G., Webber, R.J., McKinney, L. and Steadman, J.R. (1990) Collagen-based prostheses for meniscal regeneration. Future directions. *Clin. Orthop. Rel. Res.*, **252**, 129–135.

Tachibana, M., Nagamatsu, G.R. and Addonizio, J.C. (1985) Ureteral replacement using collagen sponge tube grafts. *J. of Urology*, **133**, 866–869.

Takaoka, K., Koezuka, M. and Nakahara, H. (1991) Telopeptide-depleted bovine skin collagen as a carrier for bone morphogenetic protein, *J. Ortho. Res.*, **9**, 902–907.

Tanner, M.G., Solt, C.W. and Vuddhakanok, S. (1988) An evaluation of new attachment formation using a microfibrillar collagen barrier. *J. Periodontol.*, **59**, 524–530.

Thomas, J.C. and Fletcher, G.C. (1979) Dynamic light scattering from collagen solutions. II. Photon correlation study of the depolarized light. *Biopolymers*, **18**, 1333–1352.

Topol, B., Haimes, H., Dubertret, L. and Bell, E. (1986) Transfer of melanosomes in a skin equivalent model in vitro, *Invest. Dermatol.*, **87**, 642–647.

Truscott, W. (1994) ISO 10993: A world standard for assessing material biocompatibility? *Medical Device & Diagnoistic Industry*, January, 176–193.

van der Rest, M. and Garrone, R. (1991) Collagen family of proteins. *FASEB*, **5**, 2814–2823.

van der Rest, M. and Bruckner, P. (1993) Collagens: diversity at the molecular and supramolecular levels. *Current Opinion in Structural Biology*, **3**, 430–436.

Vanderveen, E.E., McCoy, J.P., Schade, W., Kapur, J.J., Hamilton, T., R., Ragsddale, C., Frekin, R.C. and Swanson, N.A. (1986) The association of HLA and immune responses to bovine collagen implants. *Arch. Dermatol.*, **122**, 650–654.

Wasserman, A.J., Doillon, C.J., Glasgold, A.I., Kato, Y.P., Christiansen, D., Rizvi, A., Wong, E., Goldstein, J. and Silver, F.H. (1988) Clinical application of electron microscopy in the analysis of collagenous biomaterial. *Scanning Microscopy International*, **2** 1635–1646.

Weadock, K., Olson, R.M. and Silver, F.H. (1984) Evaluation of collagen crosslinking techniques. *Biomat. Med. Dev. Art. Org.*, **11**, 293–318.

Weadock, K., Silver, F.H. and Wolff, D. (1986) [125] I-calmodulin through collagen membranes: Effect of source concentration and membrane swelling ratio. *Biomaterials*, **7**, 263–276.

Weadock, K. S., Wolff, D. and Silver, F.H. (1987) Diffusivity of [125] I-labelled macromolecules through collagen: mechanism of diffusion and effect of adsorption. *Biomaterials*, **8**, 105–112.

Weinberg, C. and Bell, E. (1986) A blood vessel model constructed from collagen and cultured vascular cell. *Science*, **231**, 397–400.

Weiner, A.l., Carpenter-Green, S.S., Soehngen, E.C., Lenk, R.P. and Popescu, M.C. (1985) Liposome-collagen gel matrix: A novel sustained drug delivery system. *J. Pharmaceutical Sciences*, **74**, 922–925.

Worst, P.K.M., Valentine, E.A. and Fusenig, N.E. (1974) Formation of epidermis after reimplantation of pure primary epidermal cell cultures from perinatal mouse skin *J. Natl. Cancer Inst.*, **53**, 1061–1064.

Yaffe, A., Ehrlich, J. and Shoshan, S. (1982) One-year follow-up for the use of collagen for biological anchoring of acrylic dental roots in dogs. *Archs. Oral Biol.*, **27**, 999–1001.

Yannas, I.V., Burke, J.F., Orgill, D.P. and Skrabut, E.M. (1982) Wound tissue can utilize a polymeric template to synthesize a functional extension of skin. *Science*, **215**, 174–176.

Yannas, I.V., Orgill, D.P., Silver, J., Norregaard, N.T., T.V., Zervas, N.T. and Schoene, W.C. (1985) Polymeric template facilitates regeneration of sciatic nerve across 15-mm gap. *Trans. Soc. Biomat.*, **9**, 175.

Yannas, I.V. and Orgill, D.P. (1986) Artificial skin: A fifth route to organ repair and replacement In: Piskin, E. and Hoffman, A.S., Eds., Polymeric Biomaterials, Dordrecht: Martinus Nijhoff.

18. FIBRINOGEN AND FIBRIN: CHARACTERIZATION, PROCESSING AND MEDICAL APPLICATIONS

JOHN W. WEISEL[1] and STEWART A. CEDERHOLM-WILLIAMS[2]

[1]*Department of Cell and Developmental Biology, University of Pennsylvania School of Medicine, Philadelphia, Pennsylvania 19104–6058, USA*
[2]*Oxford Bioresearch Laboratory, The Magdalen Center, Oxford Science Park, Oxford OX4 4GA, England*

INTRODUCTION

Fibrinogen is essential for hemostasis. It is a glycoprotein normally present in human blood plasma at a concentration of about 2.5 g/L. It is a soluble macromolecule, but forms an insoluble gel on conversion to fibrin by the action of the serine proteolytic enzyme, thrombin, which itself is activated by a cascade of enzymatic reactions triggered by injury or a foreign surface (Figure 1). Fibrinogen is also necessary for the aggregation of blood platelets, an initial step in blood clotting. Each end of a fibrinogen molecule can bind with high affinity to the integrin receptor on activated platelets, $\alpha_{IIb}\beta_3$, so that the bi-functional fibrinogen molecules act as bridges to link platelets. In its various functions as a clotting and adhesive protein, the fibrinogen molecule is involved in many intermolecular interactions and has specific binding sites for several proteins and cells (Table 1).

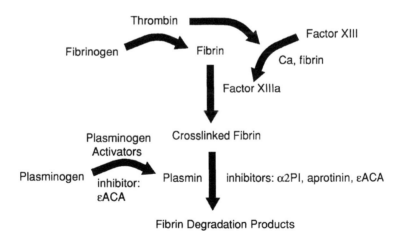

Figure 1 Some of the chemical reactions involved in fibrin clotting and degradation. Thrombin cleaves the fibrinopeptides from fibrinogen to generate a fibrin clot. Thrombin also activates the transglutaminase, Factor XIII, which covalently crosslinks fibrin to stabilize the clot. Plasminogen activators convert plasminogen to active plasmin in the presence of fibrin. As a part of normal wound healing, plasmin cleaves fibrin into fibrin degradation products to dissolve the clot. Various inhibitors, such as aprotinin, ϵ-amino caproic acid (ϵACA), and α_2-plasmin inhibitor (α2PI), can block plasminogen activation and/or fibrin degradation.

Table 1 Functional Sites on Fibrinogen

Sites related to blood coagulation and fibrinolysis
calcium binding
thrombin cleavage
polymerization
ligation
plasmin cleavage

Protein binding sites
thrombin
Factor XIII
plasmin(ogen)
plasminogen activators
α_2-antiplasmin
fibronectin
thrombospondin
collagen

Cell binding sites
platelets
fibroblasts
endothelial cells
monocytes
macrophages
erythrocytes
Streptococcal bacteria
Staphylococcal bacteria

Fibrin clots are dissolved by another series of enzymatic reactions termed the fibrinolytic system (Figure 1). The proenzyme, plasminogen, is activated to plasmin by a specific proteolytic enzyme, typically tissue-type plasminogen activator or urokinase plasminogen activator. Plasmin then cleaves fibrin at certain unique locations to dissolve the clot. The activation of the fibrinolytic system is greatly enhanced by taking place on the surface of fibrin. Thus, these reactions are highly specific for cleavage of the insoluble fibrin clot, rather than circulating fibrinogen.

There is a dynamic equilibrium between clotting and fibrinolysis. Any imbalance can result in either bleeding or thrombosis, causing blockage of the flow of blood through a vessel. Thrombosis, often accompanying atherosclerosis or other pathological processes, is the most common cause of myocardial infarction and stroke. As a consequence, various fibrinolytic enzymes that activate plasminogen are now commonly used clinically to treat these conditions.

Fibrinogen was first isolated in large quantities more than fifty years ago (Cohn *et al.*, 1946). A few of the many different purification methods will be mentioned here. Since it was first purified, fibrinogen has been used in various medical applications. However, recent technical developments have accelerated the pace of discovery of potential applications and diminished some of the hazards and side effects of its use.

An increasingly important clinical application of fibrinogen is its use as a biodegradable adhesive or sealant (Schlag & Redl, 1988; Gibble & Ness, 1990; Lerner & Binur, 1990; Sierra, 1993; Atrah, 1994). Because of the nature of the coagulation process, fibrinogen and thrombin can be used in a manner analogous to a two-part epoxy resin, with application either simultaneously or sequentially to the repair site by squirting or spraying with a special syringe or related device. Both the extrinsic and intrinsic mechanisms of coagulation are bypassed, but the final common pathway is replicated by such methods. The solutions can be applied precisely in a very localized site or broadly over a wide area to any biological tissue. The clot will adhere to a moist or even wet surface. After polymerization, the clot becomes a semi-rigid, fluid-tight sealant which can hold tissue together and stop bleeding. The strength and biological properties can be adjusted by varying the concentrations of fibrinogen and thrombin and other proteins that are added. The fibrinolytic system and the cells and other proteins normally present during wound healing will degrade the clot in a physiological fashion.

Most of the aspects of the biology and biochemistry of fibrinogen and fibrin considered briefly here have been reviewed more extensively in a variety of contexts (Doolittle, 1984; Budzynski, 1986; Henschen & McDonagh, 1986; Shafer & Higgins, 1988; Mosesson, 1990; Blombäck, 1991; Hantgan *et al.*, 1994).

METHODS OF ISOLATION

Fibrinogen and the other components of the clotting system are normally isolated from blood plasma, but there are many variations in purification methods. In addition, preparations may be either large-scale using pooled plasma from multiple donors or single donor (autologous). Since fibrinogen is one of the least soluble plasma proteins, purification methods have generally been based on its low solubility in various solvents or its isoelectric point. Fibrinogen is particularly insoluble at low temperatures, a fact utilized in the classical precipitations by cold ethanol (Cohn *et al.*, 1946) or glycine (Blombäck & Blombäck, 1956). Precipitations by ammonium sulfate (Laki, 1951), polyethylene glycol (Longas *et al.*, 1980), low pH (Hafter *et al.*, 1978), or freeze-thawing (Ware *et al.*, 1947) have also been described. Several examples illustrating the range of methodology in commercial and clinical laboratory settings will be given.

Tissucol or Tisseel is marketed in Europe as a two-component kit with lyophilized human fibrinogen/Factor XIII and human thrombin (bovine thrombin was used until recently). The fibrinogen is reconstituted in an antifibrinolytic solution, while the thrombin is reconstituted with calcium chloride. The fibrinogen component is made from pooled human donor plasma by cryoprecipitation followed by washing with a citrate buffer to extract the cold-soluble proteins. The extract is then lyophilized in a buffer containing various inhibitors and pasteurized by steam. On reconstitution, the solution contains about 110–130 mg/mL fibrinogen, 2–9 mg/mL fibronectin, 10–50 units Factor XIII, 3000 KIU/mL aprotinin, and 70–110 mg/ml plasminogen.

Fibronectin, Factor XIII and plasminogen are all proteins normally present in blood, and some of their functions will be described below. Aprotinin is a small peptide (6200 molecular mass), derived from bovine lung tissue, that is added to many sealant formulations to prolong the *in vivo* lifetime of the sealant. Aprotinin inhibits plasmin by combining almost irreversibly with the active center of the enzyme. The benefit of adding aprotinin to fibrin sealants is not wholly proven and does present certain clinical safety issues since, being a foreign protein, it is known to produce hypersensitivity reactions and even fatal anaphylaxis in some individuals. In addition, aprotinin is a potential vehicle for the transmission of infectious agents, e.g. bovine spongiform encephalopathy.

Cryoprecipitation is a commonly used method of preparing single donor autologous fibrin sealant. While citrated plasma is stored for at least one hour at $-20°C$ or less, fibronectin acts as a nidus for fibrinogen, thrombospondin, von Willebrand factor and plasminogen to aggregate and become insoluble (Gestring & Lerner, 1983). Then, the preparation is slowly thawed overnight between 2 and 8°C and the precipitate is isolated by centrifugation. Repeated cycles of freezing and thawing can yield fibrinogen concentrations of 40–60 mg/mL. Preparations can be stored frozen with no appreciable degradation. These methods necessitate the use of a skilled technician and require some forethought.

Ethanol precipitation has been a commonly used method to isolate fibrinogen since it was developed by Cohn and colleagues (Cohn *et al.*, 1946). A self-contained system for preparation of autologous fibrinogen by the ethanol precipitation method has been developed (Weis-Fogh, 1988; Dahlstrøm *et al.*, 1992). Blood is collected from the patient with a syringe containing anticoagulants and centrifuged. Using a system of sterile syringes and bags connected by tubing, the plasma supernatant is treated with ethanol and centrifuged again. After dissolving the precipitate in plasma, the protein concentrate is ready for use. However, the home-made single donor or autologous preparations made using ethanol can give a product with variable fibrinogen levels, which is a criticism also directed to single donor cryoprecipitation procedures.

All of the preceding methods are based on the principle of the addition of thrombin to fibrinogen to make a gel. Recently, a different approach has been introduced that is based on the solubility of fibrin monomer in dilute acid. Clots made by the addition of thrombin to fibrinogen in plasma can be dissolved in dilute acetic acid; on return to neutral pH, the fibrin monomer reassembles to yield a typical fibrin clot (Belitser *et al.*, 1965). In a recent application of this technique, a specially developed apparatus is used to remove cells and prepare fibrin monomer from a patient's blood (Amery *et al.*, 1995). A snake venom enzyme (batroxobin) that preferentially cleaves the A fibrinopeptides and is not particularly antigenic in humans converts fibrinogen in the plasma to fibrin, which is solubilized and concentrated by dissolution in a small volume of dilute acetic acid. This method yields 4–5 ml of desA fibrin solution at 25 ± 4 mg/ml. The batroxobin is removed, and then a clot is produced by rapidly mixing the fibrin monomer with one fifth volume of pH 10 bicarbonate buffer.

CHARACTERIZATION OF FIBRINOGEN

Fibrinogen is made up of three pairs of polypeptide chains, designated Aα, Bβ and γ, with molecular masses of 66,500, 52,000, and 46,500, respectively (Table 2 and Figure 2). The post-translational addition of asparagine-linked carbohydrate to the Bβ and γ chains brings the total molecular mass to about 340,000. The nomenclature for fibrinogen arises from the designation of the small peptides that are cleaved from fibrinogen by thrombin to yield fibrin as fibrinopeptides A and B and the parent chains, without the fibrinopeptides, as α and β. No peptides are cleaved from the γ chains by thrombin. The physicochemical characteristics of fibrinogen are listed in Table 3.

Table 2 Fibrinogen's Polypeptide Chains

	Aα	Bβ	γ
number of residues	610	461	411
amino terminal residue	alanine	pyroglutamic acid	tyrosine
carboxyl terminal residue	valine	glutamine	valine
high affinity calcium binding sites			311–336
carbohydrate attachment site		364	52
thrombin cleavage sites	arg16-gly17	arg14-gly15	
factor XIIIa ligation sites			
acceptor (glutamine)	328, 366		398
donor (lysine)	508-584		406

Table 3 Physicochemical Characteristics of Fibrinogen

Molecular mass	340,000
Sedimentation coefficient ($S_{20,w}$)	7.8×10^{-13} s
Translational diffusion coefficient ($D_{20,w}$)	1.9×10^{-7}
Rotary diffusion coefficient ($O_{20,w}$)	40,000 sec^{-1}
Frictional ratio (f/f_0)	2.34
Partial specific volume	0.72 cm^3/g
Molecular volume	3.7×10^3 nm^3
Extinction coefficient (A_{280})	16.3
Intrinsic viscosity (η)	0.25 dl/g
Degree of hydration (g/g of protein)	6
α-helix content	33%
Isoelectric point	5.5

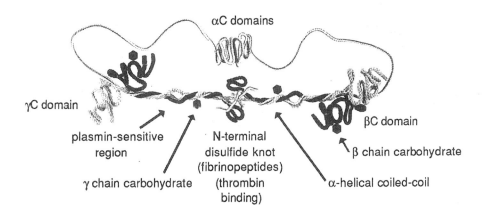

Figure 2 Schematic model for the fibrinogen molecule. Fibrinogen is made up of three pairs of polypeptide chains: Aα–striped; Bβ–black; γ–grey. All six chains are linked together in the central region, called the N-terminal disulfide knot, which contains the pairs of A and B fibrinopeptides and the thrombin binding sites. Three-chain, α-helical coiled-coils connect the central region with the C-terminal ends of the Bβ chain (βC domains), γ chain (γC domains), and Aα chain (αC domain and connector) A non-helical stretch, where plasmin cleavage occurs, interrupts each of the coiled-coils. Carbohydrate moieties on the Bβ and γ chains are represented as black hexagons.

The entire amino acid sequence of all polypeptide chains of human fibrinogen has been determined by the methods of protein chemistry and later deduced from the nucleotide sequences of the cDNAs coding for the polypeptide chains (Doolittle, 1984; Henschen & McDonagh, 1986). The amino acid sequences of the three chains are homologous, but important differences exist, giving rise to specific functions for certain molecular domains. Many structural variants exist, including a splice variant of the γ chain (Hantgan et al., 1994) and a variant of the Aα chain with an extra domain at the carboxyl terminal end (Fu et al., 1992; Fu & Grieninger, 1994).

All six chains are held together by 29 disulfide bonds (Henschen & McDonagh, 1986). The amino termini of all six chains are held together by 11 disulfide bonds in a region termed the central domain or the N-terminal disulfide knot (Figure 2). The A and B fibrinopeptides are also located at the amino terminal ends of the α and β chains, respectively. Unusual Cys-Pro-X-X-Cys sequences occurring twice in each chain are involved in disulfide ring structures in which all three chains are linked together at each end of a three-chain α-helical coiled-coil (Doolittle, 1984). Amino acid sequence predictions suggest that this coiled-coil structure is interrupted in the middle in a region known to be susceptible to plasmin cleavage.

The central region contains the two pairs of A and B fibrinopeptides and binds thrombin. The distal end nodule is the carboxyl terminal γ chain and is made up of two domains, while the proximal end nodule is the carboxyl terminal Bβ chain, also made up of two domains (Weisel et al., 1985; Medved', 1990; Rao et al., 1991). The carboxyl terminal ends of the two Aα chains extend from the molecular ends and interact with the central domains in fibrinogen, but there is a conformational change in this part of the molecule on release of the B fibrinopeptides (Veklich et al., 1993; Gorkun et al., 1994).

FIBRIN POLYMERIZATION AND CHARACTERIZATION OF THE CLOT

Fibrin polymerization is initiated by the enzymatic cleavage of the fibrinopeptides, converting fibrinogen to fibrin monomer. Then, several non-enzymatic reactions yield an orderly sequence of macromolecular assembly steps. Several other plasma proteins bind specifically to the resulting fibrin network (Table 1). Finally, the clot is stabilized by covalent ligation of specific amino acids by a transglutaminase, Factor XIIIa.

Thrombin, which is produced upon proteolytic cleavage of prothrombin by Factor Xa in the presence of Factor V, is a serine protease with a particular specificity for fibrinogen's fibrinopeptides. This specificity arises in part because of hydrophobic interactions between the fibrinopeptides and thrombin's catalytic site, as well as aspects of the three-dimensional structure of the fibrinopeptides (Stubbs & Bode, 1993). In addition, thrombin contains an anion-binding exosite which binds via specific ionic interactions to a region in the central domain of fibrinogen. The A fibrinopeptides are cleaved by thrombin more rapidly than the B peptides, so that usually the B fibrinopeptides are removed after polymerization begins. Hirudin, which is produced by the medicinal leech, is a potent inhibitor of thrombin.

Fibrinogen is about 47 nm long and consists of a globular region in the center connected to several nodules at each end via the rod-like coiled-coils (Hall & Slayter, 1959; Weisel et al., 1981; Erickson & Fowler, 1983; Williams, 1983; Weisel et al., 1985)

(Figure 3). Cleavage of the A fibrinopeptides exposes binding sites in the central domain, called "A", that are complementary to sites, called "a", always exposed at the ends of the molecules (Doolittle, 1984; Budzynski, 1986). The "A" sites consist in part of the newly exposed amino terminus of fibrin's α chain, Gly-Pro-Arg, but also include part of the β chain. The "a" sites include the carboxyl terminus of the γ chain, in the region of residues 356–411 or 303–405. There are also "B" sites exposed in the middle of the molecule on cleavage of the B fibrinopeptides after polymerization begins; these sites are complementary to "b" sites in the carboxyl terminal β chain at the ends of fibrin (Shainoff & Dardik, 1983; Medved' et al., 1993).

Specific interactions between these complementary binding sites produce aggregates in which the fibrin monomers are half-staggered, since the central domain of one molecule binds to the end of the adjacent molecule (Figure 4) (Weisel, 1987). Initially, a dimer is formed and then additional molecules are added to give a structure called the two-stranded protofibril (Figure 3) (Fowler et al., 1981; Medved' et al., 1990).

Once protofibrils reach a sufficient length (usually about 600–800 nm), they aggregate laterally to form fibers (Hantgan & Hermans, 1979; Hantgan et al., 1980). The intermolecular interactions that occur in lateral aggregation are specific so that the fibers have a repeat of 22.5 nm, or about half the molecular length, and a distinctive band pattern, as observed by electron microscopy of negatively contrasted specimens (Cohen et al., 1963; Weisel, 1986) or X-ray fiber diffraction (Stryer et al., 1963) (Figure 3).

The process of lateral aggregation of protofibrils to form fibers is enhanced by the action of the αC domains (carboxyl terminal α chains), which are released from the central domain on cleavage of the B fibrinopeptides after protofibril formation (Figure 4) (Veklich et al., 1993; Gorkun et al., 1994). The αC domains interact intramolecularly in fibrinogen and then intermolecularly in desAB fibrin to produce thicker fibers (Weisel, 1986; Mosesson et al., 1987; Weisel et al., 1993). Lateral growth of fibers seems to be limited because protofibrils in the fibers are twisted, so that as the fiber diameter increases, they must be stretched to traverse an increasingly greater path length (Weisel et al., 1987). With this model, fiber growth stops when the energy required to stretch an added protofibril exceeds the energy of bonding.

The fibrils making up the fibers branch, leading to a three-dimensional network (Figure 3). This property of branching is essential for the properties of fibrin since it leads to the production of a space-filling gel. Such a gel can be formed even with very low fibrinogen concentrations (<0.01% protein), and even clots made from 50 mg/ml fibrinogen are 95% liquid. Most branch points in a clot are simply points at which two parallel strands diverge from each other (Hantgan & Hermans, 1979; Baradet et al., 1995). The physical and mechanical characteristics of clot networks vary greatly depending on the conditions of polymerization.

The clot is stabilized by the formation of covalent bonds introduced by a transglutaminase, Factor XIIIa (Lorand et al., 1980; Henschen & McDonagh, 1986). The active enzyme, Factor XIIIa, is generated from its precursor, Factor XIII, by the action of thrombin (Figure 1). The activity of Factor XIIIa is dependent on both calcium ion concentration and fibrin(ogen). As many as six isopeptide bonds are formed between the side chains of lysine (donor) and glutamine (acceptor) residues. The γ chains of fibrinogen are ligated first, followed by the carboxyl terminal α chains. The formation of these covalent bonds renders the clot mechanically

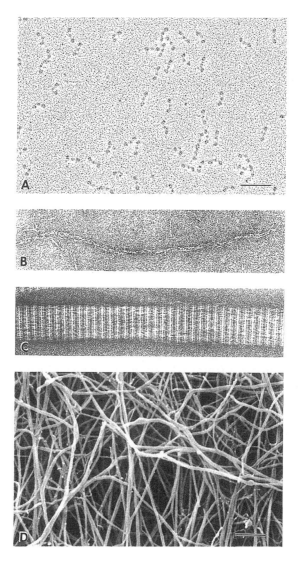

Figure 3 Electron micrographs of some structures of fibrin polymerization, including fibrinogen, fibrin monomer, protofibrils, fiber, clot. A, transmission electron micrograph of individual, rotary shadowed fibrinogen molecules. Each molecule is about 47 nm long and is made up of a nodule in the center and two nodules at each end. B, Image of negatively contrasted fibrin protofibril. With this technique the protein appears bright against a darker background. Individual fibrin monomers are trinodular, with basically the same appearance as in A, but they appear thinner without the coat of metal. Fibrin monomers aggregate in a half-staggered manner to yield a twisted, two-stranded protofibril. C, Negatively contrasted fibrin fiber. Again, areas with more protein exclude stain and appear bright while stain penetrates areas with less stain so that they appear darker. Protofibrils aggregate laterally to yield such fibers, and the band pattern and staining density is a direct reflection of the molecular packing. D, Scanning electron micrograph of a fibrin clot. Fibers grow in length and branch to produce a three-dimensional network or gel.

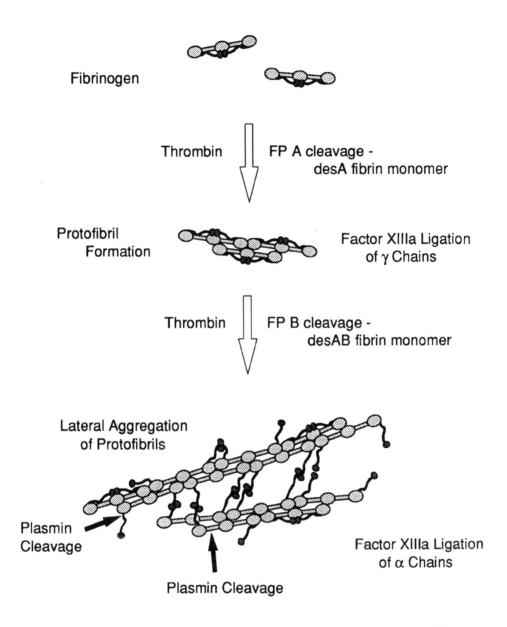

Figure 4 Schematic representation of the polymerization of fibrin and fibrinolysis. Fibrinogen molecules are soluble in the blood until thrombin cleaves the A fibrinopeptides, yielding desA fibrin monomer, which assemble in a half-staggered fashion to make two-stranded protofibrils. Factor XIIIa rapidly crosslinks the adjacent γ chains. The B fibrinopeptides are cleaved more slowly, mostly after polymerization begins. On cleavage of the B fibrinopeptides, the αC domains are released, bringing protofibrils together more efficiently to produce thicker fibers. Factor XIIIa crosslinks the α chains more slowly, to stabilize the clot more fully. Plasminogen is converted to plasmin by tissue plasminogen activator on the fibrin surface; plasmin cleaves specific peptide bonds (arrows) to dissolve the clot.

stronger and more resistant to both chemical and enzymatic dissolution. In addition, other proteins, notably $\alpha2$–plasmin inhibitor and fibronectin are also covalently ligated to fibrin by Factor XIIIa.

Some proteins, such as fibronectin and albumin, interact with fibrin, altering clot structure and properties, though the former becomes crosslinked to fibrin but the latter does not. Albumin has significant effects on the extent of lateral aggregation (Galanakis et al., 1987). Fibronectin is a large glycoprotein normally present in blood and connective tissue that is made up of a number of different domains that serve as binding sites to various cells via specific receptors called integrins and to a number of proteins, including fibrinogen, collagen and gelatin (Ruoslahti, 1988). Fibronectin is important in cell adhesion and may affect wound healing, though definitive clinical trials have not established this point. As a result of these and other interactions, fibrin clots formed in plasma have very different properties than those made with purified proteins (Shah et al., 1987; Carr, 1988; Blombäck et al., 1994). The presence of platelets or cells can also have a dramatic effect on clot structure and mechanical properties.

The thrombin concentration can have an effect on the clot structure, especially if the rate of generation of fibrin monomer by cleavage of fibrinopeptides becomes limiting, relative to the rates of polymerization (Blombäck et al., 1989; Blombäck et al., 1992; Blombäck et al., 1994). In fact, it appears that many factors, including pH, ionic strength, calcium ion concentration, and other plasma proteins, exert an influence on clot structure by affecting the rates of various steps in the polymerization process (Carr et al., 1985; Carr et al., 1986; Carr & Hardin, 1987; Weisel & Nagaswami, 1992). Fibrinogen's carbohydrate moieties also have striking effects on polymerization (Martinez et al., 1977; Langer et al., 1988; Dang et al., 1989).

The mechanical properties have been measured for many different types of clots under various conditions, often with the use of a torsion pendulum under either dynamic or static conditions (Ferry, 1988). Fine clots, made at high ionic strength and pH, have a lower storage and loss moduli than coarse clots, made at physiological pH and salt concentration. The storage modulus is a reflection of the stiffness or elasticity of the clot, whereas the loss modulus is related to dissipative or non-elastic processes. Clots ligated with Factor XIIIa, tend to have a higher storage modulus and a lower loss modulus. Some aspects of the mechanical properties of commercial sealants have been determined. For example, it has been shown that the ultimate tensile strength and Young's modulus increase proportional to the fibrinogen and the thrombin concentrations (Nowotny et al., 1982). In addition, breaking strengths of purified fibrinogen and cryoprecipitate at different concentrations have been measured (Marx & Blankenfeld, 1993).

Fibrin clot formation is a normal part of the wound healing process, sealing the injury and preventing bleeding. Fibrin(ogen) binds specifically to integrin receptors on platelets, endothelial cells and many other cells (Table 1) and plays a vital role in platelet aggregation and many other adhesive interactions. The fibrin clot serves as a scaffold for the migration of various cells, but less is known about this process. Fibroblasts migrate through the gel and deposit collagen, while macrophages and elements of the fibrinolytic system degrade the clot. Fibrin degradation products in turn affect several cellular functions. Thrombin also plays an important part in cellular aspects of wound healing in addition to its roles in blood coagulation. Although fibrin is a normal part of this granulation tissue, the effects of additional high concentrations of fibrin and thrombin have not been well defined.

Some of the adhesive characteristics of fibrin sealant used to attach two pieces of tissue have also been measured. For example, a comparison of Tissucol fibrin sealant with sutures for incisional wounds in rodent dorsal skin indicated that the fibrin sealant repair had a higher bonding strength than the suture controls until four days post-operatively, when both groups were equivalent (Jorgensen *et al.*, 1987). With these two methods there was a difference in the development of mechanical strength over time.

FIBRINOLYSIS

One advantage of fibrin sealant as a biodegradable polymer is that the clot is meant to be a temporary plug for hemostasis or wound healing, so there are natural mechanisms in the body for the efficient removal of fibrin. Various proteolytic enzymes and cells can dissolve fibrin depending on the circumstances, but the most specific mechanism involves the fibrinolytic system. The dissolution of fibrin clots under physiological conditions involves the binding of circulating plasminogen to fibrin, and the activation of plasminogen to the active protease, plasmin, by tissue-type plasminogen activator, also bound to fibrin. Highly efficient, specific fibrinolysis requires the ternary complex, fibrin-plasmin(ogen)-tissue plasminogen activator, and activation of plasminogen is further stimulated by the initial cleavage of fibrin (Wiman & Collen, 1978; Thorsen, 1992; Hantgan *et al.*, 1994). Plasminogen appears to bind at the end-to-end junction of two fibrin molecules, which accounts for the specificity of its binding to polymerizing fibrin rather than fibrinogen (Weisel *et al.*, 1994).

Plasmin then cleaves fibrin at specific sites, yielding certain identifiable, soluble fragments (Figure 4) (Marder *et al.*, 1969; Hantgan *et al.*, 1994). The carboxyl terminal ends of the α chains and the amino terminal ends of the β chains are first removed, yielding fragment X. Then, all three polypeptide chains are cut at plasmin-sensitive sites in the middle of the coiled-coil. If fibrin is not ligated by Factor XIIIa, these cleavages produce from each fibrin monomer: one fragment E, consisting of the central domains and the proximal coiled-coils; two D fragments, each consisting of the distal coiled-coil and the carboxyl terminal ends of the β and γ chains; two αC fragments, consisting of the carboxyl terminal α chains. However, under typical conditions, the D fragments are ligated to produce D dimer and the α chains are ligated to produce αC multimers. Non-covalent interactions in the clot result in the formation of DD/E complexes as the result of plasmin cleavage, although larger complexes are also seen, depending on the extent of lysis (Marder *et al.*, 1969; Weisel *et al.*, 1993).

Antifibrinolytic compounds can block the conversion of plasminogen to plasmin or directly bind to the active site of plasmin to inhibit fibrinolysis. The plasma protein, α_2-macroglobulin, is a primary physiological inhibitor of plasmin. Plasmin released from fibrin is also very rapidly inactivated by α_2-antiplasmin, which plays a role in the regulation of the fibrinolytic process (Aoki & Harpel, 1984). α_2-antiplasmin inactivates plasmin in a very rapid reaction, interferes with plasminogen binding to fibrin and is ligated to fibrin by Factor XIIIa. After α_2-antiplasmin is covalently linked to fibrin's carboxyl terminal α chain, it retains it ability to inhibit plasmin, a function which helps to prevent early clot lysis.

Various other compounds are used clinically as antifibrinolytics. As mentioned above, aprotinin, a small polypeptide that inhibits plasmin, trypsin and kallikrein, is often used in coagulation therapy and in fibrin sealant formulations (Trautschold et al., 1967). In the latter applications, the aprotinin serves to delay the fibrinolytic action of plasmin until the fibrin has fulfilled its function. A monoamino carboxylic acid, ε-amino caproic acid, is similar in structure to lysine and interacts with the active kringles or lysine-binding sites of plasminogen and plasminogen activators to inhibit their binding to fibrin and thus their fibrinolytic actions (Thorsen, 1992). This compound is also used in some clinical applications.

CLINICAL APPLICATIONS

History

Various derivatives of fibrinogen have been used in clinical applications at least since the early part of this century (Bergel, 1909). For example, fibrin sheets or foam were used to assist hemostasis in some kinds of surgery (Grey, 1915; Harvey, 1916). The first published use of a fibrinogen solution with added thrombin as a tissue adhesive appears to be that of Young and Medawar (Young & Medawar, 1940), who clotted plasma to aid peripheral nerve anastomosis. Fibrinogen was first purified in large quantities as part of a program of large-scale plasma fractionation during World War II. At that time, several types of preparations were developed for use on the battlefield, including fibrin film for replacement of the dura mater over the brain in head injuries and fibrin foam used for hemostasis (Bering, 1944; Ferry & Morrison, 1944; Morrison & Singer, 1944; Ferry, 1983).

More recent clinical applications of fibrin sealant all rely on the enhanced strength and other properties of a product that contains elevated levels of fibrinogen. High concentrations of thrombin cause gelation in well under a minute. Cryoprecipitate, a plasma derivative used in the treatment of hemophilia A which also contains elevated levels of fibrinogen, was clotted with thrombin to make an adhesive for peripheral nerve anastomoses (Matras et al., 1972). Many variants of these methods have been developed to produce fibrin sealants. Most of these preparations contain a variety of other proteins that also have significant effects on clot formation and properties. For example, Factor XIIIa covalently ligates fibrin's γ and α chains and thereby stabilizes the clot. Residual amounts of plasminogen may be important for the eventual degradation of the clot.

Examples

For more than fifteen years, fibrin glues have been extensively used in surgical repairs as a hemostatic agent, as a sealant, as a barrier to fluid and as a biodegradable space-filling matrix (Schlag & Redl, 1988; Gibble & Ness, 1990; Lerner & Binur, 1990; Sierra, 1993; Atrah, 1994). A few examples will be outlined. As mentioned above, some of the first applications of fibrin glue were in nerve repair, and applications in neurosurgery and ophthalmic surgery have continued (Matras, 1985). Fibrin sealant has been used for peripheral nerve reattachment and sealing of dura mater to prevent cerebrospinal fluid leakage. In many such applications, the need for suturing is eliminated or reduced and nerve growth and healing may be improved. Ophthalmic applications include sealing of lens perforations and other emergency procedures, lacrimal and conjunctival reconstructions, and corneal ulcer repair (Lagoutte et al., 1989).

However, the majority of uses of these products has been in cardiothoracic surgery, including sealing anastomoses and other suture points and as a hemostatic agent on heart, pericardial, pleural or mediastinal surfaces oozing blood (Wolner, 1982; Tawes et al., 1994; Spotnitz, 1995). Fibrin sealant is used for attachment of vascular grafts and heart valves and for preventing leakage from porous, synthetic graft materials. Fibrin glue can also be used for arresting bleeding of the liver or spleen or external bleeding in patients with hemophilia (Kram et al., 1984; Chisolm et al., 1989).

Fibrin sealant has been used extensively for skin grafting and in craniofacial reconstructions (Marchac & Reiner, 1990; Dahlstrøm et al., 1992). The use of fibrin glue seems to decrease the likelihood of seroma formation in mastectomy (Lindsey et al., 1990). In several microsurgical applications, especially in otorhinolaryngology, fibrin glue is ideal for topical application to secure small but functionally important structures, as in ossicular chain reconstructions and for repair of the tympanic membrane or oval window (Staindl, 1979; Silverstein et al., 1988). Fibrin sealant has played a role in surgery of the nasal septum and vocal cords.

Orthopedic surgeons have used fibrin sealants in a variety of tissue and bone repair operations, including tendon repair, filling of bone defects, and cartilage healing (Schlag & Redl, 1988). There have also been a variety of uses in obstetrics and gynecology (Adamyan et al., 1991). Pulmonary air and fluid leaks have been repaired with fibrin adhesives (Jesson & Sharma, 1985). Antibiotics and other drugs have been incorporated into clots for prolonged delivery to a specific area of the body, since the drugs appear to be released only as the clot is degraded (Kram et al., 1991). Such combinations of treatments may aid in the prevention of adhesions (de Virgilio et al., 1990).

Fibrin Delivery Systems

Since almost all of the formulations of fibrin sealant are two-component systems, involving the combination of fibrinogen and other components with thrombin (or alternatively, fibrin monomer at acidic pH with buffer), various devices have been produced for efficient mixing. One of the most common devices consists of two syringes held together along their barrels and at the plungers; the two solutions are mixed either after exiting the needles or in the hub just prior to exiting. Other devices are designed to produce an aerosol or to spray in a variable pattern depending on the application.

Safety Issues

There are several different safety issues involved in the use of fibrin sealants. As with any blood product, there is concern over viral contamination, especially with products that are prepared from pooled blood. During the late 1970's, human fibrinogen was withdrawn from the market for fear of transmission of pathogenic viruses, and fibrin sealants have not been licensed for clinical use in the United States since that time, although they have been widely used in Europe and Japan. However, improvements in viral cleansing techniques for blood products have been made since that period, such that the risk of disease transmission has been greatly reduced, though not totally eliminated. The methods developed for dealing with viral contamination of fibrin sealants include solvent detergent disruption of lipid enveloped viruses, pasteurization of freeze-dried components, nanofiltration (<70 kd), and chemical and ultraviolet cleaning (Horowitz *et al.*, 1988; Burnouf-Radosevich *et al.*, 1990; Marx *et al.*, 1996). Only autologous fibrinogen preparations from the patient's own plasma remain free of all risk of viral transmission to the patient but most of these methods are relatively labor intensive and time consuming (Cederholm-Williams, 1994).

Commonly, bovine thrombin has been used in fibrin glues, and immune reactions occasionally occur (Zehndre & Leung, 1990; Rapaport *et al.*, 1992). The use of human thrombin avoids this problem, but there may still be problems of infectivity and thrombogenicity, especially from platelet activation. These problems may be solved through the recent introduction of rapid and easy methods for the production of fibrin monomer from the patient, as described above. With any of these fibrin sealants, there are generally few concerns of toxicity, biodegradability or biocompatability since the materials involved are normally involved in human hemostasis and wound healing.

Clinical Trials

A recently developed bibliographic database of papers on fibrin sealant revealed that more than 700 animal studies including about 25,000 individual animal procedures have been published (Fairbrother, 1995). Controlled clinical trials are described in 85 papers, but the majority of the 1400 clinical papers describe case reports and anecdotal, uncontrolled studies. One fibrin sealant, Tissucol or Tisseel, has been utilized in applications involving more than 2 million patients, 41% in reconstructive surgery and 14% in cardiothoracic surgery, and studies using it have been reported in more than 2000 papers.

The great majority of clinical trials reported with statistical analysis show a benefit in favor of the use of fibrin sealant in a range of clinical applications (Fairbrother, 1995). The major usage of fibrin sealants is for the reduction of blood loss during cardiothoracic surgery, and it is believed that the use of sealants may reduce post-operative recovery time. Controlled trials have been conducted in lymphatic lung, esophagus, ulcer, ureter, and eye. Trials involving more patients and using objective endpoints are necessary to resolve issues of the benefits and safety of various fibrin sealant formulations.

REFERENCES

Adamyan, L., Myinbayev, O.A. and Kulakov, V.I. (1991) Use of fibrin sealant in obstetrics and gynaecology: a review of the literature. *Int. J. Fertil.*, **36**, 76–77, 81–88.

Amery, M.J., Edwardson, P.A.D., Fairbrother, J.E., Hollingsbee, D.A., Holm, N.-E., Westwood, B. and

Cederholm-Williams, S.A. (1995) The first totally autologous fibrin sealant which requires no added thrombin. *Thromb. Haemost.*, **73**, 1463a.

Aoki, N. and Harpel, P.C. (1984) Inhibitors of the fibrinolytic enzyme system. *Semin. Thromb. Hemost.*, **10**, 24–41.

Atrah, H.I. (1994) Fibrin glue. *Brit. Med J.*, **308**, 933–934.

Baradet, T.C., Haselgrove, J.C. and Weisel, J.W. (1995) Three-dimensional reconstruction of fibrin clot networks from stereoscopic intermediate voltage electron microscope images and analysis of branching. *Biophys. J.*, **68**, 1551–1560.

Belitser, V.A., Varetskaia, T.V. and Tarasenko, L.A. (1965) [Polymerization of fibrin-monomer and its relation to pH]. *Ukr Biokhim Zh*, **37**, 665–70.

Bergel, S. (1909) Uber wirkugen des fibrins. *Dtsch. Med. Wochenschr.*, **35**, 633–665.

Bering, E.A. (1944) Chemical, clinical and immunological studies on the products of human plasma fractionation. XX. The development of fibrin foam as a hemostatic agent and for use in conjunction with human thrombin. *J. Clin. Invest.*, **23**, 586–589.

Blombäck, B. (1991) Fibrinogen and Fibrin Formation and Its Role in Fibrinolysis. In *Biotechnology of Blood* (J. Goldstein, ed.), pp. 225–279, Butterworth-Heinemann, Boston.

Blombäck, B. and Blombäck, M. (1956) Purification of human and bovine fibrinogen. *Arkiv Kemi*, **10**, 415–443.

Blombäck, B., Blombäck, M., Carlsson, K., Fatah, K., Hamsten, A. and Hessel, B. (1992) Fibrin gel structure in health and disease. In *Fibrinogen: A New Cardiovascular Risk Factor* (E. Ernst, W. Koenig, G.D.O. Lowe and T.W. Meade, eds.), pp. 11–18, Blackwell-MZW, London.

Blombäck, B., Carlsson, K., Fatah, K., Hessel, B. and Procyk, R. (1994) Fibrin in human plasma: gel architectures governed by rate and nature of fibrinogen activation. *Thromb. Res.*, **75**, 521–538.

Blombäck, B., Carlsson, K., Hessel, B., Liljeborg, A., Procyk, R. and Aslund, N. (1989) Native fibrin gel networks observed by 3D microscopy, permeation and turbidity. *Biochim. Biophys. Acta*, **997**, 96–110.

Budzynski, A.Z. (1986) Fibrinogen and fibrin: biochemistry and pathophysiology. *Crit. Rev. Oncol.-Hematol.*, **6**, 97–146.

Burnouf-Radosevich, M., Burnof, T. and Huart, J.J. (1990) Biochemical and physical properties of a solvent-detergent treated fibrin glue. *Vox Sang.*, **58**, 77–84.

Carr, M.E. (1988) Fibrin formed in plasma is composed of fibers more massive than those formed from purified fibrinogen. *Thromb. Haemost.*, **59**, 535–539.

Carr, M.E. and Hardin, C.L. (1987) Fibrin has larger pores when formed in the presence of erythrocytes. *Am. J. Physiol.*, **253**, H1069–H1073.

Carr, M.E., Kaminski, M., McDonagh, J. and Gabriel, D.A. (1985) Influence of ionic strength, peptide release, and calcium ion on the structure of reptilase and thrombin-derived gels. *Thromb. Haemost.*, **54**, 159–165.

Carr, M.J., Gabriel, D.A. and McDonagh, J. (1986) Influence of Ca2+ on the structure of reptilase-derived and thrombin-derived fibrin gels. *Biochem. J.*, **239**, 513–516.

Cederholm-Williams, S.A. (1994) Fibrin glue. *Brit. Med. J.*, **308**, 1570.

Chisolm, R.A., Jones, S.N. and Lees, W.R. (1989) Fibrin sealant as a plug for post liver biopsy needle track. *Clin. Radiol.*, **40**, 627–628.

Cohen, C., Revel, J.-P. and Kucera, J. (1963) Paracrystalline forms of fibrinogen. *Science*, **141**, 436–438.

Cohn, E.J., Strong, L.E., Hughes, W.L., Mulford, D.J., Ashworth, J.N., Melin, M. and Taylor, H.L. (1946) *J. Am. Chem. Soc.*, **68**, 459–475.

Dahlstrøm, M.D., Weis-Fogh, U.S., Medgyesi, S., Rostgaard, J. and Sørensen, H. (1992) The use of autologous fibrin adhesive in skin transplantation. *Plast. Reconstr. Surg.*, **89**, 968–972.

Dang, C.V., Shin, C.K., Bell, W.R., Nagaswami, C. and Weisel, J.W. (1989) Fibrinogen sialic acid residues are low affinity calcium-binding sites that influence fibrin assembly. *J. Biol. Chem.*, **264**, 15104–15108.

de Virgilio, C., Dubrow, T. and Sheppard, B.B. (1990) Fibrin glue inhibits intra-abdominal adhesion formation. *Arch. Surg.*, **125**, 1378–1381.

Doolittle, R.F. (1984) Fibrinogen and fibrin. *Annu. Rev. Biochem.*, **53**, 195–229.

Erickson, H.P. and Fowler, W.E. (1983) Electron microscopy of fibrinogen, its plasmic fragments and small polymers. *Ann. N. Y. Acad. Sci.*, **408**, 146–163.

Fairbrother, J.E. (1995) A global publication profile of fibrin sealants. *Thromb. Haemost.*, **73**, 1470a.

Ferry, J.D. (1983) The conversion of fibrinogen to fibrin: Events and recollections from 1942 to 1982. *Ann. N.Y. Acad. Sci.*, **408**, 1–10.

Ferry, J.D. (1988) Structure and rheology of fibrin networks. In *Biological and Synthetic Polymer Networks* (O. Kramer, ed.), pp. 41–55, Elsevier, Amsterdam.

Ferry, J.D. and Morrison, P.R. (1944) Chemical, clinical and immunological studies on the products of human plasma fractionation. XVI. Fibrin clots, fibrin films and fibrinogen plastics. *J. Clin. Invest.*, **23**, 566–572.

Fowler, W.E., Hantgan, R.R., Hermans, J. and Erickson, H.P. (1981) Structure of the fibrin protofibril. *Proc Natl Acad Sci U S A*, **78**, 4872–6.

Fu, Y. and Grieninger, G. (1994) Fib420: A normal human variant of fibrinogen with two extended alpha chains. *Proc. Natl. Acad. Sci. USA*, **91**, 2625–2628.

Fu, Y., Weissbach, L., Plant, P.W., Oddoux, C., Cao, Y., Liang, J., Roy, S.N., Redman, C.M. and Grieninger, G. (1992) Carboxy-terminal-extended variant of the human fibrinogen alpha subunit: a novel exon conferring marked homology to beta and gamma subunits. *Biochem.*, **31**, 11968–11972.

Galanakis, D.K., Lane, B.P. and Simon, S.R. (1987) Albumin modulates lateral assembly of fibrin polymers: evidence of enhanced fine fibril formation and of unique synergism with fibrinogen. *Biochem.*, **26**, 2389–2400.

Gestring, G.F. and Lerner, R. (1983) Autologous fibrinogen for tissue-adhesion, hemostasis, and embolization. *Vasc. Surg.*, **17**, 294–304.

Gibble, J.W. and Ness, P.M. (1990) Fibrin glue: the perfect operative sealant? *Transfusion*, **30**, 741–747.

Gorkun, O.V., Veklich, Y.I., Medved', L.V., Henschen, A.H. and Weisel, J.W. (1994) Role of the αC domains in fibrin formation. *Biochemistry*, **33**, 6986–6997.

Grey, E.G. (1915) Fibrin as a hemostatic in cerebral surgery. *Surg. Gynecol. Obstet.*, **21**, 452–454.

Hafter, R., von Hugo, R. and Graeff, H. (1978) Acid precipitation of fibrinogen. A technique for isolation of fibrinogen and its derivatives from highly diluted solutions. *Hoppe–Seyler's Z. Physiol. Chem.*, **359**, 759–763.

Hall, C.E. and Slayter, H.S. (1959) The fibrinogen molecule: Its size, shape and mode of polymerization. *J. Biophys. Biochem. Cytol.*, **5**, 11–16.

Hantgan, R., Fowler, W., Erickson, H. and Hermans, J. (1980) Fibrin assembly: a comparison of electron microscopic and light scattering results. *Thromb Haemost*, **44**, 119–124.

Hantgan, R.R., Francis, C.W. and Marder, V.J. (1994) Fibrinogen structure and physiology. In *Hemostasis and Thrombosis: Basic Principles and Clinical Practice* (R. W. Colman, J. Hirsh, V. J. Marder and E. W. Salzman, eds.), pp. 277–300, J.B. Lippincott, Philadelphia.

Hantgan, R.R. and Hermans, J. (1979) Assembly of fibrin. A light scattering study. *J. Biol. Chem.*, **254**, 11272–11281.

Harvey, S.C. (1916) The use of fibrin papers and foams in surgery. *Boston Med. Surg. J.*, **174**, 659–662.

Henschen, A. and McDonagh, J. (1986) Fibrinogen, fibrin and factor XIII. In *Blood Coagulation* (R.F.A. Zwaal and H.C. Hemker, eds.), pp. 171–234, Elsevier Science, Amsterdam.

Horowitz, M.S., Rooks, C., Horowitz, B. and Hilgartner, M.W. (1988) Virus safety of solvent/detergent-treated antihemophilic factor concentrate. *Lancet*, **ii**, 186–188.

Jesson, C. and Sharma, P. (1985) Use fibrin glue in thoracic surgery. *Ann. Thorac. Surg.*, **39**, 521–524.

Jorgensen, P.H., Jensen, K.H., Andreassen, B. and Andreassen, T.T. (1987) Mechanical strength in rat skin incisional wounds treated with fibrin sealant. *J. Surg. Res.*, **42**, 237–241.

Kram, H.B., Bansal, M., Timberlake, O. and Shoemaker, W.C. (1991) Antibacterial effects of fibrin glue-antibiotics mixture. *J. Surg. Res.*, **50**, 175–178.

Kram, H.B., Shoemaker, W.C., Hino, S.T. and Harley, D.P. (1984) Splenic salvage using biological glue. *Arch. Surg.*, **119**, 1309–1311.

Lagoutte, F.M., Gautier, L. and Comte, P.R. (1989) A fibrin sealant for perforated and pre-perforated corneal ulcers. *Brit. J. Ophthalmol.*, **73**, 757–761.

Laki, K. (1951) The polymerization of proteins: The action of thrombin on fibrinogen. *Arch. Biochem. Biophys.*, **32**, 317–324.

Langer, B.G., Weisel, J.W., Dinauer, P.A., Nagaswami, C. and Bell, W.R. (1988) Deglycosylation of fibrinogen accelerates polymerization and increases lateral aggregation of fibrin fibers. *J. Biol. Chem.*, **263**, 15056–15063.

Lerner, R. and Binur, N.S. (1990) Current status of surgical adhesives. *J. Surg. Res.*, **48**, 165–181.

Lindsey, W.H., Masterson, T.M., Spotnitz, W.D., Wilhelm, M.C. and Morgan, R.F. (1990) Seroma prevention using fibrin glue in a rat mastectomy model. *Arch. Surg.*, **125**, 305–307.

Longas, M.O., Newman, J. and Johnson, A.J. (1980) An improved method for the purification of human fibrinogen. *Int. J. Biochem.*, **11**, 559–564.

Lorand, L., Losowsky, M.S. and Miloszewski, K. (1980) Human Factor XIII: Fibrin stabilizing factor. In *Progress in Hemostasis and Thrombosis* (T. H. Spaet, ed.), pp. 245–290, Grune and Stratton, New York.

Marchac, D. and Reiner, D. (1990) Fibrin glue in craniofacial surgery. *J. Craniofac. Surg.*, **1**, 32–34.

Marder, V.J., Shulman, N.R. and Carroll, W.R. (1969) High molecular weight derivatives of human fibrinogen produced by plasmin. *J. Biol. Chem.*, **244**, 2111–2119.

Martinez, J., Palascak, J.E. and Peters, C. (1977) Functional and metabolic properties of human asialofibrinogen. *J. Lab. Clin. Med.*, **89**, 367–377.

Marx, G. and Blankenfeld, A. (1993) Kinetic and mechanical parameters of pure and cryoprecipitate fibrin. *Blood Coagul. Fibrinolysis*, **4**, 73–78.

Marx, G., Mou, X., Freed, R., Ben-Hur, E., Yang, C. and Horowitz, B. (1996) Protecting fibrinogen with rutin during UVC irradiation for viral inactivation. *Photochem. Photobiol.*, **63**, 541.

Matras, H. (1985) Fibrin sealant: the state of the art. *J. Oral Maxillofacial Surg.*, **43**, 605–611.

Matras, H., Dinges, B. and Manoli, B. (1972) Zur nahtlosen interfaszikularen nerventransplantation im tierexperiment. *Wien Med. Woschtr.*, **122**, 517–523.

Medved', L.V., Ugarova, T., Veklich, Y., Lukinova, N. and Weisel, J. (1990) Electron microscope investigation of the early stages of fibrin assembly. Twisted protofibrils and fibers. *J. Mol. Biol.*, **216**, 503–509.

Medved', L.V. (1990) Relationship between exons and domains in the fibrinogen molecule. *Blood Coag. Fibrinolysis*, **1**, 439–442.

Medved', L.V., Litvinovich, S.V., Ugarova, T.P., Lukinova, N.I., Kalikhevich, V.N. and Ardemasova, Z.A. (1993) Localization of a fibrin polymerization site complementary to Gly-His-Arg sequence. *FEBS Letters*, **320**, 239–242.

Morrison, P.R. and Singer, M. (1944) Chemical, clinical and immunological studies on the products of human plasma fractionation. XVIII. A note on adsorption rates of fibrin films in tissue. *J. Clin. Invest.*, **23**, 573–575.

Mosesson, M.W. (1990) Fibrin polymerization and its regulatory role in hemostasis. *J. Lab. Clin. Med.*, **116**, 8–17.

Mosesson, M.W., DiOrio, J.P., Müller, M.F., Shainoff, J.R., Siebenlist, K.R., Amrani, D.L., Homandberg, G.A., Soria, J., Soria, C. and Samama, M. (1987) Studies on the ultrastructure of fibrin lacking fibrinopeptide B (beta-fibrin) *Blood,*, **69**, 1073–1081.

Nowotny, R., Chalupka, A., Nowotny, C. and Bosch, P. (1982) Mechanical properties of fibrinogen adhesive material. In *Biomaterial 1980* (G.D. Winter, G.F. Gibbons and H. Plenk, eds.), John Wiley and Sons, London.

Rao, S., Poojary, M., Elliott, B., Melanson, L., Oriel, B. and Cohen, C. (1991) Fibrinogen structure in projection at 18 Å resolution. Electron density by co-ordinated cryo-electron microscopy and X-ray crystallography. *J. Mol. Biol.*, **222**, 89–98.

Rapaport, S., Zivelin, A., Minow, R., Hunter, C. and Donnelly, K. (1992) Clinical significance of antibodies to bovine and human thrombin and Factor V after surgical use of bovine thrombin. *Am. J. Clin. Pathol.*, **97**, 84–91.

Ruoslahti, E. (1988) Fibronectin and its receptors. *Annu. Rev. Biochem.*, **57**, 375–413.

Schlag, G. and Redl, H. (1988) Fibrin sealant in orthopedic surgery. *Clin. Orthopedics Related Res.*, **227**, 269–285.

Shafer, J.A. and Higgins, D.L. (1988) Human Fibrinogen. *CRC Critical Rev. in Clin. Lab. Sci.*, **26**, 1–41.

Shah, G.A., Nair, C.H. and Dhall, D.P. (1987) Comparison of fibrin networks in plasma and fibrinogen solution. *Thromb. Res.*, **45**, 257–264.

Shainoff, J.R. and Dardik, B.N. (1983) Fibrinopeptide B in fibrin assembly and metabolism: physiologic significance in delayed release of the peptide. *Ann. N.Y. Acad. Sci.*, **408**, 254–268.

Sierra, D.H. (1993) Fibrin sealant adhesive systems: a review of their chemistry, material properties and clinical applications. *J. Biomater. Appl.*, **7**, 309–352.

Silverstein, L.E., Williams, L.J., Hughlett, M.A., Magee, D.A. and Weisman, R.A. (1988) An autologous fibrinogen-based adhesive for use in otologic surgery. *Transfusion*, **28**, 319–321.

Spotnitz, W.D. (1995) Fibrin sealant in the United States: clinical use at the University of Virginia. *Thromb. Haemost.*, **74**, 482–485.

Staindl, O. (1979) Tissue adhesion with highly concentrated human fibrinogen in otolaryngology. *Ann. Otol. Rhinol. Laryngol.*, **88**, 413–418.

Stryer, L., Cohen, C. and Langridge, R. (1963) Axial period of fibrinogen and fibrin. *Nature*, **197**, 793–794.

Stubbs, M.T. and Bode, W. (1993) A model for the specificity of fibrinogen cleavage by thrombin. *Semin. Thromb. Hemost.*, **19**, 344–351.

Tawes, R.L., Sydorak, G.R. and DuVall, T.B. (1994) Autologous fibrin glue: the last step in operative hemostasis. *Amer. J. Surg.*, **168**, 120–122.

Thorsen, S. (1992) The mechanism of plasminogen activation and the variability of the fibrin effector during tissue-type plasminogen activator-mediated fibrinolysis. *Ann. N.Y. Acad. Sci.*, **667**, 52–63.

Trautschold, I., Werle, E. and Zickgraf-Rudel, G. (1967) Trayslol. *Biochem. Pharm.*, **16**, 59–72.

Veklich, Y.I., Gorkun, O.V., Medved, L.V., Nieuwenhuizen, W. and Weisel, J.W. (1993) Carboxyl terminal portions of the alpha chains of fibrinogen and fibrin: Localization by electron microscopy and the effects of isolated αC fragments on polymerization. *J. Biol. Chem.*, **268**, 13577–13585.

Ware, A.G., Guest, M.M. and Seegers, W.H. (1947) *Arch. Biochem. Biophys.*, **13**, 231–236.

Weis-Fogh, U.S. (1988) Fibrinogen prepared from small blood samples for autologous use in a tissue adhesive system. *Eur. Surg. Res.*, **20**, 381–389.

Weisel, J.W. (1986) The electron microscope band pattern of human fibrin: various stains, lateral order, and carbohydrate localization. *J. Ultrastruct. Mol. Struct. Res.*, **96**, 176–188.

Weisel, J.W. (1986) Fibrin assembly. Lateral aggregation and the role of the two pairs of fibrinopeptides. *Biophys. J.*, **50**, 1079–1093.

Weisel, J.W. (1987) Molecular symmetry and binding sites in fibrin assembly. *Thromb. Res.*, **48**, 615–617.

Weisel, J.W., Francis, C.W., Nagaswami, C. and Marder, V.J. (1993) Determination of the topology of factor XIIIa-induced fibrin gamma-chain cross-links by electron microscopy of ligated fragments. *J. Biol. Chem.*, **268**, 26618–26624.

Weisel, J.W. and Nagaswami, C. (1992) Computer modeling of fibrin polymerization kinetics correlated with electron microscope and turbidity observations: clot structure and assembly are kinetically controlled. *Biophys. J.*, **63**, 111–128.

Weisel, J.W., Nagaswami, C., Korsholm, B., Petersen, L.C. and Suenson, E. (1994) Interactions of plasminogen and polymerizing fibrin and its derivatives monitored with a photoaffinity cross-linker and electron microscopy. *J. Mol. Biol.*, **235**, 1117–1135.

Weisel, J.W., Nagaswami, C. and Makowski, L. (1987) Twisting of fibrin fibers limits their radial growth. *Proc. Natl. Acad. Sci. USA*, **84**, 8991–8995.

Weisel, J.W., Phillips, G.J. and Cohen, C. (1981) A model from electron microscopy for the molecular structure of fibrinogen and fibrin. *Nature*, **289**, 263–267.

Weisel, J.W., Stauffacher, C.V., Bullitt, E. and Cohen, C. (1985) A model for fibrinogen: domains and sequence. *Science*, **230**, 1388–1391.

Weisel, J.W., Veklich, Y.I. and Gorkun, O.V. (1993) The sequence of cleavage of fibrinopeptides from fibrinogen is important for protofibril formation and enhancement of lateral aggregation in fibrin clots. *J. Mol. Biol.*, **232**, 285–297.

Williams, R.C. (1983) Morphology of fibrinogen monomers and of fibrin protofibrils. *Ann. N.Y. Acad. Sci.*, **408**, 180–93.

Wiman, B. and Collen, D. (1978) Molecular mechanism of physiological fibrinolysis. *Nature*, **272**, 549–550.

Wolner, E. (1982) Fibrin gluing in cardiovascular surgery. *Thorac. Cardiovasc. Surg.*, **30**, 236–237.

Young, J.Z. and Medawar, P.B. (1940) Fibrin suture of peripheral nerves. *Lancet*, **ii**, 126–132.

Zehndre, J. and Leung, L. (1990) Development of antibodies to thrombin and Factor V with recurrent bleeding in a patient exposed to topical bovine thrombin. *Blood*, **76**, 2011–2016.

COMMERCIAL SOURCES

Name	Type of Preparation	Principal Components	Viral Inactivation	Source
Tissucol or Tisseel	pooled donor plasma	fibrinogen (110–130 mg/mL) Factor XIII (10–50 U/mL) plasminogen (70–110 mg/mL) aprotinin (3000 KIU/mL) human thrombin (500 U/mL)	vapor heat treatment of fibrinogen & thrombin	Immuno AG Industriestrasse Vienna Austria
Beriplast	pooled donor plasma	fibrinogen (65–115 mg/mL) Factor XIII (40–80 U/mL) plasminogen (detectable) aprotinin (1000 KIU/mL) human thrombin (400–600 U/mL)	pasteurization (60°C for 10 h) of fibrinogen & thrombin	Behringwerke AG Postfach 1140 Marburg 1 Germany
Biocoll	pooled donor plasma	fibrinogen (>75 mg/mL) Factor XIII (<10 U/mL) plasminogen (31 mg/mL) aprotinin (30 mg/mL) human thrombin (>500 U/mL)	solvent detergent treatment of thrombin & fibrinogen; nanofiltration of thrombin	Centre de Transfusion Sanguine Lille, France
Bolheal	pooled donor plasma	fibrinogen (80 mg/mL) Factor XIII (75 U/mL) plasminogen (detectable) aprotinin (1000 KIU/mL) human thrombin (250 U/mL)	pasteurization (65°C): fibrinogen (144 h) & thrombin (96 h)	Kaketsu-ken 668 Okubo Shimizu Kumamoto 860 Japan
Vivostat	single autologous donor	fibrin monomer in acetic acid (25 mg/mL) Factor XIII (detectable) plasminogen (trace) bicarbonate buffer, pH 10	autologous	ConvaTec Skillman New Jersey USA
MelGlu	pooled donor plasma	fibrinogen (≈75 mg/mL) Factor XIII (<10 U/mL) plasminogen (trace) human thrombin (200 U/mL)	solvent detergent and UVC	Melville Biologics Melville New York USA

19. TRANSDUCTIONAL ELASTIC AND PLASTIC PROTEIN-BASED POLYMERS AS POTENTIAL MEDICAL DEVICES

DAN W. URRY[1], ASIMA PATTANAIK[1,2], MARY ANN ACCAVITTI[3],
CHI-XIANG LUAN[1], DAVID T. MCPHERSON[4], JIE XU[1,2],
D. CHANNE GOWDA[1,2], TIMOTHY M. PARKER[2],
CYNTHIA M. HARRIS[2] and NAIJIE JING[1]

[1]*Laboratory of Molecular Biophysics, School of Medicine,*
The University of Alabama at Birmingham,
VH 300, Birmingham, AL 35294-0019
[2]*Bioelastics Research, Ltd., 1075 South 13th Street,*
Birmingham, AL 35205
[3]*Department of Medicine, School of Medicine,*
The University of Alabama at Birmingham,
THT 466 Birmingham, AL 35294
[4]*The UAB AIDS Center, The University of Alabama at Birmingham,*
BBRB 346, Birmingham,AL 35294-2170

INTRODUCTION

Protein-based polymers are polymers comprised of repeating peptide sequences; and transductional polymers exhibit transitional behavior of a nature enabling them to function in free energy transductional processes. The transitional behavior of the present polymers is due to an inverse temperature transition of hydrophobic folding and assembly as the temperature is raised from below to above the onset temperature, designated as T_t, or alternatively as the transition temperature, T_t, is lowered from above to below an operating temperature. There are innumerable ways of changing the value of T_t (Urry, 1993a), each of which could be useful in medical applications (Urry, 1993b; Urry *et al.*, 1995a, 1995b, 1996, 1997).

At a temperature below that of the transition, the polymers are soluble in water in all proportions. As the temperature is raised above T_t or as T_t is lowered from above to below the operating temperature (e.g., room temperature or physiological temperature), the soluble polymers undergo a phase separation with the formation of a more-dense state which is fixed for a given polymer composition but which may vary, depending on polymer sequence, from little or no water to more than 60% water by weight. If the polymers are cross-linked, the resulting matrices form hydrogels at temperatures below T_t, but on raising the temperature above T_t or on lowering T_t from above to below the working temperature, the hydrogels convert either to an elastic matrix or to a plastic state depending on the sequence of the repeating peptide.

*Correspondence: Dan W. Urry, Ph.D., Director, Laboratory of Molecular Biophysics, School of Medicine, The University of Alabama at Birmingham, 1670 University Boulevard, VH 300, Birmingham, AL 35294–0019. Tel: (205) 934-4177; Fax: (205) 934-4256; E-mail: Mobi006@uabdpo.dpo.uab.edu

This capacity to convert from one state to another is central to the transductional property, but it also appears to be central to the biodegradability of the polymers. The available implant data to date indicate that the elastic or plastic states are not readily biodegradable, but that conversion of either to the hydrogel state allows biodegradation to occur such that the half-life in the peritoneal cavity is of the order of a month.

In this brief review of the elastic and plastic protein-based polymers, the chemical and microbial syntheses of these polymers are noted; several of the more commonly used physical characterizations of these polymers are described; the important biological characterizations of biocompatibility (toxicity), immunogenicity, and biodegradability are considered, and the applications of drug delivery and tissue reconstruction are discussed.

SYNTHESES

One of the great advantages of protein-based polymers is that their preparation can be either by chemical or by biological means. Our general approach has been to synthesize a family of polymers chemically, to characterize the polymers physically in order to determine those polymers which have the most preferred physical properties, and then to evaluate their biological properties. Once a particular polymer has been determined to be of interest for more extended studies, the genes are designed, constructed, and placed in a desired expression system. Fermentations for the production of the desired polymer are carried out and the purification schemes established for a given composition.

Chemical Syntheses

Over the past two decades and more, several thousand protein-based polymers have been prepared in our laboratories primarily by classical solution methods but also by solid phase methodologies. The details of these syntheses have been reported in many original publications and in reviews (Prasad *et al.*, 1985; Urry and Prasad, 1985; Gowda *et al.*, 1994). In the chemical syntheses, extraordinary care is required to achieve adequate purity. The requirement for purity in the preparation of protein-based polymers focuses on the prevention of racemization, which dictates that preparation of the basic repeating unit be achieved without impurities and without racemization and that the polymerization of the basic repeat be carried out in such a way that the amino acid at the activated carboxyl end not be able to racemize during polymerization. This issue, which is central to successful chemical syntheses, has also been treated in detail previously (Urry *et al.*, 1996). Emphasis in this review will be on the status of gene constructions and microbial fermentations to produce the elastic and plastic protein-based polymers which, with avoidance of extreme conditions during purification to separate from other cellular constituents, become the standard for the absence of racemization.

Genetic Engineering and Microbial Biosyntheses

Synthetic Gene Constructions and Expression in E. coli

By using standard molecular biology approaches (Perbal, 1988; Ausubel, 1989; Sambrook, 1989), the protein-based polymers are produced in *E. coli* from genes that are themselves polymers, or concatemers, of a basic monomer gene (McPherson *et al.*, 1992; Urry, *et al.*, 1995a; McPherson *et al.*, 1996). This monomer gene, or catemer, is constructed utilizing a combined organic synthesis and enzymatic approach. First, two single-stranded oligonucleotides are chemically synthesized, each representing opposite strands of adjacent regions (halves) of the double-stranded catemer gene, and each having a short segment of overlapping complementarity at their 3' ends. Then, the two oligonucleotides are annealed through that segment of overlapping complementarity, and extended from their 3' ends with a DNA polymerase and free deoxynucleotides to give the full-length double-stranded monomer gene.

Included at the termini of the basic monomer gene are the restriction endonuclease recognition sites required for initial cloning and subsequent polymerization, or concatenation, of the catemer gene. The catemer gene is isolated and propagated by "cloning" it into a plasmid vector. The plasmid is a circular double-stranded DNA molecule that can be inserted into an *E coli* cell where it can replicate to many copies. The catemer gene fragment is inserted into the plasmid by cleaving each with a restriction endonuclease that linearizes the circular plasmid and leaves both the plasmid and the gene fragment with compatible cohesive ends. These ends are then joined using a DNA ligase enzyme, resulting in a re-circularized plasmid that contains the catemer gene fragment.

To construct the concatemer gene encoding multiple copies of the basic repeat to form the protein-based polymer, it is first necessary to prepare a large amount of the catemer gene fragment. This is done by growing a culture of the *E. coli* cells which contain the plasmid molecules carrying the catemer gene, thereby greatly amplifying the amount of the plasmid, followed by purifying the plasmid DNA from the cells. The catemer gene fragment is then released from the plasmid by cleavage with a restriction endonuclease, the recognition site for which was engineered into the gene sequence. A large amount of this catemer gene fragment is then purified and mixed in solution with DNA ligase to concatenate, or polymerize, the genes through their cohesive ends into multiple repeats. The above restriction endonuclease is one that leaves cohesive termini that allow the gene fragments to join in only a head-to-tail fashion thereby maintaining a uniform gene polarity throughout the concatemer. In addition, synthetic double-stranded oligonucleotide cloning adaptors are added to the ligation mix at much lower amounts relative to the gene fragments; these have the same cohesive ends as the gene fragments and provide terminal restriction endonuclease sites needed to insert the concatemer genes into the expression plasmids.

The above use of DNA ligase for concatenation has been well-described in the literature; the conditions for enhancing linear polymerization over cyclization have been described (Dugaiczyk *et al.*, 1975). The ligase reaction was used (Doel *et al.*, 1980) to prepare $(Asp-Phe)_{150}$ starting with chemically prepared dodecadeoxynucleotides encoding for Asp-Phe-Asp-Phe, and Hartley and Gregori (1981) demonstrated "a general method for the construction of plasmids containing many tandem copies of a DNA segment." The general proposition that genetic engineering be used for the preparation of protein-based polymers of the type considered here appeared in the patent literature a few years later (Urry and Long, filed October 31, 1985 and issued September 15, 1987), followed by a more detailed description using many of the above referenced approaches (Ferrari, *et al.*, filed October 29, 1987 and issued September 7, 1993). There are now a number of examples in the scientific literature wherein genetic engineering has been used to prepare designed protein-based polymers (Goldberg *et al.*, 1989; Kaplan *et al.*, 1991; Capello, 1992; Deguchi *et al.*, 1993; Murata *et al.*, 1993; Salerno and Goldberg, 1993).

Production of the protein-based polymers in *E. coli* relies upon insertion of the concatemer genes adjacent to a strong promoter region on an expression plasmid. This promoter region is a DNA sequence that contains information required to drive the expression of the concatemer gene in the cell. In the present case, the expression plasmid system used to produce the protein-based polymers is one that utilizes the *E. coli* phage T7 promoter (Studier *et al.*, 1990) with an appropriate host strain. This system is one that is designed to provide regulatable control over gene expression; promoter function is largely repressed until the addition of a chemical inducer to the growing culture, thereby "switching on" the production of protein. However, for protein that is seemingly innocuous to the *E. coli* cell, as these protein-based polymers seem to be, it is possible to use this expression system in a manner that achieves high level production without the use of inducer. In fact, examples of expression have been observed wherein 80% of the cell volume was inclusion bodies of the protein-based polymer (Guda *et al.*, 1995).

Examples of Prepared Basic Monomeric Sequences for Elastic and Plastic Protein-based Polymers

Examples of a number of basic monomeric sequences that have been genetically engineered in our laboratories are given in Table 1. These polypeptide sequences, some of which are seen as repeats at the polypeptide level, contain no repeats at the nucleotide sequence level and as such are referred to as the basic gene, or catemer, sequences. For example, $(GVGVP)_{10}$, $(GVGIP)_{10}$, and $(AVGVP)_{10}$ each repeat 10 times at the polypeptide sequence level, but within the entire sequence of 150 bases required to encode for each sequence of 50 amino acid residues, there is no repeating nucleotide sequence due to the appropriate use of codon degeneracy.

The basic gene sequences, IV through VII, were designed to have one Glu(E) residue per 30 mer but to have a systematic change in the number of Val(V) residues replaced by the more hydrophobic Phe(F) residue. As has been shown with the chemically synthesized polymers, there is a systematic non-linear increase in the pKa of the carboxyl of the Glu residue with increasing number of Phe(F) residues. While this is of substantial basic interest with respect to the nature of hydrophobic-induced pKa shifts and efficiencies of energy conversion, it becomes of particular interest when designing transductional protein-based polymers for controlled release of cationic drugs. A few examples of the latter will be given below.

Examples of Concatemerized Genes, i.e., Series of Achieved Chain Lengths

As described above, the basic monomer gene sequence, designed with the desired restriction site at each end, is then concatemerized to obtain oligomers of the basic gene sequence. The result is a series of protein-based polymers of different chain lengths. The concatemer genes that have been obtained for the basic gene sequence, listed as IV in Table 1, has been obtained with many different repeats of the basic 30 mer, namely with 1, 2, 3, 4, 6, 8, 14, 17, 32, and 42 repeats. There

Table 1 Microbial Biosynthesis: Basic Monomer Gene Sequences

I.	(GVGVP)$_{10}$
II.	(GVGIP)$_{10}$
III.	(AVGVP)$_{10}$
IV.	GVGVP GVGFP GEGFP GVGVP GVGFP GFGFP
V.	GVGVP GVGFP GEGFP GVGVP GVGFP GVGFP
VI.	GVGVP GVGVP GEGVP GVGVP GVGFP GFGFP
VII.	GVGVP GVGFP GEGFP GVGVP GVGVP GVGVP
VIII.	GVGVP GVGFP GKGFP GVGVP GVGFP GFGFP
IX.	GVGVP GVGFP GKGFP GVGVP GVGFP GVGFP
X.	GVGVP GVGVP GKGVP GVGVP GVGFP GFGFP
XI.	GVGVP GVGFP GKGFP GVGVP GVGVP GVGVP
XII.	GVGVP GVGFP GDGFP GVGVP GVGFP GFGFP
XIII.	GVGVP GVGFP GHGFP GVGVP GVGFP GFGFP
XIV.	GVGVP GVGFP GRGFP GVGVP GVGFP GFGFP
XV.	GVGVP GVGFP GEGFP GVGVP GVGVP GKGVP
XVI.	GVGVP GVGFP GEGFP GVGVP GVGFP GKGVP
XVII.	GVGVP GVGKP GEGFP GVGVP GVGVP GFGVP
XVIII.	GVGVP GVGKP GEGFP GVGVP GVGFP GFGVP
XIX.	GVGIP GFGEP GEGFP GVGVP GFGFP GFGIP
	GVGIP GFGEP GEGFP GVGVP GFGFP GFGIP
XX.	GVGIP GFGEP GEGFP GVGVP GFGFP GFGIP
	GVGIP GFGEP GEGFP GVGVP GFGFP
	GFGIP GVGVP GVGRGYSLG VPGV
XXI.	(GVGIP)$_{10}$-GVGVPGRGDSP-(GVGIP)$_{10}$
XXII.	(GVGVP)$_{10}$-GVGVPGRGDSP-(GVGVP)$_{10}$
XXIII.	(GVGVP)$_{10}$-GVGVPGRGDSP-GVGIP)$_{10}$

is also a carboxyl terminal (GVGVP) which derives from the oligo adaptor used in the concatemerization reaction and which provides the restriction site for insertion into the expression vector. Additional examples are (GVGVP)$_n$ with n = 41, 121, 141, and 251; (AVGVP)$_{n+1}$ with n = 20, 40, 70, and 290, and (GVGIP)$_n$(GVGVP) with n = 40, 70, 150, and 290. To have such an exact series of chain lengths presents an extraordinary opportunity for the physical chemist interested the chain length dependent properties of polymers, but again it provides for a series of polymers with a useful gradation of properties for a number of applications including drug delivery. It allows, for example, for the design of a polymer with an optimal pKa for the ion-pair loading and release of drug.

Purification

Convenient purification of microbially produced transductional protein-based polymers, e.g. poly(GVGVP), from the cell lysate is based on a methodology which utilizes the fundamental inverse temperature transitional properties (Urry *et al.*, 1995b; McPherson *et al.*, 1996). First the bacterial cells are separated from the growth medium either by centrifugation or filtration and resuspened in Tris-HCl buffer, 50 mM, pH 8.0. Then the cells are lysed by ultrasonic disruption or French press to release the cell contents. The cell lysate is cooled to 4°C and centrifuged at high speed (10,000 × g) to remove the cold insoluble materials. The supernatant fraction is then heated to 37°C to cause the protein-based polymers to form massive aggregates and to separate out of solution (coacervate) so that the more dense phase can readily be recovered by centrifugation (5,000 × g, 37°C). The recovered protein-based polymer phase is brought back into solution by adding pre-cooled buffer followed by centrifugation (10,000 × g) to remove again the cold insoluble materials. Repeating this warm-cold process several times results in protein-based polymer with essentially all bacterial proteins and membrane contaminants removed.

Related strategies can also be used based on more complex compositions of the protein-based polymers. If, for example, the polypeptide contains a repeating ionizable function, instead of heating and cooling, a change of pH or salt concentration can be used to control the phase transition between soluble and insoluble states, and, thereby, to enable their separation from bacterial proteins by centrifugation.

PHYSICAL AND BIOLOGICAL CHARACTERIZATIONS

Of the many important physical characterizations of these transductional protein-based polymers, the temperature at which the thermally elicited transition occurs is perhaps most fundamental (Urry, 1993a); the verification of sequence by two dimensional nuclear magnetic resonance provides a most satisfying demonstration of structure, and the mechanical properties of elastic modulus of the cross-linked matrices and of the extrusibility through fine hypodermic needles are fundamental to many applications. As such, these will be briefly noted here. As to the many significant biological characterizations, the battery of biocompatibility tests will be noted for three different polymers which represent three different physical states of the polymers, those states being plastic, elastic and hydrogel. In addition, the extensive efforts to obtain monoclonal antibodies to poly(GVGVP) and poly(AVGVP) are reported.

Inverse Temperature Transitional Properties

As indicated in the INTRODUCTION, these polymers derive their capacity to be transductional by exhibiting transitional behavior. When the temperature is sufficiently low, they are soluble in water in all proportions, but on raising the temperature they aggregate and phase separate to form a more-dense state of 40% or more polymer by weight. The temperature for the onset of the transition is designated as T_t and it is determined experimentally by plotting the onset of turbidity as the temperature is raised, as shown in Figure 1A. Also apparent in Figure 1A is a systematic dependence of T_t on the composition where the effect of four residues per 100 residues is seen to be quite significant. More hydrophobic residues lower T_t whereas introduction of less hydrophobic residues raises T_t. This dependence of T_t on composition is the basis of a hydrophobicity scale, as derived in Figure 1B, which is based for the first time on the hydrophobic folding and assembly process of interest (Urry, 1993a).

The largest changes in T_t result from the formation of charged species as on the ionization of a carboxyl or, most dramatically, the phosphorylation of a neutral residue such as serine. By raising the value of T_t above the operating temperature,

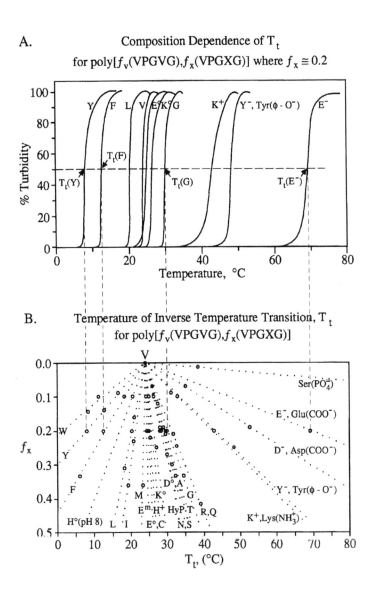

Figure 1 A. Temperature versus normalized turbidity curves for a series of guest amino acid residues occurring at the frequency of 4 guest residues per 100 residues, i.e., $f_X = 0.2$. The value of T_t, the temperature for the onset of the inverse temperature transition for hydrophobic folding and assembly, is defined as the temperature for 50% of maximal turbidity. Note that more hydrophobic guest residues lower the value of T_t and less hydrophobic, more polar residues raise the value of T_t.

B. Plots of T_t versus f_X, the mole fraction of pentamers containing guest residues, are essentially linear to $f_X = 0.5$. Extrapolation to the value of T_t at $f_X = 1$ is used to provide an index of the relative hydrophobicities of the amino acid residues. This provides for a hydrophobicity scale which is based on the hydrophobic folding and assembly of interest. Adapted with permission from Urry, 1993a.

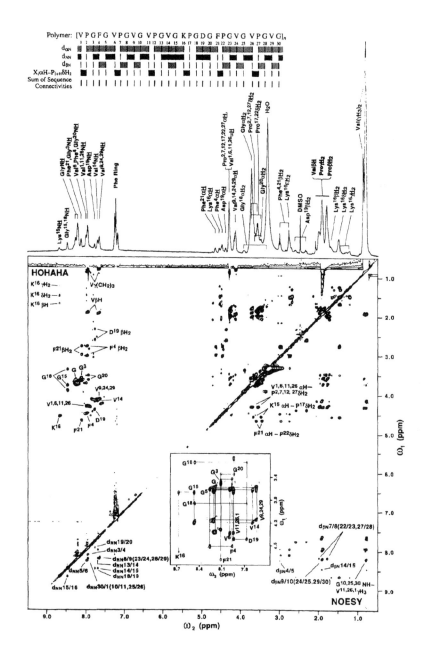

Figure 2 Proton nuclear magnetic resonance spectra (600 Mhz) of poly(VPGFG VPGVG VPGVG KPGDG FPGVG VPGVG). At the top are the connectivities that provide the sequence information and that derive from the information of the HOHAHA map (upper left side) which identifies the resonances of each of the amino acid residues and from the NOESY map (lower right side) which provides the through-space contacts for verifying sequence. Above the two-dimensional maps is the one-dimensional spectrum with resonance assignments indicated and the bar plots at top showing the sequence continuity.

the polymers can be caused to go into solution, or by lowering T_t below the working temperature, the polymers can be caused to hydrophobically fold and assemble. For cross-linked matrices, lowering T_t drives contraction and raising T_t causes the matrix to swell. One dramatic way to lower T_t is to ion-pair, for example, to pair a carboxylate of the polymer with a cation. As will be seen below, the pairing of the Glu carboxylate with a cationic drug results in a transductional process that can be the basis for controlled loading and release of drugs.

Nuclear Magnetic Resonance Verification of Structure

Nuclear magnetic resonance is a remarkable technique making it possible to identify every atom and its bonding in a large biomolecule and even to determine the folding of the molecule. Most generally, it is the protons that are observed and it is possible with proton spectra alone to verify sequence and integrity of the residues. Figure 2 reports the one-dimensional spectrum (above), the two-dimensional HOHAHA (the upper left of the map), and the two-dimensional NOESY(the lower right half of the map) of poly(VPGFG VPGVG VPGVG KPGDG FPGVG VPGVG) which is the chemically synthesized aspartic acid analogue of sequence XVII of Table 1. The bars at the top of Figure 2 indicate the occurrence of cross peaks, that is, proton-proton contacts, in the maps that can be used to verify the sequence. The sum of connectivities that allow verification of sequence are complete from the single pair of maps as shown for the entire 30 residues except for the step connecting residues 17 and 18, that is, the nuclear Overhauser enhancement, through space interactions between the NH of Gly[18] and the aCH (the d_{aN}) and the bCH$_2$ (the d_{bN}) of Pro[17] are not seen in the NOESY map as reported. While it may be that changing the parameters for collecting the data or for computing the NOESY map would bring up these contacts, the data as reported in Figure 2 clearly show the sequence to be correct.

The fidelity of the protein-based polymer expression from a prepared gene is extraordinary when compared to the chemically prepared product. Even so, it is important to use purification procedures that limit alteration of the structure and to establish that proteolytic or other enzymatic activity is sufficiently limited that the desired properties of the expressed protein-based polymer are retained. An important method with which to achieve adequate verification of the gene product again utilizes multi-dimensional nuclear magnetic resonance.

Mechanical Properties

The mechanical properties of the polymers determine in large measure their potential applications. The two physical properties of note here are the elastic modulus of the cross-linked matrices and the extrusibility of viscoelastic solutions of the polymers through fine gauged hypodermic needles as required for injection.

Elastic Modulus of Cross-Linked Matrices

Presently the most convenient means of achieving cross-linking is by γ-irradiation. An effective dose is 20 Mrads, and the cross-linked matrix of poly(GVGVP) so formed is designated as X^{20}–poly(GVGVP), or more explicitly for example for the microbially prepared sample as X^{20}–(GVGVP)$_{251}$. The elastic moduli vary depending on composition from 10^4 to 10^9 dynes/cm^2. By way of comparison the elastic modulus of the femoral artery of the dog is about 4 to 6×10^6 dynes/cm^2 and tendon is of the order of 10^9 dynes/cm^2. Specifically, the elastic moduli in dynes/cm^2 for a series of protein-based polymers are: 2×10^6 for X^{20}–(GVGVP)$_{251}$, 7×10^6 for X^{20}–poly(GVGIP), 2×10^8 for X^{20}–poly(AVGVP), and less than 10^5 for X^{20}–poly(GGAP). Combinations can give intermediate values. Accordingly, a very wide range of useful elastic moduli are possible, for example, for application to a range of tissue reconstruction efforts.

Extrusibility of Polymers

Extrusibility depends on composition and chain length, and may differ between chemically and microbially prepared polymers because the occurrence of racemization disrupts spiral structure formation and results in greater chain entanglements with the consequence of greater viscosity and poorer extrusibility. The chemically synthesized high polymers of molecular weights greater than 50,000 kDa – poly[0.2(GVGVP), 0.5(GAGVP), 0.3(GFGVP)], poly[0.75(GGVP), 0.25(GGFP)], poly[0.9(GGAP), 0.1(GGFP)] and poly(GGAP), as well as the microbially prepared (GVGVP)$_{251}$ — have each been extruded at 37°C through a 27 gauge hypodermic needle at the high concentrations of 500 mg/ml. On the other hand, chemically synthesized poly(GVGVP) of about 150,000 kDa molecular weight at a concentration of 400 mg/ml will not extrude through a 22 gauge needle. The above noted extrusibilities open the door for soft tissue augmentation applications.

Programmed Biodegradation of Matrices Due to Presence of Chemical Clocks

Whether *in vitro* in phosphate buffered saline or *in vivo*, the carboxamide of the natural amino acids, asparagine (Asn, N) and glutamine (Gln, Q), will hydrolyze to form aspartic acid (Asp, D) and glutamic acid (Glu, E), respectively. In general, this represents the conversion of an uncharged carboxamide, $CONH_2$, to a charged carboxylate, COO^-, and the charged carboxylate raises the value of T_t such that cross-linked matrices containing the carboxamides swell, the uncross-linked coacervate states dissolves, and non-cross-linked plastic constructs also dissolve. As has been shown by Robinson (1974), the half-life for the breakdown of carboxamide to carboxylate depends on the sequence and can vary from a few days to a decade. These carboxamides are referred to as chemical clocks as they can be used to trigger desired changes leading to degradation.

Intraperitoneal Implantation of the Elastic State Containing Chemical Clocks

Poly[0.89(GVGVP),0.11(GNGVP)] was γ-irradiation cross-linked using a dose of 20 Mrads to form an elastomeric sheet, designated as X^{20}-poly[0.89(GVGVP), 0.11(GNGVP)]. The sheet was cut into 1 cm^2 pieces. One such piece was then implanted in the peritoneal cavity of each of ten rats. After residing in the peritoneal cavity for one month, the rat abdominal cavities were then opened in search of the 1 cm^2 pieces or in search of that fraction of the pieces that did remain. In terms of cross-sectional areas, of the total of 10 cm^2 of material that had been implanted just less than one-half of the material was recovered. The data is available in the report from the independent testing laboratories of NAmSA[-] and is directly accessible in their archival files. These results are to be compared to an earlier set of identical studies in which the cross-linked matrix, X^{20}-poly(GVGVP), had been implanted, and 9 of the 10 implanted 1 cm^2 squares were recovered in the clear, transparent state in which they had been implanted. It was presumed that the tenth square was also there as implanted but was missed because the 9 recovered, transparent squares had been difficult to find. In all cases, the materials appeared to be biocompatible in this setting. The conclusion of these studies is that these matrices do not readily biodegrade when in the contracted state but do biodegrade once in the swollen, hydrogel state.

In Vitro Swelling of Plastic and Elastic States due to Chemical Clocks

Discs of the plastic state, X^{20}-poly(AVGVP), and the elastic states, X^{20}-poly(GVGVP) and X^{20}-poly(GVGIP), when exposed to phosphate buffered saline at pH 7.4 and 37°C, remain intact and unchanged, indefinitely. On the other hand, the introduction of asparagine, Asn(N), as in X^{20}-poly[0.94(AVGVP),0.06(GNGVP)], X^{20}-poly[0.9(GVGVP),0.1(GNGVP)], and X^{20}-poly[0.9(GVGIP),0.1(GNGVP)] causes the discs to swell linearly with time but at different rates with the more hydrophobic Ile(I)-containing polymer showing the slowest rate of swelling and with the plastic discs having been seen to disintegrate after about 100 days. The essential point is that the rate of swelling can be varied by control of composition. Since it is the swollen state that appears to be biodegradable in vivo, it appears that the rate of degradation in vivo should be controllable.

Biocompatibility (Toxicity) of Representative States

Transductional protein-based polymers can occur in three different physical states: elastic, plastic and hydrogel. The representative polymers for those three states under physiological conditions, respectively, are poly(GVGVP), poly(AVGVP), and poly(GGAP). These three polymers and their cross-linked matrices have been extensively studied for their biocompatibility using the set of eleven tests recommended for materials in contact with tissues, tissue fluids and blood. The studies were carried out by the independent testing laboratories of North American Science Associates, Inc., NAmSA. The results are published (Urry *et al.*, 1991; Urry *et al.*, 1995a, 1995b) and can be found in the archival files of NAmSA. In brief, poly(GGAP) and poly(GVGVP) and the γ-irradiation cross-linked matrices exhibit extraordinary biocompatibility (Urry *et al.*, 1991; Urry *et al.*, 1995b), and poly(AVGVP) and its (-irradiation cross-linked matrices exhibit good biocompatibility (Urry *et al.*, 1995a). Of course, as each protein-based polymer is designed for a specific function by varying the composition, it will require the complete set of tests. The results to date for sterile, low endotoxin polymers, however, are very encouraging.

Immunogenicity of Representative Elastic and Plastic Protein-Based Polymers

Synthetic Polypeptides as Antigens

Synthetic polypeptide polymers differ from most naturally occurring protein antigens in that they carry a limited assortment of antigenic determinants. In the early 1970s, experiments conducted by McDevitt and colleagues (McDevitt *et al.*, 1972) elegantly demonstrated that immune responses to synthetic polypeptide antigens in inbred mouse strains are controlled by genes that map to the major histocompatibility or H-2 complex. The antigens used in these studies consisted of a linear poly(lysine) chain having multiple poly(alanine) sequences attached as side chains, a structure represented as A-L. To complete the structure of the polymers either histidine and glutamic acid (H,G) or tyrosine and glutamic acid (T,G) were attached to the amino ends of the poly(alanine) side chains, to give polymers represented as (H,G)A-L, or (T,G)A-L, respectively. Upon injection of these polymers into inbred strains of mice, varying degrees of responsiveness were observed. CBA mice (H-2^k) demonstrated high antibody responses to (H,G)A-L but poor responses to (T,G)A-L, whereas the C57BL strain (H-2^b) exhibited the opposite response pattern. Crosses between high and low responder strains resulted in F_1 hybrids that exhibited an intermediate response as compared to the two parental strains. Additional backcross experiments indicated that responsiveness to each of the two synthetic polymers was controlled by a single codominant gene referred to as an immune response (Ir) gene.

Due to numerous advances in the field of immunology, it is now clear that Ir genes encode class II major histocompatibility complex (MHC) proteins found on the surface of antigen presenting cells (such as macrophages and B cells). Class II MHC molecules (also called Ia molecules) are heterodimers containing noncovalently bound polypeptides of approximately 25,000 and 33,000 daltons. These molecules play a pivotal role in the initiation of immune responses to protein antigens. Processing of protein antigens begins following their phagocytosis by antigen presenting cells (APC). Within the lysozome of the APC the protein is partially degraded by lysosomal enzymes into relatively small (13–17 amino acid) fragments. The protein fragments eventually make their way to the surface of the APC where they are found in a complex with class II protein molecules. The results of X-ray crystallographic studies illustrate that the peptide sits in a cleft formed by the interaction of the 2 subunits of the class II MHC molecule. Amino acid sequence data confirms that both the larger (alpha) and smaller (beta) units of the heterodimer contain polymorphic regions. Many of the polymorphic regions of the alpha and beta subunits are situated in the cleft area. Although it is surmised that the cleft of most class II MHC molecules can accommodate a wide range of structures, it is clear from the previously described experiments that some strains of mice lack class II molecules with affinity for certain synthetic peptides. Since amino acid co-polymers also represent antigens with a restricted number of antigenic determinants, the production of monoclonal antibodies (Mabs) to these substances may require the immunization of a number of different mouse strains with varying genetic backgrounds.

Hybridoma Production

The immunogenicity of amino acid co-polymers, like that of other protein antigens, is enhanced by suspension in complete Freund's adjuvant (CFA). For the production of monoclonal antibodies to the protein-based polymers, poly(AVGVP) and poly(GVGVP), primary immunization of Balb/c mice consisted of 200 μg of each polymer separately emulsified in CFA. The emulsion was injected subcutaneously into several sites over the abdominal region. Approximately 28 days, weeks after primary immunization, mice were bled and their sera tested for antibody activity. An ELISA assay (described below) performed on primary antisera from 2 mice immunized with poly(AVGVP) revealed that both mice exhibited a modest (titer approximately 1:2560) antibody response to poly(AVGVP). However, these sera also demonstrated a titer of approximately 1:320 to poly(GVGVP). In contrast, 2 mice immunized with poly(GVGVP) exhibited relatively low titers (1:40 to 1:80) to this antigen by ELISA. No activity to poly(AVGVP) was detected in sera from these mice. These findings demonstrate that both poly(AVGVP) and poly(GVGVP) are weakly immunogenic with poly(GVGVP) being the poorer antigen. Both sets of mice were given secondary immunizations consisting of 200 μg of polymer emulsified in incomplete Freund's adjuvant (IFA). Sera collected 28 days post-secondary immunization demonstrated slightly increased titers to their respective antigens; however, a significant amount of cross-reactivity between the 2 polymers was now apparent in both sets of sera. The mice were boosted a final time with 200 μg of polymer in IFA. Three days later mice were sacrificed and their spleens removed for fusion. Splenocytes were fused with the mouse myeloma line P3X63 Ag8.653 using standard techniques (Berzofsky *et al.*, 1993). Fused cells were plated into 24 well plates in complete medium consisting of RPMI 1640 containing 15% fetal calf serum, 5×10^{-5} M 2-mercaptoethanol, 2mM glutamine, penicillin-streptomycin at 100 U/ml and 100 μg/ml, respectively, and HAT supplement (Gibco-BRL) diluted 1:100. Fusion wells were fed on days 7, 10, and 12 after fusion, with the final feeding consisting of a complete change of media. Approximately 14 days after fusion, supernatants were screened for antibody activity by ELISA. Positive lines were cloned by limiting dilution into wells containing peritoneal exudate feeder cells.

Screening Assays

Mouse antisera and hybridoma supernatants were tested for reactivity with poly(GVGVP) or poly(AVGVP) by ELISA. For these assays, polymers were suspended in cold (4°C) borate buffered saline (BBS) at a concentration of 10 μg/ml. ELISA wells were coated with 100 μl of antigen solution overnight at 4°C. After coating, wells were washed 3 times with BBS and blocked by the addition of 1% BSA in BBS for 1 hour at room temperature (RT). After washing, serum dilutions or hybridoma supernatants were added to the wells and incubated overnight at 4°C. Plates were then washed 6 times with BBS and incubated with phosphatase-labeled goat anti-mouse immunoglobulin (Jackson Laboratories), diluted 1:4000 in blocking buffer, for 90 minutes at RT. Following extensive washing, plates were developed by the addition of p-nitrophenyl phosphate (PNPP) at 1 mg/ml diluted in 10 mM diethanolamine (pH 9.5) containing 0.5 mM MgCl$_2$. After 15–30 minutes, the optical density, OD, at 405 nm was determined spectrophotometrically.

Fusion of splenocytes from a mouse immunized with poly(AVGVP) resulted in the production of three antigen-specific hybridoma lines. Two of these lines were strongly reactive with the immunizing antigen, poly(AVGVP) and did not cross react with poly(GVGVP). Interestingly, one line obtained in this fusion reacted weakly with poly(AVGVP) but showed significant reactivity with poly(GVGVP). Two fusions performed with splenocytes from mice immunized with poly(GVGVP) failed to produce antigen-specific hybridomas. The failure of these fusions is probably related to the low antibody titers found in Balb/c animals immunized with this polymer. Immunization of two additional mouse strains (DBA and CB6 F_1) with poly(GVGVP) did not result in improved antibody responses.

The preceding results suggest that certain protein-based polymers, such as the elastic poly(GVGVP), appear to be incapable of eliciting a significant immune response. On the basis of the comparison of the biocompatibility studies noted above for the elastic poly(GVGVP), the plastic poly(AVGVP) and the hydrogel poly(GGAP), the hydrogel, poly(GGAP), would also be expected to elicit little or no antibody response. These findings are particularly encouraging for the consideration of medical devices comprised of, or coated with, these elastic and hydrogel protein-based polymers.

APPLICATIONS

Of a rather long list of applications that are under consideration for these polymers, only two will be discussed here. One is drug delivery because this application can take full advantage of the rich transitional properties of these transductional elastic and plastic protein-based polymers (Urry *et al.*, 1997), and the other is tissue reconstruction because this application can simultaneously take advantage of the range of elastic moduli that are possible with these polymers, the capacity to introduce cell attachment sequences, and their ultimate biodegradability (Urry, 1993b).

Drug Delivery

Just as there are innumerable ways of varying T_t, there are innumerable ways of achieving controlled release of drug. One specific means of controlled release using transductional elastic and plastic protein-based polymers involves ion-pairing.

As seen in Figure 1, protonation of a carboxylate side chain of a transductional protein-based polymer dramatically lowers the value of T_t. This means that at pH 7.4 the polymer, polymer VI of Table 1 which has a pKa of 5.4, is soluble in water at 37°C, but on lowering the pH to 4.5, the polymer will aggregate to form a more dense coacervate phase. If the polymer is cross-linked to form a matrix, the matrix will be swollen at pH 7.4, but will contract on lowering the pH to 5.4 as the protons neutralize the carboxylates. Increasing the hydrophobicity of the polymer raises the pKa of the carboxyl, i.e., increases the affinity for protons.

Analogously, other cations and even cationic drugs can neutralize the charge and drive assembly to form the drug-laden coacervate state or drive contraction of the swollen matrix to result in a drug-laden contracted state. Also, increasing the hydrophobicity of the polymer increases the affinity of the cationic drug for the pKa shifted carboxylic site.

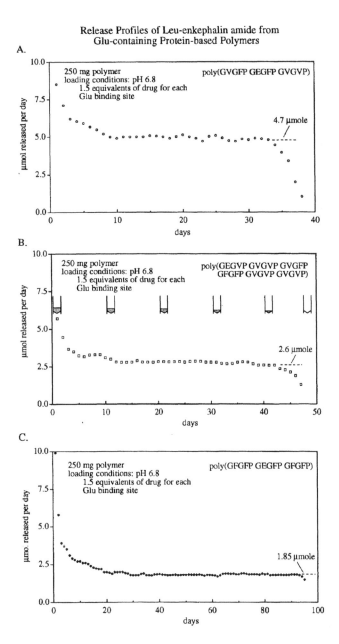

Figure 3 Release profiles of Leu-enkephalin amide from Glu-containing protein-based polymers. The loading conditions in each case were the addition of 1.5 equivalents of drug for each Glu binding site of 250 mg of polymer at pH 6.8, and the release data is for a constant surface area of 0.55 cm². **A.** Poly(GVGFP GEGFP GVGVP), giving a near constant release of 4.7 μmoles/day for one month. **B.** Poly(GEGVP GVGVP GVGFP GFGFP GVGVP GVGVP), giving a near constant release of approximately 2.7 μmoles/day for a period of more than 40 days. The included sketches are cross sections of the straight walled tubes utilized showing that the polymer disburses as the drug is released. **C.** Poly(GFGFP GEGFP GFGFP), giving a near constant release of 1.8 μmoles/day for a period of 3 months.

The association of cationic drug with anionic site to form a "phase separated" or "contracted" drug-laden state is analogous to the association of ions to form an insoluble salt, the dissociation of which is described by a solubility product constant, K_{sp}. Accordingly, the drug-laden coacervate will dissociate with the release of drug and polymer until the overlying solution is saturated. If there is a continuous removal of the surrounding medium, then there will occur a constant release of drug for a constant surface area. Importantly, the density of the drug in the vehicle is determined by the number of carboxylates and the rate of release is controlled by the hydrophobicity as shown in Figure 3 for the cationic opioid peptide, Leu-enkephalin amide. In this example, a release rate of 1.8 μmole/day is sustained for over 3 months or for a different composition a release rate of 4.7 μmole/day is observed. In both cases the release rate is constant for a constant surface area of 0.55 cm^2 and the release rate can be maintained for an approximate volume of 0.5 ml.

Tissue Reconstruction

Tissue reconstruction has as its objective the regeneration of lost or damaged tissues with restoration of normal function. Strategies for tissue reconstruction include i) cell transplantation, ii) introduction of growth factors as inducers, iii) implantation of cell-attached matrices, and iv) introduction of temporary functional scaffoldings which, due to appropriate design, would be remodeled into natural functional tissues. Although implantation of cell-attached matrices and the cellular remod-elling of these functional scaffoldings hold great promise, the biocompatibility of the material for reconstruction is, of course, essential to avoid adverse host reactions.

Polymers such as polyglycolic acid have been studied as a biomaterial for reconstruction of ureteral organs. Isolated urothelial cells have been grown on a polyglycolic acid mesh which was then implanted into animals, and multilayering of urothelial cells on the polymer was observed after 20 days (Atala *et al.*, 1992; Atala *et al.*, 1993; Langer and Vacanti, 1993). The limitations of polyglycolic acid, however, include: i) a significant reaction to degradation as ester hydrolysis results in acid release whereas hydrolysis of the peptide bond results in no release of acid, ii) the difficulty of including cell attachment sequences, and iii) the absence of elasticity such that attached cells will not readily sense the tensional forces to which the cells in the natural tissue are subjected.

The Principle of Tensegrity

In the principle of tensegrity (Ingber, 1991), cells respond to the tensional forces required of their tissues by turning on the genes required to produce an extracellular matrix sufficient to sustain those forces (Leung *et al.*, 1977). Thus, to our knowledge the only biomaterials, currently under development that exhibit the requisite set of biological and biophysical properties, are the elastic protein-based polymers, often referred to as bioelastic materials. As described above, these are biocompatible, biodegradable, and elastic polymers that can be prepared with the desired elastic modulus and cell attachment sequences as an integral part of their primary structure, as for the catemer sequences, XXI, XXII, and XXIII of Table 1.

Consideration of Urological Prostheses

In the urinary tract, smooth muscle cells and uroepithelial cells are regularly exposed to a changing mechanical strain environment. Just how cells transduce the force changes to which they are subjected to functional responses is not yet clear. Several *in vitro* studies have addressed the modulation of smooth muscle and epithelial cells by the mechanical forces (Leung *et al.*, 1977; Brunette, 1984; Ingber, 1991; Karim *et al.*, 1992; Baskin *et al.*, 1993; Barbee *et al.*, 1994; Lyall *et al.*, 1994). Baskin *et al.* (1993) specifically addressed the alteration of extracellular matrix protein synthesis in response to mechanical stimuli. Others have reported that changes in tension increase in the rate of DNA synthesis, suggesting an increase in cell growth (Karim *et al.*, 1992). Brunette found that mechanical strain affects the proliferation and the ultrastructure of epithelial cells (Brunette, 1984). In their study, a 4.2% increase in length increased the number of desmosomes, which suggested an increase in cell-cell interaction. In the preliminary study noted below, urothelial cells are subjected to approximately 100% extensions in a realistic mimicking of the filling and emptying of a functional bladder.

Accordingly, the design of the appropriate matrix for urological tissue reconstruction should consider the design of a bioelastic material with the physical properties of the respective urological tissue. For example, the material for reconstruction of a urological bladder should exhibit the same elastic modulus as the normal bladder; and the composition should be able to contain peptide sequences that support cell attachment, growth to confluence and other cellular functions.

The elastic modulus of strips of the human bladder have been reported to be about 1.9×10^5 N/m^2 (Van Mastright and Van Duyl, 1981). This is a value reasonably close to that of X^{20}-(GVGVP)$_{251}$ which has been determined to be 1.6×10^6 dynes/cm^2 or 1.6×10^5 N/m^2. The elastic modulus for X^{20}-poly(GVGIP) is in the range of 4 to 8×10^5 N/m^2 such that a functional range of elastic matrices are available.

In our preliminary studies, we have designed, synthesized and cross-linked protein-based polymers to form bioelastic matrices containing the cell attachment sequence, GRGDSP, and have studied urothelial cell outgrowth onto the matrices from human ureter explants. The urothelial cells grew out onto the bioelastic matrices and multiplied in the normal fashion as seen in Figure 4A in a manner comparable to the outgrowth onto a collagen substrate, as shown in Figure 4B.

The next step was to design a computerized apparatus for simulated bladder function which slowly fills with increasing pressure over a 3-hr period and which releases the volume and associated pressure in 23 seconds. In this bladder filling and emptying simulation, the bioelastic matrix substrate to which the cells are initially attached and over which the developing extracellular matrix of the cells spread stretches slowly and relaxes rapidly. *This system mimics, both temporally and in terms of the magnitude of the tensional force changes with 100% extensions, the normal filling and micturition of the bladder, as can only be achieved with an elastic cell-attachment matrix having an elastic modulus similar to that of the natural bladder* (Urry and Pattanaik, 1997).

In the bladder simulation study, uroepithelial cells grew out onto the bioelastic matrix, X^{20}-(GVGVP)$_{251}$ adsorbed with GRGDSP-containing fibronectin; the attached cells and matrix were subjected to the stretch/relaxation cycles over several days, and the result was a formation of a dense layer of cells. Figures 4C and D and 4E and F at higher magnification were, respectively, before and after being exposed to

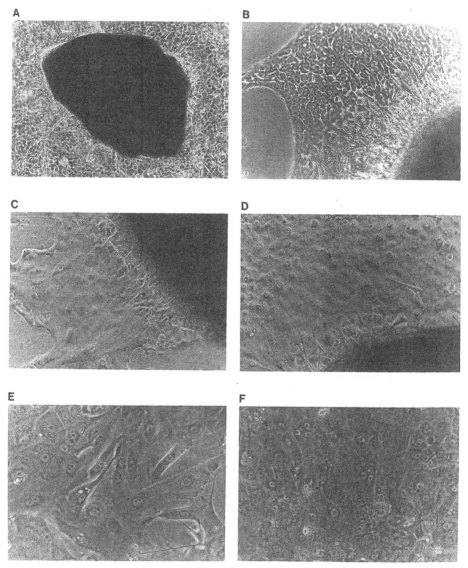

Figure 4 Comparison of urothelial cell growth from human explants onto collagen and bioelastic matrices containing the GRGDSP cell attachment sequence with and without tensional force changes mimicking bladder filling and emptying.

A. Outgrowth from human ureteral explant (dark area) onto static bioelastic matrix, cross-linked poly[40(GVGVP),(GRGDSP)].

B. Outgrowth from explant onto collagen coated surface.

C. Outgrowth from explant onto *static* cross-linked $(GVGVP)_{251}$ coated with GRGDSP-containing human fibronectin.

D. Outgrowth from explant onto *dynamic* cross-linked $(GVGVP)_{251}$ coated with GRGDSP-containing human fibronectin resulting from three days of simulating bladder function.

E. and F. Higher magnification of C. and D., respectively. This shows the richer, more developed urothelial cell and extra cellular matrix coverage resulting from the changing tensional forces in support of the principle of tensegrity. Reproduced with permission from Urry and Pattanaik, 1997.

several days of stretch/relaxation cycles. Qualitatively, an increase in cell density, in cell-cell interaction, and in extracellular matrix has been observed.

ACKNOWLEDGMENTS

The authors wish to acknowledge the following support: the Office of Naval Research under the grant No. N00014-89-J-1970; the National Institutes of Health with support of the contract 1 R43 DA09511-01 from the National Institute of Drug Abuse; N43-DK-4-2209 from National Institute of Diabetes and Digestive and Kidney Diseases; and of the U. S. Army Natick Research Development and Engineering Center DAAK60-93-C-0094.

REFERENCES

Atala, A., Freeman, M.R., Vacanti, J.P., Shepard, J. and Retik, A.B. (1993) "Implantation *in vivo* and retrieval of artificial structures consisting of rabbit and human urothelium and human bladder muscle, *J. Urol.*, **150**, 608–612.

Atala, A., Vacanti, J.P., Peters, C.P., Mandell, J.M., Retik, A.B. and Freeman, M.R. (1992) "Formation of urothelial structures *in vivo* from dissociated cells attached to biodegradable polymer scaffolds *in vitro*," *J. Urol.*, **148**, 658.

Ausubel, F.M. (1989) *Current Protocols in Molecular Biology*, Vols. 1&2, Greene Publishing Associates and Wiley-Interscience, John Wiley & Sons, NY.

Barbee, K.A., Macarak, E.J. and Thibault, L.E. (1994) "Strain measurements in cultured vascular smooth muscle cells subjected to mechanical deformation," *Annals of Biomedical Engineering*, **22**, 14–22.

Baskin, L., Howard, P.S. and Macarak, E. (1993) "Effect of physical forces on bladder smooth muscle and urothelium," *J. of Urol*, **150**, 601–607.

Berzofsky, J.A., Berkower, I.J. and Epstein, S.L. (1993) "Antigen-antibody interactions and monoclonal antibodies" in: *Fundamental Immunology*, William H. Paul, Ed., Raven Press, New York pp. 421–465.

Brunette, D.M., "Mechanical stretching increases the number of epithelial cells synthesizing DNA in culture," *J. Cell Sci.*, **69**, 35–45.

Capello, J. (1992) "Protein engineering for biomaterials applications," *Curr. Opin. Struct. Biol.*, **2**, 582–586.

Deguchi, Y., Krejchi, M.T., Borbely, J., Fournier, M.J., Mason,. T.L., Tirrell, D.A. (1993) *Mat. Res. Soc. Symp. Proc.*, **292**, 205–210.

Doel, M.T., Eaton, M., Cook, E.A., Lewis, H., Patel, T. and Carey, N.H. (1980) "The expression in E coli of synthetic repeating polymeric genes coding for poly(L-aspartyl-L-phenylalanine)," *Nucleic Acids Research*, **8**, 4575–4592.

Dugaiczyk, A., Boyder, H., and Goodman, H.M. (1975) "Ligation of EcoRI endonuclease-generated DNA fragments into linear and circular structures," *J. Mol Biol*, **96**, 174.

Ferrari, F.A., Richardson, C., Chambers, J., and Causey, S.C. (1993) Patent No. 5,243,038 (U.S.), "Construction of synthetic DNA and its use in large polypeptide synthesis," filed October 29, 1987, issued September 7, 1993.

Goldberg, I., Salerno, A.J., Patterson, T. and Williams, J.I. (1989) "Cloning and expression of a collagen-analog-encoding synthetic gene in *Escherichia coli*," *Gene*, **80**(2), 305–314.

Gowda, D. Channe, Parker, Timothy M., Harris, R. Dean and Urry, Dan W. (1994) "Synthesis, characterizations and medical applications of bioelastic materials," in *Peptides: Design, Synthesis, and Biological Activity* (Channa Basava and G. M. Anantharamaiah, Eds.), Birkhäuser, Boston, MA, pp. 81–111.

Guda, C., Zhang, X., McPherson, D.T., Xu, Jie, Cherry, J.H., Urry, D.W. and Daniell, H. (1995) "Hyper expression of an environmentally friendly synthetic polymer gene," *Biotechnology Letters*, **17**, 745–750.

Hartley, James L. and Gregori, Tanara J. (1981) "Cloning multiple copies of a DNA segment." *Gene*, **13**, 347–353.

Ingber, D. (1991) "Integrins as mechanochemical transducers," *Current Opinion in Cell Biology*, **3**, 841–848.

Kaplan, D.L., Lombardi, S.J., Muller, W.S. and Fossey, S.A. (1991) *Biomaterials*, 3–53.

Karim, O.M.A., Pienta, K., Seki, N. and Mostwin, J.L. (1992) "Stretch-mediated visceral smooth muscle growth *in vitro*," *Am. J. Physiol.*, **262**, R895–R900.

Langer, R. and Vacanti, J.P. (1993) "Tissue engineering," *Science,* **260**, 920–926.

Leung, D.Y.M., Glagov, S. and Mathews, M.B. (1977) "A new *in vitro* system for studying cell response to mechanical stimulation," *Exp. Cell Res.,* **109**, 285–298.

Lyall, F., Deehan, M.R., Greer, I.A., Boswell, F., Brown, W.C. and McInnes, G.T. (1994) Mechanical stretch increases proto-oncogene expression and phosphoinositide turnover in vascular smooth muscle cells," *J. of Hypertension,* **12**, 1139–1145.

McDevitt, H.O., Deak, B.D., Shreffler, D.C., Klein, J., Stimpfling, J.H., and Snell, G.D. (1972) "Genetic control of the immune response. Mapping to the Ir-1 locus," *J. Exp. Med,* **135**, 1259–1278.

McPherson, David T., Morrow, Casey, Minehan, Daniel S., Wu, Jianguo, Hunter, Eric and Urry, Dan W. (1992) "Production and purification of a recombinant elastomeric polypeptide, G-(VPGVG)$_{19}$-VPGV, from *Escherichia coli.*" *Biotechnol. Prog.,* **8**, 347–352.

McPherson, David T., Xu, Jie and Urry, Dan, W. (1996) "Product purification by reversible phase transition following *E. coli* expression of genes encoding up to 251 repeats of the elastomeric pentapeptide GVGVP," *Protein Expression and Purification,* **1**, 51–57.

Murata, T., Horinouchi, S. and T. Beppu, (1993) *J. of Biotech.,* **28**, 301–312.

Perbal, Bernard (1988) *A Practical Guide to Molecular Cloning,* 2nd Edition, A Wiley-Interscience Publication, John Wiley & Sons, NY.

Prasad, K.U., Iqbal, M.A. and Urry, D.W. (1985) "Utilization of 1–hydroxybenzotriazole in mixed anhydride coupling reactions," *Int. J. Pept. and Protein Res.,* **25**, 408–413.

Robinson, A.B. (1974) "Evolution and the distribution of glutaminyl and asparaginyl residues in proteins," *Proc. Natl. Acad. Sci. USA,* **71**, 885–888.

Salerno, A.J. and Goldberg, I. (1993) *Mater. Res. Soc. Symp. Proc.,* **292**, 99–104.

Sambrook, J., Fritsch, E.F. and Maniatis, T. (1989) *Molecular Cloning: A Laboratory Manual.* 3 volumes, Cold Spring Harbor Laboratory Press; CSH, NY.

Studier, F.W., Rosenberg, A.H., Dunn, J.J., and Dubendorff, J.W. (1990) "Use of T7 polymerase to direct expression of cloned genes" in *Methods in Enzymology: Gene Expression Technology* (Goeddel, D.V., Ed.), **185**, 60–89, Academic Press, San Diego.

Urry, Dan W. (1993a) "Molecular machines: How motion and other functions of living organisms can result from reversible chemical changes," *Angew. Chem. (German),* **105**, 859–883; *Angew. Chem. Int. Ed. Engl.,* **32**, 819–841.

Urry, Dan W. (1993b) "Bioelastic materials as matrices for tissue reconstruction," in *Tissue Engineering: Current Perspectives* (Eugene Bell, Ed.), Birkhäuser Boston, Div. Springer-Verlag, New York, NY, pp. 199–206.

Urry, D.W., Harris, Cynthia M., Luan, Chi Xiang, Luan, Chi-Hao, Gowda, D. Channe, Parker, Timothy M., Peng, ShaoQing and Xu, Jie (1997) Chapter entitled "Transductional protein-based polymers as new controlled release vehicles," *Part VI: New Biomaterials for Drug Delivery, in Controlled Drug Delivery: The Next Generation,* (Kinam Park, ed.) Am. Chem. Soc. Professional Reference Book, (In press).

Urry, D.W. and Long, M.M. (1987) Patent No. 4,693,718 (U.S.), "Stimulation of chemotaxis by chemotactic peptides (nonapeptide)," filed October 31, 1985, issued September 15, 1987.

Urry, Dan W., McPherson, D.T., Xu, J., Daniell, H., Guda, C., Gowda, D.C., Jing, Naijie and Parker, T.M. (1996) "Protein-based polymeric materials: Syntheses and properties" in *The Polymeric Materials Encyclopedia: Synthesis, Properties and Applications,* CRC Press, Boca Raton. pp. 7263–7279.

Urry, Dan W., McPherson, David T., Xu, Jie, Gowda, D. Channe and Parker, Timothy M. (1995a) "Elastic and plastic protein-based polymers: Potential for industrial uses," (Am. Chem. Soc.) Div. Polym. Mat.: Sci. & Engr., *Industrial Biotechnological Polymers,* Washington, D.C., 259–281.

Urry, Dan W., Nicol, Alastair, McPherson, David T., Xu, Jie, Shewry, Peter R., Harris, Cynthia M., Parker, Timothy M. and D. Channe Gowda (1995b) "Properties, preparations and applications of bioelastic materials," in *Handbook of Biomaterials and Applications,* Marcel Dekker, Inc., New York, pp. 2645–2699.

Urry, Dan W., Parker, Timothy M., Reid, Michael C. and Gowda, D. Channe (1991) "Biocompatibility of the bioelastic materials, poly(GVGVP) and its γ-irradiation cross-linked matrix: Summary of generic biological test results," *J. Bioactive Compatible Polym.,* **6**, 263–282.

Urry, D.W. and Prasad, K.U. (1985) "Syntheses, characterizations and medical uses of the polypentapeptide of elastin and its analogs," in *Biocompatibility of Tissue Analogues,* (D.F. Williams, Ed.), CRC Press, Inc., Boca Raton, Florida, 89–116.

Urry, D.W. and Pattanaik, A. (1997) NY Acad. Sci. (in press).

Van Mastright, R., Coosaet, B.L.R.A. and Van Duyl, N.A. (1981) "First results of stepwise straining of the human urinary bladder and human bladder strips," *Investigative Urology,* **19**, 58–61.

20. GENETICALLY ENGINEERED PROTEIN POLYMERS

JOSEPH CAPPELLO

Protein Polymer Technologies, Inc., 10655 Sorrento Valley Rd.,
San Diego, CA 92121, USA

DESIGN, SYNTHESIS AND PURIFICATION OF BIOLOGICALLY PRODUCED PROTEIN POLYMERS

Our laboratory has developed methods for the production of synthetically designed protein polymers consisting of repeated blocks of amino acid sequence (Ferrari *et al.*, 1986; Cappello and Ferrari, 1994). Protein polymers are sequential polypeptides with complex structural repeats. Through a combination of chemical and biological methods, protein polymers are produced using gene template directed synthesis. This technology has been used by ourselves and investigators in other laboratories to produce an increasing number and variety of distinct protein polymer compositions. While the products of this technology, sequential polypeptides, may be used for any purpose, so far, they have almost exclusively been designed for use as materials or as surface modifiers of materials.

The development of a new protein polymer involves the design, chemical synthesis, and polymerization of a gene template that encodes the amino acid sequence of the desired protein. For each new composition, the construction of the gene template occurs only once. Thereafter, it is preserved and can be reproduced faithfully and propagated indefinitely by introducing it into a microorganism. Through the construction of synthetic genes, one can specify the precise amino acid sequence of protein blocks (the unit of repetition of a protein polymer) up to several hundred amino acids in length, almost an order of magnitude greater than the limit of sequence control through chemical synthesis. Our experience suggests that except for a few examples any number and combination of the twenty natural amino acids can be used, and the complexity of the amino acid sequence has no direct bearing on the efficiency of producing the product. One can produce protein chains greater than one thousand amino acids in length (approximately 100,000 daltons), and, depending on the gene template selected for production, a discrete molecular weight, monodisperse product can be obtained. Currently, production of protein polymers occurs in microorganisms under controlled fermentation processes that can be scaled to industrial levels. The process utilizes simple renewable resources. Because master stocks of production strains can be cryopreserved, identical product can be produced from different batches at different times.

Our work has focused on the production of synthetic protein analogs of the four best known structural protein materials, silk, elastin, collagen, and keratin (Figure 1). Initially we set out to prove that synthetically designed protein polymers, which consisted of repeating structural blocks from natural structural proteins, could be produced by microbial synthesis. We demonstrated that these synthetic proteins would reproduce defined properties of their natural counterparts and that we

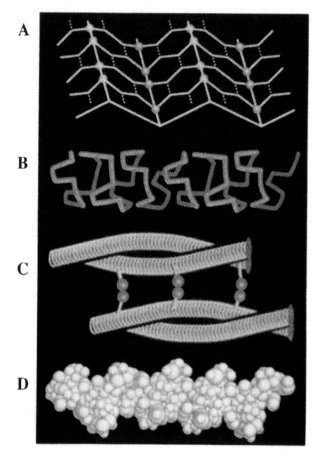

Figure 1 Diagrammatic depiction of the protein chain conformations of protein structural materials. The major, characteristic protein chain conformations of silk fibroin (A, hydrogen bonded beta strands forming a sheet), elastin (B, recurring beta turns linked by suspended flexible segments), keratin (C, alpha helical coiled ropes with interstrand dissulfide bonding), and collagen (D, triple helix consisting of three polypeptide chains each consisting of a glycineXY repeat) are displayed.

could combine structural blocks from different natural proteins and obtain new properties found in neither the synthetic homoblock polymers nor the natural proteins (Table 1). Finally, we showed that blocks of unique amino acid sequence could be introduced into a protein polymer, providing novel chemical or biological activities without disrupting the physical properties of the structural blocks. These capabilities enable the production of unique protein polymer-based materials for specific biomedical applications; materials whose properties have significant advantages over currently available materials.

David Tirrell and his collaborators at the University of Massachusetts, Amherst have produced protein polymers designed to address fundamental questions of polymer science (McGrath *et al.*, 1992; Creel *et al.*, 1991; Tirrell *et al.*, 1991).

Table 1 Primary Protein Structural Building Blocks

Structure	Example	Block Sequence	Properties
Beta sheet	Silk	$(GAGAGS)_n$	Strength, crystallinity
Reverse turn	Elastin	$(GVGVP)_n$	Flexibility, elasticity
Triple helix	Collagen	$(GX^aY^b)_n$	Soft matrix
Alpha helix	Keratin	$(AKLK/E^cLAE)_n$	Hard matrix

n can be any number greater than or equal to 2
[a]X refers to any amino acid, often proline
[b]Y refers to any amino acid, often 4-hydroxyproline
[c]Charged amino acid positions can be interchanged with acidic or basic amino acids

Using the precision of gene template directed biosynthesis, they have produced chain folded lamellae of exact molecular dimensions. These molecular films may be used to position functional groups in precise locations on the lamellar planes. Tirrell *et al.* (1991) have also demonstrated that under special circumstances certain unnatural amino acids with chemically interesting side chains can be incorporated into protein polymer chains. This further increases the potential diversity of compositions and properties that may be obtained through protein polymers.

Structural Components Of Natural Proteins

Natural proteins derive their structures from their amino acid sequences which position chemical groups throughout the chain. Designing new proteins requires knowing how to locate amino acids in a linear chain such that a desirable structure is obtained. Over the last forty years, the accumulation of protein crystal data has provided a growing database from which structural biologists have established general patterns of protein structure. Though proteins may perform very different functions, they share common structural elements (Brandon and Tooze, 1991). The four most common elements found in structural proteins are alpha helices, beta strands, reverse turns, and collagen helices (Figure 1 and Schulz *et al.*, 1979). Even when the pendant side chains vary greatly, the backbone atoms of secondary structural elements from different proteins can often be superimposed. In many cases, certain sets of amino acids can be used interchangeably at specific locations in the sequence of a protein (Chou and Fasman 1974; Garnier *et al.*, 1978, and Schulz 1988).

The use of this information in conjunction with computer modelling allows one to design amino acid sequences which conform to a particular secondary structural geometry. The treatment of proteins with solvents can also be used to drive a protein chain to a particular conformation.

Biological Production Methods

We have developed a generic production system that through simple manipulation can be used to produce a wide variety of polymer genes in sufficient quantities for chemical and physical characterization. New protein polymer compositions can be rapidly purified and analyzed to verify that the product has the intended properties. When a product is ready to enter commercial development, it can be easily transferred into a system that can be optimized for large scale production.

We and others have produced protein polymers from over 50 different amino acids sequence designs. Many of these have been purified in multigram quantities. We have demonstrated the scaleability of our process by producing two protein polymers in kilogram quantities.

Polymer Gene Construction

The elements of protein polymer gene construction and the expression of the products they encode can be described in three steps (Cappello, 1992). In the first step, the amino acid sequence of the repeating peptide block is defined. Since this sequence must ultimately be encoded by synthetic DNA, the length of the sequence is the first constraint. Typical automated oligonucleotide synthesis can produce single strands of DNA up to about ninety nucleotides in length. Under optimized conditions, we have produced DNA monomers of over 265 nucleotides from a single synthesis. Therefore, the maximum amino acid coding potential of each contiguous segment of DNA can be about 80 amino acids. If a polymer is to consist of a block of amino acids five residues long, for example, sixteen contiguous blocks can be encoded by one DNA segment. If the block of amino acids exceeds 100 residues, multiple DNA segments must be synthesized and assembled to produce the gene monomer.

After synthesis, the monomer gene segment is cloned into a DNA plasmid that has been engineered to perform suitably as an acceptor plasmid (ie., it contains a unique restriction endonuclease site with asymmetric overhanging ends which are compatible with the ends of the gene monomer). The main function of the acceptor plasmid is to readily accept the gene monomer during cloning and to allow its characterization by nucleotide sequencing.

In the second step, the cloned gene monomer is purified in sufficient quantity to be polymerized by a reaction that allows the asymmetric but complementary ends of each gene monomer segment to link in a head-to-tail fashion. The enzyme T4 DNA ligase is added to the reaction to covalently bond the linked segments. A collection of polymer genes are produced, all varying in size by the precise increment of one gene monomer segment. The products of the polymerization are introduced into a population of bacteria that are receptive to foreign DNA. Bacterial cultures containing polymer genes are analyzed to identify those containing polymer genes encoding protein polymers of 50 to 150 kilodaltons in molecular weight. Individual bacterial cultures containing an appropriately sized gene are selected for production and testing.

In the third step, the polymer gene is transferred to a DNA plasmid capable of efficiently expressing the encoded product. The expression of the product can be controlled by a variety of chemical or thermal means. A heat-inducible, lambda phage promoter system is just one example of several expression systems that have been used to produce protein polymers. In this case, growth of the organism can be conducted in the absence of product expression (at a low culture temperature). At an optimal time, expression of the product is induced by elevating the temperature at which point a substantial part of the cell's resources are expended in making product.

PROTEIN POLYMER PROPERTIES AND CHARACTERIZATION

Crystalline BetaSilk Protein Polymers

Examples of the BetaSilk, SLP (silk-like protein), polymer class were designed to adopt a crystalline beta sheet structure characteristic of silk fibroin (Pauling and Corey 1953, Marsh *et al.*, 1955, Lucas *et al.*, 1956, and Lucas *et al.*, 1957). The SLP4 polymer is comprised of 160 repetitions of the six amino acid sequence glycine-alanine-glycine-alanine-glycine-serine (GAGAGS). This repeating hexamer is prevalent in silk fibroin and may be responsible for the stability of silk fiber protein (Fraser *et al.*, 1965, and Lucas *et al.*, 1958) and the high strength of silk fiber. Protein polymers composed of this sequence also display a high degree of chemical and thermal stability. The SLP3 polymer has a melting temperature in the range of 300.5–313.06°C and a heat of fusion of 299 J/g. Above this temperature, the protein chain decomposes.

A prerequisite for this stability is the formation of hydrogen bonded and oriented beta sheets. Synthetic peptides containing the GAGAGS sequence have been analyzed in their crystalline states (Lucas *et al.*, 1958). X-ray diffraction studies of poly(alanylglycine) have provided model unit cell crystal dimensions for *beta* sheet structure (Fraser *et al.*, 1965). Films created from manually stroked evaporating solutions of these peptides were also studied using FTIR dichroism (Fraser *et al.*, 1965). Using these data for comparison, it has been possible to demonstrate the physical similarity between BetaSilk protein polymers and these model peptides (Cappello *et al.*, 1990).

The unprocessed, lyophilized powders of the four protein polymers, SLP3 and 4 and SELP1 and 3, were studied by X-ray diffraction. The X-ray analysis qualitatively confirmed the crystallinity of the polymers (Cappello *et al.*, 1990). As deduced from wide angle X-ray scattering data, the SLP3 powder contained crystal unit cell dimensions of 9.38 Å (a), 6.94 Å (b), and 8.99 Å (c). The unit cell of native silk fibroin has been measured at 9.40 Å (a), 6.97 Å (b), and 9.20 Å (c) (Fraser *et al.*, 1966). The crystalline arrangement of chains in all four polymers is similar as indicated by the similarity in position and relative intensity of diffraction rings. Since all four contain the same GAGAGS basic crystalline repeat sequence, we would expect the crystallized blocks of each polymer to be similar in structure. Of the four polymers, SLP3 and SELP3 gave more diffuse scattering rings than SLP4 and SELP1 indicating a relatively lower degree of crystallinity. This is consistent with the fact that SELP3 contains the highest composition of noncrystallizable blocks within this set.

Our data demonstrate that BetaSilk polymers produce organized crystalline structures similar to the *beta* sheet structures found in natural silk. The degree of crystallinity is affected by periodically disrupting the crystalline blocks with blocks of amino acids which cannot be accommodated within the crystal packing dimensions observed for alanylglycine sequential polypeptides. For example, the inclusion of the pentapeptide sequence, GAAGY, containing the bulky tyrosine side chain, in SLP3 decreases the crystallinity of the poly(GAGAGS) blocks present in SLP4 as determined by increased diffusivity of X-ray powder diffraction rings.

Table 2 Compositional Variation in ProLastin Monomer Sequences

Polymer (MW)	PolymerBlock Sequence[a]	Domain Abbr.[b]	E/S[c]	%S[d]
SELP0 (80,502)	$[(GVGVP)_8 (GAGAGS)_2]_{18}$	E8S2	4.0	21.9
SELP8 (69,934)	$[(GVGVP)_8 (GAGAGS)_4]_{13}$	E8S4	2.0	35.3
SELP7 (80,338)	$[(GVGVP)_8 (GAGAGS)_6]_{13}$	E8S6	1.33	45.1
SELP3 (84,267)	$[(GVGVP)_8 (GAGAGS)_8]_{12}$	E8S8	1.0	51.9
SELP4 (79,574)	$[(GVGVP)_{12} (GAGAGS)_8]_9$	E12S8	1.5	42.2
SELP5 (84,557)	$[(GVGVP)_{16} (GAGAGS)_8]_8$	E16S8	2.0	35.7

[a]The first and last block domain of each polymer is split within the silklike blocks such that both parts sum to a whole domain. All polymers also contain an additional head and tail sequence which constitutes approximately 6% of the total amino acids.
[b]Designates the number of consecutive blocks per repeating domain (E = elastinlike block, S = silklike block)
[c]Ratio of blocks per monomer
[d]% of total amino acids in polymer contributed by silklike blocks

Prolastin Polymers

A family of protein polymers, ProLastins, were produced consisting of silklike (S) and elastinlike (E) peptide blocks in various block lengths and compositional ratios (Ferrari et al., 1986, and Cappello et al., 1990). The elastinlike block consists of the five amino acid sequence, GVGVP (Urry 1984, Urry et al., 1976, and Urry 1988). As a family, silk-elastinlike copolymers (SELP's) consist of exact periodic alternation of silk and elastinlike domains. Individual members of the family vary in the number of silk or elastin blocks within the repeating domains. Table 2 displays the compositional structures of some ProLastin monomers.

The insertion of non-*beta* sheet, flexible sequence segments modeled after a repeating oligopeptide of mammalian elastin was done to decrease the overall crystallinity of the chains and impart flexibility to the structure. The nature of the elastinlike blocks, their length and position within the SELP monomers, influences the molecular chain properties as evident from the changes in solubility of the polymers in water. The BetaSilk polymers, SLP3 and SLP4, are completely insoluble in aqueous solutions. The disruption of the crystalline silklike blocks with ELP blocks, as in SELP1 and SELP3, increases their water solubility. After several days in solution, SELP1, and later, SELP3 form gels even at relatively low protein concentration (1% w/v). Decreasing the length of the SLP block domains while maintaining the length of ELP block domains further increases water solubility and the concentration at which gelation occurs from solution. SELP0, SELP8, SELP7, and SELP3 contain two, four, six, and eight SLP blocks per monomer, respectively. SELP0 is completely soluble in water at all concentrations up to at least 33% w/v. This increase in water solubility is accomplished without substantially changing the chemical properties of the SELP protein chain. If anything, the overall hydrophobicity of the SELP copolymers increases due to their increased composition of valine and their relative decrease in serine content. This is consistent with the hypothesis that the ELP blocks effect the chain properties of the SELP copolymer by disrupting the crystallization of the silklike structure.

Upon drying from solution or treatment with a nonsolvent such as ethanol most SELP polymer materials are converted to water stable products. Presumably, once induced to crystallize, the cohesion of the silklike blocks dominates the solubility properties of the block copolymer. This property of ProLastin polymers allows them to be processed or formulated with other compounds in aqueous, physiological solution and then converted to a water stable material by simply drying, lyophilizing, or treating the formulation with alcohol.

ProNectin Protein Polymers

ProNectin polymers consist of SLP backbones into which peptide segments conferring biological recognition have been introduced. The first example, ProNectin F, was designed using two oligopeptide blocks, the six amino acid SLP block and a seventeen amino acid block from human fibronectin (Pierschbacher and Rouslahti 1984, and Cappello and Crissman 1990). The SLP block provides structural stability, thermal and chemical resistance, and the ability to adsorb to hydrophobic surfaces. The fibronectin block provides biologically recognized cell adhesion activity (Hynes 1987). The blocks are configured into a repeating gene monomer such that one cell adhesion block occurs after every nine silklike structural blocks. Specifically, this was accomplished by inserting an oligonucleotide encoding the fibronectin peptide shown below into the preexisting BetaSilk, SLP3, gene monomer. The resulting 71 amino acid monomer is repeated approximately thirteen times to yield a protein polymer chain of 980 amino acids in length with a theoretical molecular weight of 72,738.

Fibronectin RGD peptide   AVTGRGDSPASA

SLP3 monomer sequence   GAGSGAGAGSGAGAGSGAGAGSGAGAGSGAGAGS
GAGAGSG**AA**G**Y**GAGAGSGAGAGSGA

SLPF monomer design
    ...*GAGAGSGAGAGSGA***AVTGRGDSPAS**<u>A</u>AGYGAGAGSGAGAGS...

The insertion site, between the two alanines in the SLP3 monomer sequence (bold), was chosen because it precedes the location in the monomer sequence where the beta strand structure might be disrupted. Computer modeling indicated that the tyrosine side chain (Y) in the SLP3 sequence could not be easily accommodated in the SLP3 crystal lattice. Furthermore, the AGYG sequence containing the tyrosine side chain could adequately adopt a beta turn conformation. Inserting peptide segments adjacent to this sequence should position them at or near reverse turns flanked by crystallizable beta strands. This location would promote exposure of the peptide sequence while maintaining a stable structure.

ProNectin F is used as a coating reagent for cell cultureware, promoting the adhesion of more than 50 animal cell types to synthetic substrates like polystyrene, polyester, and Teflon onto which it can be deposited. Consistent with its design, ProNectin F has the following features making it a useful cell culture coating:

1. withstands sterilization by autoclaving at $120°C$ without loss of activity
2. adheres to plastic surfaces such as polystyrene without denaturation
3. is optically clear permitting microscopic visualization of cells

4. is insoluble in aqueous solutions maintaining its adsorption to plastic in culture media at 37°C
5. efficiently presents a natural cell adhesion sequence of human fibronectin

The conformation of crystallized ProNectin F has been studied by wide angle X-ray scattering (WAXS), transmission electron microscopy (TEM), and selected area electron diffraction (SAED) (Anderson *et al.* 1994). Results indicate that ProNectin F crystallizes into molecular "tiles" with area dimensions of 6.1 nm × 11.8 nm. Depending on the mode in which the polymer solution is deposited (ie. equilibrium adsorption, dipped, or sprayed), the ProNectin F concentration, and the speed of drying or crystallization, the molecular "tiles" have various ways in which they can assemble. At low concentrations with very slow or no drying, an apparent monolayer will cover a hydrophobic surface. At higher concentration stacking will occur depositing more than a single monolayer. If sprayed from formic acid solution and allowed to dry slowly, fibrils composed of "tiles" are formed. Under more rapid drying, sprayed droplets of ProNectin F produce two dimensional lateral assemblies.

From WAXS and SAED data of both ProNectin F lyophilized powder and sprayed fibrils, the current model indicates that ProNectin F crystallizes into a chain folded pleated sheet of beta strands (Anderson *et al.*1994). The strands are oriented antiparallel. The beta strands are not fully extended, but have a more compressed "crankshaft" conformation. This conformation agrees with the predicted conformation of unoriented silk fibroin protein, the "Silk I" structure (Lotz and Keith 1971). The crystal dimension in the "c" direction (along the peptide backbone) is consistent with a theoretical length of 11.6 nm for nine SLP blocks (54 amino acids) in this conformation. This predicts that the width of the ProNectin F "tile" is controlled at least in part by the number of amino acids in the silklike block domains.

The conformation of the fibronectin blocks was not determined from these analyses. Since the fibronectin blocks are located between SLP block domains in the sequence of ProNectin F, it must be presumed that they occur at the turns between the crystallized SLP blocks. Their conformation is probably variable, but the fact that they are highly active as cell adhesion ligands, indicates that at least some of them are accessible and are exposed at the exterior of the crystalline region of the "tiles".

Three additional protein polymers, ProNectin L's, have been designed and produced using the ProNectin strategy. Peptides from different regions of human laminin which have been reported to promote the adhesion of tumor cells and the outgrowth of neurites from neural cells (Tashiro *et al.*, 1989, Sakamoto *et al.*, 1991, and Charonis *et al.*, 1988) were incorporated into the ProNectin designs.

Laminin IKVAV peptide   PGASIKVAVSAGPS
Laminin F9 peptide   RYVVLPRPVCFEKA
Laminin YIGSR peptide   VCEPGYIGSRCD

As for ProNectin F, each of these peptides was inserted within the monomer block sequence of SLP3 to produce the following monomers:

SLPL3.0 monomer design

... *GAGAGSGAGAGSGA*APG**ASIKVAVSA**GPS*AGYGAGAGSGAGAGS*...

SLPL-F9 monomer design

   ... *GAGAGSGAGAGSGA*<u>A</u>**RYVVLPRPVCFEK** *AGYGAGAGSGAGAGS...*

SLPL1 monomer design

   ...*GAGAGSGAGAGSGA*<u>VC</u>**EPGYIGSR**<u>CD</u>*AGYGAGAGSGAGAGS...*

All three of these additional ProNectin monomers were polymerized to create polymer genes for production and testing. The SLPL3.0 polymer is 70,000 molecular weight and contains 12 repeats of the SLPL3.0 monomer. The polymer was produced using standard extraction and protein separation techniques. Purity of the final lyophilized product was determined by amino acid compositional analysis and microchemical elemental analysis to be 94.6% by weight. About 5% of the mass was water.

SLPL3.0 polymer was evaluated for its ability to promote the attachment of several cell lines, HT1080 (human fibrosarcoma), RD (embryonal rhabdomyosarcoma), and PC12 (rat pheochromocytoma). The purified product was dissolved in a solution of 4.5 M $LiClO_4$ and diluted in phosphate buffered saline (PBS) to concentrations ranging from 100 to 0.14 μg/ml. 0.1 ml of the diluted polymer solution was dispensed to individual wells of a tissue culture polystyrene multiwell dish. The solution was left in contact with the surface of the dish for 2 hours then the dish was rinsed with PBS several times and incubated with freshly harvested HT1080 cells in serum-free medium. After one hour, nonadherent cells were removed by rinsing in PBS and adherent cells were fixed and stained with naphthol blue black dye. The stained cells were quantified by solubilizing the dye in 10% sodium dodecyl-sulfate and determining its solution absorbance by spectrometry at a wavelength of 595 nm.

The adhesion activity of SLPL3.0 was compared to fibronectin, laminin, and a synthetic peptide containing the laminin sequence (RKQAASIKVAVS). HT1080 and RD cells adhered to SLPL3.0 coated wells in a dose dependent fashion (useful range of 30 to 100 μg/ml). Growth of cells was not affected by the coating. Cells reached normal confluency in all cases. The cell morphology of HT1080 and RD cells was improved when grown on SLPL3.0 coated plasticware. These cells appeared to be better spread and elongated than those grown on polylysine or laminin. PC12 cells grown on SLPL3.0 coated plasticware in media containing nerve growth factor responded after 2 to 5 days by producing elongated processes. While the effect was not as pronounced as that observed on laminin coated dishes, the ProNectin L3.0 coating outperformed uncoated dishes and those coated with polylysine, collagen type I, and fibronectin.

Collagenlike Protein (CLP) Polymers

The production of synthetic collagen analogues has long been a desire of structural biologists. Collagen is a major component of most of the structural and mechanical components of animal bodies. While taking many macroscopic forms, cornea, tendon, and intervertebral disk, the underlying molecular structure of all collagen materials is uniformly the collagen triple helix (Eyre 1980). Many medical devices are produced from or contain collagen protein which has been harvested from animals. In most applications, collagen is chemically crosslinked in order to increase its mechanical strength, reduce its immunogenicity, or increase its durability after implantation. A collagenlike polymer that could be designed with selected biological and physical properties such as gelation at predetermined temperatures, rapid hemostatic properties, excellent biocompatibility in wound healing, and controllable resorption is of high interest.

Early attempts to create protein polymer analogues of collagen were relatively unsuccessful, illustrating one of the limitations of using biological synthesis for production of proteins (Goldberg et al., 1989, and Salerno and Goldberg 1992). Not all protein compositions are expressed in microorganisms at equal efficiencies. We learned that E. coli is especially reticent to express proteins with high contents of the amino acid proline. Natural collagen chains contain about 22% proline or hydroxyproline (Bornstein and Piez, 1964). In order to be configured as triple helices, collagen polypeptide chains require amino acid sequences in which glycine occupies every third position. Many amino acids appear in the other two positions, however proline and hydroxyproline are highly preferred (Hulmes et al., 1973). We have been successful in expressing protein polymers whose repeating amino acid blocks are segments of human collagen.

Conformational studies of synthetic peptides modelled on collagen have shown that in order to adopt a triple helix conformation, proline contents between 33% and 66% are preferred (Kobayashi et al., 1970, Brown et al., 1972, Doyle et al., 1971, and Cabrol et al., 1981). We have shown that one of our polymers, CLP-CB, which contains 39% proline (see table) undergoes reversible gelation upon cooling at a concentration of 0.5% (w/v) in water. Circular dichroism (CD) of more dilute solutions indicated that a conformational transition occurs with decreasing temperature. CD spectra consistent with triple helix formation were obtained for CLP-CB solutions at 5–50°C (Figure 2). Heating above this temperature caused a change in the spectrum corresponding to a random polypeptide conformation. We have not studied all of our CLP polymers by circular dichroism. However, all of the collagenlike protein polymers that we have tested undergo thermoreversible gelation upon cooling. We observe that the temperature at which they gel correlates with their proline contents (CLP3.1<CLP6<CLP3.7).

The collagenlike polymers that we have produced to date are designed to be relatively deficient in charged amino acids. The repeating monomers of CLP3.1 and CLP3.7 contain no amino acids with ionizable side chains. This feature forces the polymer chains to associate on the basis of hydrogen bonding and hydrophobic interactions which although weak individually, collectively can promote the formation of stable collagen triple helices. Charged residues participate in strong ionic bonds which might disrupt this process.

Except for CLP-CB which has a molecular weight of 37,000, all of our CLP polymers have molecular weights between 72,000 and 92,000. They are all extremely soluble in water. Solutions greater than 50 wt% are easily produced and are quite fluid. We have fabricated dense films from CLP3.1 which have sufficient strength to be handled and manipulated into articles. The dry films have an ultimate tensile strength of approximately 22 MPa or 3,160 psi, an elongation to break of 4.7%, and an elastic modulus of 495 MPa or 71,000 psi.

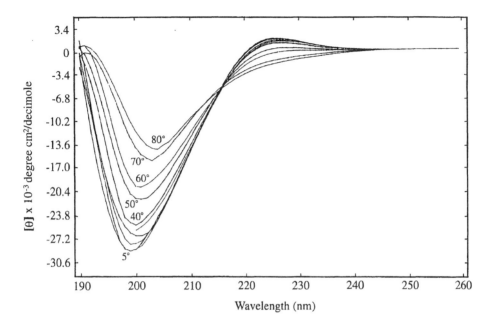

Figure 2 Circular dichroic spectrum of collagenlike protein polymer, CLP-CB. Spectra were obtained using an Aviv Associates Model 60DS spectropolarimeter calibrated in the UV range at wavelengths of 221 nm and 291 nm with D-pantolactone and D-10-camphorsulfonic acid, respectively. The temperature of the optical cell was controlled by circulating water from a Neslab Model RTE-4DD refrigerated bath. The initial spectrum was taken of a solution that had been stored at 4°C for at least 16 hours and had been equilibrated at 5°C in the optical cell. Spectra were obtained after increasing the temperature of the cell by 5°C up to 30°C and then by 10°C up to 80°C. All spectra were collected after a stable 220nm ellipticity was obtained. Data was collected every 0.5nm with a bandwidth of 1.5nm and a time average constant of 5 seconds. The polymer concentration was 0.02 mg/ml in 0.1 M sodium phosphate (pH 7.0).

Chemical Characterization Of Protein Polymers

Protein polymers are analyzed by various methods to confirm their identity and to determine their quality (Hugli 1989, and Copeland 1994). Amino acid analysis, polyacrylamide gel electrophoresis (PAGE), western blot immunoreactivity, and mass spectrometry can be used to verify the identity of the protein polymer (Beavis *et al.*, 1992). Because most protein polymers contain highly atypical amino acid compositions usually consisting primarily of a select few amino acids, amino acid analysis becomes a useful tool for identification of the product. The degree to which actual amino acid ratios in a product agree with the theoretical amino acid ratios as determined by the gene sequence will distinguish even closely related protein polymers. Since protein polymers are essentially monodisperse, they will electrophorese typically as a single band whose extent of migration through the gel can be correlated to standard molecular weight markers. We have developed antibody reagents specific for various amino acid sequence blocks contained in most of our products. After electrophoresis, the protein bands can be transferred to filters which are reacted with various antibodies (western blot analysis). ProNectin F, for example, will react with anti-SLP and anti-FCB (fibronectin cell binding peptide) antibody but will not react with anti-ELP antibody. On the other hand, ProLastins will react with anti-SLP and anti-ELP but not anti-FCB antibodies. With some protein polymers, it may be possible to obtain a molecular ion whose mass can be precisely measured by matrix assisted laser desorption mass spectrometry and compared to its theoretical mass.

Some of these same techniques as well as other standard methods can be used to determine the purity of the protein polymer preparations. PAGE, high performance liquid chromatography (HPLC), and moisture analysis are used to analyze the entire sample in bulk. The amount of non-product species which are detected can be used to estimate a maximum purity. Specific contaminants such as polysaccharides, triglycerides, nucleic acids, and endotoxin are detected by specific tests. Assays for host cell derived DNA and antigens is performed with specific reagents developed for these agents.

MATERIALS PROCESSING AND PROPERTIES

Protein Polymer Solutions

Depending on their specific compositions and the procedures by which they are produced, protein polymers may be soluble or insoluble in aqueous solution. The BetaSilk and ProNectin polymers are insoluble in water. From a lyophilized powder they can be dissolved in concentrated chaotropic solvents such as aqueous formic acid or lithium halide salt solutions. For example, SLP3 is soluble in 88% formic acid at 10 wt% or greater. If the formic acid is either diluted by addition of water or if its pH is neutralized, SLP3 will precipitate. SLPF is soluble in 4.5 M lithium perchlorate solution at 1 wt%. Once dissolved the solution can be diluted with water or phosphate buffer to 0.1 wt% or less. It will remain in solution temporarily depending on its concentration (0.01 wt% solution will precipitate in about 8 hours).

Some ProLastin polymers are soluble directly in water (SELP0 and SELP8). The solubility of ProLastins with SLP domains containing eight or more silklike blocks are generally not directly soluble in water. This is probably dependent on the degree to which the silk blocks have crystallized prior to production of the dry product. For example, SELP3 is not soluble in water while SELP5 is soluble in water if lyophilized from dilute solution. Aqueous solutions of SELPs can be produced by initially dissolving the polymer in formic acid or urea solutions and then either diluting with water and neutralizing the pH or dialyzing the solution against water or buffer. SELP3 solutions greater than about 1 wt% will ultimately gel. The other SELP solutions are stable for longer periods. SELP0 can be kept in solution at room temperature indefinitely.

Our protein polymers are generally not soluble in organic solvents. In fact, common organic solvents such as methanol and acetone can be used as nonsolvents which will induce the crystallization and precipitation of polymers containing silklike blocks. Except for SELP0, once crystallization is induced they are no longer soluble in water. These properties are used in the preparation of water stable materials from ProLastin polymers.

ProLastin Gels

ProLastins can form aqueous stable gels using procedures which take advantage of their immediate water solubility and the propensity of their silk blocks to crystalize. For ProLastins that are readily soluble in water such as SELP8 (and SELP7 and SELP5, depending on their lyophilization conditions), the polymer is dissolved in water or buffer, mixed thoroughly, and allowed to set at room temperature or 37°C. In minutes to hours, a solid hydrogel is formed which has not been contacted by organic solvents, resists dissolution in water, is thermally stable, and is mechanically stable to vigorous agitation.

ProLastins that are not immediately water soluble such as SELP3 can be formed into gels by first dissolving the lyophilized protein in formic acid solution at 2–6 wt% or greater. A coagulant solution consisting of 5 parts 100% ethanol, 2 parts 30% ammonium hydroxide, and 2 parts deionized water is slowly added such that a mixture of 9 parts of the coagulant solution to 1 part of the protein polymer solution is produced. The pH of the final mixture is approximately neutral. The mixture is sealed in a container and allowed to set at room temperature. Once set, the gel can be equilibrated against phosphate buffered saline (PBS) or other solutions in order to remove salts and alcohol. Gels of SELP3 can also be produced by solvent exchange methods. SELP3 is dissolved in 6–10 M urea solution at 10 mg/ml or greater. The solution is then exchanged with water or buffer to produce a gel.

ProLastin Dense Films

Transparent and flexible thin sheets of ProLastin polymer can be produced by solvent film casting methods. Films of all ProLastins have been produced using a common method. A polymer solution containing 1–5 wt% protein in 88% formic acid or water is filtered and degassed. The solution is poured into a casting tray whose lower surface is siliconized glass, polystyrene, or polyethylene and the solution is dried under vacuum. Drying times can be reduced by slowly purging the vacuum chamber with air or nitrogen and heating up to 60°C. The dry film can be removed from the dish directly or floated off with methanol or ethanol. Films produced from all of the ProLastins except SELP0 are completely stable in water and can be washed and equilibrated with buffer. SELP0 films will swell in water and over the course of 24 hours will dissolve.

ProLastin films are mechanically stable and flexible in their dry states. Mechanical properties of some dry SELP and CLP films are shown in Table 3. Those films which are stable in water soften and become quite flexible upon wetting. The tensile strength of wet SELP8 films decreases to about 1–2 MPa as compared to almost 20 MPa for the dry film and the elastic modulus decreases by about 5 fold. These properties should be considered minimums since the mechanical testing of film specimens is extremely sensitive to specimen defects.

Protein Polymer Fibers

A number of solvents and coagulants have been used for wet spinning fibers composed of BetaSilk and ProLastin polymers. Solvents include 98% formic acid,

Table 3 Mechanical Properties of Protein Polymer Films (Dry)

Polymer	Thickness (cm)	Tensile Strength (MPa)	Elongation (%)	Elastic Modulus (MPa)
SELP3	0.075	30.1 ± 4.9	4.46 ± 0.89	627 ± 101
SELP4	0.075	21.0 ± 8.1	6.27 ± 0.79	423 ± 225
SELP7	0.025	18.9 ± 0.8	2.86 ± 0.13	629 ± 51
CLP3.1	0.075	21.8 ± 2.0	4.69 ± 0.34	495 ± 70

50% phosphoric acid, formic acid containing 5–15% lithium chloride or lithium bromide, 9M aqueous lithium bromide, and hexafluoroisopropanol (Cappello and McGrath 1994, and Lock 1992). Methanol, acetone, acetic acid, and 75% (w/v) aqueous ammonium sulfate are effective coagulants.

Fibers have been produced by extruding 40% (w/v) polymer dopes in 9 M LiBr from a 26–gauge needle through a coagulant bath containing acetic acid or acetic acid/acetone mixtures. Fiber diameters ranged from 10 μm for SELP1, 20 μm for SLP3 and SLP4, to 40 μm for SELP3. The fibers showed obvious inhomogeneous internal morphology when observed by light microscopy. In the cases of SELP3 and SLP3, this roughness could represent microcrystalline defects in the fiber due to incomplete or nonuniform coagulation. When fibers of all four compositions were placed in water to evaluate their stability, no changes occurred in their appearance, diameter, length, or internal morphology.

Fibers of all four polymers were analyzed by wide-angle, X-ray scattering in order to determine their crystallinity and their degree of orientation (Cappello and McGrath, 1994). All samples showed a strong reflection at approximately 4.6 Å. This reflection gave clear evidence of substantial amounts of beta sheet structure in the fibers. However, no evidence of molecular alignment or orientation was seen in any of the samples.

Recently, fiber spinning and mechanical properties of fibers produced from protein polymers were reported (Lock, 1992). Fibers were produced as continuous filaments from our protein polymer compositions, SLP4, SELP1, SELP3, and SLPF. The filaments were drawn while wet at various draw ratios to improve mechanical properties. The properties obtained for these protein polymer fibers are displayed in Table 4. A range of fiber properties can be obtained from a single protein polymer composition depending on a number of processing variables including solvent, protein weight percent in the dope, time of coagulation, draw ratio, drying parameters, etc. As expected, tenacity and initial modulus of these fibers are improved by drawing. Trapping the drawn configuration of the fiber by drying it at a fixed elongation gives the greatest improvement in strength. Drawing and allowing the fiber to shrink during drying improves elongation.

ProLastin Fibrous Meshes

SELP polymers were fabricated as nonwoven fibrous meshes to produce fibrillar mats which were flexible, had good drapability, and were stable in wet environments. Fibrous meshes were produced from SELP5, SELP7, and SELPF by extruding the polymer dope containing 20–33% (w/v) protein into a high velocity gas stream.

Table 4 Protein Polymer Fiber Properties

Polymer	Draw Ratio	Diameter[a] (microns)	Tensile Strength[b] (MPa)	Elongation[c] (%)	Modulus[d] (MPa)
SLP4	1x	111	48	2	3200
SLP4[e,g]	2.5x	74	97	13	3900
SLP4	3x	66	180	26	5300
SELP1[e]	4x	68	190	8	6000
SELP1[f]	4x	79	130	45	5500
SELP3	2x	28	120	17	3200
SELP3[e,g]	4x	70	97	16	4600
SELP3[e]	6x	25	290	11	7100
SLPF	1x	111	48	2	3000
SLPF[e]	4x	61	170	19	6300
SLPF[f]	4x	63	170	40	5000

[a]Calculated from denier, assuming fiber density of 1.1 g/cm^3, [b]Ultimate tensile strength at break, [c]% of resting length, [d]Initial modulus, [e] filaments were fixed at constant length during drying, [f] filaments were allowed to shrink while drying, [g]Dope solvent was formic acid containing 10% (w/v) LiCl. All other filaments were spun from hexafluoroisopropanol dopes.

Fine filaments were collected on a circular, metal wire loop forming a web of suspended filaments in the center. Webs were removed from the loop and pressed into fibrous meshes. The meshes were stabilized by immersing them in either methanol or ethanol and dried under ambient conditions. The meshes were sterilized by electron beam irradiation at a dose of 2.5 MRads.

The meshes consisted of fine filaments which varied in diameter from 0.1 to 10 μm. The meshes were stable when placed in saline for more than 24 hours.

ProLastin Porous Sponges

SELP polymers have been produced as three dimensional, porous sponges to serve as implantable materials that might support cell and tissue ingrowth. A 3.0% (w/v) aqueous solution of SELP5 was poured into sterile plastic Petri dishes (35 mm diameter), covered, placed on a small plastic tray and frozen. After freezing, the Petri dishes were placed into a lyophilization flask and lyophilized to dryness. The sponges were removed from their Petri dishes and immersed in methanol at room temperature for 5 minutes. Methanol was removed from the sponges by washing extensively with water. The sponges were returned to the 35 mm diameter Petri dishes, frozen, and again lyophilized. The lyophilized sponges were sterilized by electron beam irradiation at 2.5 Mrads.

SELP5 sponges were dimensionally stable when immersed in saline or water. Minimal swelling was observed.

Coatings

Some proteins such as fibrinogen, serum albumin, fibronectin and collagen have been used to coat the surfaces of synthetic materials to improve their biocompatibility. A protein's inherent ability to stably adsorb to hydrophobic surfaces is due to its surfactant properties. Since proteins normally exist in an aqueous environment, their external surfaces are hydrophilic and their internal surfaces are relatively more hydrophobic. When exposed to a hydrophobic surface, a protein may invert itself exposing its hydrophobic core such that a tight, multiple contact hydrophobic bond occurs between the hydrophobic material and the protein. While the hydrophilic surfaces of the protein may still be available for exposure to the aqueous environment, considerable conformational distortion can occur during this rearrangement possibly altering the protein's biological activities.

BetaSilk and ProNectin polymers avoid this problem because their biological and chemical properties are robust and distinct from their surfactant properties. The crystalline silklike blocks have the ability to expose both hydrophilic and hydrophobic faces depending on the orientation of the beta strands within the sheet. In this sense, the protein is a planar surfactant. When exposed to a hydrophobic surface, a crystalline SLP sheet containing mostly alanine side chains (methyl groups) adsorbs to the surface. The serine side chains (containing hydroxyl groups) may rotate away creating a fairly hydrophilic surface. The coating of polystyrene or polypropylene surfaces with SLP3 increases their wettability. Depending on the coating conditions and the degree of surface coverage, advancing contact angles on polystyrene surfaces can be decreased from $110°$ to $40°$, demonstrating a significant increase in immediate wetting upon coating with SLP3. The coatings are stable to washing in water or isopropanol, resist further adsorption of proteins, and are not readily displaced by low molecular weight surfactants.

The RGD cell adhesion blocks in ProNectin F occur at positions outside the crystallizable silklike blocks. Since they do not participate in crystallization, they also are not disturbed by it. If ProNectin F adsorbs onto hydrophobic plastic surfaces like polystyrene in a mode similar to SLP3, then once adsorbed, the cell adhesion blocks should be fully exposed to the aqueous environment available for interaction with cellular receptors. ProNectin F promotes the attachment of a wide variety of cell types at coating concentrations as low as 160 ng/cm^2. With some of these cells, ProNectin F exceeds the activity of natural cell-attachment proteins, even exceeding the activity of fibronectin, from which the ProNectin F cell-attachment sequence block is modeled. At room temperature, the polymer coating is optically clear and has a dry shelf life of at least one year at room temperature.

ProNectin's chemical stability has allowed us to develop fine aerosol spraying methods for covering flat surfaces with minute amounts of the cell adhesion polymer. Original spray methods for ProNectin F employed concentrated formic acid, one of the few solvents in which the protein was completely soluble. However, a more friendly process was developed when ProNectin F was rendered water soluble by chemically modifying the protein. The chemically modified ProNectin F can be sprayed from a solution of polymer in deionized water allowing for easy solvent handling. Once dried on the surface, the chemically modified ProNectin F is not removed by water and the sprayed surface promotes cellular adhesion just like original ProNectin F. Presumably, neither the planar surfactant qualities nor the biological recognition of the polymer were altered by the chemical modification (Stedronsky, unpublished).

Surface Modification By Co-Injection Molding

The BetaSilk backbone on which ProNectin F is built renders the molecule considerably chemically and thermally resistant to denaturation and loss of biological activity. Thermogravimetric analysis of ProNectin F revealed that except for the loss of bound water at about 93°C little, if any, weight loss occurred up to 200°C. Differential scanning calorimetry confirmed that this weight loss corresponded to loss of water of hydration in the amount of 5–7% by weight. DSC analysis revealed no other thermodynamic transitions, consistent with a proposed structure consisting of highly hydrogen bonded, compact, crystalline sheets. This degree of thermal stability allowed us to demonstrate that ProNectin F can be injection molded directly with plastics which have melting temperatures up to 180°C (Stedronsky et al., 1994). The molded parts contain biologically active ProNectin F at their surfaces, promoting cellular adhesion. In fact, under appropriate conditions, the surface concentration of ProNectin F can be 100 times that expected for an additive distributed homogeneously throughout the bulk of the plastic.

BIOLOGICAL PROPERTIES

In-Vivo Resorption

The rate at which an implanted material resorbs or biodegrades within the body can be a major factor in determining its utility as a biomaterial. A limited number of materials have been shown to be suitable for the fabrication of medical devices which must degrade and resorb in the body without detrimental consequences. Synthetic polymers have demonstrated their ability to vary their resorption characteristics by compositional changes. However, this variability is accomplished at the expense of mechanical properties. Change in composition of the polymer also effects the mechanical properties of the material. Rapidly resorbing compositions are often soft and weak. Slow resorbing compositions are stiff and strong.

ProLastin polymers are a family of protein-based materials whose resorption rate *in vivo* can be controlled by adjusting the sequence and not just the composition of the polymer (Cappello et al., 1995). These adjustments can be made so as to cause little change in the formulation characteristics of the materials, their physical forms, or their mechanical properties. They have good mechanical integrity with no need for chemical crosslinking. They degrade by enzymatic proteolysis and are presumed to resorb by surface erosion. Their breakdown products are peptides or amino acids which are electroneutral at physiological pH and cause no undue inflammation or tissue response.

ProLastin film specimens for implantation were produced from five SELP compositions and denatured collagen protein (DCP) using identical solvent evaporation methods. Bovine collagen was obtained from Colla-Tec, Inc. (Plainsboro, New Jersey, fibrillar form, lot number 921101). It was completely solubilized in 88% formic acid producing a clear but viscous solution. After drying, the films were laminated onto clean polyethylene and 1.3 cm diameter discs were punched out. The specimens were sterilized by electron beam irradiation at 2.5 Mrads.

Each disc was implanted subcutaneously in the back of a rat such that the protein film was in direct contact with the muscle tissue. The specimens remained in the animals for 1 week, 4 weeks, and 7 weeks post implantation. At each time interval 6 specimens per polymer group were retrieved for protein analysis. Additional specimens from each group were evaluated for tissue reaction by histology.

Table 5 Resorption of ProLastin Polymer Films: Film mass remaining as a percent of nonimplanted initial specimen mass

Time	DCP	SELP0	SELP8	SELP3	SELP4	SELP5
0 Week	100%	100%	100%	100%	100%	100%
1 Week	2.3%	4.3%	132%	99%	96%	91%
4 Weeks	3.9%	2.2%	98%	104%	112%	88%
7 Weeks	1.1%	0.8%	18%	58%	105%	103%

Note: SELP7 film resorption results originally reported in this study (Cappello *et al.*, 1995) are not included in this data because the film thickness of those specimens was less than half that of all other films. Film thickness has a strong influence on ultimate resorption time.

Nonimplanted and retrieved specimens were analyzed to determine the mass of SELP film contained per specimen. By amino acid analysis (Henrickson and Meredith 1984), the amino acid contribution of the SELP protein was estimated based on the total content of the amino acids glycine, alanine, serine, valine and proline which for the pure polymers is >95%.

Table 5 displays values for the mass of protein film contained on specimens after implantation as a percent of the initial weight prior to implantation as determined by the mean mass of the nonimplanted specimens. Each value is the mean of at least 5 specimens. The results indicate that upon implantation, SELP0 and DCP (denatured collagen protein) were substantially resorbed by 1 week, falling below 5% of their non-implanted masses. SELP8 and SELP3 were resorbing by 7 weeks with mean values of 18.1% and 58.2% remaining, respectively. SELP4 and SELP5 films showed no evidence of resorption by 7 weeks. (Note: SELP7 film specimens included in the original report of this data (Cappello *et al.*, 1995) is being reevaluated because, although they were completely resorbed by 4 weeks, their thickness was less than half of all other film specimens in this study. The thickness of the film is an important variable in determining ultimate resorption time.)

The different rates of resorption observed for different SELP films is apparently due to the difference in composition of the polymers. SELP0 films resorbed by 1 week, SELP8 films were almost completely resorbed at 7 weeks, SELP3 films were 60% resorbed by 7 weeks, and neither SELP4 or SELP5 films showed any resorption at 7 weeks. Faster resorption correlated with compositions containing domains of silklike blocks less than eight. SELP0 and 8 have repeating domains of 2 and 4 silklike blocks. SELP3, 4, and 5 all consist of domains containing 8 silklike blocks. The average content of silklike blocks in the copolymer composition did not correlate with resorption rate. SELP8 and SELP5 have identical compositional ratios of elastin to silk blocks (2.0), yet SELP8 resorbed substantially in 7 weeks and SELP5 showed no signs of resorption.

These data indicate that the resorption rate of SELP copolymers can be changed by precisely controlling the number or the length of silklike and elastinlike blocks contained per repeating segment. This control cannot be obtained simply by controlling the compositional ratio of these blocks in the final copolymer. Faster resorption of ProLastins can be accomplished by decreasing the number of silklike blocks per repeating domain to less than 8. Slower resorption can be obtained by increasing the number of elastinlike blocks per repeating domain to greater than 8. It may be possible to obtain equivalent resorption and different material or biological properties by adjusting both block structures.

Histological Evaluation

ProLastin films

The SELP films analyzed for resorption were also evaluated histologically to determine the nature of the tissue response and the consequence of resorption on the surrounding tissue. The tissue responses of four implanted specimen types, blank polyethylene discs and discs laminated on one side with protein films composed of SELP0, SELP5, and denatured bovine type I collagen (DCP), were compared after being implanted for 1 week (Figure 3). The protein sides of the polyethylene discs on which the protein films were laminated were implanted adjacent to the muscle in subcutaneous pockets on the back of rats. The resorption studies indicated that the DCP and SELP0 films resorbed completely at 1 week and that the SELP5 film showed no evidence of resorption through 7 weeks.

The blank polyethylene implants elicited a mild inflammatory response which at 1 week was characterized by the presence of densely packed monocytes and macrophages (Figure 3A). Some fibroblasts were present at the outer perimeter of the cellular infiltrate which surrounded the implants. No significant collagen deposition had occurred. The total width of the inflammatory zone averaged about 10–15 cells in thickness.

The protein side of the DCP specimens elicited a more pronounced cellular infiltrate which was approximately 40 cells thick, consisted mostly of monocytes and macrophages, and was much more diffuse than the blank polyethylene (Figure 3B). The cells adjacent to the implant formed a dense, two cell thick layer of macrophages which were more or less rounded. Farther from the implant, the macrophage infiltration became more diffuse. A few thin layers of fibroblasts were present at the outer perimeter of the inflammatory zone.

Protein specimens containing SELP0 films elicited a mild response which was comparable in thickness to the collagen specimens but was remarkably more diffuse (Figure 3C). There was a single layer of macrophages adjacent to the implant which were quite flattened parallel to the implant surface. Beyond this inner layer, the zone consisted of a mixture of macrophages and fibroblasts and some loosely packed collagen fibrils.

Specimens containing the nonresorbing SELP5 films elicited a mild response at 1 week after implantation, but the response was peculiar in that the features of the zone were those characteristic of a healing response beyond the 1 week stage as compared with the blank polyethylene, the DCP, and the SELP0 film specimens (Figure 3D). The zone was narrow, comparable in thickness to the blank polyethylene, and consisted primarily of fibroblasts with only occasional, small pockets of macrophages. Collagen deposition was extensive and well organized with fibrils parallel to the implant surface. The cell layer adjacent to the implant was a single cell thick and consisted of highly flattened cells whose identity could not be distinguished as macrophage or fibroblast.

Overall, the tissue responses elicited by these specimens at 1 week were mild. The protein film specimens were distinguished from each other by the types of cells contained in the zone of cellular infiltration, their density, and their degree of advancement through the healing response. The nonresorbing SELP5 film specimens exhibited the most advanced cellular zone with a well organized and fairly condensed collagen capsule. The SELP0 film which had resorbed

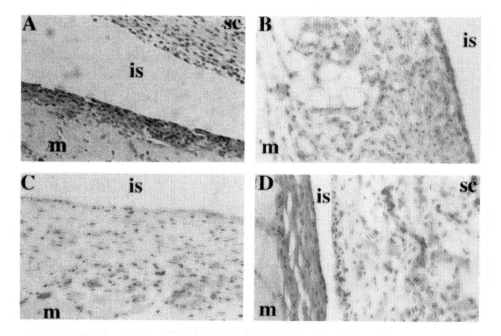

Figure 3 Histology of implanted ProLastin films at one week. Film specimens were surgically implanted in male Fisher rats weighing approximately 150 grams by creating four subcutaneous pockets on the back of each animal symmetrically across the spine, two anteriorly between the shoulders and two posteriorly over the lumbar area. One cm diameter polyethylene discs with and without protein films laminated on one side (total thickness was 2–3 mm) were inserted protein face toward the muscle into each pocket about 1.5 cm from the incision along the spine. The incisions were closed with stainless steel clips. At one week after implantation, tissue blocks surrounding each implant were excised intact and prepared for histology. Fixed tissue was dissected by hand to expose an implant containing cross section. In most cases, the implant popped free during this procedure. Tissue sections were embedded in paraffin and microtomed to 5 ÿE6m thin sections, mounted and stained with hematoxylin/eosin. Photomicrographs were taken at 80x magnification. Blank polyethylene specimen (A), and specimens to which denatured collagen film (B), SELP0 film (C), and SELP5 film (D) were laminated are displayed. Muscle tissue (m), the implant space (is), and the subcutaneous tissue (sc) are marked in each image.

by 1 week had a zone diffusely populated with macrophages and fibroblasts and unorganized collagen fibrils. This was in contrast to the collagen film specimens which contained the widest zone of cellular infiltration which was populated primarily with macrophage and monocytes. Since it is likely that macrophage cells are involved in degradation of the protein films, and that their presence would persist as long as protein were actively being degraded, their presence next to the collagen and SELP0 specimens at 1 week can be explained. Since the SELP5 film was not subject to degradation at this time, perhaps the tissue response proceeded expediently from the early inflammatory phase to maturation phase by 1 week.

This hypothesis, however, does not explain why the nonresorbing polyethylene blank disc or the opposite face of all of the protein film containing specimens, which

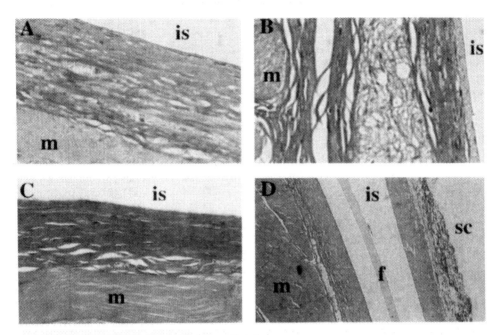

Figure 4 Histology of implanted prolastin films at four weeks. Film specimens implanted as described for Figure 3 were explanted after 4 weeks and processed for histological examination. Tissue sections were stained with Masson's trichrome stain. Panels A-C were photographed at 160x magnification and panel D was photographed at 8x. Blank polyethylene specimen (A), and specimens to which denatured collagen film (B), SELP0 film (C), and SELP4 film (D) were laminated are displayed. Muscle tissue (m), the implant space (is), and the subcutaneous tissue (sc) are marked in each image. The protein film in the SELP4 specimen remained on the slide (f).

also were polyethylene, did not progress to this same level of healing. The cellular zone adjacent to polyethylene at 1 week consisted almost exclusively of inflammatory cells. In-fact, the polyethylene side of the SELP5 specimen, facing the skin, showed a typical 1 week macrophage inflammatory profile and little collagen deposition (Figure 3D).

These data indicate that SELP protein films are biocompatible when implanted subcutaneously in rats. Their resorption, per se, has little effect on the tissue response associated with the implanted material. After 4 weeks of implantation the healing responses surrounding all implants had progressed to a common level, exhibiting comparable fibrous capsule formation as would be expected from stable polyethylene implants (Figure 4). The tissue responses observed were directed exclusively at the implant and there were no apparent effects on the surrounding muscle or subcutaneous tissues.

ProLastin sponges

The ProLastin, SELP5, and bovine type I collagen were prepared as porous sponges of similar density, pore size, and geometry by lyophilization. Sponge specimens were implanted subcutaneously in rats and evaluated histological at 3, 7, 14, 21, 28, and 56 days. Both materials elicited a mild inflammatory response which initiated upon implantation and resolved itself rapidly progressing through a healing response from 3 to 21 days. Particularly with respect to the SELP5 sponges, there was no evidence of edema or hemorrhage and no prolonged neutrophil or polymorphonuclear phagocyte infiltration. This indicated that there was no persistence of the acute inflammatory phase reflecting good surgical technique, the lack of infection and, importantly, the overall biocompatibility of the materials.

There was no progression of the inflammatory response beyond the mid-stage macrophage foreign body reaction. Spatially and temporally, the response spread into the sponge from the exterior, and the density of cells did not change over what it was at the perimeter of the implant. Very few, if any, foreign body giant cells were present. There were no indications that the response was progressing toward chronic inflammation or that the response was causing a destruction of adjacent tissue. Through 28 days, there was no evidence of large numbers of monocytes or lymphocytes which would be indicators of chronic irritation to the surrounding tissues due either to low level toxicity or immunogenicity of the material.

Two distinguishing characteristics of the SELP5 sponges as compared to the collagen sponges were their greater stability and their more rapid and complete infiltration by cells as early as 14 days. The SELP5 sponges remained essentially stable for at least 28 days. The SELP5 sponges allowed the migration of cells into their internal structures. Considerable infiltration had occurred at 14 and 21 days, although it was somewhat inconsistent. At 28 days, SELP5 sponges were almost completely infiltrated. This was in contrast to the collagen sponges which did not infiltrate even at 21 days. In those areas of the SELP5 sponges where infiltration was complete, a new tissue formed consisting of cells, sponge, and new collagen. There were very few open spaces. Blood vessels had formed. Fibroblasts were present and new connective tissue had been laid down, evident by the "veins" of collagen that marbled through the more open areas of the sponge.

Biocompatibility

Fibrous meshes produced from SELP7 and SELP5 were applied to 2 x 2 cm partial and full thickness dermal wounds in pigs in order to investigate their biocompatibility with healing tissue. The edges of the meshes could be pulled across the tissue allowing the mesh to be spread and/or arranged over the wound. The wounds were covered and examined every two days for signs of bioincompatibility. No adverse effects were observed in wounds containing SELP fibrous meshes. After 14 days, the wounds were completely epithelialized. Histological examination of tissues from wounds to which SELPF fibrous webs had been applied showed that some of the filaments had been incorporated into the healing tissue.

SELP5 sponges were similarly evaluated. $2 \times 2 \times 0.4$ cm specimens were applied to 2×2 cm full thickness, dermal wounds in pigs and compared to $2 \times 2 \times 0.3$ cm specimens of Helistat (collagen hemostatic sponge, Marion Laboratories, Inc., Kansas City, MO.). After bleeding was controlled and the wound flushed with saline, the specimens were laid into the tissue void such that they would firmly contact the wound bed. The sponges adhered well to the wound bed and resisted removal under mild tension.

All wounds were covered with petrolatum gauze pads and bandaged. After 7 days, the wounds were undressed and observed to determine the extent of healing. Wounds containing SELP5 sponges had progressed normally through the healing process as compared to wounds to which no material was applied. No evidence of excessive inflammation was observed. Epithelialization of the wound was in progress.

Immunogenicity

SLPF and several SELP polymers have been evaluated for immunogenicity in rabbits. Five rabbits were injected with 1 ml of a 10 mg/ml polymer solution (or suspension) in saline. After 6 weeks and 8 weeks, the same animals received booster injections of 1 ml of 0.5 mg/ml polymer solution. Serum was drawn prior to the initial injection, just prior to the first booster injection, and one week after the last booster injection. The sera were analyzed for immunoreactivity by ELISA. ELISA plates were either coated with the injected polymer or with synthetic peptides containing the sequences of the various repeating blocks of the polymers. Antipeptide hyperimmune serum generated against SLP peptide ($[GAGAGS]_8$), ELP peptide ($[VPGVG]_8$), or FCB peptide (human fibronectin cell binding sequence, YTITVYAVTGRGDSPASSKPISINYC) were used as positive controls. The degree of immunoreactivity was expressed as the maximum dilution of serum giving twice the background color intensity of negative control wells which contained reagents only and no serum as compared to positive control dilutions.

The rabbit study results indicated that as compared to hyperimmune anti-peptide sera, minimal reactivities directed at the injected polymers were found with sera from all animals injected with SELP0, and SLPF (<100, Table 6). Only one of five animals injected with SELP3 had a slightly elevated reactivity in the final serum (180). The other four animals had reactivities of 60 or below giving an average reactivity for all five animals of 53.2. In all of these assays, positive control hyperimmune anti-SLP and anti-FCB sera gave reactivities from 1000 to 4000 and anti-ELP sera gave reactivities from 400 to 800. Three of the five animals injected with SELPF had reactivities in their final sera of 120, 320, and 640. The other two animals had scores of 12, and 32. The five animal average for SELPF was 225. The level of reactivity in the final serum samples either remained unchanged or was only slightly elevated as compared to the six week serum samples (prebooster). This indicated that little lymphocyte clonal propagation had occurred upon initial injection of the polymers and that the successive booster injections did not lead to hyperimmunity.

By coating the ELISA wells with individual peptides containing the sequences of the SLP, ELP and FCB blocks, we were able to determine where the immunoreactivity of the SELPF polymer was directed. The final sera from each of the animals exhibiting elevated reactivities reacted only with the SLP peptide with a reactivity score of 108. No significant reactivity was directed at the ELP or FCB peptides.

Immunogenicity of a protein product depends on the exact composition of the product. Changes in the amino acid sequence can effect the nature and magnitude of the immunological response obtained. The lack of significant immunoreactivity of products composed of SLP, ELP and FCB peptide blocks gives us confidence that such materials may be used safely in the body. However, continued testing of all new products in multiple animal species and, ultimately, in humans is needed to assure that a product is not immunogenic.

Table 6 Immunogenicity of SELP0, SELP3, SELPF and SLPF Polymers

Injected Antigen	ELISA Antigen	Prebooster Serum Titer	Final Serum Titer
SELP0	SELP0	<2	14.4
	SLP peptide	NT	28
	ELP peptide	NT	<2
SELP3	SELP3	13.8	53.2
	SLP peptide	NT	84
	ELP peptide	NT	<2
SELPF	SELPF	202	225
	SLP peptide	NT	108
	ELP peptide	NT	37.6
	FCB peptide	NT	<2
SLPF	SLPF	17.8	24.1
	SLP peptide	NT	60
	FCB peptide	NT	4.7
SLP peptide*	SLP peptide	NT	4,370
ELP peptide*	ELP peptide	NT	480
FCB peptide*	FCB peptide	NT	1,450

Values indicate the maximum dilutions of serum which gave reactivities at least twice the background reactivity of negative control wells in ELISA assays. Each value represents the average of three to five animal sera tested per polymer composition except for positive control sera which came from a single animal. NT = not tested.

*Positive control hyperimmune sera were generated in rabbits by conjugating peptides to keyhole limpet hemocyanin or bovine serum albumin, mixing with Freund's complete adjuvant, and injecting 10 mg of conjugate followed by four successive 10 mg booster injections at weeks 2, 3, 5, and 7. Final sera were obtained at 8 weeks.

ACKNOWLEDGEMENTS

The author thanks Dr. Franco Ferrari and the entire crew of Protein Polymer Technologies, Inc. for their dedication and perseverance in the successful development of a new technology. Special acknowledgments go to Dr. Erwin Stedronsky for his accomplishments in protein polymer materials fabrication and characterization, Dr. James Anderson at Case Western Reserve University for advice in design and evaluation of implant studies, Dr. David Martin at the University of Michigan for molecular characterization of ProNectin F, Dr. David Tirrell at the University of Massachusetts, Amherst for polymer science and characterization, and Dr. Alan Bergold at the University of Iowa for use of his CD spectrometer.

REFERENCES

Anderson, J.P., Cappello, J. and Martin, D.C. (1994) Morphology and primary crystal structure of a silk-like protein polymer synthesized by genetically engineered *Escherichia coli* bacteria. *Biopolymers*, **34**(8), 1049–1058.

Beavis, R.C., Chait, B.T., Creel, H.S., Fournier, M.J., Mason, T.L. and Tirrell, D.A. (1992) Analysis of artificial proteins by matrix-assisted laser desorption mass spectrometry. *J. Am. Chem. Soc.*, **114**, 7584–7585.

Bornstein, P. and Piez, K.A. (1964) A biochemical study of human skin collagen and the relation between intra- and intermolecular cross-linking. *J. Clin. Invest.*, **43**, 9, 1813–1823.

Brandon, C. and Tooze, J. (1991) *Introduction to Protein Structure*, Carl Garland Publishing, Inc., New York.

Brown, F.R., DiCorato, A., Lorenzi, G.P. and Blout, E.R. (1972) Synthesis and structural studies of two collagen analogues: poly (L-prolyl-L-seryl-glycyl) and poly (L-prolyl-L-alanyl-glycyl). *J. Mol. Biol.* **63**, 85–99.

Cabrol, D., Broch, H. and Vasilescu, E.D. (1981) Sequences tripeptidiques typiques du collagene: etude conformationelle quantique. *Biochimie,* **63**, 851–855.

Cappello, J., Textor, G. and Bauerle, B. (1995) Bioresorption of implanted protein polymer films controlled by adjustment of their silk/elastin block lengths. In Gebelein, C.G., Carraher, C.E. Jr., (eds.), *Industrial Biotechnological Polymers*, Technomic Publishing Co., Lancaster, PA, pp. 249–256.

Cappello, J. and McGrath, K.P. (1994) Spinning of Protein Polymer Fibers. In Kaplan, D., Adams, W.W., Farmer, B., Viney, C., (eds.), *Silk Polymers: Materials Science and Biotechnology*, ACS Symposium Series, **544**, American Chemical Society, Washington, DC, pp. 137–147.

Cappello, J. and Crissman, J.W. (1990) The Design and Production of Bioactive Protein Polymers for Biomedical Applications. *Polymer Preprints,* **31**, 193.

Cappello, J., Crissman, J., Dorman, M., Mikolajczak, M., Textor, G., Marquet, M. and Ferrari, F. (1990) Genetic engineering of structural protein polymers. *Biotechnol. Prog.,* **6**, 198–202.

Cappello, J. and Ferrari, F (1994) Microbial production of structural protein polymers. In Mobley, D.P., (edr.), *Plastics from Microbes*, Hanser Publishers, Munich, pp. 35–92.

Cappello, J. (1992) Genetic production of synthetic protein polymers. *MRS Bulletin,* **17**, 48–53.

Charonis, A.S., Skubitz, A.P.N., Koliakos, G.G., Reger, L.A., Dege, J., Vogel, A.M., Wohlhueter, R. and Furcht, L.T. (1988) A novel synthetic peptide from the B1 chain of laminin with heparin-binding and cell adhesion-promoting activities. *J. Cell Biol.,* **107**, 1253–1260.

Chou, P.Y. and Fasman, D.A.D. (1974) Prediction of protein conformation. *Biochemistry,* **13**, 222–244.

Copeland, R.A. (1994) *Methods for Protein Analysis*, Chapman and Hall, New York, NY.

Creel, H.S., Fournier, M.J., Mason, T.L. and Tirrell, D.A. (1991) Genetically directed syntheses of new polymeric materials. Efficient expression of a monodisperse copolypeptide containing fourteen tandemly repeated – (AlaGly)4[subscript]ProGluGly – elements. *Macromolecules,* **24**, 1213–1314.

Doyle, B.B., Traub, W., Lorenzi, G.P. and Blout, E.R. (1971) Conformational investigations on the polypeptide and oligopeptides with the repeating seqeunce L-alanyl-L-prolylglycine. *Biochemistry,* **10**, 3052–3060.

Eyre, D.R. (1980) Collagen molecular diversity in the body's protein scaffold. *Science,* **207**, 1315–1322.

Ferrari, F.A., Richardson, C., Chambers, J., Causey, J. and Pollock, S.C. (1986) Construction of synthetic DNA and its use in large polypeptide synthesis. WPO Patent Application No. WO88/03533.

Fraser, R.D.B., MacRae, T.P. and Stewart, F.H. (1966) Poly-L-alanylglycyl-L-alnylglycyl-L-serylglycine: a model for the crystalline regions of silk fibroin. *J. Mol. Biol.,* **19**, 580–582.

Fraser, R.D.B., MacRae, T.P., Stewart, F.H. and Suzuki, E. (1965) Poly-L-alanylglycine. *J. Mol. Biol.,* 11, 706–712.

Garnier, J., Osguthorpe, D.J. and Robson, B. (1978) Analysis of the accuracy and implications of simple methods for predicting the secondary structure of globular proteins. *J. Mol. Biol.,* **120**, 97–120.

Goldberg, I., Salerno, A.J., Patterson, T. and Williams, J.I. (1989) Cloning and expression of a collagen-analog-encoding synthetic gene in *Escherichia coli., Gene,* **80**, 305–314.

Henrickson, R.L. and Meredith, S.C. (1984) Amino acid analysis by reverse-phase high-performance liquid chromatography: precolumn derivatization with phenylisothiocyanate. *Anal. Biochem.,* **136**, 65–74.

Hugli, T.E. (1989) *Techniques in Protein Chemistry*, Academic Press, San Diego, USA.

Hulmes, D.J.S., Miller, A., Parry, D.A.D., Piez, K.A. and Woodhouse-Galloway, J. (1973) Analysis of primary structure of collagen for the origins of molecular packing. *J. Mol Biol.,* **79**, 137–148.

Hynes, R.O. (1987) Integrins: a family of cell surface receptors. *Cell,* **48**, 549–554.

Kobayashi, Y., Sakai, R., Kakiuchi, K. and Isemura, T. (1970) Physicochemical analysis of (Pro-Pro-Gly)$_n$ with defined molecular weight–temperature dependence of molecular weight in aqueous solution. *Biopolymers,* **9**, 415–425.

Lock, R.L. (1992) A process for spinning polypeptide fibers. European Patent Office, Publication number 0 488 687 A2.

Lotz, R.L. and Keith, H.D. (1971) Crystal structure of poly (L-Ala-Gly) II. *J. Mol. Biol.,* **61**, 201–215.

Lucas, F., Shaw, J.T.B. and Smith, S.G. (1958) The silk fibroins. *Adv. Protein Chem.,* **13**, 107–242.

Lucas, F., Shaw, J.T.B. and Smith, S.G. (1957) The amino acid sequence in a fraction of the fibroin of Bombyx mori. *Biochem. J.,* **66**, 468–479.

Lucas, F., Shaw, J.T.B. and Smith, S.G. (1958) Amino-acid sequence in a fraction of Bombyx silk fibroin. *Nature*, **178**, 861.

Marsh, R.E., Corey, R.B. and Pauling, L. (1955) An investigation of the structure of silk fibroin. *Biochim. biophys. Acta*, **16**, 1–34.

McGrath, K.P., Fournier, M.J., Mason, T.L. and Tirrell, D.A. (1992) Genetically directed syntheses of new polymeric materials. Expression of artificial genes encoding proteins with repeating (AlaGly)$_3$ProGluGly elements. *J. Am. Chem. Soc.*, **114**, 727–733.

Pauling, L. and Corey, R.B. (1953) Two pleated-sheet configurations of polypeptide chanis involving both cis and trans amide groups. *Proc. Nat. Acad. Sci.*, **39**, 247–256.

Pierschbacher, M.D. and Rouslahti, E. (1984) Cell attachment activity of fibronectin can be duplicated by small synthetic fragments of the molecule. *Nature*, **309**, 30–33.

Sakamoto, N., Iwahana, M., Tanaka, N.G. and Osada, Y. (1991) Inhibition of angiogenesis and tumor growth by a synthetic laminin peptide, CDPGYIGSR-NH$_2$. *Cancer Research*, **51**, 903–906.

Salerno, A.J. and Goldberg, I. (1992) Expression and maintenance of synthetic genes encoding analogs of natural proteins in *Escherichia coli*. *J. Cell Biochem. Suppl.*, **16**, F, 123.

Schultz, G. (1988) Critical evaluation of methods for prediction of protein secondary structure. *Annu. Rev. Biophys. Biophys. Chem.*, **17**, 1–21.

Schulz, G.E. and Shirmer, R.H. (1979) *Principals of Protein Structure*, Springer-Verlag, New York.

Stedronsky, E.R., Cappello, J., David, S., Donofrio, D.M., McArthur, T., McGrath, K., Panaro, M.A., Putnam, D., Spencer, W. and Wallis, O. (1994) Injection molding of ProNectin F dispersed in polystyrene for the fabrication of plasticware activated towards attachment of mammalian cells. In Alper, M., Bayley, H., Kaplan, D., Navia, M., (eds.) *Biomolecular Materials By Design*, Symposium Proceedings Series, 330, Materials Research Society, Boston, MA, pp. 157–164.

Tashiro, K., Sephel, G.C., Weeks, B., Sasaki, M., Martin, G.R., Kleinman, H.K. and Yamada, Y. (1989) A synthetic peptide containing the IKVAV sequence from the A chain of laminin mediates cell attachment, migration, and neurite outgrowth. *J. Biol. Chem.*, **264**, 16174–16182.

Tirrell, D.A., Fournier, M.J. and Mason, T.L. (1991) Genetic engineering of polymeric materials. *MRS Bulletin*, **18**(7), 23–28.

Urry, D.W. (1984) Protein elasticity based on conformations of sequential polypeptides: the biological elastic fiber. *J. Prot. Chem.*, **3**, 403–437.

Urry, D.W. (1988) Entropic elastic processes in protein mechanisms. I. Elastic structure due to an inverse temperature transition and elasticity due to internal chain dynamics. *J. Prot. Chem.*, **7**, 1–34.

Urry, D.W., Okamoto, K., Harris, R.D. and Hendrix, C.F. (1976) Synthetic, cross-linked polypentapeptide of tropoelastin: an anisotropic fibrillar elastomer. *Biochemistry*, **15**, 4083–4089.

Polymer Composition Table: Source of all polymers is Protein Polymer Technologies, Inc.

Each protein polymer is produced as a repetition of the amino acid sequence blocks as shown below under "Polymer Structure". Additionally, each polymer chain contains a unique "Head" and "Tail" sequence which is contiguous with the repetitive portion of the chain. All of the head and tail sequences are derived from the *E. coli* expression gene to which each polymer gene is fused. Variations in the head and tail sequences are due to DNA changes which facilitate the fusion of the various polymer genes to the expression gene. If desired, all head and tail sequences have been designed such that they can be efficiently cleaved and removed from the polymer chain by treatment with cyanogen bromide which cleaves peptide bonds immediately adjacent to methionine amino acids (M).

Polymer Name	Polymer Structure	Chain Length	Molecular Weight
BetaSilks			
SLP4	Head[1] (GAGAGS)$_{167}$ Tail[1]		
		1101 amino acids	76,194
SLP3	Head[2] [(GAGAGS)$_9$ GAAGY]$_{18}$ Tail[2]		
		1178 amino acids	82,958
SLPC	Head[3] [(GAGAGS)$_9$ GAAGAGCGDPGKGCCVAGAGY]$_{16}$ Tail[2]		
		1332 amino acids	97,077
ProNectins			
SLPF	Head[4] [(GAGAGS)$_9$ GAAVTGRGDSPASAAGY]$_{12}$ Tail[2]		
		980 amino acids	72,738
SLPL1	Head[5] [(GAGAGS)$_9$ GAAVCEPGYIGSRCDAGY]$_{13}$ Tail[2]		
		1065 amino acids	81,763
SLPL3.0	Head[6] [(GAGAGS)$_9$ GAAPGASIKVAVSAGPSAGY]$_{12}$ Tail[2]		
		1019 amino acids	75,639
SLPL-F9	Head[7] [(GAGAGS)$_9$ GAARYVVLPRPVCFEKAAGY]$_{11}$ Tail[2]		
		945 amino acids	75,561
ProLastins			
SELP0	Head[8] [(GVGVP)$_8$ (GAGAGS)$_2$]$_{18}$ Tail[3]		
		997 amino acids	80,502
SELP1	Head[9] [GAA(GVGVP)$_4$ VAAGY(GAGAGS)$_9$]$_{13}$ Tail[4]		
		1205 amino acids	89,000
SELP3	Head[10] [(GVGVP)$_8$ (GAGAGS)$_8$]$_{11}$ Tail[5]		
		1113 amino acids	84,267
SELP4	Head[10] [(GVGVP)$_{12}$ (GAGAGS)$_8$]$_8$ Tail[6]		
		1029 amino acids	79,574
SELP5	Head[10] [(GVGVP)$_{16}$ (GAGAGS)$_8$]$_7$ Tail[7]		
		1081 amino acids	84,557
SELP7	Head[10] [(GVGVP)$_8$ (GAGAGS)$_6$]$_{12}$ Tail[8]		
		1045 amino acids	80,338
SELP8	Head[10] [(GVGVP)$_8$ (GAGAGS)$_4$]$_{12}$ Tail[9]		
		889 amino acids	69,934
SELPF	Head[10] [(GVGVP)$_8$ (GAGAGS)$_{12}$ GAAVTGRGDSPASAAGY (GAGAGS)$_5$]$_6$ Tail[10]		
		1011 amino acids	75,957
Collagenlike Polymers			
CLP-CB	Head[11] [[GAP(GPP)$_4$]$_2$ GLPGPKGDRGDAGPKGADGSPGPA GPVGSP]$_6$ Tail[11]		
		417 amino acids	36,958
CLP6	Head[11] [(GAHGPAGPK)$_2$ (GAQGPAGPG)$_{24}$ (GAHGPAGPK)$_2$]$_4$ Tail[12]		
		1065 amino acids	85,386
CLP3.1	Head[11] (GAPGAPGSQGAPGLQ)$_{68}$ Tail[12]		
		1077 amino acids	91,266
CLP3.7	Head[11] (GAPGTPGPQGLPGSP)$_{52}$ Tail[12]		
		838 amino acids	72,637

Head[1] = fMDPVVLQRRDWENPGVTQLNRLAAHPPFASDPMGAGS
Tail[1] = GAGAMDPGRYQLSAGRYHYQLVWCQK
Head[2] = fMDPVVLQRRDWENPGVTQLNRLAAHPPFASDPMGAGS(GAGAGS)$_6$GAAGY
Tail[2] = (GAGAGS)$_2$GAGAMDPGRYQLSAGRYHYQLVWCQK
Head[3] = fMDPVVLQRRDWENPGVTQLNRLAAHPPFASDPMGAGS(GAGAGS)$_6$GAAGA
GCGDPGKGCCVAGAGY
Head[4] = fMDPVVLQRRDWENPGVTQLNRLAAHPPFASDPMGAGS(GAGAGS)$_6$GAAVT
GRGDSPASAAGY
Head[5] = fMDPVVLQRRDWENPGVTQLNRLAAHPPFASDPMGAGS(GAGAGS)$_6$GAAVC
EPGYIGSRCDAGY
Head[6] = fMDPVVLQRRDWENPGVTQLNRLAAHPPFASDPMGAGS(GAGAGS)$_6$GAAPG
ASIKVAVSAGPSAGY
Head[7] = fMDPVVLQRRDWENPGVTQLNRLAAHPPFASDPMGAGS(GAGAGS)$_6$GAARY
VVLPRPVCFEKAAGY

Head[8] = fMDPVVLQRRDWENPGVTQLNRLAAHPPFASERFCMGS
Tail[3] = MCYRAHGYQLSAGRYHYQLVWCQK
Head[9] = fMDPVVLQRRDWENPGVTQLNRLAAHPPFASDPMGAGS(GAGAGS)$_6$
Tail[4] = GAA(GVGVP)$_4$VAAGY(GAGAGS)$_2$GAGAMDPGRYQLSAGRYHYQLVWCQK
Head[10] = fMDPVVLQRRDWENPGVTQLNRLAAHPPFASDPMGAGS(GAGAGS)$_2$
Tail[5] = (GVGVP)$_8$(GAGAGS)$_5$GAGAMDPGRYQLSAGRYHYQLVWCQK
Tail[6] = (GVGVP)$_{12}$(GAGAGS)$_5$GAGAMDPGRYQLSAGRYHYQLVWCQK
Tail[7] = (GVGVP)$_{16}$(GAGAGS)$_5$GAGAMDPGRYQLSAGRYHYQLVWCQK
Tail[8] = (GVGVP)$_8$(GAGAGS)$_3$GAGAMDPGRYQLSAGRYHYQLVWCQK
Tail[9] = (GVGVP)$_8$(GAGAGS)$_2$GAGAMDPGRYQLSAGRYHYQLVWCQK
Tail[10] = (GAGAGS)$_2$GAGAMDPGRYQLSAGRYHYQLVWCQK
Head[11] = fMDPVVLQRRDWENPGVTQLNRLAAHPPFASDPM
Tail[11] = GAMCAHRYQLSAGRYHYQLVWCQK
Tail[12] = GAMDPGRYQLSAGRYHYQLVWCQK

Single amino acid code: A, alanine; C, cysteine; D, aspartic acid; E, glutamic acid; F, phenylalanine; G, glycine; H, histidine; I, isoleucine; K, lysine; L, leucine; M, methionine; N, asparagine; P, proline; Q, glutamine; R, arginine; S, serine; T, threonine; V, valine; W, tryptophan; Y, tyrosine.

SECTION 3:
GENERAL PROPERTIES OF POLYMERS

21. SURFACE CHARACTERISATION OF BIOERODIBLE POLYMERS USING XPS, SIMS AND AFM

ALEX G. SHARD, KEVIN M. SHAKESHEFF, CLIVE J. ROBERTS,
SAUL J. B. TENDLER and MARTYN C. DAVIES

*Laboratory of Biophysics and Surface Analysis, Department of Pharmaceutical Sciences,
The University of Nottingham, Nottingham NG7 2RD, UK*

INTRODUCTION

Examining the interfacial chemistry of drug delivery devices which are based upon bioerodible polymers has been recognised as being one of the important routes to understanding the *in vivo* operation of these rather complex systems. When any foreign material is placed in intimate contact with biological fluids a vast number of physicochemical processes may occur including protein adsorption, cell attachment, diffusion of chemical species between biofluid and material, and chemical and morphological alteration of the foreign material. These processes are often defined by the nature of the interface between the material and the biofluid and the potential utility of surface analytical techniques to characterise biomaterial surfaces was recognised about twenty years or so ago, an early review on the subject was published by Ratner (1983). Much of this early research was driven by the realisation that the properties of surfaces are not easily extrapolated from bulk material data. To fully understand the reactions that are elicited from a biological system in contact with a material surface it is first necessary to be able to define the exact nature of the surface.

One of the most common problems encountered by workers in this area is the ubiquity of surface contamination. All materials, even those in the artificial environment of an ultra high vacuum (UHV) chamber, procure a hydrocarbon surface layer. In addition to this ever present contamination, it is found that low level impurities in polymers will often preferentially reside at the surface, typical examples of this form of contamination being poly (dimethyl siloxane), Ratner *et al.* (1993) and the plasticiser di-octyl phthalate. A layer of any type of surface contaminant can completely alter the observed reactivity of the polymer and ultimately lead to flawed conclusions regarding it's performance as a biomaterial.

Even when contaminant free surfaces are obtained there is still no reliable way of predicting surface properties from the bulk properties of the material. At an interface, a polymer has a greater degree of freedom than that found in the bulk. It was demonstrated by Keddie *et al.* (1994), for instance, that the glass transition temperature of polymers decrease close to an interface. Polymer orientation, copolymer stoichiometry and blend composition are all highly liable to alter in the proximity of a surface.

In this chapter we concentrate on a specific type of biomaterial, bioerodible polymers. The term bioerodible is usually reserved for systems where polymer erosion occurs in the same time scale as drug release. In most cases the polymer degrades via a hydrolytic mechanism and may proceed either homogenously throughout the polymer bulk, or heterogenously at the surface of the polymer. Polymers which specifically erode at the surface have potential for producing devices which can give zero-order drug release. In these cases polymer degradation and erosion are confined to the interfacial region between polymer and biofluid. To gain some insight into the nature of these reactions and the resulting changes in polymer chemistry and morphology it is vital to employ surface analytical techniques.

A number of techniques are available to interrogate material surfaces. The development of instrumentation suitable for this type of examination has occured relatively recently, in the last quarter of a century or so. Some of these methods, however, are simply not suitable for polymer analysis, (for example, Auger electron spectroscopy (AES), where the electron beam used as an excitation source is too energetic to avoid damage to organic materials) and all of them have limitations in terms of the information they can provide. By combining techniques which give complementary information one can obtain a detailed description of a polymeric interface.

Within this chapter we describe the principles of three of the most important techniques employed to unravel the interfacial properties of polymers. It is not our intention to provide an exhaustive theoretical and historical perspective on the methods described and only an overview will be given, interested readers should refer to some of the excellent texts which are cited in each section if they wish a more thorough grasp of these techniques.

SURFACE ANALYTICAL TECHNIQUES

X-ray Photoelectron Spectroscopy (XPS)

XPS, also called electron spectroscopy for chemical analysis (ESCA), requires specialised and rather expensive equipment. This is because XPS relies upon both the photoelectron emission and Auger electron emission processes. A beam of monoenergetic X-rays illuminates the sample, typically the Mg K_α (1253.6 eV) or the Al K_α (1486.6 eV) X-ray lines, and this causes electrons to be emitted from the sample. By measuring the kinetic energy of electrons which depart from the sample surface it is possible to produce an XPS spectrum, an example is shown in Figure 1. Analysis of these electrons require that they can travel from the sample, through an energy discriminating analyser and to a detector, without elastic collisions occuring between them and atoms and molecules in their flight path. This requirement, and the reason of minimising surface contamination, means that a pressure of much less than 10^{-6} mbar has to be employed for the XPS technique, typically XPS equipment is operated at around 10^{-9} mbar, in the UHV regime.

Using this technique it is possible to identify elements and determine their surface concentration, with the exception of hydrogen and helium. Photoelectrons are produced by direct emission of an electron from an atom that has been excited by a photon, as depicted in Figure 2(a). The kinetic energy of the photoelectron is determined by the energy of the photon and the binding energy of the electron to the atom. The electron energy may be expressed as a kinetic energy or, more commonly, as a binding energy since this value is independent of the X-ray source energy. The two are related by the simple equation;

$$E_K = h\nu - E_B - \phi \qquad (1)$$

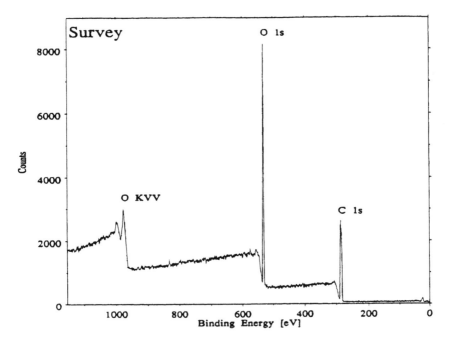

Figure 1 XPS wide scan of PLA showing oxygen and carbon 1s photoelectron peaks and oxygen KVV Auger peaks.

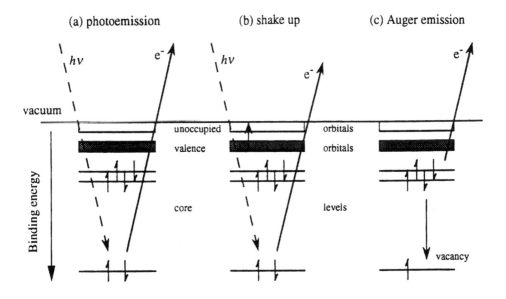

Figure 2 Electron emission processes observed in XPS spectra. (a) Photoemission. (b) Shake up during photoemission. (c) Auger emission.

where E_K and E_B are respectively the kinetic and binding energies, $h\nu$ is the energy of the X-ray photon and ϕ is the work function (a small value which depends upon both sample and spectrometer). Each elemental core level orbital has a distinctive binding energy, and thus it is possible to identify elements present within a sample surface. Since each core level orbital within each type of atom has a well defined cross section for ionisation it is possible, with the use of appropriate sensitivity factors, to determine the elemental composition of the material surface. Hydrogen and helium possess no core level electrons and cannot be identified using XPS, the inability of this technique to quantify the hydrogen content of a surface is one of the limitations of this technique as applied to polymer surfaces.

There are some complicating factors involved in the simple picture of photo-emission given above. The appearance of satellite peaks to an atomic core level orbital are sometimes observed, all of these features have a specific cause and some may provide important information about the sample itself. Equation (1) implies that when an X-ray interacts with an atom only one electron is affected. This is very often the case, especially for organic materials. Some materials, however, demonstrate a two electron transition which is represented in Figure 2(b). During the process of core level photoemission a valence electron can be promoted to an unoccupied state. This event gives rise to a spectral peak which appears on the low kinetic energy (high binding energy) side of the main photoelectron line and is known as a "shake up satellite". Such satellites arise in organic systems from the presence of some form of unsaturated bond where a $\pi \rightarrow \pi^*$ transition may occur in addition to photoemission. These are most evident in systems with conjugated and aromatic functionalities (for example, polystyrene and polyethylene terephthalate).

The presence of Auger lines in an XPS spectrum is common, these are not often used in polymer surface analysis, sometimes they may be useful in identification of an unexpected element, or a hindrance if they overlap with the photoelectron lines of the sample. When a core level electron has been ejected, the vacant "hole" is filled by an electron resident in an orbital higher in energy. The difference in energy between the two states can be lost in two ways, either by the emission of a photon (this is termed X-ray fluorescence) or by the ejection of another electron with kinetic energy given by equation (2). E_B represents the binding energy of the initial vacancy, E_C and E_D are respectively the binding energies of the atomic energy levels from which the electron falling to fill the vacancy and the ejected electron come. This process is called Auger emission and is represented schematically in Figure 2(c). Auger peaks in an XPS spectrum can be identified by virtue of the fact that their kinetic energies are not affected by changing the energy of the X-ray source.

$$E_{Auger} = E_B - E_C - E_D \qquad (2)$$

XPS may be used to non-destructively depth profile a surface. This is an important ability since the most common depth profiling method used in surface techniques is acheived by sputtering surface atoms from the material with the use of a high flux ion gun, such an approach would be inappropriate for analysing polymers due to the extreme levels of damage which would occur. The surface sensitivity of XPS derives from the low inelastic mean free path of electrons in solids. This means that photoelectrons emitted from deep in the sample have a very small probability of being emitted and detected with their original velocity. Only those electrons which come from close to the surface can be expected to retain their kinetic energy. The inelastic mean free path of an electron is denoted by the symbol λ. The actual value of λ within a solid is dependent upon the type of material and the kinetic energy of the electron. For electrons deriving from the C 1s core level shell of a polymer (kinetic energy ~969 eV with Mg Ka source) a typical value of inelastic mean free path within the solid is 3 nm. The term "sampling depth" is often used to indicate the depth from which 95% of the electrons which have not undergone elastic collisions are collected from, it can be calculated from equation (3) where θ represents the angle between the sample surface and the direction to the electron analyser.

$$\text{sampling depth} = 3 \, \lambda \, \sin \, \theta \qquad (3)$$

As can be seen from this equation, the sampling depth of XPS can be varied by changing the orientation of the sample with respect to the entrance of the analyser. This is shown schematically in Figure 3 where, at grazing angle, electrons from atoms at a depth within the sample that would normally be accessible to XPS cannot reach the detector because of the extra distance they have to travel through the sample. Depth profiling in this manner is often useful in complex systems like polymer blends and copolymers in order to gain an understanding of concentrations in the near surface region. It is important to note that determining a concentration depth profile from variable take off angle XPS is not straightforward and algorithms capable of performing this transformation have been investigated, Tyler *et al.* (1989) for example.

Most importantly, with regard to polymer surface analysis, XPS also provides chemical state information; particularly in the C 1s region. Figure 4 shows a high resolution scan of the C 1s spectrum of poly lactic acid (PLA), and each of the three separate chemical environments within the polymer gives rise to a peak in the C 1s envelope. The figure also demonstrates a deconvolution of the envelope using Gaussian peaks and it is found that the three peaks have identical areas, reflecting the concentrations of these environments in the polymer.

Chemical environment influences the binding energy position of elemental core levels. The change in binding energy away from the position typical of the pure element is known as chemical shift. Shifts of this nature can be thought of as arising from partial (or whole) charges existing on atoms, in the case of PLA carbon atoms bound to electronegative oxygen atoms are positively charged with respect to the methyl carbon atom. Electrons exiting from positive charge centres require a greater amount of energy to escape, this means that their kinetic energy is lowered and thus they have a higher binding energy. All of the C 1s orbitals in the sample have the same cross section for ionisation regardless of their chemical state and this means that the area under each peak in a deconvoluted spectrum is proportional to the number of carbon atoms in one particular environment.

Chemical shifts can be observed not just for C 1s electrons, but for all elements detectable by XPS, although for polymers the C 1s envelope does tend to contain most of the important chemical information. Sensible deconvolutions of XPS core level envelopes are often quite difficult to acheive even with the best software packages available. Familiarity with the technique and experience in the subtleties of XPS data is necessary to avoid inconsistent interpretations.

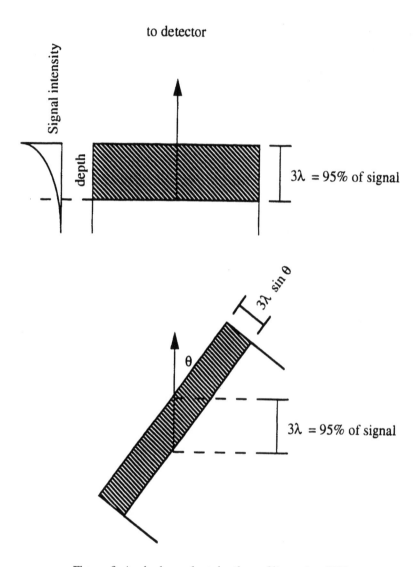

Figure 3 Angle dependent depth profiling using XPS.

Readers who are interested in delving a little deeper into polymer surface analysis by XPS are guided to the following works. Miller *et al.* (1986) produced a review of biomaterial applications for XPS with a heavy emphasis upon the theoretical aspects of the technique. A useful handbook for all aspects of XPS and Auger spectroscopies was edited by Briggs and Seah (1990).

Static Secondary Ion Mass Spectrometry (SSIMS)

SSIMS is another UHV technique. It relies upon the emission of charged fragments from a surface following a high energy particle impact. When a rapidly moving

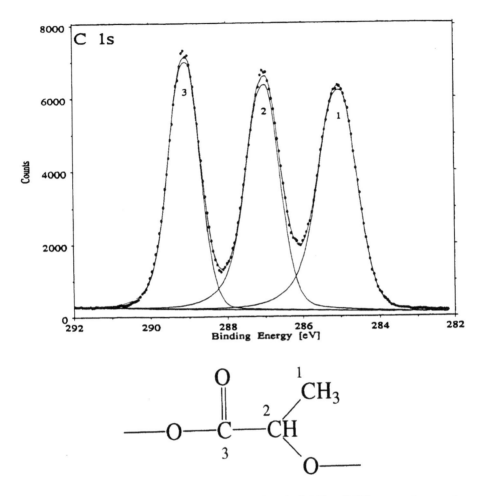

Figure 4 C 1s spectrum of PLA demonstrating chemical shifts of different carbon environments.

ion or atom collides with a solid material it will most often penetrate the surface and by colliding with atoms in the bulk of the material will rapidly lose energy. Some of the energy that is transferred to the solid results in the emission of electrons, neutral and charged fragments from the material surface. The sputtering of material from a surface is outlined in Figure 5. Of the fragmentary species that leave the surface the majority, for polymers at least, are uncharged, but those that are charged can be mass analysed.

The term *static* SIMS is applied to experiments in which mass spectra are collected from undamaged sample surfaces. This is very important for polymer surface analysis because such materials degrade very rapidly under atom or ion bombardment. In such experiments the flux of primary, high energy particles at the sample must be low enough that the probability of any primary ion striking an area damaged by a previous ion impact is remote. Studies of the rate of polymer damage indicate that

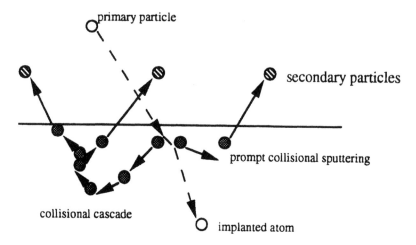

Figure 5 Collisional processes during SIMS.

the threshold doses for static SIMS are well below 10^{14} ions cm^{-2}. Typically spectra are acquired using a dose of 10^{12} ions cm^{-2} for quadrupole based instruments and 10^{10} to 10^{11} ions cm^{-2} for instruments employing time of flight (ToF) mass analysers. Using these regimes it is possible to detect fragments from polymer surfaces which contain a great deal of structural information.

The method of mass analysing these fragments is very important, most polymeric materials do not tend to produce ions with a mass to charge ratio higher than m/z 300, but in the case of oligomeric materials and very thin films on metallic substrates much higher mass ions can be detected. Although quadrupole instruments tend to have a mass range of up to m/z 1,000 and the instrument's sensitivity decreases with increasing mass these machines are adequate for most SSIMS applications. In the quadrupole instruments only one specific mass may be analysed at any one time and in the collection of a spectrum much information is lost. In contrast, ToF-SIMS instruments collect all masses simultaneously and have no mass dependent sensitivity limitations. The use of ToF-SIMS to detect high mass ions has been demonstrated, but it often requires extremely artificial conditions to produce such ions from polymeric materials. There are other advantages to using ToF-SIMS, for instance, the high sensitivity of such instruments enables the mass resolution to be increased to such an extent that ions of the same nominal mass but different chemical composition can be distinguished.

The removal of fragments from the sample into vacuum is termed sputtering. Simulations of argon ions impacting copper indicate that atoms leaving the copper surface come from a mean depth of less than 1 nm. Briggs (1989) reported that comparative SSIMS and XPS studies indicate that this is the typical depth of analysis for SSIMS of polymer surfaces also. Although quite large primary particle kinetic energies are employed for SSIMS (usually between 1 and 10 keV) it is found that many secondary ions detected retain much of the structural identity of the polymer and have low kinetic energies, they are therefore thought to have been desorbed via relatively low energy pathways.

The identification of surface species purely from SSIMS spectra is fraught with problems, it is almost always necessary to combine the data with some other source of information. This may be simply a prior knowledge of the chemical species present in the system being investigated, or it may be another analytical technique, such as XPS. When examining polymeric systems by SSIMS, spectral interpretation is based upon normal mass spectrometric rules. Quite often dominant ions contain the intact monomer repeat unit, or fragments derived from it through scission of main chain bonds. It is often found that fragments deriving from a side chain of a polymer are observed in high intensity. If the surface is a mixed system (e.g. a blend or copolymer) SSIMS data is not currently acceptable as a quantitative technique to describe the surface expression of each of the components in the mixture. Recently there have been investigations to determine in which cases it may be possible to use SSIMS data in this manner and some of these are presented within this chapter. At the very least SSIMS can identify which components of the blend are present or absent from the surface and give an estimation of their relative concentrations. The positive ion SSIMS spectra of PLA is given in Figure 6, structural assignments to ions are made later in the text.

There are many excellent reviews on the use of SSIMS to examine surfaces in general, Benninghoven (1994); polymeric surfaces, Briggs (1986) and biomaterials, Davies *et al.* (1990b), to name a few. Useful reference handbooks are Briggs and Seah (1992) and Briggs *et al.* (1989).

Scanning Probe Microscopes

The scanning probe microscopes are a growing number of closely related techniques which obtain structural information from the surfaces of samples (Quate, 1994; Roberts *et al.*, 1994). The scanning tunneling microscope (STM) was the first type to be designed (Binnig *et al.*, 1982) and it is a measure of the impact of the instrument that within 5 years of the first published description the inventors were presented the 1986 Nobel Prize in Physics. The STM obtains images of surface topography by monitoring the tunneling current flowing between a sharp metallic probe and the sample as the probe is scanned across the sample surface. The use of piezoceramic scanners and the sensitivity of the instrument to changes in topography allows the STM to achieve atomic resolution on flat samples (Hamers *et al.*, 1986). However, the imaging mechanism requires the sample to conduct electrons and as a consequence its application in biomedical science has concentrated on visualizing the molecular structure of biomolecules (Roberts *et al.*, 1994) which have been deposited onto conductive substrates or coated with thin metallic films.

For the analysis of bioerodible polymers, the atomic force microscope (AFM) is more suited than the STM because its imaging mechanism is independent of electron conduction and hence insulating organic materials can be directly visualized (Burnham and Colton, 1993). The basic components of an AFM are shown in Figure 7. The sample is mounted onto a piezoceramic scanner device capable of moving the sample in all three-dimensions. Above the sample surface is the AFM probe unit, which is composed of a sharp probe positioned on the end of a flexible cantilever. The cantilever is microfabricated in a V-shape with the two arms of the V attached to a stationary support.

ALEX G. SHARD *et al.*

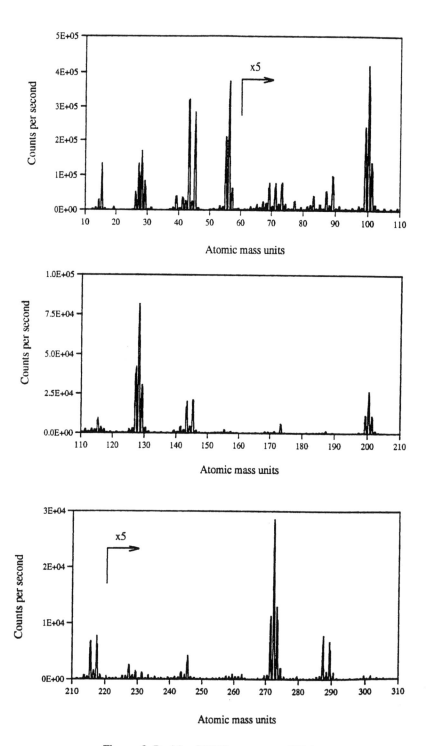

Figure 6 Positive SSIMS spectrum of PLA.

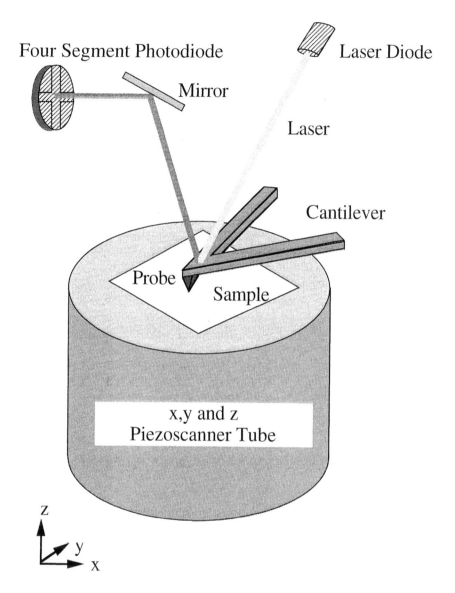

Figure 7 Schematic diagram of an Atomic Force Microscope.

To obtain an image the sample is raised, using the piezoceramic scanner, until the surface just contacts with the tip of the AFM probe. This contact generates a repulsive force due to the overlap of the electron shells of the probe and sample surfaces. Because the cantilever acts as a soft spring this repulsive force causes the AFM probe unit to deflect upwards. This deflection is measured by monitoring the angle of reflection of a laser from the upper surface of the cantilever using a four-segment photodetector. As the AFM probe moves across the sample surface any changes in sample topography will change the probe-sample repulsive force. If, for example, the AFM probe scans over a protruding surface feature, the cantilever will be deflected further upwards (i.e. there is an increased repulsion). Therefore, by plotting the angle of deflection of the cantilever (or the repulsive force) against the lateral position of the AFM probe (controlled by the piezoceramic scanner) a topographical representation of the sample is generated. Clearly, the contact between the AFM probe and the surface could potentially damage the sample and therefore it is common practise to operate the AFM in constant force mode. In this mode, at each sampling point on the surface, any deflection of the AFM probe unit is compensated by a vertical movement of the piezoceramic. Hence, for our protruding surface feature, having detected the increase in the angle of deflection of the probe unit, the AFM control unit would instruct the piezoceramic to decrease the height of the sample until a preset deflection was obtained. Using constant force mode it is possible to ensure that the repulsive force is minimized and very soft samples, such as individual protein molecules, can be repeatedly imaged (Yang *et al.*, 1994). Recently, it has become possible to further decrease the chance of sample damage and increase the resolution of the AFM using non-contact imaging modes in which the AFM probe unit is vibrated just above the surface of the sample. This mode of imaging is becoming increasingly important and has been the subject of a number of publications (Burnham and Colton, 1993; Hansma *et al.*, 1994).

The potential of the AFM to enhance the characterization of bioerodible polymers, and biomaterials in general, stems from ability of the instrument to obtain very high resolution images from uncoated polymer samples within aqueous environments. Therefore, it has become possible to image dynamic processes like surface erosion and protein adsorption which are fundamental to the mechanisms of drug delivery and biocompatibility.

SURFACE SPECTROSCOPY OF BIOERODIBLE HOMOPOLYMERS

Table 1 lists much of the important work carried out on the surface characterisation of bioerodible polymer systems. In addition, high quality XPS spectra of many of these compounds may be found in the Scienta ESCA300 database, Beamson *et al.* (1992). The cited examples of XPS and SSIMS data have been acquired, to the best of our knowledge, on clean polymer surfaces and the results obtained may be used as a basis for further investigations such as studies of protein and cellular adhesion and polymer degradation. We briefly describe some of the main features of SSIMS and XPS spectra from typical bioerodible polymers. Illustrating the depth of elemental and molecular information gleaned from such techniques and the different advantages of each technique.

Poly (esters)

The XPS C 1s spectrum for PLA is given in Figure 4, and demonstrates all the main features observed in this type of polymer. Almost all of the poly esters employed as biodegradable systems contain three types of carbon environment, hydrocarbon, ester ($\underline{C}$-O) and ester ($\underline{C}$OO). Since the latter two environments are always equal

Table 1 Typical C 1s chemical shifts for common polymer functionalities

Chemical functionality	Binding energy (eV)
-$\underline{C}H_3$, -$\underline{C}H_2$-, =$\underline{C}H$-	+0.0 (285.0)
$\underline{C}R_3$-CO-OR, $\underline{C}R_3$-C(-OCR$_3$)$_3$	+0.6
$\underline{C}R_3$-OH, $\underline{C}R_3$-O-CO-R, $\underline{C}R_3$-O-$\underline{C}R_3$	+1.6
CR_3-$\underline{C}O$-CR_3, CR_3-O-$\underline{C}R_2$-O-CR_3	+2.9
R-$\underline{C}O$-OR, (CR$_3$-O-)$_3\underline{C}$R	+4.2
RO-$\underline{C}O$-OR	+5.4
R$_3\underline{C}$-NR$_2$	+1.0
R$\underline{C}$ONR$_2$	+3.1
R$_2$N-$\underline{C}O$-OR	+4.6
R$_3\underline{C}$-Cl	+2.0
R-$\underline{C}F_2$-R	+5.9

in intensity, the main difference between polymers of this class is to be observed in the size of the hydrocarbon peak at 285 eV. For instance in poly (glycolic acid) (PGA) it is negligable, in PLA it is equal to the other two peaks, in poly (3-hydroxy butyrate) (PHB) it is twice as intense as the other peaks, and so on. The O 1s spectra of poly (esters) typically demonstrates two peaks of equal intensity due to the C = O environment (~532.2 eV binding energy) and the C-O-C environment (~533.6 eV). Identification of a poly (ester) solely from XPS spectra is thus quite difficult, for instance PHB and poly (ethylene adipate) have virtually identical core level spectra, the only differences appearing in the rather weak valence band region (~0–40 eV binding energy). The valence band region may be used to "fingerprint" polymers, but more striking differences between isomeric polymers are often evident in the SSIMS spectra.

Table 2 Literature examples of surface characterisation of bioerodible polymers

Polymers	XPS	SSIMS
Poly (esters) poly (lactide), poly (caprolactone) *etc.*	Davies *et al.* (1989) Davies *et al.* (1990a) Clark *et al.* (1991) Brinen *et al.* (1991)	Davies *et al.* (1989) Davies *et al.* (1990a) Brinen *et al.* (1991)
Poly (orthoesters)	Davies *et al.* (1991b) Davies *et al.* (1991c)	Davies *et al.* (1991c)
Poly (anhydrides)	Davies *et al.* (1991a)	Davies *et al.* (1991a)
Poly (phosphazines)	Leadley (1994)	Leadley (1994)
Collagen	Radu *et al.* (1993)	
Gelatin		Batts *et al.* (1994)
Hyaluronic acids	Barbucci *et al.* (1993)	Shard *et al.* (1994)

Table 3 Masses of prominent ions observed in SSIMS of poly (hydroxy acids) commonly used as bioerodible systems. Weak ion intensities are noted by (w), – denotes ions too weak to be observed

Ion	Poly (glycolic acid)	Poly (lactic acid)	Poly (3-hydroxy butyrate)	Poly (caprolactone)
Positive ions				
[M-OH]	–	55	69	97(w)
[M-O+H]	43	57(w)	71(w)	–
[M-H]	57	71	85	113(w)
[M+H]	59(w)	73	87	115
[2M-OH]	99(w)	127	155	–
[2M-O+H]	101	129	157(w)	–
[2M-H]	115	143	171	–
[2M+H]	117	145	173	–
Negative ions				
[M-H]	57(w)	71	85	113
[M+H]	59	73	87	115
[M+O-H]	73	87	101	129
[M+OH]	75	89	103	131
[2M-H]	115(w)	143	171	–
[2M+H]	117	145	173	–
[2M+O-H]	131	159	187	–
[2M+OH]	133	161	189	–

In poly (hydroxy acids) analysed by SSIMS a diagnostic series of ions exist in the positive and negative ion spectra which reflect the polymer structure. These are provided in Table 3 for the three polymers mentioned above and poly (caprolactone) (PCL). These ions result from scission of the polymer either at the ester linkage or in sites adjacent to it. Assignments of structures to some these ions are given below based on the discussion of Shard *et al.* (1996).

The ions [nM-OH]$^+$ and [nM-H]$^-$ are not shown above, in the case of all polymers except PGA the structure of these ions are respectively similar to the [nM-O+H]$^+$ and [nM+H]$^-$ ions, with loss of the marked hydrogen atom, and one on an adjacent methyl or methylene unit, to form a double bond. PGA does not possess such an adjacent group and this explains why these series of ions are particularly weak in PGA polyester.

In the positive ion spectra of PLA and PGA a number of radical cations are also observed. These are even electron species, which are quite common in normal mass spectrometry, McLafferty *et al.* (1993), but observed rarely in SSIMS spectra. For fragments which contain only carbon, hydrogen and oxygen, radical cations appear at an even mass. The radical cations observed in PLA and PGA can be assigned to the [nM-O]$^{\cdot+}$ and the [nM-CO2]$^{\cdot+}$ fragments. The latter sequence of ions are only visible for the n = 1 and n = 2 cases whereas the former produces ions for n = 1 to at least n = 5 and the observation of higher mass ions of this series may be limited only by the fact that most published data has relied on the use of quadrupole instruments with their intrinsic mass limitations. The fact that PCL, PHB and most other poly (ester)s do not produce ions of this type indicates that they are a product of the 2-hydroxy acid linkage and the structures given below were assigned to them

$$H_2R-C\equiv O^+ \qquad O=R-C\equiv O^+ \qquad HO-RH-C\equiv O^+$$

$$[M\text{-}O+H]^+ \qquad\qquad [M\text{-}H]^+ \qquad\qquad [M+H]^+$$

$$H_2R-C\overset{O}{\underset{O^-}{\diagdown}} \qquad O=R-C\overset{O}{\underset{O^-}{\diagdown}} \qquad HO-RH-C\overset{O}{\underset{O^-}{\diagdown}}$$

$$[M+H]^- \qquad\qquad [M+O\text{-}H]^- \qquad\qquad [M+OH]^-$$

$$RH = \qquad -CHH- \qquad \overset{CH_3}{\underset{|}{-CH-}} \qquad \overset{CH_3}{\underset{|}{-CH-CH_2-}}$$

$$\qquad\qquad\qquad PGA \qquad\quad PLA \qquad\qquad PHB$$

$$-CHH-CH_2-CH_2-CH_2-CH_2-$$

$$PCL$$

by Shard *et al.* (in press). It was reasoned that the stability of these radical cations derived from a resonant stabilisation of the radical centre. Two canonical forms for each set of ions are given below.

Poly (β-malic acid) (PMA) and it's benzyl and butyl derivatives have been analysed using XPS and ToF-SIMS by Leadley *et al.* (in preparation b). The structure of this polymer is given below. In the XPS of this polymer there are only the three main environments described above. In the case of the benzyl derivative a shake up satellite due to $\pi - \pi^*$ transitions in the phenyl ring is also observed $\sim$7 eV higher in binding energy than the hydrocarbon peak. The free acid and butyl derivative do not display this transition.

$$\overset{R}{\underset{|}{HC}}\left[\overset{O}{\underset{||}{C}}-\overset{R}{\underset{|}{CH}}\right]_{n-1}C\equiv O^+ \quad\rightleftharpoons\quad \overset{R}{\underset{|}{HC}}=\overset{O}{\underset{||}{C}}-\overset{R}{\underset{|}{CH}}\text{\textouterspace}$$

$$[nM\text{-}O]^{\bullet+} : R = Me\ (PLA) : R = H\ (PGA)$$

$$R-HC-C\overset{O}{\underset{\underset{+}{O}}{\diagdown}}CH-R \quad\rightleftharpoons\quad R-HC=C\overset{O}{\underset{\underset{+}{O}}{\diagdown}}CH-R$$

$$[2M\text{-}CO_2]^{\bullet+}$$

Poly (malic acid); R = H

Benzyl derivative; R =

Butyl derivative; R = $-CH_2-CH_2-CH_2-CH_3$

In the negative ion SSIMS spectra of these polymers, the most notable ions are the [M-H]$^-$ appearing at m/z 115, 171 and 205 for the free acid, butyl and benzyl derivatives respectively and the [M+OH]$^-$ cation at m/z 133, 189 and 223, as before. The [M+H]$^-$ and [M+O-H]$^-$ ions are very weak for these polymers and the reasons for this are obscure. There is also a strong ion appearing at m/z 107 for the benzyl derivative which is due to the benzyl oxide anion.

The positive ion spectra of the three PMA derivatives show little difference between each other below m/z = 100, with the exception that the benzyl derivative has an intense peak at m/z 91 which is assigned to the $C_7H_7^+$ tropyllium ion commonly observed in SSIMS spectra of compounds containing benzyl groups. The only readily assignable ions of interest observed at higher mass are the [M+H]$^+$ ions observed for the butyl and free acid, and the [M-OH]$^+$ and [M+O-H]$^+$ for the butyl derivative. The benzyl derivative does not display any of these ions, and none of the polymers demonstrated strong radical cations.

Poly (ortho esters)

Two homopolymers of poly (ortho esters) based upon the diketene acetal 3,9 – diethylidene – 2, 4, 8, 10 – tetraoxaspiro [5.5] undecane (DETOSU) and one of two diols, either hexanediol (HD) or *trans*-cyclohexanedimethanol (t-CDM), the structures of which are given below, were examined using XPS by Davies *et al.* (1991b), and using ToF-SIMS by Davies *et al.* (1991c). The ortho ester R-C(-OR)$_3$ functionality has a similar chemical shift in XPS to that of an acid or ester carbon, +4.2 eV, see Table 1. In addition a β-shifted carbon atom appears at ±0.5 eV for those carbons next to the ortho ester functions, these are also outlined in Table 1. The carbon atom singly bonded to oxygen also has a shift typical of this chemical environment, +1.6 eV, and is found to be three times as intense as the ortho ester functionality. A C 1s peak fit is given in Figure 8. Only a single oxygen environment is observed, determined by Beamson *et al.* (1992) to have a binding energy of 533.0 eV. All of the binding energies were referenced to the hydrocarbon environment in the polymers.

De Matteis *et al.* (1993) used high resolution XPS to study poly (ortho ester)s based upon DETOSU with either N-methyl-diethanolamine (MDE) or N-phenyl-diethanolamine (PDE) as the diol in the polymer, R = $CH_2CH_2N(Me)CH_2CH_2$ and $CH_2CH_2N(Ph)CH_2CH_2$ respectively. The XPS binding energy data was compared with ab-initio molecular orbital calculations to demonstrate that the PDE polymer contains a planar nitrogen and the MDE polymer has a pyramidal structure around the nitrogen atom.

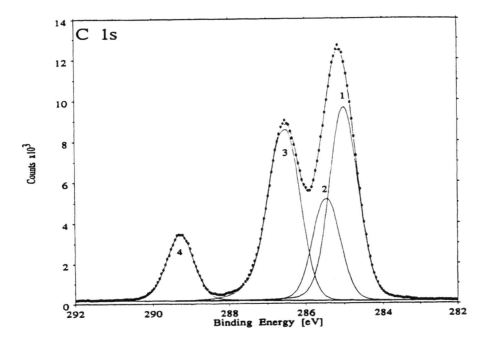

Figure 8 XPS C 1s peak fit of poly (orthoester) DETOSU – HD.

The use of ToF-SIMS on these polymers allowed the observation of ions with masses of over 1,000 amu. Ions below m/z ~300 were derived from fragments of either the DETOSU or the diol units. Typical ions diagnostic of the DETOSU functionality are given below. The positive charge centre in all these ions can be distributed to one of two oxygen heteroatoms, only one canonical form is shown in the structures. For the HD and t-CDM polymers, the characteristic low mass diol fragments were found to be $[R-H]^+$, $[R+OH]^+$ and $[R+OH_3]^+$, which were observed in relatively low intensity, except for the t-CDM $[R-H]^+$ ion at m/z 109. It should be noted that the $[R+OH]^+$ ion of t-CDM appears at m/z 127 and is thus obscured by one of the DETOSU peaks.

Hexanediol : $R = (CH_2)_6$

trans - cyclohexanedimethanol : $R = H_2C$—⟨ ⟩—CH_2

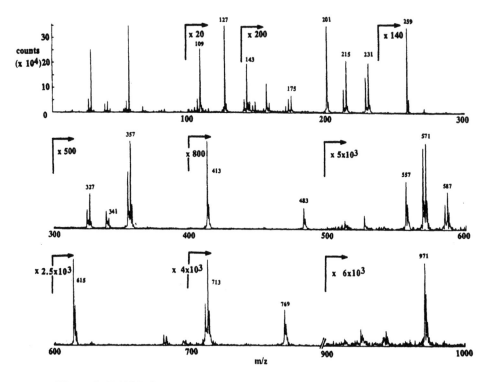

Figure 9 ToF-SIMS spectrum of DETOSU – t-CDM (Davies *et al.*, 1991c).

The positive ion spectrum for the t-CDM polymer from mass 0 to 1,000 is shown in Figure 9. Several series of ions may be observed at higher mass which incorporate one or more of both the DETOSU and diol units. These are given in Table 4 along with the masses the ions appear at for both the t-CDM and HD diols. The most intense ions at high mass are those ions which have a 259 amu moiety added. This corresponds to a DETOSU unit + 28 amu, which could either be CO or C_2H_4. No structures have been assigned to these fragments.

Poly (anhydrides)

The anhydride group appears slightly higher in chemical shift to that of the acid functionality given in Table 1. It is determined by Beamson *et al.* (1992) as being 4.46 eV above the hydrocarbon peak for poly (sebacic anhydride) (PSA), and by Davies *et al.* (1991a) at 4.5 eV for a range of aliphatic poly (anhydrides), $[-CO-(CH_2)_x-CO-O-]_n$, where x = 4, poly (adipic anhydride) (PA); x = 6, poly (suberic anhydride) (PSU); x = 7, poly (azelaic anhydride) (PAZ) and x = 8, PSA. Two peaks have to be used to fit the hydrocarbon environment, the extra one is used to account for the secondary shift on methylene groups adjacent to the anhydride functionalities. The O 1s spectra can be fitted with two peaks, at ~532.8 and ~534.0 eV. The lower binding energy peak is assigned to the carbonyl C=O oxygen atom and

Table 4 Prominent high mass cations observed in ToF-SIMS of DETOSU poly (ortho esters) – very weak or not observed

Assignment	Masses for DETOSU t-CDM n = 1, 2, 3, 4	Masses for DETOSU HD n = 1, 2, 3, 4
$(M_{detosu} + M_{diol})_n$ - CH_2CH_3	327, 683, 1039, –	301, 631, –, –
$(M_{detosu} + M_{diol})_n$ - OH	339, 695, 1051, –	313, 643, –, –
$(M_{detosu} + M_{diol})_n$ + H	357, 713, 1069, 1425	331, 661, 991, 1321
$(M_{detosu} + M_{diol})_n$ + COC_2H_5	413, 769, 1125, 1481	387, 717, 1047, 1377
$(M_{detosu} + M_{diol})_n$ + [259]	615, 971, 1327, –	589, 919, 1249, –

$$m/z = 231$$

$$m/z = 175$$

$$m/z = 215$$

$$m/z = 143$$

$$m/z = 201$$

$$m/z = 127$$

has approximately twice the intensity of the other peak which is due to the CO-O-CO oxygen, as expected. Example spectra of PSA are given in Figure 10.

ToF-SIMS spectra of the four poly (anhydrides) described above were collected by Davies *et al.* (1991a). Apart from fragment ions typical of the hydrocarbon section at masses lower than m/z 100 in the positive ion spectra, all of the polymers displayed prominent $(M+H)^+$ ions. The radical cations $M^{\cdot+}$ noted by the authors appear at one mass unit less than the $(M+H)^+$ ions but are of weak intensity in comparison to them. Also apparent are ions of the general formula $(M-OH)^+$, which appear to have higher intensity with decreasing repeat unit molecular weight, for PSA this ion is barely visible. Above the molecular mass of the repeat unit, the $(M+HCO)^+$ ion is found for all of the polymers, but only PA has an ion assignable as $(2M-OH)^+$ at m/z 239.

In the negative ion spectra all of the anhydrides produced an ion at m/z 71 assigned to the acrylate anion, $CH_2=CH\text{-}COO^-$. At higher mass, strong ions were

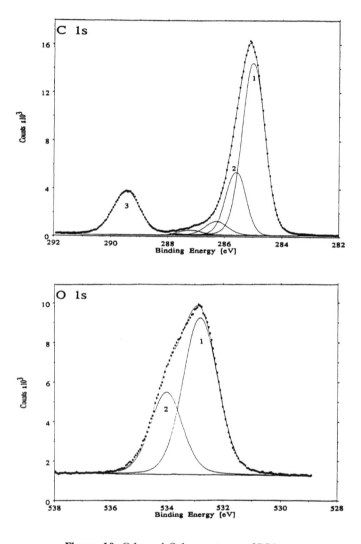

Figure 10 C 1s and O 1s spectrum of PSA.

found which could be assigned to the (M-H)⁻, (M+OH)⁻ and (M+CO₂H)⁻ anions. Some of the ions described here and in the preceding paragraph are assigned the structures shown below.

SURFACE SPECTROSCOPIES OF BIOERODIBLE POLYMER BLENDS AND COPOLYMERS

Poly (ester) copolymers

A brief report of the XPS and SIMS analysis of copolymers of lactic and glycolic acids (PLGA) was made by Davies *et al.* (1988). Both XPS and SSIMS confirmed that

		PA (n=4)	PSU (n=6)	PAZ (n=7)	PSA (n=8)
$^+O\equiv C-(CH_2)_{n-1}-CH=C=O$	$(M-OH)^+$	111	139	153	167
$^+O\equiv C-(CH_2)_n-C(=O)(OH)$	$(M+H)^+$	129	157	171	185
$^+O\equiv C-(CH_2)_n-C(=O)(O-CHO)$	$(M+HCO)^+$	157	185	199	213
$^-O-C(=O)-(CH_2)_{n-1}-CH=C=O$	$(M-H)^-$	127	155	169	183
$^-O-C(=O)-(CH_2)_n-C(=O)(OH)$	$(M+OH)^-$	145	173	187	201
$^-O-C(=O)-(CH_2)_n-C(=O)(O-CHO)$	$(M+CO_2H)^-$	173	201	215	229

the surface compositions of the copolymers were similar to the bulk compositions by elemental ratios and the presence of diagnostic ions, although the XPS data appeared to suggest a level of surface hydrocarbon contamination. It was also reported that a peptide (LH-RH) could be detected at the surface of PLGA loaded with this drug model.

A more detailed analysis of the information given by SSIMS into these systems was undertaken by Shard *et al.* (in press). Ions containing mixtures of the two monomeric units were identified, and the intensities of the radical cations were studied in detail. It was found that for the radical cations of general formulae $(nM-O)^{·+}$ in the PLA homopolymer, n+1 ions were observed in the copolymer. So, for n = 3, ions assignable to $(3L-O)^{·+}$, $(2LG-O)^{·+}$, $(2GL-O)^{·+}$ and $(3G-O)^{·+}$ were observed (at m/z = 200, 186, 172 and 158 respectively), where L is a repeat unit of lactic acid and G is a repeat unit of glycolic acid. Furthermore, the intensities of these ions reflected both the composition and short range structure of the copolymer. A similar approach was used by Briggs *et al.* (1984) to demonstrate the random nature of methacrylate copolymers.

Copolymers synthesised by two methods were investigated. The routes employed were either a polycondensation method using lactic acid and glycolic acid as comonomers, or a ring opening polymerisation of lactide and glycolide using stannous octoate as catalyst. This latter method could be expected to produce a polymer with a "random dimeric" structure, however, at the reaction temperatures generally employed, trans-esterification reactions are expected to scramble this short

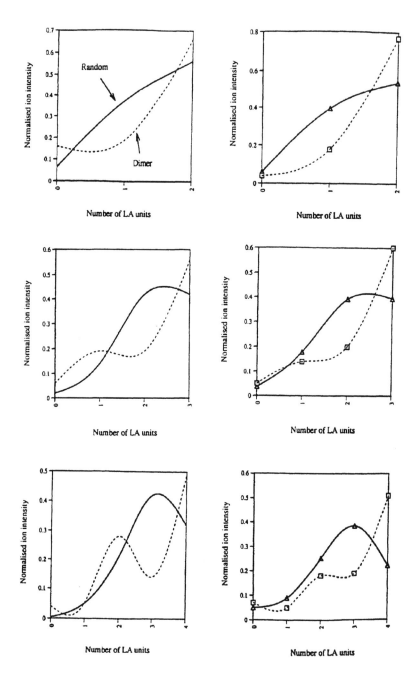

Figure 11 Comparison of theoretical and experimental intensities for 75% LA PLGA. Left hand side, theory for random and dimeric structure, right hand side experimental results from copolymers produced from monomeric acids (triangles) and dimeric lactide and glycolide (squares). Top, $[2M-O]^{+}$ series, middle, $[3M-O]^{+}$ series, bottom, $[4M-O]^{+}$ series (Shard *et al.*, 1996).

range order. It is found by NMR that a degree of non-randomness exists in polymers produced by this second route, but it cannot provide direct evidence of the expected random dimeric structure.

Comparisons of the SSIMS radical cation intensities with statistical models based upon the probabilities of finding monomer units adjacent to each other in the polymer for both random and random dimeric structures gave interesting results. Figure 11 shows these comparisons, the two polymers given both contained 75% lactic acid and were produced through the different routes given above. This not only confirms that the surface chemistry of these polymers is similar to the bulk, but gives direct evidence that dimeric structure is retained in the synthesis of PLGA from lactide and glycolide.

Copolymers of glycolide (GA) and trimethylene carbonate (TMC) (repeat structure -CH2-CH2-CH2-O-CO-O-) have been examined using ToF-SIMS by Brinen *et al.* (1993). The polymers examined were either statistical, segmented or block copolymers of the two comonomers. In the study ions were identified which arose from adjacent G and TMC units. The intensities of these ions decreased as the block lengths of the G and TMC segments increased (in the order statistical, segmented and block). This, with the example of PLGA copolymers is an illustration of the fact that fragment ions in SSIMS retain the structure of the parent copolymer and complex rearrangements are unobserved.

Poly (β-malic acid) copolymers comprising of partially hydrolysed butyl and benzyl derivatives have been investigated using ToF-SIMS by Leadley *et al.* (1995b). Effectively these are random copolymers of PMA and the PMA derivatives. Ions specific to each of the derivatised functionalities have been identified previously. By plotting the intensity of these ions against the bulk copolymer composition it is possible to demonstrate that the surface chemistry of these polymers is similar to the bulk composition.

Poly (ortho ester) copolymers

In their study of DETOSU polymerised with HD and t-CDM copolymers Davies *et al.* (1991c) also studied two copolymers containing 32.5 mol % and 17.5 mol % HD in the copolymers, 50% being DETOSU and the remainder being t-CDM. The XPS data were consistent with the compositions of the copolymers and ToF-SIMS demonstrated an almost linear relationship between ions diagnostic of the molar diol content and the copolymer composition as shown in Figure 12. At higher mass ions were observed in which contained both HD and t-CDM units. These were present in quite large intensity in comparison to the equivalent ions which contained purely one type of diol. This is direct evidence that the copoly (ortho esters) were statistical copolymers and had many adjacent HD and t-CDM units.

A number of copolymers of MDE and PDE based DETOSU poly (ortho esters) were examined with XPS by De Matteis *et al.* (1993) and the surface elemental compositions were found to be in accord with the bulk compositions of the copolymers.

Poly (anhydride) copolymers

Leadley *et al.* (in preparation a) studied the surface chemistry of copolymers containing PSA and ricinoleic acid maleate (RAM), the general structure is shown

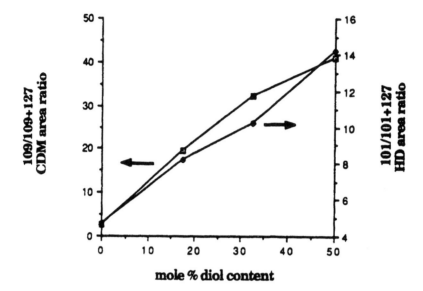

Figure 12 Graph of the area ratios of signals diagnostic of HD (m/z 101), t-CDM (m/z 109) and DETOSU (m/z 127) as a function of poly (ortho ester) bulk composition (Davies *et al.*, 1991c).

below and the copolymers contained between 0 and 50 mol. % RAM. Subtle differences were observed in the XPS of these copolymers, the elemental compositions were consistent with the bulk. In the C 1s spectra a peak attributable to the C-O environment in the RAM segment of the copolymer was observed and correlated with the presence of this monomer unit at the surface.

ToF-SIMS of these copolymers demonstrated all of the ions diagnostic of PSA. Several ions in both the positive and negative ion spectra were attributed to the RAM segment of the polymer. An ion at m/z 115 in the negative spectra, and another at m/z 263 in the positive ion spectra were assigned to the RAM segment of the copolymer. The intensities of these ions have been compared with PSA derived ions and a linear relationship found between the bulk copolymer composition and the RAM ion intensities which implies that the surface composition of these copolyanhydrides is similar to the bulk.

Blends of PLA and PSA

The mixing of polymers which have different physical properties and degradation kinetics may result in materials with novel and controllable bioerodible qualities. Davies *et al.* (1996) performed combined XPS, SSIMS and AFM investigations upon blends of PLA and PSA. It has been reported by Domb (1993) that these polymers are miscible if the PLA component has a low molecular weight (below 3 kDa), but otherwise are immiscible. In the surface analytical investigations two molecular weights of PLA were used, namely 2 kDa and 50 kDa. It was found that both the miscible (2kDa) and immiscible (50kDa) blends presented a surface excess of PLA, see Figure 13(a) and (b). A plot of oxygen content against weight % PLA in the blend should produce a straight line if the surface composition was identical to the blend bulk. It was found that the chemical environments in the C 1s spectra were a straight combination of PLA and PSA chemical environments for the miscible blends, and the surface molar concentration of PLA relative to PSA could be calculated from the C-O environment relative intensity in the C 1s spectra. Figure 13(c) shows a plot of the surface and bulk molar concentrations for the miscible systems.

The C 1s spectra of immiscible blends demonstrated some unusual features. In many cases it was found that the spectrum was broad and some of the components had shifted to higher binding energy. This effect is attributed to a differential charging phenomenon and could be corrected with a charge compensating flood of electrons as demonstrated in Figure 14. The charge compensated spectrum is very similar to C 1s spectra of the miscible blends and could be fitted with peaks in the same manner as those mixtures. It was found that the spectra with charging features could also be fitted if the PLA components of the fit were shifted in binding energy by +1.3 eV. This implies that the two blend components are electrically isolated from each other and charge to different extents under X-ray irradiation. Since it is known that the PLA (50 kDa) and PSA polymers are immiscible, this phenomenon was taken as evidence that there was a phase separation of the two components in the near surface region. Further evidence of this surface phase separation was provided by AFM analysis.

In the SSIMS spectra of these blends all of the ions diagnostic of PLA and PSA were observed. It was demonstrable that there was a surface excess of PLA in all of the blends by examining the relative intensities of each set of ions in accordance with the XPS results. However, even in high weight % PLA blends it was possible to observe PSA derived ions and this may result from the extension of PSA domains to the top 10 nm of the surface, or due to a small but detectable concentration of PSA in the PLA domains.

AFM CHARACTERIZATION OF BIOERODIBLE POLYMERS

When bioerodible polymers are employed in the body their degradation behaviour within the aqueous environment will be determined by the chemical structure and the morphology of the polymer. The surface chemistry sensitive techniques of XPS and SIMS can analyse the nature of chemical units which will be presented by the polymer to this aqueous environment. These chemical units will determine the hydrolytic sensitivity of the polymer backbone and the thermodynamics of the

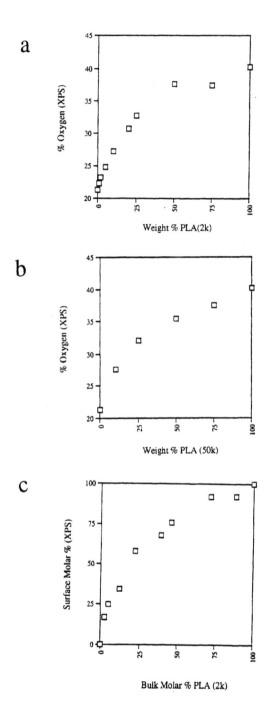

Figure 13 (a) Plot of surface oxygen concentration against weight % PLA in miscible (2 kDa) blends with PSA. (b) Plot of surface oxygen concentration against weight % PLA in immiscible (50 kDa) blends with PSA. (c) Comparison of bulk and surface molar concentrations of PLA in miscible blends with PSA (Davies *et al.*, 1996).

polymer/liquid interactions. However, to fully characterize the dynamic interactions which underlie surface erosion it is also necessary to analyse the surface morphology of the polymer. For example, the kinetics of surface erosion of PSA will be highly influenced by the degree of crystallinity of the polymer at the polymer/water interface because the tightly packed polymer chains within crystalline areas of a surface will degrade significantly slower than amorphous regions (Göpferich and Langer, 1993; Mathiowitz et al., 1993).

The AFM has proved to be a valuable tool with which to study the influence of surface morphology on degradation. Its value is derived from the ability of the instrument to continually visualize the changes in surface morphology resulting from erosion. This ability was first demonstrated on samples of PSA where it was possible to visualize the initial stages of surface erosion (Shakesheff et al., 1994). An example of the AFM imaged erosion of a PSA film is shown in Figure 15. These images show the effect of the preferential loss of amorphous PSA from around crystalline fibres which produces a etching effect resulting in the increase in height of fibres above the surface. Further experiments on PSA samples have highlighted that the AFM can be used to determine the influence of fabrication parameters on the proportion of crystalline material at the surface and hence the initial behaviour of these systems during polymer degradation (Shakesheff et al 1995). These studies indicate that the AFM can rapidly characterize the influence of manufacturing procedures on the suitability of controlled delivery systems.

A major merit of using in situ AFM to characterize the degradation of bioerodible polymers is the ease of interpretation of the patterns of degradation of complex systems such as polymer blends. An example of the characterization of such a system is shown in Figure 16 where a blend of PSA and poly(DL-lactic acid) (PLA) has been visualized during surface erosion (Shakesheff et al., 1994). These two polymers are immiscible at high molecular weight and therefore they phase separate during fabrication into solid devices. This phase separation causes problems in predicting the kinetics of erosion of such blends. The PLA component degrades at a slower rate than the PSA component. Therefore, if the phase separation generates a surface morphology dominated by the PLA, the overall degradation kinetics of the blend would be slower than predicted for a miscible blend. The images in Figure 16 show how the surface morphology of the blends becomes dominated by the underlying PLA morphology as the PSA material is lost through degradation. This PLA morphology is shown to be dependent on concentration of the polymer in the initial polymer solution from which the film was cast.

When bioerodible polymers are employed as controlled drug delivery a principle concern is the relationship between polymer erosion and drug release at the polymer/water interface. Using the AFM it is possible to visualize the release of drug from eroding polymer films (Shakesheff et al., 1995a). This has been demonstrated with a system composed of a protein (bovine serum albumin) embedded in a poly(ortho ester) film. Initial characterization of the surfaces of these systems showed the presence of granules of the protein surrounded by a relatively smooth polymer morphology. On exposure to a pH 6 aqueous environment the polymer morphology became rougher and, as that occurred, the protein granules appeared to shrink indicating the dissolution of the protein into the environmental water. An example of the change in surface topography during such an experiment is shown in Figure 17.

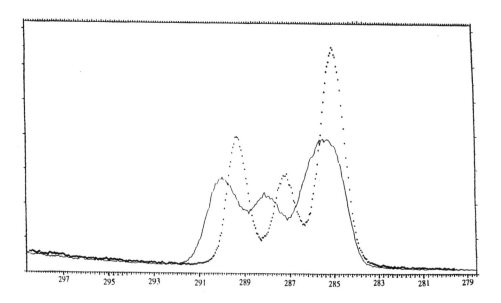

Figure 14 C 1s spectra of a 50 % weight blend of PLA (50 kDa) and PSA (solid line) and with electron flood gun charge compensation (crosses) (Davies *et al.*, 1996).

A further interesting aspect of the AFM characterization of the release of protein from a polymer film is the ability to use the data to study the relative kinetics of polymer degradation and protein dissolution (Shakesheff *et al.*, 1995a). This ability is derived from the three-dimensional nature of the AFM data. Using computational methods it is possible to measure the relative changes in volume of the polymer matrix to the embedded protein particles. This analysis has indicated that some protein particles begin dissolution immediately after exposure to the aqueous environment indicating a very thin coverage of polymer at most. Other protein particles retain the same volume for more than 30 minutes before dissolution commences indicating thicker polymer coverage.

This ability to follow the kinetics of surface erosion has been extended with the combination of the AFM with surface plasmon resonance analysis (SPR) (Chen *et al* 1995). SPR is a technique for measuring nanometric changes in the thickness of films achieved through the measurement of changes in refractive index of the film. SPR has been extensively employed in the quantification of antibody/antigen interactions and is a promising technique for analysing the adsorption of proteins to surface (Davies 1994). The combination of SPR with an AFM has allowed the simultaneous visualization and quantification of the surface erosion of thin films of bioerodible polymers (Chen *et al.*, 1995; Shakesheff *et al.*, 1996). At present, this new technique has been employed to films of a poly(ortho ester) (Chen *et al.*, 1995) and to blends of PSA/PLA (Shakesheff *et al.*, 1996). These studies have demonstrated that the combination of AFM and SPR allows changes in polymer surface morphology to be directly related to the kinetics of biodegradation.

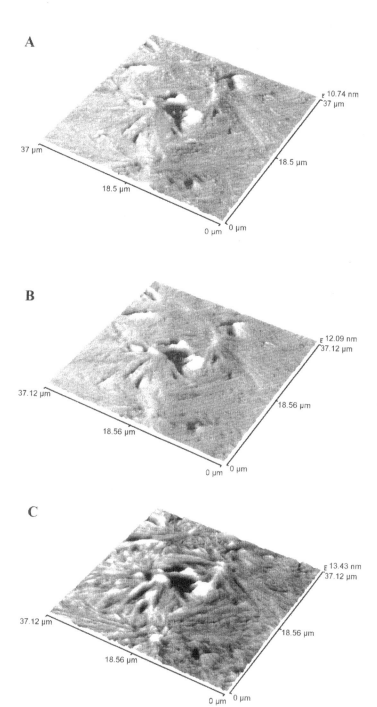

Figure 15 *In situ* AFM imaging of the degradation of a melt crystallised PSA sample after (A) 0 min. (B) 20 min. (C) 60 min. exposure to a pH 12.5 buffer (Shakesheff *et al.*, 1994).

A

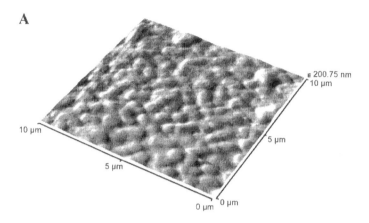

B

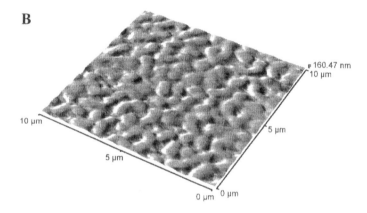

C

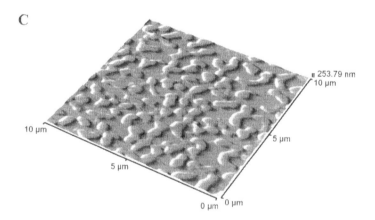

Figure 16 AFM visualisation of the erosion of an immiscible biodegradable blend containing 50% PSA and 50% PLA (A) 0 min. (B) 5 min. (C) 10 min. exposure to a pH 12.5 buffer (Shakesheff *et al.*, 1994).

A

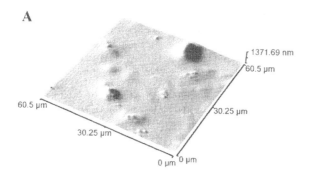

B

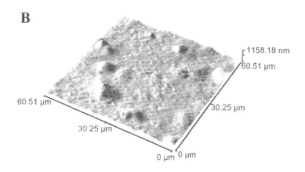

C

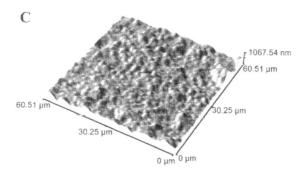

Figure 17 The release of bovine serum albumin from a poly (ortho-ester) film. The *in situ* AFM images show the dissolution of protein from the eroding polymer film after (A) 0 min. (B) 45 min. (C) 90 min. exposure to a pH 6.0 solution (Shakesheff *et al.*, 1994).

CONCLUDING REMARKS

This review has outlined the use of surface analytical techniques in the study of bioerodible polymers. Both the chemical information obtainable from XPS and SIMS, and the analysis of topographical changes provided by AFM can supply profound insights into the surface mediated processes occuring in these systems. A large amount of research is still required before complete understanding of the interfacial chemistry and physics involved can be achieved and several issues have yet to be addressed. In particular, the surface spectroscopies cannot be applied in aqueous environments with the ease of AFM and SPR because of the vacuum requirements of these techniques. Removal and analysis of polymers which are undergoing degradation is fraught with problems, not least being the liability of polymer surfaces to restructure on going from a hydrated to dry environment. The development of cryogenic techniques to 'freeze' organic samples in a hydrated state can potentially address this issue, Lin *et al.* (1993).

The development of high mass resolution ToF-SIMS capable of determining the elemental composition of secondary ions allows the SIMS technique to be capable of studying highly complex polymer systems and blends, Shard *et al.* (1995). In addition, the imaging capabilities of SSIMS (for example, Short *et al.* (1992)) has hardly been explored in the analysis of bioerodible polymers. The employment of imaging could be relevant in assessing the surface concentrations and homogeneity of drugs, polymer end groups and contaminants in drug delivery devices.

REFERENCES

Barbucci, R., Magnani, A., Baszkin, A., Dacosta, M. L., Bauser, H., Hellwig, G., Martuscelli, E. and Cimmino S. (1993) *Journal of Biomaterial Science, Polymer Edition,* **4**, 245.

Batts, G.N. and Paul, A.J. (1994) A study of the competitive adsorption of a fluorosurfactant at the gelatin-air interface using time-of-flight mass spectrometry. *Langmuir,* **10**, 218–224.

Beamson, G. and Briggs, D. (1992) *High Resolution XPS of Organic Polymers: The Scienta ESCA300 Database,* Wiley, Chichester.

Benninghoven, A. (1994) Surface analysis by secondary ion mass spectrometry. *Surface Science,* **299/300**, 246–260.

Binnig, G., Rohrer, H., Gerber, Ch. and Weibel, E. (1982) Surface studies by scanning tunneling microscopy. *Physical Review Letters,* **49**, 57.

Briggs, D., Hearn, M.J. and Ratner, B.D. (1984) Analysis of polymer surfaces by SIMS. 4. A study of some acrylic homo- and co-polymers. *Surface and Interface Analysis,* **6**, 184–193.

Briggs, D. (1986) SIMS for the study of polymer surfaces: a review. *Surface and Interface Analysis,* **9**, 391–404.

Briggs, D. (1989) Recent advances in secondary ion mass spectrometry (SIMS) for polymer surface analysis. *British Polymer Journal,* **21**, 3–15.

Briggs, D., Brown, A. and Vickerman, J.C. (1989) *Handbook of Secondary Ion Mass Spectrometry,* Wiley, Chichester.

Briggs, D. and Seah, M.P. (1990) *Practical Surface Analysis Volume 1, Auger and X-ray Photoelectron Spectroscopy,* 2nd edition, Wiley, Chichester.

Briggs, D. and Seah, M.P. (1992) *Practical Surface Analysis Volume 2, Ion and Neutral Spectroscopy,* 2nd edition, Wiley, Chichester.

Brinen, J.S., Greenhouse, S. and Jarrett, P.K. (1991) XPS and SIMS studies of biodegradable suture materials. *Surface and Interface Analysis,* **17**, 259–266.

Brinen, J.S., Rosati, L., Chakel, J. and Lindley, P. (1993) The effect of polymer architecture on the SIMS spectra of glycolide / trimethylene carbonate copolymers. *Surface and Interface Analysis,* **20**, 1055–1060.

Burnham, N.A. and Colton, R.J. (1993) Scanning Tunneling Microscopy and Spectroscopy Theory, Techniques and Applications. *Force microscopy,* ed. Bonnell D.A., New York, VCH Publishers Inc.

Chen, X., Shakesheff, K.M., Davies, M.C., Heller, J., Roberts, C.J., Tendler, S.J.B. and Williams, P.M. (1995) The degradation of a thin polymer film studied by simultaneous in situ atomic force microscopy and surface plasmon resonance analysis. *Journal of Physical Chemistry,* **11**, 2547.

Clark, M.B., Burkhardt, C.A. and Gardella, J.A. (1991) Surface studies of polymer blends. 4. An ESCA, IR and DSC study of the effect of homopolymer molecular weight on crystallinity and miscibility of poly (ϵ-caprolactone) / poly (vinyl chloride) homopolymer blends. *Macromolecules,* **24**, 799–805.

Davies, J. (1994) Surface plasmon resonance - The technique and its applications to biomaterial processes. *Nanobiology*, **3**, 5.

Davies, M.C., Khan, M.A., Domb, A., Langer, R., Watts, J.F. and Paul, A.J. (1991a) The analysis of the surface chemical structure of biomedical aliphatic polyanhydrides using XPS and ToF-SIMS. *Journal of Applied Polymer Science*, **42**, 1597–1605.

Davies, M.C., Khan, M.A., Lynn, R.A.P., Heller, J. and Watts, J.F. (1991b) X-ray photoelectron spectroscopy of the surface chemical structure of some biodegradable poly (orthoesters). *Biomaterials*, **12**, 305–308.

Davies, M.C., Khan, M.A., Short, R.D., Akhtar, S., Pouton, C. and Watts, J.F. (1990a) XPS and SSIMS analysis of the surface chemical structure of poly (caprolactone) and poly (b-hydroxybutyrate - b-hydroxyvalerate) copolymers. *Biomaterials*, **11**, 228–234.

Davies, M.C. and Lynn, R.A.P. (1990b) Static secondary ion mass spectrometry of polymeric biomaterials. *Critical Reviews in Biocompatibility*, **5**, 4, 297–341.

Davies, M.C., Lynn, R.A.P., Watts, J.F., Paul, A.J., Vickerman, J.C. and Heller, J. (1991c) ToF-SIMS and XPS analysis of the surface chemical structure of some linear poly (orthoesters). *Macromolecules*, **24**, 5508–5514.

Davies, M.C., Shakesheff, K.M., Shard, A.G., Domb, A., Roberts, C.J., Tendler, S.J.B. and Williams, P.M. (1996) Surface analysis of biodegradable polymer blends of poly (sebacic anhydride) and poly (DL-lactic acid). *Macromolecules*, **29**, 2205–2212.

Davies, M.C. and Short, R.D. (1988) The surface characterisation of PLA-PGA and LH-RH loaded PLA-PGA copolymer systems by XPS and SIMS. *Proceedings of the International Symposium of Controlled Release of Bioactive Material*, **15**, 318–319.

Davies, M.C., Short, R.D., Khan, M.A., Watts, J.F., Brown, A., Eccles, A.J., Humphrey, P., Vickerman, J.C. and Vert, M. (1989) An XPS and SIMS analysis of biodegradable biomedical polyesters. *Surface and Interface Analysis*, **14**, 115–120.

De Matteis, C.I., Davies, M.C., Leadley, S., Jackson, D.E., Beamson, G., Briggs, D., Heller, J. and Fransom, N.M. (1993) High resolution monochromated X-ray photoelectron spectroscopy (XPS) studies on novel biodegradable poly (ortho esters) prepared using N-methyl- and N-phenyl-diethanolamine, correlated with binding energies obtained from ab-initio molecular orbital methods. *Journal of Electron Spectroscopy and Related Phenomena*, **63**, 221–238.

Domb, A.J. (1993) Degradable polymer blends: 1. Screening of miscible polymers. *Journal of Polymer Science. A. Polymer Chemistry*, **31**, 1973.

Göpferich, A. and Langer, R. (1993) The influence of microstructure and monomer properties on the erosion mechanism of a class of polyanhydrides. *Journal of Polymer Science: Part A: Polymer Chemistry*, **31**, 2445.

Hamers, R.J., Tromp, R.M. and Demuth J.E. (1986) Surface electronic structure of Si(111)-(7x7) resolved in real space. *Physical Review Letters*, **56**, 1972.

Hansma, P.K., Cleveland, J.P., Radmacher, M., Walters, D.A., Hillier, P.E., Bezanilla, M., Fritz, M., Vie, D., Hansma, H.G., Prater, C.B., Massie, J., Fukunaga, L., Gurley, J. and Elings, V. (1994) Tapping Mode atomic force microscopy in liquids. *Applied Physics Letters*, **64**, 1738.

Keddie, J.L., Jones, R.A.L. and Cory, R.A. (1994) Interface and surface effects on the glass-transition temperature in thin polymer films. *Faraday Discussions*, **98**, 219–230.

Leadley, S.R. (1994), Ph.D. thesis, University of Nottingham.

Leadley, S.R., Davies, M.C., Domb, A., Nudelman, R., Paul, A.J. and Beamson, G. (in preparation a) The analysis of the surface chemical structure of anhydride copolymers of sebacic acid and ricinoleic acid maleate using XPS and ToF-SIMS. In preparation.

Leadley, S.R., Davies, M.C., Vert, M., Braud, C., Paul, A.J., Shard, A.G. and Watts, J.F. (in preparation b) Probing the surface chemical structure of the novel biodegradable polymer poly (β-malic acid) using ToF-SIMS and XPS. In preparation.

Lin, H.B., Lewis, K.B., Leachscampavia, D., Ratner, B.D. and Cooper, S.L. (1993) Surface properties of RGD-peptide grafted polyurethane block copolymers — variable take off angle and cold stage ESCA studies. *Journal of Biomaterial Science — Polymer Edition*, **4**, 183–198.

Mathiowitz, E., Jacob, J., Pekarek, K. and Chickering, D. (1993) Morphological characterization of bioerodible polymers. 3. Characterization of the erosion and intact zones in polyanhydrides using scanning electron microscopy. *Macromolecules*, **26**, 6756.

McLafferty, F.W. and Turecek, F. (1993) *Interpretation of Mass Spectra, Fourth Edition*. University Science Books, Mill Valley.

Quate C.F. (1994) The AFM as a tool for surface imaging. *Surface Science*, **299/300**, 980.

Radu, G.-L. and Baiulescu, G.L. (1993) Surface analysis of collagen membranes by X-ray photoelectron spectroscopy. *Journal of Molecular Structure*, **293**, 265–268.

Ratner, B.D. (1983) Surface characterisation of biomaterials by electron spectraoscopy for chemical analysis. *Annals of Biomedical Engineering*, **11**, 313–336.

Ratner, B.D., Leach-Scampavia, D. and Castner, D.G. (1993) ESCA surface characterisation of four IUPAC reference polymers. *Biomaterials*, **14**, 148–152.

Roberts, C.J., Williams, P.M., Davies, M.C., Jackson, D.E. and Tendler, S.J.B. (1994) Atomic force microscopy and scanning tunneling microscopy: Refining techniques for studying biomolecules, *Trends in Biotechnology*, **12**, 127.

Shakesheff, K.M., Chen, X., Davies, M.C., Domb, A., Roberts, C.J., Tendler. S.J.B. and Williams, P.M. (1995) Relating the phase morphology of a biodegradable polymer blend to erosion kinetics using simultaneous in situ atomic force microscopy and surface plasmon resonance analysis., *Langmuir*, **11**, 3921–3927.

Shakesheff, K.M., Davies, M.C., Domb, A., Roberts, C.J., Tendler, S.J.B. and Williams, P.M. (1995) In situ AFM visualization of the degradation of melt crystallized poly(sebacic anhydride). *Macromolecules*, **28**, 1108.

Shakesheff, K.M., Davies, M.C., Heller, J., Roberts, C.J., Tendler, S.J.B. and Williams, P.M. (1995a) The release of protein from a poly(ortho ester) film during surface erosion studied by in situ atomic force microscopy. *Langmuir*, **11**, 2547.

Shakesheff, K.M., Davies, M.C., Roberts, C.J., Tendler, S.J.B., Shard, A.G. and Domb, A. (1994) In situ imaging of polymer degradation in an aqueous environment. *Langmuir*, **10**, 4417.

Shard, A.G., Davies, M.C., Tendler, S.J., Jackson, D.E., Bennedetti, L., Paul, A.J., Beamson, G. and Purbrick, M.D. (1994) Surface chemistry of hyaluronic acid derivatives investigated by high resolution XPS and time of flight SIMS. *Abstracts of the 11th European conference on biomaterials*, Pisa.

Shard, A.G., Sartore, L., Davies, M.C., Ferruti, P., Paul, A.J. and Beamson, G. (1995) Investigation of the surface chemical structure of some biomedical poly (amidoamine)s using high resolution X-ray photoelectron spectroscopy and time of flight secondary ion mass spectrometry. *Macromolecules*, **28**, 8259–8271.

Shard, A.G., Volland, C., Davies, M.C. and Kissel, T., Information on the monomer sequence of poly lactic acid and random copolymers of lactic acid and glycolic acids by examination of SSIMS ion intensities. *Macromolecules*.

Short, R.D., Ameen, A.P., Jackson, S.T., Pawson, D.J., O'Toole, L. and Ward, A.J. (1992) C-R-Burch prize. ToF SIMS in polymer surface studies. *Vacuum*, **11–12**, 1143–1160.

Tyler, B.J., Castner, D.G. and Ratner, B.D. (1989) Regularisation — a stable and accurate method for generating depth profiles from angle dependent XPS data. *Surface and Interface Analysis*, **14**, 443–450.

Yang, J., Mou, J. and Shao, Z. (1994) Structure and stability of pertussis toxin studied by in situ atomic force microscopy. *FEBS Letters*, **338**, 89.

22. MECHANISMS OF POLYMER DEGRADATION AND ELIMINATION

ACHIM GÖPFERICH

Department of Pharmaceutical Technology, University of Erlangen-Nürnberg, Cauerstraße 4, 91058 Erlangen, Germany

INTRODUCTION

Polymers were originally synthesized because of their excellent physical and chemical stability to serve as durable materials. During the last 25 years, however, applications have emerged that utilize the degradability of polymers. Degradable polymeric packing materials, for example, facilitate waste management (Vert *et al.*, 1992) and protect the marine fauna from ingestion and entanglement, which is a well known problem with non-degradable polymer waste (Doi *et al.*, 1992). In contrast to waste management where the breakdown of a polymer structure is the primary objective, the mechanism of degradation and erosion is crucial for some other applications. Many examples can be found in the medical, biomedical and pharmaceutical field where degradation and erosion can bring about some desired effect. In medicine, for example, biodegradable polymers are used as resorbable suture (Miller & Williams, 1984), resorbable plates and screws (Leenslang *et al.*, 1987), plasma volume expanders (Beetham *et al.*, 1987), and polymer stents that maintain the lumen of heart arteries open after surgery (Agrawal *et al.*, 1992). In biomedical research, degradable polymers are used as scaffolds for the growth and repair of tissue and organs (Langer & Vacanti). In pharmaceutical applications, degradable polymers are used as drug carriers for the delivery of drugs (Langer, 1990). Examples are the development of vaccines (Singh *et al.*, 1991), the local treatment of cancer and infectious diseases (Mauduit *et al.*, 1993; Brem *et al.*, 1993) and drug targeting (Kopecec, 1984).

All applications mentioned above depend markedly on the physical chemical processes of polymer degradation. It is, therefore, essential to understand polymer degradation and erosion when using a degradable polymer for such an application because these events are critical for the successful polymer performance. Due to the large number of degradable polymers it is, however, impossible to review degradation and elimination processes in detail for each individual polymer. The objective of this chapter is rather to summarize the most important features of polymer degradation, erosion and elimination. Examples are given to illustrate major principles of these processes. Finally, approaches to modeling degradation and erosion on the basis of theoretical models are briefly reviewed.

Correspondence: Tel: 49-9131-859557; Fax: 49-9131-859545; e-mail: goepf@damian.pharmtech.uni-erlangen.de

TERMINOLOGY

The Definition of Degradation, Erosion, Biodegradation, Bioerosion and Elimination

There is still no agreement on a single definition for erosion and degradation. It is, therefore, necessary, to define how these terms will be used in this review. The following definitions were adapted from the literature (Vert *et al.*, 1992; Göpferich 1996a): **Degradation** is the process of polymer chain scission by the cleavage of bonds between the monomers in the polymer backbone. Accordingly, degradation leads to a size reduction of the polymer chains. Erosion, in contrast, designates the breakdown of a polymer in a broader sense. **Erosion** is the mass loss of a polymer matrix which can be due to the loss of monomers, oligomers or even pieces of non-degraded polymer. Erosion can be the result of biological, chemical or physical effects. From the definitions above it is obvious, that polymer degradation is part of polymer erosion. In the biomedical field as well as in waste management **biodegradation** and **bioerosion** are of significance. In contrast to many other definitions both terms will be used here in a broad sense and imply that degradation and erosion are at least mediated by a biological system. This means that the biological environment in which a polymer degrades or erodes is at least contributing to degradation and erosion (Vert *et al.*, 1992). In this review, polymer **elimination** is defined as the excretion and metabolism of polymer and erosion products from mammals.

The Definition of a Degradable Polymer

With the definitions above we should be able to define what a degradable polymer is, but in fact we cannot. The reason is simple: all polymers degrade one way or another and, therefore, all polymers would have to be considered degradable. As we do not feel comfortable with such a result we have to introduce some additional criteria that are met by degradable polymers but not by "non-degradable" ones. A solution is found by measuring degradation not in terms of absolute time but relative to the duration of an application or in relation to our human life time. If a polymer degrades not within the human life time it is usually not considered degradable. A distinction between degradable and non-degradable polymers could be made by defining a Deborah number (Reiner, 1964). Deborah numbers are dimensionless and have originally been developed to classify viscoelastic materials into more viscous materials and into more elastic ones (Metzner *et al.*, 1967) and are also useful for the characterization of polymer swelling. A Deborah number to distinguish degradable from non-degradable polymers could be defined as shown in Equation 1.

$$D = \frac{\text{time of degradation}}{\text{human lifetime}} \quad (1)$$

Degradable polymers would then have small values for D (D → 0) and non-degradable polymers have large ones (D → ∞).

A SURVEY ON MAJOR MECHANISMS

Potential Polymer Degradation Mechanisms

There are 4 major modes of polymer degradation: photo-, mechanical-, thermal- and chemical degradation (Banford & Tipper, 1972). All of them might be important for biodegradable polymers as well. Photodegradation can occur during the exposure of polymers to UV or gamma radiation. This is relevant when trying to decrease the risk of bacterial contamination during polymer processing by UV irradiation or when subjecting polymer samples to γ-sterilization (Sepälä *et al.*, 1991). Mechanical as well as thermal degradation might occur during polymer processing. During extrusion, for example, polymers are exposed to elevated temperatures and high shear forces which can lead to a loss of molecular weight (Gelovoy *et al.*, 1989). Mechanical degradation is important when polymers are used for providing a mechanical function where they are subject to stress (Miller *et al.*, 1984). Of all degradation mechanisms chemical degradation is the most important one for biodegradable polymers. By introducing hydrolyzable functional groups into the polymer backbone, the polymer chains become labile to an aqueous environment and thus, chemical degradation initiates polymer erosion. A brief list of different types of chemically degradable polymer bonds is given in Table 1. These bonds differ amongst other properties mainly by the velocity at which they hydrolyze.

The Erosion of Biodegradable Polymers

The erosion of hydrolytically degradable polymers starts with the diffusion of water into the polymer. As the polymer chains are hydrated, the functional groups hydrolyze and absorb part of the water. During degradation the polymer is broken down into oligomers and monomers. These compounds are transported from the polymer bulk, which can be controlled by diffusion (Göpferich & Langer, 1995a). The release of degradation products leads to the mass loss which is characteristic for erosion. Degradation is the most important part of erosion.

THE CHEMICAL DEGRADATION OF BIODEGRADABLE POLYMERS

Factors Affecting the Velocity of Passive Hydrolysis

The Type of Chemical Bond

The type of functional group in a degradable polymer backbone has the highest impact on the degradation velocity. This becomes obvious when comparing the velocity at which polymers lose molecular weight during degradation, which can be a matter of hours for poly(anhydrides) (D'Emanuelle *et al.*, 1992), but may be a matter of weeks for poly(α-hydroxy-esters) (Shah *et al.*, 1992). There have been many approaches to classify degradable polymers based upon the reactivity of their functional groups (Park *et al.*, 1993; Baker, 1987). An example for selected polymers is given in Table 2.

Table 1 Functional groups contained in degradable polymers

poly(cyanoacrylates)   poly(anhydrides)   poly(ketals)

poly(ortho esters)   poly(acetals)   poly(α-hydroxy-esters)

poly(ε-caprolactone)   poly(phosphazenes)   poly(β-hydroxy-esters)

poly(imino-carbonates)   polypeptides   poly(carbonates)

poly(phosphate esters)

Table 2 Half-life of some classes of degradable polymers according to Park *et al.*, 1993

Polymer class	half-life
poly(anhydrides)	.1 h
poly(ortho esters)	4 h
poly(esters)	3.3 years
poly(amides)	83,000 years

Poly(amides), for example, hydrolyze almost 10 orders of magnitudes slower than poly(anhydrides). The rates given in Table 2 are, however, subject to variations. One factor that can affect these rates is catalysis. Most of the bonds shown in Table 1 can be subject to acid and base catalysis. The hydrolysis rate of ester bonds can, for example, change by orders of magnitude depending on pH. When the same functional groups are inserted into the polymer backbone of degradable polymers, the rate of hydrolysis also has been shown to be altered by acid-base catalysis (Nguyen *et al.*, 1986; Leong *et al.*, 1985). Changes in hydrolysis rates were also observed in the presence of ions. The hydrolysis of acetic anhydride for example can be catalyzed by acetate and formate ions. Besides catalysis, steric effects modify hydrolysis rates. The higher reactivity of poly(glycolic acid) compared to poly(lactic acid) for example appears to be due to steric effects because the methyl group hinders the attack of water (Vert *et al.*, 1992). Another factor are electronic effects during hydrolysis. Introducing electronegative substituents in the α-position of esters for example increases ester reactivity (Kirby, 1972). In summary it can be concluded that the hydrolysis rate of degradable polymers depends mostly on the nature of the functional group, but is affected substantially by its neighborhood and the chemical environment in which hydrolysis takes place.

Water Uptake

The hydrolysis of the polymer backbone requires water and can be considered a bimolecular reaction. The reactivity of such processes can be increased by raising the concentration of either reaction partner (Pitt *et al.*, 1981). The water uptake of degradable polymers can be controlled by altering the lipophilicity of the system. Lipophilic polymers have a reduced tendency to take up water and decrease, thereby their hydrolysis rate (Leong *et al.*, 1985). The degradation rates of poly(imino carbonates), for example, were found to correlate with water uptake (Pulapura *et al.*, 1990). That water uptake affects degradation rates was also proven with physical blends of polymers such as poly(lactic-co-glycolic acid) with poly(vinyl alcohol) (Pitt *et al.*, 1992) as well as by introducing hydrophilic monomers into copolymers (Heller *et al.*, 1987). In either case the degradation rates increased when raising the content of the hydrophilic component. The degree to which water uptake can be used to decrease degradation seems, however, to be limited as water probably cannot be completely prevented from diffusing into polymers (Heller, 1986; Crank, 1968).

Crystallinity and Molecular Weight

Polymer crystallinity has a direct impact on degradation. Comparing the degradation rate of poly(D,L-lactic) acid with that of poly(L-lactic acid) might serve as a good example. The latter is partially crystalline and degrades substantially slower than the amorphous product (Baker, 1987) which proves that crystalline polymer regions degrade slower than amorphous ones. The same has been observed for other polymers such as poly(anhydrides) (Göpferich and Langer, 1993a). Also liquid crystalline regions in polymers were reported to have an increased resistance against degradation (Mathiowitz *et al.*, 1993). The impact of molecular weight on degradation is complicated as it acts in a direct as well as an indirect way. An example for the indirect effect of molecular weight on degradation is the glass transition temperature which depends on the molecular weight of the polymer (Hiemenz, 1984). Higher molecular weight causes an increase in the glass transition temperature and leads to slower degradation because glassy polymers degrade slower than rubbery ones (Baker, 1987). The direct effect of molecular weight is mediated through the chain length of the polymer. Higher molecular weight increases the chain length and, therefore, more bonds have to be cleaved in order to generate water soluble oligomers or monomers to allow erosion to proceed. Degradation takes, therefore, more time with increasing molecular weight which can be seen from the shift of the onset of weight loss from poly(lactic acid) matrices to higher times (Asano *et al.*, 1990).

Interesting is the effect of degradation on crystallinity and molecular weight of a polymer matrix. By the preferential degradation of amorphous polymer areas, an increase in total crystallinity has been observed during the degradation of partially crystalline polymers such as poly(L-lactic acid) in aqueous media (Pistner *et al.*, 1993). The effect of degradation on the molecular weight loss of polymer matrices is different for fast and slow degrading polymers. Poly(anhydride) and poly(ortho-ester) matrix discs for example loose their molecular weight very rapidly (D'Emanuelle *et al.*, 1992; Nguyen *et al.*, 1986) while poly(lactic acid) matrices are substantially more stable (Göpferich, 1996a). In the case of poly(lactic acid) and poly(lactic-co-glycolic acid) the slow degradation causes the loss of molecular weight over the polymer matrix cross-section which follow first order kinetics (Shah *et al.*, 1992). Fast eroding polymer matrices in contrast seem to degrade faster on their surface compared to their core, which leads to a non-homogenous degradation that cannot be described by first order kinetics any more (D'Emanuelle *et al.*, 1992).

pH

pH is one of the most important factors of hydrolytic polymer degradation. pH changes can modify hydrolysis rates by orders of magnitude (Kirby, 1972; Leong *et al.*, 1985). In addition, the degradation products of many degradable polymers change pH by their acid functionality (Göpferich & Langer, 1993a). Poly(esters) are a quite versatile example for how pH might affect degradation. Their hydrolysis can be either acid or base catalyzed. Lactic acid, a degradation product of poly(lactic acid) and poly(lactic-co-glycolic acid), for example, has a very high water solubility. Due to the α-hydroxy group the pKa of 3.8 is lower than that of unsubstituted aliphatic carboxylic acids which usually have pKa values of about 4.5. The low pKa in combination with high water solubility give lactic and glycolic acid the potential to decrease the pH inside an eroding polymer matrix substantially. The decreased pH and the generated carboxylic acids cause autocatalytic effects (Vert *et al.*, 1991) leading to the faster erosion inside poly(α-hydroxy acids) compared to their surface (Li *et al.*, 1990a; Göpferich, 1996a). Inside eroding rods of these materials, pH values as low as 2 were measured although they were eroded in a pH 7.4 phosphate buffer (Göpferich, 1996a). Substantial pH changes were also found during the erosion of other polymers. Poly(anhydrides) such as poly(1,3 bis[p-carboxy phenoxypropane]-co-sebacic acid) 20:80 were reported to form a network of pores in which a pH of 4.5 was measured using embedded glass electrodes (Göpferich, 1996a). This explains that pH values measured on the surface of eroding poly(anhydride) discs using confocal microscopy were one unit lower than in the surrounding pH 7.4 buffer solution (Göpferich & Langer, 1993a). Interesting is the effect of pH on the degradation of poly(ortho esters). Their degradation can only be increased by lowering pH, as hydroxide ions are not able to attack the carbon of the ortho ester bond (Heller, 1986). Adding magnesium hydroxide to poly(ortho ester) matrices, therefore, has been used to decrease their degradation rate while the addition of carboxylic acid anhydrides served the opposite purpose (Shih *et al.*, 1984; Heller, 1985). Attention has to be paid to drugs and other substances with an acid or base functionality when they are incorporated into degradable polymers, because they might alter degradation rates as well (Yoshioka *et al.*, 1991).

Copolymer Composition

Copolymers have properties that are different from the properties of the corresponding homopolymers. Crystallinity and glass transition temperature, for example, change tremendously upon copolymerization (Hiemenz, 1984). Poly(α-hydroxy acids) might serve again as a good example. Poly(L-lactic acid) and poly(glycolic acid) are crystalline while poly(D,L-lactic acid) or poly(lactic-co-glycolic acid) are amorphous (Gilding & Reed, 1979). Changes in the hydrophilic/lipophilic balance and in crystallinity as well as steric effects (Vert *et al.*, 1992) were made responsible for changes in the degradation rate of poly(lactic-co-glycolic acid) when increasing the glycolic acid content in the polymer backbone (Li *et al.*, 1990a). The presence of a variety of functional groups in a copolymer might also affect degradation rates. A polymer built from two monomers A and B, for example, contains 4 types of bonds: A-A, B-B, A-B and B-A. This has some significance, when these bonds have different hydrolysis rates. In poly(anhydrides) such as poly(1,3-bis[p-carboxyphenoxy]propane-co-sebacic acid), the degradation rate was found to decrease substantially with the content of aromatic monomer for which the increasing number of slow reacting bonds between the aromatic monomers were made responsible. In summary, copolymers must be considered new polymers with degradation and erosion mechanisms different from the corresponding homopolymers.

Enzymatic Degradation

Biodegradable polymers can be hydrolyzed either passively or actively via enzymatic catalysis. For enzymatically degradable polymers, both mechanisms compete against each other and the fastest process controls the overall degradation mechanism. Enzymatic degradation is mainly effective for naturally occurring polymers such as polysaccharides and polypeptides. Consequently mainly natural polymers such as collagen, fibrin, chitin, albumin and hyaluronic acid have been used as enzymatically degradable polymers. However, even with non-hydrolyzable, water insoluble polymers such as poly(ethylene), enzymatic degradation might occur. Measuring respiratory $^{14}CO_2$ during degradation experiments with microbes even poly(ethylene) was found to be enzymatically degradable (Albertsson, 1980), however, at a low rate. Synthetic polymers with functional groups have higher chances of non-specific enzymatic degradation. A list of enzymatically degradable synthetic polymers can be found in Park *et al.*, 1993. The competition between mechanisms of enzymatic and passive hydrolysis were observed when comparing the *in vivo* degradation of poly(ε-caprolactone) with cross-linked poly(ε-caprolactone). The cross-linked amorphous polymer was enzymatically degradable on its surface most likely due to the high mobility of its polymer chains, whereas the crystalline homopolymer hydrolyzed only passively (Pitt *et al.*, 1984). Increasing the degree of cross-linking decreased the rate of enzymatic poly(ε-caprolactone) degradation which was also found for other polymers such as albumin (Shalaby *et al.*, 1990). Poly(isobutyl cyanoacrylate) nanoparticles have been reported to be mainly enzymatically degraded *in vitro* when rat liver microsomes were present (Lenaerts *et al.*, 1984). The combination of enzymatically degradable and non-degradable materials has been investigated by cross-linking *N*-(2-hydroxypropyl)methacrylamide copolymers with oligopeptide side chains which are stable in plasma but degrade in the presence of lysosomal enzymes (Rejmanová *et al.*, 1985), an approach that might be useful for drug targeting and gene therapy. Bacteria have also a substantial effect on polymer degradation because of their potential to degrade polymers enzymatically. This has been studied extensively especially for those polymers that are in use or are intended for use as packing materials or for agricultural applications (Lenz, 1993). The enzymatic degradation of polymers by microorganisms introduces even more factors into degradation and elimination processes. Factors such as the colonization of polymer surfaces (Nishida & Tokiwa, 1991), the microbial environment (Lenz, 1993) or even catalyst residues (DiBenedetto *et al.*, 1988) affect degradation.

POLYMER EROSION

Homogenous and Heterogeneous Erosion

The diffusion of water into the polymer bulk and polymer degradation compete against each other during polymer erosion. If degradation is fast, as shown in Figure 1A diffusing water is absorbed quickly by hydrolysis and hindered from penetrating deep into the polymer bulk. Erosion is confined to the surface of the polymer in this case, a phenomenon referred to as heterogeneous or surface erosion. This does not require, however, that the thickness of the surface layer in which erosion takes place is ultimately thin as it is often assumed. This erosion behavior changes if degradation is slower than water diffusion. In that case water cannot be absorbed quickly enough to be hindered from reaching deep layers of the polymer bulk and the polymer degrades all over its cross-section, a behavior which has been termed homogenous- or bulk erosion. In this case the complete polymer cross-section is subject to erosion as shown in Figure 1B. Surface and bulk erosion have some significance for the performance of polymers. An example is drug delivery. There are three potential mechanisms by which drug release from polymers can be controlled: Polymer swelling, polymer erosion, and drug diffusion out of the polymer. If polymer erosion is intended to be the release controlling mechanism, it has to be faster than the other two processes, otherwise, drug release might be controlled by swelling and diffusion too as shown in Figure 2. Therefore, only fast eroding polymers are suitable candidates to control drug release by erosion only. From the previous considerations it is obvious, that only those polymers are surface eroding, which contain fast hydrolyzing bonds such as poly(anhydrides) and

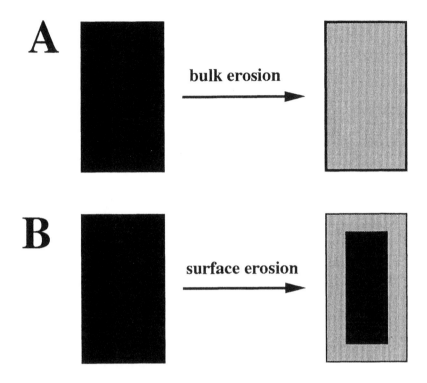

Figure 1 Schematic illustration of surface and bulk erosion. (▨ eroding polymer, ▮ non-eroding polymer).
(A) bulk erosion, (B) surface erosion.

poly(ortho esters) for example. It must be stressed, however, that surface and bulk erosion are two extremes and that the erosion mechanism of a degradable polymer shows usually characteristics of both. In addition to diffusion and degradation, other factors such as water uptake which depends on the hydrophilicity of the polymer affect the erosion behavior of polymers substantially.

Changes in Polymer Structure During Erosion

Morphological Changes

Early changes that are observed during erosion affect the polymer surface. Surface defects were found on biodegradable sutures made of poly(glycolic acid) and poly(D,L-lactic acid) after a couple of weeks of degradation *in vivo* (Rudermann *et al.*, 1973). An increasing surface roughness was detectable on poly(ortho ester) matrices using atomic force microscopy (Shakesheff *et al.*, 1994). More severe changes affect some poly(anhydrides) that tend to form cracks on their surface which has been investigated by scanning fluorescence confocal microscopy (Göpferich & Langer, 1993a). Consequently, these poly(anhydrides) become porous.

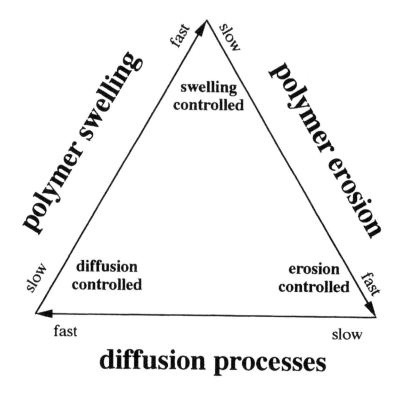

Figure 2 Control of drug release depending on the velocities of drug diffusion, polymer swelling and polymer erosion.

Mercury intrusion porosimetry was used to determine pore sizes. Poly(1,3–bis[p-carboxyphenoxy]propane-co-sebacic acid) 20:80, a partially crystalline polymer, was found to form two kinds of pores: macropores with a diameter of 100 μm which stem from surface cracks and micropores with a diameter of approximately 100 nm that stem from the preferential erosion of amorphous polymer areas (Göpferich, 1996a). Pore formation has also been reported for other types of degradable polymers such as poly(lactic-co-glycolic acid) (Shah *et al.*, 1992). Interesting are the cross-sections of degradable polymers during erosion which have been investigated by light microscopy as well as scanning electron microscopy (Mathiowitz *et al.*, 1993; Göpferich & Langer, 1993a). Fast degrading polymers such as poly(anhydrides) and poly(orthoesters) show surface erosion behavior which is visible from the appearance of so-called erosion zones or reaction zones (Heller, 1990). Such zones were found, for example, in poly(1,3-bis[p-carboxyphenoxy]propane-co-sebacic acid) (Mathiowitz *et al.*, 1989) or poly(fatty acid dimer-co-sebacic acid) (Shieh *et al.*, 1994). Surprisingly they are also found during the erosion of poly(lactic acid) and poly(lactic-co-glycolic acid) matrices (Göpferich, 1996a) which are originally bulk eroding polymers. In this case erosion fronts move inversely from the inside to the outside of the polymer matrix and are created by the autocatalytically accelerated degradation inside these poly(esters) (Li *et al.*, 1990a).

Changes in Composition and Changes of Physico-chemical Properties

At an early stage of the erosion process a drop in the glass transition temperature due to polymer swelling and relaxation (Göpferich, 1996a) and a drop of molecular weight due to the degradation process (Baker, 1987) can be observed. The weight loss lags behind the loss of molecular weight because a certain degree of degradation has to be reached before the degradation products become water soluble. For poly(lactic acid), oligomers containing only a few monomer units have been reported to be water soluble (Maniar *et al.*, 1991). Amorphous polymer regions are more vulnerable to erosion than crystalline ones, since the former are more accessible to water and, therefore, faster degradable (Li *et al.*, 1990b; Pistner *et al.*, 1993). This was verified for a number of degradable polymers such as poly(anhydrides) (Göpferich & Langer, 1993a), poly(β hydroxy acids) (Nishida & Tokiwa, 1993) and poly(α hydroxy acids) (Pistner *et al.*, 1993) for which an increase in crystallinity during erosion was observed. Changes in crystallinity can also root from the recrystallization of polymer. Intrinsically amorphous polymers have been reported to recrystallize during erosion (Li *et al.*, 1990b). Degradation products such as oligomers (Vert *et al.*, 1992) and monomers (Göpferich & Langer, 1993a) have been reported to crystallize inside eroding polymer matrices. Oligomers crystallizing from eroding poly(D,L-lactic acid) have been identified as a poly(D-lactic acid)/poly(L-lactic acid) stereocomplex (Ikada *et al.*, 1987). Sebacic acid and 1,3-bis[p-carboxyphenoxy]propane, two monomers used for the manufacture of poly(anhydrides), were reported to crystallize inside eroding poly(anhydrides) such as poly(sebacic acid) and poly(1,3-bis[p-carboxy phenoxy]propane-co-sebacic acid) (Göpferich & Langer, 1993a).

The Release of Monomers and Drugs from Degradable Polymers During Erosion

Polymer erosion is typically accompanied by a massive loss of monomers and oligomers. The release of monomers from fast degrading polymers such as poly-(anhydrides) or poly(ortho esters) starts from the very beginning of degradation (D'Emanuelle *et al.*, 1992; Nguyen *et al.*, 1986). Polymers containing less reactive functional groups such as poly(α-hydroxy esters) and poly(β-hydroxy esters) tend to show lag periods of a couple of days before the weight loss due to the release of monomers and oligomers sets in (Asano *et al.*, 1990; Shah *et al.*, 1992). The release of monomers from eroding polymer is pH dependent (Göpferich & Langer, 1996b). 1,3-bis[p-carboxyphenoxy]propane release from poly(1,3-bis[p-carboxyphenoxy]propane-co-sebacic acid), for example, has been reported to depend on the pH inside pores. The pH in pores is controlled by sebacic acid because its solubility is much greater than that of 1,3-bis[p-carboxyphenoxy]propane (Göpferich & Langer, 1993a). The release rate of the latter increases only after sebacic acid has diffused out of the eroding matrix which causes the pH in the pores to rise followed by an increase in the solubility of the poorly soluble organic acid. Drug and monomer release from degradable polymers also depends strongly on the structure of the eroding polymer matrix. Erosion zones and layers of monomers deposited on the surface of eroding poly(anhydrides), for example, act as diffusion barriers to released substances (Göpferich & Langer, 1996b; Shieh *et al.*, 1994). Therefore, diffusivity, solubility and dissolution rate of the drug become important

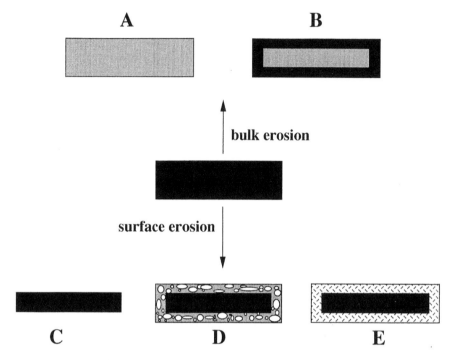

Figure 3 Potential appearance of polymer cross-sections during erosion. (▨▨▨ eroding polymer, ▬▬ non-eroding or slowly eroding polymer, ▨▨ monomer, ▨◌◯ eroding polymer containing pores):
(A) bulk eroding polymer,
(B) bulk eroding polymer with autocatalytically accelerated degradation,
(C) perfectly surface eroding polymer,
(D) surface eroding polymer with porous erosion zone,
(E) surface eroding polymer with monomer deposits on the surface.

for the release kinetics. The release of indometacin from p(CPP-SA), for example, is slower than the release of sebacic acid monomer due to dissolution and diffusion effects (Göpferich *et al.*, 1996b). Drug release from degradable polymers is not necessarily always controlled by erosion. For example, drug release from poly(β-hydroxy butyrate-co-β-hydroxy valerate) 73:27 devices has been reported to be completed by diffusion after 2 days, while the copolymer half-life is more than 100 weeks (Akhtar *et al.*, 1991). In general it can be concluded that the release of low molecular weight substances such as oligomers, monomers and incorporated drugs depends on a number of factors. Most important is the erosion mechanism of the polymer. Equally important are the physicochemical properties of the released drug molecules.

A Brief Summary on Potential Erosion Mechanisms

Surface and bulk erosion are ideal cases. The erosion of most polymers does not follow one of these two mechanisms unequivocally. This becomes obvious if

A 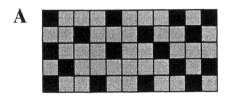 B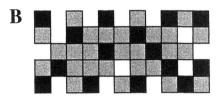

Figure 4 Representation of a poly(anhydride) cylinder cross-section (black pixels: drug-free crystalline polymer, gray pixels: drug loaded amorphous polymer, white pixels: pores). (A) prior to erosion, (B) during erosion.

one compares the cross-sections of eroding polymers. In Figure 3 the potential appearance of such sections during erosion is shown schematically. When water diffusion is fast compared to degradation, the polymers are bulk eroding as shown in Figure 3A. If the degradation products accelerate degradation autocatalytically inside a bulk eroding polymer, erosion is faster than on its surface as shown in Figure 3B. Poly(lactic-co-glycolic acid) serves here as an example (Göpferich, 1996a; Li *et al.*, 1990a). Poly(orthoesters) have been reported to be almost ideally surface eroding under certain conditions and their cross-sections appear as the one shown in Figure 3C (Heller, 1986). Polyanhydrides behave as described in Figure 3D and E. They tend to form either porous erosion zones such as poly(sebacic acid) and poly(1,3-bis[p-carboxyphenoxy]propane-co-sebacic acid) (Göpferich & Langer, 1993a; Göpferich, 1996a), or accumulate monomer on their surface such as poly(fatty acid dimer-co-sebacic acid) does (Shieh *et al.*, 1994).

POLYMER ELIMINATION

The degradation and erosion of a biodegradable polymer during an *in vivo* application might be very different from *in vitro* conditions. For poly(anhydrides) for example, degradation was reported to be slower *in vivo* than *in vitro*, while for other polymers the opposite was found. This makes clear that biodegradation and bioerosion are influenced by a number of additional parameters compared to degradation and erosion alone. When a degradable polymer is implanted a cascade of events occur (Anderson, 1994). The injury of tissue is followed by an inflammation reaction which attracts phagocytosing cells such as macrophages and monocytes to the location of the implant. Later the wound healing process sets in, which is characterized by the proliferation of endothelial cells and fibroblasts. The alternative to wound healing is the foreign body reaction, where macrophages cover the surface of the foreign material followed by the encapsulation with fibroblasts referred to as fibrosis. Using a cage implant system, it was possible to determine the composition of the extracellular fluid around the implant (Marchant *et al.*, 1983) from which a number of enzymes was identified. *In vivo* conditions at the implantation site are subject to a continuous change. This has been proven by implanting surgical suture material into rats for 10 weeks. When nylon suture was transferred weekly from one rat to another, the tensile strength was higher than when the material rested in the same animal during the experiment. The opposite was found for silk (Yui *et al.*, 1993). In addition to enzymatic degradation, particles of a few microns or less might by phagocytosed by cells of the reticulo endothelial system. In that case the polymer particles are subject to direct lysosomal digestion which exposes them to pH values of 5–5.5, and a number of lysosomal enzymes (Duncan, 1986). The lysosomal degradation is the major pathway of polymer elimination from the blood for polymers that cannot be excreted directly via the kidney. Some polymers have been synthesized in such a way, that the lysosomal degradation cleaves drugs from the polymer backbone (Rejmanová et al., 1985), an approach that might be useful for drug targeting. Long circulating polymers are lysosomally degraded on their way out of the body. This can be followed by radiolabeling the polymer with ^{14}C and measuring the distribution and excretion of radioactive matter. A number of interesting results on polymer biodegradation and elimination in vivo have been obtained in this way. ^{14}C-labeled poly(styrene), poly(ethylene) and poly(methyl methacrylate) were implanted into rats. Feces, urine and respiratory CO_2 were monitored for radioactivity for more than a year (Williams, 1982). Polystyrene degraded fastest with radioactivity detectable in the urine after 21 weeks, followed by poly(ethylene) after 26 weeks and poly(methyl methacrylate) after 54 weeks. No radioactivity was detected in the respiratory CO_2 or in the surrounding tissue after the removal of the polymer. The results prove that even polymers that are considered non-degradable can be degraded in vivo to some extent, and that their biodegradation products are readily excreted. The difference to degradable polymers becomes apparent by comparing these results with the biodegradation and elimination of poly(D,L-lactide-co-glycolide), for example, where no radioactivity was detectable at the injection site of microspheres after 56 days in rats (Visscher et al., 1985), indicating the completeness of the biodegradation process. After intravenous administration of poly(lactic acid) nanoparticles to rats, 30% of the administered amount of ^{14}C was eliminated during the first day but then slowed down substantially indicating that lactic acid might have been incorporated into endogenous compounds (Bazile et al., 1992). After oral administration of poly(D,L-lactic acid) nanospheres to rats 17.8% of the radioactivity were recovered in urine, pulmonary excretions and the blood, which indicates that nanospheres cross the intestinal barrier (Ropert et al., 1993).

APPROACHES TO POLYMER EROSION MODELING

The predictability of polymer erosion with theoretical models would be beneficial in several ways. For pharmaceutical applications such information could be used to assess the potential of a polymer to release drugs by an erosion-controlled mechanism and to predict drug release. In waste management it would be favorable to predict the polymer half-life in various environments.

Unfortunately, there have been only few attempts to model the erosion of degradable polymers. Most of them focused on the description of erosion for fast degrading polymers that tend to be surface eroding. In early approaches, linearly moving erosion fronts, separating eroding from non-eroded polymer have been assumed to move at constant speed. This allowed the effect of polymer matrix geometry on the erosion of surface eroding polymers to be investigated (Cooney, 1972; Hopfenberg, 1976). For slow degrading polymers such as poly(α-hydroxy esters), the degradation of polymer was assumed to follow first-order or second-order kinetics. Other models introduced diffusion theory that either described the diffusion of water into the polymer as a steady-state process and regarded degradation as a first-order process (Baker & Lonsdale, 1976; Heller & Baker, 1980) or described the degradation by erosion fronts that move at constant speed (Thombre & Himmelstein, 1984). At present there are two major approaches to erosion modeling. One uses sets of partial differential equations, the second uses simplified discrete models.

Approaches using differential equations are usually based on the second law of diffusion to describe the diffusion of water into the polymer (Joshi & Himmelstein, 1991). As water diffuses, however, not unhindered through degradable polymers because it is consumed by hydrolyis, the basic kinetics is a process of diffusion coupled with reaction (Cussler, 1989). Such models have successfully been applied to describe the degradation of poly(ortho esters) and drug release from these polymers (Thombre & Himmelstein, 1985). Equation 2 is the basic partial differential equation that describes the diffusion controlled mass transfer in one dimension coupled with a chemical reaction.

$$\frac{\partial c_i}{\partial t} = \frac{\partial}{\partial x} D_i \frac{\partial c_i}{\partial t} + r_i \tag{1}$$

c designates the concentration of species i (e.g. water, degradation products etc.) and the reaction terms r_i denotes the velocity at which the ith species is generated. The advantage of these models is their ability to account for the diffusion of water, degradation products and drugs as well as for polymer degradation. They fail, however, to describe changes in microstructure caused by the preferential erosion of amorphous compared to crystalline polymer areas for example which in some instances affects erosion substantially.

To describe changes in polymer microstructure, a spatial aspect has to be introduced into modeling approaches for which two-dimensional discrete models are suitable. Originally, these models have been developed to describe drug release from compressed cylindrical matrices consisting of a drug, a polymer and a filler (Zygourakis, 1990). The release of drug was assumed to be initiated upon contact with water. Using Equation 2, it was possible to simulate effects such as drug loading and filler content on drug release:

$$\frac{dV_d}{dt} = v_d \cdot S_d \tag{2}$$

Vd is the volume of dispersed drug, v_d a dissolution rate constant and S_d the drug/solvent interfacial area. A different two-dimensional approach was used later to describe the erosion of poly(anhydrides) (Göpferich & Langer, 1993b). First polymer cross-sections were covered with a two-dimensional computational grid as the one shown in Figure 4A. Depending on polymer crystallinity, each pixel on the grid was assigned the quality "crystalline" or "amorphous". As the hydrolytic degradation of polymer bonds usually follows pseudo first-order kinetics (Vert et al., 1992), degradation can be regarded a random process with Poisson kinetics. This was taken into account by randomly choosing the life-time of polymer pixels after water contact from a zero order Erlang distribution via Monte Carlo techniques. Equation 3 illustrates how the life-time is calculated:

$$t = \frac{1}{\lambda \cdot n} \cdot \ln(1 - \varepsilon) \tag{3}$$

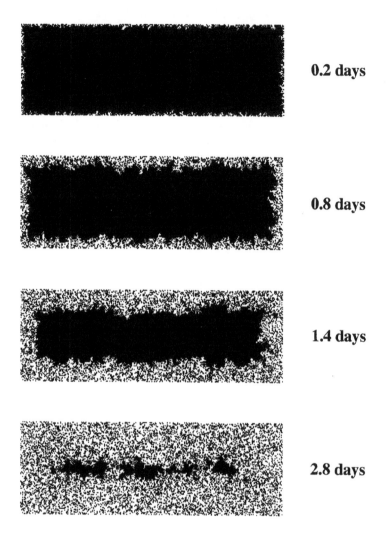

Figure 5 Erosion time series for a cylindrical polyanhydride matrix disc, complete cross-section (black pixels = non-eroded polymer, white pixels = eroded polymer). (Reproduced with permission of the American Institute of Chemical Engineers from (Göpferich & Langer, 1995b). Copyright © 1995 AIChE. All rights reserved).

t is the life-time of a pixel after contact to the degradation medium, λ a rate constant, n the size of the computational grid and ε a random number equally distributed between 0 and 1. The crystallinity of polymers is accounted for by using distributions that yield on the average longer lifetimes for "crystalline" pixels compared to "amorphous" ones. This is achieved by using different λ values for the two species of pixels. To simulate how erosion proceeds, pixels are removed from the grid in the order of their lifetimes. The algorithm proceeds until all pixels have been removed from the grid. During such simulations porous structures are obtained as the one shown in Figure 4B. With these models, the erosion of polyanhydrides, for example, can be predicted precisely (Göpferich & Langer, 1993b). An example for an erosion simulation is shown in Figure 5. The cross-sections obtained by these simulations agree very well with experimental findings by scanning electron microscopy (Göpferich & Langer, 1993a). This modeling approach has the advantage that structural changes such as porosity, which is decisive for modeling the release of drugs from porous systems, can be predicted. Unfortunately, these discrete models cannot describe the release of drugs from eroding polymers except when these substances have a high solubility inside pores and diffuse very fast so that mass transport phenomena are not rate limiting (Göpferich, 1996b). To overcome this limitation, they have to be combined with diffusion models which allow the diffusion of low-molecular-weight substances such as monomers from eroding polymers to be simulated, an approach that has successfully been applied to describe the release of monomers from eroding poly(anhydrides) (Göpferich & Langer, 1995a). Equation 4 describes the diffusion of monomer through the pores of eroded polymer:

$$\frac{\partial}{\partial t} c \cdot \varepsilon = \frac{\partial}{\partial x} D \cdot \varepsilon \cdot \frac{\partial c}{\partial x} + \frac{\partial S}{\partial t} \tag{4}$$

c is the concentration of diffusing monomer, D the effective diffusion coefficient, and $\partial S / \partial t$ a rection term that accounts for the precipitation and dissolution of monomer. The porosity ε was calculated from Monte Carlo simulations.

In general it can be concluded, that erosion modeling has, so far, been only successful for some degradable polymers. More research in this area will be necessary to develop models that predict the behavior of polymers based on their chemical structure, or models that are at least applicable to a variety of different types of polymers.

SUMMARY

Polymers degrade by four potential mechanisms: photo-, mechanical-, thermal- and chemical-degradation. Degradation is characterized by a loss of molecular weight and initiates polymer erosion, which proceeds with the loss of newly-formed monomers and oligomers. A degradable polymer has degradation times that are under ambient conditions smaller than our human life time. For degradable polymers, chemical degradation in the form of hydrolysis is the most important mode of degradation. These polymers differ by their hydrolyzable functional groups which cause major differences in degradation velocity. Besides the functional groups, the chemical environment in which the process takes place affect degradation. The pH of the degradation medium, for example, can change degradation rates by orders of magnitude. Erosion, which comprises all processes leading to the mass loss of a degradable polymer, changes the microstructure of polymers tremendously. These changes include processes such as swelling, surface cracking and increases in porosity to mention just a few. During biodegradation and bioerosion a biological system affects the processes of degradation and erosion. After implantation into the human body, for example, cells of the immune system influence the erosion process. The same might occur when polymers are eroded in the presence of microbes, which is important when disposing polymers in landfills. The multitude of parameters makes the theoretical modeling of erosion complicated. Some progress has however been made to describe the erosion of selected classes of polymers using partial differential equations or Monte Carlo models. Modeling is substantially complicated by the fact that degradable polymers show an individual erosion behavior, which can be substantially different even for members of the same polymer group. In summary, one can conclude that tremendous progress has been made during recent years towards a better understanding of polymer erosion. More research is, however, necessary for a complete understanding of erosion and the changes in polymer properties during erosion.

ACKNOWLEDGMENTS

Part of the presented work was sponsored by the Deutsche Forschungsgemeinschaft with research grant Go 565/3-1 and by NATO with collaborative research grant CRG 940721.

REFERENCES

Agrawal, C.M., Haas, K.F., Leopold, D.A. and Clark, H.G. (1992) Evaluation of poly(L-lactic acid) as material for intravascular polymeric stents, *Biomaterials*, **13**, 176–182.

Akhtar, S., Pouton, C.W. and Notarianni, L.J. (1991) The influence of crystalline morphology and copolymer composition on drug release from solution cast and melt-processed p(HB-HV) copolymer matrices, *J. Controlled Release*, **17**, 225–234.

Albertsson, A.-C. (1980) The shape of the biodegradation curve for low and high density polyethenes in prolonged series of experiments, *Eur. Pol. J.*, **16**, 623–630

Anderson, J.M. (1994) *In vivo* biocompatibility of implantable drug delivery systems. *Eur. J. Pharm. Biopharm.*, **40**,1–8.

Asano, M., Fukuzaki, H., Yoshida, M., Kumakura, M., Mashimo, T., Yuasa, H., Imai, K. and Yamanaka, H. (1990) Application of poly D,L-lactic acids of varying molecular weight in drug delivery applications, *Drug Design and Delivery*, **5**, 301–320.

Baker, R.W. and Lonsdale, H.K. (1976) Erodible controlled release systems, *Am. Chem. Soc. Div. Org. Coat. Plast. Chem. Prepr.*, **3**, 229.

Baker, R. (1987) *Controlled Release of Biologically Active Agents.* John Wiley & Sons, New York, pp. 84–131.

Banford, C.H. and Tipper, C.F.H. (1972) *Comprehensive Chemical Kinetics*, Volume 14: Degradation of Polymers, Elsevier, New York.

Bazile, D.V., Ropert, C., Huve, P., Verecchia, T., Marlard, M., Frydman, A., Veilard, M. and Spenlehauer, G. (1992) Body distribution of fully biodegradable [^{14}C]-poly(lactic acid) nanoparticles coated with albumin after parenteral administration into rats. *Biomaterials*, **13**, 1093–1102.

Beetham, R., Dawnay, A. and Cattell, W. (1987) The effect of a synthetic polypeptide on the renal handling of protein in man. *Clinical Science*, **72**, 245–249.

Brem, H., Walter, K.A. and Langer, R. (1993) Polymers as controlled drug delivery devices for the treatment of malignant brain tumors. *Eur. J. Pharm. Biopharm.*, **39**, 2–7.

Cooney, D.O. (1972) Effect of geometry on the dissolution of pharmaceutical tablets and other solids: surface detachment kinetics controlling. *AIChE Journal*, **18**, 446–449.

Crank, J. and Park, G.S. (1968) *Diffusion in Polymers*, Academic Press, London.

Cussler, E. (1989) *Diffusion.* Cambridge University Press, Cambridge.

D'Emanuelle, A., Hill J., Tamada, J.A., Domb, A.J. and Langer, R. (1992) Molecular weight changes in polymer erosion. *Pharm. Res.*, **9**, 1279–1283.

DiBenedetto, L., Cameron, J.A. and Huang, S.J. (1988) Biodegradation of hydroxylated polymers. *Pol. Sci. Technol.*, **38**, 61–74.

Doi, Y., Kanesawa, Y. and Tanahashi, N. (1992) Biodegradation of microbial polyesters in the marine environment. *Pol. Degrad. Stab.*, **36**, 173–177.

Duncan, R. (1986) Lysosomal degradation of polymers used as drug carriers, in: *CRC Critical reviews an Biocompatibility, Vol. 2.2,* CRC Press, Boca Raton, 127–145.

Gelovoy, A., Cheung, M.F. and Zimbo, M. (1989) Molecular weight degradation of polyacrylate during processing of polyacrylate/poly(phenylene sulfide) blends, *Polym. Comm.,* **30,** 322–324.

Gilding, D.K. and Reed, A.M. (1979) Biodegradable polymers for use in surgery — polyglycolic/poly(lactic acid) homo- and copolymers: 1. *Polymer,* **20,** 1459–1464.

Göpferich, A. and Langer, R. (1993a) The influence of microstructure and monomer properties on the erosion mechanism of a class of polyanhydrides. *J. Polym. Sci.,* **31,** 2445–2458.

Göpferich, A. and Langer, R. (1993b) Modeling of polymer erosion. *Macromolecules,* **26,** 4105–4112.

Göpferich, A. and Langer, R. (1995a) Modeling monomer release from bioerodible polymers. *J. Controlled Release,* **33,** 55–69.

Göpferich, A. and Langer, R. (1995b). Modeling of Polymer Erosion in Three Dimensions – Rotationally Symmetric Devices. *AIChE Journal,* **41,** 2292–2299.

Göpferich, A. (1996a) Mechanisms of Polymer Degradation and Erosion, *Biomaterials,* **17,** 103–114.

Göpferich, A. (1996b) Polymer Degradation and Erosion: Mechanisms and Applications. *Eur. J. Pharm. Biopharm.,* **42,** 1–11.

Heller, J. and Baker, R.W. (1980) Theory and practice of controlled drug delivery from bioerodible polymers. *In Controlled release of bioactive materials,* Baker, R.W. (Ed.), Academic Press, New York.

Heller, J. (1985) Controlled drug release from poly(ortho esters). A surface eroding polymer. *J. Controlled Release,* **2,** 167–177.

Heller J. (1986) Control of surface erosion by the use of excipients. *Pol. Sci. Tech.,* **34,** 357–368.

Heller, J., Penhale, D.W., Fritzinger, B.K. and Ng, S.Y. (1987) The effect of copolymerized 9,10-dihydroxystearic acid on erosion rates of poly(ortho esters) and its use in the delivery of levonorgestrel. *J. Controlled Release,* **5,** 173–177.

Heller, J. (1990) Development of poly(ortho esters): a historical overview. *Biomaterials,* **11,** 659–665.

Hiemenz, P.C. (1984) *Polymer Chemistry,* Marcel Dekker, New York.

Hopfenberg, H.B. (1976) Controlled Release from Bioerodible Slabs, Cylinders and Spheres. In: *Controlled Release Polymeric Formulations,* ACS Symp. Ser. No. 33, Paul D. R., Harris F.W., Edts., American Chemical Society, Washington D.C., 26–32.

Ikada, Y., Jamshidi, K., Tsuji, H. and Hyon, S.-H. (1987) Stereocomplex formation between enantiomeric poly(lactides). *Macromolecules,* **20,** 904–806.

Joshi, A. and Himmelstein, K. (1991) Dynamics of controlled release from bioerodible polymers, *J. Controlled Release,* **15,** 95–104.

Kirby, A.J. (1972) Hydrolysis and formation of esters of organic acids, in: *Comprehensive Chemical Kinetics. Volume 10: Ester Formation and Hydrolysis and Related Reactions,* Banford C.H. and Tipper C.F.H. (Eds.), Elsevier, Amsterdam, 57–202.

Kopecec, J. (1984) Controlled biodegradability of polymers – a key to drug delivery systems. *Biomaterials,* **5,** 19–25.

Langer, R. (1990) New Methods of Drug Delivery, *Science,* **249,** 1527–1532.

Langer, R. and Vacanti, J.P. (1993) Tissue Engineering. *Science,* **260,** 920–926.

Leenslang, J. W., Pennings, A. J., Ruud, R.M., Rozema, F.R. and Boering, G. (1987) Resorbable materials of poly(L-lactide). VI. Plates and screws for internal fracture fixation. *Biomaterials,* **8,** 70–73.

Lenaerts, V., Couvreur, P., Christiaens-Leyh, D., Joiris, E., Roland, M., Rollmann, B. and Speiser, P. (1984) Degradation of poly(isobutyl cyanoacrylate) nanoparticles. *Biomaterials,* **5,** 65–69.

Lenz, R. (1993) Biodegaradable polymers. In: *Advances in Polymer Science, No. 107, Biopolymers I.,* Peppas, N., Langer, R. (Eds.), Springer Verlag, Berlin, pp. 1–40.

Leong, K.W., Brott, B.C. and Langer, R. (1985) Bioerodible polyanhydrides as drug-carrier matrices. I: characterization, degradation and release characteristics, *J. Biomed. Mat. Res.,* **19,** 941–955.

Li, S.M., Garreau, H. and Vert, M. (1990a) Structure-property relationships in the case of the degradation of massive poly(α-hydroxy acids) in aqueous media part 2: degradation of lactide-glycolide copolymers: PLA37.5 GA25 and PLA75GA25, *J. Mater. Sci. Mat. Med.,* **1,** 131–139.

Li, S.M., Garreau, H. and Vert, M. (1990b) Structure-property relationships in the case of the degradation of massive poly(α-hydroxy acids) in aqueous media, Part 3: influence of the morphology of poly(L-lactic acid). *J. Mater. Sci. Mat. Med.,* **1,** 198–206.

Maniar, M.L., Kalonia, D.S. and Simonelli, A.P. (1991) Determination of specific rate constants of specific oligomers during polyester hydrolysis. *J. Pharm. Sci.,* **80,** 778–782.

Marchant, R., Hiltner, A., Hamlin, C., Rabinovitch, A., Slobodkin, R. and Anderson, J.M. (1983) *In vivo* biocompatibility studies I. The cage implant system and a biodegradable hydrogel. *J. Biomed. Mat. Res.*, **17**, 301–325.

Mathiowitz, E., Jacob, J., Pekarek, K. and Chickering III, D. (1993) Morphological characerization of bioerodible polymers. 3. characterization of the erosion and intact zones in polyanhydrides using scanning electron microscopy. *Macromolecules*, **26**, 6756–6765.

Mathiowitz, E., Ron, E., Mathiowitz, G., Amato, C. and Langer, R. (1989) Surface morphology of bioerodible polyanhydrides. Polymer Prepr., **30**, 460–461.

Mauduit, J., Bukh, N. and Vert, M. (1993) Gentamycin/ poly(lactic acid) blends aimed at sustained release local antibiotic therapy administered per-operatively. I. The release of gentamycin base and gentamycin sulfate in poly(D,L-lactic acid) oligomers. *J. Controlled Release*, **23**, 209–220.

Metzner, A.B., White, J.L. and Denn, M.M. (1967) Constitutive equations for viscoelastic fluids for short deformation periods and for rapidly changing flows: significance of the deborah number, *AIChE Journal*, **12**, 863–866.

Miller, N.D. and Williams, D.F. (1984) The *in vivo* and *in vitro* degradation of PGA suture material as a function of applied strain. *Biomaterials*, **5**, 365–368.

Nguyen, T.H., Himmelstein, K.J. and Higuchi, T. (1986) Erosion of poly(ortho ester) matrices in buffered aqueous solutions, *J. Controlled Release*, **4**, 9–16.

Nishida, H. and Tokiwa, Y. (1991) Effects of higher-order structure of poly(3-hydroxybutyrate) on its biodegradation. I. Effects of heat treatment on microbial degradation. *J. Appl. Pol. Sci.*, **46**, 1467–1476.

Nishida, H. and Tokiwa, Y. (1993) Effects of higher-order structure of poly(3-hydroxybutyrate) on its biodegradation. II. effects of crystal structure on microbial degradation. *J. Environmental Polymer Degradation*, **1**, 65–80.

Park, K., Shalaby, W.S.W. and Park, H. (1993) *Biodegradable Hydrogels for Drug Delivery*. Technomic Publ., Lancaster.

Pistner, H., Bendix, D.R., Mühlig, J. and Reuther, J.F. (1993) Poly(L-lactide): a long-term degradation study *in vivo*. Part III. Analytical Characterization, *Biomaterials*, **14**, 291–298.

Pitt, C.G., Chasalow, F.I., Hibionada, Y.M. and Klimas, D.M. (1981) Schindler, A., Aliphatic polyesters. I. The degradation of poly(ε-caprolactone) *In Vivo*, *J. Appl. Pol. Sci.*, **26**, 3779–3787.

Pitt, C.G., Hendren R.W. and Schindler A. (1984) The enzymatic surface erosion of aliphatic polyesters. *J. Controlled Release*, **1**, 3–14.

Pitt, C.G., Cha, Y., Shah, S.S. and Zhu, K.J. (1992) Blends of PVA and PGLA: control of the permeability and degradability of hydrogels by blending. *J. Controlled Rel.* **19**, 189–200.

Pulapura, S., Li, C. and Kohn, J. (1990) Structure-property relationships for the design of polyiminocarbonates. *Biomaterials*, **11**, 666–678.

Reiner, M. (1964) The Deborah Number, *Physics Today*, **1**, 62.

Rejmanová, P., Kopecek, J., Duncan, R. and Lloyd, J.B. (1985) Stability in rat plasma and serum of lysosomally degradable oligopeptide sequences in N-(2hydroxypropyl)methacrylamide copolymers. *Biomaterials*, **6**, 45–48.

Ropert, C., Bazile, D., Bredenbach, J., Marlard, M., Veillard, M. and Spenlehauer, G. (1993) Fate of ^{14}C radiolabelled poly(D,L-lactic acid) nanoparticles following oral administration to rats, *Colloids and Surfaces: B: Biointerfaces*, **1**, 233–239.

Ruderman, R.J., Bernstein, E., Kairinen, E. and Hegyeli, A.F. (1973) Scanning electron microscopic study of surface changes on biodegradable Sutures. *J. Biomed. Mat. Res.*, **7**, 215–229.

Seppälä, J., Linko, Y.-Y. and Su, T. (1991) Photo- and biodegradation of high volume thermoplastics, *Acta Polytech. Scand.*, **198**, 10–12.

Shah, S.S., Cha, Y. and Pitt, C.G. (1992) Poly(glycolic acid-co-D,L-lactic acid): diffusion or degradation controlled drug delivery?, *J. Controlled Release*, **18**, 261–270.

Shakesheff, K.M., Davies, M.C., Domb, A., Glasbey, T.O., Jackson, D.E., Heller, J., Roberts, C.J., Shard, A.G., Tendler, S.J.B. and Williams, P.M. (1994) Visualizing the degradation of polymer surfaces with an atomic force microscope. *Proc. Int. Symp. Contr. Rel. Bioact. Mater.*, **21**, 618–619.

Shalaby, S.W.S., Blevins, W.E. and Park, K. (1990) Enzyme-induced degradation behavior of albumin-crosslinked hydrogels. *Pol. Prepr.*, **31**, 169–170.

Shieh, L., Tamada, J., Chen, I., Pang, J., Domb, A. and Langer R. (1994) Erosion of a new family of biodegradable polyanhydrides, *J. Biomed. Mater. Res.*, **28**, 1465–75.

Shih, C., Higuchi, T. and Himmelstein, K.J. (1984) Drug delivery from catalysed erodible polymeric matrices of poly(ortho ester)s. *Biomaterials*, **5**, 237–240.

Singh, M., Singh, A. and Talwar, G.P. (1991) Controlled delivery of diphtheria toxoid using biodegradable poly(D,L-Lactide) microcapsules. *Pharm. Res.*, **8**, 958–961.

Thombre, A.G. and Himmelstein, K.J. (1984) Modelling of drug release kinetics from laminated device having an erodible drug reservoir, *Biomaterials*, **5**, 250–254.

Thombre, A.G. and Himmelstein, K.J. (1985) A simultaneous transport-reaction model for controlled drug delivery from catalyzed bioerodible polymer matrices, *AIChE Journal*, **31**, 759–766.

Vert, M., Li, S. and Garreau, H. (1991) More about the degradation of LA/GA-derived matrices in aqueous media. *J. Controlled Release*, **16**, 15–26.

Vert, M., Feijen, J., Albertson, A., Scott, G. and Chiellini, E. (1992) *Degradable Polymers and Plastics*, Redwood Press Ltd., England.

Visscher, G.E., Robinson, R.L., Maulding, H.V., Fong, J.W., Pearson, J.E. and Argentieri, G.J. (1985) Biodegradation of and tissue reaction to 50:50 poly(DL-lactide-co-glycolide) microcapsules, *J. Biomed. Mat. Res.*, **19**, 349–366.

Williams, D.F. (1982) Biodegradation of surgical polymers. *J. Mat. Sci.*, **17**, 1233–1246.

Yoshioka, S., Kishida, A., Izumikawa, S., Aso, Y. and Takeda, Y. (1991) Base-induced polymer hydrolysis in poly(β-hydroxybutyrate/β-hydroxyvalerate) matrices. *J. Controlled Release*, **16**, 341–348.

Yui, N., Nihira, J., Okano, T. and Sakurai, Y. (1993) Regulated release of drug microspheres from inflammation responsive degradable matrices of crosslinked hyaluronic acid. *J. Controlled Release*, **25**, 133–143.

Zygourakis, K. (1990) Development and temporal evolution of erosion fronts in bioerodible controlled release devices. *Chem. Eng. Sci.*, **45**, 2359–2366.

23. NON-MEDICAL BIODEGRADABLE POLYMERS
Environmentally Degradable Polymers

GRAHAM SWIFT

Rohm and Haas Company, Norristown Road, Spring House, PA 19477, USA

INTRODUCTION

To relate this chapter to the rest of the book, environmentally degradable polymers designed as a waste-management option for commodity and specialty polymers, and *in vivo* degradable polymers designed for medical applications have many common elements and some very key differences. In both cases, polymers are exposed in living environments in which degradation may be promoted either by biocatalytic processes involving enzymes or by chemical processes such an hydrolysis, oxidation, and irradiation. Adverse environmental responses in both cases must be negligible for the polymers to be acceptable; they should be degradable in some controlled and predictable manner, such that their life time may be estimated; and no harmful or toxic persistent residues should be produced. Because of these similarities, it should not be surprising that there are common synthesis chemistries that may be applied in the two fields. However, the value placed on polymers for applications in the two fields, medicine, and commodity and specialty industrial polymers, is vastly different. Cost/performance is a much more sensitive driver for commodity and specialty polymers, where competition from existing polymers such as polyethylene, polypropylene, polystyrene, polyacrylates, poly(vinyl alcohol), poly(alkylene oxides), poly(vinyl chloride), poly(ethylene terephthalate), etc. demands excellent performance at low cost. This imposes a very severe synthesis limitation on the design and development of environmentally degradable polymers relative to those for use in the medical field. Nevertheless, basic synthesis concepts are similar in the two fields and one should be familiar with both to be successful in either one.

Interest in environmentally degradable polymers began more than thirty years ago, when it was first recognized that the commonly used commodity packaging plastics such as polyolefins, poly(vinyl chloride, polystyrene, and poly(ethylene terephthalate) were accumulating in the environments in which they were discarded, after use. Since these polymers were developed for their resistance properties, it should not have been surprising that they were recalcitrant in landfills and as litter when disposed of in a negligent manner.

More recently, the problem of polymer accumulation in the environment has been recognized as more general than packaging plastics, extending to recalcitrant water-soluble and other specialty polymers and plastics such as poly(acrylic acid), poly(vinyl alcohol), polyacrylamide, poly(alkylene oxides), and even some modified natural polymers, for example cellulosics. These polymers are widely used in coatings, pigment dispersants, temporary coatings, mining, detergents, water treatment, etc.; and all are potential contributors to environmental problems and are, therefore, targets for replacement with environmentally degradable substitutes.

To avoid confusion in the reader's mind, the term polymer will be used in a general sense throughout this chapter to include water-soluble polymers and plastics, the two major polymer types under discussion. Where differentiation is necessary, reference will be to the individual categories.

This review will attempt to cover the major synthetic approaches to polymers that are designed to degrade in the environment by any of the accepted degradation pathways, photodegradation, biodegradation, and chemical degradation (which is hydrolytic or oxidative degradation). Of these degradation pathways, biodegradation is recognized as the only one capable of completely transforming a polymer into naturally occurring products through assimilation by microorganisms into biomass and residual inorganic compounds such as salts and carbon dioxide; therefore, it has received an appropriately higher degree of attention than the others. The other approaches are more realistically described as environmental deterioration or disintegration, since their degradation products are left in the environment, unless they are biodegradable.

With the broadened perspective of research on environmentally degradable polymers to include water-soluble polymers, there has also been an acceptance that such polymers are only one of several waste-management options for polymers in the environment. Alternative solutions exist and are recognized such as recycle of polymers, recycle of polymers to monomers and alternate feedstocks, incineration, and continued landfilling. Each of the disposal method has advantages and disadvantages and acceptance will depend on many factors including available processing facilities, land space, local customs and expectations, ease of collection, cost relative to virgin polymers, and physical and chemical property requirements for the polymer. Consequently, research on environmentally degradable polymers is becoming more focused targeting applications where they offer unique advantages in disposal and properties over their competitive alternatives. Examples, as we shall see, include agricultural films where photo/biodegradable plastics are competitive with the cost of current films plus the associated collection and recycle or disposal costs; compostable plastics for fast-food packaging which eliminates separation from food waste; water-soluble polymers and plastics that are difficult to recover for other disposal options, after use; and personal hygiene products.

For wide-spread acceptance, environmentally degradable polymers must overcome some significant obstacles not recognized earlier. Environmental safety assessments must be completed to ensure that any claimed environmentally degradable polymer leaves no potentially harmful product in the environment. Hence developments in this area require multidiscipline teams involving polymer scientists, biologists, microbiologists, environmental chemists, analytical chemists, etc. to ensure a balanced polymer is developed that not only has desirable properties but also is environmentally acceptable. In addition to the science, there are, also, emotional issues that must be reconciled with the general public, legislators, and lawyers before wide spread acceptance of environmentally degradable polymers. In order to reach this level of acceptance, it will be necessary to achieve some level of international consensus on definitions and laboratory testing protocols that correlate with real world exposure for these polymers. The environment can not be compromised, on that we all agree. Consequently, this chapter will include sections on the current definitions of the common environmental degradation pathways;

the inter-relationship of these pathways to one another and to environmental safety which dictates a severe restriction on the acceptability of environmentally degradable polymers; opportunities for environmentally degradable polymers; test protocols; synthetic approaches; currently available polymers; and, finally, some projections for the future.

DEFINITIONS FOR ENVIRONMENTALLY DEGRADABLE POLYMERS

There have been and continue to be numerous attempts to define environmentally degradable plastics and polymers in a manner acceptable by everyone. All encompass the same broad general concepts but are slightly different in phraseology, depending on the author's perspective and discipline, chemist, biochemist, layman, lawyer, legislator, etc. This indicates a broad understanding of the problem but it also implies that we will never reduce to words an acceptable definition which has world-wide consensus. Nevertheless, definitions are important because they are indicative of expectations for environmentally degradable polymers and of the types of testing protocols that are needed to establish the acceptability of the polymers designed for environmental degradation. At this time, the definitions developed by the American Society for Testing and Materials (ASTM D 883–93) for degradable, biodegradable, hydrolytically degradable, and oxidatively degradable plastics indicated below are probably the most widely accepted as written or in some slightly modified form. They are equally applicable to polymers in general simply by substitution of the term polymer for plastic.

Degradable plastic, a plastic designed to undergo a significant change in its chemical structure under specific environmental conditions resulting in a loss of some properties that may vary as measured by standard test methods appropriate to the plastic and the application in a period of time that determines its classification.

Biodegradable plastic, a degradable plastic in which the degradation results from the action of naturally-occurring micro-organisms such as bacteria, fungi, and algae.

Hydrolytically degradable plastic, a degradable plastic in which the degradation results from hydrolysis.

Oxidatively degradable plastic, a degradable plastic in which the degradation results from oxidation.

Photodegradable plastic, a degradable plastic in which the degradation results from the action of natural daylight.

It is apparent that the definitions do not quantify the extent of degradation, they are only indicative of the mechanism that is operating to promote degradation. While this is acceptable in a scientific sense to define the chemical process, they do not define environmentally acceptable polymers (Swift, 1992) which in the minds of legislators and lay people is the key issue. For environmentally degradable polymers and plastics to be acceptable as a waste-management option, definitions have to be more practical to reflect the goal of acceptability as synonymous with the assurance of no harmful residues after degradation. The above ASTM definitions, therefore, require elaboration as recommended by Swift (1993) to address this deficiency. Environmental degradation processes are interrelated as shown schematically below in Figure 1.

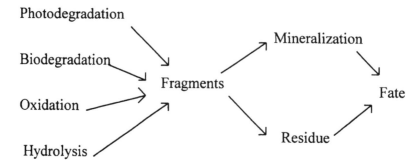

Figure 1 Inter-relationships for Environmentally Degradable Polymers.

Biodegradation, oxidation, hydrolysis and photodegradation initially all give intermediate products or fragments which may biodegrade further to some other residue;
biodegrade completely and be removed from the environment entirely (being
converted into biomass and carbon dioxide or methane, depending on aerobic or
anaerobic system), and ultimately mineralized as indicated in Figure 1; or remain
unchanged in the environment. In degradations where residues remain, these
must be established as harmless in the environment by suitably rigorous fate and
effect evaluations. Clearly, only biodegradation has the potential to remove plastic
and polymers completely from the environment. This should be recognized when
developing and designing polymers and plastics for degradation in the environment
by any of the pathways, the final stage should preferably be complete biodegradation
and removal from the environment. This ensures environmental acceptance and the
polymers may be considered to be recycled through nature into microbial cells,
plants and higher animals (Narayan, 1990), and potentially back into chemical
feedstocks. Therefore, as proposed by Swift (1993), **an acceptable environmentally
degradable plastic or polymer, may be defined as one which degrades by any of
the above defined mechanisms, biodegradation, photodegradation, oxidation, or
hydrolysis to leave no harmful residues in the environment**. This definition does
not limit the degree of degradation for a particular polymer but does require
sufficient testing of fragments and degradation products that are incompletely
removed from the environment to ensure no long-term damage or adverse effects
to the ecological system. Polymers and plastics meeting this definition should be
completely acceptable for disposal in the appropriate environment anywhere in the
world.

POTENTIAL USES FOR ENVIRONMENTALLY DEGRADABLE POLYMERS

Polymer applications are the drivers for research into environmentally degradable
plastics and polymers. They indicate property requirements, disposal methods and
identify the testing protocols that must be established to evaluate their environmental degradation under laboratory simulated environmental exposure conditions.
Drivers for environmentally degradable polymers and plastics are waste-management

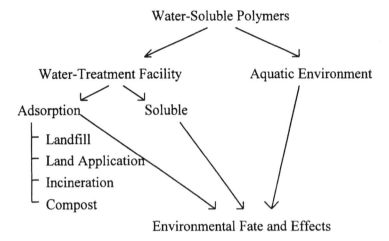

Figure 2 Environmental Disposal of Water-Soluble Polymers

programs and the desire to restrict the uses for currently available non-degradable polymers and plastics in the environment, wherever possible. Figures 2 and 3 indicate likely disposal pathways of water-soluble polymers and plastics, respectively.

Water-soluble polymers are usually disposed of as dilute aqueous solutions through waste-water treatment facilities or sometimes directly into the aquatic environment. On entering a waste-water treatment plant, a polymer may pass straight through into streams, rivers, lakes and other aquatic environments or it may be adsorbed onto the suspended solids. If the polymer passes straight through the wastewater treatment plant, it is no different from direct disposal into aquatic environments, both raise similar questions as to their fate and effects. Adsorption of a polymer onto sewage sludge results in the possibility of the polymer being landfilled, incinerated, composted, or land applied as fertilizer or for soil amendment, depending on the local options. In all cases, environmental fate and effects need to be addressed; it must be established how these polymers move in their new environmental compartments, and what are their incineration byproducts, if that is the method of disposal. Therefore, there is an obvious and distinct advantage for water-soluble polymers to be environmentally degradable. There is a preference for biodegradable water-soluble polymers as it is unlikely that the other degradation paths would be applicable in dilute aqueous solutions. Water-soluble polymers should be designed to be completely biodegradable and removed in the disposal environment, generally the sewage treatment facility, since they may move rapidly and without difficulty throughout the aquatic environment. Hence, if biodegradation of a polymer is not complete in the disposal environment, it must be assessed in subsequent environmental compartments that it may enter. Once the complete biodegradation of a polymer has been confirmed no uncertainty remains as to its fate and effects in any subsequent environmental compartments. Complete biodegradation can be established with a high degree of certainty with the appropriate test methods, whereas the assessment of environmental fate and effect of any residue is always a risk assessment based on a limited study with a few aquatic species.

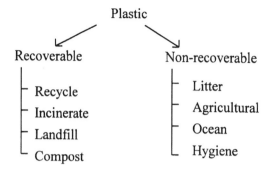

Figure 3 Environmental Disposal of Plastics

Plastics have more disposal options than water-soluble polymers because they are usually solid, handleable materials and are generally recoverable after use. Disposal options include landfilling, recycling, incineration, and composting. Composting, which is predominantly biodegradation with the possibility of oxidation and hydrolysis, is an opportunity for environmentally degradable plastics which are used in food applications such as wrappers and utensils. In these uses plastics are contaminated with food residues and the mix is ideally suitable for composting without separation. There are opportunities for environmentally degradable plastics where recovery of current plastics is not economically feasible, viable, or controllable such as litter; discarding at sea from naval vessels; farm and agricultural applications such as pre-emergence plant protection with sheets, and mulch; and hygienic applications such as diapers, sanitary napkins and hospital garments and swabs, etc. These opportunities are the drivers for the development of environmentally acceptable degradable plastics and polymers.

In summary, it is apparent that there are selective opportunities for environmentally degradable polymers and plastics. Notably for water-soluble polymers where recovery is unlikely. There are niche opportunities for plastics such as fast food wrappers, fast food utensils, where separation is avoided and the plastic is already in a food matrix that is suitable for biodegradation; litter; ocean dumping; agricultural applications where recovery is unlikely and expensive for films and mulch; and hygienic products.

A major advantage that is conducive for the development of environmentally degradable water-soluble polymers is the well established disposal infrastructure; almost all centers of population in the industrial world have access to waste-water treatment plants. Whereas environmentally degradable plastics being developed for composting, for example, have a major impediment in that they will require the development of a composting infrastructure for their disposal and this will certainly influence their rate of acceptance.

TEST PROTOCOLS FOR ENVIRONMENTALLY DEGRADABLE POLYMERS

Environmentally degradable polymers fall into the categories of biodegradation, photodegradation, oxidation, and hydrolysis. Biodegradation has been the most actively researched, followed by photodegradation. Chemically promoted hydrolysis is frequently followed by the biodegradation of the initially produced fragments, and oxidation often operates in tandem and is inseparable from photodegradation.

Laboratory test protocols are usually evaluations of environmental degradation under simulated real world conditions to which a particular polymer or plastic will be exposed on disposal. However, it is occasionally desirable to test under the most favorable conditions to establish whether environmental degradation is at all possible. Regardless of the test, laboratory results only indicate the rate and extent of degradation under the test conditions. Correlation with real world exposure must be established as the ultimate goal of laboratory testing protocols is to be predictable of environmental response to new polymers and plastics. The goal is more difficult for biodegradation than for photodegradation, since the environments for biodegradation differ widely in microbial composition, pH, temperature, moisture, etc. and are not readily reproduced in the laboratory. In addition, once an environment sample has been placed within the confinements of a laboratory vessel, it can no longer interact with the greater environment in response to an added xenobiotic and results are frequently difficult to reproduce and may not always be representative of the response in real world exposures.

Test protocols to be discussed are essentially those developed by the ASTM which have been extensively tested and modified by many testing organizations world-wide. The protocols are either classified as 'test methods' or 'practices'. Methods are designed to give a measurable value or change, and practices are conditioning for a subsequent test method to measure an effect (ASTM D883–93).

Test Protocols for Photodegradable / Oxidatively Degradable Plastics

Photodegradation and oxidation test protocols are practices for exposing a plastic to some form of radiation and subsequently measuring property loss in an appropriate test method, for example a tensile strength loss (ASTM D 882–83), impact resistance loss (ASTM D 1709–85), tear strength loss (ASTM D 1922–67), molecular weight loss, friability, disintegration, brittle point, etc. Several standard test practices have recently been developed within ASTM (1993) for the plastic exposure part of the experiment and are listed in Table 1, subsequent testing in any of the standard ASTM test methods, including those above, may be done at various time intervals throughout the exposure to assess the rate of degradation.

Table 1 ASTM Standard Practices for Photodegradation

Test Number	Standard Test Practice
D 3826–91	Degradation End Points Using a Tensile Test
D 5071–91	Operation of a Xenon Arc ARC-Type Exposure Apparatus
D 5208–91	Operation of a Fluorescent Ultraviolet (UV) and Condensation Apparatus
D 5272–92	Outdoor Exposure Testing of Photodegradable Plastics
D 5437–93	Marine Weathering, Floating
D 5510–94	Heat-aging of Oxidative Degradation of Polymers

Table 2 Biodegradation Ratings in Fungal and Bacterial Growth Tests

Rating	Growth
0	no visible growth
1	< 10% of surface with growth
2	10–30% surface with growth
3	30–60% surface with growth
4	60–100% surface with growth

Test Protocols for Biodegradable Polymers

In the early years, the only tests conducted to establish biodegradability were related to microbial growth, weight loss, tensile changes and other physical property losses. These are all indirect measurements of biodegradation and often lead to results that were difficult to reproduce from laboratory to laboratory, giving rise to confusion on the susceptibility to biodegradation of a given polymer. They have been reviewed in many articles (ASTM. 1993; Potts, 1973, 1978; Andrady 1994). Two of these tests were 'growth ratings' based on ASTM Tests G 21–70 and ASTM G 22–76 which are really tests developed for assessing the resistance of plastics to fungal and bacterial growth, respectively. Fungal organisms such as *Aspergillus niger, Aspergillus flavus, Chaetomium globosum, and Penicillium funiculosum* and bacterial standards such as *Pseudomononas aeruginosa* are suggested in the test protocol, though it is not limited to these, are evaluated for growth on suitable plastics. After a suitable time period, growth is assessed by a subjective numerical rating, shown in Table 2, in which higher numbers are considered to correlate with the susceptibility of the plastic to biodegradation.

The advantage of these tests is that they are quick and easy to do and give an indication of biodegradation potential. Conversely, they are not conclusive for biodegradation of the plastic or polymer since any impurities such as plasticizers and solvents, may interfere by promoting growth and give false positive results.

Other simple tests include the soil burial test used widely by Potts and Clendinning (1984) to demonstrate the biodegradability of polycaprolactone, following its disappearance as a function of time; and the clear zone method which indicates biodegradation by the formation of a clear zone in an agar medium of the test polymer or plastic as it is consumed (Fields *et al.*, 1973). The burial test is still used as a confirmatory test method in the real world environment after quantitative laboratory methods indicate the degree of biodegradation.

In the last few years, the need to develop better tests, both qualitative and quantitative, has become apparent and is generating a great deal of activity in both plastics and water-soluble polymers. Many of the quantitative test protocols for both plastics and polymer are based on the tests developed over the last 25 years in the detergent industry for water-soluble organic compounds, particularly surfactants, which have been scrutinized for biodegradability since the late 1960's. Many of these tests are summarized in Swisher's excellent book (1987) on surfactant biodegradation and are formalized in publications by the Environmental Protection Agency (EPA) (1982) and the Organization for Economic and Cooperative Development (OECD) (1981) . This earlier work showed the value of choosing the environment applicable to the disposal method, with detergents the waste-water treatment plant, rivers, aquifers, and soil, the conditions for running laboratory tests with temperature control, addressing toxicity issues, acclimation potential, etc. Above all, the value of quantitative measurements of the products of biodegradation as the only reasonable means of assessing biodegradation. The necessity of applying similar rigor to plastic biodegradation was pointed out by Swift (1993).

Chemical equations describing the biodegradation of a hydrocarbon polymer in aerobic and anaerobic environments are indicated below. The equations may be readily modified to include other elements that may be present in a particular polymer which will appear in the oxidized or reduced form depending on the environment, aerobic or anaerobic, respectively. Most of the testing reported in the literature over the years has been with aerobic biodegradation conditions probably because this is easier to do in the laboratory and because polymers and plastics are usually discarded into these environments. However, anaerobic degradation, which is particularly pertinent to water-soluble polymers which may enter anaerobic digestors in sewage treatment facilities, is receiving more interest and more information will be developed in the future on this condition. In this article, unless specified, the term biodegradation should be understood to convey degradation in an aerobic environment.

Aerobic Environment

$$Polymer + O_2 \rightarrow CO_2 + H_2O + Biomass + Residue$$

Anaerobic Environment

$$Polymer \rightarrow CO_2 \, / \, CH_4 + H_2O + Biomass + Residue$$

To quantitatively assess the degree of biodegradation, analytical techniques are needed for all of the reactants and products; the polymer; oxygen uptake, known as biochemical oxygen demand (BOD), and the residue. The more rigorous the analysis, the more reliable the measurement of the extent of biodegradation, and, of course the acceptability of the data and the conclusions drawn therefrom. For total biodegradation, there should be no residue remaining in the environment.

Qualitative assessment of biodegradation where changes such as weight loss, tensile strength loss, disintegration, etc. are measured is a useful metric for plastics in that it gives an indication of the loss of properties that are a useful guideline as to the physical break down of the plastic and its decomposition in various environments such as compost, landfill, etc. and particularly the intended disposal environment. To differentiate biodegradation and abiotic degradation such as oxidation and hydrolysis, it is usually necessary to do a simultaneous control test with a killed inoculum (cyanide or mercury salts are acceptable) in which no degradation should be observed, if only biodegradation is involved in the plastic degradation. This test, obviously, would also serve as a test method for hydrolytic or oxidative degradation should degradation occur in the abiotic environment.

There have been numerous recent communications on the subject of biodegradation test methods and practices, including aerobic compost (Gross, 1992), anaerobic bioreactor (Gross, 1992), general methodology and future directions (Pettigrew *et al.* 1992; Swift 1993, 1994) and a fine review article by Andrady (1994). ASTM (1993) and

Table 3 Biodegradation Test Protocols

Test Number	Environment	Measurement
ASTM D 5209–92	Aerobic sewage sludge	CO_2
ASTM D 5210–92	Anaerobic sewage sludge	CO_2/CH_4
ASTM D 5247–92	Aerobic specific microorganisms	Molecular weight
ASTM D 5271–93	Aerobic activated sewage sludge	O_2/CO_2
ASTM D 5338–92	Aerobic controlled composting	CO_2
ASTM D 5437–93	Marine floating conditions	Physical properties
ASTM D 5509–94	Simulated compost	Physical properties
ASTM D 5511–94	Anaerobic biodegradation	CO_2/CH_4
ASTM D 5512–94	Simulated compost	Physical properties
ASTM D 5525–94	Simulated landfill	Physical properties
ASTM D 5526–94	Accelerated landfill	CO_2/CH_4
MITI Test	Mixed microbial	O_2

Masuda (1994) have also set forth standard testing protocols for plastics as shown in Table 3, whereas OECD test methods(1981) are more suited for water-soluble polymers.

Although the current state of biodegradation testing is greatly improved, the test protocols developed at this time should be recognized as only screening tests for readily biodegradable polymers and plastics. Failure in these test does not exclude biodegradation, it merely indicates that under the environmental conditions evaluated or the time frame of the test there is no biodegradation. Repeated tests, particularly in other environments, are recommended before accepting non-biodegradability. The possibility of toxicity of the polymer and the need for lower concentration testing should also be explored, in the latter case it may be necessary to resort to isotopic labeling in order measure biodegradation by monitoring low concentrations of carbon dioxide evolution. In many cases with synthetic polymers it may also be important to allow acclimation to a given environment so that enzymes may be induced that will biodegrade the polymer.

In addition to attention to improved biodegradation test method development, there is still emphasis needed on establishing the fate and effect of residues and degradation fragments in the environment where biodegradation is incomplete, whether from biodegradable polymers and plastics or from any of the other environmental degradation pathways. The acceptability of environmentally degradable polymers will depend on these methodologies. Some efforts are being made in this area with the work of Scholz in water-soluble polymers (1991) and ASTM Standard Practice D 5152–91 (1993) for extracting aqueous solubles from the solid fragments produced by the environmental degradation of plastics for testing in standard aquatic toxicity protocols. In general, though, it has not yet become accepted and understood in that the degree of biodegradation of polymer or plastic in itself has no value, it is only part of the much more important Environmental Safety Assessment (Swift, 1993). Biodegradation is related to the environmental concentration of the polymer or fragments and is therefore related to an environmental safety assessment (ESA) by equations (3), (4), and (5):

$$\text{Environmental concentration} = f \text{ (rate and degree of biodegradation)} \quad (3)$$

$$\text{ESA} = f \text{ (environmental concentration)} \quad (4)$$

therefore:

$$\text{ESA} = f \text{ (rate and degree of biodegradation)} \quad (5)$$

SYNTHESIS OF ENVIRONMENTALLY DEGRADABLE POLYMERS

There are several excellent reviews on environmentally degradable polymers covering all the aspects of the subject (Heap, 1968; Rodriquez, 1970; Cocarelli, 1972; Potts, 1978; Kuster, 1979; Kumar, *et al.*, 1987; Potts, Jopski, 1993; David, 1994; Swift, 1990; Huang, 1990; Amirabhavi *et al.*, 1990; Yabbana *et al.*, 1993; Narayan, 1993; Nakamura, 1994; Udipi *et al.*, 1993; Satyananaya *et al.*, 1993; Lenz *et al.*, 1993; Swift, 1990), plus others mentioned in this article, which are recommended reading. The two major degradation pathways, photodegradation and biodegradation, will be treated separately for convenience. Polymers degrading by oxidation and hydrolysis, as mentioned, generally are difficult to separate from photodegradation and biodegradation, respectively and will not be given separate attention.

Photodegradable and Oxidative Degradable Polymers

Photodegradable polymers are designed to degrade by chain scission promoted by natural daylight (and usually oxygen) to low molecular weight fragments that are more susceptible to biodegradation than the original polymer. The polymers are generally structurally similar to currently used environmentally stable polymers, such as polyolefins and polyesters, but have been modified during synthesis or post-treatment to insert photochemically active groups. The addition of carbonyl functionality (Guillet, 1968, 1970, 1971) into the polymer main or side chain, or by adding external photosensitizers and pro-oxidants such as metal salts (Griffin, 1973, 1976, Scott, 1964; Sipinen, 1993) benzophenone (Swanhol, 1975) ketones (Taylor, 1977), ethers (Taylor, 1977) mercaptans (Taylor, 1977), and polyunsaturated compounds (White, 1974) are representative examples. The mechanism of fragmentation is well established and there are commercial products based on ethylene/carbon monoxide copolymers for use in six-pack holders and agricultural film. The ultimate fate of the fragments produced is not yet fully established, in most cases the argument is put forward that if the molecular weight of the degradation products is low enough then they will biodegrade. Only recently has there has been good scientific evidence that appears to support this expectation (David, 1995), and more work is underway in the same laboratory to confirm this.

The synthesis of photodegradable copolymers of olefins with carbon monoxide and ketones are shown schematically below, when $R = C_6H_5$ the monomer is styrene, and $R = H$ the monomer is ethylene (Scott, 1973; Cooney, 1981; Guillet, 1990).

When exposed to ultraviolet radiation, the activated ketone functionalities may fragment by two different mechanisms, known as Norrish types I and II. Polymers with the carbonyl functionality in the backbone of the polymer cleave the main

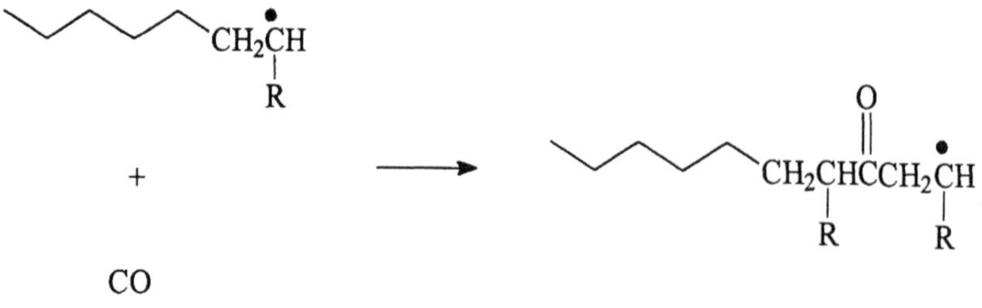

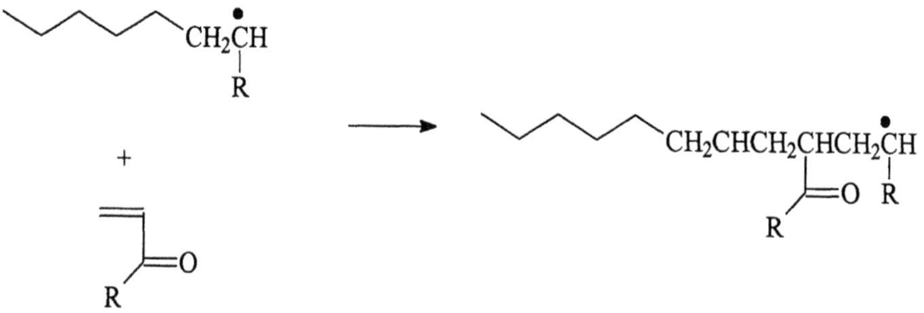

$$\text{CH}_2\overset{\bullet}{\text{CH}}$$
$$\underset{\text{R}}{|}$$

+

CO

$\longrightarrow$

$$\overset{\text{O}}{\overset{||}{\text{CH}_2\text{CHCCH}_2\overset{\bullet}{\text{CH}}}}$$
$$\underset{\text{R}}{|}\qquad\underset{\text{R}}{|}$$

$$\text{CH}_2\overset{\bullet}{\text{CH}}$$
$$\underset{\text{R}}{|}$$

+

$$\underset{\text{R}}{\overset{||}{\text{C}=\text{O}}}$$

$\longrightarrow$

$$\text{CH}_2\text{CHCH}_2\text{CHCH}_2\overset{\bullet}{\text{CH}}$$
$$\underset{\underset{\text{R}}{\overset{|}{\text{C}}}=\text{O}}{|}\quad\underset{\text{R}}{|}$$

chain by both mechanisms, but when the carbonyl is in the polymer side chain, only the Norrish type II degradation mechanism produces main chain scission (Guillet, 1970; Kato, 1973). The fragmentation reactions are illustrated below for a carbon monoxide copolymers with main chain carbonyl and for an alkyl vinyl ketone copolymer with a side chain carbonyl functionality.

Norrish type I chemistry is claimed to be responsible for about 15% of the chain scission of ethylene/carbon monoxide polymers at room temperature, whereas at 120°C, it promotes 59% of the degradation. Norrish I reactions are independent of temperature and oxygen concentration at temperatures above the Tg of the polymer (Reich, 1971).

Norrish Type I Reaction for Backbone Carbonyl Functionality

$$\overset{\text{O}}{\overset{||}{\text{wwCH}_2\text{CH}_2\text{CCH}_2\text{CH}_2\text{CH}_2\text{CH}_2\text{ww}}}$$

$\longrightarrow$

$$\overset{\text{O}}{\overset{||}{\text{wwCH}_2\text{CH}_2\text{C}\cdot}} \;+\; \cdot\text{CH}_2\text{CH}_2\text{CH}_2\text{CH}_2\text{ww}$$

$\longrightarrow$

$$\overset{\text{O}}{\overset{||}{\text{wwCH}_2\text{CH}_2\text{COH}}} \quad + \quad \text{HOCH}_2\text{CH}_2\text{CH}_2\text{CH}_2\text{ww}$$

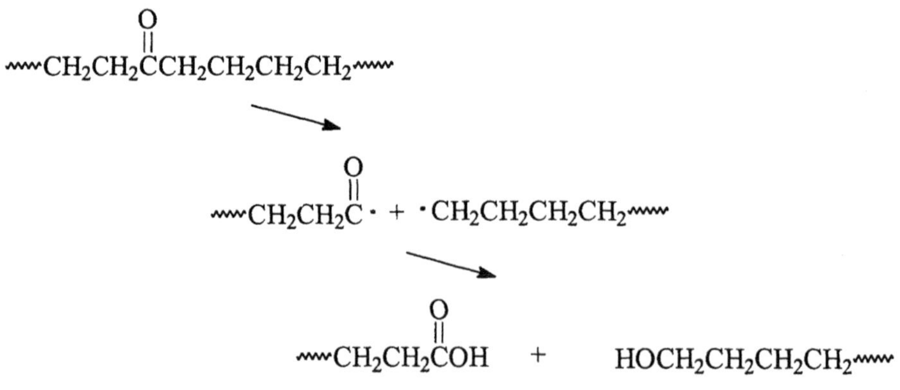

Norrish Type II for Backbone Carbonyl Functionality

$$\text{wwwCH}_2\text{CH}_2\overset{\displaystyle\overset{\text{O}}{\|}}{\text{C}}\text{CH}_2\text{CH}_2\text{CH}_2\text{CH}_2\text{www}$$

$$\text{wwwCH}_2\text{CH}_2\overset{\displaystyle\overset{\text{O}}{\|}}{\text{C}}\text{CH}_3 \quad + \quad \text{H}_2\text{C}=\text{CHCH}_2\text{CH}_2\text{www}$$

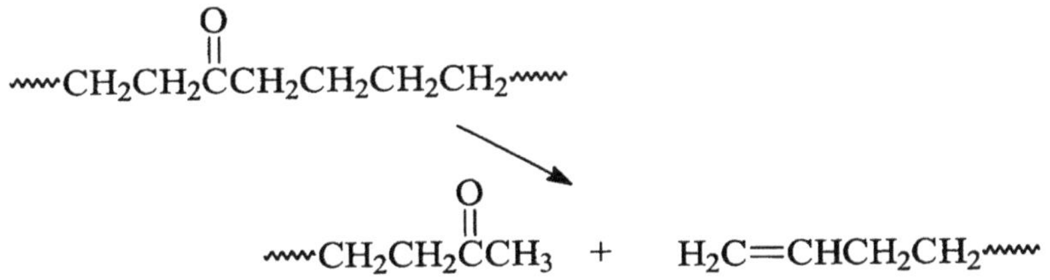

Norrish Type I for Side-Chain Carbonyl Functionality

$$\text{wwwCH}_2\text{CH}_2\underset{\underset{\text{R}}{\overset{|}{\underset{|}{\text{C}=\text{O}}}}}{\text{CH}}\text{CH}_2\text{CH}_2\text{CH}_2\text{CH}_2\text{www}$$

$$\text{wwwCH}_2\text{CH}_2\overset{\displaystyle\cdot}{\text{C}}\text{HCH}_2\text{CH}_2\text{CH}_2\text{CH}_2\text{www} \quad + \quad \text{R}\overset{\displaystyle\cdot}{\text{C}}=\text{O}$$

$$\text{wwwCH}_2\text{CH}_2\underset{\underset{\text{OH}}{|}}{\text{CH}}\text{CH}_2\text{CH}_2\text{CH}_2\text{www} \quad + \quad \text{RCO}_2\text{H}$$

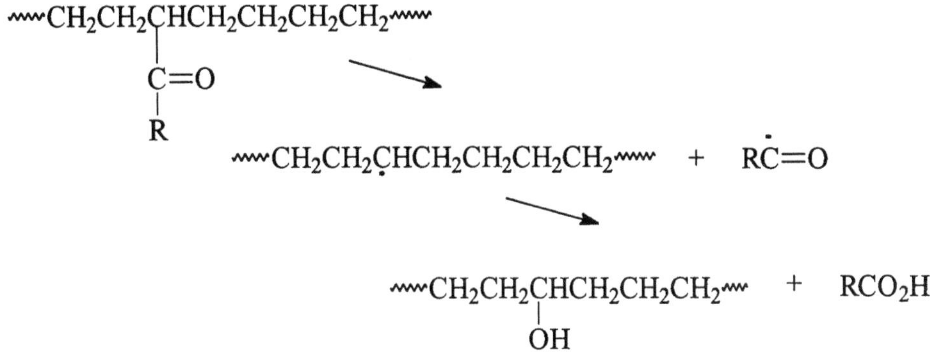

Norrish Type II for Side-Chain Carbonyl Functionality

$$\text{wwwCH}_2\text{CH}_2\underset{\underset{\text{R}}{\overset{|}{\underset{|}{\text{C}=\text{O}}}}}{\text{CH}}\text{CH}_2\text{CH}_2\text{CH}_2\text{CH}_2\text{www}$$

$$\text{wwwCH}_2\text{CH}_2\underset{\underset{\text{R}}{\overset{|}{\underset{|}{\text{C}=\text{O}}}}}{\text{CH}}_2 \quad + \quad \text{CH}_2=\text{CH}_2\text{CH}_2\text{CH}_2\text{www}$$

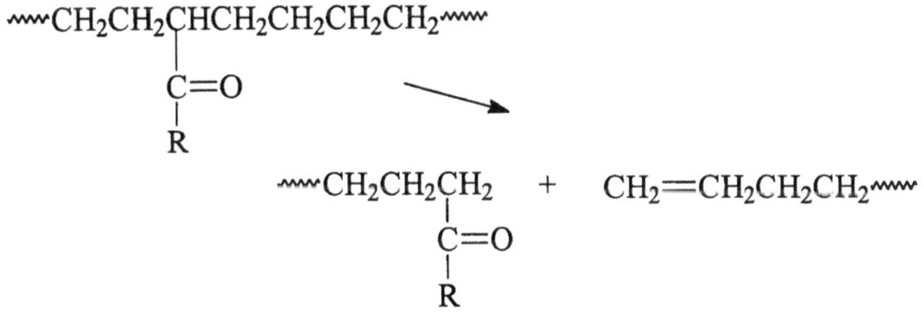

    Degradation of polyolefins such as polyethylene, polypropylene, polybutylene, and polybutadiene promoted by metals and other pro-oxidants occurs via a combination photo-oxidative mechanism. A general mechanism common to all these reactions is that shown below. The reactant radical may be produced by any suitable mechanism from the interaction of air/oxygen with polyolefins (Scott, 1964; Sipinen et al., 1993) to form peroxides which subsequently decompose by the action ultraviolet radiation. These decomposition intermediates may undergo many reactions such as abstraction of hydrogen atoms from the polymer until the backbone ultimately converted into a polymer with ketone functionalities which is degraded by a Norrish mechanism discussed above. A recent review by Scott (1995) elegantly discusses this subject.

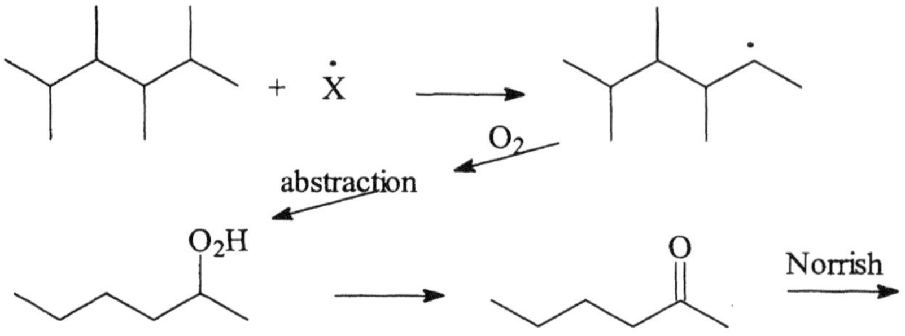

Research on photodegradable polymers is still very active as an end in itself and also in combination with biodegradable polymers to promote more rapid biodegradation by rapidly decreasing primary polymer molecular weight. Austin of Exxon (1992) and Chang of Quantum (1991) have recently received patents for polyolefin/polyester/carbon monoxide compositions prepared by totally different routes. The former is by copolymerization of ethylene, carbon monoxide, and 2-methylene-1,3-dioxapane. The use of dioxapanes is based on W.J. Bailey's research in which he introduced an ester linkage into polyolefins during free radical polymerization such that they become susceptible to biodegradation, and will be discussed later. Research by Exxon scientists is an equally clever combination of Bailey's work with a known photodegradable product to enhance biodegradation of the fragments. The polymer will have the structural elements shown below in a concentration related to the degradation response required and controlled by the synthesis variables.

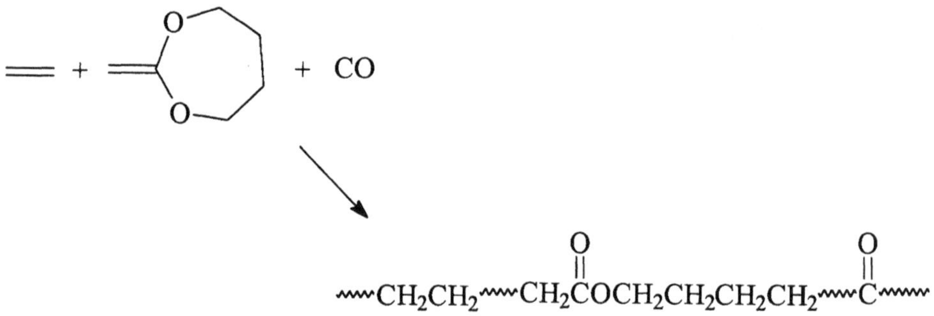

Quantum, on the other hand, converted an ethylene/carbon monoxide polymer into a biodegradable polyester acidic with hydrogen peroxide, the Baeyer-Villiger reaction. Depending on the degree of conversion to polyester, the polymer will be totally or partially degraded by a biological mechanism.

Other recent patents include copolymers of vinyl ketones with acrylates, methacrylates, and styrene (O'Brien, 1993); an ethylene/carbon monoxide (1–7 wt%) blend as a photo initiator in polycaprolactone/polyethylene blends (Hirsoe, 1992); ethylene/carbon monoxide for degradable golf tees (Akimoto); a vinyl ketone analog of Exxon's carbon monoxide/dioxapane/ethylene (Priddy, 1992); a photodegradable food wrapper based on blends of a polyolefin/starch and photo activators for the

$$\text{wwwCH}_2\text{CH}_2\text{CH}_2\text{CH}_2\text{www} \overset{\overset{\displaystyle O}{\|}}{C}\text{CH}_2\text{CH}_2\text{www}$$

$$\text{H}^+ \overset{\text{H}_2\text{O}_2}{\searrow}$$

$$\text{wwwCH}_2\text{CH}_2\text{CH}_2\text{CH}_2\text{www} \overset{\overset{\displaystyle O}{\|}}{C}\text{OCH}_2\text{CH}_2\text{www}$$

polyolefin degradation (Kurata, 1992); and a carboxylated polyethylene/carbon monoxide/norbornene-2,3-dicarboxylic acid (Dent, 1991).

Photodegradation chemistry has evolved into a highly practical state over the last few years to where commercial products are available and others are being evaluated. The degradation mechanisms are understood to the point of property loss for the plastics. The gap on environmental acceptability still needs attention, it is not sufficient to expect low molecular weight fragments to be biodegradable, this must be demonstrated.

Biodegradable Polymers

For ease of discussion, this section is divided into three broad classifications – natural, synthetic, and modified natural based biodegradable polymers and plastics. Natural polymers indicate no modification of isolated polymer, synthetic polymers include carbon chain and heteroatom chain polymers, and modified natural polymers encompass grafts and blends, and chemical modifications such as oxidations and esterifications.

Natural polymers or biopolymers are produced by all living organisms. As a class, they are accepted as biodegradable since they are produced naturally, even though with polymers such as lignin, biodegradation may be very slow, and even not measurable by laboratory test protocol. The most wide spread natural polymers are the polysaccharides, such as celluloses and starch. Other important classes of natural polymers include polyesters such as polyhydroxyalkanoates; proteins like silk and poly(γ-glutamic acid); and hydrocarbons such as natural rubber. An excellent description of many biopolymers is given in a recent book edited by Byrom (1991) with chapters on silk proteins (Kaplan, Lombardi, Muller, and Fossey); collagen (Gorham); polyhydroxyalkanoates (Steinbuchel); microbial polysaccharides (Linton, Ash, and Huybrechts); microbial cellulose (Byrom); hyaluronic acid (Swann and Kuo); alginates (Sutherland); and miscellaneous biomaterials (Byrom). Other information sources include Lenz's review on biodegradable polymers mentioned earlier (1993), and the proceedings of the NATO Advanced Research Workshops on New Biodegradable Microbial Polymers (Dawes, 1990; Vert, 1992). With a few notable exceptions, natural polymers are not suitable polymers for practical applications at this time either because they lack the property requirements or are too expensive for other than specialty high value niche markets such as biomedical applications, which are discussed in other chapters of this book. There are, however, opportunities in blends and for chemically modified polymers such as starch, celluloses, and proteins, which will be discussed later in this chapter.

Polysaccharides are largely limited to starch and cellulose derivatives for practical applications either in plastics or as water-soluble polymers. Both these polymers are composed of thousands of D-glycopyranoside repeat units to very high molecular weight. They differ in that starch is poly-1,4-αD-glucopyranoside, and cellulose is poly-1,4-βD-glucopyranoside, as illustrated below. This difference in structure influences biodegradation rates and properties of the polymers.

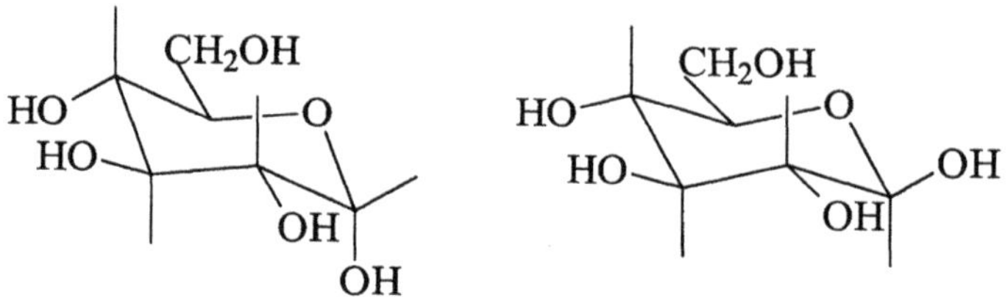

CH$_2$OH

HO
HO
OH
OH

(Starch)

CH$_2$OH

HO
HO
OH
OH

(Cellulose)

Complex carbohydrates such as microbially produced xanthan, curdlan, pullulan, hyaluronic acid, alginates, carageenan, and guar are accepted as biodegradable and are finding uses where cost is not an impediment. Xanthan is the predominant microbial polysaccharide on the market, ca 10,000 tonnes world-wide (Byrom, 1991), and finds use in the food industry and as a thickener in many industrial applications. It is foreseeable that the others will gain acceptance in specialty areas where biodegradability is essential.

Proteins have not yet found widespread use as plastic materials since they are difficult to process, they are not fusible without decomposition or soluble in practical solvents so they have to be used as found in nature. Examples are silk used as a fiber, gelatin(collagen), which is used as an encapsulant in the pharmaceutical and food industries, and wool also used as a fiber. The structure of proteins is an extended chain of amino acids joined through amide linkages which are readily degraded by enzymes, particularly proteases. Recent activity in poly(γ-glutamic acid) (Gross, 1996) with control of stereochemistry by the inclusion of manganese ions may have import for future developments in biodegradable water-soluble polymers with carboxyl functionality, which is an intensely researched and desirable goal for detergent and other applications, as will be seen later in this chapter.

Polyesters are produced by many bacteria as intracellular reserve materials for use as a food source during periods of environmental stress. They have received a great deal of attention in the last two decades because they are biodegradable, can be processed as plastic materials, are produced from renewable resources and are produced by many bacteria in a range of compositions. These thermoplastic polymers have properties that vary from soft elastomers to rigid brittle plastics depending on the structure of the pendant side-chain of the polyester. The general structure of this class of compounds is shown below.

They have been comprehensively reviewed in a recent book by Doi (1990) and by others (Dawes, 1990; Lenz et al., 1990). All the polyesters are 100% optically pure and are 100% isotactic. Where R is CH$_3$, poly-β-hydroxybutyrate (PHB), the polymer

$$\left[OCHCH_2\overset{\overset{\displaystyle O}{\|}}{C} \right]_n$$
$$\underset{R}{|}$$

$$R = -\left(CH2\right)_m CH_3 \qquad n => 100; \ m = 0 \ \text{or higher}$$

is highly crystalline with a melting point of 180°C and a glass transition temperature, Tg, of 5°C (Marchessault, 1988). Because of this combination of high Tg and high crystallinity polymer films and plastics are very brittle and plasticization is necessary for processing and to improve properties. This is accomplished in the commercially available BIOPOL from Zeneca (formerly ICI) by using bacteria to producing a copolymer containing β-hydroxyvalerate ($R = C_2H_5$). By feeding the bacteria, *Alcaligenes eutrophus*, a mixed feed of propionic acid and glucose (Cox 1992) a random copolymer is produced with some control over the gross composition. These copolymers have better mechanical properties and are produced on a relatively large scale for this new polymer fermentation technology (a few hundred tonnes per year). Their acceptance would certainly be more wide spread if the price were closer to the synthetic commodities with which they compete, since the fact that they are produced from renewable resources and are of natural origin makes them appealing in an environmental age. Attempts to reduce costs are underway in a number of laboratories: the cost of the processing with the current polymers is excessive due to the isolation steps which include clean up of bacterial debris. If these polymers could be produced by isolated enzymatic processes this would be avoided. Another intriguing possibility is to produce these natural polymers in plants rather than by bacteria by transferring the bacterial genes to suitable plants. Some work by Somerville and Dennis (1993), and Byrom at Zeneca (1992, 1993), in mustard plants has succeeded in the production of minute quantities of polyhydroxyalkanoates throughout the plant. More recently rape seed plant has also been evaluated in this biotechnology approach (Anonymous 1995). The day when polyhydroxyalkanoates are produced in the volumes and at the price of starch may herald a new age for plastics and polymers, but they are not yet here. Some of the opportunities for these materials are discussed by Byrom (1993).

The longer side-chain polyesters, where m is 3–6, are produced by a variety of bacteria, usually as copolymers and with low crystallinity, low melting points and low glass transition temperatures. These are elastomeric and have excellent toughness and strength (Lenz, 1990). They are inherently biodegradable but as the chain length is increased the biodegradation rate is greatly reduced, indicating that hydrophilic/hydrophobic balance of polymer plays a major roll in biodegradation (69), even for naturally occurring polymers. Other biodegradation studies are underway to evaluate mechanisms (Saito, 1993) using ^{13}C labeled polyhydroxybutyrate, and effect of environment (Brandl, 1992) on the rate of biodegradation in lake Lugano, Switzerland.

All these polyesters are produced by bacteria in stressed conditions in which they are deprived of some essential component for their normal metabolic processes. Under normal conditions of balanced growth the bacteria would utilize any substrate for energy and growth, whereas under stressed conditions they utilize any suitable substrate to produce polyesters as reserve material. When the bacteria can no longer subsist on the organic substrate due to depletion, they consume the reserve for energy and food for survival. Similarly, upon removal of the stress, the reserve is consumed and normal activities resumed. This cycle is utilized to produce the polymers which are harvested at maximum cell yield. This process is outlined schematically below(Lenz), and in more detail in the paper by Steinbuchel (1991) on the mechanism of biosynthesis of polyhydroxyalkanoates.

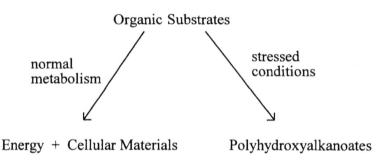

Synthetic polymers are well established in many applications where their environmental resistance properties are highly valued, as was mentioned in the introduction. With the advent of environmental awareness and polymer waste-disposal problems, there has been increasing activity to develop biodegradable analogs of these polymers, particularly water-soluble polymers and plastics which are used in packaging and other areas of opportunity discussed earlier. Natural polymers have been produced in nature for eons and are ultimately degraded and consumed in nature in a continuous recycling of resources. However, the plethora of enzymes available in nature for degrading natural polymers are not generally useful for synthetic polymers. The search for synthetic polymeric structures that can be biodegraded has progressed from minor modification of the non-degradables now in use, all the way to structures that mimic nature, where more success is being achieved. In spite of the fact that biodegradation testing, until very recently, has been very unreliable and the interpretation of many reported data in the literature are questionable, general guidelines have emerged for predicting the biodegradability of synthetic polymers (Swift, 1990; Cooke, 1990; Huang, 1989). These include: hydrophilic/hydrophobic balance; higher ratio is better for biodegradation; carbon chain polymers are unlikely to biodegrade; chain branching is deleterious to biodegradation; condensation polymers are more likely to biodegrade; lower molecular weight polymers are more susceptible to biodegradation; and crystallinity slows biodegradation. The guidelines are useful for approaching polymer synthesis.

Carbon chain backbone polymers are represented by the general structure below and may be considered derivatives of polyethylene, where n is the degree of polymerization and R is a functional group such as hydrogen (polyethylene), methyl(polypropylene), carboxylic[poly(acrylic acid)], chlorine[poly(vinyl chloride)], aromatic(polystyrene) hydroxyl[poly(vinyl alcohol)], ester[poly(vinyl acetate)], nitrile(polyacrylonitrile), vinylic(polybutadiene) etc. The functional group and the molecular weight of the polymers, control properties such as hydrophile / hydrophobe balance, solubility characteristics, glass transition temperature, and crystallinity.

$$\left(\text{CH}_2\underset{R}{\text{CH}}\right)_n$$

The biodegradation of low molecular weight polyethylenes were studied very extensively by Potts (1973, 1974, 1978) and Albertsson (1976, 1978, 1979, 1980, 1987, 1988, 1993) has had long term experiments with radiolabeled high molecular weights running for many years. Potts' work was based on fungal and bacterial growth test and indicated that polyethylene and other high molecular weight carbon chain polymers did not support growth, a few anomalous results were attributed to plasticizer or low molecular weight impurities which he had already shown to be biodegradable in similar tests with pyrolyzed polyethylene and simple hydrocarbons. In this work, Potts found that branching of hydrocarbon chains limits biodegradation and the molecular weight cut-off for the biodegradation of linear molecules appears to be in the 500 dalton range. However, more rigorous testing is needed for confirmation of the molecular weight phenomenon. Some representative data are shown in Tables 4 and 5.

The increase in degradation with lower molecular weight may be due to many factors, for example transportation of polymer across cell walls being more likely at lower molecular weight, and the mechanism of exocellular biodegradation, random or chain-end cleavage prior to entering the cell. Chain end, exo biodegradation, would explain slower biodegradation at higher molecular weight where there are fewer chain ends than at lower molecular weight.

It is worth noting that the terminal groups found in oxidized and photodegraded polyethylene are oxygen containing and these should predictably expedite biodegradation via a β-oxidation mechanism, lending support to the claims that these fragments are biodegradable.

Table 4 Hydrocarbon Branching and Molecular Weight Effects on Biodegradability

Compound	MW	No. of Branches	Growth Test
dodecane	170	0	4
2,6,11-trimethyldodecane	212	3	0
hexadecane	226	0	4
2,6,11,15-hexadecane	282	4	0
tetracosane	338	0	4
squalene	422	6	0
hexatriacontane	506	0	0
tetratetracontane	618	0	0

Table 5 Biodegradability of Low MW Pyrolysis Products of
HDPE and LDPE

Pyrolysis Temperature C	MW	Growth Rating
Control (HDPE)	123000	0
450	8000	1
500	3200	3
535	1000	3
Control (LDPE)	56000	0
450	12000	1
500	2100	2
535	1000	3

Albertsson's results on experiments that are now several years along continue to suggest that polyethylene is slowly biodegraded and is slightly accelerated by pre-treatment with surfactants or oxidation to permit better surface wetting with enzymes. This meticulous work has identified many of the degradation products of polyethylene, many of which are oxygen containing consistent with a degradation mechanism promoted initially by oxidation with subsequent enzymatic activity.

Other high molecular weight hydrocarbon polymers do not appear to be biodegradable, but, as with polyethylene, T. Suzuki found that oligomers of cis-1,4-isoprene (1979), butadiene (1978), styrene (1977), are degradable; he also confirmed the work of Potts with oligomeric ethylene (1980).

Functional derivatives of polyethylene, particularly poly(vinyl alcohol) and poly-(acrylic acid) and their derivatives have received attention because of their water-solubility and disposal into the aqueous environment. Poly(vinyl alcohol) is used in a wide variety of applications, including textiles, paper, plastic films, etc.; and poly(acrylic acid) is used in detergents as a builder, as a super-absorbent in diapers and feminine hygiene products, water treatment, thickeners, pigment dispersant, etc.

Poly(vinyl alcohol), obtained by the hydrolysis of poly(vinyl acetate) is the only synthetic carbon chain polymer accepted as fully biodegradable. The biodegradation mechanism is established as a random chain cleavage of 1,3 diketones formed by an enzyme catalyzed oxidation of the secondary alcohol functional groups in the polymer backbone. Biodegradation was first observed by Yamamoto (1966) as a reduction in the aqueous viscosity of the polymer in the presence of soil bacteria. Subsequently, T. Suzuki (1973) identified a *Pseudomonas* species as the soil bacteria responsible for the degradation over a degree of polymerization range of 500 to 2000. Utilizing the polymer as a sole carbon source in an aqueous polymer solution at a concentration of 2700 ppm, it was reduced to 250–300 ppm concentration in 7–10 days at pH 7.5–8.5 and 35–45°C. An oxidative endo mechanism was proposed and later substantiated (Suzuki, T. 1979) by quantifying the oxygen uptake at one mole for every mole of hydrogen peroxide produced and identifying the degradation products as ketones and carboxylic acids, as shown in the flow diagram below. Included in the diagram is the alternative mechanism proposed by Watanabe (1975, 1976) in which the products were identified as an alcohol and a carboxylic acid. This was subsequently proved to be in error and the Suzuki mechanism is now widely accepted. It is also supported by the rapid biodegradation of chemically oxidized poly(vinyl alcohol) (Huang, S.J. 1982; Huang, J.C. 1990; Endo, 1991).

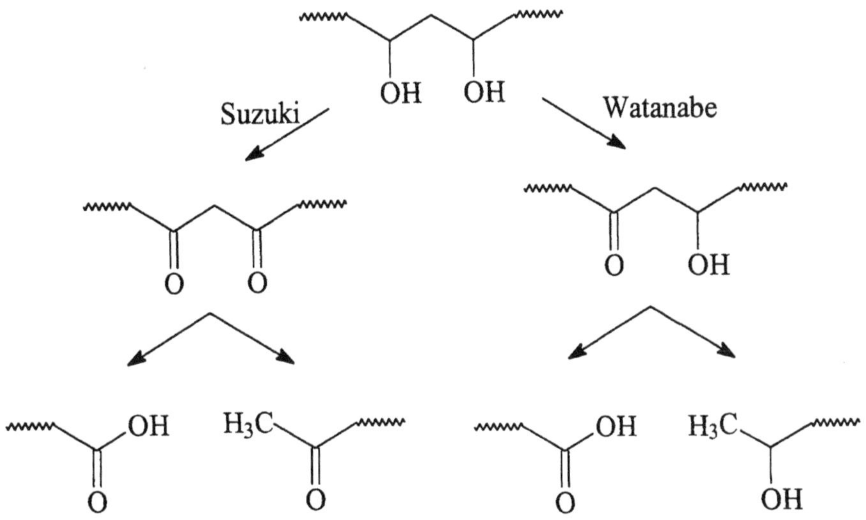

Other bacterial strains identified as biodegrading poly(vinyl alcohol) include *Flavobacterium* (Fukunaga, 1977) and *Acinetobacter* (Fukunaga, 1976), as well as fungi, molds, and yeasts (Kato, 1990). Industrial evaluations at DuPont (Casey, 1976) and Air Products (Wheatley, 1976) indicate that over 90% of poly(vinyl alcohol) entering waste water treatment plants is removed and hence no environmental pollution is likely.

Poly(vinyl acetate), the precursor for poly(vinyl alcohol), hydrolyzed to less than 70% is claimed to be non-biodegradable under conditions similar to those that biodegrade the fully hydrolyzed polymer (Matsumura 1988).

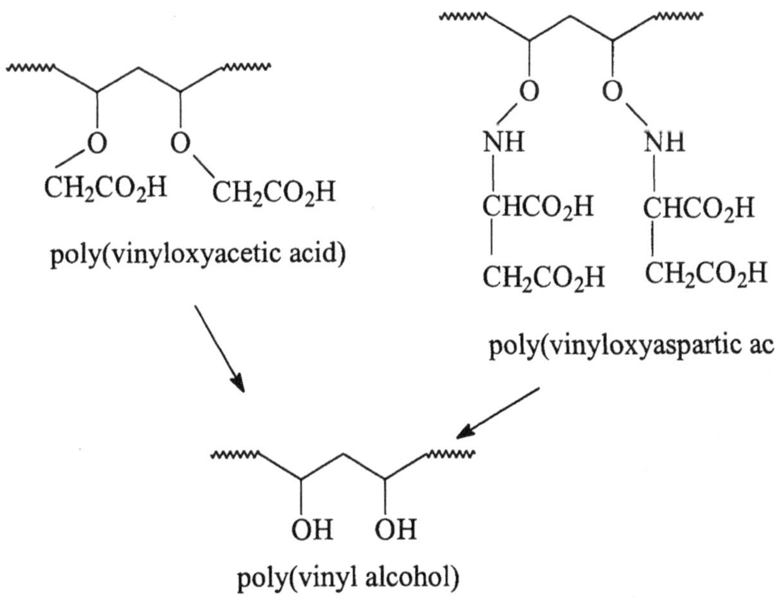

poly(vinyloxyacetic acid)

poly(vinyloxyaspartic ac

poly(vinyl alcohol)

Carboxylate derivatives of poly(vinyl alcohol) are biodegradable and functional in detergents as cobuilders, although too costly to be practical at this time. Matsumura polymerized vinyloxyacetic acid (Matsumura, 1988) and Lever has a patent to polymers based on vinyl carbamates obtained from the reaction of vinyl chloroformates and amino acids such as aspartic and glutamic acids (Garafalo, 1991). Both hydrolyze in a suitable environment to poly(vinyl alcohol) and then biodegrade.

Copolymers of vinyl alcohol with acrylic and/or maleic acid have been evaluated in detergents as potentially biodegradable cobuilders by a number of laboratories (Matsumura 1993, Swift 1993), but the results were not encouraging for balancing biodegradation and performance. Higher than 80 mole % of vinyl alcohol is required for high levels of biodegradation, and less than 20 mole % for acceptable performance.

The use of poly(carboxylic acids) in detergents is well established and has been well reviewed (Lester 1988). Their lack of biodegradability at preferred performance molecular weights, ca. 5000 dalton for poly(acrylic acid) and 70000 daltons for copoly(acrylic/maleic acids), even though there is ample data to indicate no harmful environmental effects, has resulted in a massive search for degradable replacements. After many efforts to radically copolymerize acrylic and maleic acids to biodegradable polymers with a whole range of vinyl monomers (Matsumura, 1980, 1981, 1984, 1985, 1986), graft substrates, including polysaccharides (Murai, 1986), it is now generally accepted that the pioneering work of Suzuki (1978) with functional oligomers and polymers, and Potts (1973, 1974, 1978) with oligomeric hydrocarbons is correct and only carbon chain oligomers are likely to be biodegradable, regardless of functionality. Recent support of this generalizations comes from Kawai (1993) and research in NSKK laboratories in Japan (1993) with acrylic acid oligomers.

Suzuki (1978) ozonized high molecular weight poly(acrylic acid), polyacrylamide, and poly(vinyl pyrrolidone) to oligomers with molecular weights less than 1000 daltons and observed a marked increase, with the exception of polyacrylamide, in their biodegradability. The work of Kawai (1993) and Tani (1993) was based on oligomers of acrylic acid obtained by chromatographic separation from low molecular weight polymers. Their results indicated that poly(acrylic acids) are not biodegradable above a degree of polymerization of about 6–8 units (a molecular weight of about 400–600 daltons).

Other efforts to use radical polymerization to synthesize biodegradable carboxylated polymers have been based on combining low molecular weight oligomers through degradable linkages and by introducing weak links into the polymer backbone. Both BASF (Baur and Boeckh *et al.* 1988) and NSKK (Irie 1993) have patented acrylic oligomers chain-branched with degradable linkages(X) in the schematic below. Grillo Werke has patented copolymers of acrylic acid and enol sugars (1988). The degradability of these polymers has not been clearly established, but the branching is likely to be a hindrance, if not a deterrent.

Several miscellaneous carbon chain backbone polymers have been claimed as biodegradable without clear evidence, these include copolymers of methyl methacrylate and vinyl pyridinium salts (Kawabata 1993), where the pyridinium salt is hypothesized as a 'magnet' for bacteria which then cleave the chain into small fragments which biodegrade completely. An ethylene/vinyl alcohol copolymer that is converted into a polyester by a Baeyer Villiger reaction (Brima 1993) (below);

and polymers of α-hydroxyacrylic acid (Mulders 1977). The former, being a polyester is probably biodegradable, but the latter study fails to differentiate adsorption of the polymer on solids in a sewage sludge test from biodegradation.

Heteroatom chain backbone polymers include polyesters, which have been widely studied from the initial period of research on biodegradable polymers, polyamides, polyethers, polyacetals and other condensation polymers. Their chemical linkages are widely found in nature and these polymers are more likely to biodegrade than the hydrocarbon-based polymers discussed in the previous section.

Potts (1974) determined that low melting, low molecular weight aliphatic polyesters were readily biodegradable using soil burial tests and ASTM bacterial and fungal growth methods. From this work, polycaprolactone is recognized as one of a select few commercially available synthetic polymers that is biodegradable. Since that time many other workers have confirmed the biodegradability of aliphatic polyesters using other test protocols (Suzuki, 1977, 1978, 1986) such as lipase hydrolysis with measurement of the rate of production of water-soluble oligomers. Results also indicate that as the aliphatic polyesters become more hydrophobic, either from acid or alcohol chain length extension, the biodegradation rate is slowed. Kendricks showed that the amorphous regions of polyesters are more readily biodegradable than crystalline regions.

During this early period, Bailey developed a very clever free radical route
to polyesters which he used to introduce weak linkages into the backbones of
hydrocarbon polymers and render them susceptible to biodegradability (Bailey
1975, 1979, 1985, 1991). Copolymerization of ketene acetals with vinyl monomers
incorporates an ester linkage into the polymer backbone by rearrangement of
the ketene acetal radical as illustrated below. The ester is a potential site for
biological attack. The chemistry has been demonstrated with polyethylene (Bailey),
poly(acrylic acid) (Bailey, 1990), and polystyrene (Tokiwa).

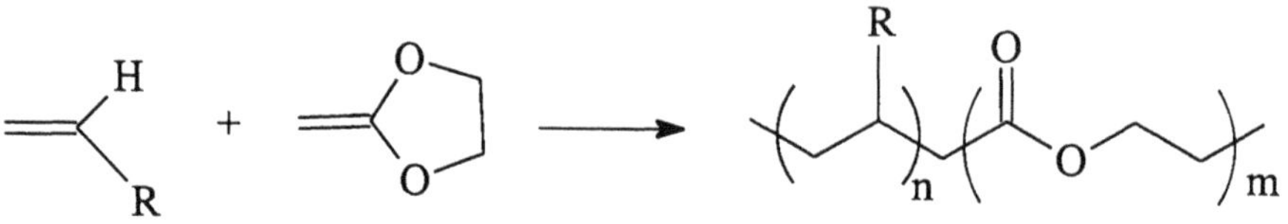

Recent activity on biodegradable polyesters has maintained interest in predominantly aliphatic polyester structures and includes ring opening of 1,5-dioxepan-2-one
(Albertsson, 1993), polyesters of aliphatic acids with glycols, some terephthalic
acid and sulphoterephthalic acid as a compostable diaper (Gallagher, 1992, 1993).
Perhaps the biggest advances in the synthetic polyester area are the close match
for BIOPOL®, the expensive bacterial polyester(vide supra), by Kobayashi (1993)
which has the chirality but lacks the molecular weight of the natural polymer; and
the new product BIONOLLE® polymer from Showa High Polymer (Takiyama 1992,
1993; Fujimaka 1994; Taniguchi, 1993) which is supposedly a property match for a
'biodegradable' polyolefin. Recent patents suggest that BIONOLLE is an aliphatic
ester coupled with a polyisocyanate to increase molecular weight and that ease of
biodegradation and properties are inversely related. Both these inventions indicate
that progress is being made on meeting the property requirements for biodegradable
polymers.

Aliphatic polyesters are also available by the chemical reaction of carbon monoxide and formaldehyde (Masuda, 1991), carbon dioxide and epoxy compounds
(Yoshida, 1992), and bis epoxies and bis carboxylic acids (Yamamoto, 1991).

Poly(lactic acid), has until very recently been known only in the medical field as
an expensive polymer for the manufacture of sutures and other biomaterials. With
cheap lactic acid becoming available from the fermentation of waste agricultural
products, there has been a surge of activity to develop new polymerization methods
and to find commercial outlets for the products. Though still more expensive than
common commodity polymers, the introduction of products for niche markets such
as agricultural films and mulch, and the fast-food industry is anticipated to be in the
near future. There are many patents issuing in the homo- and copolymer synthesis
and process areas, from principally just a few corporate players, Cargill (Benson,
1992, 1993), DuPont (Ford, 1992, 1993), Mitsui Toatsu (Kitamura, 1992, 1993), and
Battelle Memorial Laboratories (Sinclair, 1992). Other players with some activity in
this area include Shimadzu, and Argonne National Laboratories.

Polyamides have received some attention and the results indicate that the
stereochemistry of the groups close to the amide linkages and the hydrophilic
nature control biodegradability (Bailey, 1975; Huang, 1976). A more general study

on polyesters, polyureas, polyurethanes and polyamides is a good fundamental early study (Huang, 1977) guiding the later work in this general area. Gonsalves (1993) has demonstrated that polyesteramides are difficult to chemically hydrolyze yet can be biodegraded rapidly by enzymes under ambient conditions in the right environment. Support for this observation comes from Tokiwa (1979) and Yamamoto (1992) in Japan. While polymers of Nylon-6 are considered non-biodegradable, oligomers and low molecular weight polymers of less than 11000 daltons will biodegrade (Baggi, 1993).

Water-soluble polyesters and polyamides containing carboxyl functionality are reported to be biodegradable detergent polymers in patents to BASF and may be obtained by condensation polymerization of monomeric polycarboxylic acids such as citric acid, butane-1, 2, 3, 4-tetracarboxylic acid, tartaric acid, and malic acid with polyols (BASF, 1992, 1993); amino compounds, including amino acids (BASF, 1993, 1994); and polysaccharides (BASF, 1992). Earlier work by Matsumura (1986) and Lenz and Vert (1978) had demonstrated the self-condensation of malic acid to biodegradable polyesters regardless of the ester structure, -α or β-linkage, formed. Procter and Gamble has patented a polyester dispersant based on succinylated poly(vinyl alcohol)(Schectman).

Polyanionics are available also from polyamino acids based on polycarboxy amino acids such as glutamic acid and aspartic acid. Though both are known and claimed as biodegradable homopolymers, aspartic acid is more amenable to a practical industrial synthesis by thermal polymerization since it has no tendency to form an internal N-anhydride. An alternative synthesis is from ammonia and maleic acid. Only the acid catalyzed condensation of L-aspartic acid yields an authenticated biodegradable polymer (Swift, 1994). The non-catalyzed process and the ammonia/maleic acid processes give partially (ca 30 wt % residue remains in the Sturm test, which measures biodegradation by carbon dioxide evolution and residual soluble organic carbon) biodegradable polymers due to the molecules being branched and resistant to enzymatic attack. The structures are shown below starting from aspartic acid, pathway 'A' is acid catalyzed thermal condensation, and 'B' is non-catalyzed thermal condensation. The polysuccinimides shown hydrolyze at the positions indicated to give copoly(α, β-aspartic acid) salts. Regardless of the stereochemistry of the starting aspartic acid, L or D, the final product is the DL racemate.

Many synthesis patents and publications from aspartic acid (Koskan, 1992, 1993; Swift, 1994; Ponce, 1992; Rao, 1993) and ammonia/maleic (Wood, 1993, 1994) processes have issued recently and the product is expected to find use in many applications including dispersants (Sikes, 1993, Koskan, 1993), and detergents (Imanaka, Rocourt, 1993). BASF has an aspartic acid copolymer patent with carbohydrates and polyols (Baur, 1994), and Procter and Gamble (Hall, 1993) has a patent for poly(glutamic acid), both for biodegradable detergent cobuilders. There is one patent for poly(methyl — glutamate) as a transparent plastic with excellent strength and biodegradability (Endo, 1991). Other opportunities for these polymers include crosslinking to form biodegradable superabsorbent materials for diapers (Choi 1995).

Polyethers have been investigated since about 1962, especially poly(ethylene glycol), which is water-soluble, is widely used in detergents and as a synthesis intermediate in polyurethanes. This whole field has been very comprehensively reviewed by Kawai (1987) who established the symbiotic nature of the degradation of poly(ethylene glycols) with molecular weights higher than 6000 daltons to be symbiotic, while below a molecular weight of 1000 daltons the polymer is biodegraded by many individual bacteria. The enzymatic exo degradation pathway described by Kawai is shown schematically below, the first stage is a dehydrogenation, the second stage is an oxidation, the third stage is an oxidation which is followed by a hydrolysis to remove a two carbon fragment as glyoxylic acid. Degradation of poly(ethylene glycols) with molecular weights of 20,000 daltons has been reported.

Anaerobically, poly(ethylene glycol) degrades slowly, although molecular weights up to 2000 daltons have been reported (Schink, 1983, 1986) to biodegrade. The biodegradation poly(alkylene glycols) is hindered by their lack of water solubility and only the low oligomers of poly(propylene glycol) are biodegradable with any certainty (Kawai, 1977, 1982, 1985), as are those of poly(tetramethylene glycol) (Kawai, 1986). A similar exo oxidation mechanism to that reported for poly(ethylene glycol) was proposed.

Poly(ether carboxylates) have been evaluated as biodegradable detergent polymer, initially by Crutchfield (1978), Procter and Gamble (1986, 1987), and Matsumura (1987). They all fit the general structure of the series made in Matsumura's extensive evaluation of anionic and cationic polymerized epoxy compounds in the molecular weight range of several hundreds to a few thousands, where X or Y may be carboxyl functionality and X or Y may be hydrogen or a substituent bearing a carboxyl functionality. Biodegradability, based on biochemical oxygen demand (BOD), is structure dependent.

$$\sim\!\!\sim\!\!\sim\!\!\sim\!\!\sim\!\!\mathrm{OCH_2CH_2OCH_2CH_2OCH_2CH_2OH}$$

$$\downarrow$$

$$\sim\!\!\sim\!\!\sim\!\!\sim\!\!\mathrm{OCH_2CH_2OCH_2CH_2OCH_2\overset{\overset{\displaystyle O}{\|}}{C}H}$$

$$\downarrow$$

$$\sim\!\!\sim\!\!\sim\!\!\sim\!\!\mathrm{OCH_2CH_2OCH_2CH_2OCH_2CO_2H}$$

$$\downarrow$$

$$\sim\!\!\sim\!\!\sim\!\!\sim\!\!\mathrm{OCH_2CH_2OCH_2CH_2O\overset{\overset{\displaystyle OH}{|}}{C}HCO_2H}$$

$$\downarrow$$

$$\sim\!\!\sim\!\!\sim\!\!\sim\!\!\mathrm{OCH_2CH_2OCH_2CH_2OH} \;+\; \mathrm{H\overset{\overset{\displaystyle O}{\|}}{C}CO_2H}$$

Water-soluble biodegradable polycarboxylates with an acetal or ketal weak link were the clever invention of Monsanto scientists in their search for biodegradable detergent polymers. However, economics prevented the polymers reaching commercial status. The polymers are based on the anionic or cationic polymerization of glyoxylic esters at low temperature (molecular weight is inversely proportional to the polymerization temperature) and subsequent hydrolysis to the salt form of the polyacid which is stable under basic conditions being a hemi acetal, or ketal if methylglyoxylic acid is used. Biodegradation results from the pH drop such a detergent polymer will experience as it leaves the alkaline laundry environment (pH ca. 10) and enters the sewage or ground water environment (pH close to neutral). The polymer is unstable and hydrolyzes to monomer which rapidly biodegrades. The chemistry is outline schematically below and is reported in many patents (Monsanto, 1979, 1980) and several publications (Gledhill, 1978, 1987).

$$\mathrm{HO}\!\!\left(\!\!\begin{array}{c}\overset{\displaystyle H}{|}\;\;\overset{\displaystyle H}{|}\\[-2pt]\mathrm{C-C-O}\\[-2pt]\underset{\displaystyle X}{|}\;\;\underset{\displaystyle Y}{|}\end{array}\!\!\right)_{\!\!n}\!\!\mathrm{H}$$

Similar polyacetals were prepared by BASF scientists from -aldehydic aliphatic carboxylic acids (Baur, 1992; Koeffer, 1993), and by the addition of polyhydroxy carboxylic acids, tartaric acid for example, to divinyl ethers (Baur 1993) as biodegradable detergent polymers.

Modified natural polymers offer a way of capitalizing on their well accepted biodegradability of natural polymers to develop environmentally acceptable polymers. This, of course, will only be true if the modification is shown to not interfere with the biodegradation process and the product meets the guidelines listed earlier for environmental acceptability: *either demonstrated to be totally biodegraded and be removed from the environment or biodegradable to the extent that no environmentally harmful residues remain.* With this in mind, the approaches that have received the major attention include blends with other natural and synthetic polymers; grafting of another polymeric composition; and chemical modification to introduce some desirable functional group by oxidation or other simple chemical reaction such as esterification or etherification.

Starch is made thermoplastic at elevated temperatures in the presence of water as a plasticizer, allowing melt processing alone or in blends with other thermoplastics (Zobel, 1984; Stepto, 1987, Thomka). Good solvents, such a water, lower the melt transition temperature of amylose, the crystalline component of starch, so that processing can be done well below the decomposition/degradation temperature.

The most important commercial application has been the blending of polyethylene with starch in the presence or absence of other additives to promote compatibility. The interest in this approach goes back to Griffin in the 70's (1971, 1977, 1987) and there is continuing activity with commercial products from several companies. There are very many other contributors to this field, Otey was also a pioneer, who developed starch/polyethylene compatibilized with ethylene/acrylic acid copolymers (1977, 1987) and ethylene/vinyl alcohol (1976). Fanta (1993) was a player from the same USDA laboratory. Later work with polyethylene capitalizing on this early research includes that of Bastioli of Novamont (also known as Butterfly in some of their patents) with starch blends containing hydroxy acids, urethanes, polyamides, and polyvinyls (1992, 1993, 1994); Warner Lambert (Novon) (Stepto, 1989, 1993; Miller, 1990); US Army with cellulose acetate (Meyer, 1994), Henkel with alkyds (Beck, 1993), Iowa State University (Jane, 1992, 1993) with proteins and oxidized polyethylene; ADM (Koutlakis, 1993); Solvay (Dehennau, 1994) with polycaprolactone; and Agritech (Wool, 1991) with a starch minimum loading of 30 wt %.

Biodegradation studies of starch blends have not been conclusive where a non-degradable synthetic polymer has been the blend component, probably biodisintegration would be a better term to describe these polymers. The major deficiencies of products based on this chemistry aside from the incomplete biodegradation are water sensitivity of manufactured articles, and the balance of this and biodegradation with the starch level in the product.

Other blends such as polyhydroxyalkanoates(PHA) with cellulose acetate (Buchanan, 1992); PHA with polycaprolactone (Mita Ind.); poly(lactic acid) with poly(ethylene glycol) (Bazile, 1992); chitosan and cellulose (Nishiyama 1994); poly (lactic acid) with inorganic fillers (Shikinami, 1993); PHA and aliphatic polyesters with inorganics (Tokiwa, 1992), polyesters with cellulosics (Buchanan, 1995), and proteins and cellulosics (Vaidya, 1995) are receiving attention. The different blending compositions seem to be limited only by the number of polymers available and the compatibility of the components. The latter blends with all natural or biodegradable components appear to be the best approach for future research as property balance and biodegradability is attempted. A recent paper, evaluates starch and additives in detail from the perspective of structure and compatibility with starch (Poteate 1994).

Starch has also been a substrate of choice for biodegradable polymers by grafting with synthetic polymers to achieve property improvement and new properties such as carboxyl functionality not available in starch with retention of as much biodegradability as possible. Thermoplastic polymers from the ionic grafting of styrene to starch was demonstrated by Narayan (Narayan, 1989), and radical grafting of acrylate esters (Dennenberg, 1978) has also been reported. The latter was recommended as a mulch as it rapidly decomposed in the presence of fungi. One must question the extent of biodegradation of both these materials as the acrylic and styrene components are known for their resistance to biodegradation.

Other grafts to natural materials are exemplified by Dordick's work (1992) in which he produced polyesters from sugars and polycarboxylates by enzyme catalysis, these polymers and the method of synthesis may well be one of the future directions of renewable resource chemistry. This is similar to some very early research by Stannet's group with cellulose condensation with polyfunctional isocyanates and optionally propylene glycol (1973). Some degradation was claimed. Meister (1992) has shown the utility and potential for lignin grafted with styrene, and it is claimed that the product is totally biodegradable due to the potency of white rot *basidiomycetes* which is a lignin degrader. Further proof is required but this is a promising lead.

Natural polymers have also received attention as graft sites for carboxylic monomers to produce detergent polymers without great success, the synthetic portion of the graft is not usually biodegradable even though in some cases attempts were made to meet the molecular weight limitations mentioned earlier(less than DP of ca 6–8). Acrylic grafts onto polysaccharides in the presence of alcohol chain transfer agent (Kim, 1993) were not completely biodegradable, nor were the ones based on initiation with Ce^{4+} (Vidal, 1992; Jost, 1992) and mercaptan (Baur, 1991; Klimmek, 1994). Protein substrates (Kroner 1992) are expected to be similar to the starch grafts, the fundamental problem is the need to control acrylic acid polymerization to the oligomers range, as indicated earlier, in order to have complete biodegradability.

Simple chemical reactions on natural polymers are well known to produce polymers such as hydroxyethyl cellulose, hydroxypropyl cellulose, carboxymethyl cellulose, cellulose acetates and propionates, and many others that have been in commerce for many years. Their biodegradability is not at all well established. Carboxy methyl cellulose, for example, has been claimed as biodegradable below a degree of substitution of about 2, which is similar to cellulose acetate. More recently, there has been attempts to more rigorously quantify biodegradation of the cellulose acetates (Stannett 1973, Buchanan 1994, Gross 1994), and to establish a structure/property/biodegradation relationship. A Rhone-Poulenc publication also indicates that cellulose acetate with a degree of substitution of about 2 is biodegradable, in agreement with the earlier references (1993). Cellulose has been discussed as a renewable resource (Arch). A recent publication (Monal 1993) on chitosan reacted with citric acid indicates that the ampholytic product is biodegradable.

Carboxylated natural polymers have been known for many years with the introduction of carboxymethyl cellulose, as noted above. This product has wide use in detergents and household cleaning formulations, even though of questionable biodegradability at the level of substitution required for performance. Nevertheless, carboxylated polysaccharides are a desirable goal for many application and the balance of biodegradation with performance has been recognized as an attractive target with a high probability of success by many people. Three approaches have been employed, esterification, oxidation or Michael addition of the hydroxyl groups to unsaturated carboxylic acids such as maleic and acrylic, with some attempts to react specifically at the primary or secondary sites.

Esterification with polycarboxylic anhydrides can be controlled to minimize diesterification and crosslinking to produce carboxylated cellulosic esters. Eastman Kodak in a recent patent claimed the succinylation of cellulose to different degrees, 1 per three anhydroglucose rings and 1 per two rings Faber 1993). Henkel (Engelskirck) also has a patent for a surfactant by the esterification of cellulose with alkenylsuccinic anhydride, presumably substitution governs the hydrophile/hydrophobe balance of the product.

Oxidation of polysaccharides is a far more attractive route to polycarboxylates, potentially cheaper and cleaner than esterification. Selectivity at the 2,3-secondary hydroxyls and the 6-primary is possible. Total biodegradation with acceptable property balance has not yet been achieved. For the most part, oxidations have been with hypochlorite/periodate under alkaline conditions, more recently catalytic oxidation has appeared as a possibility, and chemical oxidations have also been developed that are specific for the 6-hydroxyl oxidation.

Matsumura (1990, 1992, 1993, 1994) has oxidized a wide range of polysaccharides, starch, xyloses, amyloses, pectins, etc. with hypochlorite/periodate. The products are either biodegradable at low oxidation levels or functional at high oxidation levels, the balance has not yet been established. Other than Matsumura, van Bekkum, at Delft University, has been the major contributor in the search to control the hypochlorite/periodate liquid phase oxidations of starches (1987, 1988) and he has been searching for catalytic processes to speed up the oxidation with hypochlorite. Hypobromite is one solution which is generated in situ from the cheap hypochlorite and bromide ion (Besemer and van Bekkum, 1993, 1992). van Bekkum (1994) has published a method for oxidizing specifically the 6-hydroxyl group(primary) of starch by using TEMPO and bromide/hypochlorite as shown below.

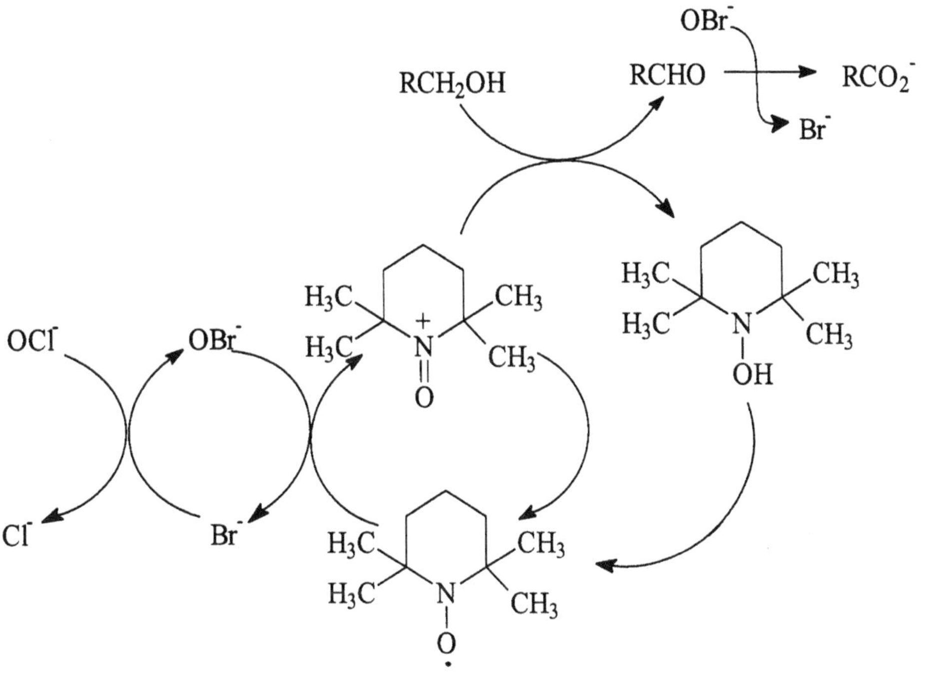

Chemical oxidation with strong acid is reportedly selective at the 6-hydroxyl, either with nitric acid/sulfuric acid/vanadium salts (DuPont 1993) which is claimed as specific for the 6-hydroxyl up to 40% conversion or with dinitrogen tetroxide in carbon tetrachloride with similar specificity up to 25% conversion (Engelskirchen 1993).

Catalytic oxidation in the presence of metals is claimed as both non-specific and specific for the 6-hydoxyl depending on the metals used and the conditions employed for the oxidation. Non-specific oxidation is achieved with silver or copper and oxygen (Sakharov, 1979, 1993), and noble metals with bismuth and oxygen (Gosset, 1991; Fleche, 1991). Specific oxidation is claimed with platinum at pH 6–10 in water in the presence of oxygen (Watanabe, 1993). Related patents to water-soluble carboxylated derivative of starch are Hoechst's on the oxidation of ethoxylated starch and another on the oxidation of sucrose to a tricarboxylic acid, all the oxidations are specific to primary hydroxyls and are with a platinum catalyst at pH near neutrality in the presence of oxygen (Dany, 1993; Friscela, 1993).

Table 6 Some Currently Available Environmentally Degradable Polymers

Polymer	Developer	Degradation
Poly(lactic acid)	Cargill, Ecochem, Biopak, Mitsui Toatsu	hydrolysis/biodegradation
Cellophane	Flexel	biodegradation
PHBV	Zeneca	biodegradation
Starch-based	Novamont	biodegradation
Polycaprolactone	Union Carbide, Solvay	biodegradation
Starch-activator	Ecostar	photo/biodegradation
Starch foam	National Starch	biodegradation
Polyolefin-activator	Plastigone	photodegradation
Polyester	Showa High Polymer	biodegradable
Polyethylene/CO	Dow	photodegradation
Poly(vinyl alcohol)	Rhone-Poulenc, Air Products Kuraray, Hoechst	biodegradation
Poly(ethylene glycol)	Union Carbide, Dow	biodegradation
Cellulosics	Rhone Poulenc, Eastman	biodegradation
Poly(aspartic acid)	Rohm and Haas	biodegradable
Poly(ethylene oxide)s	Planet Technologies	biodegradable

Polysaccharides as raw materials in the detergent industry were reviewed in a presentation by Swift (1993) at the Chicago ACS Meeting in 1993 and recently published.

AVAILABLE ENVIRONMENTALLY DEGRADABLE POLYMERS

There are indications that the use of environmentally degradable polymers and plastics is expanding. As the market begins to realize the availability of these new materials, it is expected that they will move into the niche opportunities mentioned earlier in this review. When this happens, production will increase and costs, the biggest barrier to acceptance at this time should begin to come down. Some of the polymers currently in production at some scale larger than laboratory are shown in Table 6.

The only product with substantial sales at this time is a photodegradable environmentally degradable ethylene/carbon monoxide polymer used as a six pack holder which are often carelessly thrown away to litter the environment. The litter will slowly disappear as it degrades into fine fragments. EPA has let it be known that although this is acceptable for now, future products should degrade by a combination of photodegradation and biodegradation to ensure complete removal from the environment.

CONCLUSIONS

There has been much progress in the last few years in environmentally degradable polymers. Standard protocols are available to determine degradation in the environment of disposal, definitions are understood and accepted in a broad sense, if not in detail (I doubt whether it is necessary to go beyond the current conceptual levels), and fate and effects issues for these new polymers are being addressed and will be resolved when appropriate test methods are developed. International agreement on the acceptability of these polymers is, therefore, close with the general understanding that preservation of the environment is the major priority.

Perhaps the major remaining issue is the cost of these polymers, most are at least 3 to 5 times that of the current products that must be displaced. There has to be some recognition that a clean environment has a price. Regardless of cost, all the major interested parties in this area realize that environmentally degradable polymers are unlikely to be a major force in the waste-management of polymers, they will be used where they are best suited and offer advantages over the competing technologies such as recycle, landfilling, and incineration. Some obvious areas of opportunity are in fast food, agriculture, sanitary articles for plastics and water-soluble polymers which are not recoverable after use. A major advantage for environmentally degradable water-soluble polymers is the ready availability of a disposal infra structure in municipal waste-water treatment plants. Composting, a promising disposal avenue for plastics, on the other hand is in need of development, and no plans are in place to do this which is hindering the growth of these plastics.

REFERENCES

Akimoto, I. *et al.* (1992) Photodegradation. *JP 04058962* (to Nippon Unicar).
Albertsson, A.C. Proc. (1976) Degradation of Polyethylene. *Int. Biodeg. Symp.*, 743.
Albertsson, A.C. (1979) Polyethylene Degradation. *J. Appl. Poly, Sci., Poly. Sci. Symp.*, **35**, 423.
Albertsson, A.C. (1978) Polyethylene Degradation. *Europ. Poly. J.*, **16**, 123.
Albertsson, A.C. (1978) Polyethylene Degradation. *J. Appl. Poly. Sci.*, **22**(11), 3419, 3435.
Albertsson, A.C. (1980) Polyethylene Degradation. *J. Appl. Poly. Sci.*, **25**(12), 1655.
Albertsson, A.C. and Karlsson, S. (1990) Polyethylene Degradation. *Poly. Mater. Sci. Eng. Preprint* pp. 60–64.
Albertsson, A.C., Andersson, S.O., Karlsson, S. (1987) Polyethylene Degradation. *Polym. Deg. Stab.*, **17**, 73.
Albertsson, A.C. and Karlsson, S. (1988) Polyethylene Degradation. *J. Appl. Polym. Sci.*, **35**, 1289.
Albertsson, A.C. (1993) Polyethylene Degradation. *J.M.S. Pure Appl. Chem.*, **A30**(9 and 10), 757–765.
Albertsson, A.C. (1993) Aliphatic polyesters. *J. M. S. Pure and Appl. Chem.*, **A30**(12), 919–931.
Amirabhavi, T.M., Balundgi, R.H. and Cassidy, P.E. (1990) *Polym. Plast.Technol. Eng.*, **29**(3), 235–62.
Andrady, A.L. (1994) *JMS-Rev. Macromol. Chem. Phys.* **C34**(1), 25–76.
Anonymous (1995) Rape seeds in PHA production. *Europ. Chem. News*, (1995) Sept. 25–Oct. 1, 39.
Arch, A. (1993) Cellulose as a renewable resource. *J. Macromol. Chem.*, **A30**(9/10), pp. 733–740.
ASTM D 883 – 93, Terminology Relating to Plastics.
ASTM (1993) *Standards on Environmentally Degradable Plastics*, ASTM Publication Code Number (PCN): 03-420093-19.
Austin, R.G. (1992) Photo/Biodegradation *US Patent 5281681 and WO 9212185–A2* (to Exxon).
Baggi, G., Andreoni, V.S., Guaita, C., Manflin, P. (1993) Biodegradation of oligomers of Nylon-6. *Int. Biodeterior. Biodegrad.*, **31**(1), pp. 41–53.
Bailey, W.J., *et al.* (1975) Hydrocarbon Biodegradation. *J. Polym. Sci., Polym. Lett. Edn.*, **13**, 193.
Bailey, W.J. (1976) Biodegradation of polyamides. *Proc. of the 3rd. Int. Biodeg. Symp (1975)*, Applied. Sci Publ., 765.
Bailey, W.J. *et al.* (1979) Hydrocarbon Biodegradation. Contemp. *Topics Polym. Sci.*, **3**, 29.
Bailey, W.J. and Gapud, B. (1985) Hydrocarbon degradation. *Polym. Stab. and Deg.*, **280**, 423.
Bailey, W.J. (1990) Biodegradable poly(acrylic acid). (1990) *US Patent 4923941* (to American Cyanamid).
Bailey, W.J., Gu, W.J., Lin, Y., Zheng, Z. (1991) *Hydrocarbon Biodegradation. Makromol. Chem. Makromol. Symp.*, **42/42**, 195.
Bastioli, C. (1992) Starch blends. *WO 9219680-A1, WO 9214782-A1, O9202363*; (1993) US Patent *5262458*; (1994) *US Patents 5286770, 5288765* (to Novamont).

Bastioli, C. (1992) Starch blends. *Spec. Publ. R. Soc. Chem.*, No. **109**, 101.

Bastioli, C., Bellotti, V., Del Guidice, L. and Gilli, G., (1993) Starch blends. *J. Environ. Polym. Degrad.*, **1**(3), 181–192.

BASF, Biodegradable polymeric carboxylic acids. *DE 3716543/4A* (1988) and *EP 291808A* (1988); and *EP 292766A* (1988) (R. Baur et al.); EP 2 *89787/8A* (1988); EP *289827A* (1988) and DE *3733480A* (1988) (D. Boeckh et al).

BASF (1992) Biodegradation of polyesters. *WO 9216493–A1, DE 4108626–A1.*

BASF (1993) Biodegradation of polyesters. *US Patent 5217642.*

BASF (1993) Polyamides. *DE 4213282–A1.*

BASF (1994) Polyamides. *DE 4225620–A1.*

BASF (1992) Polyesters based on polysaccharides. *DE 4108626–A1, DE 4034334–A1.*

Baur, R. *et al.* (1991) Acrylics grafted to starch in the presence of mercaptan. *DE 4003172* (to BASF).

Baur, R. *et al.* (1992) Biodegradable polyacetals. *DE 4106354-A!, WO 9215629–A1* (to BASF).

Baur, R. (1993) Biodegradable polyacetals. *DE 4142130-A1* (to BASF).

Baur, R. *et al.* (1994) Poly(aspartic acid) copolymers. *DE 4221875-A1* (to BASF).

Bazile, D. (1992) PHA blends with poly(ethylene oxide). *EP's 520888, 520889* (to Rhone Poulenc).

Beck, M. and Ritter, W. (1993) Starch Blends. *DE 4209095–A*(to Henkel).

Bensen, R.D. *et al.* (1992, 1993) Poly(lactic acid). *US Patents 5142023, 5247058, 5247059, 5258488* (to Cargill).

Besemer, A.C. Thesis (Delft), (1993) and *EP 4273459–A2, WO 9117189.*

Besemer, A.C. and van Bekkum, H. (1994) Oxidation of polysaccharides. *Starch-Starke*, **46**, pp. 95–100; pp. 101–106.

Brandl, H. and Pucchner P. (1992) *Biodegradation*, **2**(4), 237–243.

Brima, T.S., Chang, B. and Kwiatek, J. (1993) US Patent, 5219930 (to Quantum Chemicals).

Buchanan, C.M. *et al.* (1992) Cellulose acetate blends with polyhydroxyalkanoates. *Macromolecules*, **25**, 7381.

Buchanan, C.M., Komanek, R., Dorschel, D., Boggs, C. and White, A.W. (1994) Measuring biodegradation of cellulosics. *J. Appl. Poly. Sci.*, **52**(10), pp. 1477–1488.

Buchanan, C.M., Gardner, R.M., White, A.W. (1995) Blends of polyesters and cellulosics. *US Patent 5446079* (to Eastman Chemical Company).

Byrom, D. (1991) *Biopolymers: Novel Materals from Biological Sources*, Stockton Press, New York, ISBN 1-56159-037-1.

Byrom, D., Bright, S.W., Fentem, P.A. (1992) *WO 9219747-A1* (to Imperial Chemical Industries).

Byrom, D. (1993) *Inter. Biodeter. Biodegrad.*, **31**(3), 199–208.

Casey, J.P. and Manley, D.G. (1976) *Proc. 3rd. International Biodeg. Symp. Appl. Sci. Publ.*, 731–741.

Chang, B.H. and Lee, L.Y. (1991) *Photo/Biodegradation.* US Patent 5064932 (to Quantum).

Choi, H.J., Yang, R. and Kunioka M. (1995) Superabsorbents. **58**, pp. 806–814.

Cooke, T.F. (1990) *J. Polym. Eng.*, **9**, 171.

Cooney, C.G., Carlsson, J.D. and Wiles, D.J. (1981) Photodegradation. *J. Appl. Poly. Sci.*, **26**, 509.

Coscarelli, W. (1972) *Polymer Stabilization*, W.L. Hawkins Edn. Wiley, N.Y.

Cox, M.K. (1992) Polyhydroxyalkanoates. *Spec. Publ. R. Soc. Chem.*, **109**, 95.

Crutchfield, M.M. (1978) Poly(ether carboxylates). *J. Amer. Oil Chem. Soc.*, **55**, 58.

Dany, F.J. *et al.* (1991 and 1993) Oxidation of ethoxylated starch. *WO 9102712 and US Patent 5223642* (to Hoechst).

David, C., DeKersel, C., LeFevre, F. and Wieland, M. (1994) *Angew. Makromol. Chem.*, **216**, 21–35.

David, C., Wieland, M. and Davis, A. (1995) Biodegradation of photodegradation fragments. *Polym. Degrad. and Stab.*, **48**, pp. 275–289.

Dawes, E. (1990) *Novel Biodegradable Microbial Polymers*, NATO ASI Series, Series E: Applied Sciences, Volume 186, Kluwer Academic Publisher Dordrecht, Boston, London.

Dawes, E.A. and Anderson, A.J. (1990) *Microbiol. Rev.*, **54**, 450.

Dehennau, C. and Depireux, T (1994) Starch Blends. *EP 580032–A1* (to Solvay SA).

Dennenberg, R.J., Bothast, R.J., Abbot, T.P. (1978) Acrylics grafted to starch. *J. Appl. Polm. Sci.*, **22**(2), 459–465.

Dent, E. (1991) Photo/Biodegradation. *US Patent 5059676* (to Shell).

Doi, Y. (1990) *Microbial Polyesters*, VCH Publishers, New York.

Dordick, J., Patil, D.R. and Rethwisch, D.G. (1992) WO 9221765 (to Univeristy Iowa, State Res. Found.).

Driemel, K., Bunthoff, K. and Nies, H. (1988) Enol-sugar copolymerization. *EP289895A* (to Grillo Werke).

Dupont, J.S. and Heinzman, S.W. (1993) Site-specific oxidation of starch. *EP 542496-A1* (to Procter and Gamble).

Endo, T. (1991) Degradation of Oxidized Poly(vinyl alcohol). *JP 03263406 and JP 03263407* (to Kuraray).

Endo, T. *et al.* (1991) Poly(glutamic acid). *EP 445923* (to Meija-Seika K.K.).

Engelskirchen, K. *et al.* (1993) Oxidation of starch. *DE 4203923–A1 and WO 9308251–A1* (to Henkel).

Engelskirchen, K. *et al.* (1988) Cellulose alkylated with alkenylsuccinic acid. *EP 254025–B* (to Henkel).

Faber, J.W.H. (1993) Succinylated cellulose. *WO 9210521-A1* (to Eastman Kodak).

Faber, J.W.H. (1993) Succinylated cellulose. *EP 560891–A1* (to Eastman Kodak).

Fanta, G.F., Swanson, C.L., Shogren, R.L. and Imam, S.H. (1993) *Polyethylene-starch blends. J. Environ. Polym. Degrad.*, **1**(2), 155–166.

Fields, R.D., Rodriquez, F. and Finn, R.K. (1973) *J. Amer. Chem. Soc., Divn. Poly. Chem.*, **14**, 2411.

Fleche, G. and Fuertes, P. (1991) Oxidation of starch. *US Patent 4985553* (to Roquette Freres).

Ford, T.M., Hyvnkook, S., Bellis, H.E. (1992, 1993) Poly(lactic acid). *US Patents 5097005, WO 9204410, 5210108* (to DuPont).

Fritschela, W. *et al.* (1993) Oxidation of ethoxylated starch. *US Patent 5238597* (to Hoechst).

Fujimaka, T. *et al.* (1994) Biodegradable polyesters. *US Patent 5310782* (to Showa High Polymer).

Fukunaga, F. *et al.* (1976) Bacterial Degradation of Poly(vinyl alcohol). *Japan Kokai 76125786.*

Fukunaga, F. *et al.* (1977) Bacterial Degradation of Poly(vinyl Alcohol). *Japan Kokai 7794471.*

Gallagher, J.F. (1992, 1992, 1993) Compostable polyesters. *US Patents* 5171308, 5171309, and 5219646, respectively (to Procter and Gamble).

Garafalo, A. and Wu, S.R. (1991) *US Patent 5062995* (to Lever).

Gledhill, W.E. and Saeger, V.W. (1987) Poly(ether carboxylates) *J. Ind. Microbiol.*, **2**(2), 97.

Gledhill, W.E. (1978) Poly(ether carboxylates). *Appl. Environ. Microbiol.*, **12**, 591.

Gonsalves, K. *et al.* (1993) Biodegradation of polyesteramides. *J. Appl. Poly. Sci.*, **50**, pp. 1999–2006.

Gosset, S. and Videau, D. (1991) Oxidation of starch. *EP 455522-A*(to Roquette Freres).

Griffin, G.J.L. (1971) Blending of polyethylene with starch. *Adv. Chem. Ser.*, **134**, 159; (1977) *US Patent* 4016117, *GB* 1487050; (1979) *GB* 1485833; and (1994) The *Chemistry and Technology of Biodegradable Polymers*, Blackie Academic and Professional Press, Chapman and Hall, 1994, Ed. G.J.L. Griffin.

Griffin, G.J.L. (1973) Pro-oxidants. *J. Amer. Chem. Soc., Divn. Org. Coat. Plast. Chem.*, **33**(2), 88.

Griffin, G.J.L. (1976) Pro-oxidants. *J. Poly. Sci.*, **57**, 281.

Gross, R., Tanna, R.J., McCarthy S.P. (1992) Aerobic Composting. *Poly. Mater. Sci. Eng.*, **67**, 230–231.

Gross, R., Gu, J.D., McCarthy, S.P., Smith, G.P., Eberiel, D. (1992) Anaerobic Bioreactors. *Polym. Mater. Sci. Eng.*, **67**, 294–295.

Gross, R.A., Gu, J.D., Ebereiel, D.T., McCarthy, S.P. (1994) Quantifying cellulosic biodegradation. *J. Environ. Polym. Degrad.*, **1**(2), pp. 143–155.

Gross, R.A., Cromlich, A.M., Bierrer, G.A. (1996) Gamma-glutamic acid. *Biotech. BioEng.*, **50**, 22.

Guillet, J.E. and Golemba, F.J. (1970) Photodegradation. *SPE J.*, **26**, 88.

Guillet, J.E. and Hartley, G.H. (1968) Photodegradation. *Macromolecules*, **1**, 165.

Guillet, J.E. and Heskins, M. (1968) Photodegradation. *Macromolecules*, **1**, 97.

Guillet, J.E. and Americk, Y. (1971) Photodegradation. *Macromolecules*, **4**, 375.

Guillet, J.E. (1990) Perspective, Issues and Opportunities. *Degradable Materials*, CRC Press, Boca Raton, pp. 55–97.

Hall, R., Wiley, A.D. and Hall, R.G., (1993) Poly(aspartic acid) copolymers. *WO 9306202* (to Procter and Gamble).

Heap, W.M. and Morell, S.H. (1968) *J. Appl. Chem.*, **18**, 189.

Hirsoe, K. (1992) Photodegradation. *US Patent 5147712* (to Nippon Unicar).

Huang, J-C., Shetty, A.S. and Wang, M.S. (1990) *Adv. Polym. Technol.*, **10**(1), 23–30.

Huang, S.J. (1976) Biodegradation of Polyamides. *Proc. of the 3rd. Int. Biodeg. Symp (1975)*, Applied Sci. Publ., 731.

Huang, S.J., Bittrito, M., Cameron, J.A., Leong, K.W., Pavlisko, J.A., Roby, M.S., Bell, J.P. and Knox, J.R. (1977) Biodegradation of condensation polymers. *Poly. Preprints*, **18**(1), 438–41.

Huang, S.J., Quingua, E. and Wang, I.F. (1982) Degradation of Oxidized Poly(vinyl alcohol). *Org. Coat. Appl. Polym. Sci. Proc.*, **46**, 345.

Huang, S.J. (1989) *Comprehensive Polymer Science*, **6**, 597, Pergamon, New York.

Huang, S.J., Shetty, A.S. and Wang, M.S. (1990) Degradation of Oxidized Poly(vinyl alcohol). *Adv. Polym. Tech.*, **10**, 23.

Imanaka, T., *et al.*, Poly(aspartic acid) in detergents. *EP 454125* (to Montedipe).

Irie, Y. (1993) Biodegradable polymeric carboxylic acids. *EP 529910A* (to NSKK)

Jane, J-L. *et al.* (1992) Starch Blends. *US Patent 5115000* (to Iowa State University).

Jane, J-L. and Lim, S. (1993) Starch Blends. *WO 9319125–A1* (to Iowa State University).

Jopski, T. (1993) *Kunststoffe*, **83**(10), 248–251.

Jost P. and Tournilhac, F. (1992) Acrylic grafts to starch. *EP 465287* (to Rhone-Poulenc).

Kato, M. and Yoneshige, Y. (1973) Photodegradation Mechanism. *Macromol. Chem.*, **164**, 159.

Kawai, F., Hanada, K., Tani, Y. and Ogata, K. (1977) Biodegradation of poly(propylene oxides). *J. Ferment. Technol.*, **55**, 89.

Kawai, F. (1982) Biodegradation of poly(propylene oxides). *J. Kobe Univ. Commerce*, **18**(1–2), 23.

Kawai, F., Okamoto, T. and Suzuki, T. (1985) Biodegradation of poly(propylene oxides). *J. Ferment. Technol.*, **63**, 239.

Kawai, F. and Yamanaka, H. (1986) Biodegradation of poly(tetramethylene oxides). *Ann. Meeting Agric. Chem. Sic. Japan, Kyoto.*

Kawai, F. (1987) Biodegradation of polyethers. *CRC Critical Reviews in Biotechnology*, C.R.C. Press, **6**, 273.

Kawai, F. (1993) Biodegradation of Poly(acrylic acid). *Appl. Microbiol. and Biotech.*, **39**(3), 382–385.

Kawabata, N. (1993) Biodegradation of vinyl pyridinium salts. *Nippon Gomu Kyokaishi*, **66**(2), 80–87.

Kendricks, *Diss. Abstr.*, 82329391.

Kim, Y.W., Park, I.H. and Park, T.S. (1993) Acrylic grafts onto starch. *WO 9302118–A1* (to Teaching Muslin Co. Ltd).

Kitamura, T., Ajioka, M., Morita, K. *et al.* (1992, 1993) Poly(lactic acid). *EP 510999–A1, JP 05339557–A1, EP 5726750–A1* (to Mitsui Toatsu).

Klimmek, H. and Krause, F. (1994) Acrylics grafted to starch in the presence of mercaptans. *WO 9401476-A1* (to Stockhausen).

Kobayashi, S. (1993) Synthetic analogues of natural polyhydroxyalkanoates. *Macromol. Chem. Rapid Commun.*, **14**, 785–790.

Koeffer, D., Lorz, P.M. and Roeper M. (1993) Biodegradable polyacetals. *DE 4204808-A1* (to BASF).

Koskan, L.P. and A.M. Atencio (1992) Poly(aspartic acid). *WO 9214753* (to Donlar Corporation)

Koskan, L.P. and Meah, A. (1993) Poly(aspartic acid). *US Patent, 5219952* (to Donlar).

Koskan, L.P. and Atencio, A.M. (1993) Poly(aspartic acid). *US Patent 5221733* (to Donlar Corporation).

Koskan, L.P. and Lowe, K.C. (1993) Poly(aspartic acid) dispersants. *US Patent 5116513* (to Donlar Corporation).

Koutlakis, G., Lane, C.C. and Lenz, R.P. (1993) Starch Blends. *US Patent* 5271766 (to ADM).

Kroner, M. *et al.* (1992) Protein grafted with acrylics. *DE 4029348* (to BASF).

Kumar, G.S. *et al.* (1981–83) *J. Macromol. Sci. Rev., Macromol. Chem. and Phys.*, **C22**(2), 225, and (1987) *Biodegradable Polymers*, Marcel Dekker, Inc., New York and Basel.

Kurata, T. *et al.* (1992). Photoactivators. *JP 04173869* (to Japan Synthetic Rubber).

Kuster, E. (1979) *J. Appl. Polym. Sci. Appl. Polym. Symp.* **35**, 395.

Lenz, R.W., Brandl, Gross, R.A. and Fuller, R.C. (1990) *Adv. Biochem. Eng./Biotech.* **41**, 77.

Lenz, R.W. and Vert, M. (1978) Malic acid condensation. *ACS Polym. Preprints*, **20**, 608.

Lenz, R.W. (1993) *Adv. Polm. Sci.* **107**, 1–40.

Lenz, R.W., Gilmore, D.F., Antoria, S., Fuller, R.C. (1993) *J. Environ. Polym. Degrad.*, **1**(4), 269–274.

Lester, J.N., Hunter, M., daMotta Marques, D.M.L., Perry, R. (1988) Degradability of Polymeric Carboxylic Acids. *Environ. Technol. Lett.*, **9**, 1–22.

Marchessault, R.H., Blilim, T.L., Deslandes, Y., Hamer, G.K., Orts, W.J., Gundararajan, P.R., Taylor, M.G., Bloembergen, S., Holden, D.A. (1988) *Makromol. Chem. Makromol. Symp.*, **19**, 235.

Masuda, T. (1991) Aliphatic Polyesters. *Kagaku to Kogyo*, **44**, 1737.

Masuda, T. (1994) *Technol. Japan*, **24**, 56.

Matsumura, S. *et al.* (1980) Degradability of Vinyl Polymers. *Yukagaku*, **30**, 31.

Matsumura, S. *et al.* (1981) Degradability of Vinyl Polymers. *Yukagaku*, **30**, 757.

Matsumura, S. *et al.* (1984) Degradability of Vinyl Polymers. *Yukagaku*, **33**, 211, 228.

Matsumura, S. *et al.* (1985) Degradability of Vinyl Polymers. *Yukagaku*, **34**, 202, 456.

Matsumura, S. *et al.* (1985) Degradability of Vinyl Polymers. *Yukagaku*, **35**, 167.

Matsumura, S. *et al.* (1986) Degradability of Vinyl Polymers. *Yukagaku*, **35**, 937.

Matsumura, S., Abe, Y. and Imai, K. (1986) Malic acid polymerization. *Yukagaku*, **35**(11), 937–944.

Matsumura, S., Hashimoto, K. and Yashikawa, S. (1987) Poly(ether carboxylates). *Yukagaku*, **36**(110), 874–881.

Matsumura, S., Maeda, S., Takahashi, J. and Yoshikawa, S. (1988) Biodegradation of Poly(vinyl acetate). *Kobunshi Robunshi*, **45**(4), 317.

Matsumura, S., Takahashi, J., Maeda, S. and Yoshikawa, S. (1988) Vinyloxyacetic acid. *Macromol. Chem. Rapid Commun.*, **9**(1), 1–5.

Matsumura, S., Takahashi, J., Maeda, S. and Yoshikawa, S. (1988) Vinyloxyacetic acid. *Kobunshi Ronbunshi*, **45**(4), 325–331.

Matsumura, S., Maeda, S. and Yoshikawa, S., (1990) Oxidation of polysaccharides. *Macromol. Chem.*, **191**(6), 1269–75.

Matsumura, S. *et al.* (1992) Oxidation of polysaccharides. *Poly Preprints Japan*, **41**(7), 2394–2396.

Matsumura, S. *et al.* (1993) Degradation of Vinyl Polymers. *J. Am. Oil. Chem. Soc.*, **70**, 659–665.

Matsumura, S., Ii, S., Shigeno, H., Tanaka, T., Okuda, F., Shimura, Y. and Toshima, T. (1993) Degradation of Vinyl Polymers. *Makromol. Chem.*, **194**(12), 3237–3246.

Matsumura, S. *et al.* (1993) Oxidation of polysaccharides. *Angew. Makromol. Chem.*, **205**, 117–129.

Matsumura, S., Amaya, K. and Yoshikawa, S. (1993) Oxidation of polysaccharides. *J. Environ. Polym. Degrad.*, **1**(1), 23–31.

Matsumura, S., Aoki, K. and Toshima, K. (1994) Oxidation of polysaccharides. *J. Amer. Oil. Chem. Soc.*, **71**(7), 749–755.

Meister, J.J., Milstein, O., Gersonde, R., Hutterman, A., Chen, M.J., (1992) Lignin- grafts. *Environ. Microbiol.*, **58**(10), 3225–3237.

Meyer, J.M. and Elion, G.R. (1994) Starch cellulose acetate blends. *US Patent 5288318* (to US Army, Natick Lab.).

Miller, B. (1990) Starch blends. *Plastic World*, **48**(3), 12.

Mita Industry PHA blend with polycaprolactone. *JP 0429261-A*.

Monal, W.A. and Covac, C.P. (1993) Chitosan modifications. *Macromol. Chem. Rapid Commun.*, **14**, 735–740.

Monsanto (1979) Poly(ether carboxylates). *US Patents*, 4144226, 4146495, 4204052; (1980) 4233422, 4233423.

Mulders, J. and Gilain, (1977) Hydroxyacrylic acid. *J. Water Res.*, **11**(7), 571–574.

Murai, K. and Oota, A. (1986) Degradation of Vinyl Grafted Polysaccharides JP 6131498 (to Sanyo Chem.)

Nakamura, S. (1994) *Kansai Res. Inst. (KRI), Report* #10, Update #1.

Narayan, R., Stacy, N., Lu, Z. J. and Chen, Z. X. (1989) Styrene grafted to starch. *Antec Confer. Proc.*, 1362–4.

Narayan, R. (1990) *Annual Meeting of the Air and Waste Management Assoc.*, June 24–29, 40.

Narayan, R. (1993) *Biotech. NIST GCR* 93–633.

Nishiyama. M. *et al.* (1994) PHA blend with chitosan and cellulose. *JP 06001881-A* (to A.I.S.T.).

O'Brien, J.J. *et al.* (1993) Photodegradation. *US Patent 5194527* (to Dow).

OECD Guidelines for Testing Methods.(1981) *Degradation and Accumulation Section*; OECD: Washington, DC); Nos. 301A-E, 302A-C, 303A, and 304A.

Otey, F.H. and Mark, A.M. (1976) Ethylene, poly(vinyl alcohol, starch blends. *US Patent 3949145* (to USDA).

Otey, F.H., Westoff, R.P. and Doane, W.M. (1987) Polyethylene starch blends. *Ind. Eng. Chem. Res.*, **26**, 1659; (1977) *Org. Coat. Plast/ Chem.*, *37*(2), 297.

Pettigrew, C.A. and Palmisano, A.C. (1992) *Bioscience*, **42**(9), 680–685.

Ponce, A. and Tournilhac, F. (1992) Poly(aspartic acid). *EP 511037–A1* (to Rhone-Poulenc).

Poteate, H., Ruecker, A. and Nartop, B. (1994) Starch compatibility with additives. *Starch*, **46**(20), 52–59.

Potts, J.E., *Encyl. Chem. Tech. Suppl.*, Kirk-Othmer, 3rd. Edn., Wiley.

Potts, J.E. *et al.* (1973) *Polymers and Ecological Problems, Plenum Press* and EPA Contract (1972), CPE-70–124.

Potts, J.E., Clendinning, R.A. and Cohen, S. (1974) *Great Plains Agric. Council Pub.*, **68**, 244.

Potts, J.E. (1978) *Aspects of Degradation and Stabilization of Polymers*, H.H.J. Jellinek ed. Elsevier, Amsterdam, 617.

Priddy, D.B. and Sikkema, K.D. (1992) Photodegradation. *US Patent 5115058* (to Dow).

Procter and Gamble (1986) Poly(ether carboxylates). *EP 192441, EP 192442; (1987) US Patents 465415931, 4663071, 4689167; EP 236007; (1988) EP. 264977.*

Rao, V.S. (1993) Poly(aspartic acid). *Makromol. Chem.*, **194**, 1095–1104.

Reich, L. and Stivala, S.S. (1971) Photodegradation Mechanism. *Elements of Polymer Degradation*, (McGraw Hill, Inc., New York), 32–35.

Rhone-Poulenc Announcement (1993) *Eur. Plastics News*, **20**, 16.

Rocourt, A.P.A. (1993) Poly(aspartic acid) in detergents. *EP 561464 and EP 561452* (to Lever).

Rodriquez, F. (1970) *Chemtech.*, **1**, 409.

Sakharov, AS.M., Skibida, I.P. and Brussani, G. (1979) and (1993) Oxidation of starch. *EP 548399–A1 and WO 9218542–A1* (to Novamont).

Saito, T., Shiraki, M. and Tatsumichi, M. (1993) *Kobunshi Robuni*, **50(10)**, 781–3.

Satyananaya, D., Chatterji, P.R. (1993) *J. Macromol. Sci.; Rev. Macromol-Chem. Phys.*, **C33**(3), 349–368.

Schechtman, L., Degenhardt, C.R. and Kozikowski, B.A. (1991, 1992) Succinylated poly(vinyl alcohol). *US 5093170 and US Patent 5093170*, (to Procter and Gamble).

Scholz, N. Tenside, (1991) Environmental toxicity of surfactants. *Tensides, Surfactants and Detergents*, **28**, 277–281.

Schink, B. and Strab, H. (1983) Anaerobic biodegradation of polyethers. *Appl. Environ. Microbiol.*, **45**, 1905.

Schink, B. and Strab, H. (1986) Anaerobic biodegradation of polyethers. *Appl. Microbiol. Biotech.*, **25**, 37.

Sikes, S. and Donachy, J. (1993) Poly(aspartic acid) dispersants. *US Patent 5260272* (to University of South Alabama).

Scott, G., Mellor, D.C. and Moir, A.B. (1973) Photodegradable. *European Poly. J.*, **9**, 219.

Scott, G., Hodsworth, J.D. and Williams, D. (1964) Oxidation. *J. Chem. Soc.*, 4692.

Scott, G. (1995) Oxidative degradation. *Polym. Degrad. and Stab.*, **48**, 315– 324.

Shikinami, Y. *et al.* (1993) PHA blend with inorganics. *JP 05237180-A* (to Takiron).

Shimao, M. and Kato, N. (1990) Degradation of Poly(vinyl alcohol). *International Symp. on Biodegradable Polymers Abstr.*, 80.

Sipinen, A.J. and Rutherford, D.R. (1993) Oxidation. J. Environ. *Polym. Degrad.*, **1**(3), 193–203.

Sinclair, R.G. and Preston, J.R. (1992) Poly(lactic acid). *WO 9204413-A* (to Battelle Mem. Inst.).

Sommerville, D.C., Dennis, D. (1993) *WO 9302187-A1* (to Univ. Mich.); [see also *Adv. Mater.* (1993), **5**(1), 30–36.].

Stannett, V.T., Kim, S. and Gilbert, R.D. (1973) Cellulosics and polyisocyanate condensations. *J. Polym. Sci. Polym. Lett. Ed.*, **11**(12), 731–735.

Steinbuchel, A. (1991) *Acta Biotechnol.*, **11**(5), 419–427.

Stepto, R., Silbiger, J. and Wexler, F.C. (1993) Starch blends. *WO 9314911–A1.* (to Warner Lambert).

Stepto, R.F. and Thomka, I. (1987) *Chimia.*, 41.

Suzuki, T. and Tokiwa, Y. (1970) Biodegradability of Aliphatic Polyesters. *Agric. Biol. Chem.*, **42**(5), 1071.

Suzuki, T., Ichihara, Y., Yamada, M. and Tonomura, K. (1973) Poly(vinyl alcohol) Degradation. *Agric Biol. Chem.*, **34**(4), 747–756.

Suzuki, T. *et al.* (1977) Styrene Oligomer Degradation. *Agric. Biol. Chem.*, **41**(12), 2417.

Suzuki, T. and Tokiwa, Y. (1977) Biodegradability of Aliphatic Polyesters. *Nature*, **270**(5632), 76.

Suzuki, T., Hukushima, K. and Suzuki, S. (1978) *Environmental Science and Technology*, **12**(10), 1180–1183.

Suzuki, T. *et al.* (1978) Butadiene Oligomers. *Agric. Biol. Chem.*, **42**(6), 1217.

Suzuki, T. *et al.* (1979) Degradation of Isoprene. *Agric. Biol. Chem.*, **43**(12), 2441.

Suzuki, T., Ichihara, Y., Yamada, M. and Tonomura, K. (1979) Poly(vinyl alcohol) Degradation. *J. Appl. Poly. Sci., Appl. Poly. Symp.*, **35**, 431–437.

Suzuki, T. *et al.* (1980) Oligomeric Ethylene Degradation. *Report of the Fermentation Inst., Japan.*

Suzuki, T. and Tokiwa, Y. (1986) Biodegradability of Aliphatic Polyesters. *Agric. Biol. Chem.*, **50**(5), 1323.

Swanholm, C.E. (1975) Photosensitizers. *US Patent 3888804* (to Biodeg. Plastics).

Swift, G. (1990) *Poly. Mater. Sci. Eng. Preprint. Amer. Chem Soc., Washington D. C.* 846–52.

Swift, G. (1990) ACS Symp. Ser. #433, *Agricultural and Synthetic Polymers, Biodegradation and Utilization*, 1–12. J. Edward Glass and Graham Swift (eds.).

Swift, G. (1992) Biodegradability of polymers in the environmental., *FEMS Microbiol. Revs.*, **103**, 339–346.

Swift, G. (1993) Directions for Environmentally Degradable Polymers. *Acc. Chem. Res.*, **26**, 105–110.

Swift, G. (1993) *International Biodegradable Polymer Workshop*, Osaka, Japan, November.

Swift, G. and Weinstein, B. (1993) Degradation of Vinyl Polymers. *US Patent 5191048* (to Rohm and Haas).

Swift, G., Paik, Y.H., Simon, E.S. (1993) Polysaccharides as raw materials for the detergent industry. *ACS Polym. Mater. Sci. and Eng. Abstr.*, **69**, 496; and (1995) Chem. & Ind., pp. 55–59.

Swift, G. (1994) *Polymer News*, **19**, 102–106.

Swift, G., Freeman, M.B., Paik, Y.H., Wolk, S.A., Yocom, K.M. (1994) Poly(aspartic acid). *6th International Conference on Polymer Supported Reactions in Organic Chemistry(POC)*, Venice, June 19–23, Abstr. pp. 21.13. and, *35th, IUPAC International Symposium on Macromolecules*, Akron, Ohio. July 11–15, Abstr. 0–4.4–13th, pp. 615.

Swift, G., Paik, Y.H. and Simon, E.S. (1994) Poly(aspartic acid). *EP 578448–A1* (to Rohm and Haas).

Swisher, R.D. (1987) *Surfactant Biodegradation*, Marcel Dekker, 2nd. Edition.

Takiyama, E. (1993) Biodegradable polyesters. *JP 05070543, '566, '571, '572, '574, '575, '576, '577, '579, and EP 572682–A1* (to Showa High Polymer).

Takeyama, E. and Fujimaki, T. (1992) *Plastics*, **43**, 87.

Tani, Y. *et al.* (1993) Biodegradation of Poly(acrylic acid). *Applied and Environ Microbiol.*, 1555–1559.

Taniguchi, W. *et al.* (1993) *Polym. Prep. J.*, **42**, 3787.

Taylor, L.J. (1977) Photosensitizers. *US Patent 4056499* (to Owens-Illinois).

Thomka, I. and Whitmer, F. *US Patent 4673438.*

Thomka, I. and Whitmer, F. *US Patent 4673438.*

Tokiwa, Y. (1979) Enzyme degradation of polyesteramides. *J. Appl. Poly. Sci.*, **24**, 1701.

Tokiwa, Y. and Koyama, M. (1992) PHA blends with polyesters and inorganics *JP's 04146952–A, 04146953–A, 04146929–A* (to A.I.S.T.).

Tokiwa, Y. (1994) Polystyrene biodegradation. *Private communication.*

Udipi, K. and Zolotor, A.M. (1993) *J. Poly. Sci.*, **75**, 109–117.

US Environmental Protection Agency (1982) *Chemical Fate Testing Guidelines, NTIS No. PB 82–233008.*

Vaidya, U.R. and Bhattacharya, M. (1995) Blends of proteins and cellulosics. *US Patent 5446078* (to University of Minnesota).

van Bekkum, H. *et al.* (1987) Oxidation of polysaccharides. *Prog. Biotech.*, **3**, 157.

van Bekkum, H. (1988) Oxidation of polysaccharides. *Starch-Starke*, 192.

van Bekkum, H., deNooy, A.E.J. and Besemer, A.C. (1994) Specific oxidation of starch. *Rec. Trav. Chim.*, **113**(3), 165–166.

Vert, M., Feijin, J., Albertsson, A., Scott, G. and Chiellini, E. (1992) *Biodegradable Polymers and Plastics: Proceedings of the Second International Scientific Workshopon Biodegradable Polymers and Plastics, November 25–27, 1991.* The Royal Society of Chemistry, London, ISBN 0–85186–207–1, Special Publication 109.

Vidal, C. and Vaslin, S. (1992) Acrylic grafts to starch. *EP 465286* (to Rhone-Poulenc).

Watanabe, Y., Morita, M., Hamada, N., Tsujisaka, Y. (1975) Poly(vinyl alcohol) Degradation. *Agric. Biol. Chem.*, **39**(12), 2447–2448.

Watanabe, Y., Morita, M., Hamada, N. and Tsujisaka, Y., (1976) Poly(vinyl alcohol) Degradation. *Arch. Biochem. Biophys.*, **174**, 575.

Watanabe, and Tsuchiyama, Y (1993) Specific oxidation of polysaccharides. *JP 05017502–A* (to Mercian Corp.).

Wheatley, O.D. and Baines, C.F. (1976) *Text. Chem. Color*, **8**(2), 28–33.

White, R.A. (1974) Photosensitizers. *US Patent 3847852* (to DE Bill and Richardson).

Wood, L.L. (1994) Poly(aspartic acid) from ammonia and maleic acid. *US Patent 5288783* (to SR Chem).

Wood, L.L. (1993) Poly(aspartic acid) from ammonia and maleic acid. *WO 9323452* (to SR Chem).

Wool, R.P., Peanasky, J.S. and Long, J.M. (1991) Starch Blends. *WO 9115542* (to Agritech).

Yabbana, A. and Bartha, R. (1993) *Soil Biol. Biochem.* 25(11), 1469–1475.

Yamamoto, T., Inagaki, H., Yagu, J. and Osumi, T. (1966) Poly(vinyl alcohol) Degradation. *Abst. Annual Meeting Agric. Chem. Soc. Japan*, 133.

Yamamoto, N. (1991) *Polym. Preprints Japan*, **41**, 2240.

Yamamoto, N. (1992) Enzyme degradations. *Preprints Ann. Meeting of Kagaku Kogaku-Kai*, 34.

Yoshida, Y. (1992) *,Fine Chemicals*, **21**, 12.

Zobel, H.F. (1984) *Starch: Chemistry and Technology.* 2nd. Ed. R.L. Whistler, J.N. Bemiller, E.F. Paschall, ed., Academic Press, 285.

INDEX